The Enzymes

VOLUME XI

OXIDATION–REDUCTION

Part A

DEHYDROGENASES (I)

ELECTRON TRANSFER (I)

Third Edition

CONTRIBUTORS

BRIAN M. AUSTEN
LEONARD J. BANASZAK
KENNETH M. BLUMENTHAL
RALPH A. BRADSHAW
CARL-IVAR BRÄNDÉN
KEITH DALZIEL
RICHARD E. DICKERSON
HANS EKLUND
BO FURUGREN
BUNJI HAGIHARA
J. JOHN HOLBROOK
HANS JÖRNVALL
ANDERS LILJAS
JOSEPH F. NYC
MICHAEL G. ROSSMANN
NOBUHIRO SATO
EMIL L. SMITH
STEVEN J. STEINDEL
RUSSELL TIMKOVICH
TATEO YAMANAKA

ADVISORY BOARD

THE ENZYMES

Edited by PAUL D. BOYER
Molecular Biology Institute and
Department of Chemistry
University of California
Los Angeles, California

Volume XI

OXIDATION-REDUCTION

Part A

DEHYDROGENASES (I)
ELECTRON TRANSFER (I)

THIRD EDITION

ACADEMIC PRESS New York San Francisco London 1975
A Subsidiary of Harcourt Brace Jovanovich, Publishers

ACADEMIC PRESS, INC.
111 Fifth Avenue, New York, New York 10003

United Kingdom Edition published by
ACADEMIC PRESS, INC. (LONDON) LTD.
24/28 Oval Road, London NW1

Library of Congress Cataloging in Publication Data
Main entry under title:

The Enzymes.

Includes bibliographical references.
CONTENTS: v. 1. Structure and control.–v. 2. Kinetics and mechanism.–v. 3. Hydrolysis: peptide bonds. [etc.]
1. Enzymes. I. Boyer, Paul D., ed. [DNLM: 1. enzymes. QU135 B791e H2vc. v. 1-10]
QP601.E523 574.1'925 75-117107
ISBN 0–12–122711–1

PRINTED IN THE UNITED STATES OF AMERICA

Contents

List of Contributors

Numbers in parentheses indicate the pages on which the authors' contributions begin

BRIAN M. AUSTEN (293), Department of Biological Chemistry, UCLA School of Medicine, and Molecular Biology Institute, University of California, Los Angeles, California

LEONARD J. BANASZAK (61, 369), Department of Biological Chemistry, Division of Biology and Biomedical Sciences, Washington University, St. Louis, Missouri

KENNETH M. BLUMENTHAL* (293), Department of Biological Chemistry, UCLA School of Medicine, and Molecular Biology Institute, University of California, Los Angeles, California

RALPH A. BRADSHAW (369), Department of Biological Chemistry, Division of Biology and Biomedical Sciences, Washington University, St. Louis, Missouri

CARL-IVAR BRÄNDÉN (61, 103), Department of Chemistry, Agricultural College of Sweden, Uppsala, Sweden

KEITH DALZIEL (1), Department of Biochemistry, University of Oxford, Oxford, England

RICHARD E. DICKERSON (397), Norman W. Church Laboratory of Chemical Biology, California Institute of Technology, Pasadena, California

HANS EKLUND (103), Department of Chemistry, Agricultural College of Sweden, Uppsala, Sweden

BO FURUGREN (103), Department of Chemistry, Agricultural College of Sweden, Uppsala, Sweden

BUNJI HAGIHARA (549), Department of Biochemistry, The School of Medicine, Osaka University, Osaka, Japan

* Present address: Department of Biochemistry, University of Florida, and C. V. Whitney Marine Laboratory, St. Augustine, Florida.

J. JOHN HOLBROOK (191), Department of Biochemistry, University of Bristol, Bristol, England

HANS JÖRNVALL (103), Department of Chemistry, Karolinska Institutet, Stockholm, Sweden

ANDERS LILJAS (61, 191), Department of Molecular Biology, Wallenberg Laboratory, University of Uppsala, Uppsala, Sweden

JOSEPH F. NYC (293), Department of Biological Chemistry, UCLA School of Medicine, and Molecular Biology Institute, University of California, Los Angeles, California

MICHAEL G. ROSSMANN (61, 191), Department of Biological Sciences, Purdue University, West Lafayette, Indiana

NOBUHIRO SATO (549), Department of Biochemistry, The School of Medicine, Osaka University, Osaka, Japan

EMIL L. SMITH (293), Department of Biological Chemistry, UCLA School of Medicine, and Molecular Biology Institute, University of California, Los Angeles, California

STEVEN J. STEINDEL* (191), Department of Biological Sciences, Purdue University, West Lafayette, Indiana

RUSSELL TIMKOVICH (397), Norman W. Church Laboratory of Chemical Biology, California Institute of Technology, Pasadena, California

TATEO YAMANAKA (549), Department of Biology, Faculty of Science, Osaka University, Osaka, Japan

* Present address: Clinical Chemistry Branch, United States Army Medical Laboratory, Fort McPherson, Georgia.

Preface

The important area of oxidation–reduction enzymes, covered in Volumes XI–XIII, will complete the third edition of "The Enzymes." Some of the most remarkable examples of biological catalysis and versatility of coenzyme and cofactor function are found within this group. As with previous volumes, the excellent quality is a result of the outstanding and well-recognized contributors. For planning the volumes on oxidation–reduction, the Editor and readers are indebted to the Advisory Board Members, namely, Professors Britton Chance, Lars Ernster, Bo Malmström, and Vincent Massey.

This volume covers the general topics of the kinetics and mechanism of the nicotinamide-nucleotide-linked dehydrogenases and of the evolutionary and structural relationships of these enzymes. These general topics are followed by detailed coverage of well-understood individual dehydrogenases, namely, alcohol dehydrogenase, lactate dehydrogenase, glutamate dehydrogenase, and malate dehydrogenase. The last but by no means least important portion of the volume is devoted to a comprehensive survey of the *c*-type and *b*-type cytochromes.

One additional dehydrogenase, glyceraldehyde-3-phosphate dehydrogenase, could have logically been covered in this volume. A chapter on this dehydrogenase as well as a chapter on transhydrogenases will appear in Volume XIII. Postponement of these chapters was necessary to meet the more important goal set for the treatise: timeliness of the contributions in a given volume. Taken as a whole, the Editor is under the impression that this goal has been met, and the promptness of publication of all contributions in "The Enzymes" is far better than for most multiauthored treatises.

The rapidly accumulating knowledge of the past decade has brought a depth of understanding to many oxidation–reduction enzymes which has yet to have its full impact on biochemistry. The basic molecular information that has been accumulated will remain vital for years to come. Many students of enzymology have followed advances from the kinetic evidence that binding sites on enzymes must exist to the detailed description of these sites made possible by X-ray techniques. To such long-time students, it is professionally rewarding and even an esthetic experience to examine a table, such as that on page 232 of this volume, in which the interactions of some 34 amino acid residues of lactate dehydrogenase with its bound coenzyme are given.

Volumes XII and XIII, covering Parts B and C of oxidation–reduction enzymes, will include chapters on members of the second great family of dehydrogenases: those linked to flavin derivatives. Coverage will also include flavin-linked electron-transferring enzymes. Other sections will cover oxygenases and oxidases, including cytochrome oxidase. Chapters on catalase and peroxidase will complete the volumes.

Again, it is a pleasure to acknowledge the help and encouragement of the staff of Academic Press and the editorial and organizational assistance of Lyda Boyer. They made the Editor's task much lighter and added considerably to the professional quality of the volume.

Paul D. Boyer

Contents of Other Volumes

Volume I: Structure and Control

Volume II: Kinetics and Mechanism

Volume III: Hydrolysis: Peptide Bonds

Volume IV: Hydrolysis: Other C–N Bonds, Phosphate Esters

Volume V: Hydrolysis (Sulfate Esters, Carboxyl Esters, Glycosides), Hydration

Volume VI: Carboxylation and Decarboxylation (Nonoxidative), Isomerization

Volume VII: Elimination and Addition, Aldol Cleavage and Condensation, Other C–C Cleavage, Phosphorolysis, Hydrolysis (Fats, Glycosides)

Volume VIII: Group Transfer, Part A: Nucleotidyl Transfer, Nucleosidyl Transfer, Acyl Transfer, Phosphoryl Transfer

Volume IX: Group Transfer, Part B: Phosphoryl Transfer, One-Carbon Group Transfer, Glycosyl Transfer, Amino Group Transfer, Other Transferases

Volume X: Protein Synthesis, DNA Synthesis and Repair, RNA Synthesis, Energy-Linked ATPases, Synthetases

The Enzymes

VOLUME XI

OXIDATION–REDUCTION

Part A

DEHYDROGENASES (I)

ELECTRON TRANSFER (I)

Third Edition

1

Kinetics and Mechanism of Nicotinamide-Nucleotide-Linked Dehydrogenases

KEITH DALZIEL

I. Introduction

A. Scope of the Chapter

Nicotinamide-nucleotide-linked dehydrogenases were among the earliest two-substrate enzymes to be subjected to detailed kinetic study by steady-state (*1–3*) and rapid reaction techniques (*4*), and provided much of the original stimulus for the necessary extension of kinetic theory already developed for one-substrate and hydrolytic enzymes (*5–8*). This was partly because of the convenience and precision with which rates can be measured by means of the light absorption or fluorescence emission (*9–11*) of the reduced coenzymes and because of the changes of these properties which accompany the binding of reduced coenzymes to many dehydrogenases (*12, 13*).

Some 250 dehydrogenases using NAD(P) as acceptor are known, and perhaps one-tenth of this number has been studied kinetically in sufficient detail to give some information about mechanism. Several are discussed individually in the following chapters of this volume, including those for which structural information has been obtained recently by X-ray crystallography. The rapidly increasing knowledge of the structures of dehydrogenases, or even of their stable compounds with coenzymes and substrate analogs, does not diminish the need for more detailed kinetic studies of the right kind, which are still among the most informative means for the investigation of mechanism. As Haldane remarked 40 years ago, "The key to the knowledge of enzymes is the study of reaction velocities, not of equilibria" (*14*). Knowledge of the static structure of an enzyme puts us almost in the same advantageous position as the chemist studying the mechanism of a reaction between simpler compounds

1. H. Theorell and R. Bonnichsen, *Acta Chem. Scand.* **5,** 1105 (1951).
2. H. Theorell, A. P. Nygaard, and R. Bonnichsen, *Acta Chem. Scand.* **9,** 1148 (1955).
3. M. T. Hakala, A. J. Glaid, and G. W. Schwert, *JBC* **221,** 191 (1956).
4. H. Theorell and B. Chance, *Acta Chem. Scand.* **5,** 1127 (1951).
5. R. A. Alberty, *JACS* **75,** 1928 (1953).
6. K. Dalziel, *Acta Chem. Scand.* **11,** 1706 (1957).
7. R. A. Alberty, *JACS* **80,** 1777 (1958).
8. W. W. Cleland, *BBA* **67,** 104 (1963).
9. H. Theorell and A. P. Nygaard, *Acta Chem. Scand.* **8,** 877 (1954).
10. K. Dalziel, *BJ* **84,** 244 (1962).
11. P. C. Engel and K. Dalziel, *BJ* **115,** 621 (1969).
12. P. D. Boyer and H. Theorell, *Acta Chem. Scand.* **10,** 447 (1956).
13. S. P. Colowick, J. van Eys, and J. H. Park, *Compr. Biochem.* **14,** 1 (1966).
14. J. B. S. Haldane, "Enzymes." Longmans, Green, New York, 1930.

of known structure. But the fundamental questions about the dynamic mechanism of the enzyme-catalyzed reaction remain. How many elementary or one-step reactions make up the overall reaction and what are the structures of the molecular intermediates? What are the structures and properties of the transition states which determine the rates of these elementary reactions? If there are two or more active centers in an enzyme molecule, are they equivalent and does the reaction occur independently at each? Or are there "interactions" between them, in substrate-binding or catalytic steps, which play a part in the catalytic mechanism or in the regulation of the enzymic activity?

It was a difficult matter to decide upon the scope and organization of this chapter. The literature pertaining to several individual dehydrogenases, let alone dehydrogenases in general, is vast, and much of the wealth of information has been obtained by kinetic studies of one kind or another. Kinetic studies can hardly be discussed without reference to other kinds of investigation, and it would be inappropriate to attempt a comprehensive account of the kinetics and mechanisms of the enzymes considered individually in other chapters. Rather the author's aims have been to outline the kinds of information that have been—or could be—derived from kinetic studies, to attempt generalizations, indicate gaps, and perhaps stimulate further studies. Naturally, a few well-studied enzymes such as the alcohol dehydrogenases and lactate dehydrogenases will figure largely in this account. It has also been necessary to amplify some aspects of the general theory of steady-state and rapid reaction kinetics, considered in Chapters 1 and 2 of Volume III of this work.

B. Some Preliminary Generalizations

Evidence of mechanisms from kinetic studies must of course be considered in relation to the results of other kinds of investigation, and it is useful to note some tentative generalizations at the start. The nicotinamide nucleotides are coenzymes, and not prosthetic groups, and can be considered as substrates from the kinetic point of view. They also form stable, reversible compounds with dehydrogenases. An oxidation reduction reaction between a dehydrogenase (excluding those with flavin prosthetic groups) and a coenzyme, in the complete absence of the other substrate, has never been demonstrated. There is no reliable evidence, therefore, for an enzyme-substitution (ping-pong) mechanism; indeed, for many dehydrogenases this mechanism has been excluded simply by the results of initial rate measurements.

It is reasonable to suppose that the enzyme–coenzyme compounds are intermediates in the overall catalytic reaction. A compulsory-order mechanism (or at least a preferred-order mechanism), with the coenzyme combining first with the enzyme and dissociating last, has been established for a few dehydrogenases. For others, there is evidence that an alternative pathway through an enzyme–substrate complex contributes significantly to the overall reaction in a random mechanism. An ordered mechanism in which the substrate combines before the coenzyme is unlikely, and has been eliminated for several dehydrogenases by isotope exchange kinetics and kinetic studies with alternative substrates.

The work of Westheimer and Vennesland and their colleagues has shown that direct transfer of hydrogen between coenzyme and substrate occurs in dehydrogenase reactions and that there is stereospecificity with respect to the prochiral C-4 of the coenzyme and the chiral or prochiral centers of the substrates (*15*). These findings imply that hydrogen transfer occurs within a ternary complex of enzyme, coenzyme, and substrate. There is evidence for the existence of these unstable intermediates from steady-state and pre-steady-state studies of several dehydrogenases and from isotopic exchange rates at equilibrium; in addition, stable ternary complexes of enzyme, coenzyme, and inactive substrate analog and of enzyme, reduced coenzyme, and reduced substrate (abortive complexes) have been demonstrated in many cases.

II. Steady-State Kinetics

Initial rate measurements, especially with alternative substrates and with a product or substrate analog as inhibitor, and measurements of the rate of isotope exchange at equilibrium, can give a great deal of information about mechanism, and in some cases allow estimates of individual velocity constants and dissociation constants. The results of such studies, which require little enzyme, are an essential basis for the proper interpretation, in relation to the overall catalytic reaction, of pre-steady-state studies and kinetic and thermodynamic studies of enzyme–coenzyme reactions in isolation.

A. Phenomenological Initial Rate Equations for Two-Substrate and Three-Substrate Reactions (*16*)

For many simple dehydrogenases which catalyze only reversible hydrogen transfer between a coenzyme and another substrate, it has been found

15. G. Popják, "The Enzymes," 3rd ed., Vol. 2, p. 115, 1970.
16. Including the coenzyme but excluding water and the hydrogen ion. Dehydrogenases with flavin prosthetic groups are not considered.

that initial rate measurements in each direction over quite wide ranges of reactant concentrations can be described by an equation of the form (*6*):

$$\frac{e}{v} = \phi_0 + \frac{\phi_A}{A} + \frac{\phi_B}{B} + \frac{\phi_{AB}}{AB} \tag{1}$$

$$= \left(\phi_0 + \frac{\phi_B}{B}\right) + \left(\phi_A + \frac{\phi_{AB}}{B}\right)\frac{1}{A} \tag{1a}$$

$$= \left(\phi_0 + \frac{\phi_A}{A}\right) + \left(\phi_B + \frac{\phi_{AB}}{A}\right)\frac{1}{B} \tag{1b}$$

where v is the initial velocity, e is the enzyme (or active center) concentration, A and B are coenzyme and second substrate concentrations, respectively, and the ϕ's are constants (kinetic coefficients) (*17*). This means that primary plots of e/v vs. 1/A (or 1/B) for several fixed values of B (or A) are linear and nonparallel, from Eqs. (1a) and (1b). Secondary plots of the slopes and intercepts against the reciprocal of the "fixed" substrate concentration are also linear. The four kinetic coefficients are estimated experimentally as the slopes and intercepts of these secondary plots (*6*).

Rate equations of this form have been derived for ordered and random ternary complex mechanisms, but not for an enzyme-substitution (ping-pong) mechanism (*5,6*). The steady-state rate equation for the latter lacks the last term of Eq. (1); that is, $\phi_{AB} = 0$ and primary plots with respect to either substrate at all fixed concentrations of the other would be parallel (*6*). On these grounds alone this mechanism can be dismissed for all dehydrogenases for which adequate initial rate studies have been made on the reaction in each direction. The conclusion that glycerol dehydrogenase operates by this mechanism (*19*) rests mainly on apparently parallel plots for the glyceraldehyde–NADPH reaction over limited substrate concentration ranges. The reverse reaction could not be studied

17. Equation (1) may also be written (*18*) in terms of K_m values and the maximum velocity V. The relationships between the two sets of constants are $\phi_0 = e/V$, $\phi_A = eK_A/V$, $\phi_B = eK_B/V$, and $\phi_{AB} = eK_{iA}K_B/V$. The system used here has the advantage that the ϕ's are the fundamental parameters independent of enzyme concentration, are the more directly estimated experimentally, and vary independently of one another with inhibitor concentration, pH, etc. (*18*). The ϕ's are also generally simpler functions of velocity constants in a mechanism than K_A, etc., especially in ordered mechanisms and for three-substrate reactions. In an ordered mechanism ϕ_A or ϕ_B is the reciprocal of a single velocity constant; the diagnostic importance of this for comparing the kinetics with alternative substrates has often been overlooked and useful data neglected.

18. W. W. Cleland, "The Enzymes," 3rd ed., Vol. 2, p. 1, 1970.

19. C. J. Toews, *BJ* **105**, 1067 (1967).

because of the unfavorable equilibrium. An interpretation more consistent with the product inhibition and other data reported is an ordered mechanism with $\phi_{AB}/\phi_B \ll \phi_A/\phi_0$. It can be seen from Eqs. (1a) and (1b) that there may then be no significant change of slope of the primary plots over a limited range of concentrations of the "fixed" substrate, although there may be a large change of intercept. In fact, this situation exists for several dehydrogenases, including liver alcohol dehydrogenase when A = NADH and B = acetaldehyde, simply because the dissociation constant of the enzyme-NADH compound (ϕ_{AB}/ϕ_B) is much smaller than K_m for NADH (ϕ_A/ϕ_0) and this makes ϕ_{AB} difficult to estimate. However, for NAD the reverse is true, and consequently the changes of slope of primary plots for the NAD–ethanol reaction are much more marked than the changes of intercept (*20*).

For glyceraldehyde-3-phosphate dehydrogenase, glutamate dehydrogenase, and the NADP-linked oxidative decarboxylases, which have three substrates in one direction, initial rate measurements with a fixed concentration of any one of the three substrates also conform to Eq. (1). This again rules out any form of enzyme-substitution mechanism in which free product is formed before all the substrates have combined with the enzyme and indicates the involvement of a quaternary enzyme complex. The appropriate generalized form of Eq. (1) is

$$\frac{e}{v} = \phi_0 + \frac{\phi_A}{A} + \frac{\phi_B}{B} + \frac{\phi_C}{C} + \frac{\phi_{AB}}{AB} + \frac{\phi_{AC}}{AC} + \frac{\phi_{BC}}{BC} + \frac{\phi_{ABC}}{ABC} \tag{2}$$

The rate equations for the several possible enzyme-substitution mechanisms all lack the last term of this equation. The kinetic coefficients in Eq. (2) can be estimated by primary, secondary, and tertiary plots (*21*).

Accurate estimates of all the kinetic coefficients in Eq. (1) or (2) are needed for the reaction in each direction, and at the same pH value, if reliable conclusions about mechanisms are to be reached. A sensitive method for measuring changes of reactant concentration is demanded because accurate estimates of v can only be made from progress curves that are linear for 30 sec or more, during which only a small fraction of the total reaction to equilibrium occurs. Moreover, K_m values and dissociation constants of binary complexes of the coenzyme are often small (10^{-6}–10^{-8} M), and the equilibrium is usually unfavorable for NAD oxidation at pH 7.0 and in studies of product inhibition. Measurements of the fluorescence of the reduced coenzymes with a simple recording fluorometer provide a more sensitive analytical method than spectrophotometry (*10,11*).

20. K. Dalziel, *JBC* **238,** 1538 (1963).
21. K. Dalziel, *BJ* **114,** 547 (1969).

B. Initial Rate Equations for Ordered and Random Mechanisms

Four mechanisms for the reaction $A + B \rightleftharpoons P + Q$ [A = NAD(P), P = NAD(P)H] consistent with initial rate measurements that conform to Eq. (1) will be described. Examination of the initial rate equations will indicate various ways in which the mechanisms may be distinguished and tested experimentally, and the results of such studies will be discussed in subsequent sections of this chapter.

1. *The Simple Ordered Mechanism*

The simplest mechanism that gives an initial rate equation of the form of Eq. (1) is

$$E + A \rightleftharpoons EA + B \rightleftharpoons Q + EP \rightleftharpoons P + E \tag{3}$$

proposed for liver alcohol dehydrogenase by Theorell and Chance (*4*). The probable existence of central ternary complexes EAB and EPQ was recognized, but they were omitted because initial rate measurements, in conjunction with kinetic and equilibrium studies of the isolated reaction $E + NADH \rightleftharpoons ENADH$, showed that they were not kinetically significant (*1,2,4*). In fact, the insertion of any number of such complexes in Eq. (3) does not alter the form of the initial rate equation, and for the reasons given earlier, therefore, the most reasonable minimal mechanism for simple dehydrogenases is

$$E + A \underset{k_{-1}}{\overset{k_1}{\rightleftharpoons}} EA + B \underset{k_{-2}}{\overset{k_2}{\rightleftharpoons}} EAB \underset{k'}{\overset{k}{\rightleftharpoons}} EPQ \underset{k_2'}{\overset{k_{-2}'}{\rightleftharpoons}} Q + EP \underset{k_1'}{\overset{k_{-1}'}{\rightleftharpoons}} P + E \tag{4}$$

The initial rate equation derived by steady-state analysis is Eq. (1) with the kinetic coefficients for the reaction from left to right as defined in Table I. (The kinetic coefficients for the reverse reaction, ϕ_0', ϕ_P, etc., are obtained by addition or deletion of primes on the rate constants.) The characteristic features of this mechanism (*6*) are that the individual velocity constants for the formation and dissociation of the enzyme–coenzyme compounds can be calculated from experimental values for the kinetic coefficients:

$$k_1 = 1/\phi_A \quad \text{and} \quad k_1' = 1/\phi_P \tag{5}$$

$$k_{-1} = \phi_{AB}/\phi_A\phi_B \quad \text{and} \quad k_{-1}' = \phi_{PQ}/\phi_P\phi_Q \tag{6}$$

The two most important tests of the applicability of this mechanism are therefore comparison of these values with direct estimates of the velocity constants from studies of the isolated enzyme–coenzyme reactions by rapid reaction techniques (Section III,B), and initial rate studies with two or more alternative substrates B (and Q), which should give the same values for ϕ_A and $\phi_{AB}/\phi_A\phi_B$, while the individual values of ϕ_B and of ϕ_{AB} may vary greatly (Section II,E).

TABLE I

THE PHYSICAL SIGNIFICANCE OF KINETIC COEFFICIENTS FOR ORDERED AND RANDOM MECHANISMS FOR TWO-SUBSTRATE REACTIONS

Mechanism	Kinetic coefficient[a]				Ref.
	ϕ_0	ϕ_A	ϕ_B	ϕ_{AB}	
Simple ordered, Eq. (4)	$\frac{1}{k_{-1}'} + \frac{1}{k_{-2}'} + \frac{A^b}{k}$	$\frac{1}{k_1}$	$\frac{Ak_{-2} + k}{kk_2}$	$\frac{k_{-1}(Ak_{-2} + k)}{k_1kk_2}$	*4,6*
Ordered with isomeric coenzyme complexes, Eq. (12)	$\frac{1}{k_{-1}'} + \frac{1}{k_{-2}'} + \frac{A}{k} + \frac{1}{k_c} + \frac{1}{k_{-c}'} + \frac{k_c'}{k_{-1}'k_{-c}'}$	$\frac{1}{k_1}\left[1 + \frac{k_{-1}}{k_c}\right]$	$\left[\frac{Ak_{-2} + k}{kk_2}\right]\left[1 + \frac{k_{-c}}{k_c}\right]$	$\frac{k_{-c}k_{-1}(Ak_2 + k)}{k_ck_1kk_2}$	*20*
Rapid equilibrium random, Eq. (13)	$\frac{1}{k}$	$\frac{K_{EBA}}{k}$	$\frac{K_{EAB}}{k}$	$\frac{K_{EA}K_{EAB}^c}{k}$	*4,14*
Preferred order, Eq. (13)	$\frac{1}{k_{-1}'} + \frac{1}{k_{-2}'} + \frac{A}{k}$	$\frac{Ak_{-4}}{kk_4} + \frac{1}{k_1}$	$\frac{Ak_{-2} + k}{kk_2}$	$\frac{k_{-1}(Ak_{-2} + k)}{k_1kk_2}$	*39*

[a] The kinetic coefficients are the constants in Eq. (1) in the text. The mechanisms and rate constants are defined in the text.

[b] $A = 1 + k'/k_{-2}'$.

[c] $K_{EA}K_{EAB} = K_{EB}K_{EBA}$. These are dissociation constants for the complexes in Eq. (13): $K_{EA} = k_{-1}/k_1$, $K_{EAB} = k_{-2}/k_2$, etc.

Because the maximum rate in either direction cannot exceed the rate of dissociation of the product coenzyme, Eq. (6) leads to the "maximum rate relations"

$$\frac{1}{\phi_0'} \leqslant \frac{\phi_{AB}}{\phi_A \phi_B} \quad \text{and} \quad \frac{1}{\phi_0} \leqslant \frac{\phi_{PQ}}{\phi_P \phi_Q} \tag{7}$$

The equalities characterize the Theorell–Chance mechanism, which is best defined as a limiting case of Eq. (4) with $k_{-1}' \ll k_{-2}'$, k and $k_{-1} \ll k_{-2}$, k', so that $\phi_0 = 1/k_{-1}'$ and $\phi_0' = 1/k_{-1}$. The steady-state concentrations of the ternary complexes will then be negligible compared with those of EA and EP. The other three kinetic coefficients in Eq. (1) are the same as for the general case of Eq. (4).

The thermodynamic dissociation constants of the enzyme–coenzyme compounds can also be calculated from the kinetic coefficients

$$K_{EA} = \phi_{AB}/\phi_B \quad \text{and} \quad K_{EP} = \phi_{PQ}/\phi_Q \tag{8}$$

However, these relations are also required by random mechanisms; agreement with direct estimates of the dissociation constants (Section III) will not distinguish between the mechanisms considered here, but is important for establishing the cause of any deviations from the maximum rate relations [Eq. (7)]. It may be noted that in an ordered mechanism neither ϕ_{AB}/ϕ_A, nor the K_m values ϕ_A/ϕ_0 and ϕ_B/ϕ_0, have any simple physical significance, for example, as dissociation constants (Table I).

The equilibrium constant of the overall catalyzed reaction is related to kinetic coefficients for the forward and reverse reactions by the Haldane relation (*5,6*)

$$K_{eq} = \frac{P \times Q}{A \times B} = \frac{\phi_{PQ}}{\phi_{AB}} \tag{9}$$

The proton involved stoichiometrically in the reaction has been neglected in these formulations, and K_{eq}/H^+ is the true, pH-independent equilibrium constant. Consideration of the physical significance of ϕ_B and ϕ_Q (Table I) shows that the equilibrium constant for the reaction of enzyme-bound coenzymes and the substrates, $EA + B \rightleftharpoons EP + Q$, is (*20*)

$$K_{eq}' = \phi_Q/\phi_B \tag{10}$$

A proton may or may not be involved stoichiometrically in this reaction involving the protein. Equation (10) is of importance in the interpretation of pH (Section III,A) and temperature (*22*) variation of kinetic coefficients, and has been somewhat neglected.

22. K. Dalziel, *Acta Chem. Scand.* **17**, S27 (1963).

From Eqs. (8), (9), and (10), which apply to all the mechanisms considered here,

$$K_{eq}' = K_{eq}K_{EA}/K_{EP} \tag{11}$$

as was first shown by Theorell and Bonnichsen (*1*).

2. *The Ordered Mechanism with Isomeric Enzyme–Coenzyme Compounds: Conformational Changes*

The *raison d'etre* for an ordered mechanism may be that binding of the coenzyme creates the binding site for the second substrate by inducing a conformational change of the enzyme (*23*). There is direct evidence for this from recent X-ray diffraction studies of lactate dehydrogenase (*24*). The conformational change will, however, be manifest kinetically only if it occurs as an elementary step separate from coenzyme binding, thus,

$$\mathrm{E} + \mathrm{A} \underset{k_{-1}}{\overset{k_1}{\rightleftharpoons}} \mathrm{EA} \underset{k_{-c}}{\overset{k_c}{\rightleftharpoons}} \mathrm{E'A} + \mathrm{B} \underset{k_{-2}}{\overset{k_2}{\rightleftharpoons}} \mathrm{E'AB} \underset{k'}{\overset{k}{\rightleftharpoons}} \mathrm{E'PQ}$$

$$\underset{k_2}{\overset{k_{-2}'}{\rightleftharpoons}} \mathrm{Q} + \mathrm{E'P} \underset{k_c'}{\overset{k_{-c}'}{\rightleftharpoons}} \mathrm{EP} \underset{k_1'}{\overset{k_{-1}'}{\rightleftharpoons}} \mathrm{P} + \mathrm{E} \tag{12}$$

The initial rate equation is again of the form of Eq. (1) with the kinetic coefficients as in Table I, which shows that the mechanism differs from the simple ordered mechanism in three important respects. First, the isomerization steps are potentially rate-limiting; evidence for such a rate-limiting step not attributable to product dissociation or the hydride-transfer step (k) has been put forward for pig heart lactate dehydrogenase (*25*). Second, Eqs. (5) and (6) no longer apply; in each case the function of kinetic coefficients will be smaller than the individual velocity constant (Table I). Third, because $\phi_{AB}/\phi_A\phi_B$ is smaller than k_{-1}', it may also be smaller than the maximum specific rate of the reverse reaction; that is, one of the maximum rate relations in Eq. (7) need not hold (*26*). This mechanism was in fact first suggested to account for anomalous maximum rate relations obtained with dehydrogenases for which there was other evidence for an ordered mechanism (*27–29*).

23. D. E. Koshland, Jr., *Proc. Nat. Acad. Sci. U.S.* **44**, 98 (1958).

24. M. J. Adams, M. Buehner, K. Chandrasekhar, G. C. Ford, M. L. Hackert, A. Liljas, M. G. Rossmann, I. E. Smiley, W. S. Allison, J. Everse, N. O. Kaplan, and S. S. Taylor, *Proc. Nat. Acad. Sci. U. S.* **70,** 1963 (1973).

25. R. S. Criddle, C. H. McMurray, and H. Gutfreund, *Nature* (*London*) **220,** 1091 (1968).

26. It can be shown that *one* of the two maximum rate relations in Eq. (7) must hold for the isomeric coenzyme compound mechanism, and this does not preclude the formation of isomeric complexes of each coenzyme (*27–29*).

27. H. R. Mahler, R. H. Baker, and V. J. Shiner, Jr., *Biochemistry* **1,** 47 (1962).

28. V. Bloomfield, L. Peller, and R. A. Alberty *JACS* **84,** 4367 (1962).

29. K. Dalziel, *JBC* **238,** 2850 (1963).

It should be noted that the relations in Eq. (8) are valid for this mechanism, however, because ϕ_{AB}/ϕ_B is equal to the apparent dissociation constant, [E][A]/{[EA] + [E′A]}, which will also be measured in any direct equilibrium study of the enzyme–coenzyme reaction (Section III,A). Moreover, the functions of kinetic coefficients in Eqs. (5) and (6), as well as those in Eq. (8), should be independent of the nature of the second substrate B or Q, as in the case of the simple ordered mechanism, since they are defined by rate constants for the enzyme–coenzyme and isomerization reactions only (Section II,E).

3. *Rapid Equilibrium Random Mechanism*

The random mechanism is

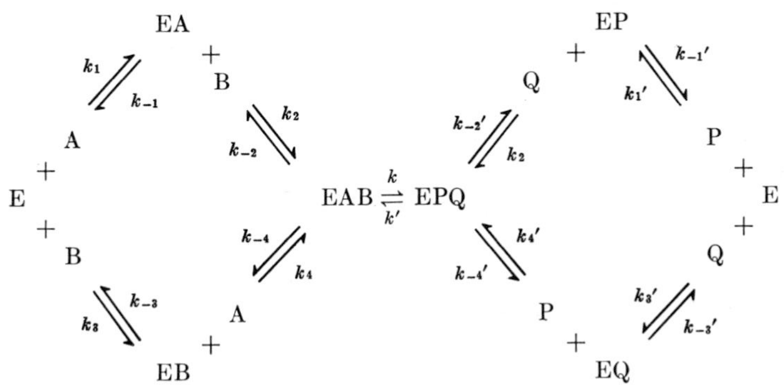

(13)

The initial rate equation derived by steady-state analysis is of the second degree in A and B (*30*). It simplifies to the form of Eq. (1) if the rates of dissociation of substrates and products from the complexes are assumed to be fast compared with the rates of interconversion of the ternary complexes (k, k'); thus, the steady-state concentrations of the complexes approximate to their equilibrium concentrations, as was first shown by Haldane (*14*). The kinetic coefficients for this rapid equilibrium random mechanism (Table I), together with the thermodynamic relations $K_{EA}K_{EAB} = K_{EB}K_{EBA}$ and $K_{EP}K_{EPQ} = K_{EQ}K_{EQP}$, suffice for the calculation of k, k' and all the dissociation constants $K_{EA} = k_{-1}/k_1$, $K_{EAB} = k_{-2}/k_2$, etc.

In this mechanism, the K_m values (ϕ_A/ϕ_0, ϕ_B/ϕ_0) are dissociation constants for the *ternary* complex. Equation (8) holds, and in addition $\phi_{AB}/\phi_A = K_{EB}$ and $\phi_{PQ}/\phi_P = K_{EQ}$. The mechanism was early rejected for beef heart lactate dehydrogenase on the grounds that enzyme–lactate and enzyme–pyruvate compounds having these dissociation constants

30. K. Dalziel, *Trans. Faraday Soc.* **54**, 124 (1959).

calculated from initial rate data could not be detected by equilibrium binding studies (*31*). It has also been conclusively disproved for liver and yeast alcohol dehydrogenases with ethanol as substrate (*32*), and for lactate (*33*) and malate (*34*) dehydrogenases by isotopic exchange studies (Section II,C). The observation of a "burst" of NADH formation in pre-steady-state studies of liver alcohol dehydrogenase with ethanol (*35*) and propan-1-ol (*36*) as substrates, and of lactate (*37*) and glutamate (*38*) dehydrogenases also precludes this mechanism, which requires that the initial formation of NADH, as the complex EPQ, be slower than all the following product dissociation steps and equal to the steady-state maximum rate. This mechanism is a possible one for liver alcohol dehydrogenase with secondary alcohols as substrates and for yeast alcohol dehydrogenase with primary alcohols other than ethanol (Section II,E).

4. *Preferred Pathway Mechanism*

The steady-state rate equation for the random mechanism will also simplify to the form of Eq. (1) if the relative values of the velocity constants are such that net reaction is largely confined to one of the alternative pathways from reactants to products, of course. It is important, however, that dissociation of the coenzymes from the reactant ternary complexes need not be excluded. Thus, considering the reaction from left to right in Eq. (13), if $k_{-2}' \gg k_{-4}'$, then product dissociation will be effectively confined to the upper pathway; this condition can be demonstrated by isotope exchange experiments (Section II,C). Further, if $k_3k_4\mathrm{B} \ll k_1k_{-3} + k_1k_4\mathrm{A}$, then the rate of net reaction through EB will be small compared with that through EA (*39*). The rate equation is then the same as that for the simple ordered mechanism, except that ϕ_A is now a function of the dissociation constant for A from the ternary complex, k_{-4}/k_4, as well as k_1 (Table I). Thus, Eqs. (5), (6), and (7) do not hold; instead, $1/\phi_\mathrm{A} < k_1$ and $\phi_\mathrm{AB}/\phi_\mathrm{A}\phi_\mathrm{B} < k_{-1}$, and this mechanism can account for anomalous maximum rate relations. In contrast to the ordered mechanism with isomeric complexes, however, the same values for these two functions of kinetic coefficients would not be expected if an alterna-

31. Y. Takenaka and G. W. Schwert, *JBC* **223**, 157 (1956).
32. E. Silverstein and P. D. Boyer, *JBC* **239**, 3908 (1964).
33. E. Silverstein and P. D. Boyer, *JBC* **239**, 3901 (1964).
34. E. Silverstein and G. Sulebele, *Biochemistry* **8**, 2543 (1969).
35. J. D. Shore and H. Gutfreund, *Biochemistry* **9**, 4655 (1970).
36. R. L. Brooks and J. D. Shore, *Biochemistry* **10**, 3855 (1971).
37. H. d'A. Heck, C. H. McMurray, and H. Gutfreund, *BJ* **108**, 793 (1968).
38. M. Iwatsubo and D. Pantaloni, *Bull. Soc. Chim. Biol.* **49**, 1563 (1967).
39. K. Dalziel and F. M. Dickinson, *BJ* **100**, 34 (1966).

tive substrate B were used, since ϕ_A is a function of k_4, k_{-4}, k, and k' as well as k_1. The relations in Eq. (8) should hold, however.

This mechanism in which reactant coenzyme may "leak" from the ternary complex was first suggested for liver alcohol dehydrogenase to reconcile the results of initial rate studies with primary and secondary alcohols as substrates and isotope exchange experiments (*39*), and it has also been proposed for yeast alcohol dehydrogenase (*40*) and α-glycerophosphate dehydrogenase from rabbit muscle (*41*).

The preferred pathway mechanism also provides one hypothesis for substrate activation or inhibition (*6,39*). It is simply a limiting case of a nonequilibrium random mechanism which can be formally reconciled with linear primary plots over wide reactant concentrations. However, with sufficiently large concentrations of B, net reaction through EB may become significant, and cause deviations from Eq. (1) toward lower or higher activity, depending upon whether $k_4 > k_1$ or $k_4 < k_1$ [Eq. (10)]. This and other mechanisms for substrate inhibition and activation will be discussed in Section II,F,1.

The preferred pathway mechanism might be regarded at present as a rather general one for dehydrogenases, in place of the simple ordered mechanism.

5. *Ordered and Random Mechanisms for Three-Substrate Reactions*

The rate equations for fully random and ordered mechanisms for three-substrate reactions are shown in Table II and can only be briefly discussed here. For the random mechanism, the rate equation derived by the rapid equilibrium assumption (*42*) contains all the terms of Eq. (2), and from experimental values for the eight kinetic coefficients for the reaction in each direction the dissociation constants for all the complexes may be calculated (*cf. 43*).

For the ordered mechanism

$$\mathrm{E} + \mathrm{A} \underset{k_{-1}}{\overset{k_1}{\rightleftharpoons}} \mathrm{EA} + \mathrm{B} \underset{k_{-2}}{\overset{k_2}{\rightleftharpoons}} \mathrm{EAB} + \mathrm{C} \underset{k_{-3}}{\overset{k_3}{\rightleftharpoons}} \mathrm{EABC} \underset{k'}{\overset{k}{\rightleftharpoons}}$$

$$\mathrm{EPQ} \underset{k_2'}{\overset{k_{-2}'}{\rightleftharpoons}} \mathrm{Q} + \mathrm{EP} \underset{k_1'}{\overset{k_{-1}'}{\rightleftharpoons}} \mathrm{P} + \mathrm{E} \quad (14)$$

the steady-state rate equation lacks the term ϕ_{AC}/AC in Eq. (2), and it is therefore possible in principle to distinguish between this mechanism and

40. F. M. Dickinson and G. P. Monger, *BJ* **131,** 261 (1973).
41. P. Bentley and F. M. Dickinson, *BJ* **143,** 11 and 19 (1974).
42. C. Frieden, *JBC* **234,** 2891 (1959).
43. P. C. Engel and K. Dalziel, *BJ* **118,** 409 (1970).

TABLE II

KINETIC COEFFICIENTS FOR ORDERED AND RANDOM MECHANISMS FOR THREE-SUBSTRATE REACTIONS

Mechanism	Kinetic coefficient[a]								Ref.
	ϕ_0	ϕ_A	ϕ_B	ϕ_C	ϕ_{AB}	ϕ_{AC}	ϕ_{BC}	ϕ_{ABC}	
Ordered, Eq. (14)	$\frac{1}{k_{-1}'} + \frac{1}{k_{-2}'} + \frac{A^b}{k}$	$\frac{1}{k_1}$	$\frac{1}{k_2}$	$\frac{B^c}{k_3}$	$\frac{k_{-1}}{k_1 k_2}$	0	$\frac{k_{-2}B^c}{k_2 k_3}$	$\frac{k_{-1}k_{-2}B^c}{k_1 k_2 k_3}$	*21,42*
Rapid equilibrium random[d]	$\frac{1}{k}$	$\frac{K_{BCA}}{k}$	$\frac{K_{ACB}}{k}$	$\frac{K_{ABC}}{k}$	$\frac{K_{ACB}K_{CA}}{k}$	$\frac{K_{ABC}K_{BA}}{k}$	$\frac{K_{ABC}K_{AB}}{k}$	$\frac{K_A K_{AB} K_{ABC}}{k}$	*42*

[a] The kinetic coefficients are the constants in Eq. (2) in the text. The rate constants are defined in Eq. (14).

[b] $A = 1 + k'/k_{-2}'$.

[c] $B = (k_{-3}k' + k_{-3}k_{-2}' + kk_{-2}')/kk_{-2}'$.

[d] K_A, K_{AB}, K_{ABC}, etc., are dissociation constants of binary, tertiary, and quaternary complexes of the substrates with the enzyme.

the random mechanism by initial rate measurements alone (*42*). The velocity constants k_1, k_{-1}, k_2, and k_{-2} can be calculated from the kinetic coefficients, and there are maximum rate relations analogous to those for the simple ordered mechanism for a two-substrate reaction. In addition, the dissociation constant of EA is given by $\phi_{ABC}/\phi_{BC} = K_{EA}$. (*21*).

Detailed initial rate studies of bovine liver glutamate dehydrogenase (*43*) showed that all the eight terms in Eq. (2) were needed to describe the reductive amination of 2-oxoglutarate, indicating a random mechanism. This has been confirmed by isotope exchange studies (*44*). With 6-phosphogluconate dehydrogenase, on the other hand, the term ϕ_{AC}/AC in Eq. (2), with A = NADPH and C = CO_2, could not be detected (*45*), suggesting an ordered mechanism in which ribulose 5-phosphate is the second substrate to combine. That CO_2 is the last substrate to combine is indicated by the tritium exchange experiments of Leinhard and Rose (*46*) which showed that labilization of a hydrogen atom in ribulose-5-P by the enzyme requires NADPH but not CO_2.

For glyceraldehyde-3-phosphate dehydrogenase, conflicting conclusions have been reached by different workers (*47–49*). From one analysis of initial rate data, for the pig muscle enzyme, it appears that with glyceraldehyde as substrate the mechanism is random (*50*), whereas with glyceraldehyde 3-phosphate there is random combination of this substrate and NAD followed by phosphate as the compulsory third substrate (*48*). On the other hand, inhibition studies with the rabbit muscle enzyme indicate an ordered mechanism in which NAD is the first and acyl acceptor the last substrate to combine (*51*). More detailed comparative initial rate studies with the several aldehydes which act as substrates (Section II,E), preferably by a fluorimetric method (*11,47*), and isotope exchange studies at equilibrium are needed for this enzyme.

C. Isotope Exchange at Equilibrium

Measurements of the rate of isotope exchange between substrate and product pairs as a function of reactant concentrations at chemical equilibrium were introduced by Silverstein and Boyer (*33*) to distinguish

44. E. Silverstein and G. Sulebele, *Biochemistry* **12,** 2164 (1973).
45. R. H. Villet and K. Dalziel, *Eur. J. Biochem.* **27,** 244 (1972).
46. G. E. Leinhard and I. A. Rose, *Biochemistry* **3,** 185 (1964).
47. C. S. Furfine and S. F. Velick, *JBC* **240,** 844 (1965).
48. T. Keleti and J. Batke, *Acta Physiol. Hung.* **28,** 195 (1965).
49. L. A. Fahien, *JBC* **241,** 4115 (1966).
50. T. Keleti, *Acta Physiol.* **28,** 19 (1965).
51. B. A. Orsi and W. W. Cleland, *Biochemistry* **11,** 102 (1972).

random and ordered mechanism and have been of great value. For the details of the method and its full potentialities, the reader is referred to the original papers and the article by Cleland (*18*).

From inspection of the ordered mechanism in Eq. (4), it is evident that if the concentrations of B and Q are increased, chemical equilibrium being maintained, the rate of isotope exchange between pools of free coenzymes A and P will increase to a maximum value, as the flux between EA and EP increases, and then fall again to zero at still higher concentrations of B and Q, as the coenzymes become effectively confined to the ternary complexes and the concentrations of EA and EP fall to zero. The rate of isotope exchange between B and Q, however, will rise to a plateau value and remain constant, the rate at high reactant concentrations being determined by the values of k_{-2}, k_{-2}', k, and k' only. Results of this kind were obtained, at pH 7.9, with rabbit muscle and bovine heart lactate dehydrogenases (*33*) and with pig heart mitochondrial malate dehydrogenase (*34*), and substantiated the conclusions from initial rate measurements that the mechanisms of these enzymes are ordered. The total inhibition of A/P exchange by high concentrations of B and Q shows that dissociation of A and P from the ternary complexes, provided for in the random mechanism in Eq. (13) and not subject to such inhibition, does not occur to a significant extent under the conditions of these experiments.

With horse liver and yeast alcohol dehydrogenases at pH 7.9 (*32*), and also with bovine heart lactate dehydrogenase at pH 9.7 (*33*) and malate dehydrogenase at pH 9.0 (*34*), isotope exchange between NAD and NADH persisted at a finite rate with very large reactant concentrations, showing that dissociation of the coenzymes from the ternary complexes can occur to form the binary complexes EB and EQ, as in the random mechanism, Eq. (13). However, the limiting rate of exchange between A and P at high reactant concentrations, determined by the values of k_{-4}, k_{-4}', k, and k', was in each case much less than that between B and Q. This finding eliminates a truly rapid equilibrium random mechanism, for which k and k' must be much smaller than k_{-4}, k_{-4}', k_2, and k_{-2}', since the two exchange rates must then be equal. In fact, the differences between the two exchange rates show that the dissociation of A and/or P from the ternary complexes must be slow compared with that of B and/or Q, and also slow relative to the interconversions of the ternary complexes (*32*). This means that in at least one direction of reaction the dissociation of products in the overall reaction is essentially ordered for all these enzymes, the coenzymes dissociating last, as in the preferred pathway mechanism (Section I,B,4). With malate, lactate, and liver alcohol dehydrogenases, the NAD/NADH exchange rate increased to a maximum and then decreased to a constant level at very large B and Q concentrations, showing that the coenzymes are more firmly bound in their reactive ternary complexes than in their binary complexes with the enzyme alone. This is not so for yeast alcohol dehydrogenase (*32*), a finding which may be correlated with recent initial rate studies of this enzyme and with the absence of substrate inhibition (Sections I,E,2 and I,F,1).

While these isotope exchange experiments establish that binary complexes EB and EQ can be formed by dissociation of coenzymes from the ternary complexes, they give no information about the net rate of the overall reaction through the lower pathway of Eq. (13) compared with that through the upper pathway. Thus, the preferred pathway mechanism will account for these results and also for conformity of initial rate data for all these enzymes to Eq. (1) over wide ranges of reactant concentrations. Certain relationships between the rate constants in the preferred pathway mechanism can also be inferred from the relative values of the rate of isotope exchange between B and Q, with saturating reactant concentrations, and the maximum initial rate of the overall reaction; for example, with liver alcohol dehydrogenase the fact that the former rate is 30 times larger than the maximum rate of ethanol oxidation implies that the rate of dissociation of E·NADH (k_{-1}') determines the maximum rate, in accordance with conclusions from initial rate and other studies (*39*).

Recent isotope exchange studies with bovine liver glutamate dehydrogenase (*44*) also showed that coenzyme dissociation from ternary complexes can occur, and in contrast to the results with other dehydrogenases just discussed, the rate of exchange between NAD(P) and NAD(P)H, with large reactant concentrations, was not much smaller than that between glutamate and 2-oxoglutarate. In this case, therefore, an essentially ordered dissociation of products cannot be inferred, and the mechanism appears to approximate more closely to a random mechanism, which may account in part for the fact that initial rate data for the oxidative deamination of glutamate (*11*) cannot be described by Eq. (1) over substantial substrate and coenzyme concentration ranges.

The formation of abortive complexes of the type EAQ and EPB, a common cause of substrate inhibition of dehydrogenases (Section II,F), can also be detected in isotope exchange experiments by inhibition of all exchanges when the concentrations of A and Q or P and B are increased together. The complexes E·NADH·malate (*34*) and E·NAD(P)H·glutamate (*44*), for example, were detected in this way, and are probably responsible for the substrate inhibition observed in initial rate studies of these enzymes (Section II,F,1). The latter complex, but not the former, had already been demonstrated by its distinctive fluorescence (*52*) and absorption (*53*) spectrum.

Perhaps the most important general conclusions from isotope exchange studies, supported by other investigations, are that the preferred pathway mechanism is common to a number of dehydrogenases and that for most, but not all, the formation of the active ternary complex with the second substrate increases the firmness of coenzyme binding, presumably by a protein conformational change.

D. Maximum Rate Relations and Haldane Relations

The Haldane relation, Eq. (9), common to all the mechanisms considered in Section II,B, is reasonably well satisfied by data for very many dehydrogenases, in some cases over a range of pH, especially when the difficulty of estimating ϕ_{PQ} is taken into account. For the Theorell–Chance mechanism there is a second Haldane relation (*5*), but this derives from the two limiting maximum rate relations which can be tested separately with greater precision.

Tests of the maximum rate relations, Eq. (7), for several dehydrogenases, are shown in Table III (*29,40,41,54–59*). The ratios $\phi_A\phi_B/\phi_{AB}\phi_0'$ and $\phi_P\phi_Q/\phi_{PQ}\phi_0$ cannot exceed unity for a simple ordered mechanism. Values close to unity indicate that dissociation of the product coenzyme determines the maximum rate, i.e., a Theorell–Chance mechanism. The results for liver alcohol dehydrogenase with several primary alcohols as substrates (*29,39*), and with mitochondrial malate dehydrogenase (*54*) over a wide range of pH, are substantial evidence for this mechanism, and for the former enzyme this conclusion is supported by direct estimates of the rate constant for dissociation of E·NADH (Section III,B). For lactate dehydrogenase also the deviations from unity are probably not outside the combined experimental errors except at pH 6.0. Values much smaller than unity for α-glycerophosphate dehydrogenase and cytoplasmic malate dehydrogenase indicate that some step other than product coenzyme dissociation is rate limiting under some conditions.

The large values of $\phi_A\phi_B/\phi_{AB}\phi_0'$ for yeast alcohol dehydrogenase with butanol as substrate, for bovine heart lactate dehydrogenase at pH 6.0,

52. G. W. Schwert and A. D. Winer, *BBA* **29,** 424 (1958).
53. R. R. Egan and K. Dalziel, *BBA* **250,** 47 (1972).
54. D. N. Raval and R. G. Wolfe, *Biochemistry* **1,** 1112 (1962).
55. M. Cassman and S. Englard, *JBC* **241,** 787 and 793 (1966).
56. A. D. Winer and G. W. Schwert, *JBC* **231,** 1065 (1958).
57. G. W. Schwert, B. R. Miller, and R. J. Peanasky, *JBC* **242,** 3245 (1967).
58. V. Zewe and H. J. Fromm, *JBC* **237,** 1668 (1962).
59. V. Zewe and H. J. Fromm, *Biochemistry* **4,** 782 (1965).

TABLE III

TESTS OF MAXIMUM RATE RELATIONS [EQ. (7)] FROM INITIAL RATE MEASUREMENTS[a]

Enzyme	pH	$\phi_{AB}/\phi_A\phi_B$ (sec^{-1})	$1/\phi_0'$ (sec^{-1})	$\frac{\phi_A\phi_B}{\phi_{AB}\phi_0'}$	$\phi_{PQ}/\phi_P\phi_Q$ (sec^{-1})	$1/\phi_0$ (sec^{-1})	$\frac{\phi_P\phi_Q}{\phi_{PQ}\phi_0}$	Ref.
Alcohol dehydrogenase[b] (horse liver)	6.0	100	125	1.3	3.5	1.6	0.5	*29*
	7.1	99	125	1.3	4.2	2.7	0.6	
	8.0	29	48	1.7	4.7	3.2	0.7	
	9.0	7.6	7.6	1.0	4.0	3.8	1.0	
Alcohol dehydrogenase (yeast)								
(i)[c]	7.0	1354	3850	2.8	500	455	0.9	*40*
(ii)[d]	7.	15.6	3450	22.0	250	25	0.1	
Malate dehydrogenase	6.0	150	133	0.9	28	50	1.8	
(pig heart mitochondria)	7.0	315	330	1.1	56	84	1.5	*54*
	8.0	620	580	0.9	152	167	1.1	
	9.0	760	630	0.8	208	267	1.3	
Malate dehydrogenase (bovine heart supernatant)	8.5	377	335	0.9	1470	46	0.03	*55*
α-Glycerophosphate dehydrogenase (rabbit muscle)	6.0	0.27	5.6	21	0.36	0.02	0.06	
	7.0	55	38	0.7	42	3.1	0.07	*41*
	8.0	545	127	0.23	102	39	0.4	
	9.0	1465	119	0.08	42	63	1.5	
Lactate dehydrogenase (bovine heart)	6.0	32	260	8.0	42	18	0.4	
	7.0	100	230	2.3	36	50	1.4	*56,57*
	8.0	400	190	0.5	25	80	3.2	
	9.0	500	180	0.4	150	100	0.7	
Lactate dehydrogenase (rabbit muscle)	8.0			1.8			0.3	*58,59*

[a] The original data have been corrected, where necessary, to express the rates as moles substrate per mole enzyme active center. $1/\phi_0$ and $1/\phi_0'$ are therefore the maximum turnover numbers.

[b] With ethanol and acetaldehyde as substrates. Similar values were obtained with propan-1-ol, butan-2-ol, and 2-methylpropan-2-ol and the corresponding aldehydes as substrates (*39*).

[c] With ethanol and acetaldehyde as substrates.

[d] With butan-1-ol and butyraldehyde as substrates.

and for α-glycerophosphate dehydrogenase at pH 6.0 are inconsistent with a simple ordered mechanism. Purified coenzymes were used in all these studies, and the discrepancies are unlikely to be caused by competing nucleotide impurities, responsible for similar but smaller deviations in earlier kinetic studies with liver alcohol dehydrogenase (*29*). The formation of isomeric enzyme–NAD compounds, either reactive (Section II,B,2) or nonproductive (*27*) is a possible explanation suggested for lactate dehydrogenase (*57*), but there is no direct evidence in its support. An alternative explanation, consistent with isotope exchange experiments, is a "leak" of NAD from the reactive ternary complex resulting in $1/\phi_A < k_1$ and $\phi_{AB}/\phi_A\phi_B < k_{-1}$ (Section I,B,4). For yeast alcohol dehydrogenase (*40*) and α-glycerophosphate dehydrogenase this is supported by initial rate measurements with alternative substrates, discussed in the next section.

E. Kinetic Studies with Alternative Substrates

Comparisons of the kinetic coefficients in Eq. (1) obtained from initial rate measurements with alternative substrates have given a considerable amount of information about reaction pathways as well as indications of the molecular basis of specificity (*60*). This approach, much used for proteolytic enzymes, has been exploited particularly with the alcohol dehydrogenases, which catalyze the oxidation of a variety of primary and secondary alcohols (*61*). While several other dehydrogenases have been studied in this way, most of the results have been reported only as apparent maximum rates and apparent K_m values for the alternative substrate, which restricts the amount of information that can be derived.

1. *Liver Alcohol Dehydrogenase*

Initial rate measurements with ethanol, propan-1-ol, butan-1-ol, 2-methylpropan-1-ol (*10,39*), and also cyclohexanol (*62*) gave the same values for ϕ_0, and for ϕ_A and ϕ_{AB}/ϕ_B, while the individual values for ϕ_{AB} and ϕ_B varied over a fiftyfold range. Analogous results were obtained for the reverse reactions. These findings are convincing evidence for an ordered mechanism of the Theorell–Chance type with all these substrates. Butanol gave the smallest ϕ_B value and is therefore the best of these substrates when the alcohol concentration is rate-limiting. From the physical significance of ϕ_B in an ordered mechanism (Table I), this suggests that the rate of reaction of the alcohol with E·NAD, or the rate of hydride transfer in the ternary complex, increases with increase of

60. F. M. Dickinson and K. Dalziel, *BJ* **104**, 165 (1967).
61. H. Sund and H. Theorell, "The Enzymes," 2nd ed., Vol. 7, p. 25, 1963.
62. K. Dalziel and F. M. Dickinson, *BJ* **100**, 491 (1966).

alkyl chain length. Brooks and Shore (*63*) showed by pre-steady-state studies that hydride transfer is indeed five times faster with propan-1-ol than with ethanol. Hydrophobic interactions between substrate and enzyme are clearly important in the formation of a ternary complex with the best configuration for catalysis.

Several secondary alcohols are also substrates and give different values for all four kinetic coefficients, larger than those for primary alcohols (*39,60*). The ratio ϕ_{AB}/ϕ_B is, however, essentially the same for all the primary and secondary alcohols and in satisfactory agreement with direct estimates of the dissociation constant for E·NAD (*64*). The results for secondary alcohols, particularly the varying ϕ_A values, are not consistent with an ordered mechanism, with or without isomeric enzyme–coenzyme compounds. The large and variable ϕ_0 values suggest that hydride transfer in the ternary complex is slower than with primary alcohols and rate-limiting with saturating substrate concentrations; this has been established by the absence of a "burst" of NADH formation in stopped-flow studies of propan-2-ol oxidation and a large deuterium isotope effect on the maximum steady-state rate (*63*).

The enzyme also catalyzes the oxidation of aldehydes to acids irreversibly (*65*). With catalytic amounts of NAD complete dismutation to equivalent amounts of acid and alcohol occurs. The mechanism has been studied in some detail (*66*). The ϕ_2 (and K_m) values for aldehydes in the aldehyde dehydrogenase reaction, measured by proton production, are large, comparable to those for secondary alcohols, as would be expected if the hydrated forms of the aldehydes are the true substrates (*39*). The maximum rates of aldehyde oxidation ($1/\phi_0$) are, however, much greater than those for the oxidation of secondary and even primary alcohols and different for different aldehydes. This also is to be expected, because the product complex E·NADH, the dissociation of which is rate limiting with primary alcohols, is immediately reoxidized by the aldehyde in the reverse alcohol dehydrogenase reaction. Hydride transfer in the E·NAD·aldehyde complex was shown to be rate limiting in the overall dismutation reaction (*66*). Not surprisingly, the reduction of acids is not catalyzed by the enzyme since the protonated acid would be the potential substrate. A similar dismutation of glyoxylate to oxalate and glycollic acid by NAD is catalyzed by lactate dehydrogenase from rabbit muscle (*67*) and pig heart (*68*).

63. R. L. Brooks and J. D. Shore, *Biochemistry* **10**, 3855 (1971).
64. H. Theorell and J. S. McKinley-McKee, *Acta Chem. Scand.* **15**, 1811 (1961).
65. R. H. Abeles and H. A. Lee, *JBC* **235**, 1499 (1960).
66. K. Dalziel and F. M. Dickinson, *Nature* (*London*) **206**, 255 (1965).
67. M. Romano and M. Cerra, *BBA* **177**, 421 (1969).
68. W. A. Warren, *JBC* **245**, 1673 (1970).

The kinetics with all the substrates and isotope exchange studies (*32*) are consistent with the preferred pathway mechanism described in Section II,B,4. With secondary alcohols, hydride transfer is slow, the steady-state concentration of the reactant ternary complex is large, and "leak" of NAD from this complex occurs; thus, $\phi_A = Ak_{-4}/kk_4$ (Table I) and is different for different alcohols. With primary alcohols as substrates, hydride transfer and aldehyde dissociation are much faster than NADH dissociation. Under initial rate conditions, therefore, the ternary complex is not present in significant steady-state concentration, and dissociation of NAD from it does not occur to an appreciable extent; thus, $\phi_A = 1/k_1$, and like ϕ_0 is the same for all primary alcohols.

Detailed initial rate studies at pH 7.0 have also been reported with ethanol and the acetylpyridine (APAD) and hypoxanthine (NHD) analogs of NAD as coenzyme (*69,70*). As would be expected for an ordered mechanism, all the kinetic coefficients are changed. With APAD, the maximum rate relations are, however, again consistent with a Theorell–Chance mechanism, but the maximum rate is significantly smaller than both the rate of dissociation of E·APADH measured directly by the stopped-flow method and the rate of hydride transfer measured from "burst" experiments. A rate-limiting isomerization of the E·APADH complex has been suggested (*71*), and is discussed in Section IV.

2. *Yeast Alcohol Dehydrogenase*

The recent results of Dickinson and Monger (*40*) with yeast alcohol dehydrogenase are in marked contrast to those for the liver enzyme. The effectiveness of primary alcohols as substrates decreases in the series ethanol, propan-1-ol, butan-2-ol, because all four kinetic coefficients increase markedly. In the reverse reactions, acetaldehyde and butyraldehyde gave the same values for ϕ_0 and for ϕ_P, but different values for ϕ_Q and ϕ_{PQ}. These results, together with direct estimates of the dissociation constants of the enzyme–coenzyme compounds by binding studies (*72,73*) and isotope exchange experiments (*32*), indicate a preferred pathway mechanism. Dickinson and Monger (*40*) concluded that with ethanol and acetaldehyde the mechanism approximates to the Theorell–Chance type, except that significant dissociation of NAD from the ternary complex results in $1/\phi_A < k_1$ and an anomalous maximum rate relation. With the

69. J. D. Shore and H. Theorell, *Eur. J. Biochem.* **2,** 32 (1963).
70. J. D. Shore and M. J. Gilleland, *JBC* **245,** 3422 (1970).
71. J. D. Shore and R. L. Brooks, *ABB* **147,** 825 (1971).
72. F. M. Dickinson, *BJ* **120,** 821 (1970).
73. F. M. Dickinson, *BJ* **126,** 133 (1972).

higher alcohols, they concluded that hydride transfer in the ternary complex is rate-limiting, and dissociation of NAD from the ternary complex is therefore pronounced (Table III). This has been confirmed by the finding of a large deuterium isotope effect on the maximum rates with propan-1-ol and butan-1-ol (*74*).

The conclusion that even with the best substrate, ethanol, dissociation of NAD occurs from the active ternary complex is consistent with the evidence from isotope exchange experiments (*32*), mentioned previously, that the dissociation of coenzyme is not greatly suppressed in the ternary complex compared with the binary complex, in contrast to liver alcohol dehydrogenase. This is also indicated by the initial rate data in another way; ϕ_A/ϕ_0 for the preferred pathway mechanism approximates to the dissociation constant for NAD from the ternary complex (Table I), and is reasonably constant for the three primary alcohols and approximately equal to the dissociation constant of E·NAD, determined independently (*40*).

3. *Other Enzymes*

Lactate dehydrogenases use several 2-hydroxy and 2 keto acids and other analogs as substrates (*75–79*). With the rabbit muscle enzyme (*77*), 2-ketobutyrate and 2-keto-4-hydroxyglutarate give smaller maximum rates than pyruvate, indicating that some step before dissociation of the product coenzyme—perhaps hydride transfer but not, of course, substrate binding (*79*)—is rate-limiting with these poorer substrates; they also have larger K_m values, and much larger ϕ_Q ($= K_m/V$) values, than pyruvate. Fluoropyruvate is almost as good a substrate as pyruvate (*78*). Unfortunately, ϕ_P values or K_m values for NADH (ϕ_P/ϕ_0) were not determined with these alternative substrates; thus, the applicability of a simple ordered mechanism cannot be tested. The apparent maximum rates and K_m values for glyoxylate as a unique substrate for both oxidation and reduction, with different pH optima of 9.3 and 6.9, respectively (*68*), are difficult to interpret in spite of a detailed kinetic study, because the initial rates for both oxidation and reduction were measured only by NADH formation or utilization. The product complexes E·NADH and E·NAD would presumably be oxidized or reduced without dissociation to varying extents at different pH values by the coupled dismutase reac-

74. C. J. Dickenson and F. M. Dickinson, *Biochem. Soc. Trans.* **1**, 1270 (1973).
75. J. S. Nesselbaum, D. E. Packer, and O. Bodansky, *JBC* **239**, 2830 (1964).
76. L. Schatz and H. L. Segal, *JBC* **244**, 4393 (1969).
77. R. S. Lane and E. E. Decker, *Biochemistry* **8**, 2958 (1969).
78. E. H. Eisman, H. A. Lee, and A. D. Winer, *Biochemistry* **4**, 606 (1965).
79. J. Everse and N. O. Kaplan, *Advan. Enzymol.* **37**, 61 (1973).

tions, and a separate study of the aldehyde dehydrogenase reaction by measurements of proton production seems to be needed.

Glyoxylate also seems to be a feeble substrate for α-glycerophosphate dehydrogenase from rabbit muscle, all four kinetic coefficients being several orders of magnitude larger than those for the "natural" substrate, although K_m (ϕ_B/ϕ_0) is nearly the same for both substrates (*41*). An ordered mechanism therefore cannot operate with both substrates. The same conclusion was drawn from studies of glutamate dehydrogenase with norvaline as an alternative substrate to glutamate; in addition, the negative rate cooperativity with respect to NAD and NADP observed with glutamate as substrate was not apparent with norvaline, suggesting that the reactive ternary complex is involved in the negative cooperativity (*11*).

Maximum rates and K_m values were recently reported for seven aldehydes as substrates of rabbit muscle glyceraldehyde-3-phosphate dehydrogenase (*51*). Values for ϕ_A, which might have provided further evidence for the conclusion that the mechanism is ordered, were not reported, however.

For mitochondrial malate dehydrogenase a reciprocating subunit mechanism was suggested by Harada and Wolfe (*80*) on the basis of kinetic studies with ketomalonate as an alternative substrate, and is of particular interest in view of the firm binding of NAD to only one of the two subunits of cytoplasmic malate dehydrogenase in the crystal of the complex (*81*), and subsequent suggestions of a similar mechanism for liver alcohol dehydrogenase (*82*). Isotope exchange experiments provide evidence against this mechanism, however (*34*). The mechanism was proposed on the basis of inhibition studies with hydroxymalonate and the finding that the maximum rate of ketomalonate reduction was only half that of oxaloacetate reduction (*80*). It was assumed that with both substrates the rate of dissociation of the product coenzyme NAD from just one subunit determines the maximum rate, and varies according to the nature of the substrate bound to the other subunit. An alternative explanation is that with ketomalonate, hydride transfer in the ternary complex is rate-limiting, and receives some support from the fact that ϕ_P, calculated from the data reported, is doubled when ketomalonate replaces oxaloacetate as substrate, which is not consistent with a simple ordered mechanism. Stopped-flow "burst" studies and deuterium isotope effects would be of value.

80. K. Harada and R. G. Wolfe, *JBC* **243,** 4131 (1968).
81. D. Tsernoglou, E. Hill, and L. J. Banaszak, *JMB* **72,** 577 (1972).
82. S. A. Bernhard, M. F. Dunn, P. L. Luisi, and P. Schack, *Biochemistry* **9,** 185 (1970).

F. Inhibition and Activation by Substrates, Substrate Analogs, and Products

The interpretation and value of reversible inhibition kinetics have been set forth clearly by Cleland (*8,18*). The uses of product inhibition studies, particularly for distinguishing random and ordered mechanisms, was suggested by Alberty (*7*) and first applied and developed by Fromm (*58,83*). Schwert and his colleagues first gave a detailed theoretical and experimental analysis of inhibition by substrate analogs in their important work on oxamate and oxalate inhibition of bovine heart lactate dehydrogenase (*84*).

1. *Substrate Inhibition and Activation*

Deviations of initial rate data from Eq. (1) toward lower or higher activity at high concentrations of substrate, particularly of nonnucleotide substrate B and Q, are common and have been studied in some detail with liver alcohol dehydrogenase, malate, lactate, and glutamate dehydrogenases and glyceraldehyde-3-phosphate dehydrogenase. The results seem to be explicable in terms of the formation of abortive complexes EPB and EAQ, together with, in some cases, diversion of reaction through EB and EQ; there seem to be no good grounds for assuming nonproductive substrate binding at the active site nor for postulating a substrate binding site distinct from the active site.

Theorell and McKinley-McKee (*85*) first suggested a kinetically significant abortive complex to explain ethanol inhibition of liver alcohol dehydrogenase. The existence of such compounds of enzyme, NADH, and reduced substrate had already been recognized for alcohol, lactate and glutamate dehydrogenases by the effects of the reduced substrates on the fluorescence emission of enzyme-bound NADH (*52,86*), and Fromm (*87*) first detected an inhibitory NAD–pyruvate compound of rabbit muscle lactate dehydrogenase by difference spectroscopy. Analogous compounds of many other dehydrogenases have now been recognized by these and other techniques, including isotope exchange at equilibrium (*34,44*).

The addition of an abortive complex to the preferred pathway mechanism, Eq. (15), seems to account adequately for the data presently

83. H. J. Fromm and D. R. Nelson, *JBC* **237,** 215 (1962).
84. W. B. Novoa, A. D. Winer, A. J. Glaid, and G. W. Schwert, *JBC* **234,** 1143 (1959).
85. H. Theorell and J. S. McKinley-McKee, *Acta Chem. Scand.* **15,** 1834 (1961).
86. A. D. Winer, W. B. Novoa, and G. W. Schwert, *JACS* **79,** 6571 (1957).
87. H. J. Fromm, *BBA* **52,** 199 (1961).

available. It is assumed that the abortive complex may dissociate the product coenzyme P to give an active complex EB.

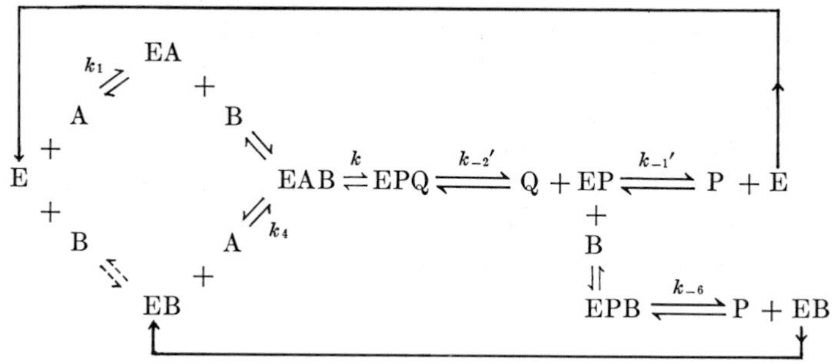

(15)

The initial rate equation is discussed in detail elsewhere (*39,60*), and a qualitative examination of the mechanism will suffice here to indicate its main features. If $k_{-6} < k_{-1}'$, k_{-2}', and k, substrate inhibition will occur at sufficiently large concentrations of B, and will be most pronounced when A is saturating. A large steady-state concentration of EP ($k_{-1}' < k$, k_{-2}') will favor substrate inhibition. If $k_{-6} > k_{-1}'$, substrate activation can occur but only if EP dissociation is the rate-limiting step. (Theorell–Chance mechanism). If $k_{-6} = 0$, EPB is a dead-end complex, and only ϕ_0 in Eq. (1) will be affected by the inhibition which will therefore be uncompetitive with respect to A and complete at high concentration of B. The inhibition will also be uncompetitive when $k_{-6} \neq 0$ provided $k_4 \simeq k_1$, so that A can react with EB as fast as with E. The inhibition will then be partial and nonlinear, however, and this can be shown most simply by a plot of ϕ_0 (or $1/v$ with saturating A) vs. B. If $k_4 < k_1$, there will be an additional source of inhibition competitive with respect to A, i.e., most marked at low concentrations of A.

The substrate inhibition of liver alcohol dehydrogenase by several primary alcohols is partial and uncompetitive with respect to NAD (*39*). The limiting initial rate at high alcohol concentrations may be identified with k_{-6}, in accordance with the smaller dissociation constant for NADH from the abortive complex than from E · NADH shown by equilibrium binding studies (*85*). The magnitude of k_{-6} decreases in the sequence ethanol, propan-1-ol, butan-1-ol, as does the alcohol concentration at which substrate inhibition begins (8–1 m*M*). Since the efficiency of the alcohols as substrates increases in this sequence (Section I,E,1) it may be presumed that the affinity of both E · NAD and E · NADH is greater for the longer alkyl chain. Secondary aliphatic alcohols are much poorer substrates and do not show substrate inhibition up to 0.4 *M*: the affinity of both the enzyme–coenzyme compounds for these alcohols is evidently small, and the steady-state concentration of EP will also be small because hydride transfer is rate-limiting. This correlation between substrate activity and substrate inhibition has also been observed with alternative substrates for glyceraldehyde-3-phosphate dehydrogenase (*51*). With cyclohexanol (*60*), which is almost as good a substrate as ethanol, there is substrate activation of alcohol dehydrogenase when the NAD concentration is also large, consistent with $k_4 > k_{-1}'$, and incidentally providing further evidence for a Theorell–Chance mechanism with this relatively good substrate. With low concentrations of NAD, there is competitive inhibition by cyclohexanol, attributed to $k_4 < k_1$. Substrate inhibition of the reverse reaction with butyraldehyde (> 1 mM) as substrate is due to the "abortive" complex EAQ which is, however, the reactive ternary complex of the aldehyde dehydrogenase reaction also catalyzed by this enzyme (*66*).

Ethanol activation of yeast alcohol dehydrogenase reported earlier (*88*) has not been confirmed in recent initial rate studies (*40*), nor was substrate inhibition observed with very large concentrations of primary alcohols and aldehydes. This may be correlated with other evidence from initial rate studies (*40*) and isotope exchange (*32*) showing that with this enzyme, in contrast with the liver enzyme, binding of substrates to form ternary complexes does not greatly strengthen the binding of coenzymes.

Substrate inhibition of rabbit muscle glyceraldehyde-3-phosphate (GAP) dehydrogenase by GAP (*47,51*), with phosphate or arsenate as acyl acceptor, is uncompetitive toward NAD and apparently complete. This was attributed by Orsi and Cleland (*51*) to the formation of a dead-end abortive complex EPB, in accordance with the deduction, from the identity of the maximum rates with arsenate and phosphate as acceptors and uncompetitive inhibition by inactive substrate analogs, that dissociation of EP is rate-limiting. With arsenate as acceptor, 3-hydroxypropionaldehyde 3-phosphate (HPAP) gave a maximum rate not much less than that for GAP and also showed uncompetitive substrate inhibition. With phosphate as acceptor, HPAP gave a much smaller maximum rate and no substrate inhibition. This indicates that some earlier step than dissociation of NADH is rate-limiting, resulting in a smaller steady-state concentration of E·NADH (*51*). There does not appear to be any direct evidence as yet for the existence of a compound E·NADH·GAP.

Substrate inhibition of bovine liver glutamate dehydrogenase has not been studied in such detail. It is most marked when the NAD(P) concentration is also large (*11*), and is relieved by ADP. This appears to be the reason why ADP activates the enzyme when large glutamate and

88. A. P. Nygaard and H. Theorell, *Acta Chem. Scand.* **9**, 1300 (1955).

NAD(P) concentrations are used (*89,90*); with smaller NAD(P) concentrations, ADP inhibits the enzyme, as does GTP. The abortive complex of enzyme, NAD(P)H and glutamate has been studied by fluorescence measurements; the reduced coenzymes are bound more firmly than in their binary complexes with the enzyme (*91–93*), largely because of a smaller dissociation velocity constant (*92,94*), but coenzyme binding in both the abortive and the active binary complexes is negatively cooperative (*93*).

The reversible inhibition of mitochondrial malate dehydrogenase by oxaloacetate (>0.25 m*M*) has been studied in some detail (*95*). It is noncompetitive toward NADH. A dead-end abortive complex, which would cause uncompetitive inhibition, and a nonproductive binary complex of enzyme and oxaloacetate, to account for the competitive element of the inhibition, were postulated. The mechanism in Eq. (15) involving an active complex EB is an alternative interpretation for these results and for the less pronounced oxaloacetate inhibition observed with the cytoplasmic enzyme (*55*). Cytoplasmic malate dehydrogenases from several tissues and species are less susceptible to oxaloacetate inhibition than the mitochondrial enzymes (*96,97*) and, in the case of the chicken heart enzymes, more susceptible to malate inhibition (*97*). Activation by high concentrations of malate has been observed with the mitochondrial enzyme from bovine heart (*98,99*).

Pyruvate inhibition of lactate dehydrogenases (*3,56*) has attracted much attention and shows unusual features. The fact that the heart isozymes from several species are more susceptible to the inhibition than muscle isozymes, *in vitro,* has been correlated with distinct functional roles for the enzymes in highly aerobic heart muscle and in glycolyzing voluntary muscle, particularly by Kaplan (*79,100*), but it has been disputed whether these differences are significant under physiological condi-

89. D. Pantaloni and M. Iwatsubo, *BBA* **132,** 217 (1967).
90. J. E. Bell, Ph.D. Thesis, Oxford University, 1974.
91. H. F. Fisher and L. L. McGregor, *BBA* **43,** 557 (1960).
92. A. di Franco and M. Iwatsubo, *Eur. J. Biochem.* **30,** 517 (1972).
93. G. Melzi D'Eril and K. Dalziel, "The Behavior of Regulatory Enzymes," p. **33.** Biochem. Soc. Monogr., 1973.
94. A. d'Albis and D. Pantaloni, *Eur. J. Biochem.* **30,** 553 (1972).
95. D. N. Raval and R. G. Wolfe, *Biochemistry* **2,** 220 (1963).
96. A. Delbrück, H. Schunassek, K. Bartsch, and T. Bücher, *Biochem. Z.* **331,** 227 (1959).
97. G. B. Kitto and N. O. Kaplan, *Biochemistry* **5,** 3966 (1966).
98. D. D. Davies and E. Kun, *BJ* **66,** 307 (1957).
99. L. Siegel and S. Englard, *BBA* **54,** 67 (1961).
100. N. O. Kaplan, J. Everse, and J. Admiraal, *Ann. N. Y. Acad. Sci.* **151,** 400 (1968).

tions (*101–103*). There is some uncertainty about the interpretation of the substrate inhibition observed in initial rate measurements. Progress curves for NADH oxidation with large pyruvate concentrations (>0.3 m*M*) show a more rapid decline in rate than can be explained by product inhibition or the reverse reaction; prior incubation of the enzymes with NAD and pyruvate increases the inhibition and results in more linear progress curves (*100,104*). Substrate inhibition is said not to be observed when initial rates are measured during the first 0.1 sec (*103,104*), nor when the pure keto form of pyruvate, the true substrate (*105*), is used in initial rate measurements (*106*). The fall off in rate of NADH oxidation with large pyruvate concentrations correlates well with the relatively slow formation of an enzyme compound characterized by an NADH-like absorption spectrum and decreased protein fluorescence (*87,100,104,107*).

This compound has been isolated and characterized (*108,109*); its crystallographic structure differs from that of the apoenzyme and its NAD compound (*24*). The pyruvate is covalently linked to NAD, and the compound appears to be formed from enolpyruvate by nucleophilic attack of pyruvate C-3 at the nicotinamide C-4 in a relatively slow enzyme-catalyzed reaction formally analogous to lactate oxidation (*106*). It is formed more rapidly—and dissociates to give active enzyme more slowly—with heart isozymes than with muscle isozymes, and the half-time for the formation of the complex from enzyme, NAD, and pyruvate unexpectedly increases with the enzyme concentration (*103*). The explanation seems to be that it is formed primarily from enzyme subunits produced by reversible dissociation of the tetramer in dilute solution (*107*). This does not preclude the existence of the complex in the tissue, and Everse and Kaplan (*79*) have suggested a hypothesis for its involvement in metabolic control.

This work on lactate dehydrogenase underlines the importance of making true initial rate measurements by means of a sufficiently sensitive

101. P. G. W. Plagemann, K. F. Gregory, and F. Wroblewski, *Biochem. Z.* **334,** 37 (1961).
102. E. S. Vesell and P. E. Pool, *Proc. Nat. Acad. Sci. U.S.* **55,** 756 (1966).
103. T. Wuntch, E. S. Vesell, and R. F. Chen, *JBC* **244,** 6100 (1969).
104. H. Gutfreund, R. Cantwell, C. H. McMurray, R. S. Criddle, and G. Hathaway, *BJ* **106,** 683 (1968).
105. F. A. Loewus, T. T. Chen, and B. Vennesland, *JBC* **212,** 787 (1955).
106. C. J. Coulson and B. R. Rabin, *FEBS* (*Fed. Eur. Biochem. Soc.*) *Lett.* **3,** 333 (1969).
107. J. H. Griffin and R. S. Criddle, *Biochemistry* **9,** 1195 (1970).
108. G. Di Sabato, *Biochemistry* **10,** 395 (1971).
109. J. Everse, R. E. Barnett, C. J. R. Thorne, and N. O. Kaplan, *ABB* **143,** 444 (1971).

method (Section II,A), and of the possible effects of tautomerization or hydration of the substrate, and the presence of inactive isomers, on kinetic studies. "Slow" abortive complexes, as distinct from rapidly reversible, noncovalent abortive complexes more analogous to reactive complexes, may be formed by other dehydrogenases, notably malate dehydrogenase (*79*). The NAD–pyrazole complex of liver alcohol dehydrogenase is also covalent (*110*). Reynolds *et al.* (*111*) have shown that the keto form of dihydroxyacetone phosphate is the true substrate of α-glycerophosphate dehydrogenase. Only the free aldehyde form of glyceraldehyde-3-phosphate is active with the dehydrogenase, and its formation from the predominant gem-diol form in aqueous solution may be rate limiting in the catalytic reaction with large enzyme concentrations (*112*). Inactive optical isomers of some substrates, for example, D-glutamate and D-malate, form abortive complexes with greater affinity than the active isomer (*113, 114*); and apparent substrate inhibition as well as undetectable competitive inhibition (*20*) may occur in kinetic studies with racemic mixtures.

2. *Inhibition by Substrate Analogs*

Kinetic studies of reversible inhibition by substrate analogs give evidence of the mode of action of the inhibitor and the types of enzyme–inhibitor complex formed, and estimates of their dissociation constants. The complexes may be isolated and sometimes crystallized. Studies of the stabilities, optical properties, and structures of ternary complexes of enzymes, coenzymes, and substrate analog in particular, as stable models of the catalytically active ternary complexes or of the transition state for hydride transfer (*61,79,109,115–117*), can only be touched upon here; there is direct evidence with several enzymes that the binding of coenzymes is firmer in such complexes than in their binary complexes (*85,93,118*), which supports the indirect, kinetic evidence already mentioned for a similar stabilization in active ternary complexes.

a. Nonnucleotide Substrate Analogs. The simplest hypothesis for reversible inhibition by an analog, I, of substrate B is that it combines

110. H. Theorell and T. Yonetani, *Biochem. Z.* **338,** 537 (1963).
111. S. J. Reynolds, D. W. Yates, and C. I. Pogson, *BJ* **122,** 285 (1971).
112. D. R. Trentham, C. H. McMurray, and C. I. Pogson, *BJ* **114,** 19 (1969).
113. M. Iwatsubo, B. Lecuyer, A. di Franco, and D. Pantaloni, *C. R. Acad. Sci.* **263,** 558 (1966).
114. H. Theorell and T. A. Langan, *Acta Chem. Scand.* **14,** 933 (1960).
115. A. D. Winer and G. W. Schwert *JBC* **234,** 1155 (1959).
116. K. Mosbach, H. Guilford, R. Ohlsson, and M. Scott, *BJ* **127,** 625 (1972).
117. J. H. Holbrook and V. A. Ingram, *BJ* **131,** 729 (1973).
118. K. Dalziel and R. R. Egan, *BJ* **126,** 975 (1972).

with the substrate binding site in EA or EP [Eq. (4)] to form dead-end complexes EAI and EPI with dissociation constants for inhibitor K_{EAI} and K_{EPI}. With the addition of these steps the initial rate equation for an ordered mechanism with or without isomeric coenzyme complexes and for a preferred order mechanism [Eq. (12)] takes the form

$$e/v = \left(\phi_0 + \frac{\mathrm{I}}{k_{-1}'K_{\mathrm{EPI}}}\right) + \frac{\phi_{\mathrm{A}}}{\mathrm{A}} + \frac{\phi_{\mathrm{B}}}{\mathrm{B}}\left(1 + \frac{\mathrm{I}}{K_{\mathrm{EAI}}}\right) + \frac{\phi_{\mathrm{AB}}}{\mathrm{AB}} \qquad (16)$$

In this equation, the four kinetic coefficients have the same physical significance as in the absence of I (Table I). Inspection of Eq. (16) shows that I will be noncompetitive toward B and uncompetitive toward A (*18*). If I combines only with EA ($K_{\mathrm{EPI}} = \infty$), it will be competitive toward B, and again uncompetitive toward A when the concentration of B is finite and, of course, without effect when B is saturating. From initial rate measurements in the presence and absence of I, K_{EAI} may be calculated but not K_{EPI} unless the mechanism is of the Theorell–Chance type for which $\phi_0 = 1/k_{-1}'$.

If the same analog is used in kinetic studies of the reverse reaction, an analogous rate equation applies with K_{EAI} and K_{EPI} interchanged; thus, the latter may then be calculated. If the maximum amount of information is to be obtained, the concentrations of A and B should be varied simultaneously and the four kinetic coefficients separately evaluated as a function of I to test whether the inhibition is linear. This has rarely been done.

The classic work of Schwert and his colleagues on bovine heart lactate dehydrogenase and of Theorell on liver alcohol dehydrogenase has been reviewed (*61,119*). For lactate dehydrogenase, the conclusions were that oxamate forms a strong complex E·NADH·I and a weaker E·NAD·I compound, and is primarily an analog of ketopyruvate, while oxalate showed the reverse behavior (*120*). Direct equilibrium studies of the ternary inhibitor complexes established their stoichiometry and gave dissociation constants in satisfactory agreement with kinetic estimates (*115, 116*), and like the studies with liver alcohol dehydrogenase (*64*) emphasize the different specificities of the oxidized and reduced coenzyme compounds of dehydrogenases. The inhibition kinetics were not completely in accord with Eq. (16) in that the inhibitors were not strictly uncompetitive toward the coenzymes, but noncompetitive, indicating a small effect on ϕ_{A} and/or ϕ_{AB}. This was tentatively attributed to the formation of binary EI compounds, from which kinetically significant EB and EQ compounds might be inferred. A similar conclusion for liver alcohol dehydrogenase was reached by Theorell and McKinley-McKee by a rather more compli-

119. G. W. Schwert and A. D. Winer, "The Enzymes," 2nd ed., Vol. 7, p. 127, 1963.
120. W. B. Novoa and G. W. Schwert, *JBC* **236,** 2150 (1961).

cated interpretation of inhibition studies with fatty acids and amides (*64*). Although there is some evidence from crystallographic data for dogfish lactate dehydrogenase that oxamate can combine at the active site in absence of coenzyme (*121*), the inhibition kinetics are susceptible to other interpretations. In particular, the assumption that the complexes E·NADH·I and E·NAD·I are dead-end and cannot dissociate coenzyme is probably not justified.

Similar but generally less detailed studies have been made with lactate dehydrogenases from other species (*59*) and with a few other dehydrogenases, and can be interpreted by Eq. (16). Inhibition of glyceraldehyde-3-phosphate dehydrogenase by α-glycerophosphate and 3-phosphoglycerate indicates that the former combines rather weakly with both oxidized and reduced coenzyme binary compounds, the latter only with the NADH compound (*51*). Nitrate is competitive toward formate and uncompetitive toward NAD in the formate dehydrogenase reaction, suggesting ternary complex formation with E·NAD but not with E·NADH (*122*). Harada and Wolfe (*80*) showed that ketomalonate is a good substrate for malate dehydrogenase, but its product hydroxymalonate is a very poor substrate in the reverse reaction. Hydroxymalonate is also a competitive inhibitor of ketomalonate and oxaloacetate reduction but an uncompetitive inhibitor of malate dehydrogenation, showing that it forms an abortive ternary complex with E·NADH but that the ternary complex E·NAD·hydroxymalonate formed in the reduction of ketomalonate dissociates practically irreversibly (*80*).

b. Coenzyme Analogs. Inhibition studies with coenzyme analogs and fragments have given valuable information about the binding sites for coenzymes on dehydrogenases and are now being correlated with crystallographic structures. If an inhibitor, I, acts simply by forming an enzyme compound with which the coenzyme, A, cannot combine, the initial rate equation for an ordered or preferred-order mechanism is

$$e/v = \phi_0 + \frac{\phi_A}{A}\left(1 + \frac{I}{K_{EI}}\right) + \frac{\phi_B}{B} + \frac{\phi_{AB}}{AB}\left(1 + \frac{I}{K_{EI}}\right) \qquad (17)$$

The apparent dissociation constant K_{EI} can be calculated from the slope changes in reciprocal plots with A variable and any fixed concentration of B; it will be smaller than the true dissociation constant for EI if the latter can combine with B to form an inactive ternary complex. This must be borne in mind in interpreting inhibition studies with coenzyme fragments of varying size, since the larger the fragment the more likely

121. M. J. Adams, A. Liljas, and M. G. Rossmann, *JMB* **76**, 519 (1973).
122. D. Peacock and D. Boulter, *BJ* **120**, 763 (1970).

it may be that EIB is formed. Direct binding studies are desirable to establish the significance of K_{EI}.

It was shown some years ago that nicotinamide mononucleotide (NMN) does not inhibit liver alcohol dehydrogenase while adenosine diphosphate ribose (ADPR) is a potent competitive inhibitor, $K_{EI} = 7\ \mu M$ in neutral solutions (*20,123,124*), apparently binding much more firmly than NAD ($K_{EA} = 266\ \mu M$). Subsequent work, including binding studies, confirmed these results and showed that adenosine, AMP, and ADP are also competitive inhibitors but bind less firmly than ADPR, adenosine binding being especially weak (*125–127*). Evidently the phosphate and second ribose contribute substantially to the binding energy, but the effect of the latter could partly result from elimination of the additional negative charge on the terminal phosphate of ADP. Adenosine diphosphate ribose is also a strong competitive inhibitor of formate dehydrogenase, $K_{EI} = 12\ \mu M$, and again binds more firmly than NAD, $K_{EA} = 42\ \mu M$ (*122*).

Analogous inhibition studies have been made with lactate dehydrogenases (*128,129*). With the dogfish enzyme, McPherson (*129*) showed that NMN does not bind to the enzyme alone, but mixed inhibition studies indicated that it would bind together with AMP, suggesting that combination of the adenine moiety of NAD causes a conformational change that creates the nicotinamide binding site. Crystallographic studies showed that binding of AMP and ADP at pH < 6, and ADPR at any pH—but not adenosine binding, which is very weak—causes conformational changes similar to those which accompany NAD binding (*129,130*). While the inhibition studies show that ADPR, $K_{EI} = 1.0$ mM, binds more firmly than AMP and ADP, the conclusion that its binding energy is much smaller than that for NAD (*129*) is incorrect because it is based on the fallacy that the K_m value for NAD is a dissociation constant. In fact, the true dissociation constant for the enzyme–NAD compound, 0.5 mM (*131*), is only slightly smaller than that of the ADPR compound. In this respect, and also with regard to the relatively weak

123. K. Dalziel, *Nature* (*London*) **191,** 1098 (1961).
124. K. Dalziel, *BJ* **84,** 241 (1962).
125. H. Theorell and T. Yonetani, *ABB* **105,** 243 (1964).
126. T.-K. Li and B. L. Vallee, *JBC* **239,** 792 (1964).
127. C. H. Reynolds, D. L. Morris, and J. S. McKinley-McKee, *Eur. J. Biochem.* **14,** 4, (1970).
128. H. Geyer, *Hoppe-Seyler's Z. Physiol. Chem.* **348,** 823 (1967).
129. A. McPherson, Jr., *JMB* **51,** 39 (1970).
130. K. Chandrasekhar, A. McPherson, Jr., M. J. Adams, and M. G. Rossmann, *JMB* **76,** 503 (1973).
131. R. A. Stinson and J. H. Holbrook, *BJ* **131,** 719 (1973).

binding of ADPR, these findings for lactate dehydrogenase are in marked contrast to those for liver alcohol dehydrogenase, and are presumably related to differences between the conformational changes which accompany coenzyme binding to the two dehydrogenases. There is no evidence that conformational changes are associated with the binding of ADPR to alcohol dehydrogenase, and the "loop" region involved in such changes with lactate dehydrogenase is absent from the subunit structure (*132*) of the former enzyme.

3. *Product Inhibition*

The main value of product inhibition studies of dehydrogenases has been to distinguish between ordered and random mechanisms and to provide additional kinetic estimates of the dissociation constants of enzyme–coenzyme compounds. On both counts the method has been especially useful for reactions that are essentially irreversible or for other reasons cannot be studied in both directions (*122,133*). It is also in such circumstances that product inhibition studies are most reliable because, as Alberty (*7*) emphasized when proposing the method, with readily reversible reactions it may be difficult to estimate true initial rates with small concentrations of substrates in the presence of a product. The reality of ternary complexes in an ordered mechanism of the Theorell–Chance type has also been demonstrated with several enzymes (*134*) by product inhibition studies.

The theory of product inhibition need not be reproduced here (*8,18,83*). For nearly all the dehydrogenases adequately studied by this method, including alcohol dehydrogenases (*134*), lactate dehydrogenases (*58*), malate dehydrogenases (*55,135*), formate dehydrogenase (*122*), glyceraldehyde-3-phosphate dehydrogenase (*51*), malic enzyme (*133*), and others the results are consistent with an ordered mechanism. For glutamate dehydrogenase (*136*) a random mechanism was indicated, consistent with other evidence (*43,44*). This is also true of NADP-linked isocitrate dehydrogenases (*137,138*), but studies of this reaction are complicated

132. C.-I. Brändén, H. Eklund, B. Nordström, T. Boiwe, G. Söderlund, E. Zeppezauer, I. Ohlsson, and Å. Åkeson, *Proc. Nat. Acad. Sci. U. S.* **70,** 2439 (1973).

132a. H. Eklund, B. Nordström, E. Zeppezauer, G. Söderlund, I. Ohlsson, T. Boiwe, and C.-I. Brändén, *FEBS* (*Fed. Eur. Biochem. Soc.*) *Lett.* **44,** 200 (1974).

133. R. Y. Hsu, H. A. Lardy, and W. W. Cleland, *JBC* **242,** 5207 and 5315 (1967).

134. C. C. Wratten and W. W. Cleland, *Biochemistry* **2,** 935 (1963).

135. D. N. Raval and R. G. Wolfe, *Biochemistry* **1,** 1112 (1962).

136. P. C. Engel, Ph.D. Thesis, Oxford University, 1968.

137. J. S. Wicken, A. E. Chung, and J. S. Franzen, *Biochemistry* **11,** 4766 (1972).

138. J. C. Londesborough and K. Dalziel, *in* "Pyridine Nucleotide-Dependent Dehydrogenases" (H. E. Sund, ed.), p. 315. Springer-Verlag, Berlin and New York, 1970.

by the fact that Mg^{2+}- or Mn^{2+}-isocitrate, and dissolved CO_2 not HCO_3^-, are the true substrates of the enzyme, as was shown by kinetic methods (*138,139*).

G. Cooperative Rate Effects

The initial rate behavior of a few dehydrogenases shows more extensive deviations from Eq. (1) than can be accounted for by substrate inhibition or activation confined to the high concentration region; perhaps the number will increase as more detailed and precise measurements are made, particularly at physiological temperature and pH. Double reciprocal plots may be concave upward or concave downward, and in the former case plots of v vs. A or B are sigmoidal. Such behavior may be called positive and negative rate cooperativity, regardless of its physical basis, since the consequences are to make the rate more sensitive and less sensitive, respectively, to changes of substrate concentration near the half-saturation level (*140*). The best-documented examples among the dehydrogenases seem to be NAD-linked isocitrate dehydrogenases (*141–144*), glyceraldehyde-3-phosphate dehydrogenases (*145–148*), and bovine liver glutamate dehydrogenase (*149*), oligomeric enzymes that have three substrates in one direction of reaction.

Since the initial rate equation for a random mechanism, Eq. (13), is not of the linear form of Eq. (1), it can account for rate cooperativity with appropriate values for the rate constants (*6,30,149*) and has been suggested for isocitrate dehydrogenases (*142,150*). It does not require that there be more than one active center in the enzyme molecule nor cooperative equilibrium binding of substrates or modifiers. An alternative to this purely kinetic explanation is that there are two or more active

139. K. Dalziel and J. C. Londesborough, *BJ* **110**, 223 (1968).
140. K. Dalziel, *Symp. Soc. Exp. Biol.* **27**, 21 (1973).
141. J. A. Hathaway and D. E. Atkinson, *JBC* **238**, 2875 (1963); G. D. Kuehn, L. D. Barnes, and D. E. Atkinson, *Biochemistry* **10**, 3945 (1971); L. D. Barnes, A. J. McGuire, and D. E. Atkinson, *ibid.* **11**, 4322 (1972).
142. B. D. Sanwal, M. W. Zink, and C. S. Stachow, *JBC* **239**, 1547 (1964); B. D. Sanwal and R. A. Cook, *Biochemistry* **5**, 886 (1966).
143. M. Klingenberg, H. Goebell, and G. Wenske, *Biochem. Z.* **341**, 199 (1965).
144. A. M. Stein, S. K. Kirkman, and J. H. Stein, *Biochemistry* **6**, 3197 (1967).
145. K. Kirschner, M. Eigen, R. Bittmann, and B. Voigt, *Proc. Nat. Acad. Sci. U. S.* **56**, 1661 (1966); R. A. Cook and D. E. Koshland, Jr., *Biochemistry* **9**, 3337 (1970).
146. A. Conway and D. E. Koshland, Jr., *Biochemistry* **7**, 4011 (1968).
147. S. F. Velick, J. P. Baggott, and J. M. Sturtevant, *Biochemistry* **10**, 779 (1971).
148. C. H. Smith and S. F. Velick, *JBC* **247**, 273 (1972).
149. K. Dalziel and P. C. Engel, *FEBS* (*Fed. Eur. Biochem. Soc.*) *Lett.* **1**, 346 (1968).
150. W. Ferdinand, *BJ* **98**, 278 (1966).

centers in the molecule, and that while reaction at each may satisfy Eq. (1), they are not identical or do not react independently of one another as was assumed in deriving the rate equations in Section II,B. Nonidentical multiple active centers will explain negative but not positive rate cooperativity; interactions between active centers will explain either. Cooperative effects in the equilibrium binding of substrates or modifiers are to be expected from this "allosteric" model, and have been found with the dehydrogenases mentioned. While equilibrium binding can be rigorously described on the basis of allosteric models, it is perhaps fair to say that no experimentally useful and at the same time theoretically sound initial rate equations have been derived for these models for a two-substrate reaction without restrictive assumptions such as the rapid equilibrium assumption (*151,152*). The latter cannot be applied to an ordered mechanism (*21*), and for this reason the conclusion from initial rate measurements that yeast isocitrate dehydrogenase operates by an ordered mechanism (*152*) is open to question.

The mitochondrial NAD-isocitrate dehydrogenases show positive rate cooperativity with respect to isocitrate. They are effectively unidirectional catalysts, unlike the NADP-linked enzymes (presumably because their K_m values for CO_2 are large); the positive rate cooperativity and the common activation by AMP or ADP and citrate, which decreases the rate cooperativity, can be rationalized as a typical control mechanism for the Krebs cycle acting at an irreversible step (*141,153*). The rate cooperativity and the effects of citrate and nucleotides have been variously interpreted. For the *Neurospora* enzyme (*142*), a kinetic interpretation with a separate activating site for isocitrate and citrate, and a change from a random to an ordered mechanism in the presence of AMP, is suggested. For the enzymes from yeast (*141*) and Ehrlich ascites carcinoma (*144*), kinetic data were interpreted by the empirical Hill equation on the basis of the concerted symmetry model of Monod *et al.* (*154*). Detailed equilibrium binding studies with substrates and activators for the yeast enzyme (*141*) generally support this interpretation, and show positively cooperative binding of isocitrate in the presence of AMP. The complexities of the isocitrate dehydrogenase reaction were mentioned earlier; in most kinetic studies the effects of chelation of the essential metal ion by isocitrate, by α-ketoglutarate in product inhibition studies, and especially by citrate, have not been adequately considered.

Glutamate dehydrogenase and rabbit muscle glyceraldehyde-3-phos-

151. C. Frieden, *JBC* **242**, 4045 (1967).
152. K. E. Kirtley and D. E. Koshland, Jr., *JBC* **242**, 4192 (1967).
153. R. F. Chen and G. W. E. Plaut, *Biochemistry* **2**, 1023 (1963).
154. J. Monod, J. Wyman, and J. P. Changeux, *JMB* **12**, 88 (1965).

phate dehydrogenase exhibit negative cooperativity. It may be significant that, in contrast to the positively cooperative NAD-isocitrate dehydrogenase, the reactions catalyzed by these enzymes, present in large concentrations in the mitochondrion and cytoplasm, respectively, are close to equilibrium in the physiological steady state in several tissues (*140,148,155*). Negative cooperativity between the six active centers of the glutamate dehydrogenase oligomer was suggested to explain the peculiar "coenzyme activation" by both NAD and NADP, manifest in initial rate studies by several discontinuites in double reciprocal plots (*149*). Because this behavior was observed only with near saturating concentrations of glutamate, and not with the next best substrate norvaline, it appeared that the active ternary complex of coenzyme and dicarboxylic acid was the site of the interactions (*11*), and this hypothesis was substantiated by evidence for negatively cooperative equilibrium binding of NAD(P) in ternary complexes with the analog glutarate (*118*). A sudden conformational change in these complexes near half-saturation of the oligomer with NAD(P) was demonstrated by CD measurements and by a change of fluorescence quantum yield of NADH acting as a tracer probe (*156*). This seems to eliminate nonidentical active centers as the source of the complexities, to establish that NAD(P) binding induces conformational changes in other, unliganded subunits of the oligomer, and to suggest that the latter may function as a dimer of trimers. The allosteric inhibitor ADP normalizes the initial rate, equilibrium binding, and conformational change data as a function of oxidized coenzyme concentration (*156*).

Evidence for negative cooperativity with rabbit muscle glyceraldehyde-3-phosphate dehydrogenase has come primarily from equilibrium binding studies with NAD (*146*). Some evidence for negative rate cooperativity in the oxidative phosphorylation reaction was also given; this was not detected in earlier, more detailed studies (*47,49*), nor was it detected by Smith and Velick (*148*) in recent initial rate studies with the rabbit muscle and rabbit liver enzymes at 37° and pH 7.4. This important work (*148*) has clarified several aspects of the complex mechanism of this enzyme and of the probable metabolic significance of negative cooperativity. For the reductive dephosphorylation reaction, initial rate measurements with a sensitive fluorimetric method show positive rate cooperativity with respect to 3-phosphoglyceroyl phosphate, especially at high NADH concentrations. NAD activates the reductive dephosphorylation reaction with low acyl phosphate concentrations by acting as a hetero-

155. H. A. Krebs and R. L. Veech, *in* "The Energy Level and Metabolic Control in Mitochondria" (S. Papa *et al.*, eds.), p. 329. Adriatica Editrice, Bari, 1969.
156. J. E. Bell and K. Dalziel, *BBA* **309,** 237 (1973).

tropic effector and combining with nonacylated active centers, thereby normalizing the rate response to acyl phosphate. At high concenrtations of the latter, NAD inhibits by competing with NADH. The binding of NAD to nonacylated sites is favored over that of NADH, but the reverse is true at acylated sites in the tetramer. These effects are attributed to both steric factors and conformational changes. The regulatory effects of NAD binding in promoting acylation and reductive dephosphorylation at other subunits in the oligomer, through conformational changes, are seen by Smith and Velick as a means by which the enzyme may function in the direction of gluconeogenesis in the aerobic hepatocyte, in the face of a large NAD/NADH ratio and aldehyde concentration and a low acyl phosphate concentration.

Yeast glyceraldehyde-3-phosphate dehydrogenase also apparently shows both positive and negative cooperativity in kinetic and equilibrium studies of NAD binding (Section III) but only slight evidence of positive rate cooperativity in the overall oxidative phosphorylation reaction at 40° and pH 8.5 (*145*). There is much evidence from thiol group reactivity and acylation studies for "half of the sites" reactivity and a reciprocating dimer mechanism of catalysis for this tetrameric enzyme, which implies negative cooperativity (*157*).

III. Equilibrium and Kinetics of Enzyme–Coenzyme Reactions

A. Dissociation Constants of Enzyme–Coenzyme Compounds

1. *Dissociation Constants from Initial Rate Measurements and Binding Studies*

For the four mechanisms described in Section II,B the dissociation constants of the binary coenzyme complexes can be calculated from kinetic coefficients for the overall reaction by Eq. (8). Such estimates are compared in Tables IV–VI (*158–162a*) with direct estimates from studies of enzyme–coenzyme equilibria at similar pH values and temperatures. In comparing these values, it must be borne in mind that ϕ_{AB} and ϕ_{PQ} are estimated from initial rates with small concentrations of coenzyme

157. W. B. Stallcup and D. E. Koshland, Jr., *JMB* **80**, 41, 63, and 77 (1973).
158. S. Anderson and G. Weber, *Biochemistry* **4**, 1948 (1965).
159. S. F. Velick, *JBC* **233**, 1455 (1958); A. D. Winer, G. W. Schwert, and B. R. Miller, *ibid.* **234**, 1149 (1959).
160. H. J. Fromm, *JBC* **238**, 2938 (1963).
161. M. Silverberg and K. Dalziel, "Methods in Enzymology, Vol. 41 (in press).
162. J. J. M. de Vjilder and E. C. Slater, *BBA* **167**, 23 (1968).
162a. J. E. Bell and K. Dalziel, *BBA* (in press) (1975).

and substrate, and that ϕ_{PQ} in particular is often subject to considerable error (Section II,A). Competing nucleotide impurities in the coenzymes, particularly NADH-like impurities and ADPR in NAD preparations, may result in underestimates of dissociation constants (*10,29*). The binding studies were mostly made by measurements of protein fluorescence quenching, NAD(P)H fluorescence enhancement, and competition of NAD with NADH. The molar emission of the compound was not always adequately established, nor was a sufficiently wide range of saturation to establish simple binding always used. These points are discussed by several authors (*93,131,158*).

With these considerations in mind, it may be concluded that for liver alcohol dehydrogenase and α-glycerophosphate dehydrogenase over a wide pH range, and for yeast alcohol dehydrogenase at pH 7.0, the kinetic and direct estimates of $K_{E \cdot NAD}$ and $K_{E \cdot NADH}$ agree very well. For most of the other enzymes except glyceraldehyde-3-phosphate dehydrogenase (Table VI) the two estimates for $K_{E \cdot NAD}$ are probably not significantly different. However, for the NADH compounds of bovine heart lactate dehydrogenase and cytoplasmic malate dehydrogenase, and the NADPH compound of malic enzyme, the few direct estimates of the dissociation constants appear to be significantly smaller than the kinetic estimates. Isomeric coenzyme compounds will not account for this discrepancy (Section II,B,2). A possible explanation is a significant contribution to the overall reaction through the substrate complex EQ, which would alter the physical significance of ϕ_Q without necessarily causing detectable deviations from linearity in reciprocal plots of initial rate data (Section II,B,4). This was suggested for malate dehydrogenase (*55*).

The simple dehydrogenases (Table IV) bind NADH much more firmly than NAD. Crystal structure determinations have not yet revealed the chemical basis for this; for liver alcohol dehydrogenase electrostatic repulsion between the positively charged nicotinamide N-1 of NAD and the active center zinc atom remains a possibility which may soon be tested (*85,132*), but in dogfish lactate dehydrogenase a glutamate residue is available to balance this charge on NAD in the covalent complex with pyruvate (*24*). This difference of affinity for NAD and NADH means that the equilibrium constant for the reduction of E·NAD to E·NADH by substrate is greater by a factor of $K_{E \cdot NAD}/K_{E \cdot NADH}$ than that for the enzyme-catalyzed reduction of NAD to NADH (*1*). The effect of pH change on this ratio is discussed below. By contrast, the NADP-linked oxidative decarboxylases (Table V) and glyceraldehyde-3-phosphate dehydrogenase (Table VI) bind their oxidized and reduced coenzymes with similar affinities. The apparent equilibrium constants at physiological pH and

TABLE IV
DISSOCIATION CONSTANTS FOR COENZYME COMPOUNDS OF NAD-LINKED DEHYDROGENASES

		From initial rates[a]			From binding studies		
Enzymes	pH	$K_{E\cdot NAD}$ (μM)	$K_{E\cdot NADH}$ (μM)	Ref.	$K_{E\cdot NAD}$ (μM)	$K_{E\cdot NADH}$ (μM)	Ref.
Alcohol dehydrogenase (horse liver)	6.0	250	0.40	*10,29*	266	0.23	*64,158*
	7.0	110	0.42		160	0.31, 0.29	
	8.0	26	0.62		51	0.41	
	9.0	9	0.88		12	0.66	
	10.0	—	—		10	5.0	
Alcohol dehydrogenase (yeast)	7.0	325	12.5	*40*	350	11.0	*72,73*
α-Glycerophosphate dehydrogenase (rabbit muscle)	6.0	7.8	0.03	*41*	9.2	0.02	*41*
	7.0	320	1.3		260	0.63	
	8.0	410	6.5		325	1.82	
	9.0	730	3.2		460	2.05	
Malate dehydrogenase (mitochondrial, pig heart)	6.0–8.0	750	5.0	*54*	280	1.0	*114*
	9.0	730	9.5				
	10.0	800	48				
Malate dehydrogenase (cytoplasmic, bovine heart)	8.5	410	10.0	*55*	—	0.5	*55*
Lactate dehydrogenase (bovine heart)	6.0–8.0	200	3	*56,57*	320	0.25–0.39	*158,159*
	9.0	200	11				
	9.8	1000	24			1.9	
Lactate dehydrogenase (pig heart)	6.0	—	—		200	0.6	*131*
	7.2	—	—		300	0.9	
	8.5	—	—		500	1.8	
	10	—	—		—	8.0	

Lactate dehydrogenase (pig muscle)	6.0	—	—		400	1.7	*131*
	7.2	—	—		500	3.3[b]	
	8.5	—	—		600	5.7	
	10.0	—	—		—	11.0	
Lactate dehydrogenase (rabbit muscle)	6.9	630	8	*59*	910	2.1–3.2	*160*
	7.2	—	—		—	3.5	*131*
Formate dehydrogenase (mung bean)	8.0	42	2.5	*122*	—	—	
Glutamate dehydrogenase (bovine liver)	7.0	425	8	*11,43*	500	—	*118*
	8.0	500	14		—	2–20[c]	*93*

[a] Calculated from Eq. (8).
[b] Similar values were obtained for rabbit, dogfish, and bovine muscle lactate dehydrogenases (*131*).
[c] A wide range of values has been reported (*92*) and the binding is negatively cooperative (*93*).

TABLE V

DISSOCIATION CONSTANTS FOR COENZYME COMPOUNDS OF NADP-LINKED DEHYDROGENASES

Enzyme	pH	From initial rates[a]			From binding studies		
		$K_{E\cdot NADP}$ (μM)	$K_{E\cdot NADPH}$ (μM)	Ref.	$K_{E\cdot NADP}$ (μM)	$K_{E\cdot NADPH}$ (μM)	Ref.
Glutamate dehydrogenase (bovine liver)	7.0	750	32	*11,43*	~2500	—	*118*
	8.0	480	4		—	2–20[b]	*92*
Malic enzyme (pigeon liver)	7.0	0.96	3[c]	*133*	0.97	0.75	*133*
Isocitrate dehydrogenase (bovine heart)	7.0	0.1	0.5	*138*	~2.0	0.2	*138*
Isocitrate dehydrogenase (*A. vinelandii*)	7.2	18	24	*137*	27	41	*137*
Phosphogluconate dehydrogenase (sheep liver)	7.0	37	9[c]	*161*		5.7	*161*

[a] Calculated from Eq. (8).

[b] A wide range of values has been reported (*92*), and the binding is negatively cooperative (*90*).

[c] From product inhibition experiments.

TABLE VI

DISSOCIATION CONSTANTS FOR COENZYME COMPOUNDS OF RABBIT MUSCLE GLYCERALDEHYDE-3-PHOSPHATE DEHYDROGENASE

		From initial rates[a]			From binding studies								
					NAD				NADH				
Temp.	pH	$K_{E \cdot NAD}$ (μM)	$K_{E \cdot NADH}$ (μM)	Ref.	K_1 (μM)	K_2 (μM)	K_3 (μM)	K_4 (μM)	K_1 (μM)	K_2 (μM)	K_3 (μM)	K_4 (μM)	Ref.
20–25°	8.2				<0.05	<0.05	4.0	35					*162*
19°	7.4	100	0.3	*47*	0.023	0.20	0.55						*147*
28°	7.4				0.37	0.37	0.37	36					*147*
25°	7.6				0.01	0.09	4.0	36	0.01	0.06	4.0	36	*162a*

[a] From product inhibition experiments.

CO_2 concentration for the oxidative decarboxylation reactions are also much larger than those for simple dehydrogenations (*163*). These facts can be related to the different redox states of NAD and NADP in the cytoplasm, and to the different metabolic roles of NADH and NADPH (*155,163*)—and teleologically to the reason for the specificity of the cytoplasmic oxidative decarboxylases for NADP, perhaps.

2. *pH Effects and the Role of Histidine*

For liver alcohol dehydrogenase, mitochondrial malate dehydrogenase, and the lactate dehydrogenases, $K_{E \cdot NADH}$ increases from pH 8 to pH 10 in a similar manner (Table IV). The suggestion that in alcohol dehydrogenase ionization of the essential thiol group may be responsible (*64*) seems to be inconsistent with the latest crystallographic data (*132a*), which shows that this group, Cys-46, is a zinc ligand and is not involved in coenzyme binding. Deprotonation of Arg-47 and fission of its salt linkage with the pyrophosphate moiety of the coenzyme may be responsible for the pH effect (*132a*). For the lactate dehydrogenases, also, both crystallographic studies of the dogfish enzyme (*24*) and chemical modification of the pig heart enzyme (*164*) show that the essential thiol is not involved in coenzyme binding. Although coenzyme binding to lactate dehydrogenase has frequently been described as essentially independent of pH in the range pH 5–10 (*57*), the decreased stability of the NADH compound above pH 8 indicated in earlier studies of the bovine heart enzyme has been confirmed in the detailed binding studies with the enzymes from several species by Stinson and Holbrook (*131*). They suggested that ionization of tyrosine-85 in the dogfish enzyme, thought to furnish a hydrogen bond to adenine N1 of the coenzyme (*24*), may be responsible. The significantly firmer binding of NADH to the pig heart enzyme than to the muscle enzymes is noteworthy (Table IV); the lack of detailed kinetic studies with a muscle enzyme is unfortunate in view of the structural data available and the anomalies in kinetic data for the bovine heart enzyme.

There is no uniformity in the effect of pH on the dissociation constants of the NAD compounds. The pronounced decrease of $K_{E \cdot NAD}$ for liver alcohol dehydrogenase from pH 6.0 to pH 10.0 was attributed to ionization of a water molecule coordinated to the active center zinc atom and Coulombic interaction with the nicotinamide N-1 (*64*). The absence of a similar pH effect with the other dehydrogenases studied so far is understandable on this basis. Structural studies of the compounds of liver alcohol dehydrogenase with ADPR and 1:10-phenanthroline, competitive

163. R. L. Veech, L. V. Eggleston, and H. A. Krebs, *BJ* **115**, 609 (1969).
164. J. H. Holbrook and R. A. Stinson, *BJ* **120**, 289 (1970).

inhibitors of the coenzyme that can combine simultaneously with the enzyme (*165*), suggest that the nicotinamide of NAD may bind near the zinc atom (*132*).

Another contrast between liver alcohol dehydrogenase and the other dehydrogenases in Table IV is the marked decrease of the ratio $K_{E \cdot NAD}/K_{E \cdot NADH}$ from pH 6.0 to pH 10.0. This means, from Eqs. (9) to (11), that the apparent equilibrium constant for the oxidation of E · NAD by substrate increases much less with increase of pH than it should if a proton were liberated stoichiometrically. It also means that there are two coenzyme-linked acid groups in the enzyme which have higher pK values in E · NADH than in E · NAD, and that the proton liberated from the substrate is largely counterbalanced by protonation of these groups in E·NADH, particularly at pH > 7.5 (*29*). From these considerations, and the changes with pH of the rate constants for the formation and dissociation of the enzyme–coenzyme compounds (Section III,B), it was suggested that these enzyme groups may be involved as a proton source and sink (*29*). However, stopped-flow "burst" studies by Brooks *et al.* (*166*) indicate that the rate constant for hydride transfer in the ternary complex E · NAD · C_2H_5OH is also controlled by ionization of a group with $pK = 6.4$ in this complex. This also seems to be true with yeast alcohol dehydrogenase, as indicated by pH effects on the maximum rates of propan-1-ol and butan-1-ol oxidation (*74*), which are determined by the rate of hydride transfer (Section II,E). Moreover, chemical modification of a histidine residue in yeast alcohol dehydrogenase does not prevent the binding of NADH, but does inactivate the enzyme and prevent the combination of acetamide with E · NADH, suggesting the need for a protonated histidine residue for ternary complex formation and catalysis (*74*). There is a histidine residue in both alcohol dehydrogenases five residues away from the reactive cysteine (*167*).

The constancy of $K_{E \cdot NAD}/K_{E \cdot NADH}$ at pH 6.0–8.5 for other dehydrogenases in Table IV, notably the heart lactate dehydrogenases, means that a proton is released stoichiometrically in the reduction of E · NAD to E · NADH by substrate, just as in the overall reaction [Eq. (11)]. A histidine residue as proton source and sink was first suggested by Schwert *et al.* (*56,57*) and can now be identified with the single essential histidine per subunit (*168*) in the pig heart enzyme and histidine-195 in the dogfish enzyme. The constancy of $K_{E \cdot NAD}/K_{E \cdot NADH}$ shows that the

165. T. Yonetani and H. Theorell, *ABB* **106,** 243 (1964).
166. R. L. Brooks, J. D. Shore, and H. Gutfreund, *JBC* **247,** 2382 (1972).
167. I. Harris, *Nature* (*London*) **203,** 30 (1964); H. Jörnvall, *Proc. Nat. Acad. Sci. U. S.* **70,** 2295 (1973).
168. C. J. Woenckaus, J. Berghäuser, R. Jeck, and E. Schättle, *Hoppe-Seyler's Z. Physiol. Chem.* **353,** 559 (1972).

mechanism cannot involve different ionization states of this histidine in E·NAD and E·NADH, as originally suggested (*57*), but is consistent with different pK values for this group in the two active ternary complexes. The proton transferred from lactate to this histidine in the hydride transfer step is presumably released when pyruvate dissociates from the ternary complex, as suggested by Holbrook and Gutfreund (*169*). Strong experimental evidence to support this scheme is the finding that protonation of a histidine group is needed for the binding of the pyruvate analog oxamate to E·NADH (*117*) and that proton release in stopped-flow "burst" studies of lactate oxidation accompanies a later, slower step than hydride transfer (*169*). It remains to be discovered whether this is a general mechanism for dehydrogenases, and what is the structural basis for the different changes of the ionization states of histidine which must accompany binding of the appropriate substrate to E·NAD and E·NADH.

3. *Cooperative Effects*

Negative cooperativity in the binding of coenzymes to glutamate dehydrogenase, and the accompanying conformational changes, were discussed earlier in relation to cooperative rate effects. With NAD(P), cooperative binding has been detected only in a ternary complex with glutarate (*118*), whereas with NAD(P)H it is apparent in both the abortive ternary complex with glutamate and the binary complex (*93*). Correlation between these findings and the complex initial rate behavior has not yet been achieved. An important question, which applies also to rabbit muscle glyceraldehyde-3-phosphate dehydrogenase and should be open to experimental test, is whether there are heterotropic effects when oxidized and reduced coenzymes are bound to the same oligomer.

The binding of NAD to rabbit muscle glyceraldehyde-3-phosphate dehydrogenase can be described by four apparent dissociation constants corresponding to sequential binding to the four subunits with decreasing affinity (Table VI). There is some disagreement between the results of different workers, particularly for the affinity for the third coenzyme molecule. Velick *et al.* (*147*) studied the binding by protein fluorescence and calorimetric measurements, obtaining good agreement by the two methods, and found that the affinities for the first three coenzyme molecules converged as the temperature was increased. They present thermodynamic parameters for the binding at each site. The large difference between these dissociation constants for NAD and that derived from

169. R. A. Stinson and J. H. Holbrook, *BJ.* **131,** 739 (1973); J. H. Holbrook and H. Gutfreund, *FEBS (Fed. Eur. Biochem. Soc.) Lett.* **31,** 157 (1973).

product inhibition studies is not surprising in view of the complexity of the reaction (*148*). The binding of NADH is also negatively cooperative (*162a*) and can be described by four dissociation constants similar in magnitude to those for NAD (Table VI). Again, these results have not yet been fully integrated with the overall reaction kinetics (*148,162*). The work of Velick *et al.* (*147,148*) on this enzyme, and of Kirschner *et al.* (*145*) on the yeast enzyme, underlines the importance of studying the kinetics and coenzyme binding properties of dehydrogenases at physiological temperature. The closer proximity of the four coenzyme binding sites in lobster glyceraldehyde-3-phosphate dehydrogenase, compared with lactate dehydrogenase, revealed in recent crystallographic studies (*170*) may be significant in relation to the negative cooperativity.

B. Kinetics of Enzyme–Coenzyme Reactions

1. *Velocity Constants from Initial Rate Measurements and Fast Reaction Studies*

Velocity constants for enzyme–coenzyme reactions calculated from initial rate data by means of Eqs. (5) and (6), assuming a simple ordered mechanism, are compared in Table VII (*171–175*) with direct estimates from stopped-flow or temperature-jump studies of the reactions under similar conditions of pH and temperature (20–28°). For liver alcohol dehydrogenase stopped-flow studies by Shore *et al.* (*35,70,172*), and more precise initial rate data, confirm within much closer limits the early conclusions of Theorell *et al.* (*1,4*) that at pH 7.0 the mechanism is a simple ordered one, and that dissociation of the product coenzyme is rate-limiting for the overall reaction in each direction. Together with the maximum rate relations, dissociation constants, and studies with alternative substrates, these results justify the interpretation of pH effects (*29*) and temperature effects (*22*) on kinetic coefficients in terms of effects on the rate constants for the enzyme–coenzyme reactions. The few direct estimates of these rate constants at other temperatures (*171,172*) also agree well with those calculated from initial rate measurements; interpretation of the latter by the absolute reaction theory indicates a large negative

170. H. C. Watson, E. Duée, and W. D. Mercer, *Nature* (*London*) *New Biol.* **240,** 130 (1972); M. Buehner, G. C. Ford, D. Moras, K. W. Olsen, and M. G. Rossmann, *JMB* **82,** 563 (1974).
171. G. Geraci and Q. H. Gibson, *JBC* **242,** 4275 (1967).
172. J. D. Shore, *Biochemistry* **8,** 1588 (1969).
173. G. Czerlinsky and G. Schreck, *Biochemistry* **3,** 89 (1963).
174. H. D.'A. Heck, *JBC* **244,** 4375 (1969).
175. R. A. Stinson and H. Gutfreund, *BJ* **121,** 235 (1971).

TABLE VII

Velocity Constants for Enzyme–Coenzyme Reactions

		From initial rates[a]							Direct estimates				
		NAD			NADH				NAD		NADH		
Enzyme	pH	k_1 (μM^{-1} sec^{-1})	k_{-1} (sec^{-1})	$1/\phi_0'$ (sec^{-1})	k_1' (μM^{-1} sec^{-1})	k_{-1}' (sec^{-1})	$1/\phi_0$ (sec^{-1})	Ref.	k_1 (μM^{-1} sec^{-1})	k_{-1} (sec^{-1})	k_1' (μM^{-1} sec^{-1})	k_{-1}' (sec^{-1})	Ref.
Alcohol dehydrogenase[b]	6.0	0.4	100	125	8.7	3.5	1.6	*29*					
(horse liver)	7.0	0.9	99	125	10.0	4.2	2.7		0.84	135[c]	17[d]	3.0[d]	*70,35*
	8.0	1.1	29	48	7.6	4.7	3.2						
	9.0	0.8	7.6	7.6	4.5	4.0	3.8						
Alcohol dehydrogenase[e] (yeast)	7.1	4.2	1354	3850	40	500	455	*40*					
α-Glycerophosphate	6.0	0.03	0.27	5.6	13	0.4	0.02	*41*	0.56	3.9	25	0.5[c]	*41*
dehydrogenase	7.0	0.17	55	38	32	42	3.1				26	15.4[c]	
(rabbit muscle)	8.0	1.33	545	127	16	102	39				52	95[c]	
	9.0	2.00	1465	119	13	45	63				94	192[c]	
Malate dehydrogenase[f]	6.0	0.2	150	133	10	28	50	*54*				5	
(pig heart,	7.0	0.42	315	330	23	56	84					45	*173*
mitochondrial)	8.0	0.84	620	580	33	152	167					167	
	9.0	1.0	760	630	28	208	267					230	
Malate dehydrogenase (bovine heart, cytoplasmic)	8.5	0.93	377	335	145	1470	46	*55*					
Lactate dehydrogenase[g]	6.0	0.13	32	260	16	42	18	*56,57*					
(bovine heart)	7.0	1.0	100	230	20	36	50						
	8.0	2.0	400	190	18	25	80						
	9.0	2.5	500	180	15	150	100						
Lactate dehydrogenase[h]	6.0						8.0	*37*			33	50, 32	*37,174*
(pig heart)	7.0						21					50	
	8.0						36				54	50,	
	8.5						41(100)[g]					50	
Lactate dehydrogenase (pig muscle)	7.2						35	*175*			63	380	*175*

[a] The velocity constants are calculated on the assumption of a simple ordered mechanism by Eqs. (*5*) and (*6*). $1/\phi_0'$ and $1/\phi_0$ are the maximum specific rates of NADH oxidation and NAD reduction, respectively, in the overall reaction.

[b] With ethanol and acetaldehyde as substrates. Similar values were obtained with propan-1-ol and butan-1-ol and the corresponding aldehydes (*29*).

[c] These values were calculated from those for k_1 or k_1' and the dissociation constants (Table IV).

[d] These values were obtained at 23°. Other values are, for k_1', 5.0 μM^{-1} sec^{-1} at 3° (*171*) and 6.9 μM^{-1} sec^{-1} at 8° (*172*), and for k_{-1}', 1.7 sec^{-1} at 8° (*172*).

[e] With ethanol and acetaldehyde as substrates.

[f] The initial rate measurements were made in Tris-acetate buffers, and the direct estimates by the temperature-jump method in phosphate buffers.

[g] In Tris-chloride buffers.

[h] In phosphate buffers.

entropy of activation for the dissociation of both coenzyme compounds, which is largely responsible for their stabilities, the association reactions having the normal activation entropy for a bimolecular reaction (*22*). More studies of the molecular kinetics and thermodynamics of enzyme–coenzyme compounds would be of value.

No direct kinetic studies of the reactions of yeast alcohol dehydrogenase with coenzymes have been reported, but the rate constants from initial rate measurements serve for comparison with the liver enzyme. Dickinson and Monger (*40*) pointed out that the much greater maximum rates for the yeast enzyme largely result from the greater rates of dissociation of the NAD and NADH compounds, and that the rate of hydride transfer from ethanol in the ternary complex, which has not been measured, must also be much greater than with the liver enzyme [130 sec^{-1} (*63*)]. As far as turnover is concerned, yeast alcohol dehydrogenase with its natural substrate is a much better enzyme than liver alcohol dehydrogenase or indeed than most other simple mammalian dehydrogenases, although its K_m values for coenzymes and substrates are much larger than those for the liver enzyme.

The direct estimates of k_1 and k_1' for α-glycerophosphate dehydrogenase confirm other evidence previously discussed that the mechanism is not a simple ordered one. The dissociation velocity constant for the NADH compound is significantly larger than the maximum rate of α-glycerophosphate oxidation at all pH values, and hydride transfer or some other step must be rate-limiting. For mitochondrial malate dehydrogenase, on the other hand, NADH dissociation is rate-limiting at pH 7.0–9.0, consistent with the evidence from maximum rate relations for a Theorell–Chance mechanism, and pH effects on kinetic coefficients have been interpreted on this basis (*54*). Unfortunately, direct studies of the kinetics of the coenzyme compounds of bovine heart lactate dehydrogenase are lacking; on the other hand, the only steady-state data for the pig heart and skeletal muscle enzymes seem to be the maximum rates of lactate oxidation. The dissociation velocity constant for the NADH compound of the pig muscle enzyme is remarkably large and much faster than the steady-state maximum rate at pH 6.0–9.0 (*175*). This is also true of the pig heart enzyme at pH 6.0–7.0. The mechanisms are not of the Theorell–Chance type, and evidence regarding the rate-limiting steps will be discussed in Section IV.

Comparing the limited number of simple dehydrogenases in Table VII, it is evident that the greater stability of the NADH compounds compared with the NAD compounds (Table IV) stems from both slower combination and faster dissociation, by a factor of ten or more, of the oxidized coenzyme. The rate constants for NADH combination are all similar, of the same order as those for oxygen combination with hemoglobin and myoglobin, and about an order of magnitude less than would be expected for a diffusion controlled process.

2. *Conformational Changes*

For the enzymes in Table VII, the kinetics of the reactions with coenzymes studied by the stopped-flow or temperature-jump techniques could be accounted for by a bimolecular association opposed by a unimolecular dissociation reaction. Evidence for a two-step process in which combination of NADH with liver alcohol dehydrogenase was followed by a relatively slow isomerization of the complex was adduced by Theorell *et al.* (*176*) from stopped-flow measurements, but their observations were not confirmed by others (*171*). For rabbit muscle lactate dehydrogenase, it has been reported that temperature-jump studies of the equilibrium with NADH indicated two relaxation times, and an isomerization step with a time constant of about 10^3 sec was suggested (*177*). The precision of the measurements was, however, restricted by large blanks arising from the temperature sensitivity of the fluorescence used to monitor the relaxation. To the present author, there seems to be no conclusive evidence of a rate-limiting isomerization step accompanying the binding of coenzyme to a simple dehydrogenase.

Temperature-jump studies of the combination of NAD with yeast glyceraldehyde-3-phosphate dehydrogenase by Kirschner *et al.* (*145,178*) monitored by the light absorption of the compound showed more clear-cut complexities. At 20°, the readjustment of the equilibrium could be described by a single relaxation time, consistent with a simple reversible association reaction. But at pH 8.5 and 40°, conditions under which the equilibrium binding is a weakly sigmoidal function of the NAD concentration, three relaxation times were required, indicating a minimum of three distinct processes. The rates of the two faster processes increased with the reactant concentrations and were similar to those for the combination of coenzymes with other dehydrogenases, while the slowest process was independent of the enzyme concentration and could be attributed to a protein conformational change with a time constant of about 1 sec. Both the kinetic and the equilibrium binding data could be satisfactorily described by the two-state model of Monod *et al.* (*154*), and with certain simplifying assumptions the rate constants for the reactions of NAD with the two conformational forms of the oligomer were calculated (Table VIII). The different affinities of the two forms of the

176. H. Theorell, A. Ehrenberg, and C. de Zalensky, *BBRC* **27**, 309 (1967).
177. G. H. Czerlinski and G. Schreck, *JBC* **239**, 913 (1964).
178. K. Kirschner and I. Schuster, *in* "Pyridine Nucleotide-Dependent Dehydrogenases" (H. Sund, ed.), p. 217. Springer-Verlag, Berlin and New York, 1970.

TABLE VIII

VELOCITY CONSTANTS FOR ENZYME–COENZYME REACTIONS OF ALLOSTERIC DEHYDROGENASES

Enzyme	Coenzyme	pH	Temp.	k_1 ($\mu M^{-1}\ sec^{-1}$)	k_{-1} (sec^{-1})	Ref.
Glyceraldehyde-3-phosphate dehydrogenase (yeast)	NAD	8.5	20°	10	100	*145,178*
	NAD	8.5	40°	11	1100	
				0.32	800	
Glyceraldehyde-3-phosphate dehydrogenase (rabbit muscle)	NAD	8.2	25°	>100	—	*162*
Glutamate dehydrogenase (bovine liver)	NADH	7.5–8.5	22°	2.0	340	*179*
	NADPH	7.5	22°	0.7	150	
	NADPH	7.5	10°	2.0	30	*92*

oligomer for NAD are entirely the result of a large difference between the association velocity constants.

The reactions of bovine liver glutamate dehydrogenase with NADH and NADPH have been studied by stopped-flow (*92*) and temperature-jump (*179*) methods. There was no evidence of complexities in the reaction, in spite of the negative cooperativity demonstrated in equilibrium binding studies (*93*). The estimates of the "on" velocity constant by the two methods agree well and are somewhat smaller than those for other dehydrogenases, but the two values for the dissociation velocity constant differ considerably (Table VIII). The smaller value from stopped-flow experiments is more consistent with the best equilibrium binding studies and is significantly larger than the maximum specific rate for the oxidative deamination of glutamate (*11*). A possible explanation of the discrepancy is that the temperature-jump studies were restricted to a relatively small range of fractional saturation of the active sites with coenzyme and also involved a pH jump as well as a temperature jump.

Kinetic studies of the reaction of NAD with rabbit muscle glyceraldehyde-3-phosphate dehydrogenase do reflect the negative interactions observed in equilibrium binding studies. It is clear from both stopped-flow (*162*) and temperature-jump (*180*) studies that binding of the first molecule of NAD to the oligomer is very fast, perhaps diffusion controlled, and that relatively slow conformational changes are involved in the binding of subsequent molecules, but a detailed interpretation of the results was not possible. The relaxation spectrum is complex, and individual relaxation times were not well resolved.

IV. Kinetics of the Transient Phase

Up to the present, it appears that studies of the transient phase of dehydrogenase reactions by the stopped-flow method have given more information than studies by relaxation methods. Transient NAD(P)H-containing enzyme compounds may be detected by their light absorption or fluorescence if a relatively large concentration of enzyme is rapidly mixed with excess NAD(P) and substrate in the stopped-flow spectrophotometer. A rapid burst of enzyme-bound reduced coenzyme preceding the steady-state turnover was first observed in this way by Iwatsubo and Pantaloni (*181*) with glutamate dehydrogenase. Similar studies of

179. A. D. B. Malcolm, *Eur. J. Biochem.* **27,** 453 (1972).
180. G. Hammes, P. J. Lilleford, and J. Simplicis, *Biochemistry* **10,** 3686 (1971).
181. M. Iwatsubo and D. Pantaloni, *Bull. Soc. Chim. Biol.* **49,** 1563 (1967).

liver alcohol dehydrogenase by Shore *et al.* (*35,36,71,182*) and of lactate dehydrogenases by Gutfreund and his collaborators (*37,175,183*) have given information about the kinetics of transients, rates of hydride transfer, and, in conjunction with steady-state studies and kinetic data for isolated reactions, some evidence for the occurrence of distinct isomerization steps in the mechanisms.

Because of the difficulty of deriving useful and exact integrated rate equations for even relatively simple, one-substrate reversible mechanisms far from equilibrium (*184*), transient kinetic studies by the stopped-flow method have usually been interpreted qualitatively or on the basis of various simplified models (*185,186*). This lack of rigor may easily lead to unjustified conclusions, and a brief theoretical discussion on the basis of an ordered mechanism before discussion of the more interesting results may not be out of place.

A. Integrated Rate Equations

For several dehydrogenases, the time course of the formation of total enzyme-bound and free NADH, ΣP, measured spectrophotometrically, can be described by the relation

$$\Sigma P_t = \mathrm{E}_0[\alpha t + \beta(1 - e^{-\lambda t})] \tag{18}$$

In this equation, E_0 is the total enzyme active site concentration, and α is the steady-state specific rate of NADH formation, approached by a single exponential "burst" of amplitude β with an apparent first-order rate constant λ. β is estimated by extrapolation of the steady-state portion of the progress curve to $t = 0$, and λ is estimated by the usual logarithmic plot:

$$\ln [\mathrm{E}_0(\beta + \alpha t) - \Sigma P_t] = \ln \mathrm{E}_0\beta - \lambda t \tag{19}$$

Some results show a biphasic exponential approach to the steady state (*187*), however, from which nonequivalence of active centers has been postulated. Recognition of the possible causes of multiphasic bursts is important, therefore.

182. J. D. Shore and H. Gutfreund, *in* "Oxidation-Reduction Enzymes" (Å. Åkeson and A. Ehrenberg, eds.), p. 755. Pergamon, Oxford, 1972.
183. N. G. Bennett and H. Gutfreund, *BJ* **135,** 81 (1973).
184. G. G. Hammes and P. R. Schimmel, "The Enzymes," 3rd ed., Vol. 2, p. 67, 1970.
185. D. R. Trentham, *BJ* **122,** 59 and 71 (1971).
186. H. Gutfreund, *Annu. Rev. Biochem.* **40,** 315 (1971).
187. S. A. Bernhard, M. F. Dunn, P. L. Luisi, and P. Schack, *Biochemistry* **9,** 185 (1970); P. L. Luisi and R. Favilla, *ibid.* **11,** 2303 (1972).

Exact integrated rate laws have been derived for an ordered mechanism, Eq. (4), for the situation that the reactant concentrations A and B remain constant (A,B $\gg$ E_0 and P,Q $\ll$ A,B), together with the essential simplifying assumption that the product release steps are irreversible. Thus, Hijazi and Laidler (*188*) obtained the general solution for the ordered mechanism in Eq. (20), from which the product ternary complex EPQ is omitted.

$$E + A \rightleftharpoons EA + B \rightleftharpoons EAB \rightarrow Q + EP \rightarrow P + E \tag{20}$$

It was shown that the time course of the formation of free product P in general follows a triphasic exponential approach to the steady-state rate, while in some circumstances a biphasic burst would be observed. Whereas this work is important with regard to the interpretation of multiphasic bursts in experiments with limiting substrate concentrations, the general application of the equations is perhaps restricted at present by the omission of EPQ, because usually only the total bound and free P can be measured. Moreover, the irreversible dissociation of P is only a reasonable assumption if the concentration of A is saturating, especially since most dehydrogenases have a much greater affinity for P than for A.

A more immediately useful exact rate law can be derived (*189*) for the mechanism including EPQ in Eq. (21):

$$EAB \underset{k'}{\overset{k}{\rightleftharpoons}} EPQ \xrightarrow{k_{-2}'} Q + EP \xrightarrow{k_{-1}'} P + E \tag{21}$$

This is simply the ordered mechanism, Eq. (4), for saturating concentrations of A and B, a condition that can be achieved experimentally as described by Shore and Gutfreund (*35*), and effectively ensures the irreversibility of the last step. The necessary assumption that the dissociation of Q is also irreversible may be a reasonable one for a limited extent of reaction. The time course of the formation of ΣP is of the form

$$\Sigma P_t = E_0\{\alpha t + \beta[1 - \gamma_1 \exp(-\lambda_1 t) + \gamma_2 \exp(-\lambda_2 t)]\} \tag{22}$$

In the general case, the steady-state maximum rate $E_0\alpha$ (*190*) will therefore be approached by a biphasic exponential burst of total amplitude βE_0, but the degree of resolution of the two phases will depend upon the relative values of the rate constants in Eq. (21). For the Theorell–Chance mechanism, with $k_{-1}' \ll k_{-2}'$ and k, Eq. (22) simplifies to Eq. (23):

$$\Sigma P_t = E_0[k_{-1}'t + (1 - e^{-kt})] \tag{23}$$

188. N. H. Hijazi and K. J. Laidler, *Can. J. Biochem.* **51**, 832 (1972); *ABB* **315**, 209 (1973).
189. N. McFerran and K. Dalziel, unpublished work.
190. $\alpha = 1/\phi_0$ is defined in terms of the velocity constants in the ordered mechanism in Table I.

A burst of enzyme-bound P, as EP, equal to the active center concentration will be obtained with a first-order rate constant k, reversibility of the hydride transfer step being prevented by the fast, irreversible dissociation of Q. This has been observed experimentally with liver alcohol dehydrogenase (*35*).

Equation (22) may also approximate to a single exponential form with other relative values of the rate constants, however (*189*). The mechanism then simplifies formally to Eq. (24) and Eq. (22) reduces to Eq. (25).

$$\mathrm{EAB} \underset{k'}{\overset{k}{\rightleftharpoons}} \mathrm{EPQ} \xrightarrow{k_{-2}'} \mathrm{E} + \mathrm{Q} + \mathrm{P} \tag{24}$$

$$\Sigma P_t = \mathrm{E}_0[\alpha t + \beta(1 - e^{-\lambda t})] \tag{25}$$

In Eq. (25), $\alpha = k\,k_{-2}'/(k + k' + k_{-2}')$, $\beta = k(k + k')/(k + k' + k_{-2}')^2$, and $\lambda = (k + k' + k_{-2}')$. Thus, the apparent first-order rate constant for the burst is complex. This equation was derived by Shafer *et al.* (*191*) and is formally analogous to that derived by Gutfreund and Sturtevant (*192*) for the chymotrypsin mechanism.

TABLE IX
APPARENT FIRST-ORDER RATE CONSTANTS FOR HYDRIDE TRANSFER TO NAD IN TERNARY COMPLEXES (20–25°)

Enzyme	Substrate	pH	k (sec^{-1})	Isotope effect	Ref.
Alcohol dehydrogenase (horse liver)	Ethanol	7.0	130	6.0	*35*
	Propan-1-ol	7.0	650	4.3	*36*
	Propan-2-ol	7.0	0.23	2.3	*36*
Alcohol dehydrogenase (yeast)	Ethanol	7.0	>455[a]		*40*
Lactate dehydrogenase (pig heart)	Lactate	6.0–8.0	>800		*37*
Lactate dehydrogenase (pig muscle)	Lactate	6.0–8.5	>1000(?)		*175*
Glutamate dehydrogenase (bovine liver)	Glutamate	7.5	>30–50	1.7	*94,191*
Glyceraldehyde-3-phosphate dehydrogenase (sturgeon)	Glyceraldehyde-3-phosphate	5.4	>400		*185*

[a] The maximum specific steady-state rate of ethanol oxidation.

191. J. A. Shafer, E. Chiancone, K. L. Yielding, and E. Antonini, *Eur. J. Biochem.* **28,** 528 (1972).
192. H. Gutfreund and J. M. Sturtevant, *BJ* **63,** 656 (1956).

Estimates of apparent rate constants from burst studies are shown in Table IX.

B. Liver Alcohol Dehydrogenase

In accordance with the Theorell–Chance mechanism established for this enzyme, Shore and Gutfreund (*35*) showed that with saturating NAD and ethanol concentrations, progress curves for the formation of total NADH, measured at the isosbestic wavelength for NADH and E·NADH, could be described by Eq. (23) from 2 msec after mixing enzyme and substrates. This was the first, and so far the only, quantitative characterization of the rate of hydride transfer in a dehydrogenase reaction, confirmed by a large deuterium isotope effect of 6 (Table IX). The identity of the burst amplitude and the active center concentration, and its description by a single first-order rate constant over 90% of its course, indicates that the two active centers of the enzyme are kinetically equivalent. Qualitatively similar results were obtained with propan-1-ol as substrate (*36*), but the hydride transfer step was much faster, confirming the importance of hydrophobic bonding of the substrate in inducing the optimum conformation of the ternary complex for hydride transfer. With isopropanol, no pre-steady-state burst of NADH formation was observed, and the maximum steady-state rate showed a significant isotope effect, indicating that hydride transfer is at least partly rate-limiting in the overall reaction (*36*).

Indirect evidence for a relatively slow isomerization step that is rate-limiting under some conditions has also been obtained (*35*). The dissociation velocity constant, k'_{-1}, for the compound E·NADH is increased threefold in the presence of sodium chloride, but the maximum rate of ethanol oxidation is only slightly increased; thus, dissociation of NADH can no longer be the sole rate-determining step. Since the fast hydride transfer step was not affected by sodium chloride, and reasonable evidence that aldehyde dissociation is also relatively fast was obtained, Shore *et al.* (*35*) concluded that a new step in the mechanism had been revealed. This could be isomerization of either the ternary product complex or the binary complex E·NADH. Evidence of a similar slow step in the oxidation of ethanol and propanol with APAD as coenzyme (*71*) was referred to in Section II,E,1.

Shore and Gutfreund (*35*) also took advantage of the firm binding of NADH by the enzyme to study the oxidation of the stoichiometric compound E·NADH by excess acetaldehyde. They observed complete reduction by a single first-order process, consistent with complete equivalence of the two active centers in the enzyme molecule, and this was further confirmed by the finding that the rate constant for oxidation was unchanged when enzyme only partially saturated with NADH was used. This conclusion is in contrast with that drawn by Bernhard *et al.* (*187*) from stopped-flow studies of the reduction of aromatic aldehydes in which biphasic bursts were observed when the substrate concentrations were larger than the enzyme concentration. The complete reduction of limited concentrations of substrate by excess enzyme and NADH also occurred in two phases of equal amplitude. These interesting results were interpreted as a reciprocating subunit mechanism, similar to that proposed for malate dehydrogenase (*80*). There are, however, other possible interpretations. The maximum steady-state rates of reduction of these aldehydes are smaller than that for aliphatic aldehydes, and the Theorell–Chance mechanism is not applicable. In these circumstances an ordered mechanism may give a biphasic exponential approach to the steady state, even when the substrate concentrations are saturating, according to Eq. (22), and Hijazi and Laidler (*188*) have given specific criticisms of the conclusions on this basis. Moreover, it is not clear whether the substrate concentrations were saturating in the burst experiments. Although the K_m values are smaller for the aromatic aldehydes than for acetaldehyde, this is largely because of the smaller maximum rates, and K_m values are not relevant in pre-steady-state studies. As the authors themselves suggested (*187*), more detailed investigations of both the transient phase and steady-state kinetics of the enzyme with aromatic aldehydes are needed.

C. Lactate Dehydrogenases

The results of stopped-flow studies of the lactate dehydrogenase reaction have proved to be more difficult to interpret than those of alcohol dehydrogenase. The dissociation velocity constants for the binary NADH compounds of the pig heart and skeletal muscle lactate dehydrogenases are much larger than that for liver alcohol dehydrogenase, and also larger than the maximum specific rates of lactate oxidation at pH 6.0–7.0 (Table VII). Some earlier step must therefore be rate-limiting.

In spectrophotometric stopped-flow studies of lactate oxidation with saturating lactate and NAD concentrations at pH 6.0 and 8.0, catalyzed by the pig heart enzyme, Heck *et al.* (*37*) observed a burst of NADH formation equivalent to the enzyme active center concentration and complete within the dead-time of the apparatus (3 msec), indicating that the rate constant for hydride transfer is greater than 800 sec^{-1}. At pH 6.0, therefore, some step after hydride transfer and before NADH dissociation from the enzyme must restrict the maximum rate to 8 sec^{-1} (Table VII).

The observed full burst in the dead-time showed that the reversible hydride transfer step is prevented from equilibrating and that the slow step must occur after pyruvate dissociation. Isomerization of E·NADH before release of the product coenzyme was suggested. Studies of the oxidation of E·NADH by pyruvate at pH 6.0 showed that the reverse hydride transfer step was indeed fast.

More recent stopped-flow studies of lactate dehydrogenation, briefly reported by Holbrook and Gutfreund (*169*), gave somewhat different results. A burst of NADH formation equivalent to the active center concentration was again observed at pH 6–8, but it could now be resolved into two components: a fast phase in the dead-time of the apparatus, of amplitude equivalent to from 10 to 33% of the enzyme active center concentration, according to the pH, was followed by an observable burst with an apparent first-order velocity constant of 275 sec^{-1}. It was suggested that the fast phase represents equilibration of the reactant and product ternary complexes by hydride transfer and the slower transient the formation of E·NADH by pyruvate dissociation. The picture is clearly more complex than that for the alcohol dehydrogenase reaction, but further resolution may be aided by evidence that in the ternary complex of enzyme, NADH, and pyruvate, the coenzyme fluorescence is largely quenched, while in the binary complex E·NADH it is enhanced (*169*).

Similar studies of the enzyme from pig skeletal muscle have been reported (*175,183*). In the earlier work, a fast burst of NADH formation in the dead-time of the apparatus was observed, equal in amplitude to the active center concentration at pH 8.0, but smaller at lower pH values. The suggestion that slow isomerization of the ternary product complex before pyruvate release may be the step responsible for the low steady-state maximum rate of lactate oxidation seems to be inconsistent with the full burst observed at pH 8.0, since it might be expected to result in partial equilibration of the reactant and product ternary complexes. Direct studies of the oxidation of E·NADH by pyruvate at pH 9.0 did indicate that reverse hydride transfer from NADH to pyruvate is indeed fast, but the absence of a deuterium isotope effect suggested that the observed rate constant of 246 sec^{-1}, equal to the maximum steady-state rate of pyruvate reduction, may reflect an isomerization of the ternary complex preceding even faster hydride transfer. More recent studies (*183*) with improved techniques, however, appear to indicate no burst of enzyme-bound NADH formation preceding the steady-state phase of lactate oxidation at pH 8.0. On the basis of stopped-flow studies of lactate oxidation in the presence of oxamate, which forms a dead-end complex with E·NADH and can serve as an indicator of the rate of formation of this product complex, it is now concluded that the equilibrium of the ternary complexes is greatly in favor of the oxidized nucleotide, and that the dissociation of pyruvate controls the rate of formation of E·NADH. There are, however, other possible interpretations of the behavior of this rather complex system.

D. Other Enzymes

A number of studies of the glutamate dehydrogenase reaction by the stopped-flow method have been reported (*94,181,191,193*). Single exponential bursts of NAD(P)H formation preceding the steady-state rate and equivalent to about half the enzyme active center concentration, with a relatively small apparent rate constant (Table IX), have been observed. The reaction is complicated by the formation of three products, and, as has been indicated, the interpretation of fractional bursts in mechanistic terms is difficult. The most detailed attempt in this direction (*191*) was based on an integrated rate equation for the reaction with nonsaturating coenzyme and substrate concentrations derived by means of an equilibrium assumption, which is difficult to justify. The conclusions drawn are therefore open to question. D'Albis and Pantaloni (*94*) compared the kinetics of proton release and NADPH production in the pre-steady state, and concluded that proton liberation occurred after hydride transfer but before release of free NADPH. They also observed that in the presence of NADPH, in amount equivalent to the enzyme active centers, the pre-steady-state burst of NADPH formation in the oxidative deamination of glutamate is eliminated. This confirms the earlier conclusions of Iwatsubo *et al.* (*181,193*) that dissociation of NADPH from the abortive ternary complex with glutamate is the rate-limiting step with saturating glutamate concentrations. No convincing evidence of negative cooperativity or nonequivalence of the active centers has yet been obtained from pre-steady-state studies.

The glyceraldehyde-3-phosphate dehydrogenase reaction has been studied in detail by Trentham (*185*), taking advantage of the possibility of examining a partial reaction, acylation of the enzyme, in the absence of acyl acceptor. This work well illustrates the value of stopped-flow studies of the transient phase of dehydrogenase reactions, and has given important information about the multiple roles of NAD in the reaction in particular. The studies were made with the lobster enzyme, which contains firmly bound NAD, and the sturgeon enzyme, which does not, and could be used to investigate the effects of pre-incubation of the enzyme

193. A. di Franco and M. Iwatsubo, *Biochimie* **53**, 153 (1971); H. F. Fisher, J. R. Bard, and R. A. Prough, *BBRC* **41**, 601 (1970).

with NAD on the reductive dephosphorylation of diphosphoglycerate. With the sturgeon enzyme at pH 8.5, a lag phase before the steady-state rate was abolished by pre-incubation with NAD. There was no pre-steady-state burst of NADH disappearance, indicating that in the presence of NAD the rate-limiting step is concomitant with hydride transfer. However, the absence of a deuterium isotope effect on the stady-state rate suggested that a conformational change after combination of NADH with the acyl enzyme may be the rate-limiting step at pH 8.5. At pH 5.3, in the presence of NAD, a burst of NADH disappearance before the steady-state rate was reached showed that a step after hydride transfer is rate-limiting; this step is also accelerated by NAD, and is identified as release of glyceraldehyde 3-phosphate. The burst of NADH disappearance in the pre-steady state could be described by a single exponential of amplitude equal to the enzyme active center concentration, showing that the four sites of the oligomer are equally reactive. From studies of the oxidative phosphorylation reaction, which showed a pre-steady-state burst of NADH formation, and of the uncoupled enzyme acylation reaction, Trentham concluded that NADH release is rate-limiting in the overall reaction at high pH, and phosphorolysis of the acyl enzyme is rate-limiting at low pH. The roles of NAD other than its oxidative function are identified as facilitation of the formation and phosphorolysis of the acyl enzyme and of the release of aldehyde.

These studies show no evidence of effects which could be related to the negative cooperativity established for the reaction of the rabbit muscle enzyme, nor of the nonequivalence of the active centers indicated by acylation studies (*194*). As Trentham pointed out, there may be differences between the lobster and rabbit enzymes, and transient phase studies with the apoenzyme from rabbit muscle are needed.

194. R. A. MacQuarrie and S. A. Bernhard, *in* "Pyridine Nucleotide-Dependent Dehydrogenases" (H. Sund, ed.), p. 187. Springer-Verlag, Berlin and New York, 1970.

2

Evolutionary and Structural Relationships among Dehydrogenases

MICHAEL G. ROSSMAN • ANDERS LILJAS •
CARL-IVAR BRÄNDÉN • LEONARD J. BANASZAK

I. Introduction

The structural similarity of myoglobin to both the α and β chains of hemoglobin suggested the hypothesis that proteins of similar function have similar tertiary structures. A particularly striking example of conservation of structure over a long evolutionary period, for the important functions of electron transport, is represented by the conservation of the cytochrome *c* structure found in eukaryotes (*1*) as well as prokaryotes (*2*). Anticipation of structural similarities among dehydrogenases has been implicit in comparisons of amino acid sequences aligned on the basis of essential cysteine residues (*3–6*) and on other physical properties (*7*). One of the motivations for the study of the lactate dehydrogenase (LDH) structure (*8*) was the possibility that a structure–function correlation might be revealed in a comparison with soluble malate dehydrogenase (s-MDH) and glyceraldehyde-3-phosphate dehydrogenase (GAPDH). Some similarity of subunit structure between liver alcohol dehydrogenase (LADH) and LDH was indeed noted by Brändén and Rossmann (*9*), by comparing the three-dimensional models based on low resolution electron density maps. This first recognition of the similarity between LADH and LDH was remarkable in that cognizance was taken of the greater similarity in one-half of the subunit and in the differing subunit associa-

1. R. E. Dickerson, T. Takano, D. Eisenberg, O. B. Kallai, L. Samson, A. Cooper, and E. Margoliash, *JBC* **246,** 1511 (1971).
2. R. Timkovich and R. E. Dickerson, *JMB* **79,** 39 (1973); F. R. Salemme, J. Kraut, and M. D. Kamen, *JBC* **248,** 7701 (1973).
3. T. P. Fondy, J. Everse, G. A. Driscoll, F. Castillo, F. E. Stolzenbach, and N. O. Kaplan, *JBC* **240,** 4219 (1965).
4. S. S. Taylor, S. S. Oxley, W. S. Allison, and N. O. Kaplan, *Proc. Nat. Acad. Sci. U. S.* **70,** 1790 (1973).
5. M. O. Dayhoff, W. C. Barker, and J. K. Hardman, *Atlas Protein Sequence Struct.* **5,** 58 (1972).
6. J. I. Harris, *Nature (London)* **203,** 30 (1964).
7. T. P. Fondy and P. D. Holohan, *J. Theor. Biol.* **31,** 229 (1971).
8. M. G. Rossmann, B. A. Jeffery, P. Main, and S. Warren, *Proc. Nat. Acad. Sci. U. S.* **57,** 515 (1967).
9. C.-I. Brändén and M. G. Rossmann, *in* "Pyridine Nucleotide-Dependent Dehydrogenases" (H. Sund, ed.), p. 133. Springer-Verlag, Berlin and New York, 1970.

tions. It is improbable that this would have been found without some expectation of a structure–function relationship among dehydrogenases.

Structures at high resolution, where individual polypeptide chains and side groups can be accurately positioned, are now available for four dehydrogenases: dogfish LDH (*10–12*), pig s-MDH (*13,14*), horse LADH (*15,16*) and lobster GAPDH (*17,18*). Detailed discussion of the relationship of these structures with respect to the enzyme properties in solution will be found in other chapters of this volume, while here an attempt has been made to compare and contrast the structural results.

The structures of those parts of the polypeptide chains whose function is to bind the NAD^+ coenzyme are remarkably similar although the complete subunit conformations are not alike. The varying position of this dinucleotide binding domain (*19*) within the complete chain for each of the above four dehydrogenases is shown in Fig. 1. The remainder of the polypeptide chains are required for substrate binding, catalysis, specificity, and formation of oligomeric structure. These parts have different three-dimensional structures.

The concept of structure–function relationships might thus be sharpened. Sophisticated enzymes such as dehydrogenases are mostly constructed out of a variety of structural domains (*20,21*), which may repre-

10. M. J. Adams, G. C. Ford, R. Koekoek, P. J. Lentz, Jr., A. McPherson, Jr., M. G. Rossmann, I. E. Smiley, R. W. Schevitz, and A. J. Wonacott, *Nature (London)* **227,** 1098 (1970).
11. M. G. Rossmann, M. J. Adams, M. Buehner, G. C. Ford, M. L. Hackert, P. J. Lentz, Jr., A. McPherson, Jr., R. W. Schevitz, and I. E. Smiley, *Cold Spring Harbor Symp. Quant. Biol.* **36,** 179 (1971).
12. M. J. Adams, M. Buehner, K. Chandrasekhar, G. C. Ford, M. L. Hackert, A. Liljas, M. G. Rossmann, I. E. Smiley, W. S. Allison, J. Everse, N. O. Kaplan, and S. S. Taylor, *Proc. Nat. Acad. Sci. U. S.* **70,** 1968 (1973).
13. E. J. Hill, D. Tsernoglou, L. E. Webb, and L. J. Banaszak, *JMB* **72,** 577 (1972).
14. L. E. Webb, E. J. Hill, and L. J. Banaszak, *Biochemistry* **12,** 5101 (1973).
15. C.-I. Brändén, H. Eklund, B. Nordström, T. Boiwe, G. Söderlund, E. Zeppezauer, I. Ohlsson, and Å. Åkeson, *Proc. Nat. Acad. Sci. U. S.* **70,** 2439 (1973).
16. H. Eklund, B. Nordström, E. Zeppezauer, G. Söderlund, I. Ohlsson, T. Boiwe, and C. I. Brändén, *FEBS (Fed. Eur. Biochem. Soc.) Lett.* **44,** 200 (1974).
17. M. Buehner, G. C. Ford, D. Moras, K. W. Olsen, and M. G. Rossmann, *Proc. Nat. Acad. Sci. U. S.* **70,** 3052 (1973).
18. M. Buehner, G. C. Ford, D. Moras, K. W. Olsen, and M. G. Rossmann, *JMB* **90,** 25 (1974).
19. In this chapter, the word "domain" is defined as the complete NAD^+ binding segment in any one dehydrogenase. Thus this usage differs to that of Rossmann and Liljas (*20*) where a domain specifies a mononucleotide binding structure only. The latter will be called a mononucleotide unit here.
20. M. G. Rossmann and A. Liljas, *JMB* **85,** 177 (1974).
21. D. Wetlaufer, *Proc. Nat. Acad. Sci. U. S.* **70,** 697 (1973).

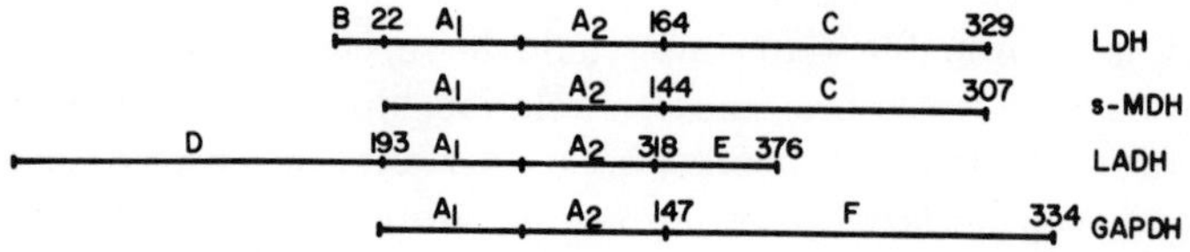

FIG. 1. Position of the two similar mononucleotide binding domains, A_1 and A_2, comprising together the NAD binding region, within the single polypeptide chains of four different dehydrogenases. The catalytic domains, C, of LDH and s-MDH are similar in structure, but different to the catalytic domains D and E in LADH and F in GAPDH. The primary function of the amino terminal arm domain, B, in LDH is to stabilize the quaternary structure.

sent gene fusion of proteins of simpler functions. The evolution of domain structures has also been studied for immunoglobulin (*22*) and for calcium binding in carp muscle protein (*23*). Similar structures are repeated in each case along the polypeptide chain.

This chapter will first compare the subunit structures of known dehydrogenases and then review the alignments arrived at by using only amino acid sequence information. This will be followed by a comparison of amino acid sequences in light of tertiary structure which, in case of homology, gives pointers to measure evolutionary distance as well as to residues important in the conservation of structure and function. Further exploration of functional properties will be considered in the development of the oligomeric structures of dehydrogenases. Finally, a larger group of enzymes is discussed which are functionally or structurally related possibly because of common nucleotide binding properties.

II. Comparison of Three-Dimensional Structures

A. KNOWN STRUCTURES

Two structures of LDH have been determined, that of apo-LDH and that of the abortive ternary complex LDH:NAD-pyruvate. In both these structures the tetrameric LDH molecule has strict 222 symmetry with all four subunits identical. In the tetrameric holo-GAPDH, on the other hand, all four subunits are crystallographically independent and might then have subtle structural differences. In the crystal structure determination of s-MDH (a dimer) there were two crystallographically independent polypeptide chains. These were differentially substituted by the

22. G. M. Edelman, *Science* **180,** 830 (1973).
23. R. H. Kretsinger, *Nature (London), New Biol.* **240,** 85 (1972).

NAD^+ coenzyme and will be referred to as "holo" and "apo" subunits. In the crystal structure of dimeric LADH the two subunits are related by strict 2-fold symmetry and do not contain NAD^+. The coenzyme binding site was determined from the complex of LADH with ADP-ribose.

B. s-MDH TO LDH COMPARISON

The first structural correlation to emerge from these studies was that the fold of the s-MDH subunit was extremely similar to that of the LDH subunit (*13,24*). This is demonstrated in comparing the NAD^+ binding regions shown in Figs. 2a and 2b and the catalytic domains seen in Figs. 3a and 3b. Apart from the first 20 residues at the amino end of LDH which are absent in s-MDH, both independent subunits of s-MDH had the same general conformation as an LDH subunit. A quantitative measure of these similarities was obtained by Rao and Rossmann (*24*). The objective of their method is to rotate and translate one structure into the other in such a way as to minimize the sum of the squared distances between all C_α atoms (*25*) assumed to have equivalence. The root mean square deviation, σ, of these distances can then be taken as a quantitative measure of similarity between structures, and the significance of conformational differences can be expressed in terms of σ.

When "apo" and "holo" s-MDH subunits were compared to each other using this procedure, a σ value of 1.8 Å was obtained for 288 equivalent C_α atoms. A comparison of one subunit of LDH to the "apo" and "holo" s-MDH subunits gave corresponding values of 2.7 and 2.9 Å, respectively, for about 250 equivalent C_α atoms. Since the substrate specificity is different, and since extensive physicochemical and kinetic investigations of these two enzymes have given no indications that they might be closely related (*7*), this high degree of structural similarity is remarkable.

C. COMPARISON OF s-MDH AND LDH WITH LADH AND GAPDH

The most exciting correlation between dehydrogenases emerged, however, when the structures of LDH and s-MDH were compared with LADH and GAPDH. There was a striking similarity in the structures of the coenzyme binding domains, whereas the catalytic domains had very different structures in all these subunits. Not only were the general folds of the NAD^+ binding domains similar but also the conformation of the bound coenzyme and its orientation and position on the protein were all similar (Fig. 2).

24. S. T. Rao and M. G. Rossmann, *JMB* **76**, 241 (1973).
25. IUPAC-IUB Commission, *JBC* **245**, 6489 (1970).

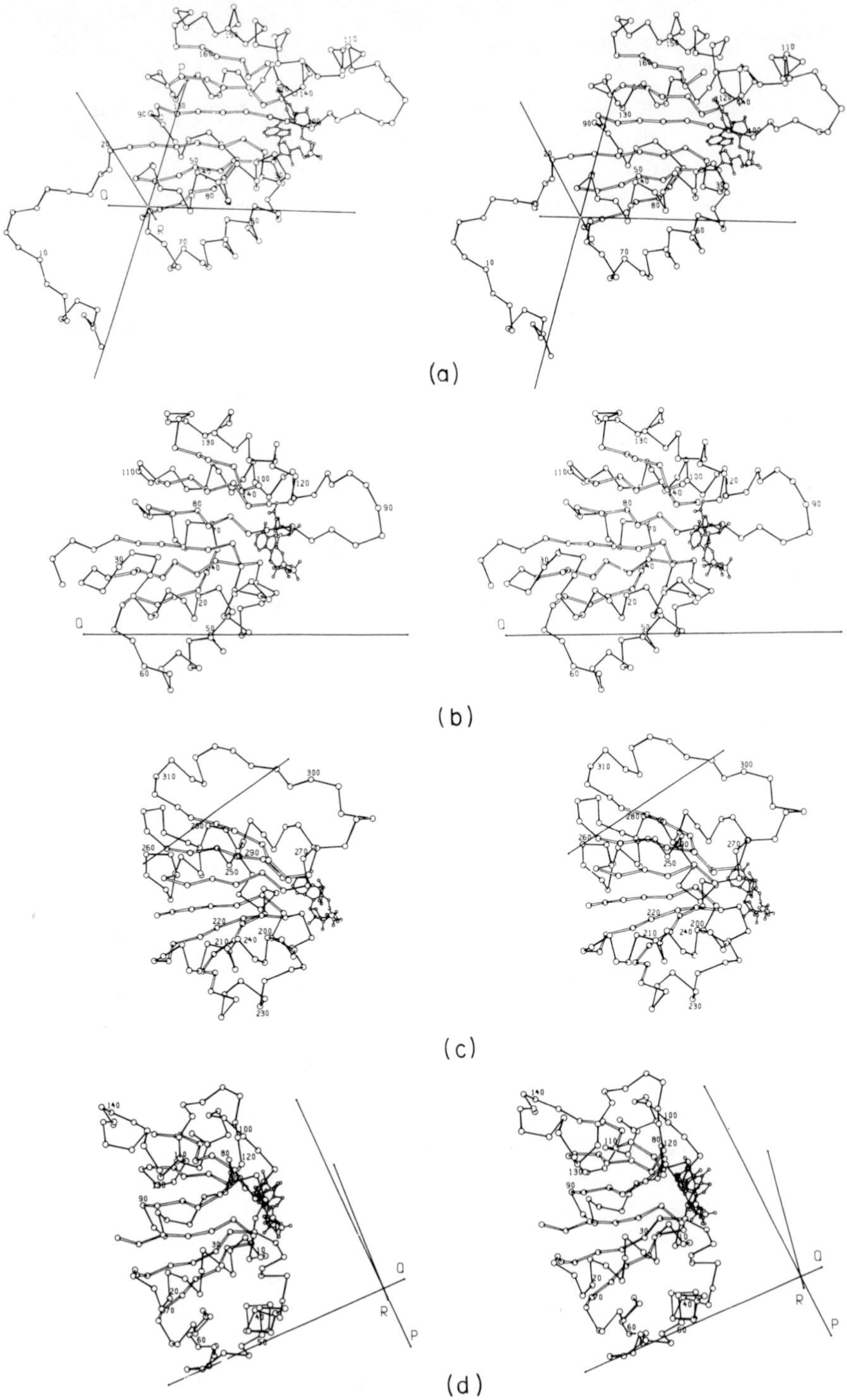

FIG. 2. The nucleotide binding domain with bound NAD⁺, viewed from a similar orientation, for (a) LDH, (b) s-MDH, (c) LADH, and (d) GAPDH. The stereo pairs show the C_α backbone.

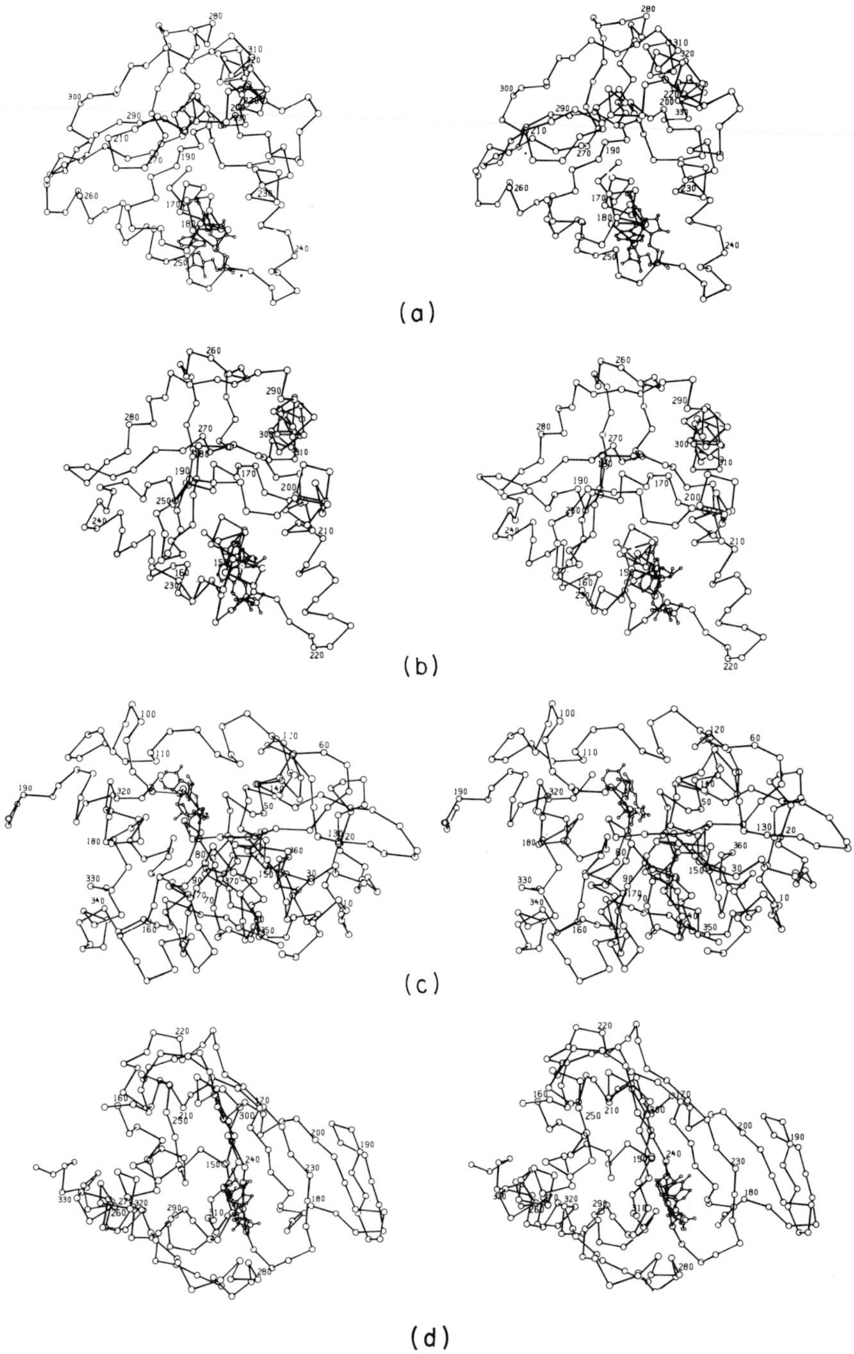

FIG. 3. The catalytic domains for (a) LDH, (b) s-MDH, (c) LADH, and (d) GAPDH. Each subunit is aligned with its NAD^+ binding region oriented as in Fig. 2 but only the bound coenzyme is shown as a reference. The stereo pairs show the C_α backbone.

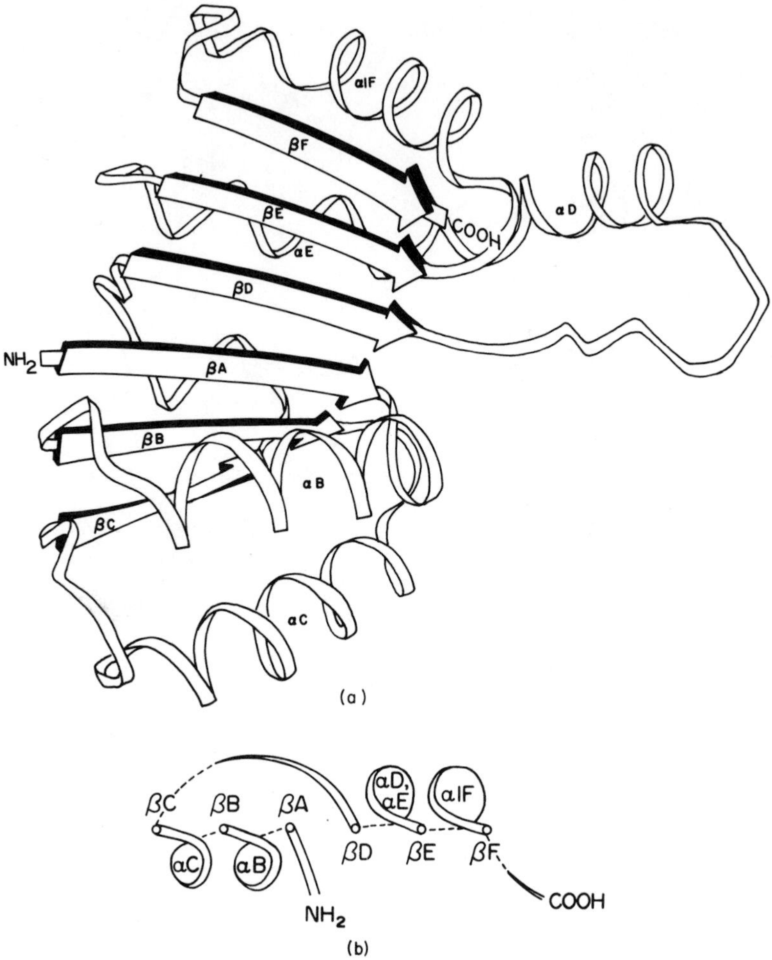

FIG. 4. Schematic drawings of the NAD^+ binding region in dehydrogenases (a) viewed from top, perpendicular to the six-stranded parallel sheet and (b) viewed from end, looking along the extended polypeptide chains in the sheet, with their amino terminal end closest to the observer. The various secondary structural elements have been labeled according to the usually accepted nomenclature.

Schematic drawings of the NAD^+ binding domain are given in Fig. 4. The nomenclature first described by Adams *et al.* (*10*) and later extended by Hill *et al.* (*13*) will be used in the description of this protein fragment (Fig. 4). The main structural elements of this domain are six strands of parallel sheet (βA, βB, βC, βD, βE, and βF) and four helices (αB, αC, αE, and α1F). There are two helices on each side of the sheet. There is a gradual, left-handed twist, from one strand to the next, amounting to about 100° between the extreme strands of the sheet (*26*).

Starting from the amino end of the polypeptide chain, the first structural element is βA, which is the third strand from the left in Fig. 4b. Helices αB and αC connect strands βA with βB and βB with βC, respectively. These two helices are on the same side of the sheet. The sequence βA, αB, βB, αC, and βC is the AMP mononucleotide binding unit of these homologous enzymes. From βC the chain passes back to the amino end of the sheet and into βD which in the molecular structure is next to βA. The sequence βD, αE, βE, α1F, βF forms the nicotinamide mononucleotide binding unit, with the helices αE and α1F on the opposite side of the sheet compared to αB and αC. The number of residues from the beginning of βA to the end of βF are 144, 149, and 127 in LDH, GAPDH, and LADH, respectively.

There are some obvious differences in this domain as found in LDH, s-MDH, GAPDH, and LADH, the functional significance of which will be discussed in later sections. There is a helix αD between βD and αE which is part of a flexible loop in LDH and s-MDH that is absent in LADH and in GAPDH. There is also an extra antiparallel sheet excursion inserted between αC and βC in GAPDH which is absent in LDH, s-MDH, and in LADH. In GAPDH the helix α1F has slipped around the edge of the sheet and is rather irregular, while in LADH there is instead a wide loop of extended chain forming part of the subunit interaction area. Thus, in general, the four central strands and two connecting helices have been better conserved than the extremities of the structure. It will be shown (Section V) that most of the important functional residues for the binding of the coenzyme are in the central, better conserved, part of this domain.

D. The Mononucleotide Binding Unit

Rao and Rossmann (*24*) showed that the coenzyme binding domain consists of two roughly identical units associated each with a mononucleotide binding area and related by an approximate 2-fold axis running parallel to the strands between βA and βD. Thus the fold (the spatial arrangement of extended polypeptide chain and interconnecting helices with due regard to their polarity) of each unit has the same unique hand. Rao and Rossmann found a σ value of 3.1 Å for 34 equivalent atoms comparing the mononucleotide binding units of LDH. The good spatial correlation between these two units is shown in Fig. 5. In this rotation βA must be compared to βD, βB with βE, βC with βF, αB with αE, and αC with α1F.

26. C. Chothia, *JMB* **75**, 295 (1973).

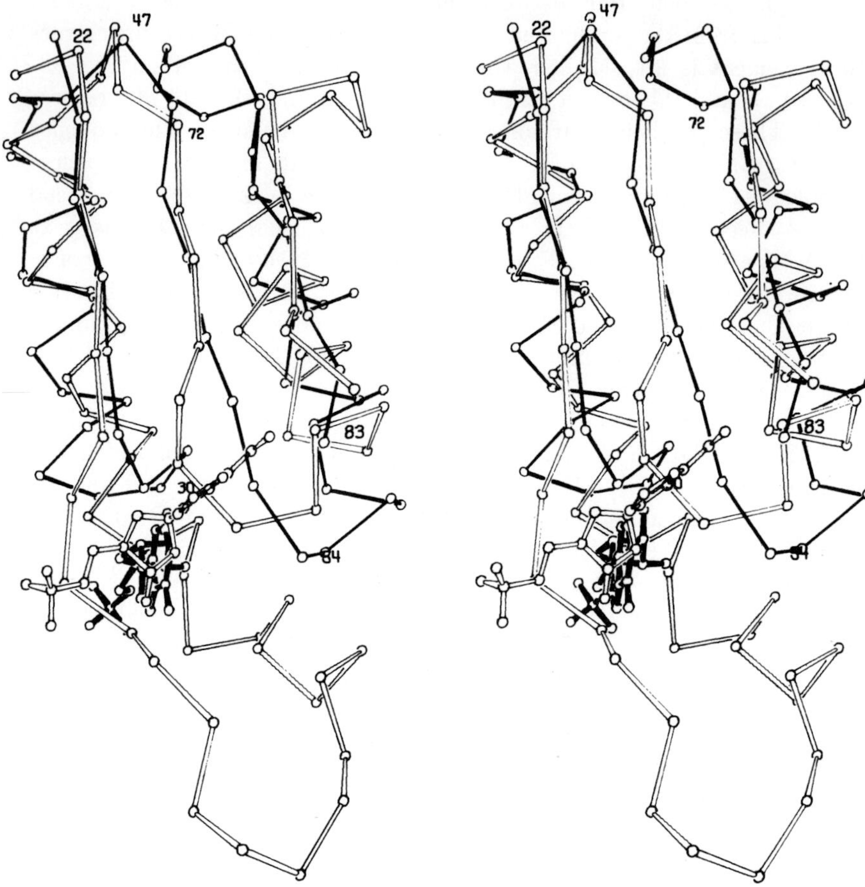

FIG. 5. Superposition of the AMP (dark and numbered bonds) and NMN nucleotide binding domains in LDH (open bonds).

E. Systematic Comparison of the NAD$^+$ Binding Structure

The methods of Rao and Rossmann have been used by Ohlsson *et al.* (*27*) to determine the degree of similarity between the whole dinucleotide binding domains in LDH, LADH, and GAPDH. In their comparisons the rotation matrices and translation vectors were defined only by the C_α atoms of those residues that take part in a common hydrogen bonding scheme within the parallel pleated sheet regions. These hydrogen bonds have been depicted in Fig. 6 for LDH, LADH, and GAPDH. When these

27. I. Ohlsson, B. Nordström, and C.-I. Brändén, *JMB* **89**, 339 (1974).

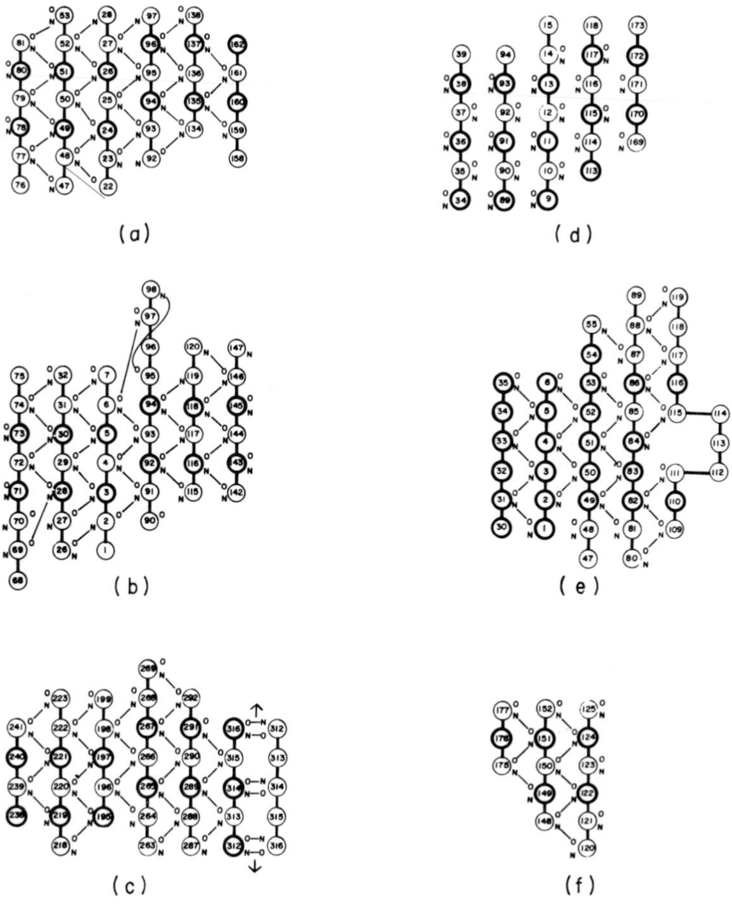

FIG. 6. The hydrogen bonding pattern of the β-pleated sheet region in (a) dogfish apo-LDH, (b) lobster holo-GAPDH, (c) horse LADH, (d) adenylate kinase, (e) *Clostridium* MP flavodoxin, and (f) BPN′ subtilisin. Residues pointing into the hydrophobic cavities between the sheet and helices have been heavily circled.

bonds are equivalently aligned, with due regard for the polarity of the peptide chains, then the orientation of the C_β atoms in each amino acid side chain must always be aligned (*28*). The residues will be pointing alternatively toward or away from the helices that link the corresponding strands (Fig. 6). The superpositioning of the hydrogen bonds is a sensitive method to determine equivalent amino acids in corresponding β-sheet regions of different dinucleotide binding domains. The distance between these equivalenced C_α atoms in the different dehydrogenases is given in

28. M. G. Rossmann, D. Moras, and K. W. Olsen, *Nature (London)* **250,** 194 (1974).

TABLE I
Distances between Equivalenced C_α Atoms Used in the Alignment of the Nucleotide Binding Regions of Different Dehydrogenases

	(1) C_α LDH	(2) C_α LADH	(3) C_α GAPDH	Distances in Å 1–2	2–3	3–1
	24	195	3	1.2	0.7	1.0
	25	196	4	0.9	1.3	0.9
βA	26	197	5	1.0	1.1	0.8
	27	198	6	0.9	0.7	1.1
	28	199	7	0.9	2.1	1.3
	49	219	28	1.0	0.8	1.0
βB	50	220	29	0.6	0.8	0.5
	51	221	30	1.6	0.8	0.7
	52	222	31	1.8	2.6	1.1
	78	238	71	1.2	2.5	2.1
βC	79	239	72	0.7	2.6	2.4
	80	240	73	0.8	2.4	2.1
	81	241	74	1.5	2.9	1.5
	92	263	90	1.4	1.5	2.1
	93	264	91	1.3	0.9	0.8
βD	94	265	92	1.2	0.9	1.0
	95	266	93	0.7	1.5	0.9
	96	267	94	0.8	1.6	1.2
	135	289	116	1.3	1.2	2.0
βE	136	290	117	1.0	2.1	1.8
	137	291	118	0.4	2.5	2.3
	138	292	119	0.4	1.4	1.9
	159	313	142	0.8	1.6	1.3
βF	160	314	143	1.3	1.6	1.0
	161	315	144	1.6	0.4	1.4
	162	316	145	1.5	0.4	1.2
Root mean square deviation of 26 equivalent pairs of atoms				1.1	1.7	1.5

Table I. The excellent agreement between these atoms demonstrates not only the correct alignment of the individual residues but also that the angle of twist between strands is essentially constant for all dehydrogenases. Differences are no greater than in the comparison of subtilisin BPN′ with subtilisin Novo (*29*) or the errors generally found in the measurement of atomic coordinates.

29. J. Drenth, W. G. J. Hol, J. N. Jansonius, and R. Koekoek, *Cold Spring Harbor Symp. Quant. Biol.* **36**, 107 (1971).

TABLE II
CORRESPONDENCE OF STRUCTURAL ELEMENTS IN LDH, LADH, AND GAPDH

Structural element	Residue numbers (1) LDH	(2) LADH	(3) GAPDH	Root mean square deviation 1–2	2–3	3–1
βA	24–28	195–199	3–7	1.0	1.3	1.0
αB	31–42	202–213	10–21	2.8	2.6	2.2
βB	49–52	219–222	28–31	1.3	1.5	0.9
αC	61–67	227–233	38–44	3.7	5.4	5.7
βC	78–81	238–241	71–74	1.1	2.5	1.3
βD	92–96	263–267	90–94	1.1	1.3	1.3
αE	122–130	274–282	104–112	4.4	3.9	5.0
βE	135–138	289–292	116–119	1.0	1.9	2.0
βF	159–162	313–316	142–145	1.3	1.1	1.2

For the analysis of the regions between the strands, Ohlsson *et al.* (*27*) excluded the local regions of striking differences described above. When these residues are excluded, the number of amino acids comprising the remaining basic segments of this region does not vary greatly between LDH, LADH, and GAPDH. Standard deviations for the distance between equivalenced atoms over each structural element are given in Table II. The central regions represented by αB and αE are clearly better conserved than the outside helix αC.

F. Recognition of Similar Structural Domains

In concluding the section on structure, some mention should be made concerning the recognition of domains and units. Frequently, common function can lead to the recognition of structural similarity. Similarly, common structures can lead to the recognition of function where none has been known. Nevertheless, the recognition of common structure is not always easy; for instance, the similarity of the two nucleotide binding units in LDH was overlooked for two years after the first solution of that structure. Rossmann and Liljas (*20*) have suggested the use of distance plots (*30*) for the structural detection of domains and units. Similarity of structure will be indicated by a similar pattern of contours.

30. T. Ooi and K. Nishikawa, *in* "Conformations of Biological Molecules and Polymers" (E. D. Bergmann and B. Pullman, eds.), p. 173. Academic Press, New York, 1973.

Deletions or insertions will not greatly affect such a contour pattern. Hence, both the separation of different structural units and similarity of domain structure can be readily recognized by eye.

An example of a distance plot is given in Fig. 13 of Chapter 4. Four structural units can easily be identified. The first two recognizable units have a similar contour pattern, each representing a mononucleotide binding domain and corresponding to A_1 and A_2 in Fig. 1. The last two units are similar to each other but different from the first two. Together they make the catalytic domain of the subunit, identified briefly as "domain C" in Fig. 1.

III. Sequence Comparisons in the Absence of Three-Dimensional Structural Information

A. Suggested Homologies or Analogies

A review of the known sequences of each of the dehydrogenases has been given in the appropriate chapters of this volume. Table III shows the available sequence information in relation to the known tertiary and quaternary structures (*4,5,10–18,30a–d,43*).

Prior to the structure determination, attempts were made to align sequences of different dehydrogenases. Harris (*6*) and Fondy *et al.* (*3*) used the essential cysteine residue as a point of alignment between LDH, GAPDH, LADH, and yeast ADH. These results were also cited by Taylor (*4*) and extended by Dayhoff *et al.* (*5*). In light of the known structures, it is now apparent that such comparisons are only valid between LDH and GAPDH, where the essential cysteine is at the end of βF and forms the junction between two different domains. In LADH the essential cysteine-46 is in the middle of the first (catalytic) domain, not at the end of βF. Indeed, in view of the association of two domains in the vicinity of the essential cysteine, it is uncertain whether the comparison beyond this residue is even valid between LDH and GAPDH. The difference in function of this cysteine residue between LDH and GAPDH is also noteworthy. During the enzymic reaction, a thioester intermediate is formed at cysteine-149 in GAPDH. Cysteine-165 of LDH, on the other hand, is termed essential because its modification blocks the active site.

30a. S. S. Taylor, private communication (1974).
30b. H. Jörnvall, *Eur. J. Biochem.* **16**, 41 (1970).
30c. H. Jörnvall, *Proc. Symp. Alc. Aldehyde Metab. Syst., 1st, 1974* (in press).
30d. G. M. T. Jones and J. I. Harris, *FEBS (Fed. Eur. Biochem. Soc.) Lett.* **22**, 185 (1972).

TABLE III
AVAILABLE STRUCTURAL INFORMATION FOR DEHYDROGENASES

Dehydrogenases	Species for which complete sequence is known	Species for which three-dimensional structure is known	Additional species for which essential thiol sequence is known[a]	Oligomeric structure
LDH	Dogfish M[b]	Dogfish M_4[c]	Bovine, chicken, pig, and bullfrog	Tetramer
s-MDH		Pig[d]		Dimer
LADH	Horse and rat E[e]	Horse E_2[f]	Horse, human liver, and *B. stearothermophilus*	Dimer
Yeast ADH			In progress	Tetramer
GAPDH	Pig, lobster, and yeast[g]	Lobster[h]	Human, rabbit, chicken, ostrich, monkey, badger, sturgeon, bovine, *E. coli*, honey bee, halibut, blue crab, *T. aquaticus*, and *B. stearothermophilus*	Tetramer
GluDH	Chicken, bovine[i]			Hexamer

[a] Data from Dayhoff *et al.* (*5*).
[b] Data from Taylor *et al.* (*4,30a*).
[c] Data from Adams *et al.* (*10–12*).
[d] Data from Hill *et al.* (*13,14*).
[e] Data from Jörnvall (*30b,30c*).
[f] Data from Brändén *et al.* (*15,16*).
[g] Data from Jones and Harris (*30d*).
[h] Data from Buehner *et al.* (*17,18*).
[i] Data from Moon *et al.* (*43*).

Possibly it occurred in an early dehydrogenase and was found advantageous in some cases like GAPDH but has not been mutated in the LDH gene.

Jörnvall (*31*) attempted to compare the sequence of LADH with that of GAPDH. His alignments match the catalytic domain of LADH with the nucleotide binding domain of GAPDH, and thus are unlikely to reflect an early common genetic origin. Similarly, Smith *et al.* (*32*) found

31. H. Jörnvall, *Eur. J. Biochem.* **16,** 25 (1970).
32. E. L. Smith, M. Landon, D. Piszkiewicz, W. J. Brattin, T. J. Langley, and M. D. Melamed, *Proc. Nat. Acad. Sci. U.S.* **67,** 724 (1970).

a dodecapeptide involving an essential lysine in glutamate dehydrogenase (GluDH) which is remarkably similar to a sequence in GAPDH. Engel (*33*) extended this sequence comparison relating two sections of GluDH (one of which contained the essential lysine) with one section of GAPDH. While it is reasonable to anticipate at least two dinucleotide binding domains in GluDH (*34*), the sequences of GluDH picked out by Engel are compared to the catalytic domain of GAPDH. It would seem most improbable that a GAPDH catalytic domain would occur twice in GluDH. In a further examination of possible sequence homologies among dehydrogenases, Engel (*35*) made a variety of suggestions relating GAPDH and LADH which do not correspond to three-dimensional homologies. However, he also suggested a gene duplication within GAPDH which corresponds essentially to the repetition of the mononucleotide binding unit. He aligned the βA, αB, and βB region with βD, αE, and βE.

B. Significance of Comparisons

The above attempts at sequence comparisons demonstrate an expectation of homology among dehydrogenases. However, the difficulty of obtaining significant results where divergence is large requires some systematic method of testing.

Many methods have been described for the measurement of evolutionary distances using amino acid sequence comparisons. In the methods used by Margoliash, Smith, and Fitch (*36,37*), amino acid sequences are first used to show homologous alignments between proteins. The mean change between sequences can then be measured by various scoring procedures such as the counts of amino acids changed, weighted counts of the amino acids according to their function (*38*), base changes of the genetic code (*39*), or on values taken from empirically constructed tables based on observing the frequency of specific changes (*40*). Although some

33. P. C. Engel, *Nature* (*London*) **241,** 118 (1973).
34. J. Krause, M. Buehner, and H. Sund, *Eur. J. Biochem.* **41,** 593 (1974).
35. P. C. Engel, *FEBS* (*Fed. Eur. Biochem. Soc.*) *Lett.* **33,** 151 (1973).
36. E. Margoliash and E. L. Smith, *in* "Evolving Genes and Proteins" (V. Bryson and H. J. Vogel, eds.), p. 221. Academic Press, New York, 1965.
37. E. Margoliash and W. M. Fitch, *Science* **155,** 279 (1967).
38. J. E. Haber and D. E. Koshland, Jr., *JMB* **50,** 617 (1970).
39. T. H. Jukes and C. R. Cantor, *in* "Mammalian Protein Metabolism" (H. N. Munro, ed.), Vol. 3, p. 22. Academic Press, New York, 1969.
40. M. O. Dayhoff, R. V. Eck, and C. M. Park, *Atlas Protein Sequence Struct.* **5,** 89 (1972).

of these techniques have implicitly built in the conservation of functional residues such as histidines or cysteines, none accounts for specific functional necessities, such as the glycine-28 in LDH, which is conserved in all known dehydrogenase structures. The scoring for deletions is also a much discussed problem, but if alignment is based on structure, then this need not be considered. Whichever method of scoring is used, a mean change (per codon or per 100 residues) can be evaluated, along with a standard deviation. This must be compared with an appropriate random change reference value (*38,41,42*).

When structure has been used to obtain alignment, then each pair of equivalent amino acids may be examined independently. Thus, the random change value (1.51 minimum base changes per codon) corresponds to the probability of changing any one amino acid into any other according to their natural frequency. When sequence is used for alignment, the reference value must be computed with respect to the frequency of the amino acids within the sequences in question. Thus, if the peptide is short this number will be appropriately lower. The reference value must also be lowered when specific features such as sheets are used for structural alignments. These features may possess an unusual distribution of amino acid residues. The average number of minimum base changes between randomly selected strands of parallel sheet in carboxypeptidase A and subtilisin has been found to be 1.48 ± 0.04.

IV. Sequence Comparisons Based on Structural Alignments

A. Alignments of LDH, LADH, and GAPDH

A β structure such as the six-stranded coenzyme binding domain in the dehydrogenases would be disrupted by insertions or deletions of amino acids (see Fig. 7 for elaboration). Hence, sequence comparisons of parallel pleated sheet regions are particularly reliable. Structural methods of alignment of sheet areas have been discussed in Section II. The corresponding amino acid comparisons are made in Table IV. For the purpose of this chapter, the present LDH numbering scheme (*4*) will be used as the generalized reference system.

In Table IV the AMP and NMN mononucleotide binding units have also been aligned below each other. The secondary structural features are shown at the top. One row indicates the functions of certain residues.

41. A. D. McLachlan, *JMB* **61**, 409 (1971).
42. S. B. Needleman and C. D. Wunsch, *JMB* **48**, 443 (1970).

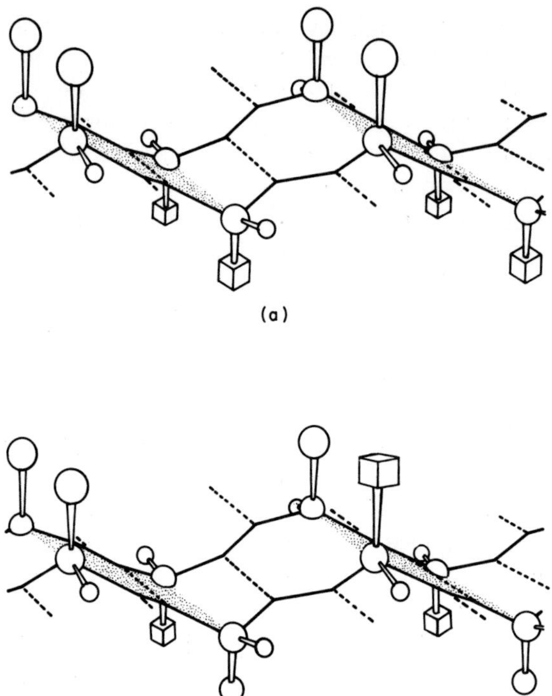

(a)

(b)

FIG. 7. A parallel pleated sheet (a) with hydrophobic residues pointing up and hydrophilic residues pointing down. If a deletion should occur in one strand it will cause a disruption in the physical properties of the protein (b), making the event of such a deletion within a sheet most improbable.

Thus, the residues heavily circled in Fig. 6, representing those sheet residues in the hydrophobic pockets facing the helices, have been labeled H. Similarly, those residues in helices which are in this pocket, and therefore face the sheet, have been labeled S. Still further, residues which are functional in the binding of coenzyme have been labeled F.

The conservation of αB is well demonstrated by the constant number of residues between the aligned strands βA and βB (see Table II). It has been necessary to assume a deletion in LADH between residues 215 and 216. A deletion in the equivalent position 24 of lobster GAPDH with respect to pig and yeast GAPDH has been observed. Thus with respect to this particular residue, lobster GAPDH is similar to LADH, whereas pig and yeast GAPDH are similar to LDH. A deletion in GAPDH of residue LDH 130 in αE has had to be assumed. Neither GAPDH nor LADH has the functional loop present in LDH between residues

99 and 120; they appear to be more like each other than like LDH in the NMN nucleotide binding unit.

B. Glutamate Dehydrogenase Comparison

A systematic search has been made of the GluDH sequences (*43*) for homologies with the nucleotide binding sequences in known dehydrogenase structures. Rossmann *et al.* (*28*) used the procedure of Jukes and Cantor (*39*) and found a region of 38 residues in bovine GluDH that corresponded to βA, αB, and βB in lobster GAPDH. Not only were the minimum number of base changes per codon similar to those observed between other dehydrogenases (Table V), but also the highly conserved glycines 28 and 33 (LDH numbering) were found in corresponding positions (see Section V) and the functional residue in LDH, aspartate-53, changed only conservatively to glutamate. In addition, the character of the other residues marked H and S have been strongly conserved. Particularly noteworthy is the striking homology between LADH and GluDH starting with βA and extending into the beginning of αB. Other regions could be found which satisfied some of the criteria. Only the region starting with bovine GluDH 245 (see Table IV) could be matched with a high degree of confidence, and very likely represents part of one of the dinucleotide binding domains of this enzyme.

Wootton (*43a*) has used the method of Chou and Fasman (*43b*) to predict secondary structural elements of *Neurospora* and beef GluDH. By combining the anticipated sequence of sheets and helices in a dinucleotide binding domain with the requirement for such functional residues as are shown in Table IV, he was able to predict the position of two such domains within the polypeptide chain. One of these corresponds to the prediction of Rossmann *et al.* (*28*).

C. Statistics of Comparisons

Table VI shows the matrix relating the number of amino acids changed per 100 residues between any two dehydrogenases when aligned as shown in Table IV. The uncertainties in correcting for back mutations and superpositions are exceedingly great for proteins that have diverged as far as these dehydrogenases. Thus no attempt has been made to determine the number of accepted point mutations. On the other hand, if the basis

43. K. Moon, D. Piszkiewicz, and E. L. Smith, *Proc. Nat. Acad. Sci. U. S.* **69,** 1380 (1972); K. Moon and E. L. Smith, *J. Biol. Chem.* **248,** 3082 (1973).
43a. J. C. Wootton, *Nature* (*London*) **252,** 542 (1974).
43b. P. Y. Chou and G. D. Fasman, *Biochemistry* **13,** 222 (1974).

TABLE IV

Secondary structure				βA								
	22							29				33
Dogfish LDH	N	K	(I)	T	(V)	V	(G)	C	B	A	V	(G)
	1							8				12
Pig GAPDH	V	K	(V)	G	(V)	D	(G)	F	G	R	I	(G)
Lobster GAPDH	S	K	(I)	G	(I)	D	(G)	F	G	R	I	(G)
Yeast GAPDH	V	R	(V)	A	(I)	D	(G)	F	G	R	I	(G)
	193							200				204
Horse LADH	S	T	(C)	A	(V)	F	(G)	L	G	G	V	(G)
Rat LADH	S	T	(C)	A	(V)	F	(G)	L	G	G	V	(G)
	245							252				256
Bovine GluDH	K	T	F	A	V	Q	G	F	G	N	V	G
	9							16				20
Pig AK	K	I	(I)	F	(V)	V	(G)	G	P	G	S	G
			1					6				10
Flavodoxin (1)[a]			(M)	K	(I)	V	Y	W	S	G	T	G
Flavodoxin (2)	M	K	(V)	N	(I)	I	Y	W	S	G	T	G
Flavodoxin (3)	P	K	(A)	L	(I)	V	Y	G	S	T	T	G
Flavodoxin (4)		M	(V)	E	(I)	V	Y	W	S	G	T	G
	120							127				131
Subtilisin[b]	D	V	(I)	N	(M)	S	L	G	G	P	S	G
Function[c]			H		H		F					S

Secondary structure							βD			
	89				92					
Dogfish LDH	A	G	S		K	L	(V)	V	(I)	T
	87				90					
Pig GAPDH	A	G	T		A	Y	(V)	V	(E)	S
Lobster GAPDH	A	G	A		E	Y	(I)	V	(E)	S
Yeast GAPDH	G	D	S		V	I	(A)	I	(R)	S
	260				263					
Horse LADH	G	G	V		D	F	S	F	E	V
Rat LADH	G	G	V		D	F	X	F	E	V
	110				113					
Pig AK	G	Q	P		T	L	(L)	L	(Y)	V
	44				47					
Flavodoxin (1)	L	N	E		D	I	(L)	I	(L)	G
Flavodoxin (2)	K	E	A		D	V	(V)	A	(F)	G
Flavodoxin (3)	F	E	G	F	D	L	(V)	L	(L)	G
Flavodoxin (4)	A	S	K		D	V	(I)	L	(L)	G
					120					
Subtilisin					D	V	(I)	N	(M)	S
Function							H		H	

Secondary structure					LOOP					
	98							105		
Dogfish LDH	A	(G)	A	R	Q	Q	—	E	G	E
	96			99						
Pig GAPDH	T	(G)	V	F						
Lobster GAPDH	T	(G)	V	F						
Yeast GAPDH	T	(G)	V	F						
	269									
Horse LADH	I	(G)								
Rat LADH	I	(G)								
	119					124				
Pig AK	D	A	G	P	E	T				
	53					62				
Flavodoxin (1)	C	S	A	M	G	D				
Flavodoxin (2)	S	P	S	M	G	S				
Flavodoxin (3)	C	S	T	W	G	D				
Flavodoxin (4)	C	P	A	M	G	S				
	126					131				
Subtilisin	L	G	G	P	S	G				
Function		F								

Secondary structure							αD						
	108							115					
Dogfish LDH	S	(R)	L	N	L	V	Q	R	N	V	N	(I)	(F)
												100	
Pig GAPDH												(T)	(T)
Lobster GAPDH												(T)	(T)
Yeast GAPDH												(T)	(E)
Horse LADH													R
Rat LADH													R
Pig AK													
Flavodoxin (1)													(F)
Flavodoxin (2)													(Z)
Flavodoxin (3)													
Flavodoxin (4)													(V)
Subtilisin													
Function		F										S	S

αB	βB		βC	
3 4	4	5	7 8 8	8
4 0	7	4	6 1 3	4
M A D Ⓐ I S Ⓥ Ⓛ M K D L A	D E Ⓥ A Ⓛ V Ⓓ	V M E D K	A K Ⓘ V Ⓢ G Ⓚ	D
1 1	2	3	6	7
3 9	6	3	9	6
R L Ⓥ Ⓣ R A Ⓐ Ⓕ N S G K V	D I Ⓥ A Ⓘ N Ⓓ	P F I D L	K A Ⓘ T Ⓘ F Ⓠ	E
R L Ⓥ Ⓛ R A Ⓐ Ⓛ S C G — V	Q V Ⓥ A Ⓥ N Ⓓ	P F I A L	K K Ⓘ T Ⓥ F Ⓝ	E
R L Ⓥ Ⓜ R I Ⓐ Ⓛ S R P B A	Z V Ⓥ A Ⓢ B Ⓑ	P F I B L	K K Ⓘ A Ⓣ Y Ⓠ	E
2 2	2	2	2	2
0 1	1	2	3	4
5 1	7	4	6	3
L S Ⓥ Ⓘ M G C K A A G — A	A R Ⓘ I Ⓖ V Ⓓ	I N K D K	G A Ⓣ E C V N	P
L S Ⓥ Ⓥ I G C K T A G — A	A K Ⓘ I Ⓐ V Ⓓ	I N K D K	G A Ⓣ D C I N	P
2 2	2	2		
5 6	6	7		
7 3	9	6		
L H S M R Y L H R F G — A	K C V A V G E	S D G S I		
2 2	8	9		
2 8	9	6		
G T Ⓠ Ⓒ E K Ⓘ Ⓥ Q K Y G Y	G F Ⓛ I D G Y	P		
1 2	2	3		
7 3	8	5		
E L Ⓘ Ⓐ K G Ⓘ Ⓘ E S G — —	K D Ⓥ N Ⓣ I Ⓝ	V S D V N		
K L Ⓘ Ⓐ E G Ⓐ Q E K G — —	A Q Ⓥ K Ⓛ L Ⓝ	V S D A K		
E T Ⓘ Ⓐ R E Ⓛ Ⓐ B A G — —	Y E Ⓥ D Ⓢ R Ⓓ	A A S V E		
N E Ⓘ Ⓔ A A Ⓥ Ⓚ A A G	E S Ⓥ R Ⓕ E Ⓓ	T N V D N		
1 1	1	1	1	1
3 3	4	5	7	7
2 8	7	4	2	9
S A Ⓐ Ⓛ K A Ⓐ Ⓥ D K A G	V V Ⓥ V Ⓐ A A	G N E G S	P S Ⓥ I Ⓐ V G	A
S S S S	H H F		H H H	

αE	βE		βF	
1 1 1	1		1 1	
2 2 2	3		5 6	
1 5 9 A B	3		8 4	
K F Ⓘ Ⓘ P N Ⓘ Ⓥ K H S P D	C I Ⓛ E Ⓛ H P		H R Ⓘ I Ⓖ S Ⓖ	— C
1 1 1	1		1	1
0 0 1	1		4	4
2 6 0	4		1	8
M E Ⓚ Ⓐ G A Ⓗ Ⓛ K — G G A	K R Ⓥ I Ⓘ S A		L K Ⓘ V Ⓢ N Ⓐ	S C
I E Ⓚ Ⓐ S A Ⓗ Ⓕ K — G G A	K K Ⓥ V Ⓘ S A		M T Ⓥ V Ⓢ N Ⓐ	S C
L D Ⓣ Ⓐ Q K Ⓗ Ⓛ K — A G A	K K Ⓥ V Ⓘ T A		L K Ⓘ V Ⓢ N Ⓐ	S C
2 2 2 2 2	2		3	3
7 7 8 8 8	8		1	1
2 6 0 3 6	7		2	9
L D Ⓣ Ⓜ V T Ⓐ Ⓛ S C C Q Y	G V Ⓢ V Ⓘ V G		R T Ⓦ K Ⓖ A Ⓘ	F G
L D Ⓣ Ⓜ A X Ⓧ Ⓛ L S C H C	G V Ⓢ V Ⓘ V G		R T Ⓦ K Ⓖ A Ⓘ	F G
1 1 1	1			
5 5 6	6			
5 9 3	8			
K A Ⓣ E P V Ⓘ Ⓐ F Y E K R	G I Ⓥ R Ⓚ V N			
			1	1
6 7 7	8		0	1
7 1 5	0		8	5
E P Ⓕ Ⓘ E E Ⓘ Ⓢ T K I S G	K K Ⓥ A Ⓛ F G		C V Ⓥ V Ⓔ T Ⓟ	L
Z P Ⓕ Ⓛ D V Ⓥ Ⓢ S I V T G	K K Ⓔ G Ⓐ F			
A Z G	R K Ⓥ A Ⓒ F G		A Z Ⓘ V Ⓩ B Ⓖ	L
E P Ⓕ Ⓕ T D Ⓛ Ⓐ P K L K G	K K Ⓥ G Ⓛ F G		A T Ⓥ I Ⓖ T Ⓐ	
1 1	1		1	1
3 3	4		7	7
2 6	7		2	9
S A Ⓐ Ⓛ K A Ⓐ Ⓥ D K A G	V V Ⓥ V Ⓐ A A		D S Ⓥ I Ⓐ V Ⓖ	A V
S S S S	H H		H H H	

[a] Flavodoxin (numbering scheme according to *Clostridium* MP): (1) *Clostridium* MP flavodoxin (*60*), (2) *Clostridium pasteurianum* flavodoxin (*64*), (3) *Desulfovibrio vulgaris* flavodoxin (*65*), and (4) *Peptostreptococcus elsdenii* flavodoxin (*63*)
[b] Subtilisin hydrogen bonding taken from Drenth *et al.* (*29*)
[c] The NMN mononucleotide binding unit has been aligned below the AMP binding unit. The secondary structural features are shown at the top. The functions H, S, and F refer to being opposite a helix or a sheet, or whether relevant to the binding function, respectively

of scoring is in terms of minimum base changes per codon (Table V), then not only is there more information per residue but also some account of the nature of the amino acid change is considered.

A comparison between the AMP nucleotide binding unit (LDH1, GAPDH1, LADH1, and GluDH1) and the NMN binding unit (LDH2, GAPDH2, and LADH2) is also given in Table V. These values are systematically higher than those from the comparison of complete dinucleotide binding domains. This supports the concept of an earlier gene dupli-

TABLE V

		Dehydrogenases									
		1	2	3	4	5	6	7	8	9	10
Dogfish LDH	1		1.17	1.17	1.08	1.21	1.19	1.31	1.25	1.20	1.24
Pig GAPDH	2			0.32	0.51	1.24	1.14	1.36	1.28	1.17	1.30
Lobster GAPDH	3				0.52	1.21	1.03	1.45	1.27	1.21	1.22
Yeast GAPDH	4					1.26	1.03	1.43	1.26	1.14	1.22
Horse LADH	5						1.14	1.44	1.37	1.39	1.31
Bovine GluDH	6							1.50	1.48	1.43	1.39
Pig AK	7								1.38	1.34	1.37
Flavodoxin (1)[b]	8									0.84	1.29
Flavodoxin (3)[b]	9										1.19
Subtilisin	10										

Dogfish LDH1	11
Pig GAPDH1	12
Lobster GAPDH1	13
Yeast GAPDH1	14
Horse LADH1	15
Bovine GluDH1	16
Pig AK1	17
Flavodoxin1 (1)	18
Flavodoxin1 (3)	19
Subtilisin1	20
Dogfish LDH2	21
Pig GAPDH2	22
Lobster GAPDH2	23
Yeast GAPDH2	24
Horse LADH2	25
Pig AK2	26
Flavodoxin2 (1)	27
Flavodoxin2 (3)	28

cation of the mononucleotide binding unit which then gave rise to the various dinucleotide binding domains observed in dehydrogenases.

V. Functional Aspects of Dinucleotide Binding Domains

Although only NAD-linked dehydrogenases have been examined so far, the structure of the NAD^+ binding region is conserved in the presence of an essential zinc atom as in LADH and where the nicotinamide is B-side specific as in GAPDH. This structure occurs in each subunit for the dimeric s-MDH and LADH molecules as well as the tetrameric LDH and GAPDH molecules. Its biological function is to bind the coenzyme NAD^+ to the enzyme in such a way as to bring the C-4 atom of the

MATRIX SHOWING RELATIONSHIPS BETWEEN MONO- AND DINUCLEOTIDE BINDING STRUCTURES IN TERMS OF MINIMUM BASE CHANGES PER CODON[a]

	Dehydrogenases 1										Dehydrogenases 2						
11	12	13	14	15	16	17	18	19	20	21	22	23	24	25	26	27	28
	1.05	1.00	0.90	1.10	1.08	1.15	1.09	0.92	1.20	1.35	1.30	1.25	1.20	1.30	1.38	1.10	1.05
		0.65	0.55	1.35	1.15	1.31	1.27	1.08	1.35	1.30	1.25	1.20	1.25	1.50	1.31	1.30	1.30
			0.65	1.35	1.08	1.31	1.09	1.08	1.30	1.30	1.35	1.35	1.35	1.50	1.46	1.40	1.35
				1.40	1.08	1.31	0.91	1.00	1.30	1.25	1.15	1.20	1.05	1.45	1.31	1.20	1.00
					1.15	1.15	1.55	1.38	1.45	1.70	1.35	1.50	1.55	1.25	1.31	1.60	1.45
						1.15	1.55	1.54	1.38	1.46	1.15	1.15	1.46	1.54	1.62	1.23	1.46
							1.45	1.46	1.23	1.23	1.46	1.46	1.69	1.00	1.00	1.54	1.54
								1.00	1.36	1.73	1.27	1.36	1.27	1.64	1.18	1.18	1.36
									1.46	1.31	1.38	1.38	1.08	1.46	1.31	1.15	1.15
										1.25	1.25	1.10	1.10	1.20	1.54	1.25	1.15
											1.20	1.30	1.15	1.40	1.31	1.40	1.10
												0.35	0.50	1.45	1.38	1.15	1.10
													0.60	1.25	1.54	1.05	1.15
														1.45	1.54	0.95	1.00
															1.15	1.30	1.45
																1.31	1.23
																	0.65

[a] Comparisons of dinucleotide fragments 1–10 are over sheet and helices as shown in Table IV. Comparison of mononucleotide fragments 11–28 is only over the sheet area.

[b] Flavodoxin (1) and (3) correspond to similar identification in Table IV.

TABLE VI
NUMBER OF AMINO ACID CHANGES PER 100 RESIDUES

		1	2	3	4	5	6	7	8	9	10
Dogfish LDH	1	—	80	81	83	85	81	84	86	82	88
Pig GAPDH	2		—	27	42	83	75	92	89	83	88
Lobster GAPDH	3			—	39	83	75	92	85	83	80
Yeast GAPDH	4				—	81	72	91	85	82	86
Horse LADH	5					—	72	86	92	89	89
Bovine GluDH	6						—	84	88	86	92
Pig AK	7							—	89	88	84
Flavodoxin (1)[a]	8								—	58	90
Flavodoxin (3)[a]	9									—	89
Subtilisin	10										—

[a] Flavodoxin (1) and (3) correspond to similar identification in Table IV.

nicotinamide moiety into a proper position with respect to the substrate and the catalytic residues of the enzyme.

A. NAD+ CONFORMATION

The conformation of the coenzyme when bound to the enzyme has been studied most thoroughly in s-MDH (*14*), but also in LDH (*44*), and to some extent in GAPDH (*18*) and LADH (*16*). Although the constraints on dihedral angles relative to the experimental electron density have been discussed for LDH (*44*) and s-MDH (*14*), results for the coenzyme conformation in GAPDH and LADH are not yet refined. Nevertheless, the structures as compared in Fig. 8 represent the major features of the coenzyme conformations. Webb *et al.* (*14*) defined five distances and an angle (Fig. 9) which they considered to characterize the overall conformation of the coenzyme. These are compared in Table VII. The adenine and nicotinamide rings are not stacked (*45,46*) but about 14Å apart. Furthermore, the adenine and nicotinamide rings are not parallel but nearly perpendicular. The adenine ribose ring was considered to be in the C-3′ endo conformation for s-MDH (*14*) and LDH (*44*) while the nicotinamide ring gave a best fit to the electron density in the C-2′ endo and the C-3′ endo conformations for s-MDH and LDH, respectively.

44. K. Chandrasekhar, A. McPherson, Jr., M. J. Adams, and M. G. Rossmann, *JMB* **76,** 503 (1973).
45. S. F. Velick, *in* "Light and Life" (W. D. McElroy and M. B. Glass, eds.), p. 108. Johns Hopkins Press, Baltimore, Maryland, 1961.
46. C. Y. Lee, R. D. Eichner, and N. O. Kaplan, *Proc. Nat. Acad. Sci. U. S.* **70,** 1593 (1973).

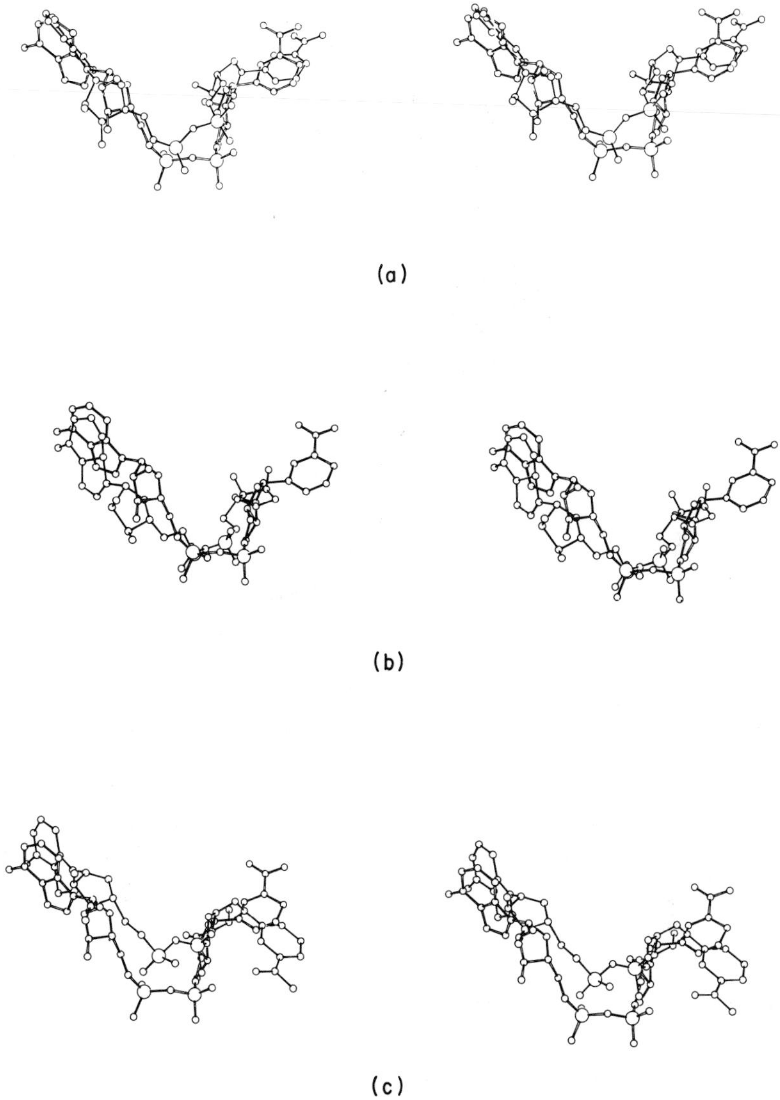

FIG. 8. Comparison of the NAD$^+$ conformation and position between (a) s-MDH and LDH, (b) LADH and LDH (only ADPR is shown for LADH), and (c) GAPDH and LDH. The structure for LDH is in double outline in every case.

The similarities of the coenzyme conformation when bound to the dehydrogenases reflect the similarities of the fold in the dinucleotide binding domains and of detailed similarities in those regions that are actually involved in binding. Unfortunately, details on the interaction between

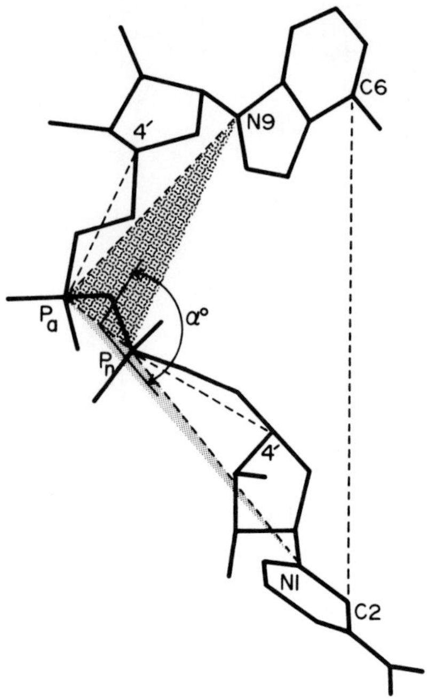

FIG. 9. Structural parameters, in Å and degrees, of bound NAD^+. The angle α, indicated by the arrow, is the dihedral angle between the planes defined by atoms N-9, P_a, P_n and N-1, P_a, P_n with the sign convention defined by Webb *et al.* (14).

NAD^+ and s-MDH are not available because of lack of sequence information, but instead a schematic view of the important interactions between GAPDH and NAD are given in Fig. 10.

TABLE VII
CONFORMATIONAL PARAMETERS OF BOUND COENZYME AS DEFINED IN FIG. 9

Parameter (Å)	s-MDH (holo)	s-MDH (apo)	LDH (ternary)	LADH (ADPR binary)	GAPDH (holo, averaged)
P_a–$N9_a$	6.9	6.9	6.9	6.5	7.1
P_a–$C4_a'$	4.0	3.9	3.9	3.6	4.3
P_n–$N1_a$	6.3	6.5	6.7	—	5.5
P_n–$C4_n'$	3.9	4.0	4.0	3.4	3.6
$C6_a$–$C2_n$	14.3	14.2	14.1	—	15.8[a]
$N1_a$–$N4_n$	16.7	16.2	16.6	—	17.6
α°[b]	−82.8	−81.3	−58.3	—	−85.6

[a] $C6_a$–$C6_n$ to compensate for the B-side specificity in GAPDH.

[b] α is the dihedral angle between the planes defined by the atoms $N9_a$, P_a, P_n and by N1, P_a, P_n.

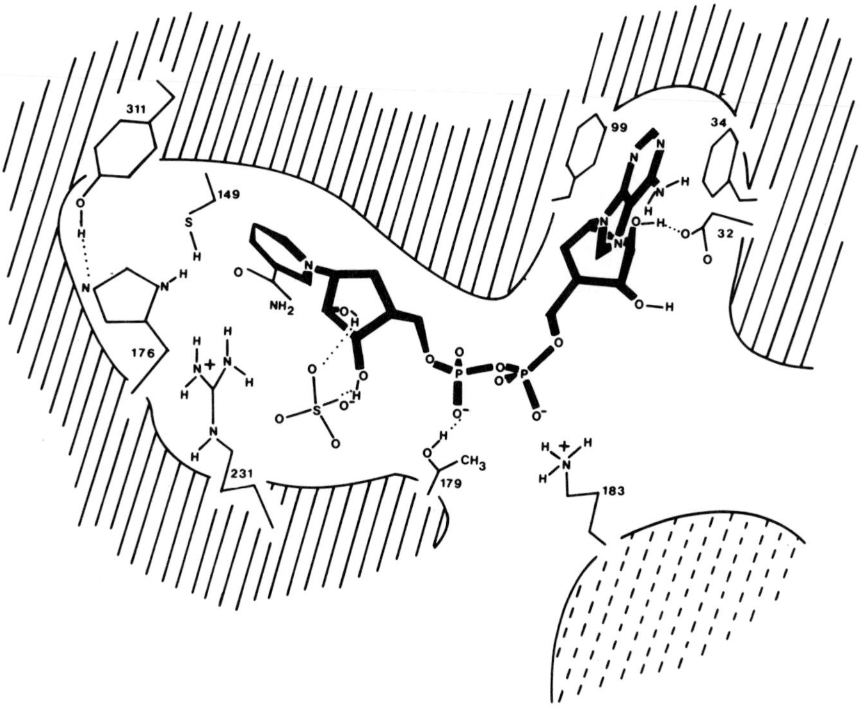

FIG. 10. Schematic view of the NAD⁺ binding to GAPDH, showing functional amino acids.

B. AMP BINDING

The adenosine moiety binds in a hydrophobic crevice, lined by residues from βA, αB, βB, and βD, at the carboxyl end of the parallel β-pleated sheet. Similar hydrophobic pockets are found in all the dehydrogenases, and equivalent residues are listed in Table VIII. The specificity of the hydrophobic pocket is for aromatic groups rather than for adenine in particular as is shown by the binding of 5-iodosalicylate to LADH (*47*) and the affinity of an iodofluorescein dye to LDH (*48*). The pocket in GAPDH is somewhat less hydrophobic in its interior because of the presence of Asp-6 and Asn-31. The adenine ring is situated between Phe-34 and Phe-99.

In all these enzymes the last residue in βA is invariably a glycine (LDH 28). Any larger residue would interfere with the adenine ribose

47. R. Einarsson, H. Eklund, E. Zeppezauer, T. Boiwe, and C.-I. Brändén, *Eur. J. Biochem.* **49,** 41 (1974).

48. P. M. Wassarman and P. J. Lentz, Jr., *JMB* **60,** 509 (1971).

TABLE VIII
RESIDUES LINING THE ADENOSINE-BINDING HYDROPHOBIC POCKET

Structure	LDH	LADH	GAPDH
βA	27 Val 28 Gly	198 Phe 199 Gly 201 Gly	6 Asn 7 Gly 9 Gly
βB	52 Val 53 Asp 54 Val 55 Met	222 Val 223 Asp 224 Ile 225 Asn	31 Asn 32 Asp 33 Pro 34 Phe 35 Ile
βC	85 Tyr		77 Met 78 Lys (main chain) NH–N1 79 Pro
βD	98 Ala	269 Ile	96 Thr
Loop			98 Val 99 Phe
αD	119 Ile		
αE	123 Ile	275 Met	

position. There may be a similar reason for the conservation of LDH Gly-99, although its relative position with respect to the nicotinamide ribose is different in LDH (which has a "loop") as opposed to LADH and GAPDH (which do not have a functional loop). A functionally important conserved residue is an aspartate (LDH 53), the last residue of βB (*12,16–18,27*). It forms a hydrogen bond with the O-2′ hydroxyl of the adenine ribose and thus is a significant constraint on coenzyme binding.

C. NMN BINDING

There is no obvious homology in the residues involved in the binding of the pyrophosphate or nicotinamide ribose moieties. However, the nicotinamide ring itself is invariably in a cavity which is hydrophobic on one side and hydrophilic on the substrate binding side.

When NAD^+ is bound to GAPDH there is a 180° rotation about the glycosidic bond linking the ribose to the nicotinamide moiety, compared to the conformation of NAD^+ in LDH, s-MDH, and by inference in LADH. The effect of this rotation is to change the available side of the nicotinamide ring, one side being blocked by hydrophobic residues on

αB (particularly LDH 32). The rotation is required in GAPDH since otherwise there would be steric hindrance between the carboxyamide group and Ala-120. The corresponding residues in LDH and LADH are Pro-139 and Gly-293, respectively. Thus, the hydrogen is transferred to the C-4 atom of the nicotinamide in either of two sterically distinguishable positions called the A (as in LDH, s-MDH, and LADH) and B (as in GAPDH) sides.

D. Some Generalizations

The structural studies have thus shown for all dehydrogenases investigated that (a) the folding of the coenzyme binding domains are similar, (b) the conformation of the coenzyme when bound to the enzyme is very similar, (c) the detailed interaction between enzyme and at least the adenosine end of the coenzyme is very similar, and (d) the difference between A and B side dehydrogenases is simply a rotation of 180° about the glycosidic bond of the pyridine ring.

Smith (*49*) has discussed the effect of evolutionary pressures on amino acid sequences with respect to the conservation of functional amino acids. The latter must retain their correct relative spatial arrangements. If the function of a protein is to be conserved, its three-dimensional structure must be maintained. Thus, it would seem probable that the above generalizations should be true for all dehydrogenases.

A characteristic feature of the mechanisms of action of most dehydrogenases is an obligatory binding order of coenzyme followed by substrate. This order is consistent with the finding that the binding of coenzyme is accompanied by a conformational change of the protein. The change has been studied most extensively for LDH (*11,12*) where the functional loop and helix αD, between βD and αE, undergo the largest structural alterations. That these changes are accomplished in different ways can be recognized by the absence of such a loop and helix in LADH and GAPDH. The presence of the functional loop in two kinases will be discussed in Section VII.

VI. Domain and Subunit Assembly

A. Gene Fusion

The structures of dehydrogenases provide an interesting example of the combination of different polypeptide sequences into a single chain

49. E. L. Smith, "The Enzymes," 3rd ed., Vol. 1, p. 267, 1970.

as a consequence of gene fusion. Such processes can be studied in bacteria (*50,51*) or observed in the repetition of amino acid sequences (*49*) in a single polypeptide chain. A particularly good example of gene multiplication is in the domains of about 110 amino acids in antibody sequences (*52*) which were subsequently found to have similar three-dimensional structures (*53,54*). The repetition of sequence over a limited range may suggest gene duplication, but this evidence alone can lead to erroneous conclusions (see Section III). The repetition of structure without conservation of function, as is the case for chymotrypsin, may suggest gene duplication. However, when a conserved structure can be associated with sequence homology (*23,55*), preferably supported by a conserved function, definite conclusions can be formed (but, see also Fitch, *55a*).

The polypeptide chains of the subunits of LDH, s-MDH, LADH, and GAPDH can be divided functionally and structurally into separate domains (Fig. 1). In contrast to the structural similarities of the coenzyme binding domains of the dehydrogenases, their catalytic domains are quite different (Fig. 3) and occupy different regions in space relative to their common dinucleotide binding domain. One attractive hypothesis is that the ancestors of each of the dehydrogenases evolved by gene fusion, one gene for the dinucleotide binding protein common to all dehydrogenases and one gene for the catalytic properties different in LDH, LADH, and GAPDH.

In the case of GAPDH sequence comparisons can be made among widely ranging species and related to structure. The NAD^+ binding region conserves 48% of its residues, while the catalytic domain conserves as many as 67% of all residues among sequences as far apart as pig and yeast. Both these mutation rates are slow (*56*) and consistent with the concept of an ancient structure produced by the fusion of two different genes evolving at different rates.

50. E. Jacob, A. Ullmann, and J. Monod, *JMB* **13,** 704 (1965).
51. J. H. Miller, W. S. Rezmkoff, A. E. Silverstone, K. Ippen, E. R. Signer, and J. R. Beckwith, *J. Bacteriol.* **104,** 1273 (1970).
52. G. M. Edelman, B. A. Cunningham, W. E. Gall, P. D. Gottlieb, V. Rutishauser, and M. J. Waxdal, *Proc. Nat. Acad. Sci. U. S.* **63,** 78 (1969).
53. R. J. Poljak, L. M. Amzel, H. P. Avey, L. N. Becka, and A. Nisonoff, *Nature (London), New Biol.* **235,** 137 (1972); R. J. Poljak, *Proc. Nat. Acad. Sci. U.S.* **70,** 3305 (1973).
54. M. Schiffer, R. L. Girling, K. R. Ely, and A. B. Edmundson, *Biochemistry* **12,** 4620 (1973).
55. A. D. McLachlan, *JMB* **64,** 417 (1972).
55a. W. M. Fitch, *Syst. Zool.* **20,** 406 (1971).
56. P. J. McLaughlin and M. O. Dayhoff, *Atlas Protein Sequence Struct.* **5,** 47 (1972).

B. Conservation of Domain Contacts within a Subunit

In all dehydrogenases investigated so far there is an α-helix (α3G in LDH and s-MDH) on the same side of the sheet as αB and αC and running roughly parallel to these and the strands in the sheet. It is situated closest to βD. However, in terms of sequence, this helix is invariably associated with the catalytic domain. Thus, in LDH it involves residues 249–263. In LADH (where it is called αA) it consists of residues 175–194, while in lobster GAPDH it forms the C-terminal helix with residues 316–334.

C. Quaternary Structure

The structural differences of the catalytic domains are also reflected in differences in subunit interactions. For LDH, s-MDH, and GAPDH these have been expressed in terms of an orthogonal, right-handed coordinate system (P,Q,R) defined by Rossmann *et al.* (*57*). In LDH they form the three mutually perpendicular molecular 2-fold axes. These axes can be defined with respect to the direction of the strands in the β-pleated sheet.

1. *Q Axis*

The Q axis is of particular interest in the comparison of dehydrogenases since it associates subunits by contacts between the NAD^+ binding domains. It generates interactions between helices αC, αB, and α3G in one subunit with helices α3G, αB, and αC in the other, respectively. In GAPDH these interactions are weak as a result of the spacing effect generated by the insertion between αC and βC. The Q axis is accurately maintained in s-MDH when compared to LDH (*24*). While diffusion of coenzyme into apo-LDH crystals causes the molecule to lose its accurate 222 symmetry, nevertheless, the Q axis is maintained (*11*). Similarly, an exact (crystallographic) Q axis is found in GAPDH crystals of holoenzyme from humans (*17,58*). Thus in LDH, s-MDH, and GAPDH, but not in LADH, there exist "Q-axis dimers."

2. *P and R Axes—Cooperativity*

Appropriately the association across the catalytic domains can be more easily perturbed. In dimeric s-MDH, there is no further association owing

57. M. G. Rossmann, M. J. Adams, M. Buehner, G. C. Ford, M. L. Hackert, A. Liljas, S. T. Rao, L. J. Banaszak, E. Hill, D. Tsernoglou, and L. Webb, *JMB* **76,** 533 (1973).
58. H. C. Watson, E. Duée, and W. D. Mercer, *Nature (London), New Biol.* **240,** 130 (1972).

to the absence of the amino terminal arm (residues 1–21 in LDH). This arm is the primary cause for the association of the two Q-axis dimers in LDH. Although the directions of the P and R axes with respect to the nucleotide binding structure are roughly the same in LDH and GAPDH, the interactions across these axes are different. If the Q-axis dimers are considered to be associated "head-to-head" in GAPDH, then they are "tail-to-tail" in LDH. Thus, while in LDH the coenzyme sites are well separated on the outside of the molecule, in GAPDH they are close to the subunit interfaces permitting direct interactions between NAD^+ sites across the R axis. The essential Lys-183 of GAPDH approaches the pyrophosphate in the adjacent active site, creating the interesting situation in which the catalytic center contains residues from two different subunits. Thus, the different association of the GAPDH subunits compared to LDH might be correlated with its cooperative phenomena *(18)*. A diagrammatic representation of these different quaternary structures is given in Fig. 11.

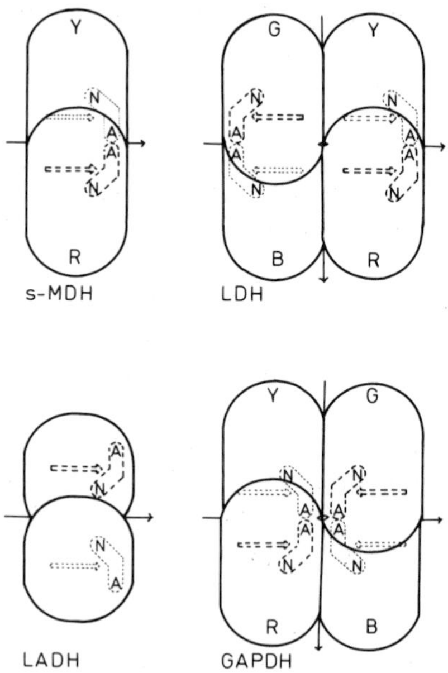

FIG. 11. Diagrammatic comparison of the association of subunits in s-MDH, LDH, GAPDH, and LADH. The Q-axis dimer in s-MDH is also present in LDH and GAPDH but not in LADH. The association of these dimers in LDH is caused by an amino terminal arm of 21 residues. In GAPDH the Q-axis dimers are associated in the opposite manner to LDH bringing the NAD sites close together which is the basis for the observed cooperative phenomena in the former.

3. *LADH Subunit Association*

In LADH the association of subunits is completely different. The molecular twofold axis runs approximately perpendicular to the plane defined by βE and βF causing βF in the associated subunits to make an antiparallel β structure with each other. Thus, in the LADH dimer there is a central 12-stranded sheet. The subunit interaction areas are mainly defined by βF and the main chain connecting βF and βE, which also makes an antiparallel association with the corresponding chain in the neighboring subunit. Because of the different arrangement of the catalytic domains relative to the coenzyme binding domains, this region is entirely on the inside of the subunit in the other dehydrogenases which would make a subunit association as in LADH impossible. Whether or not a similar βF to βF subunit association exists in tetrameric yeast ADH will be of considerable interest, and can probably be established by amino acid sequence comparisons alone.

D. QUATERNARY STRUCTURES—EVOLUTION

The differences in quaternary structure as opposed to the preservation of tertiary structure of the NAD^+ binding domains suggest that subunit associations are of more recent origin (*9*). Buehner *et al.* (*17*) used the preservation of the *Q* axis in GAPDH, LDH, and s-MDH, and the close resemblance of the dimer of s-MDH to one-half of the LDH molecule, to suggest an evolutionary tree (Fig. 12). Also included in Fig. 12 is the relationship of an FMN mononucleotide binding unit in flavodoxin (*24*) to the dehydrogenases.

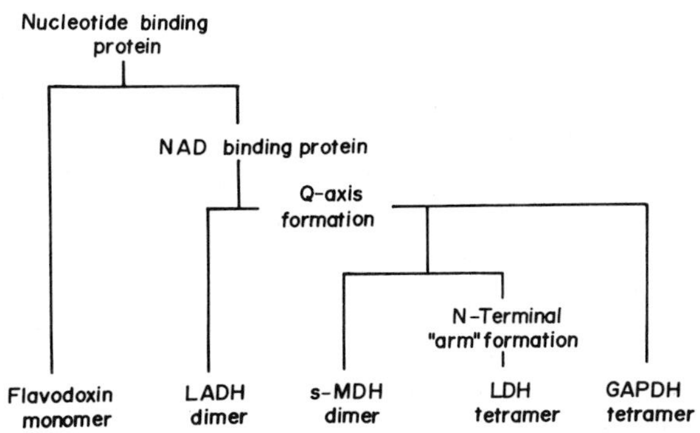

FIG. 12. Evolutionary tree representing possible divergence of present day dehydrogenases and flavodoxin according to Buehner *et al.* (*17*).

VII. The Bigger Family

A. General

The presence of structures similar to the mono- or dinucleotide binding units observed in the dehydrogenases has now been found in a variety of other proteins (Fig. 13). In every case there is evidence that this structure is associated with the binding of nucleotide or aromatic dyes in a manner exceedingly similar to that in dehydrogenases.

B. Members of the Family

1. *Flavodoxin*

The structure of the FMN-linked electron carrier, flavodoxin, has been determined from *Clostridium* MP by Ludwig and co-workers (*59,60*) and the corresponding protein from *Desulfovibrio vulgaris* by Jensen and co-workers (*61*). Rao and Rossmann (*24*) showed that, with the coordinates available to them, the standard deviation, σ, between these two structures this structure can be compared with the AMP mononucleotide binding unit of LDH and found that $\sigma = 2.4$ Å for 39 equivalent atoms. When was 2.1 Å for 102 equivalent C_α atoms. They also showed that part of these two quite different proteins were so aligned, it became clear that FMN binds to flavodoxin at a site similar to where AMP binds to LDH. Not only is it unlikely that this is an accidental structural relationship, but furthermore Baltscheffsky (*62*) has supported a common evolutionary origin for flavin and NAD^+-linked enzymes on the basis of a comparison of their redox potentials.

Since the two mononucleotide binding units in dehydrogenases are very similar, it is also possible to align the FMN binding unit in flavodoxin with the NMN binding unit. This then superimposes virtually the whole structure of flavodoxin on the NAD^+ binding region (Fig. 14a). This alignment also emphasizes the correspondence of the flavin with the functional properties of the nicotinamide moieties.

59. R. D. Anderson, P. A. Apgar, R. M. Burnett, G. D. Darling, M. E. LeQuesne, S. G. Mayhew, and M. L. Ludwig, *Proc. Nat. Acad. Sci. U. S.* **69,** 3189 (1972).
60. R. M. Burnett, G. D. Darling, D. S. Kendall, M. E. LeQuesne, S. G. Mayhew, W. W. Smith, and M. L. Ludwig, *JBC* **249,** 4383 (1974).
61. K. D. Watenpaugh, L. C. Sieker, L. H. Jensen, T. LeGall, and M. Dubourdieu, *Proc. Nat. Acad. Sci. U. S.* **69,** 3185 (1972).
62. H. Baltscheffsky, *in* "Dynamics of Energy Transducing Membranes" (L. Ernster, R. W. Estabrook, and E. C. Slater, eds.). Elsevier, Amsterdam, 1974 (in press).

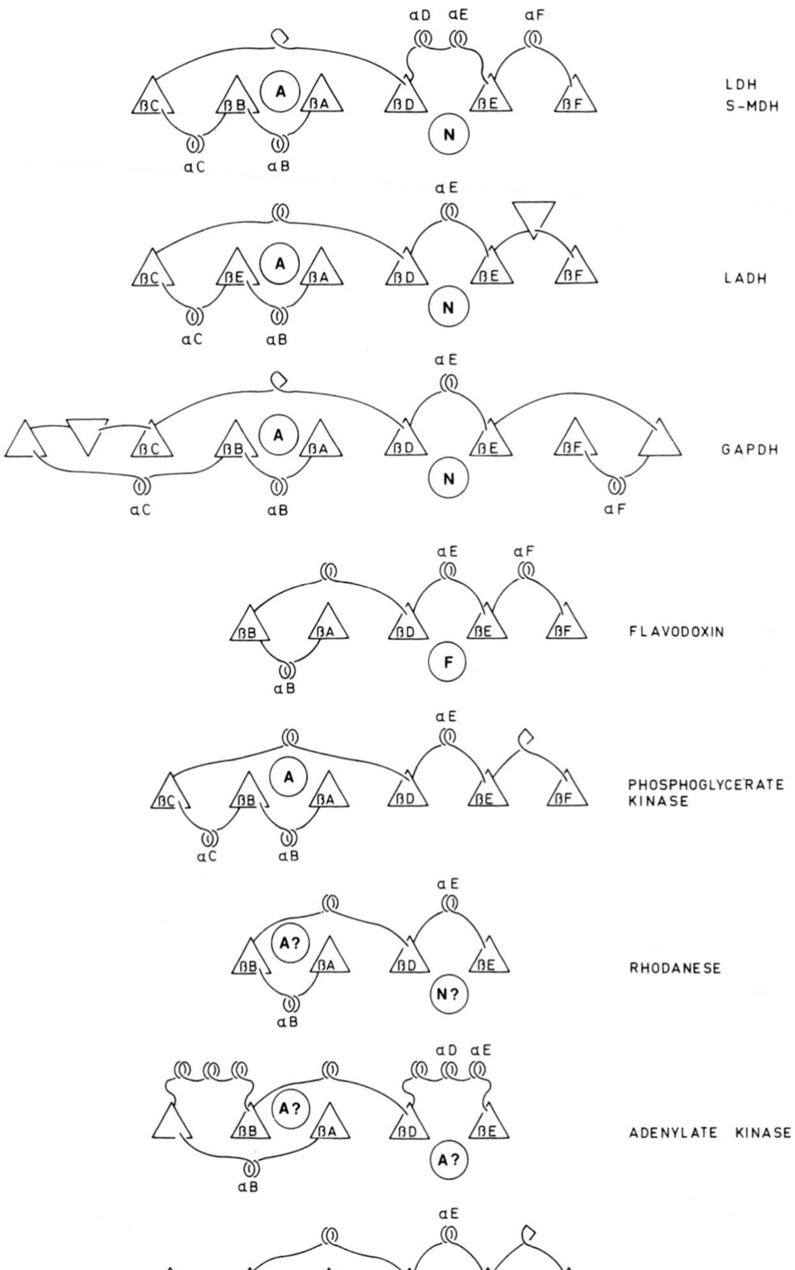

FIG. 13. Diagrammatic representation of nucleotide binding domains, and other similar structures, as found in a variety of proteins. Triangles indicate strands within a β-pleated sheet and loops indicate helices. The amino terminus of the sheets should be considered closest to the viewer.

The hydrogen bonding scheme (Fig. 6) can then be used for alignment with other dehydrogenases for the purposes of sequence comparisons. Three different amino acid sequences for flavodoxins are available (*63–65*) and these are compared, according to the structural alignments, with the dehydrogenases in Table IV. Corresponding minimum base changes per codon are given in Table V. The lack of homology of a residue corresponding to LDH Gly-28 in flavodoxin is possibly a consequence of the absence of a nucleotide binding requirement in this part of the structure. Similarly, there is no binding region corresponding to the hydrophobic adenine pocket. However, the highly conserved nature of some of the "helix" and "sheet" residues such as LDH Gly-33 and LDH residues 44, 49, 93–95, 135, and 160–161 is quite remarkable. Thus, assuming an evolutionary relationship, the earlier separation of the flavodoxin structure from the NAD^+ binding domain present in the dehydrogenases is evident in the data of Table V.

2. *Kinases*

Phosphoglycerate kinase (PGK) is a monomer with a polypeptide chain that folds into two distinct lobes (*66,67*). Blake and Evans (*68*) and also Watson *et al.* (*69*) have shown that one of these lobes is associated with ADP binding while the other is the catalytic domain. It is most probable that the ADP binding lobe has essentially the same structure as the dinucleotide binding fragment in LDH (*68*). There is, however, an exaggerated loop which appears to be subject to conformational changes in the presence of ADP. The adenosine part of ADP binds to the structure in a position corresponding to the adenosine part of NAD in the dehydrogenases (Fig. 14c).

The structure of adenylate kinase (AK) has also been determined (*70,71*). In this case a complete sequence is available (*72*). The enzyme

63. M. Tanaka, M. Hanice, K. T. Yasunobu, J. Mayhew, and V. Massey, *BBRC* **44,** 886 (1971).
64. J. L. Fox, S. S. Smith, and J. R. Brown, *Z. Naturforsch. B* **27,** 1096 (1972).
65. M. Dubourdieu, J. LeGall, and J. L. Fox, *BBRC* **52,** 1418 (1973).
66. C. C. F. Blake, P. R. Evans, and R. K. Scopes, *Nature* (*London*), *New Biol.* **235,** 195 (1972).
67. P. L. Wendell, T. N. Bryant, and H. C. Watson, *Nature* (*London*), *New Biol.* **240,** 134 (1972).
68. C. C. F. Blake and P. R. Evans, *JMB* **84,** 585 (1974).
69. T. N. Bryant, H. C. Watson, and P. L. Wendell, *Nature* (*London*) **247,** 14 (1974).
70. G. E. Schulz, M. Elzinga, F. Marx, and R. H. Schirmer, *Nature* (*London*) **250,** 120 (1974).
71. G. E. Schulz and R. H. Schirmer, *Nature* (*London*) **250,** 142 (1974).
72. A. Heil, G. Müller, L. H. Noda, T. Pinder, I. Schirmer, R. H. Schirmer, and I. von Zabern, *Eur. J. Biochem.* **43,** 131 (1974).

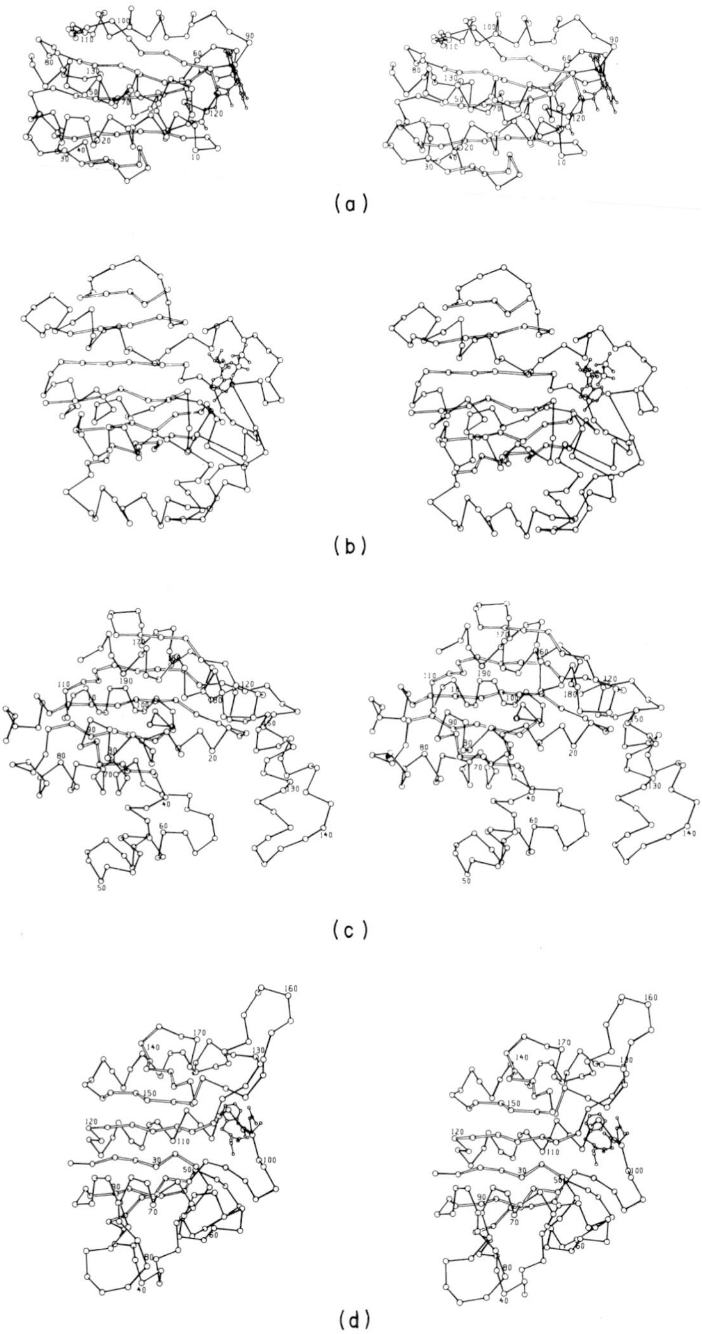

FIG. 14. Stereo views of various structures related to dehydrogenases by having similar nucleotide binding function and/or similar structures. Views are oriented as in Fig. 2 for the dehydrogenases, showing (a) flavodoxin, (b) adenylate kinase, (c) the ADP binding lobe of phosphoglycerate kinase, and (d) the central portion of subtilisin.

requires two independent binding sites for ADP as it catalyzes the interconversion of two ADP molecules to ATP and AMP in the presence of a magnesium ion. Its structure has been found to consist of five parallel strands with helices on either side (Fig. 14b) corresponding to the dinucleotide binding domains of the dehydrogenases. The positions of bound nucleotides relative to the protein have not yet been definitely established. However, comparisons of the amino acid sequence (Table IV) aligned on the basis of the hydrogen bonding pattern (Fig. 6) suggest considerable homology. It is of particular interest that the two invariant glycines (LDH 28 and 33) are homologous. Minimum base changes per codon show that, like flavodoxins, AK is not as closely related to the dehydrogenases as these are to each other.

A bilobal structure for each of the two monomers of hexokinase has also been reported (*73*). The similarity of hexokinase and PGK lobe structure suggests that these two enzymes have some similarity.

3. *Rhodanese*

Although the physiological function of rhodanese is not fully understood, it catalyzes the overall reaction:

$$S_2O_3^{2-} + CN^- \leftrightarrows SCN^- + SO_3^{2-}$$

The nucleotides NAD^+, NADH, FAD, and FMN are all known to be inhibitors. The structure of this enzyme was solved (*74*) and was found to have two repeating structural domains along a (probably) single polypeptide chain. Each domain contained a fold reminiscent of the nucleotide binding domains in the dehydrogenases. It remains to be seen whether nucleotides bind in a manner suggested by the structural similarity.

4. *Subtilisin*

The comparison of a part of subtilisin (Fig. 14d) with the mononucleotide binding unit in dehydrogenases produced unexpected results (*24*). The sequence comparison, based on the hydrogen bonding patterns in Fig. 6, is shown in Table IV and the corresponding minimum base changes per codon in Table V. It is remarkable how well some of the residues correspond. LDH Gly-33 is invariant as in every other case and residues corresponding to LDH 24, 26, 49, and 50 are all homologous. Rao and Rossmann (*24*) showed $\sigma = 4.49$ Å for 29 equivalent atoms. The nucleotide binding pockets of the dehydrogenases correspond to the

73. T. A. Steitz, R. J. Fletterick, and K. J. Hwang, *JMB* **78,** 551 (1973).
74. J. D. G. Smit, Ph.D. Thesis, University of Gröningen, Holland, 1973.

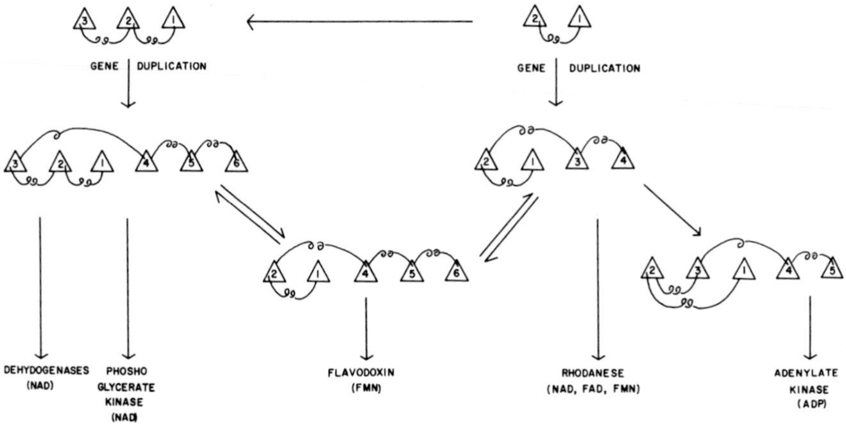

FIG. 15. Possible evolutionary relationships among various nucleotide binding proteins, as suggested by their structures.

aromatic specificity pocket of subtilisin. Kraut (*75*) pointed out that various dyes and nucleotides can competitively inhibit subtilisin, presumably by preferential binding into this pocket.

C. Structural Relationships among the Family

A conjecture, based on available structural comparisons, relating the development of the known or suspected nucleotide binding proteins has been shown in Fig. 15. If this scheme is right, then the primordial nucleotide binding fragment might be represented by two parallel strands connected by a helix.

D. A Time Scale

A relative estimate of time between the different events of gene duplication can be obtained (*28*) from the study of base changes per codon. These are summarized in Table IX. Although it is clear that these figures show a definite trend, and that the dehydrogenases are most closely related to each other among the bigger family, some caution must be given: (1) In a comparison of this kind it is assumed that the rate of accepted mutations has been roughly constant throughout the evolutionary process, and (2) the changes are not linearly related because of the absence of any correction for back mutations and superpositions. Thus, progressively increasing time intervals might be expected for equal increments in base changes per codon.

75. J. Kraut, "The Enzymes," 3rd ed., Vol. 3, p. 165, 1971.

Rossmann *et al.* (*28*) have suggested a rough time scale (Fig. 16) for the events in Fig. 12. The time of divergence of different genera is well established (nodes 1, 2, and 3 in Table IX), but less certain is the time of earlier events. The development of a mononucleotide binding unit

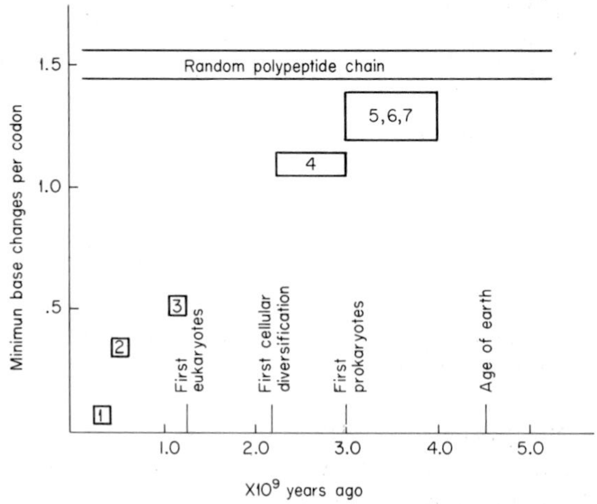

FIG. 16. Relationship between a probable time scale and the averaged minimum base changes per codon observed between different nucleotide binding structures. Numbers refer to correspondingly numbered events in Table IX.

TABLE IX

SUMMARY OF MONO- AND DINUCLEOTIDE BINDING PROTEIN COMPARISONS

Comparison	*a*	*b*	*c*	*d*
1. Bovine GluDH — chicken GluDH	1	0.03	0.03	0.03
2. Pig GAPDH — lobster GAPDH	1	0.32	0.32	0.32
3. Yeast GAPDH — muscle GAPDH	2	0.52	0.52	0.51
4. Dinucleotide binding domains in dehydrogenases	12	1.16	1.26	1.03
5. Dinucleotide binding domain in pig AK — dinucleotide binding domains in dehydrogenases	6	1.42	1.50	1.31
6. Nucleotide binding domain in flavodoxins — dinucleotide binding domain in dehydrogenases	12	1.29	1.48	1.14
7. AMP binding unit — NMN binding unit in dehydrogenases	30	1.33	1.70	1 05
8. Random comparisons: (a) in general			1.57	
(b) in sheets			1.48	

[a] Number of comparisons.

[b] Mean minimum base change per codon over all comparisons.

[c] Comparison with the largest minimum base change per codon.

[d] Comparison with the smallest minimum base change per codon.

probably preceded the formation of the first prokaryotes 3.2×10^9 years ago (*76*). Since the age of the earth is around 4.5×10^9 years ago (*77*), there is a period of 10^9 years during which the mononucleotide binding unit could have developed. The rectangles on the graph (Fig. 16) indicate degrees of uncertainty in both absolute and relative time. Also shown is the random reference level.

The significance of the evolutionary tree in Fig. 12 lies in the time scale covered. The enzymes compared and related here were already necessary in the most essential metabolic pathways (such as glycolysis) of the most primitive cells. In contrast, phylogenetic trees based on amino acid sequences or fossil data mostly relate to the development of species.

E. The Primordial Mononucleotide Binding Proteins

Inasmuch as any gradually divergent trends in the property of biological material can be taken as evidence for a single common ancestor, the progressive structural changes shown in Figs. 12, 15, and 16 are consistent with the occurrence of a common primordial mononucleotide binding protein. The data presented here are consistent with the occurrence of such a protein prior to the formation of the first biological cells.

It is now firmly established that this basic structural unit is present twice in the NAD-binding domain of dehydrogenases, and it seems very probable that the same domain was incorporated into the kinases for ATP binding. Based on these results Blake and Evans (*68*) made the interesting suggestion that this domain is adenosine specific and that the presence of adenosine in the four important cofactors, NAD^+, ATP, FAD, as well as in coenzyme A, may be a consequence of a mutual evolutionary history. The presence of the mononucleotide binding unit in flavodoxin for FMN binding and also in rhodanese and subtilisin is, however, an indication that the basic nucleotide binding unit is of wider generality.

It might thus be expected that this structure has been incorporated in many other proteins wherever there was a need for its characteristic function. A requirement of binding nucleotides to proteins occurs in energy transfer systems and in the molecular reproduction of nucleic acids. Consequently, it is possible that this molecular fossil may be found in such diverse proteins as tRNA synthetase, ribosomal proteins, and virus coat proteins. The recognition of its structure by sequence homology or from X-ray structure determinations may also give guidance as to function where none is known.

76. S. E. Barghorn, *Sci. Amer.* **225,** 30 (1971).
77. D. York and R. M. Farguhar, "The Earth's Age and Geochronology." Pergamon, Oxford, 1972.

Acknowledgments

The authors are exceedingly grateful for technical assistance in the preparation of this manuscript to Sharon Wilder; to Bo Furugren, Hannu Ukonen, and Connie Braun in the artistic preparation of some of the diagrams; to Dr. Dino Moras and Dr. Ken Olsen for preparation of Fig. 5; to Agneta Borell for preparation of Tables V and VI and other technical assistance; and to Hannu Ukonen and William Boyle for photography. They also wish to express their appreciation for help and cooperation given by Dr. Georg Schulz, Dr. Heiner Schirmer, Dr. Colin Blake, Dr. Martha Ludwig, and Dr. Susan Taylor in allowing them to work with unpublished results and to Dr. Walter Fitch for helpful comments on this manuscript, particularly in relationship to the flavodoxin comparisons.

3

Alcohol Dehydrogenases

CARL-IVAR BRÄNDÉN • HANS JÖRNVALL •
HANS EKLUND • BO FURUGREN

I. Alcohol Dehydrogenases*

A. Introduction

Since the chapter on alcohol dehydrogenase appeared in "The Enzymes" in 1963 (*1*) there has been a considerable advance in our knowledge, especially of the liver enzyme. The main advances have been the following: (a) Determination of the complete amino acid sequences of the horse and, except for a few positions, rat liver enzymes; partial sequences of the protein from other sources have also been obtained, permitting an estimation of functional and evolutionary connections between different alcohol dehydrogenases. (b) Determination of the X-ray structure of the horse liver enzyme; this structure determination provided one of the keys to the structural, functional, and evolutionary relations among different dehydrogenases that are discussed in a special chapter of this volume (*2*). Knowledge of the tertiary structure has provided a firm ground for future detailed investigations of the reaction mechanism. (c) Elucidation of the genetic and molecular basis of the isozyme pattern of LADH from several species has been achieved. (d) Inhibitor studies have provided a battery of different inhibitors specific for different steps in the catalytic mechanism. (e) Rapid kinetic studies have provided new ideas about the mechanism of action. (f) Discoveries of substrates of possible physiological significance for the liver enzyme have been made.

In this chapter we will describe these advances with an emphasis on the structures of the alcohol dehydrogenases and the relationship between structure and function. Although various aspects of alcohol dehydrogenases have been treated in some review articles since 1963 (*3–6*), none has appeared after this structural knowledge became available. Only a brief discussion of the kinetic studies will be included here since a special chapter (*7*) is devoted to the kinetics of the NAD-requiring enzymes.

* Abbreviations used: LADH, horse liver alcohol dehydrogenase (EC 1.1.1.1); YADH, yeast alcohol dehydrogenase (EC 1.1.1.1); LDH, lactate dehydrogenase (EC 1.1.1.27); s-MDH, soluble malate dehydrogenase (EC 1.1.1.37); and GAPDH, glyceraldehydephosphate dehydrogenase (EC 1.2.1.12).

1. H. Sund and H. Theorell, "The Enzymes," 2nd ed., Vol. 7, p. 25, 1963.
2. M. G. Rossmann, A. Liljas, C.-I. Brändén, and L. J. Banaszak, Chapter 2, this volume.
3. S. P. Colowick, J. van Eys, and J. H. Park, *Compr. Biochem.* **14**, 1 (1966).
4. A. S. Mildvan, "The Enzymes," 3rd ed., Vol. 2, p. 445, 1970.
5. D. D. Ulmer and B. L. Vallee, *Advan. Enzymol.* **27**, 37 (1965).
6. H. Sund, *in* "Biological Oxidations" (T. P. Singer, ed.), p. 641. Wiley, New York, 1968.
7. K. Dalziel, Chapter 1, this volume.

B. Functional Aspects

Yeast and mammalian alcohol dehydrogenase differ in substrate specificity and rate of catalytic activity (*1*). The classic yeast enzyme is more specific for acetaldehyde and ethanol, which is consistent with its recognized physiological significance to participate in alcohol fermentation at the end of the glycolytic pathway. Enzyme forms with other functions and properties also occur in yeast (Section III,A). The mammalian enzymes have a broad substrate specificity and, even with primary alcohols, the maximum activity is not observed with ethanol. Mammalian alcohol dehydrogenase has, therefore, attracted great interest; first, regarding its true physiological function, especially in view of the great amount of alcohol dehydrogenase present in liver (*1*), and, second, regarding its role in alcohol metabolism in man.

1. *Physiological Substrates*

Previously known substrates have been reviewed earlier (*1*) as well as the widespread occurrence of alcohol dehydrogenase in nature. Alcohols, including ethanol, produced in the intestinal tracts mainly by bacterial actions, are found in the portal vein. One physiological function of liver alcohol dehydrogenase may be to metabolize these products (*8*).

The discovery of the steroid-active isozymes of horse liver alcohol dehydrogenase (Sections II,A,1,a and II,A,1,d) also established that 3-keto- and 3β-hydroxysteroids are substrates, but the functional significance of this is not clear. Since this activity also seems to be present in the rat enzyme (*9*) and in all isozymes of the human enzyme (*10*), it may be more important than previously realized when the horse "ethanol-active" isozyme was the most studied form. It may also be noted that mammalian livers contain many different specific steroid dehydrogenases (*11*). Structural studies have established that mammalian alcohol dehydrogenases have a distant evolutionary link to both the yeast (*12*) and bacterial enzymes (*13*).

Recently discovered substrates of physiological interest in mammals

8. H. A. Krebs and J. R. Perkins, *BJ* **118,** 635 (1970).
9. O. Markovič, H. Theorell, and S. Rao, *Acta Chem. Scand.* **25,** 195 (1971).
10. R. Pietruszko, H. Theorell, and C. de Zalenski, *ABB* **153,** 279 (1972).
11. R. I. Dorfman and F. Ungar, "Metabolism of Steroid Hormones," p. 394. Academic Press, New York, 1965.
12. H. Jörnvall, *Proc. Nat. Acad. Sci. U. S.* **70,** 2295 (1973).
13. J. Bridgen, E. Kolb, and J. I. Harris, *FEBS (Fed. Eur. Biochem. Soc.) Lett.* **33,** 1 (1973).

include 5β-cholestane-3α,7α,12α,26-tetrol (*14,15*), which is an intermediate in bile acid biosynthesis (*16*). These, as well as ω-hydroxylated fatty acids which are intermediates in ω degradation of fatty acids (*17*), have been found to be substrates (*18*) for both types of homogeneous isozymes of horse liver alcohol dehydrogenase (*19,20*).

2. *Activity in Ethanol Metabolism*

Ingested alcohol is metabolized to acetaldehyde mainly by the action of liver alcohol dehydrogenase. Catalase (*21,22*), the "microsomal ethanol oxidizing system (MEOS)" (*23–25*), and extrahepatic pathways have also been considered as ethanol metabolizers, but these systems probably play only a minor role in most cases (for a detailed discussion of ethanol metabolism, see review by Hawkins and Kalant, *26*). Ethanol metabolism produces an increase in the NADH/NAD^+ ratio in the liver (*27*) and is inhibited *in vivo* by the alcohol dehydrogenase inhibitors pyrazole (*28,29*) and 4-methyl-pyrazole (*30*).

The multiplicity of isozyme patterns of alcohol dehydrogenase in human liver (Section II,A,2), together with racial and acquired differences in ethanol metabolism (*31,32*), have attracted great interest. No clear correlation has, however, so far been found between patterns of alcohol dehydrogenase isozymes and ethanol metabolism (*33*) or alcoholism.

14. G. Waller, H. Theorell, and J. Sjövall, *ABB* **111**, 671 (1965).
15. K. Okuda and N. Takigawa, *BBA* **220**, 141 (1970).
16. H. Danielsson, *Acta Chem. Scand.* **14**, 348 (1960).
17. K. C. Robbins, *Fed. Proc., Fed. Amer. Soc. Exp. Biol.* **20**, 272 (1961).
18. I. Björkhem, *Eur. J. Biochem.* **30**, 441 (1972).
19. I. Björkhem, H. Jörnvall, and E. Zeppezauer, *BBRC* **52**, 413 (1973).
20. I. Björkhem, H. Jörnvall, and Å. Åkeson, *BBRC* **57**, 870 (1974).
21. D. Keilin and E. F. Hartree, *Proc. Roy. Soc., Ser. B* **119**, 141 (1936).
22. D. Keilin and E. F. Hartree, *BJ* **39**, 293 (1945).
23. W. H. Orme-Johnson and D. M. Ziegler, *BBRC* **21**, 78 (1965).
24. C. S. Lieber and L. M. DeCarli, *J. Clin. Invest.* **47**, 62a (1968).
25. C. S. Lieber and L. M. DeCarli, *JBC* **245**, 2505 (1970).
26. R. D. Hawkins and H. Kalant, *Pharmacol. Rev.* **24**, 67 (1972).
27. O. Forsander, N. Räihä, and H. Suomalainen, *Hoppe-Seyler's Z. Physiol. Chem.* **312**, 243 (1958).
28. L. Goldberg and U. Rydberg, *Biochem. Pharmacol.* **18**, 1749 (1969).
29. D. Lester, W. Z. Keokosky, and F. Felzenberg, *Quart. J. Stud. Alc., Part A* **29**, 449 (1968).
30. R. Blomstrand and H. Theorell, *Life Sci., Part II* **9**, 631 (1970).
31. D. Fenna, L. Mix, O. Schaefer, and J. A. L. Gilbert, *Can. Med. Ass. J.* **105**, 472 (1971).
32. C. S. Lieber, *Ann. Intern. Med.* **76**, 326 (1972).
33. J. A. Edwards and D. A. Price Evans, *Clin. Pharmacol. Ther.* **8**, 824 (1967).

II. Liver Alcohol Dehydrogenases

A. Multiple Molecular Forms

1. *Horse Liver Alcohol Dehydrogenase*

Horse liver alcohol dehydrogenase was first crystallized by Bonnichsen and Wassén in 1948 (*34*). An acidic minor component was isolated by Dalziel (*35*), and different forms were later shown to exist (*36,37*). Neither of these studies revealed the true isozyme pattern of horse liver alcohol dehydrogenase, and an increasing number of different molecular forms have since been characterized. The multiplicity is a result of the synthesis of different types of subunits as well as of the occurrence of secondary modified forms.

a. Different Types of Subunits. Three main isozymes are formed by the dimeric combination of two different types of protein chains. This has been established by dissociation–reassociation experiments (*38–40*), structural analysis (*41*), immunological (*42,43*) and other studies of the purified isozymes (*44–46*). The subunits are not active in the monomeric state (*45*). Details of the subunit interactions are discussed in Sections II,C and II,I. Based on a difference in substrate specificity (Section II,A,1,d) the two types of protein chains have been called E (for "ethanol-active") and S (for "steroid-active") by Theorell (*39,44,45*), and the main isozymes are thus EE, ES, and SS. It should be noticed, however, that SS also has ethanol activity (although lower than ES and still lower than EE) and that EE has some activity toward certain steroid

34. R. K. Bonnichsen and A. M. Wassén, *ABB* **18**, 361 (1948).
35. K. Dalziel, *Acta Chem. Scand.* **12**, 459 (1958).
36. J. S. McKinley-McKee and D. W. Moss, *BJ* **94**, 16P (1965); **96**, 583 (1965).
37. D. H. Treble, *BJ* **82**, 129 (1962).
38. R. Pietruszko, H. J. Ringold, T.-K. Li, B. L. Vallee, Å. Åkeson, and H. Theorell, *Nature* (*London*) **221**, 440 (1969).
39. R. Pietruszko and H. Theorell, *ABB* **131**, 288 (1969).
40. U. M. Lutstorf and J. P. von Wartburg, *Fed. Eur. Biochem. Soc., Proc. Meet.* **6**, 92 (1969); *FEBS* (*Fed. Eur. Biochem. Soc.*) *Lett.* **5**, 202 (1969).
41. H. Jörnvall, *Eur. J. Biochem.* **16**, 41 (1970).
42. R. Pietruszko and H. J. Ringold, *BBRC* **33**, 497 (1968).
43. R. Pietruszko, H. J. Ringold, N. Kaplan, and J. Everse, *BBRC* **33**, 503 (1968).
44. T. Theorell, *in* "Nobel Symposium 11" (A. Engström *et al.*, eds.), p. 145. Almqvist & Wiksell, Stockholm, 1968.
45. H. Theorell, *in* "Pyridine Nucleotide-Dependent Dehydrogenases" (H. Sund, ed.), p. 121. Springer-Verlag, Berlin and New York, 1970.
46. U. M. Lutstorf, P. M. Schürch, and J. P. von Wartburg, *Eur. J. Biochem.* **17**, 497 (1970).

side chain hydroxyl groups (*15,19*). The E/S nomenclature will be used in the present article, but other lettering or numbering systems (*38,40,42,46–48*) have also been introduced and compared (*45*). The isozymes may be identified by electrophoresis (*38–40,44–47*): SS is the most basic, ES the intermediate, and EE the least basic form, with pI's of >10, 9.3, and 8.7, respectively (*46,49*), as determined by isoelectric focusing. Similar values were estimated from the electrophoretic mobilities (*50*), but lower values are obtained for EE in phosphate buffers (*35*). Six amino acid exchanges between the primary structures of the E and S subunits have been identified (Section II,B,1). The two protein chains are apparently synthesized from two different autosomal gene loci, related by a comparatively recent gene duplication (*41*).

b. Modified Forms. Several subfractions, related to the main forms by nongenetic differences, also occur. The subfractions are more acidic than the corresponding main forms, and are, in this nomenclature, identified as "primed" fractions (*39,44,45*), e.g., EE′, EE″, EE‴, following decreasing cathodic electrophoretic mobility. In horse liver, subfractions are usually most intense and numerous in the EE group and the intensity of the bands decreases with decreasing electrophoretic mobility (*40,44,45*). In total, at least 12 different forms of the enzyme have been demonstrated by electrophoresis and chromatography (*46*).

Structural interrelationships among the subfractions are not established and conversions in one (*45*) or both (*40*) directions have been reported. Binding of extrinsic molecules (*36*) as well as different conformational states ("conformers") (*40*), deamidations (*45*), and oxidations of SH groups (*51,52*) have been suggested as possible causes for these subfractions. Combined effects or alternative explanations in different cases may not be excluded.

c. Purification. EE is the main component of the originally crystallized enzyme (*34*), but all three isozymes and some subfractions may be prepared by ion-exchange chromatography (*35,40,46,48,53*). The ES (*46,48*) and SS isozymes (*54*) have also been crystallized. Recently, the enzyme has been purified, and the EE and SS isozymes separated by affinity chromatography on an immobilized AMP analog (*51*). Commercially

47. R. Pietruszko, A. Clark, J. M. H. Graves, and H. J. Ringold, *BBRC* **23**, 526 (1966).
48. H. Theorell, S. Taniguchi, Å. Åkeson, and L. Skurský, *BBRC* **24**, 603 (1966).
49. J. P. von Wartburg, P. M. Kopp, and U. M. Lutstorf, *FEBS Symp.* **18**, 195 (1970).
50. H. Jörnvall, *BBRC* **35**, 542 (1969).
51. L. Andersson, H. Jörnvall, Å. Åkeson, and K. Mosbach, *BBA* **364**, 1 (1974).
52. H. Jörnvall, *BBRC* **53**, 1096 (1973).
53. K. Dalziel, *BJ* **80**, 440 (1961).
54. Å. Åkeson and G. Lundquist, reported in Jörnvall (*41*).

available preparations contain the EE group and small amounts [2–5% (*47*)] of the ES family.

d. Substrate Specificities. Pietruszko *et al.* (*47*) established that the steroid activity previously detected in some alcohol dehydrogenase preparations (*14,55–57*) was not attributable to the original EE form. 3β-Hydroxyl groups in steroids are reduced only by isozymes containing the S subunit (*47,48,58*) which has a specific activity with ethanol lower than that of the E type (*20,39*). Hydroxyl groups in many other substrates are also oxidized by all isozymes (*46*) which therefore exhibit in common the broad substrate specificity previously known for the enzyme (*1*). Divergent kinetic behavior and differences in the dependence of the specific activities on pH (*46*) occur. Neither is coenzyme binding identical since NADH is bound more firmly to SS than to EE and with different pH dependence (*45*).

With ω-hydroxylated fatty acids as substrates, E and S subunits have specific activities of similar magnitude (*20*). On the other hand, with an ω-hydroxylated steroid, 5β-cholestane-3α,7α,12α,26-tetrol, the specific activity is more than 50 times higher for the SS isozyme than for the EE isozyme (*20*). For certain steroid hydroxyl groups there is thus an absolute difference between the subunits regarding one end (3β-OH) of the substrate molecule and a relative difference regarding another end (26-OH). The activity of the S subunit toward differently positioned steroid hydroxyl groups indicates a large hydrophobic substrate binding site (*20*) slightly different from that of the E chain (Section II,D,1,a). Differences must however be small, since 3α-steroids are efficient inhibitors (*58*) for both types of subunits.

2. *Human Liver Alcohol Dehydrogenase*

Human liver alcohol dehydrogenase was purified by von Wartburg *et al.* (*59*) and crystallized by Mourad and Woronick (*60*). Chromatographically (*61*) and electrophoretically (*62,63*) distinct forms were demon-

55. F. Ungar, *Univ. Minn. Med. Bull.* **31,** 226 (1960).
56. F. Ungar, M. Goldstein, and C.-M. Kao, *Steroids, Suppl.* **1,** 131 (1965).
57. J. M. H. Graves, A. Clark, and H. J. Ringold, *Proc. Pan-Amer. Congr. Endocrinol., 6th, 1965* p. E86 (1966).
58. M. Reynier, H. Theorell, and J. Sjövall, *Acta Chem. Scand.* **23,** 1130 (1969).
59. J. P. von Wartburg, J. L. Bethune, and B. L. Vallee, *Biochemistry* **3,** 1775 (1964).
60. N. Mourad and C. L. Woronick, *ABB* **121,** 431 (1967).
61. A. H. Blair and B. L. Vallee, *Biochemistry* **5,** 2026 (1966).
62. K. Moser, J. Papenberg, and J. P. von Wartburg, *Enzymol. Biol. Clin.* **9,** 447 (1968).
63. P. Pikkarainen and N. C. R. Räihä, *Nature* (*London*) **222,** 563 (1969).

strated in other studies. Forms with similar electrophoretic patterns but different kinetic properties ("atypical enzyme") were detected in some livers by von Wartburg and co-workers (*64,65*). Individual variations in the multiplicity (*10,65–68*) as well as changes during development (*63,66,68–70*) were also shown. Human alcohol dehydrogenase is a dimer (*10,66,67,71–74*) with subunit structures homologous to those of the horse enzyme (*73,75*) with which hybrid enzymes may be formed (*71,72*). A zinc content of 2–2.5 atoms per enzyme molecule has been found (*59,61*) but, as discussed in Section II,D,1,b, each subunit in all probability contains two firmly bound zinc atoms *in vivo*. Variations mainly result from synthesis of different types of protein chains from separate gene loci as well as from allelic genes (*66,67,74*), but the occurrence of secondary modified forms has also been demonstrated (*10,71,72*).

a. Different Types of Subunits. Genetic analyses by Smith *et al.* (*66,67,74,76*) clearly show the presence of three different autosomal gene loci (ADH_1, ADH_2, and ADH_3, coding for polypeptides α, β, and γ, respectively) for alcohol dehydrogenase. The genes are active to various extents in different tissues and at different times during development. Two common alleles occur at each of the ADH_2 and ADH_3 loci, coding for different forms of the corresponding peptides; there may also be a second allele at the ADH_1 locus (*68*). Complex mixtures of dimeric enzyme forms containing identical or nonidentical subunits are thus present. Most of the separate forms may be resolved by electrophoresis (*66,67,72,74,76*), but some are almost indistinguishable and their relative separation is dependent on pH (*10*) and type of zone electrophoresis (*68*). Different forms have been purified by ion-exchange chromatography (*10,72,74,75*).

Results from purification of isozymes (*10,72*) and structural studies (*73,75*) are compatible with the genetic analyses and also establish the

64. J. P. von Wartburg, J. Papenberg, and H. Aebi, *Can. J. Biochem.* **43**, 889 (1965).
65. J. P. von Wartburg, and P. M. Schürch, *Ann. N.Y. Acad. Sci.* **151**, 936 (1968).
66. M. Smith, D. A. Hopkinson, and H. Harris, *Ann. Hum. Genet.* **34**, 251 (1971).
67. M. Smith, D. A. Hopkinson, and H. Harris, *Ann. Hum. Genet.* **35**, 243 (1972).
68. R. F. Murray, Jr. and P. H. Price, *Ann. N.Y. Acad. Sci.* **197**, 68 (1972).
69. R. F. Murray, Jr. and A. G. Motulsky, *Science* **171**, 71 (1971).
70. P. Pikkarainen and N. C. R. Räihä, *Pediat. Res.* **1**, 165 (1967).
71. T. M. Schenker and J. P. von Wartburg, *Experientia* **26**, 687 (1970).
72. T. M. Schenker, L. J. Teeple, and J. P. von Wartburg, *Eur. J. Biochem.* **24**, 271 (1971).
73. H. Jörnvall and R. Pietruszko, *Eur. J. Biochem.* **25**, 283 (1972).
74. M. Smith, D. A. Hopkinson, and H. Harris, *Ann. Hum. Genet.* **36**, 401 (1973).
75. D. Berger, M. Berger, and J. P. von Wartburg, *Eur. J. Biochem.* **50**, 215 (1974).
76. M. Smith, D. A. Hopkinson, and H. Harris, *Ann. Hum. Genet.* **37**, 49 (1973).

presence of different types of subunits. Exact correlations between the isozymes in all reports are difficult to establish since phenotypic variations were not characterized in the chemical studies. The known amino acid exchange between two different types of subunits in the "nonatypical" enzymes (Section II,B) may therefore not be strictly correlated with the genetic analysis. The most common isozyme in adult livers is among those with highest cathodic mobility (*10,66,72*). The "atypical" enzyme probably results from a mutant allele (*72*) at the ADH_2 locus (*66,74*) and thus contains one β subunit with a known amino acid substitution (*75*). This variation is almost indistinguishable upon electrophoresis (*72*) but gives a slight separation (*66,68*).

b. Modified Forms. Isolated isozymes are converted into fractions with less electrophoretic mobility upon storage or incubation at basic pH. Reversions upon treatment with mercaptoethanol have been reported (*10,72*). These properties are similar to those of the horse (Section II,A,1,b) and rat (Section II,A,3,a) enzymes and the same explanations have been suggested (*10,71,72*).

c. Enzymic Properties. In contrast to the horse isozymes all isolated isozymes of human liver alcohol dehydrogenase appear to be active with a 3β-hydroxysteroid with an activity about 5% of that of ethanol (*10*). All isozymes of the human enzyme have a broad substrate specificity (*64,76*), but variations in relative activities (*10,64,76*) are present and sensitivities toward inhibitors are different.

Isozymes containing the atypical subunit have a pH optimum around pH 8.5–8.8 with ethanol as substrate, whereas the other isozymes have an optimum at pH 10.8–11.5 (*72,76*), agreeing with the properties of the atypical enzyme in the liver (*64,65*). The small structural difference, resulting from the mutant allele at the ADH_2 locus, thus contributes a considerable effect on the enzymic activity.

Trichloroethanol is a potent inhibitor for the ADH_1 isozymes, isobutyramide for the ADH_3 isozymes, and pyrazole and thiourea for the atypical ADH_2 isozymes (*76*). The differential effect of thiourea, pyrazole, and other inhibitors was detected early (*64,65*). The isozymes also differ in stability. The atypical ADH_2 isozymes are the most labile; ADH_1 isozymes are also less stable than the other isozymes (*76*).

3. *Rat Liver Alcohol Dehydrogenase*

Partial purifications (*15,77*) as well as preparations of pure enzyme (*9,78*) have been reported. The protein was crystallized by Markovič

77. M. Reynier, *Acta Chem. Scand.* **23,** 1119 (1969).
78. M. J. Arslanian, E. Pascoe, and J. G. Reinhold, *BJ* **125,** 1039 (1971).

et al. (*9*) who furthermore demonstrated the presence of electrophoretically separable forms. These were further studied by Jörnvall (*52*) who also showed that secondary modifications may occur. Multiplicities were observed in other studies too (*79–81*) but not in all cases (*15,78*). Various electrophoretic patterns are evidently obtained depending on the experimental conditions. This is compatible with the known differences. No evidence for different types of subunits has been detected in any of these studies or in a developmental study (*82*) of the protein. Structural work on the enzyme (*80,83,84*) has hitherto revealed only one type of subunit. The presence of other subunits, especially in low amounts or with a preponderance in tissues outside the liver, should not be excluded in view of the multiplicity of rat liver alcohol dehydrogenase (*52*) and the complex pattern of the human enzyme (Section II,A,2,a).

a. Modified Forms. Four (*9*) or more (*52*) electrophoretically different enzyme forms were found to be convertible to more cathodic forms by treatment with thiols. Differential carboxymethylation with [^{14}C]iodoacetate before and after conversion suggested (*52*) that all cysteine thiol groups were present in the most cathodic subfraction. Different combinations of a few half-cystine residues might explain the occurrence of various numbers of less cathodic forms. Three residues (Cys-97, Cys-100, and Cys-103, Section II,B) were differentially reactive, indicating if SS interchanges are excluded that these residues are most likely to participate in the conversions accompanying the subfractionation.

Similar electrophoretic properties of subfractions have also been observed in the horse (*51*) and human (*72*) enzyme forms, but it seems that the rat alcohol dehydrogenase is most sensitive to conversions. This may be accounted for by the fact that the rat enzyme has the greatest total cysteine content (*80,84*) and, in particular, an extra cysteine residue at position 112 (Section II,B,3). This is close to the four cysteine residues which are ligands to the second zinc atom (Section II,C,3,b) and which are common to all three mammalian proteins (Section II,B). These residues are therefore in close proximity, compatible with formation of disulfide bridges without great structural or functional alterations. In addition, this region, from the tertiary structure (Section II,C,3,b), seems less essential for catalytic activity.

79. O. A. Forsander and P. Saarinen (1966), quoted in Räihä *et al.* (*82*).
80. H. Jörnvall and O. Markovič. *Eur. J. Biochem.* **29**, 167 (1972).
81. A. L. Koen and C. R. Shaw, *BBA* **128**, 48 (1966).
82. N. C. R. Räihä, M. Koskinen, and P. Pikkarainen, *BJ* **103**, 623 (1967).
83. H. Jörnvall, *FEBS* (*Fed. Eur. Biochem. Soc.*) *Lett.* **28**, 32 (1972).
84. H. Jörnvall, *in* "Alcohol and Aldehyde Metabolizing Systems" (R. G. Thurman *et al.*, eds.), p. 23. Academic Press, New York, 1974.

B. Primary Structures

1. *Horse Liver Alcohol Dehydrogenase*

a. The Ethanol-Active Subunit. The sequence of a fragment containing a reactive cysteine residue, selectively ^{14}C-labeled by carboxymethylation (*85–87*), was reported by Harris (*86*), and part of this structure was also determined by Li and Vallee (*87*). Attempts to identify a protein N-terminal residue (*88,89*) were negative, but the presence in each subunit of an acetyl group, probably blocking the N-terminus, was demonstrated (*90*). The C-terminus and adjacent residues were also determined (*89*).

Jörnvall and Harris (*91*) presented data for the structures around all of the 14 cysteine residues in each protein chain. Analysis by Jörnvall (*92,93*) of different peptide mixtures obtained after treatment of the protein with trypsin (before or after maleylation), chymotrypsin, pepsin, cyanogen bromide, or thermolysin yielded amino acid sequence information for all parts of the subunit and the primary structure of the whole protein chain was deduced (*93*). It was found to contain 374 residues and is shown in Table I. An acetylated serine residue is at the N-terminus and the reactive cysteine residue is at position 46. Some residues are unevenly distributed (*93*). Six of the seven histidine residues are in the N-terminal half of the molecule, the two tryptophan residues are in either terminal region, the four tyrosine residues are in the middle of the primary structure, and none of the 14 cysteine residues occur in the C-terminal quarter of the molecule. A characteristic distribution of hydrophobic residues was also noticed (*93*), which may now be partly correlated with the presence of large hydrophobic cores in the tertiary structure of the protein (Section II,C,3). Most regions of the primary structure were analyzed in many different overlapping peptides (*92–94*) with a corresponding increase in reliability. The structure is in excellent agreement with the total composition determined by acid hydrolysis (*93*). It is compatible with independently determined partial structures of

85. T.-K. Li and B. L. Vallee, *BBRC* **12,** 44 (1963).
86. I. Harris, *Nature* (*London*) **203,** 30 (1964).
87. T.-K. Li and B. L. Vallee, *Biochemistry* **3,** 869 (1964).
88. H. Jörnvall, *Acta Chem. Scand.* **19,** 1483 (1965).
89. R. D. Hamburg, Dissertation, University of California, Berkeley, 1966.
90. H. Jörnvall, *Acta Chem. Scand.* **21,** 1805 (1967).
91. H. Jörnvall and J. I. Harris, *Eur. J. Biochem.* **13,** 565 (1970).
92. H. Jörnvall, *Eur. J. Biochem.* **14,** 521 (1970).
93. H. Jörnvall, *Eur. J. Biochem.* **16,** 25 (1970).
94. H. Jörnvall, Dissertation, Karolinska Institutet, Stockholm, 1970.

TABLE I

PRIMARY STRUCTURES OF LIVER ALCOHOL DEHYDROGENASES

```
                  1                  5                   10                    15                   20
Horse[a]   Acetyl-Ser-Thr-Ala-Gly-Lys -Val-Ile-Lys -Cys-Lys -Ala-Ala -Val-Leu-Trp-Glu-Glu-Lys -Lys -Pro -Phe-Ser -Ile-Glu-
                                                                            Gln
Rat[b]     Acetyl-Ser-Thr-Ala-Gly-Lys -Val-Ile-Lys -Cys-Lys -Ala-Ala-Val-Leu-Trp-Glu-Pro-His -Lys -Pro-Phe-Thr-Ile-Glu-
Human[c]   Acetyl-Ser-Thr-Ala-Gly-Lys/Val-Ile-Lys/Cys-Lys/                                    /Lys/Pro (Phe,Ser, Ile,Glx,

 25              30                 35                  40                   45                  50
-Glu-Val-Glu-Val-Ala-Pro-Pro-Lys -Ala-His -Glu-Val-Arg -Ile-Lys -Met-Val-Ala-Thr-Gly-Ile -Cys-Arg -Ser-Asp-Asp-His-Val-Val-Ser-
-Asp-Ile -Glu-Val-Ala-Pro-Pro-Lys -Ala-His-Glu-Val-Arg -Ile-Lys -Met-Val-Ala-Thr-Gly-Ser-Cys-Arg -Ser-Asp-Asp-His-Ala-Val-Ser-
,Asx ,Val,Glx ,Val,Ala,Pro,Gly)Lys/Ala(Tyr,Glx,Val)Arg/Ile-Lys/Met-Val-Ala-Val -Gly-Ile -Cys-Arg/Ser(Asx, Asx,His, Val, Val,Ser,
                                                                   Ala

 55               60                65                   70                  75                  80
-Gly-Thr-Leu-Val -Thr-Pro-Leu-Pro-Val-Ile -Ala -Gly-His-Glu-Ala-Ala-Gly-Ile-Val-Glu-Ser-Ile-Gly-Glu-Gly-Val-Thr-Thr-Val-Arg-
-Gly-Ser -Leu-Phe-Thr-Pro-Leu-Pro-Ala-Val-Leu-Gly-His-Glu-Gly-Ala-Gly-Ile-Val-Glu-Ser-Ile-Gly-Glu-Gly-Val-Thr-Cys-Val-Lys-
-Gly, Thr ,Leu,Val ,Thr,Pro ,Leu,Pro,Val ,Leu,Ala ,Gly,His,Glx ,Ala ,Ala,Gly,Ile ,Val,Glx,Ser,Ile ,Gly,Glx,Gly ,Val,Thr,Thr,Va ,Lys,

 85              90                 95                   100                 105
-Pro-Gly-Asp-Lys -Val-Ile-Pro-Leu-Phe-Thr-Pro-Gln-Cys-Gly-Lys-Cys-Arg -Val-Cys-Lys-His-Pro-Glu-Gly-Asn-
                                      Ile                       Ser
-Pro-Gly-Asp-Lys -Val-Ile-Pro-Leu-Phe-Ser -Pro-Gln-Cys-Gly-Lys-Cys-Arg -Ile -Cys-Lys-His-Pro-Glu-Ser -Asn-
,Pro,Gly,Asx)Lys/Val-Ile-Pro-Leu-Phe-Thr-Pro-Gln-Cys-Gly-Lys-Cys-Arg/

 110                   115                120                 125                 130
-Phe-Cys-  –  -Leu-Lys -Asn-Asp-Leu-Ser -Met-Pro-Arg -Gly-Thr -Met-Gln-Asp-Gly-Thr-Ser-Arg -Phe-Thr-Cys-Arg -Gly-
 Leu                        (Ser)
-Leu-Cys-Cys-Gln-Thr -Lys -Asn -Leu-Thr-Gln -Pro-Lys -Gly-Ala -Leu -Leu-Asp-Gly-Thr-Ser-Arg -Phe-Ser -Cys-Arg -Gly-
                     /Asn(Asx ,Leu,Ser,  Met,Pro)Arg/                                 /Phe(Thr,Cys)Arg/Gly-

 135            140                 145                 150                 155                 160
-Lys -Pro-Ile-His-His-Phe-Leu-Gly-Thr-Ser-Thr-Phe-Ser-Gln-Tyr-Thr-Val-Val-Asp-Glu-Ile-Ser-Val-Ala-Lys -Ile-Asp-Ala-Ala-Ser-
-Lys -Pro-Ile-His-His-Phe-                        -Ser-Gln-Tyr-Thr-Val-Val-Asp-Asp-Ile-Ala-Val-Ala-Lys-Ile-Asp-Ala-Ala-Ala-
-Lys/                                                                                          /Ile(Asx ,Val,Ala,Ser,

 165             170                175                 180                 185
-Pro-Leu-Glu-Lys -Val-Cys-Leu-Ile-Gly-Cys-Gly-Phe-Ser-Thr-Gly-Tyr-Gly-Ser-Ala-Val-Lys -Val-Ala-Lys -Val-
-Pro-Leu-Asp-Lys -Val-Cys-Leu-Ile-Gly-Cys-Gly-Phe-Ser-Thr-Gly-Tyr-Gly-Ser-Ala-Val-Gln -Val-Ala-Lys -Val-
,Pro,Leu,Glx)Lys/                                                                   /Val-Ala-Lys/
```

190 195 200 205 210
-Thr-Gln-Gly-Ser-Thr-Cys-Ala-Val-Phe-Gly-Leu-Gly-Gly-Val-Gly-Leu-Ser-Val-Ile - Met-Gly-Cys-Lys -Ala -Ala-
-Thr-Pro-Gly-Ser-Thr-Cys-Ala-Val-Phe-Gly-Leu-Gly-Gly-Val-Gly-Leu-Ser-Val-Val- Ile -Gly -Cys-Lys -Thr-Ala-
/Ala -Ala-

215 220 225 230 235
-Gly-Ala-Ala-Arg -Ile-Ile-Gly-Val-Asp-Ile-Asn-Lys -Asp-Lys -Phe-Ala-Lys -Ala-Lys -Glu-Val -Gly-Ala-Thr-Glu-
-Gly-Ala-Ala-Lys -Ile-Ile-Ala-Val-Asp-Ile-Asn-Lys -Asp-Lys -Phe-Ala-Lys -Ala-Lys -Glu-Leu-Gly-Ala-Thr-Asp-
-Gly-Ala-Ala-Arg/ /Asp-Lys/Phe-Ala-Lys/Ala-Lys/Glu-Leu-Gly-Ala-Thr-Glx -
Pro

240 245 250 255 260
-Cys-Val-Asn-Pro-Gln-Asp-Tyr-Lys -Lys-Pro-Ile-Gln-Glu-Val-Leu-Thr -Glu-Met-Ser-Asn-Gly-Gly-Val-Asp-Phe-
-Cys-Ile -Asn-Pro-Gln-Asp-Tyr-Thr -Lys-Pro-Ile-Gln-Glu-Val-Leu-Thr -Glu-Met-Ser-Asn-Gly-Gly-Val-Asp-Phe-
-Cys-Ile -Asx-Pro-Glx-Asx-Tyr-Lys/Lys-Pro-Ile-Gln-Glu-Val-Leu-Lys/

265 270 275 280 285 290
-Ser-Phe-Glu-Val-Ile-Gly-Arg -Leu-Asp-Thr-Met-Val-Thr-Ala-Leu-Ser -Cys-Cys-Gln-Glu-Ala-Tyr-Gly-Val-Ser- Val-Ile-Val-Gly-Val-
-Phe-Glu-Val-Ile-Gly-Arg -Leu -Asp-Thr-Met-Ala- -Leu-Leu-Ser -Cys-His-Ser -Ala-Cys-Gly-Val-Ser-Val-Ile-Val-Gly- Val-
/Leu(Asx, Thr, Met, Ile, Thr, Ala, Leu, ,Cys, Cys, Glx, Glx, Ala, Tyr, Gly, Val, Ser, Val, Ile, Val, Gly, Val,

295 300 305 310 315
-Pro-Pro-Asp-Ser -Gln-Asn-Leu-Ser-Met-Asn-Pro-Met-Leu-Leu-Leu-Ser -Gly -Arg -Thr-Trp-Lys -Gly-Ala-Ile-Phe-
-Pro-Pro-Val -Ala-Gln-Ser -Leu-Ser-Val -Asn-Pro-Met- -Leu-Gly -Arg -Thr-Trp-Lys -Gly-Ala-Ile-Phe-
,Pro, Pro, Asx, ,Glx, ,Leu, Ser, Met, Asx, Pro, Met, Leu, Leu, Leu, ,Gly)Arg/Thr-Trp-Lys/

320 325 330 335 340
-Gly-Gly-Phe-Lys -Ser-Lys -Asp-Ser -Val-Pro-Lys-Leu -Val-Ala-Asp-Phe-Met-Ala-Lys -Lys -Phe-Ala-Leu-Asp-Pro-
-Gly-Gly-Phe-Lys -Ser-Lys -Asp-Ala-Val-Pro-Lys-Leu-Val-Ala-Asp-Phe-Met-Ala-Lys -Lys -Phe-Pro-Leu-Glu-Pro-
/Ser-Lys/ Leu(Val, Ala, Asx, Phe, Met, Ala)Lys/Lys/

345 350 355 360 365 370 374
-Leu-Ile-Thr-His-Val-Leu-Pro-Phe-Glu-Lys-Ile-Asn-Glu-Gly-Phe-Asp-Leu-Leu-Arg-Ser-Gly-Glu-Ser-Ile-Arg -Thr-Ile -Leu-Thr-Phe
Lys
-Ile-Thr-His-Val-Leu-Pro-Phe-Glu-Lys-Ile-Asn-Glu-Ala-Phe-Asp-Leu-Leu-Arg-Ser-Gly-Lys-Ser-Ile-Arg -Thr-Val-Leu-Thr-Phe
/Thr-Val-Leu-Thr-Phe

---

[a] The E subunit, from Jörnvall (*93*), for which a complete structure has been determined, in top line. Numbering system refers to this sequence. Known exchanges in S subunit (*41*), detected by peptide mapping, are at positions 17, 94, 101, 110, 115, and 366, and are given immediately below corresponding E-chain residues.

[b] From Jörnvall (*84*). Except for a few positions not analyzed (empty spaces) and for some peptides ordered by homology, the structure is complete. An insertion, compared to the horse protein, is between positions 111 and 112 in the latter.

[c] From Jörnvall and Pietruszko (*73*) for sequence analysis and from Berger *et al.* (*75*) for total compositions of tryptic peptides. In the latter case residues have been ordered to yield maximum homology with the other structures. Peptides were similarily positioned in both studies. Empty spaces not analyzed. Known subunit differences are the bottom line alternatives at position 43 (*73*) affecting one set of subunits, and at 230 (*75*) affecting another set (the atypical enzyme).

homologous enzymes from other species (below), and it fits the electron density maps obtained by X-ray crystallographic analysis (Section II,C).

b. The Steroid-Active Subunit. The degree of structural identity between the E and S subunits was determined by peptide mapping of tryptic fragments and by analysis of those peptides which were found to differ. An initial study (*50*) in which mainly the EE and ES isozymes were compared demonstrated a high degree of similarity and some tentative amino acid exchanges. In later studies (*41,95*) in which the SS isozyme was also available, S-chain peptides were produced in higher yield and six amino acid exchanges were detected. They are shown in Table I and make the S chain three units more positively charged, compatible with the electrophoretic mobilities of the isozymes (*41*). The structural implications of these observed differences are discussed in Section II,D,1,a.

2. *Human Liver Alcohol Dehydrogenase*

Peptide mapping and sequence analysis of about one-fifth of all residues in the subunit of human liver ADH showed that the human and horse enzymes are identical to about 90% (*73*). A similar estimate was also obtained by analysis of the compositions of tryptic peptides from about half of the molecule (*75*). Known segments of the structure of the human enzyme are summarized in Table I.

Subunit differences have also been characterized. One of the amino acid differences between two types of subunits is a Val/Ala exchange (*73*) at position 43 (with the numbering of the horse enzyme, Table I). The subunit responsible for the atypical human alcohol dehydrogenase has been reported to have Pro instead of Ala (*75*) at position 230. This position is, in the horse enzyme, in αC in the coenzyme binding domain (Fig. 5) close to the reactive lysine residue 228 (Section II,E,1.b).

3. *Rat Liver Alcohol Dehydrogenase*

The rat and horse enzymes were found to be identical in primary structure to about 80% (*80*). The amino acid sequences around all 16 cysteine residues have been determined (*83*): Further studies of peptides obtained after treatment with trypsin, chymotrypsin, cyanogen bromide, or thermolysin (*84*) permitted identification of nearly all residues in the primary structure of the protein. Overlapping fragments were not recovered from all regions and peptides were in some places ordered utilizing the great homology with the horse enzyme. In this way a tentative amino acid sequence for the whole subunit of rat liver alcohol dehydrogenase has been obtained (*84*). It is shown in Table I. Long regions are

95. H. Jörnvall, *Nature* (*London*) **225,** 1133 (1970).

completely identical between the rat and horse proteins, and the characteristic distribution of hydrophobic residues in the amino acid sequence of the horse enzyme (Section II,B,1,a) is also present in the sequence of the rat enzyme (*84*). Therefore, it is concluded that the tertiary structure of rat liver alcohol dehydrogenase is very similar to that of the horse enzyme. At some regions, however, differences in the primary structure are extensive. One such region comprises residues 108–124 where, furthermore, an insertion in the rat enzyme (or deletion in the horse) of a cysteine residue is present. This residue is adjacent to one of the cysteine residues which is thought to be associated with the conversions between subfractions of the rat enzyme (Section II,A,3,a). At positions 276–286, differences are also extensive between the rat and horse proteins.

C. Tertiary Structure

1. *Crystallization and Preliminary X-Ray Studies*

Crystallization conditions for the preliminary X-ray work were described by Yonetani and Theorell (*96–98*) for binary complexes with NADH, ADP-ribose and 1,10-phenanthroline as well as for the ternary complexes E·NAD^+·pyrazole, E·NADH·isobutyramide, and E·ADP-ribose·1,10-phenanthroline. Crystals of both the apoenzyme and these complexes suitable for detailed X-ray diffraction studies were prepared using pure isozyme EE by precipitation with ethanol (*99*) or 2-methyl-2,4-pentane-diol (*100*).

The apoenzyme crystallizes in an orthorhombic space group with one subunit in the asymmetric unit (*99*). Crystals prepared by precipitation with alcohol are isomorphous to those obtained from weak phosphate buffer (*101*). Binary complexes with inhibitors such as ADP-ribose, 1,10-phenanthroline, imidazole, and salicylate crystallize isomorphously with the apoenzyme (*99,102*). However, all complexes containing coen-

96. T. Yonetani and H. Theorell, *ABB* **100,** 554 (1963).
97. T. Yonetani, *Biochem. Z.* **338,** 300 (1963).
98. H. Theorell and T. Yonetani, *Biochem. Z.* **338,** 537 (1963).
99. E. Zeppezauer, B.-O. Söderberg, C.-I. Brändén, Å. Åkeson, and H. Theorell, *Acta Chem. Scand.* **21,** 1099 (1967).
100. C.-I. Brändén, E. Zeppezauer, B.-O. Söderberg, T. Boiwe, B. Nordström, G. Söderlund, M. Zeppezauer, P-E. Werner, and Å. Åkeson, *in* "Structure and Function of Oxidation-Reduction Enzymes" (Å. Åkeson and A. Ehrenberg, eds.), p. 93. Pergamon, Oxford, 1972.
101. C.-I. Brändén, L.-M. Larsson, I. Lindqvist, H. Theorell, and T. Yonetani, *ABB* **109,** 195 (1965).
102. R. Einarsson, H. Eklund, E. Zeppezauer, T. Boiwe, and C.-I. Brändén, *Eur. J. Biochem.* **49,** 41 (1974).

TABLE II
CRYSTAL FORMS OF COMPLEXES WITH LADH

Orthorhombic form		Monoclinic form	Triclinic form
Cell dimensions	a = 56.0 Å	a = 51 Å	a = 51.8 Å
	b = 75.2 Å	b = 44 Å	b = 44.6 Å
	c = 181.6 Å	c = 182 Å	c = 94.3 Å
Space group	$C222_1$	γ = 108°	α = 104.4°
		Space group $P2_1$	β = 101.9°
			γ = 70.9°
1 subunit/asymmetric unit		2 subunits/asymmetric unit	
Complexes known to crystallize in this form		Complexes known to crystallize in both these forms	
LADH		LADH·NADH	
LADH·1,10-phenanthroline		LADH·1,4,5,6-tetrahydro-NAD	
LADH·AMP		LADH·NADH·isobutyramide	
LADH·ADP-ribose		LADH·NAD^+·pyrazole	
LADH·ADP-ribose·1,10-phenanthroline		LADH·NAD^+·dodecanoate	
LADH·salicylate		LADH·NAD^+·isobutyramide	
LADH·imidazole		LADH·NADH·dimethylsulfoxide	
LADH·modified with iodoacetic acid		LADH·3-I pyridine AD^+·pyrazole	

zyme give two different crystal forms, triclinic and monoclinic, depending on slight variations of crystallization conditions. Both these crystal forms contain the whole dimeric molecule in the asymmetric unit. The relationship between the symmetry and cell dimensions of these crystal forms has been discussed (*103*). The loss of crystallographic symmetry upon coenzyme binding has been interpreted (*103*) in terms of a coenzyme-induced conformational change and discussed as possible evidence for half-site reactivity (*104,105*). Table II lists the various complexes investigated and the cell dimensions obtained.

2. *Electron Density Maps*

The electron density maps published so far have been based on the orthorhombic crystal form. The anions $Pt(CN)_4^{2-}$ and $Au\ (CN)_2^-$ and the mercury complex ethyl mercurythiosalicylate were used to prepare heavy atom derivatives (*106,107*). The overall dimensions of the molecule

103. C.-I. Brändén, *ABB* **112,** 215 (1965).

104. B. W. Matthews and S. A. Bernhard, *Annu. Rev. Biophys. Bioeng.* **2,** 257 (1973).

105. M. Lazdunski, C. Petitclerc, D. Chappelet, and C. Lazdunski, *Eur. J. Biochem.* **20,** 124 (1971).

106. B.-O. Söderberg, E. Zeppezauer, T. Boiwe, B. Nordström, and C.-I. Brändén, *Acta Chem. Scand.* **24,** 3567 (1970).

107. H. Eklund, B. Nordström, E. Zeppezauer, G. Söderlund, I. Ohlsson, T. Boiwe, and C.-I. Brändén, *FEBS (Fed. Eur. Biochem. Soc.) Lett.* **44,** 200 (1974) and *JMB* (in press) (1975).

TABLE III
INHIBITOR COMPLEXES WITH LADH STUDIED BY X-RAY METHODS

Compound	Resolution of difference Fourier map (Å)	Ref.
ADP-Ribose	2.9	*111*
8-BrADP-ribose	4.5	*111*
1,10-Phenanthroline	4.5	*112*
Imidazole	3.7	*112*
5-Iodosalicylate	4.5	*102*
LADH modified with iodoacetic acid	4.5	*130*

and the location of one zinc atom per subunit were derived from an electron density map at 6 Å resolution (*108*). A comparison (*109*) of an LADH model from this map to a model of lactic dehydrogenase at 5 Å resolution revealed some of the basic resemblances between the subunit structures that have later been studied at high resolution (*2,110*). A subsequent electron density map to 5 Å resolution of LADH (*100*) in combination with inhibitor studies revealed the coenzyme binding site. The general folding of the main chain and the position of the second zinc atom were obtained from a map at 2.9 Å resolution (*110*). This map also revealed the presence of a unique structure for coenzyme binding within the family of dehydrogenases by comparison with the structure of LDH (*2,110*). The final electron density map was based on diffractometer data for four heavy atom derivatives to a resolution of 2.4 Å (*107*). Knowledge of the primary structure permitted Brändén and coworkers to position the side chains of the molecule within this map. All structural conclusions thus rely on the combined interpretation of crystallographic analysis and work on primary structure. The binding of different inhibitor molecules to the apoenzyme structure has been studied by difference Fourier techniques (*102,111,112*). Table III lists these compounds and the resolution of the corresponding studies.

108. C.-I. Brändén, E. Zeppezauer, T. Boiwe, G. Söderlund, B.-O. Söderberg, and B. Nordström, *in* "Pyridine Nucleotide-Dependent Dehydrogenases" (H. Sund, ed.), p. 129. Springer-Verlag, Berlin and New York, 1970.
109. C.-I. Brändén and M. G. Rossmann, *in* "Pyridine Nucleotide-Dependent Dehydrogenases" (H. Sund, ed.), p. 133. Springer-Verlag, Berlin and New York, 1970.
110. C.-I. Brändén, H. Eklund, B. Nordström, T. Boiwe, G. Söderlund, E. Zeppezauer, I. Ohlsson, and Å. Åkeson, *Proc. Nat. Acad. Sci. U. S.* **70**, 2439 (1973).
111. M. A. Abdallah, J.-F. Biellmann, B. Nordström, and C.-I. Brändén, *Eur. J. Biochem.* **50**, 475 (1975).
112. T. Boiwe and C.-I. Brändén, *Eur. J. Biochem.* (in press).

3. *Three-Dimensional Structure*

The general shape of the dimeric molecule is an approximate prolate ellipsoid of dimensions 45 × 60 × 110 Å. It is wider at the ends of the long axis than in the central region, giving the molecule a dumbbell-shaped appearance. A stereo diagram of the 748 α-carbon positions and the four zinc atoms of the molecule as well as of two bound ADP-ribose molecules is given in Fig. 1. The folding of the polypeptide chain within the subunit is illustrated in Fig. 2. Table IV lists the various regions of secondary structural elements of the subunit.

Each subunit is clearly divided into two domains joined by a narrow neck region and separated by the deep active site cleft. One of these domains, called the coenzyme binding domain, binds the coenzyme. The two zinc atoms are bound within the second domain, called the catalytic domain. The two domains are unequal in size: the catalytic domain is larger and comprises 231 residues, whereas the coenzyme binding domain is built up from 143 residues. The two subunits of the dimeric molecule

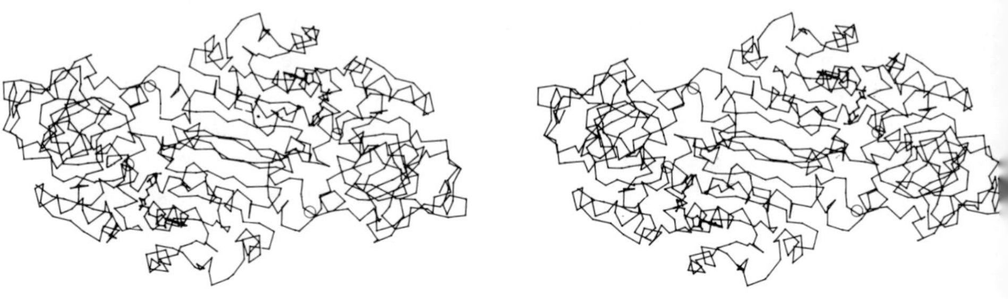

FIG. 1. Stereo diagram of the 748 α-carbon atoms and the four zinc atoms of the LADH molecule. In addition, the two bound ADP-ribose molecules are depicted.

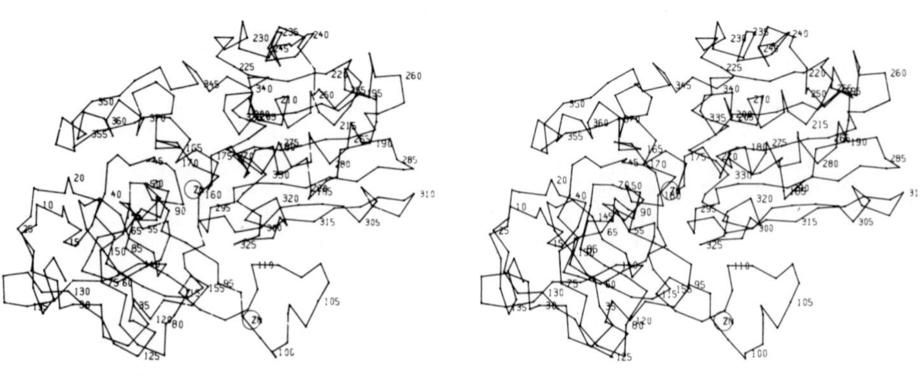

FIG. 2. Stereo diagram of the α-carbon and zinc atoms of one subunit of LADH.

TABLE IV
REGIONS OF SECONDARY STRUCTURAL ELEMENTS[a] IN THE SUBUNIT OF LADH

Residues	Element of secondary structure	Comments
		Residues 1–175 and 319–374 are within the catalytic domain.
9–14	βIII:1	
22–29	βIII:2	
32–35	R1	
33-40	βII:1	
41-45	βI:1	
46–53	α1	Cys-46 is a ligand to the catalytic zinc.
62–65	βIII:3	
68–71	β1:2	His-67 is a ligand to the catalytic zinc.
72–78	βII:2	
80–83	R2	
84–87	R3	
86–87	βII:3	
88–92	βI:3	This strand is distorted by Pro-91.
95–113		Residues 95–113 form a lobe at the outside of the molecule that binds the second zinc atom through residues Cys-97, Cys-100, Cys-103, and Cys-111.
123–126	R4	
129–132	βIII:4	
132–135	R5	
135–138	βIII:5	
140–143	R6	
145–146	βIII:6	
148–152	βII:4	
153–156	R7	
156–160	βI:4	
165–168	R8	
168–180	α2	Helix α2 bridges the two domains and provides one ligand, Cys-174, to the catalytic zinc. This helix is highly distorted and has turns of both α, π and 3^{10} types.
		Residues 176–318 comprise the coenzyme binding domain.
180–188	αA	There is a break between the helices at residue 180. The new helix αA is also distorted.
190–193	R9	
193–199	βA	
201–215	αB	The two last turns are of 3^{10} type.
218–223	βB	
229–236	αC	
238–243	βC	
244–247	R10	
250–258	αCD	The loop connecting βC and βD has a two-turn helix in LADH in contrast to the other dehydrogenases.

TABLE IV (*Continued*)

Residues	Element of secondary structure	Comments
258–261	R11	
263–269	βD	
271–280	αE	
283–286	R12	
287–293	βE	
295–298	R13	
299–304	βS	βS corresponds in space to α_1F of LDH.
304–311	3^{10}S	βS, 3^{10}S, and βF are involved in the subunit interaction.
312–318	βF	
319–322	R14	
324–338	α3	Deformed in the beginning by Pro 329.
338–341	R15	
344–347	R16	
347–352	βI:5	
351–354	R17	
355–365	α4	
369–374	βI:6	

[a] βI, βII, and βIII are the three pleated sheet regions of the catalytic domain schematically illustrated in Fig. 10. The individual strands of these sheets are numbered sequentially from the amino terminal. βA . . . βF are the strands of the parallel pleated sheet region in the coenzyme binding domain schematically illustrated in Fig. 5. The helices are labeled α1 . . . α4 in the catalytic domain and αA . . . αE in the coenzyme binding domain. R1 . . . R17 are reverse bends defined according to Venkatachalam (*117*).

are joined together by homologous interactions (*113*) within regions of the coenzyme binding domains. The whole molecule has an approximate helical content of 28% of the residues whereas 34% are in pleated sheet regions.

a. Coenzyme Binding Domain. (*i*) *Secondary structure and folding.* The coenzyme binding domain comprises residues 176–318. A stereo diagram of the α-carbon positions of this domain is given in Fig. 3. The corresponding main chain hydrogen bonding pattern is given in Fig. 4. The domain is built up from six parallel strands of pleated sheet (βA — βF) flanked by helices (αA, αB, αC, αCD, αE, and 3^{10}S) in a regular pattern. The nomenclature used in describing the helices and the β structure within this domain is given in Fig. 5. The similarities and differences in sequence and structure between the coenzyme binding

113. J. Monod, J. P. Changeux, and F. Jacob, *JMB* **6**, 306 (1963).

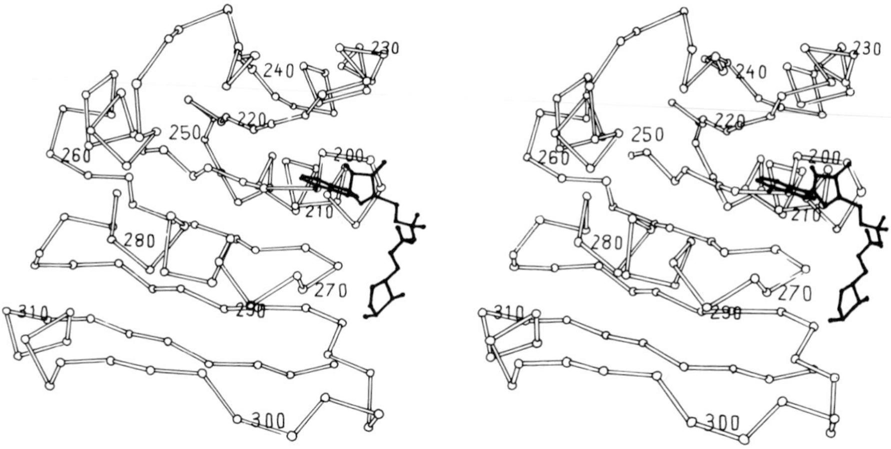

FIG. 3. Stereo diagram of the positions of the α-carbon atoms of the coenzyme binding domain in LADH and the bound ADP-ribose molecule.

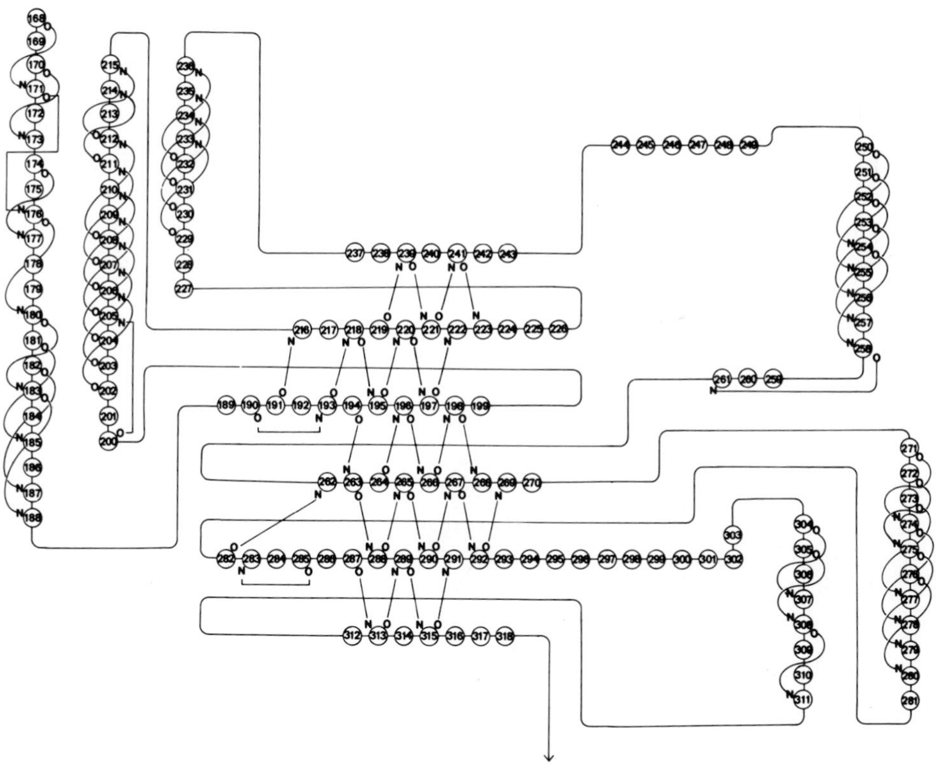

FIG. 4. The main chain hydrogen bonding pattern of the residues involved in the coenzyme binding domain of LADH.

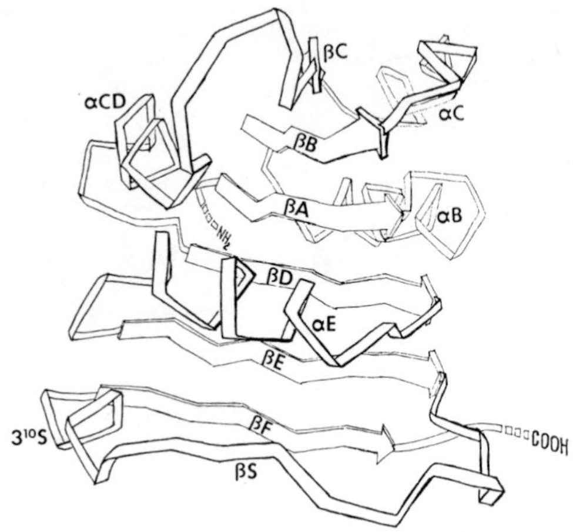

FIG. 5. Schematic diagram illustrating the fold of the polypeptide chain within the coenzyme binding domain and the nomenclature of the elements of secondary structure.

domains in the dehydrogenases have been studied (*114,115*) and are discussed in detail in a special chapter of this volume (*2*).

There are three hydrophobic regions in this domain. One is involved in the subunit interaction and is discussed in Section II,C,3,c. The other two are important for the folding of this domain since they form hydrophobic cores between the helices and the parallel pleated sheet (*116*). Table V lists the residues involved. The importance of these hydrophobic cores for the proper folding is realized from the fact that almost all residues in LDH and GAPDH which are structurally equivalent to those of LADH listed in Table V are also hydrophobic.

The amount of secondary structure within this domain is considerable (see Table IV): 44% of the residues are helical, mainly α-helices; 32% are in the pleated sheet structure; and 13% in reverse bends (*117,118*). Thus, only about 10% of the residues have no regular secondary structure and no continuous stretch of irregular structure is longer than 4 residues.

(*ii*) *Coenzyme binding.* Two coenzyme molecules are bound per enzyme molecule independently of each other (*1,119*). Neither the binding

114. M. G. Rossmann, D. Moras, and K. W. Olsen, *Nature* (*London*) **250,** 194 (1974).
115. I. Ohlsson, B. Nordström, and C.-I. Brändén, *JMB* **89,** 339 (1974).
116. K. Nagano, *JMB* **84,** 337 (1974).
117. C. M. Venkatachalam, *Biopolymers* **6,** 1425 (1968).
118. J. L. Crawford, W. N. Lipscomb, and C. G. Schellman, *Proc. Nat. Acad. Sci U. S.* **70,** 538 (1973).
119. S. Taniguchi, H. Theorell, and Å. Åkeson, *Acta Chem. Scand.* **21,** 1903 (1967).

TABLE V
RESIDUES IN THE COENZYME BINDING DOMAIN WHICH FORM HYDROPHOBIC CORES[a]

Core N1			Core N2	
Ala-183	Gly-204	Ala-237	Ala-196	Leu-254
Val-186	Val-207	Phe-264	Phe-198	Val-262
Ala-187	Ile-208	Phe-266	Ile-220	Ala-278
Val-189	Cys-211	Val-268	Val-222	Cys-281
Cys-195	Ala-216	Val-288	Ile-250	Cys-282
Val-197	Ile-219	Val-290		
Leu-200	Gly-221	Val-292		
Val-203	Ala-232	Ala-317		

[a] Core N1 is formed by residues from αA, αB, αC, and the strands of the parallel pleated sheet. Core N2 is formed by residues from αCD, αE, and the strands.

of the full coenzyme nor the coenzyme-induced conformational change has yet been studied by X-ray methods since crystals of complexes with coenzyme are not isomorphous to those of apoenzyme. However, the binding of the coenzyme analog ADP-ribose to the apoenzyme has been studied in detail (*111*). This analog is a coenzyme-competitive inhibitor (*97,120*).

It was found in the X-ray study that one ADP-ribose molecule binds to each subunit in the coenzyme binding domain. This binding is similar to the binding of corresponding parts of NAD to LDH, s-MDH, and GAPDH (*2*). It is thus obvious that there is one and only one specific, productive coenzyme binding site per subunit. This site is illustrated in Fig. 3. Investigations claiming several molecules of specifically bound coenzyme molecules or coenzyme analogs per subunit (*121,122*) are thus not compatible with the X-ray studies.

The unique folding of the nucleotide binding domain creates a specific crevice for binding of a dinucleotide molecule. The coenzyme is bound in the central region of the carboxyl end of the parallel pleated sheet. Residues from βA, βB, βD, αB, αE and the loops connecting βA with αB and βD with αE are involved in this binding. The middle strand βA is shorter at the carboxyl end than its neighboring strands βB and βD. Furthermore, the loop which connects βA with αB turns to the left

120. T. Yonetani, *Acta Chem. Scand.* **17,** S96 (1963).
121. H. Weiner, *Biochemistry* **8,** 526 (1969).
122. H. Weiner, I. Iweibo, and P. L. Coleman, *in* "Structure and Function of Oxidation-Reduction Enzymes" (Å. Åkeson and A. Ehrenberg, eds.), p. 619. Pergamon, Oxford, 1972.

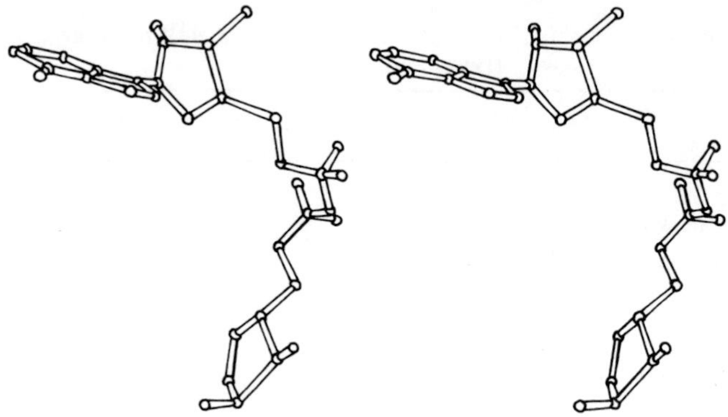

FIG. 6. Stereo diagram illustrating the conformation of ADP-ribose when bound to LADH.

TABLE VI
DIHEDRAL ANGLES[a] (IN DEGREES) FOR THE CONFORMATION OF LADH-BOUND ADP-RIBOSE

Adenosine part	Terminal ribose part
χ_a 78	ϕ_n 112
ξ_a 159	ψ_n 51
θ_a 167	θ_n 152
ψ_a 177	ξ_n 154
ϕ_a 39	Ribose$_n$ C-3′ endo
Ribose$_a$ C-3′ endo	

[a] The conventions of Arnott and Hukins (*135*) have been used to define these angles.

whereas the loop below βA connecting βD with αE turns to the right. By this arrangement a crevice is formed outside the carboxyl end of βA below αB and above αE as seen in Figs. 3 and 5. The coenzyme molecule binds in this crevice with the AMP part in the region above helix αE and the NMN part in the region below αB (see Fig. 3). A symmetrical arrangement of both the coenzyme binding domain and the bound coenzyme molecule is thus obtained as was noted by Rao and Rossmann for LDH (*123*).

The conformation of the bound ADP-ribose molecule is illustrated in Fig. 6. The corresponding torsional angles are given in Table VI. This conformation is very similar to that found for the ADP-ribose parts of the coenzyme molecules bound to LDH, s-MDH, and GAPDH (*2*).

123. S. T. Rao and M. G. Rossmann, *JMB* **76**, 241 (1973).

TABLE VII
OBSERVED INTERACTIONS BETWEEN LADH AND ADP-RIBOSE

Moiety	Residue	Comments
Adenine	Phe-198	All side chains lining the hydrophobic adenine binding pocket have been included here, although some are not in actual contact with the adenine moiety.
	Val-222	
	Ile-224	
	Pro-243	
	Ile-250	
	Ile-269	
	Thr-274	
	Thr-277	
	Arg-271	At the surface of the pocket.
	Asp-273	
Adenosine ribose	Asp-223	Hydrogen bond to O-2′.
	Gly-199	
	Ile-269	
	Asn-225	O-3′ close to these residues. Bonds not established.
	Lys-228	
Pyrophosphate	Arg-47	Charge interaction.
	Ile-269	
Nicotinamide ribose	Gly-293	
	C = O from Ile-269	Hydrogen bond to O-3′.
	C = O from Gly-293	Hydrogen bond to O-2′.

All observed interactions between ADP-ribose and the protein are listed in Table VII. Figure 7 shows a schematic diagram of the bound nucleotide and the surrounding protein.

The adenine moiety is bound in a hydrophobic pocket at one end of the coenzyme binding crevice. The NH_2 group points out from this pocket into the solution. This agrees with affinity chromatography studies (*124,125*) of specific binding of dehydrogenases to coenzyme analogs anchored to the solid support through this group. There are no specific interactions which are unique for adenine, only a general hydrophobic cleft for the binding of aromatic molecules, consistent with studies of the binding of analogs and inhibitor molecules (Sections II,G and II,H).

The adenosine ribose moiety is bound to one of the carboxyl oxygens of Asp-223 by a hydrogen bond to the O-2′ hydroxyl. The invariance of Asp-223 in other dehydrogenases (*2,115*) suggests that this hydrogen bond is an important feature of coenzyme binding. This conclusion is reinforced by a similar invariance of Gly-199. A β-carbon in this position would in all dehydrogenases prevent the ribose from forming the hydro-

124. K. Mosbach, H. Guilford, R. Ohlsson, and M. Scott, *BJ* **127**, 625 (1972).
125. M. Lindberg, P.-O. Larsson, and K. Mosbach, *Eur. J. Biochem.* **40**, 187 (1973).

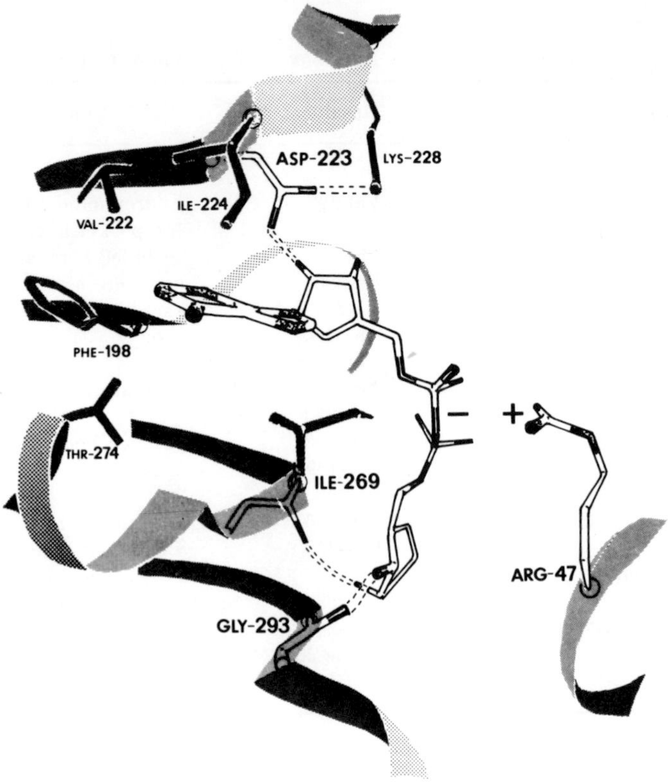

FIG. 7. Schematic diagram illustrating the interactions between ADP-ribose and LADH.

gen bond between O-2′ and the conserved aspartic acid. The specificity of LADH for NAD in contrast to NADP is a consequence of this arrangement (Section II,G). Coenzyme competitive inhibitors such as salicylate also form a hydrogen bond to this conserved aspartic acid (Section II,H,1,a).

The pyrophosphate moiety is bent over the fork of the side chain of Ile-269. The guanidinium group of Arg-47 provides a positive charge for an ionic interaction with the pyrophosphate. This site has been shown to be a general anion binding site in LADH (Section II,H,1a).

The nicotinamide ribose is bound by two hydrogen bonds from O-3′ and O-2′ to the main chain carbonyl oxygens of Ile-269 and Gly-293. A β-carbon at position 293 would overlap with this ribose and prevent its proper binding. These bonds must be essential for dinucleotide binding since ADP-ribose is bound more strongly than ADP (*120*).

Similar interactions have been found in LDH (*126*) where these hydroxyls form hydrogen bonds to the main chain carbonyl oxygen of Ala-98 and amino nitrogen of Glu-140. These residues are structurally equivalent to Ile-269 and Val-294 in LADH.

An assumed position of the nicotinamide moiety in LADH has been deduced from the similarities of coenzyme conformation and binding within the dehydrogenases by rotating the LDH-bound NAD molecule into the LADH structure so that the ADP-ribose parts overlap. No steric hindrance from the LADH protein was found for building the nicotinamide moiety onto the ADP-ribose portion using the conformation of NAD found in LDH. The nicotinamide fits nicely into a pocket lined by the residues Thr-178, Val-203, Val-268, Val-292, Gly-293, and Ile-318. Carbon atom C-4 of the nicotinamide is then 4.5 Å from the catalytic zinc atom with the A side (*127*) facing zinc. It was impossible to build in the nicotinamide with the B side facing zinc because of the steric hindrance between the side chain of Val-203 and the amide portion of the nicotinamide. This agrees with the known stereochemistry of hydrogen transfer for LADH (*127*).

There are no amino acid residues in the nicotinamide binding pocket that can be equivalent in function to Glu-140 or His-195 in LDH (*126*). The higher affinity at neutral pH for NADH compared to NAD^+ can be correlated to the hydrophobic environment in this pocket. Such an environment would be expected to favor binding of the neutral nicotinamide moiety of the reduced coenzyme. The importance of the pK_a shift of the water molecule bound to zinc for the stabilization of the binding of oxidized coenzyme at higher pH is discussed in Section II,I,1.

b. Catalytic Domain. (*i*) *Secondary structure and folding.* The catalytic domain comprises residues 1–175 and 319–374. Both ends of the polypeptide chain are thus within this region. The two zinc atoms of the subunit are bound to ligands from this domain. A stereo diagram of the α-carbon positions and the zinc atoms of the catalytic domain is given in Fig. 8. The corresponding main chain hydrogen bonding pattern is given in Fig. 9.

There are no apparent similarities between the structure of this domain and the structures of the catalytic domains of LDH (*126*) or GAPDH (*128*). The catalytic domain in LADH is built up by a complicated network of β structures. There are three main regions of pleated sheet which have been called βI, βII, and βIII. Schematic diagrams of

126. J. Holbrook, A. Liljas, S. J. Steindel, and M. G. Rossmann, Chapter 4, this volume.
127. G. Popják, "The Enzymes," 3rd ed., Vol. 2, p. 115, 1970.
128. M. Buehner, G. C. Ford, D. Moras, K. W. Olsen, and M. G. Rossmann, *JMB* **90,** 25 (1975).

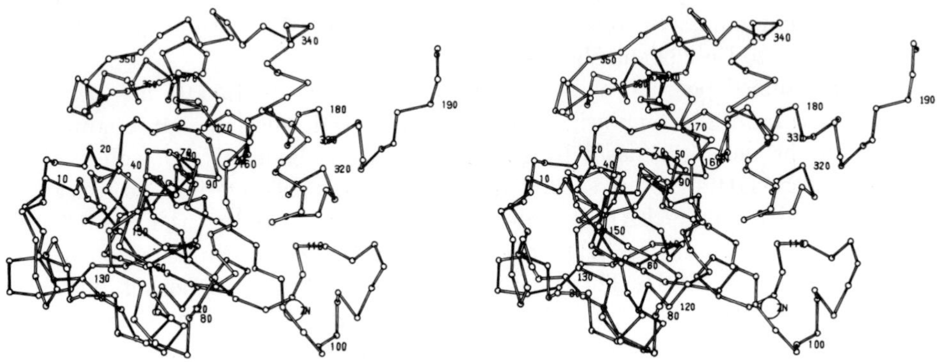

FIG. 8. Stereo diagram of the α-carbon and zinc atoms of the catalytic domain.

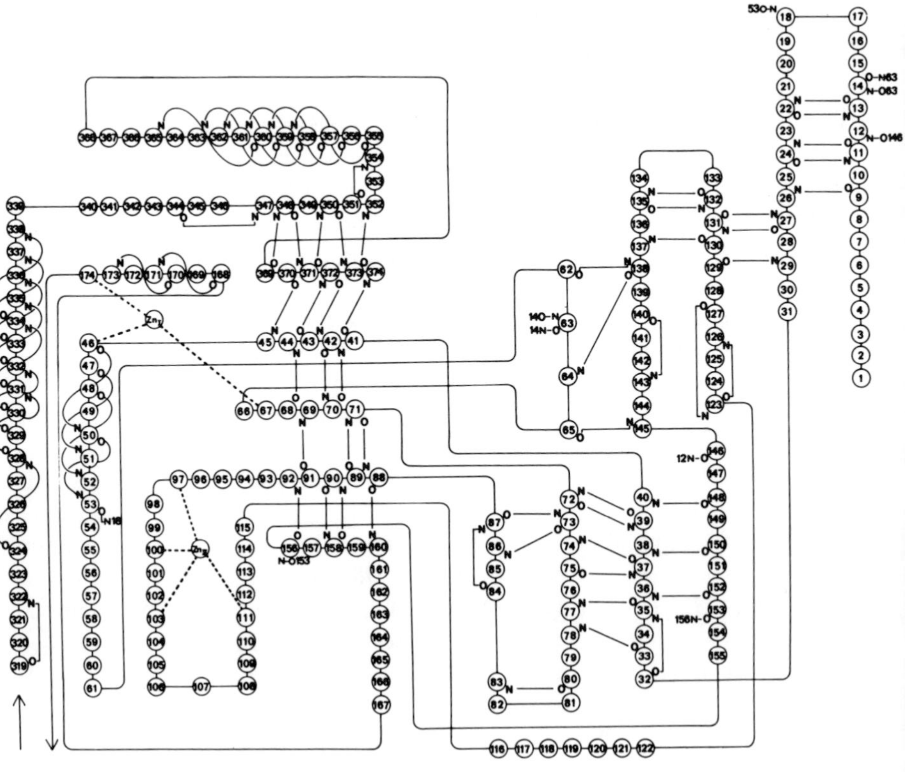

FIG. 9. The main chain hydrogen bonding pattern of the residues involved in the catalytic domain.

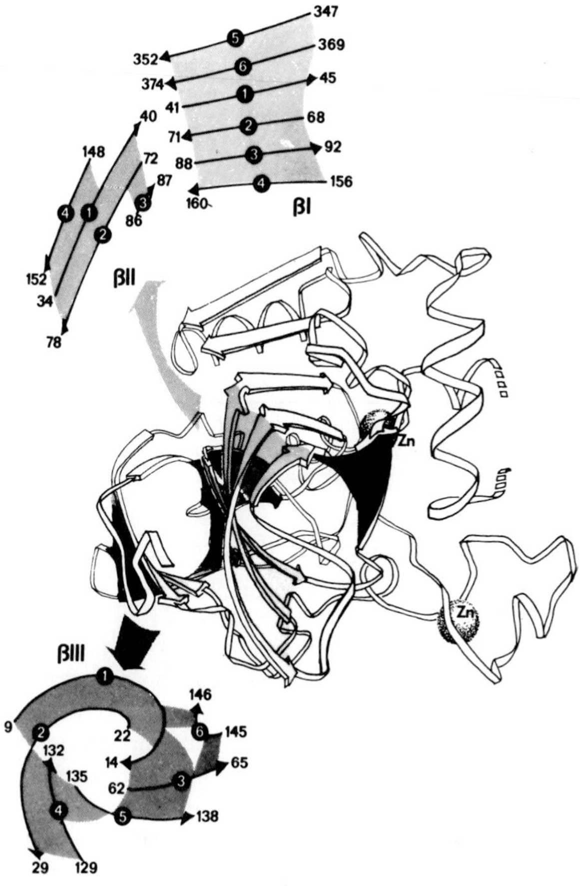

FIG. 10. Schematic diagram of the three pleated sheet regions in the catalytic domain.

these three regions are given in Fig. 10. The strands of these sheets are mainly antiparallel. The correspondence between sequence and elements of secondary structure is given in Table IV.

Region βI comprises six strands which form a wall through the top half of this domain as shown in Figs. 8 and 10. On one side of this wall are mainly residues from the C-terminal part of the chain. The two last strands of βI are formed from the last 30 residues of this part. These are parallel and joined by a helix in a manner which strongly resembles part of the structure that builds up the nucleotide binding domain. These structures even have the same unique hand (*2*).

On the other side of the wall formed by the strands of βI are the remaining residues of the amino terminal half of the chain. These can be

TABLE VIII
RESIDUES IN THE CATALYTIC DOMAIN WHICH FORM HYDROPHOBIC CORES[a]

Core K1		Core K2	Core K3		Core K4	
Val-36	Pro-91	Ala-11	Ala-12	Phe-140	Ala-70	Phe-176
Ile-38	Met-123	Val-13	Leu-14	Phe-146	Ile-90	Val-328
Met-40	Thr-143	Val-26	Phe-21	Phe-352	Ile-160	Leu-331
Ala-69	Thr-145	Ile-64	Ile-45	Ile-355	Pro-165	Val-332
Val-73	Thr-150	Phe-130	Thr-59	Phe-359	Leu-166	Phe-335
Val-80	Val-151	Ile-137	Val-63	Leu-362	Val-169	Met-336
Val-83	Val-152		Ala-65	Thr-370	Leu-171	Phe-340
Val-89	Val-157		Gly-66	Leu-372	Ile-172	Leu-342
				Phe-374	Gly-173	

[a] Core K1 comprises residues from the pleated sheet regions βI and βII; core K2 is at the interior of the cylinder formed by βIII; core K3 comprises residues from βI, βIII, and α4; and core K4 from βI, α2, and α3.

divided into three separate regions: the two remaining pleated sheet regions βII and βIII and a lobe comprising residues 95–113 which bind the second zinc atom.

Several of the strands of βII are a direct continuation of strands in βI. There is, however, a bend of approximately 90 degrees in the chains between the two sheet regions. These sheets are, furthermore, twisted in such a way that the strands of βII in combination with some of the strands of βI enclose an approximate cylinder. Inside this cylinder is a tightly packed core, K1, of 16 hydrophobic side chains, half of which are valine. The third pleated sheet region βIII comprises six strands linked in a rather irregular fashion. These six strands form a closed set in that each strand is hydrogen bonded to two of its neighbors. Similar cylindrical arrangements have also been observed in other proteins (*129*). Residues in the interior of this cylinder form a second core, K2, of hydrophobic residues. The residues that contribute side chains to these hydrophobic cores and to two additional internal hydrophobic regions in this domain are listed in Table VIII.

There are only four helical segments in the catalytic domain comprising 18% of the residues. In contrast, 35% of the residues are in pleated sheet regions and 14% in reverse bends. Thus, a large number of residues, 33%, have no regular secondary structure. However, a considerable portion of these residues, 95–113, form the lobe which binds the second zinc atom. Apart from this lobe there is only one long continuous string of residues, 114–128, which have no apparent secondary structure.

129. D. M. Blow, "The Enzymes," 3rd ed., Vol. 3, p. 185, 1971.

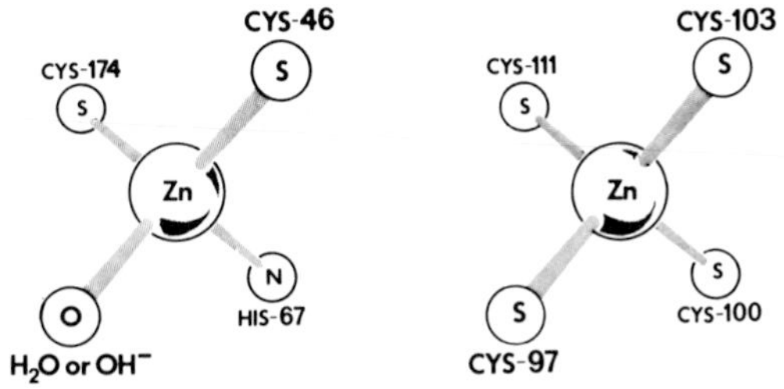

FIG. 11. Schematic diagram of the ligands to the two different zinc atoms of the subunit.

(*ii*) *The catalytic zinc atom and the substrate binding pocket.* The two domains of the LADH subunit are separated by a pocket lined almost entirely by hydrophobic residues. The catalytic zinc atom is bound at the bottom of this pocket in the middle of the subunit by three protein ligands; two sulfur atoms from Cys-46 and Cys-174 and one nitrogen atom from His-67 (see Figs. 2 and 11). A water molecule or hydroxyl ion, depending on the pH, completes a distorted tetrahedral coordination. Stereo diagrams of the active site pocket are given in Fig. 12. This pocket is about 25 Å deep from the surface to the zinc atom.

The binding of 1,10-phenanthroline has been studied by X-ray methods (*112*) in order to correlate a large number of inhibitor studies in solution with the structure (Section II,H,1,b). This chelating agent binds to the zinc atom in the active site displacing the water molecule. The metal is, thus, five coordinated in the complex. The aromatic ring system of the inhibitor is positioned in the active site pocket partly overlapping the assumed nicotinamide position.

The reactive cysteine, Cys-46, which can be selectively carboxymethylated by iodoacetate (Section II,E,1,a) is one of the zinc ligands.

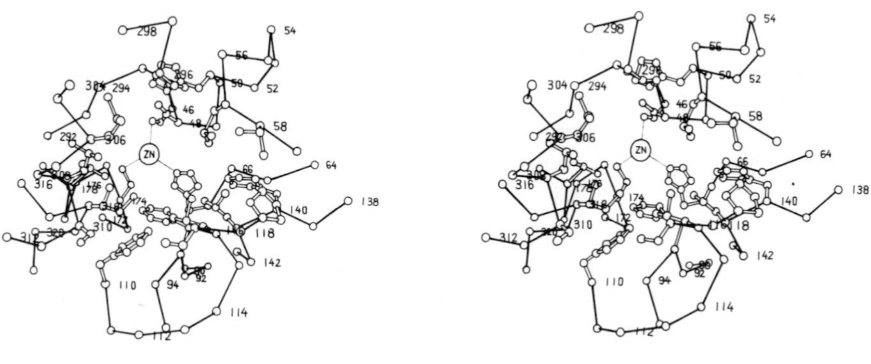

FIG. 12. Stereo diagram of the active site pocket with the catalytic zinc atom.

TABLE IX
RESIDUES LINING THE ACTIVE SITE POCKET[a]

Subunit A		Subunit B
Ser-48	Ser-117	Met-306
Leu-57	Leu-141	Leu-309
Val-58	Thr-178	Ser-310
Phe-93	Pro-296	
Phe-110	Ile-318	
Leu-116		

[a] The letters A and B refer to the two different subunits of the LADH dimer.

The importance of the general anion binding site, including Arg-47, for this reaction and the results of an X-ray study (*130*) of the carboxymethylated enzyme are discussed in Section II,H,1,a.

The zinc-bound water molecule is involved in a system of hydrogen bonds which include Ser-48 and His-51 (see Fig. 17). The possible importance of these hydrogen bonds for proton release in the catalytic mechanism is discussed in Section II,I,1.

The residues lining the hydrophobic active site pocket are listed in Table IX. There are only two polar residues, Ser-48 and Thr-178, in the part of this pocket where the catalytic reaction occurs. The other residues create a hydrophobic environment for the zinc atom, the nicotinamide moiety and the bound substrate. No obvious role for Thr-178 in the catalytic mechanism can be deduced from the structure.

Both subunits contribute residues to each active site pocket. The bottom part of the pocket which binds zinc, nicotinamide, and the reactive part of the substrate is contained entirely within each subunit. The second half of the pocket, closer to the surface, however, is lined by residues from the catalytic domain of one subunit and from the region 3^{10}S of the coenzyme binding domain of the other subunit.

(*iii*) *The second zinc atom.* Residues 95–113 form a lobe which binds the second zinc atom of the subunit. This lobe projects out from the catalytic domain (Fig. 8) having few side chain interactions with the remaining parts of the subunit which agrees with the variability of its sequence in different species (Section II,D). This zinc is liganded in a distorted tetrahedral arrangement by four sulfur atoms from cysteine residues 97, 100, 103, and 111. A schematic diagram illustrating this coordination is given in Fig. 11. Stereo diagrams of this lobe with its bound zinc are given in Fig. 13.

130. E. Zeppezauer, H. Jörnvall, and I. Ohlsson, *Eur. J. Biochem.* (in press).

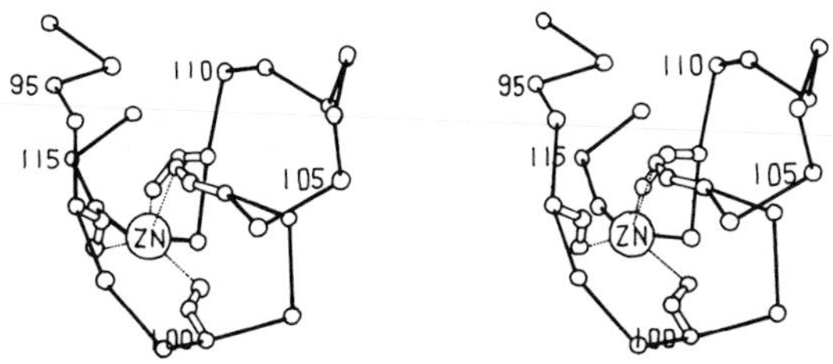

FIG. 13. Stereo diagram of the residues 95–113 and the bound zinc atom.

The coordination of this zinc atom in LADH is similar to the tetrahedral arrangement of four cysteine ligands around iron in rubredoxin (*131*). Even more striking is the similarity with the four cysteine ligands around the iron clusters in the bacterial ferredoxins (*132*). Three of the cysteines in both ferredoxin and LADH are separated by two other residues and the fourth ligand is some distance away in the sequence. The similarities extend even further. Looking down an axis defined by the metal and one of the ligands, the sequential arrangement of the other three ligands has the same unique hand in these two structures. These similarities may be the result of an energetically favorable structural arrangement or they may have evolutionary implications.

In rat liver ADH the presence of multiple molecular forms has been correlated to disulfide bridges involving the ligands to this zinc atom (Section II,B,3). These forms are active in ethanol oxidation (*52*). The lobe region which binds zinc is thus, in all probability, not essential for the catalytic action of alcohol oxidation. It has been suggested (*133,134*) that the extra zinc atom is essential for the structural stability of the enzyme. There is no evidence in the structure that this lobe region is necessary either for tertiary or quaternary structure stabilization. From the structural point of view, this region looks much more like a second catalytic center. The zinc atom is situated in one side of an obvious cleft into which the lone pair electrons of the sulfur atom of Cys-97 project.

c. The Dimer. The two subunits are joined together into the dimeric molecule through interactions within the coenzyme binding domains (see Fig. 1). A crystallographic twofold axis relates the subunits in the apoenzyme structure. This crystallographic symmetry axis is not present in

131. L. H. Jensen, *in* "Iron-Sulfur Proteins" (W. Lovenberg, ed.), Vol. 2, p. 163. Academic Press, 1974.
132. E. T. Adman, L. C. Sieker, and L. H. Jensen, *JBC* **248,** 3987 (1973).
133. D. E. Drum and B. L. Vallee, *Biochemistry* **9,** 4078 (1970).
134. Å. Åkeson, *BBRC* **17,** 211 (1964).
135. S. Arnott and D. W. L. Hukins, *Nature* (*London*) **224,** 886 (1969).

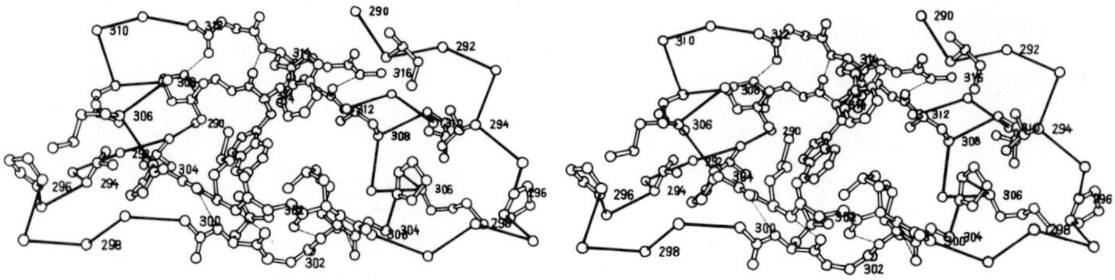

FIG. 14. Stereo diagram of the interaction area between the two subunits of the LADH dimeric molecule.

crystals of holoenzyme (*99*), indicating a coenzyme-induced asymmetry of the two subunits.

In the apoenzyme structure the βF strands of each subunit run in opposite directions perpendicular to the twofold axis and are joined together by hydrogen bonds forming two strands of antiparallel β structure. The pleated sheet structure in the coenzyme binding domains thus extends through the whole dimeric molecule and comprises 12 strands arranged in two pairs of six parallel strands in each subunit. These pairs are joined together by antiparallel β binding through the twofold symmetry axis. There is a twist of about 200 degrees from the first to the twelfth strand. Furthermore, residues 301–303 of the extended chain βS are also joined by antiparallel β binding to corresponding residues in the other subunit. Between these two regions of hydrogen bonds there is an hydrophobic core involving side chains mainly from βE, βF, βS, and 3^{10}S. Figure 14 illustrates these interactions and the subunit contacts are listed in Table X.

The two catalytic zinc atoms of the molecule are separated by 47 Å. Thus, there is no direct interaction between the two catalytic centers. Indirect interaction mediated by the conformational change induced by the coenzyme is, however, quite possible, especially since the subunits are bound together through their coenzyme binding domains.

D. Comparisons of the Structures of Different Alcohol Dehydrogenases

In this section we will correlate some of the structural information described in Sections II,B, II,C, III,B, and IV,B and discuss some functional and evolutionary implications of this structural knowledge.

1. *Structural and Functional Aspects*

a. Differences between the E and S Chains of the Horse Liver Enzyme. Three of the six observed amino acid differences between the E and S

TABLE X
INTERACTIONS BETWEEN THE TWO SUBUNITS A AND B OF LADH DIMER

Interaction type	Residue in subunit A	Residue in subunit B
Main chain hydrogen bonds	O of Arg-312	N of Gly-316
	O of Trp-314	N of Trp-314
	N of Trp-314	O of Trp-314
	N of Gly-316	O of Arg-312
	N of Leu-301	O of Met-303
	O of Leu-301	N of Met-303
	N of Met-303	O of Leu-301
	O of Met-303	N of Leu-301
Nonpolar contact between side chains	Met-275	Pro-305
	Ile-291	Leu-308
	Val-294	Pro-305, Met-306, Leu-309
	Pro-296	Met-306
	Leu-301	Met-303
	Met-303	Leu-301, Trp-314
	Pro-305	Met-275, Val-294
	Met-306	Val-294, Pro-296
	Leu-308	Ile-291, Trp-314
	Leu-309	Val-294
	Trp-314	Met-303, Leu-308, Trp-314

chains (Section II,B,1,b) are possible candidates as being responsible for the difference in substrate specificity. These are Phe-110, Asp-115, and Thr-94 which are leucine, serine or a gap, and isoleucine, respectively, in the S chain. The side chain of Phe-110 lines the active site pocket in the apoenzyme structure. Only small conformational changes are required to bring the other two side chains into the pocket, close to the surface of the molecule and far from the catalytic zinc atom.

The other three residues which are exchanged are all in different regions of the catalytic domain. No exchanges have been found in the coenzyme binding domain although there are differences in coenzyme binding between the EE and SS isozymes (*45*). The known exchanges were, however, detected by analysis based on peptide mapping and additional differences may therefore exist.

b. Comparison of ADH from Different Species. Partial sequences of yeast (Section III,B) and a bacillar (Section IV,B) ADH are known. They show that these enzymes are evolutionarily linked to the liver enzymes. From the regions where sequence comparisons can be made between the yeast and liver enzymes (Table XIV), we have extracted those residues for which the side chains are important for structure or function of the liver enzyme, as deduced from the X-ray structure. These are listed in Table XI together with the known corresponding residues from the bacillar enzyme.

Among these amino acids are the zinc ligands and residues which are in the four hydrophobic cores of the catalytic domain. This comparison shows that the three-dimensional structures of these enzymes must be very similar in the catalytic domains. Although the general sequence homology between the yeast and liver enzymes is less than 40%, all the structurally important residues are either homologous or conservatively substituted. Main differences in the folding of the polypeptide chains would have resulted in at least some substitutions for charged residues because of differences in environment.

A detailed analysis is even more convincing. Residues 11–14 (all numbering refers to the LADH sequence) must all be nonpolar since there are hydrophobic cores on both sides of this strand. Residues 36, 38, and 40 which point into a hydrophobic core are conservative, whereas the adjacent residues 37 and 41 which point into solution are nonconservative. Residues 69–71 all have small nonpolar side chains for steric reasons. Residues 66 and 77 are invariant glycines because of lack of space for a side chain in hydrophobic cores, and residue 86 is an invariant glycine which occurs in the only type II reverse bend *(117)* in this region.

Furthermore, all ligands to both zinc atoms in LADH are conserved. Since yeast ADH has been reported to have only one zinc per subunit (Section III,A), it was surprising to find that all four cysteine residues which are ligands to the second zinc in LADH are also conserved. In contrast, almost all other residues in this lobe from 95 to 113 vary (Table XIV). We can thus conclude that yeast ADH in all probability has two firmly bound zinc atoms per subunit or, alternatively, S–S bridges in this lobe region.

Incidentally, the same arguments apply to the zinc content of the human liver enzyme. The high degree of sequence homology between the human and horse enzymes (Section II,B,2) shows that the human enzyme also contains the two different binding sites for zinc in each subunit.

There are some interesting differences between the liver and yeast enzymes among the residues which are in the active site pocket region. Residues 58 and 93 are both substituted by tryptophans in the yeast enzyme which narrows the pocket. Furthermore, the two prolines which are inserted on both sides of residue 58 in the yeast enzyme could diminish the pocket even further. A smaller active site pocket in the yeast enzyme is consistent with the substrate specificity of the two enzymes (Section III,E) as well as the ability to bind big chelating molecules like 1,10-phenanthroline (Section III,D).

The first eight residues of the liver enzyme form a loosely organized

TABLE XI

SEQUENCE COMPARISONS IN DIFFERENT ALCOHOL DEHYDROGENASES OF THOSE RESIDUES THAT ARE STRUCTURALLY OR FUNCTIONALLY IMPORTANT

Residues in				Residues in		
Core No.	Horse liver	Yeast	Bacillus	Core No.	Horse liver	Yeast
1. Hydrophobic cores						
K2	11 Ala	Gly	Ala	K4	70 Ala	Ala
K3	12 Ala	Val	Ala	K1	73 Val	Val
K2	13 Val	Ile	Val	K1	80 Val	Val
K3	14 Leu	Phe	Val	K1	83 Val	Trp
K3	21 Phe	Leu	Leu	K1	89 Val	Ala
K2	26 Val	Ile	Val	K4	90 Ile	Gly
K1	36 Val	Leu	Val	K1	91 Pro	Ile
K1	38 Ile	Ile	Val	K4	169 Val	Val
K1	40 Met	Val	Ile	K4	171 Leu	Pro
K1	43 Thr	Ser	Cys	K4	172 Ile	Val
K3	45 Ile	Val	Val	K4	173 Gly	Leu
K3	59 Thr	Thr		K4	176 Phe	Gly
K3	63 Val	Leu		N1	183 Ala	Ala
K2	64 Ile	Val		N1	186 Val	Ser
K3	65 Ala	Gly		N1	187 Ala	Ala
K1	69 Ala	Gly		N1	189 Val	Leu

Horse liver	Yeast	Bacillus	Horse liver	Yeast
2. Active site pocket				
48 Ser	Thr	Thr	110 Phe	Asn
57 Leu	Leu		116 Leu	Leu
58 Val	Trp		178 Thr	Thr
93 Phe	Trp			
3. Zinc ligands				
46 Cys	Cys	Cys	97 Cys	Cys
67 His	His		100 Cys	Cys
174 Cys	Cys		103 Cys	Cys
			111 Cys	Cys
4. Structurally important glycines				
66 Gly	Gly		86 Gly	Gly
77 Gly	Gly			

loop on the outside of the molecule. The sequences of the compared species start at different positions in the NH_2-terminal regions. It is significant that the bacillar enzyme which is shorter starts at the residue corresponding to number nine of the liver enzyme. This is the first residue which participates in building up the pleated sheet regions. Furthermore, in this enzyme Bridgen *et al.* (*13*) were surprised to find cysteine in position 43 since thermophilic enzymes have in general fewer cysteine residues than other species. This side chain is, however, deeply buried between cores K2 and K3 in a hydrophobic region and not accessible to oxidation. They also found that Cys-46 could not be selectively carboxymethylated. The model of LADH shows that fairly small structural modifications are required to protect this ligand.

No sequence comparisons can be made in the pleated sheet region which builds up the nucleotide binding domain in LADH. However, these domains are similar even in dehydrogenases of different function (*2*), in spite of different structures for the catalytic domain. The general structure of the nucleotide binding domain is thus, in all probability, also present in yeast alcohol dehydrogenase. It is also possible to predict, although with less confidence, that the two domains are oriented relative to each other in a similar fashion in the yeast and liver enzymes. A comparison of the sequences in the region 167–190, which comprise the helices $\alpha 2$ and αA which join the two domains, show striking conservation for those residues that participate in two hydrophobic cores, K4 and N1, one in each domain.

In conclusion, it seems very probable that the three-dimensional structures of the subunits of the yeast and liver alcohol dehydrogenases, including the presence of two zinc atoms, are similar. There must be differences in the subunit contacts since the yeast enzyme is tetrameric and the liver enzyme dimeric. There are also differences in the size of the active site pocket of the subunit and at the carboxyl terminal region of the chain since the yeast enzyme is about 40 residues shorter.

2. *Evolutionary Aspects*

The rate of evolutionary changes in the alcohol dehydrogenases as estimated from the known primary structures is rapid in comparison with GAPDH (*80*). Thus, in spite of the homologies in the structurally and functionally important residues, yeast and mammalian ADH which have an identity of about 40% in the most similar regions show almost no homology in some regions and differ in subunit size and quaternary structure (*12*). This rapid mutation rate within a single group of dehydrogenases may explain why no clear sequence homologies have been noticed between dehydrogenases of different specificities in spite of similarities in tertiary structure (*2*).

It is of interest to compare the isozymes among the mammalian alcohol dehydrogenases. The known amino acid exchanges between the subunits of the horse isozymes (*41*) are different from those between the subunits of the human isozymes (*73,75*), while no isozymes have been detected in the rat enzyme (*80*). There is, furthermore, evidence for three different gene loci for alcohol dehydrogenase in man (Section II,A,2,a), two in horse (Section II,A,1,a), but only one in rat (Section II,A,3). The gene duplications probably occurred independently in these three species since the structural differences are found at nonequivalent positions along the molecules (*80*). Several different gene duplications are therefore apparent during the comparatively recent ancestral separation of these three species, which also suggests that the present rate of evolutionary changes in liver alcohol dehydrogenase is high.

The rate of evolution of the coenzyme binding domain may be estimated separately and compared to that of the catalytic domain. When only the known mammalian enzymes are considered (Table I), no great differences are noticed between the rates of change in the separate domains. Amino acid differences are distributed along the entire molecules (*84*), and one of the two variable regions (Section II,B,3) is inside each domain. Comparisons with the distantly related yeast enzyme, however, reveal evolutionary differences between the two domains. Thus, known segments of the junctional region between the two domains as well as long segments of the catalytic domain show homologies with identities in primary structure around 30% (Table XIV) and even 40% in a shorter segment [residues 25–87 (*12*)]. Known regions (*12,136,137*) of the coenzyme binding domain, although hitherto only partly characterized, do not, however, reveal a similar degree of identity. A similar difference in evolutionary rate between the two domains has also been observed in glyceraldehyde-3-phosphate dehydrogenase (*128*).

E. Studies on Modified Enzymes

1. *Chemical Modifications*

a. Cysteine Residues. The horse enzyme has 14 free SH groups per subunit (*53,93,138,139*) and is inhibited by mercurials and other thiol reagents (*1*). One of the cysteine residues was early found (*85–87*) to be especially reactive with iodoacetate; it was later discovered (*137*) that another cysteine residue may be selectively modified with a reactive coenzyme analog. Both these residues are now known to be ligands to the functional zinc atom (Section II,C) at the active site of the enzyme.

136. H. Jörnvall, *FEBS* (*Fed. Eur. Biochem. Soc.*) *Lett.* **38**, 329 (1974).
137. H. Jörnvall, C. Woenckhaus, and G. Johnscher, *Eur. J. Biochem.* **53**, 71 (1975).
138. A. Witter, *Acta Chem. Scand.* **14**, 1717 (1960).
139. T. Yonetani and H. Theorell, *ABB* **99**, 433 (1962).

Their positions in the primary structure have been determined as Cys-46 and Cys-174, respectively (Section II,B). Reactive cysteine residues have also been alkylated with iodoacetamide (*140*), modified with spin-labeled reagents (*141*) and with a mercury substituted salicylate molecule (*107*). In the latter case the X-ray analysis has shown that two molecules are bound to Cys-132 and Cys-240, respectively. In the homologous rat enzyme other cysteine residues, now known to be ligands to the second zinc atom (Section II,C,3,b), have been differentially carboxymethylated (*52*).

(*i*) *Specific modification of Cys-46.* Li and Vallee (*85,87*) and Harris (*86*) found that one cysteine residue per subunit may be selectively carboxymethylated with iodoacetate. The modified enzyme is inactivated and this cysteine residue, Cys-46 (*92*), was suggested to be at the active site of the enzyme. The same residue in the S subunit is also especially reactive (*20,94*). The modification is preceded by anion binding of the iodoacetate and stimulated by the presence of imidazole (*140,142,143*). By using these facts and working with the crystalline enzyme, it is possible to achieve a highly specific and complete modification (*130*). X-ray studies of the carboxymethylated enzyme and the reaction mechanism of this modification are described in Section II,H. The carboxymethylation has been used to establish that both the EE (*19*) and SS (*20*) isozymes are active in ω oxidations of fatty acids.

The inactivated carboxymethylated enzyme still binds NAD (*85*) and also has a small residual enzymic activity (*144*). Reactive cysteine residues also occur in other dehydrogenases and have been compared (*145*), but the properties are different and these comparisons have no functional or evolutionary significance (*2,73*).

(*ii*) *Specific modification of Cys-174.* Reactive bromoacetyl derivatives of NAD analogs inactivate LADH by covalent modification (*146–148*) (Table XVI).

140. N. Evans and B. R. Rabin, *Eur. J. Biochem.* **4,** 548 (1968).
141. J. E. Spallholz and L. H. Piette, *ABB* **148,** 596 (1972).
142. C. H. Reynolds and J. S. McKinley-McKee, *Eur. J. Biochem.* **10,** 474 (1969).
143. C. H. Reynolds, D. L. Morris, and J. S. McKinley-McKee, *Eur. J. Biochem.* **14,** 14 (1970).
144. C. H. Reynolds and J. S. McKinley-McKee, *BJ* **119,** 801 (1970).
145. J. J. Holbrook, G. Pfleiderer, K. Mella, M. Volz, W. Leskowac, and R. Jeckel, *Eur. J. Biochem.* **1,** 476 (1967).
146. C. Woenckhaus, M. Zoltobrocki, and J. Berghäuser, *Hoppe-Seyler's Z. Physiol. Chem.* **351,** 1441 (1970).
147. C. Woenckhaus and R. Jeck, *Hoppe-Seyler's Z. Physiol. Chem.* **352,** 1417 (1971).
148. R. Jeck, P. Zumpe, and C. Woenckhaus, *Justus Liebigs Ann. Chem.* p. 961 (1973).

Nicotinamide-[5-bromoacetyl-4-methyl-imidazole] dinucleotide was prepared by Woenckhaus and co-workers and found to be active as hydrogen acceptor in the enzymic reaction (*147*). It also causes complete inactivation after incorporation in a 1:1 ratio of the protein subunits. The enzyme is, furthermore, protected against inactivation by NAD and NADH. These properties suggest that the inactivator alkylates a residue at the active site of the enzyme. This residue has been identified (*137*) as Cys-174 which thus is labeled by the inactivator without modification of Cys-46 or other residues. This result suggests that Cys-174 and Cys-46 are close together at the active site which agrees with the crystallographic data (Section II,C,3,b). The modification of Cys-174 by an inactivator with a bromoacetyl group on the adenine side of the molecule furthermore suggests that the inactivator, at least in the case of those molecules which alkylated the protein, was anomalously bound (*137*) before reaction.

b. Lysine Residues. Reaction of horse liver alcohol dehydrogenase with methyl picolinimidate was found by Plapp to enhance the enzymic activity (*149,150*). Most of the lysine ε-amino groups were modified, but the activation is the result of substitution of only one amino group per subunit of enzyme (*151,152*). This group is specifically protected by NAD and pyrazole or NADH and isobutyramide (*150*). Its location was suggested to be near the binding site for the nicotinamide ring of the coenzyme (*150*). By differential acetimidylation the lysine residue involved has been preliminarily identified as Lys-228 (*152,153*) in the protein. A number of modifications of the amino groups at the active sites of the enzyme have been further studied (*154*). The effects of the net charge of the substituted amino group and of the size of the substituent on the enzymic properties were then determined. The position of Lys-228 in the apoenzyme structure is schematically depicted in Fig. 7. The ε-amino group forms a hydrogen bond to one of the carboxyl oxygen atoms of Asp-223. This residue is involved in the binding of adenosine ribose. Combined with the results of the chemical studies, this indicates that the side chain of Lys-228 has changed its conformation in the modified enzyme.

Lysine residues have also been modified by reaction with pyridoxal-phosphate in the presence of sodium borohydride which inactivated the enzyme (*155*). It was concluded from this study that a lysine residue

149. B. V. Plapp, *Fed. Proc., Fed. Amer. Soc. Exp. Biol.* **28,** 601 (1969).
150. B. V. Plapp, *JBC* **245,** 1727 (1970).
151. B. V. Plapp, R. L. Brooks, and J. D. Shore, *JBC* **248,** 3470 (1973).
152. B. V. Plapp, *in* "Alcohol and Aldehyde Metabolizing Systems" (R. G. Thurman *et al.*, eds.), p. 91. Academic Press, New York, 1974.
153. R. Dworschak, G. Tarr, and B. V. Plapp, *Biochemistry* **14,** 200 (1975).
154. M. Zoltobrocki, J. C. Kim, and B. V. Plapp, *Biochemistry* **13,** 899 (1974).
155. J. S. McKinley-McKee and D. L. Morris, *Eur. J. Biochem.* **28,** 1 (1972).

is at the active center close to the phosphate group of the AMP part of the coenzyme. A second site was also suggested (*155*) and several amino groups were always modified. Methylation of lysine residues by reaction with formaldehyde and sodium borohydride was found to activate the enzyme (*156,157*); the methylated enzyme is partly protected from modification by the coenzymes. Differential labeling did not permit a strict functional correlation of reactive residues, but some residues were identified as particularly sensitive to modification (*156*).

Thus, modification of lysine residues with imido esters (*150*) by reductive methylation (*156,157*) or with other reagents (*154*) is a useful way of introducing radioactive labels into the protein without destroying the enzymic activity.

c. Histidine Residues. Histidine residues in horse liver alcohol dehydrogenase may be preferentially destroyed by photooxidation (*158,159*). The enzyme is inactivated but the degree of activity loss varies as measured with different coenzyme analogs (*159*). The affected residues were not identified.

Histidine residues in the horse enzyme have also been modified by carbethoxylation with diethylpyrocarbonate. This was found to result in a rapid activation followed by a slower inactivation (*160*). Substitution of the enzyme was measured, but affected histidine residues were not identified. They were, however, suggested to be outside the active site.

d. Tyrosine Residues. Iodination of tyrosine residues of horse liver alcohol dehydrogenase has been reported (*161*). By simultaneous oxidation of protein SH groups (*161*), the enzyme is easily inactivated. However, a mild iodine treatment of the protein in the crystalline state can produce an active enzyme with partly labeled tyrosine residues (*161*). Iodination may, therefore, be an alternative way to introduce radioactive label into the protein without destroying the enzymic activity. No evidence for a tyrosine residue with particular structural or functional significance has been obtained from this study. This agrees with the tertiary structure of the enzyme (Section II,C).

e. Arginine Residues. Arginine residues in the human and horse enzymes have been modified with loss of enzymic activity by treatment

156. H. Jörnvall, *BBRC* **51,** 1069 (1973).
157. C. S. Tsai, Y.-H. Tsai, G. Lauzon, and S. T. Cheng, *Biochemistry* **13,** 440 (1974).
158. D. Robinson and D. Stollar, *Fed. Proc., Fed. Amer. Soc. Exp. Biol.* **21,** 232 (1962).
159. D. Robinson, D. Stollar, S. White, and N. O. Kaplan, *Biochemistry* **2,** 486 (1963).
160. D. L. Morris and J. S. McKinley-McKee, *Eur. J. Biochem.* **29,** 515 (1972).
161. H. Jörnvall and M. Zeppezauer, *BBRC* **46,** 1951 (1972).

with butanedione or phenylglyoxal (*162*). Analysis of total substitution indicates that two arginine residues may be associated with the activity loss, but the residues have not been identified. NADH protects the enzyme from both loss of activity and arginine modification. The modified protein does not bind the coenzyme. Arginine residues have therefore been suggested (*162*) to participate in NADH binding, perhaps through interaction with the coenzyme's pyrophosphate bridge (*162*). This agrees with the results of the crystallographic analysis (Section II,C).

f. Uncharacterized Residues. Apart from the above-mentioned nicotinamide-substituted imidazole dinucleotide (Section II,E,1,a), other reactive NAD analogs (*148,163*) have been shown to inactivate horse liver alcohol dehydrogenase by alkylation, but the modified residues have not been identified. Reactive coenzyme analogs have been further studied with yeast alcohol dehydrogenase, a summary of these studies is found in Table XVI.

2. *Changes of Metal Content*

The presence of two zinc atoms per subunit of LADH was first established by Åkeson (*134*). He also showed that one zinc was essential for activity and suggested that the second zinc may have a structural role. There is no difference in zinc content between the multiple forms of the EE isozymes (*164*). Attempts to differentiate the function and chemical reactivities of the two zinc atoms have been described in a series of papers (*133,165–167*) which have been reviewed (*168*). One zinc atom per subunit could be selectively exchanged and removed by dialysis. The modified enzyme containing one zinc atom per subunit was catalytically inactive and did not bind 1,10-phenanthroline. From this information, combined with the X-ray data which show that the catalytic zinc atom binds 1,10-phenanthroline, it is evident that the catalytic zinc atom is first removed during dialysis under these conditions (*167*). The second zinc atom can be selectively removed, in preference to the catalytic zinc, by carboxymethylation (*165*). From sedimentation experiments of zinc-free

162. L. G. Lange, J. F. Riordan, and B. L. Vallee, *Biochemistry* **13,** 4361 (1974).
163. J.-F. Biellmann, G. Branlant, B. Y. Foucaud, and M. J. Jung, *FEBS* (*Fed. Eur. Biochem. Soc.*) *Lett.* **40,** 29 (1974).
164. N. Sandler and R. H. McKay, *BBRC* **35,** 151 (1969).
165. D. E. Drum, J. H. Harrison, T. K. Li, J. L. Bethune, and B. L. Vallee, *Proc. Nat. Acad. Sci. U. S.* **57,** 1434 (1967).
166. D. E. Drum, T.-K. Li, and B. L. Vallee, *Biochemistry* **8,** 3783 (1969).
167. D. E. Drum, T.-K. Li, and B. L. Vallee, *Biochemistry* **8,** 3792 (1969).
168. D. D. Ulmer and B. L. Vallee, *Advan. Chem. Ser.* **100,** 187 (1971).

enzyme (*165,169*), it was concluded that zinc is not necessary for the subunit association.

Weiner and co-workers have shown that zinc-free LADH binds coenzyme (*121,122,170–172*) and also a spin labeled coenzyme analog (*4,173*). They concluded that zinc is not necessary for coenzyme binding, which agrees with the X-ray results.

However, the findings (*171*) that binary complexes of zinc-free enzyme and coenzyme bind substrates and substrate competitive inhibitors such as isobutyramide cannot be taken as evidence that zinc does not participate in the catalytic action. Several SH groups are probably oxidized in the zinc-free enzyme (*170*). Evidence has also been presented (*172*) of other structural differences compared to the catalytically active enzyme. Thus, artificial binding to this enzyme with no relevance to the catalytic action is not unlikely.

Experiments which substitute cobalt or other metal atoms for zinc while retaining activity have been difficult to perform for LADH and have only recently succeeded (*174–176*). Extreme caution in keeping a reduced atmosphere is required, presumably because of oxidation of the zinc cysteine ligands to S–S groups. Exposure to air after zinc removal oxidizes SH groups in LADH (*177*).

The coordination around zinc in LADH is unusual compared to the other zinc-containing enzymes which have been investigated. In LADH there are two cysteine and one histidine ligands to the catalytic zinc and four cysteine ligands to the second zinc (Fig. 11). In carbonic anhydrase (*178*) there are three histidines and in carboxypeptidase (*179*) and thermolysin (*180*) there are two histidines and one glutamic acid as zinc ligands. Thiol groups as ligands to zinc in LADH were implied by Drum and Vallee (*174*) from an analysis of the absorption spectrum of cobalt- and cadmium-substituted LADH and actually proposed by Williams and

169. R. W. Green and R. H. McKay, *JBC* **244,** 5034 (1969).
170. C. W. Hoagstrom, I. Iweibo, and H. Weiner, *JBC* **244,** 5967 (1969).
171. I. Iweibo and H. Weiner, *Biochemistry* **11,** 1003 (1972).
172. P. L. Coleman, I. Iweibo, and H. Weiner, *Biochemistry* **11,** 1010 (1972).
173. A. S. Mildvan and H. Weiner, *JBC* **244,** 2465 (1969).
174. D. E. Drum and B. L. Vallee, *BBRC* **41,** 33 (1970).
175. J. M. Young and J. H. Wang, *JBC* **246,** 2815 (1971).
176. M. Takahashi and R. A. Harvey, *Biochemistry* **12,** 4743 (1973).
177. H. L. Oppenheimer, R. W. Green, and R. H. McKay, *ABB* **119,** 552 (1967).
178. S. Lindskog, L. E. Hendersson, K. K. Kannan, A. Liljas, P. O. Nyman, and B. Strandberg, "The Enzymes," 3rd ed., Vol. 5, p. 587, 1971.
179. J. A. Hartsuck and W. N. Lipscomb, "The Enzymes," 3rd ed., Vol. 3 p. 1, 1971.
180. P. M. Colman, J. N. Jansonius, and B. W. Matthews, *JMB* **70,** 701 (1972).

co-workers (*181*), based on the presence of a charge transfer band in the cobalt-LADH spectrum.

Young and Wang (*175*) have also prepared a hybrid enzyme containing both cobalt and zinc. Binding of azide, pyrazole, or 1,10-phenanthroline caused no change in the spectrum of this enzyme, indicating that these substrate competitive inhibitors (Section II,H) did not bind to cobalt. A similar hybrid enzyme was used by Takahashi and Harvey (*176*) to study energy transfer from bound dyes and coenzyme to the cobalt atom. They calculated a minimal distance of 19 Å between the nicotinamide ring and the nearest cobalt atom. These experiments cannot be interpreted in terms of a substitution of cobalt for the catalytic zinc atom. Such an interpretation is contrary to all other experimental data on the position and function of the catalytic zinc atom including the X-ray structure.

3. *Denaturation Studies*

The stability of LADH and its denaturation has been studied under a variety of conditions such as acid pH, different concentrations of urea, guanidine hydrochloride and detergents as well as high temperature.

a. Acid Denaturation. LADH loses activity and zinc at pH 5 while still in the dimeric state (*177,182*). At lower pH dissociation occurs into subunits (*182–184*) and there are drastic changes in the protein fluorescence spectrum (*185–187*) and the fluorescence polarization spectrum (*182*). Different time dependences for the changes of the tyrosine and tryptophan difference fluorescence peaks are observed (*187*), which is consistent with a slower quenching of the buried Trp-314 (Section II,C,3,c) compared to the more exposed tyrosines. This interpretation implies that a partial unfolding of the tertiary structure occurs prior to the dissociation into subunits at acid pH.

b. Denaturation by Urea, Guanidine HCl and Detergents. LADH dissociates into two subunits in 7 to 8 *M* urea (*89, 188*) in 5 to 6 *M* guani-

181. K. Garbett, G. W. Partridge, and R. J. P. Williams, *Bioinorg. Chem.* **1,** 309 (1972).
182. R. H. Mc Kay, *ABB* **135,** 218 (1969).
183. L.-Y. Cheng and J. S. McKinley-McKee, *BBRC* **31,** 755 (1968).
184. L.-Y. Cheng, J. S. McKinley-McKee, C. T. Greenwood, and D. J. Hourston, *BBRC* **31,** 761 (1968).
185. L. Brand, J. Everse, and N. O. Kaplan, *Biochemistry* **1,** 423 (1962).
186. J. R. Heitz and L. Brand, *ABB* **144,** 286 (1971).
187. C. H. Blomquist, *ABB* **122,** 24 (1967).
188. J. A. Koepke, Å. Åkeson, and R. Pietruszko, *Enzyme* **13,** 177 (1972).

dine HCl (*169,189*), and in 5 m*M* dodecyl sulfate (*190*). The dissociation is accompanied by considerable changes in tertiary structure (*188*). It has been reported (*183,184,191*) that dissociation in 8 *M* urea after treatment with mercaptoethanol produces subspecies of molecular weight around 20,000. It is not impossible that these are the result of breaks in the polypeptide chain leading to a separation of the two domains of the subunits (Section II,C). On the other hand, it has been claimed (*188*) that the chromatography pattern of the products of LADH dissociation in urea in the presence of mercaptoethanol is indistinguishable from those in urea alone.

c. Stability as a Function of Temperature. Theorell and co-workers have shown (*139,192*) that both NAD^+ and NADH protect LADH against thermal inactivation. The protection was concentration-dependent affording almost complete protection for 15 min at 80° at high concentrations of NADH. The ternary complex with isobutyramide was even more stable (*139*). On the other hand, opposite effects have also been found (*193*) using slightly different conditions.

F. Protein Fluorescence and Phosphorescence

Quenching of the protein fluorescence of LADH has been observed by bound aromatic inhibitors and NADH (*194–198*). Possible causes of this quenching have been discussed mainly in terms of two alternative processes: either a change in the environment of the tryptophans or an excitation energy transfer process from tryptophan to the nicotinamide ring of the coenzyme. The first alternative is highly unlikely since 1,10-phenanthroline, which binds far from the tryptophans (Fig. 15) and which does not induce any conformational change (Section II,C), causes a quenching similar to that caused by NADH (*196*).

It is thus very probable that the majority of the quenching is caused by excitation energy transfer. Such a process is also strongly favored by Theorell and Tatemoto (*196*). They found in titration experiments

189. F. J. Castellino and R. Barker, *Biochemistry* **7**, 2207 (1968).
190. C. H. Blomquist, D. A. Smith, and A. M. Martinez, *ABB* **122**, 248 (1967).
191. D. B. Pho and J. L. Bethune, *BBRC* **47**, 419 (1972).
192. H. Theorell and K. Tatemoto, *ABB* **143**, 354 (1971).
193. A. Wiseman and N. J. Williams, *BBA* **250**, 1 (1971).
194. A. D. Winer and H. Theorell, *Acta Chem. Scand.* **14**, 1729 (1960).
195. G. Geraci and Q. H. Gibson, *JBC* **242**, 4275 (1967).
196. H. Theorell and K. Tatemoto, *ABB* **142**, 69 (1971).
197. P. L. Luisi and R. Favilla, *Eur. J. Biochem.* **17**, 91 (1970).
198. J. J. Holbrook, *BJ* **128**, 921 (1972).

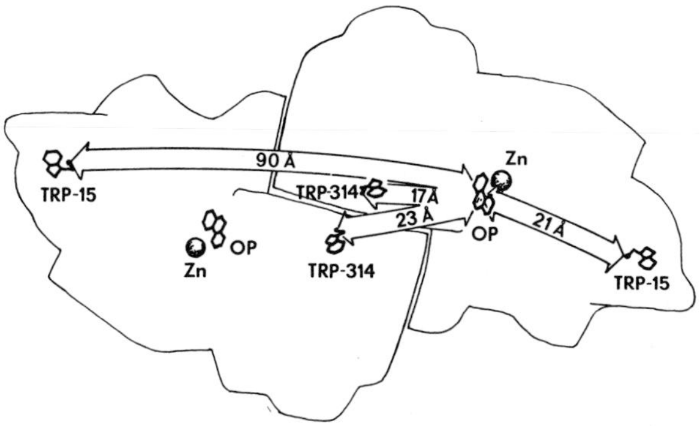

FIG. 15. Schematic diagram of the LADH molecule showing the positions of the four tryptophans in relation to the two active sites and the distances involved.

with NADH that the increase of the NADH fluorescence band parallels the decrease of the protein fluorescence band when the complex is excited at 297 nm. Only the aromatic residues of the protein are excited at this wavelength, but energy transfer can occur to the coenzyme since the absorption band of NADH overlaps the fluorescence emission band of the protein. It was also found in these experiments (*196*) that the quenching of protein fluorescence is nonlinear as a function of ligand concentration, both for NADH and for 1,10-phenanthroline as quenchers. Binding of the first NADH molecule to the LADH dimer causes a larger quenching of the protein fluorescence than the binding of the second one.

In contrast, linear curves are obtained for the enhancement of NADH fluorescence when excited at 330 nm where no excitation transfer from the protein is possible. The tertiary structure provides a possible and simple explanation for these experiments in terms of the geometrical arrangement of the four tryptophans and the two active sites of the whole molecule as illustrated in Fig. 15. This interpretation is quite different from the structural implications of the "geometric quenching" discussed by Holbrook *et al.* (*199*).

The distances from the active site in one subunit, represented by the midpoint of 1,10-phenanthroline, to Trp-15 in the same subunit and to the two Trp-314 of both subunits are between 17 and 23 Å (Fig. 15). These distances are well within the limits expected from theoretical calculations for energy transfer (*200*). In contrast, the distance to the fourth

199. J. J. Holbrook, D. W. Yates, S. J. Reynolds, R. W. Evans, C. Greenwood, and M. G. Gore, *BJ* **128**, 933 (1972).
200. G. Karreman, R. Steele, and A. Szent-Györgyi, *Proc. Nat. Acad. Sci. U. S.* **44**, 140 (1958).

tryptophan of the molecule, Trp-15 in the second subunit, is about 90 Å. Thus, in simple terms, a quencher bound only to one active site at the beginning of the titration experiment can quench the fluorescence from three of the four tryptophans. This leaves mainly the fluorescence from only one tryptophan for the next quencher molecule when it binds to the second active site.

Purkey and Galley (*201*) have shown that the phosphorescence band in LADH around 400 nm is split at low temperatures into separate peaks at 405 and 410 nm. From the time dependence of the bromide quenching, they concluded that the 410 peak arises from a tryptophan in a more hydrophobic environment than that responsible for the 405 nm peak. The 410 nm peak can, thus, in all probability, be assigned to Trp-314 which is buried deep in the hydrophobic subunit interaction area, whereas Trp-15 is exposed to the solvent (Section II,C).

G. Studies of Coenzyme Analogs

A large number of coenzyme analogs modified at various positions of the NAD molecule have been prepared and tested for coenzyme activity with LADH and YADH. The results are generally consistent with the X-ray studies of the binding of coenzyme analogs to LADH (Section II,C,3). Some analogs discussed in Sections II,E and III,C have been used to modify specific amino acid residues of the enzymes.

a. Adenine Moiety. Removal of the adenine moiety causes a drastic decrease in V_{max} and an increase in K_m values (*202,203*) presumably because of a weak binding of the analog compared to NAD. Substitution of adenine for other aromatic groups (*202–212*) causes much smaller changes in kinetic parameters. These findings are in agreement with the

201. R. M. Purkey and W. C. Galley, *Biochemistry* **9**, 3569 (1970).
202. G. Pfleiderer, E. Sann, and F. Ortanderl, *BBA* **73**, 39 (1963).
203. C. Woenckhaus and P. Zumpe, *Z. Naturforsch. B* **23**, 484 (1968).
204. C. P. Fawcett and N. O. Kaplan, *JBC* **237**, 1709 (1962).
205. G. Pfleiderer, C. Woenckhaus, K. Scholz, and H. Feller, *Justus Liebigs Ann. Chem.* **675**, 205 (1964).
206. C. Woenckhaus and G. Pfleiderer, *Biochem. Z.* **341**, 495 (1965).
207. C. Woenckhaus and M. H. Volz, *Chem. Ber.* **99**, 1712 (1966).
208. C. Woenckhaus and D. Scherr, *Z. Naturforsch B* **26**, 106 (1971).
209. D. C. Ward, T. Horn, and E. Reich, *JBC* **247**, 4014 (1972).
210. J. R. Barrio, J. A. Secrist, III, and N. J. Leonard, *Proc. Nat. Acad. Sci. U. S.* **69**, 2039 (1972).
211. D. Scherr, R. Jeck, J. Berghäuser, and C. Woenckhaus, *Z. Naturforsch.* **28c**, 247 (1973).
212. C. Y. Lee and J. Everse, *ABB* **157**, 83 (1973).

observed nonspecificity of the hydrophobic adenine binding pocket (Section II,C,3,a).

Since the adenine moiety is bound with its NH_2 group pointing out from the protein and with N-1 pointing into the hydrophobic pocket, substitutions at these two nitrogens have different effects (*213*). The N^6-hydroxyethyl-substituted analog is almost as active as NAD^+ whereas the corresponding N-1 substituted analog possesses lower enzymic activity.

b. Adenosine Ribose Moiety. Proper orientation of the adenosine ribose moiety, mediated by hydrogen bonding between O-2 and Asp-223, is an important feature of coenzyme binding (Section II,C,3,a). The importance of O-2′ was first realized by Fawcett and Kaplan (*204*) who showed that the 2′-deoxyadenosine analog had lower activity than NAD because of weaker binding. Furthermore, NADP, which has an extra phosphate bound to O-2′, has a low activity compared to NAD (*1,202,204*). The adenosine ribose of NADP cannot be oriented in the same way as NAD because of interference between the protein and the extra phosphate as well as of charge repulsion between phosphate and Asp-223.

c. Nicotinamide Ribose Moiety. The structure determination of the LADH–ADP-ribose complex indicated that the two hydrogen bonds from O-2′ and O-3′ of the nicotinamide ribose to main chain carbonyl oxygen atoms (Section II,C,3,a) are important for coenzyme binding. This conclusion is supported by studies in solution showing that ADP-ribose binds more strongly than ADP (*97,214*) and that the corresponding 2′,3′-dideoxy-NAD analog has very low activity (*215*). Furthermore, analogs in which ribose is replaced by a butyl or a propyl group show little or no activity (*146,148*).

d. Nicotinamide Moiety. A large number of analogs modified at the 3 position of the nicotinamide ring have been prepared and studied for coenzyme activity. Previous studies have been reviewed extensively (*1,3,6*) and will not be further discussed here. Recent studies include 3-CN pyridine AD^+ (*216*), which has been reported to be slightly active toward different isozymes of LADH (*217*), 3-aminopyridine AD^+ (*218*) and 3-aldoxime pyridine AD^+ (*217*) which are inhibitors showing no activity, and 3-I pyridine AD^+, which is surprisingly active (*217*).

213. H. G. Windmueller and N. O. Kaplan, *JBC* **236,** 2716 (1961).
214. B. M. Anderson, M. L. Reynolds, and C. D. Anderson, *ABB* **111,** 202 (1965).
215. C. Woenckhaus and R. Jeck, *Justus Liebigs Ann. Chem.* **736,** 126 (1970).
216. J.-F. Biellmann and M. J. Jung, *FEBS (Fed. Eur. Biochem. Soc.) Lett.* **7,** 199 (1970).
217. J.-P. Samama, Thèse Doctorat de 3ᵉ Cycle, Strasbourg, 1973.
218. T. L. Fisher, S. V. Vercellotti, and B. M. Anderson, *JBC* **248,** 4293 (1973).

Analogs with an acetyl, thioacetamide, or cyano group in the 3 position of the pyridine ring have all been shown (*219*) to exhibit the same stereochemistry for hydrogen transfer as NAD^+. It was concluded (*219*) that interactions between enzyme and coenzyme were solely responsible for the A-side stereo specificity of NAD^+ in LADH which agrees with the X-ray results (Section II,C,3,a).

Analogs having substitutions at other positions of the nicotinamide ring have also been prepared recently (*217,220*). 2-Methyl and 6-methyl NAD have both been reported (*217,220*) to exhibit different activities toward different isozymes of LADH.

H. Inhibitor Binding

LADH provides an excellent example of the generally recognized notion (*221*) that strong reversible binding of small molecules takes place predominantly in areas overlapping the binding sites for substrates, coenzymes, and prosthetic groups. There is evidence for all molecules, where the binding to LADH has been studied in detail, that they bind either at the coenzyme binding site or the substrate binding site. Since the productive substrate binding site is induced by coenzyme binding, there is one class of compounds which binds in the abortive substrate cleft of the apoenzyme but does not bind in the presence of coenzyme. Other molecules bind at the productive substrate binding site forming ternary complexes with coenzyme but do not bind to the apoenzyme. The inhibitors which have been studied are arranged here according to the most likely mode of binding.

1. *Binary Complexes*

a. Coenzyme Competitive Inhibitors. Inactive coenzyme analogs are the most obvious examples of these inhibitors. These are discussed in Section II,G. There are three main binding areas for inhibitors at the coenzyme binding site: the adenosine binding cleft, the anion binding site where the pyrophosphate group of the coenzyme binds, and the nicotinamide binding region.

It has been found (*102*) that aromatic coenzyme competitive molecules with a negative charge attached to the ring system will bind at the adenosine binding site. Examples of such molecules which have been studied

219. J.-F. Biellmann, C. G. Hirth, M. J. Jung, N. Rosenheimer, and A. D. Wrixon, *Eur. J. Biochem.* **41**, 517 (1974).
220. J.-F. Biellmann and J.-P. Samama, *FEBS (Fed. Eur. Biochem. Soc.) Lett.* **38**, 175 (1974).
221. A. N. Glazer, *Proc. Nat. Acad. Sci. U. S.* **65**, 1057 (1970).

TABLE XII
DISSOCIATION CONSTANTS FOR VARIOUS MOLECULES THAT FORM BINARY COMPLEXES WITH LADH

Compound	Ref.	$K_I(\mu M)$
Salicylate	*222*	1,250
1,5-ANS	*225*	370
1,7-ANS	*225*	53
1,8-ANS	*225*	125
TNS	*227*	13–17
Rose Bengal	*225*	3
Triiodothyroacetic acid	*229*	1.3–1.9
Triiodothyronine	*231*	6.7
1,10-Phenanthroline	*97*	8
2,2-Bipyridine	*249*	400
Hydroxymethylferrocene	*253*	140–180
Benzamide	*249*	110,000
Auramine O	*287*	16
Berberine	*255*	36

are salicylic acid (*102,222–224*), anilinonaphthalene sulfonate (ANS) (*102,225,226*), 2,6-toluidinonaphthalene sulfonate (TNS) (*227,228*), Rose Bengal (*225*), and triiodothyroacetic acid (*229*). Dissociation constants for these inhibitors are given in Table XII.

A complex between LADH and an iodo-substituted salicylic acid has been studied by X-ray methods (*102*). It was found that salicylate simulates binding of the adenosine part of the coenzyme although the molecules are structurally very different. The phenyl ring binds in the adenine binding region, the hydroxyl group is hydrogen bonded to Asp-223 which in the ADP-ribose complex forms a hydrogen bond to O-2′ of the adenosine ribose, and the carboxyl group forms a salt linkage to Arg-271. Furthermore, it has been shown by fluorimetric methods that ANS binds competitively to salicylic acid (*102*) and that Rose Bengal (*225*),

222. P. D. Dawkins, B. J. Gould, J. A. Sturman, and M. J. H. Smith, *J. Pharm. Pharmacol.* **19**, 355 (1967).
223. S. Grisolia, I. Santos, and J. Mendelson, *Nature (London)* **219**, 1252 (1968).
224. S. Grisolia, J. Mendelson, and D. Diederich, *FEBS (Fed. Eur. Biochem. Soc.) Lett.* **11**, 140 (1970).
225. L. Brand, J. R. Gohlke, and D. S. Rao, *Biochemistry* **6**, 3510 (1967).
226. D. C. Turner and L. Brand, *Biochemistry* **7**, 3381 (1968).
227. J.-C. Mani, J. Dornand, M. Mousseron-Canet, and F. Vial-Reveillon, *Biochimie* **53**, 355 (1971).
228. J. Dornand, J.-C. Mani, M. Mousseron-Canet, and F. Vial-Reveillon, *Biochimie* **53**, 1181 (1971).
229. K. McCarthy and W. Lovenberg, *JBC* **244**, 3760 (1969).

TNS (*227*), and triiodothyroacetic acid (*229*) bind competitively to ANS. Triiodothyronine (*230,231*) and thyroxine (*59,230*) have also been shown to bind to LADH and to interfere with coenzyme binding. There is, however, conflicting evidence regarding the detailed mode of inhibitory action.

The X-ray studies of LADH have shown (*107*) that the anion $Pt(CN)_4^{2-}$ binds at the same site on the enzyme as the pyrophosphate group of the coenzyme. Arginine-47 is involved in this general anion binding site. The relation between the anion binding site, Arg-47, the catalytic zinc atom and the reactive Cys-46 is shown in Fig. 16. Early kinetic studies (*1*) and optical rotatory experiments (*232,233*) showed that LADH was inhibited by anions like halides, perchlorate, and thiosulfate and that this inhibition was coenzyme-competitive. Recent kinetic studies (*234*) have confirmed this competition. Furthermore, inhibition studies (*235*) show that $Pt(CN)_4^{2-}$ and $Au(CN)_2^-$ bind in strict competition to coenzyme, ADP-ribose, AMP, and chloride ions. Kinetic studies (*235*) have also shown that there is a second binding site for $Au(CN)_2^-$ which probably coincides with the main binding site for this anion observed in the X-ray analysis (*107*). This site is inside the adenine binding pocket. Using ^{35}Cl nuclear magnetic resonance methods (*236,237*), it was found that chloride ions bind competitively to coenzyme and, furthermore (*238*), that this binding is competitive to $Pt(CN)_4^{2-}$ and independent of 1,10-phenanthroline. The anions $Pt(CN)_4^{2-}$, $Au(CN)_2^-$, and Cl^- thus all bind at the general anion binding site where the phosphate groups of the coenzyme are bound. Thus, anions like chloride ions do not bind to zinc in LADH in contrast to the situation in carbonic anhydrase (*178*).

This general anion binding site is especially interesting because of its involvement in the chemical modification of Cys-46 (Section II,E,1,a). The carboxymethylation of this residue is preceded by a reversible binding of iodoacetate to the enzyme (*142,143*). It has furthermore been shown that coenzyme, ADP-ribose, ADP, AMP, $Pt(CN)_4^{2-}$ and chloride ions but not 1,10-phenanthroline protects competitively against this reversible binding (*130,140,142,143*). An inspection of the LADH model

230. K. McCarthy, W. Lovenberg, and A. Sjoerdsma, *JBC* **243**, 2754 (1968).
231. M. J. Gilleland and J. D. Shore, *JBC* **244**, 5357 (1969).
232. T. K. Li and B. L. Vallee, *JBC* **239**, 792 (1963).
233. T. K. Li, D. D. Ulmer, and B. L. Vallee, *Biochemistry* **2**, 482 (1963).
234. P. L. Coleman and H. Weiner, *Biochemistry* **12**, 1705 (1973).
235. P.-O. Gunnarsson, G. Pettersson, and M. Zeppezauer, *Eur. J. Biochem.* **43**, 479 (1974).
236. R. L. Ward and J. A. Happe, *BBRC* **45**, 1444 (1971).
237. B. Lindman, M. Zeppezauer, and Å. Åkeson, *BBA* **257**, 173 (1972).
238. J.-E. Norne, T. E. Bull, R. Einarsson, B. Lindman, and M. Zeppezauer, *Chem. Scr.* **3**, 142 (1973).

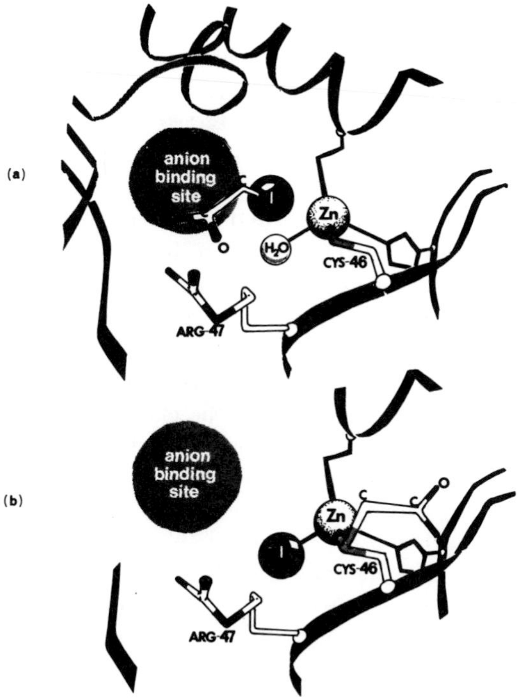

FIG. 16. Schematic diagram illustrating (a) the assumed reversible binding of iodoacetate and (b) the observed irreversible binding of iodoacetate.

showed (Fig. 16) that the sulfur atom of Cys-46 is coordinated to zinc in a position so that one of its lone pair electrons is directed toward the anion binding site. The distance from this site to the sulfur atom of Cys-46 is about 8 Å. Thus, iodoacetic acid can be positioned by reversible binding to this anion binding site for chemical attach on Cys-46, a typical example of an active-site-directed irreversible inhibitor.

An X-ray study of the carboxymethylated enzyme (*130*) shows that the carboxyl group of the reagent is positioned in a direction away from both the active site pocket and the coenzyme binding site (Fig. 16). The position of the sulfur atom of Cys-46 remains almost unchanged. The iodide ion released in the reaction is coordinated to zinc. Thus, the crevice that binds coenzyme is unchanged by the modification, which is consistent with studies in solution (*85,239*) showing that the carboxymethylated enzyme still binds coenzyme. The decreased activity of the carboxymethylated enzyme may result from changed electronic configuration of the zinc atom, the presence of bound iodide ion, interference with the necessary conformational change, or a combination of these factors.

239. T.-K. Li and B. L. Vallee, *Biochemistry* 4, 1195 (1965).

b. Substrate Competitive Inhibitors. The binding of chelating agents like 2,2-bipyridine and 1,10-phenanthroline to LADH gave the first evidence that zinc is involved in the catalytic mechanism (*240*) and has since been studied extensively (*97,110,112,120,133,241,* and references in *1,168,242*). These inhibitors bind independently of ADP-ribose and smaller coenzyme fragments, are competitive with NAD and NADH, partially competitive with ethanol, and show complex kinetic behavior with acetaldehyde (*97,120,240,243,244*). It has been shown (*245*) that these kinetic patterns can be predicted for an ordered mechanism in which the inhibitor competes with the coenzyme. It was not possible to establish conclusively from kinetic data whether the chelating agents were competitive with the substrate. However, the X-ray study (*112*) of the 1,10-phenanthroline complex has shown that the chelating groups bind to the catalytic zinc and displaces the zinc-bound water molecule at the presumed substrate binding site (Section II,C,3,b) and that the aromatic groups partly overlap the nicotinamide moiety of the coenzyme.

These chelating agents have frequently been used as probes for the interactions taking place at the catalytic site. Vallee and co-workers (*133,232,233,246–248*) have utilized the optical rotatory dispersion curves of the complexes, whereas Yonetani (*97,120*) and Sigman (*249*) have studied the changes in the absorption spectrum. By the latter method dissociation constants have been measured for a number of binary substrate and inhibitor complexes as well as some ternary inhibitor complexes. The interaction of LADH and a spin-labeled analog of 1,10-phenanthroline has also been studied (*141*).

Heterocyclic nitrogen bases such as imidazole have been found (*244,250–252*) to form both binary complexes with LADH and ternary

240. B. L. Vallee and F. L. Hoch, *JBC* **225,** 185 (1957).
241. R. B. Loftfield and E. A. Eigner, *Science* **164,** 305 (1969).
242. B. L. Vallee, "The Enzymes," 2nd ed., Vol. 3, Part B, p. 225, 1960.
243. B. L. Vallee, R. J. P. Williams, and F. L. Hoch, *JBC* **234,** 2621 (1959).
244. R. A. Plane and H. Theorell, *Acta Chem. Scand.* **15,** 1866 (1961).
245. K. Dalziel, *Nature (London)* **197,** 462 (1963).
246. D. D. Ulmer, T.-K. Li, and B. L. Vallee, *Proc. Nat. Acad. Sci. U. S.* **47,** 1155 (1961).
247. T.-K. Li, D. D. Ulmer, and B. L. Vallee, *Biochemistry* **1,** 114 (1962).
248. B. L. Vallee, T. L. Coombs, and R. J. P. Williams, *JACS* **80,** 397 (1958).
249. D. S. Sigman, *JBC* **242,** 3815 (1967).
250. C. H. Reynolds and J. S. McKinley-McKee, *FEBS (Fed. Eur. Biochem. Soc.) Lett.* **21,** 297 (1972).
251. H. Theorell and J. S. McKinley-McKee, *Acta Chem. Scand.* **15,** 1797 (1961).
252. H. Theorell and J. S. McKinley-McKee, *Acta Chem. Scand.* **15,** 1811 and 1834 (1961).

complexes in the presence of coenzyme. An X-ray study of the binary LADH·imidazole complex (*112*) shows that imidazole, like 1,10-phenanthroline, binds to the catalytic zinc atom, displacing the zinc-bound water molecule. This mode of binding was suggested by Theorell and co-workers (*244,252*) on the basis of kinetic data and the pH dependence of NAD^+ binding to this complex (Section II,I,1).

A direct demonstration of binding of alcohol to LADH in the absence of coenzyme has been made (*253*) by utilizing the spectroscopic changes and protein fluorescence quenching that occur when the chromophoric substrate hydroxymethylferrocene binds. It was not definitively established, however, that the alcohol binds in the substrate binding pocket.

The alkaloid berberine has been shown to inhibit LADH competitively with both coenzyme and substrates (*254,255*). Since berberin is displaced by 1,10-phenanthroline it probably binds in the substrate binding pocket but not necessarily as a ligand to zinc.

c. Other Reversible Inhibitors. A large number of different compounds, mainly drugs, have been tested for inhibitory actions of LADH without further characterization of the mode of inhibition (*1*). Examples of these are primaquine (*256*) and other quinoline derivatives (*256,257*), amphetamines (*258*), propranolol (*259*), folic acid (*260*), and flavensomycin (*261*). It has also been shown that metronidazole, used in treatment of alcoholism, inhibits LADH *in vitro* (*262–265*), although it has been pointed out (*266,267*) that this inhibition might result from the formation of a complex between NAD^+ and the inhibitor. Doxapram HCl has been claimed (*268*) to inhibit rabbit LADH, stimulate rat LADH, and have no effect on horse LADH.

253. R. Einarsson, L. Wallén, and M. Zeppezauer, *Chem. Scr.* **2,** 84 (1972).
254. L. Skurský and J. Kovář, *in* "Structure and Function of Oxidation-Reduction Enzymes" (Å. Åkeson and A. Ehrenberg, eds.), p. 653. Pergamon, Oxford, 1972.
255. J. Kovář and L. Skurský, *Eur. J. Biochem.* **40,** 233 (1973).
256. T.-K. Li and L. J. Magnes, *Biochem. Pharmacol.* **21,** 17 (1972).
257. R. Einarsson and M. Zeppezauer, *Acta Chem. Scand.* **24,** 1098 (1970).
258. D. V. Siva Sankar, *Res. Commun. Chem. Pathol. Pharmacol.* **1,** 460 (1970).
259. R. J. S. Duncan, *Mol. Pharmacol.* **9,** 191 (1973).
260. R. Snyder, W. Vogel, and M. P. Schulman, *JBC* **240,** 471 (1965).
261. D. Gottlieb and Y. Inoue, *J. Bacteriol.* **94,** 844 (1967).
262. R. Fried and L. W. Fried, *Biochem. Pharmacol.* **15,** 1890 (1966).
263. J. A. Edwards and J. Price, *Nature (London)* **214,** 190 (1967).
264. J. A. Edwards and J. Price, *Biochem. Pharmacol.* **16,** 2026 (1967).
265. E. Paltrinieri, *Farmaco, Ed. Sci.* **22,** 1054 (1967).
266. R. Fried and L. W. Fried, *Experientia* **24,** 56 (1968).
267. N. K. Gupta, C. L. Woodley, and R. Fried, *Biochem. Pharmacol.* **19,** 2805 (1970).
268. J. P. da Vanzo, E. S. Kline, and L. S. Kang, *Eur. J. Pharmacol.* **6,** 152 (1969).

TABLE XIII
DISSOCIATION CONSTANTS FOR A SELECTION OF INHIBITOR MOLECULES THAT FORM TERNARY COMPLEXES WITH LADH AND COENZYME

Compound	Ref.	$K_I(\mu M) = K_{EO,I}$
1. Complexes with LADH-NAD$^+$		
Isobutyramide	*279*	7900
C_{10} Fatty acid	*194*	5.0
Pyrazole	*277*	0.22
4-Methyl pyrazole	*277*	0.013
4-Propyl pyrazole	*277*	0.004
4-Pentylpyrazole	*277*	0.0008
Auramine O	*287*	7.0
Compound	**Ref.**	$K_I(\mu M) = K_{ER,I}$
2. Complexes with LADH-NADH		
Isobutyramide	*251*	140
Benzamide	*281*	530
p-Methyl benzamide	*281*	54
m-Methyl benzamide	*281*	33
N-Methyl benzamide	*281*	14450
p-Hydroxybenzamide	*281*	294
m-Hydroxybenzamide	*281*	1260
Auramine O	*287*	9

2. *Ternary Complexes*

a. Fatty Acids and Pyrazole. Long chain fatty acids (*194,269*) and pyrazole (*98*) were shown to form specific ternary complexes with LADH-NAD$^+$ but not with LADH-NADH. These early studies have been reviewed (*1*). Since pyrazole is very strongly bound (Table XIII), it was proposed (*98*) that the heterocyclic ring forms a bridge through the two nitrogen atoms between the catalytic zinc atom and C-4 of the nicotinamide moiety. Both kinetic studies of this binding (*270*) using coenzyme analogs and the X-ray model are consistent with this picture.

Intense interest in these derivatives developed when it was found (*271*) that 4-substituted pyrazole derivatives formed very tight complexes which seemed to be specific for LADH from several mammalian sources (*77*) including man (*272*). A large number of studies have been devoted

269. A. D. Winer and H. Theorell, *Acta Chem. Scand.* **13**, 1038 (1959).
270. J. D. Shore and M. J. Gilleland, *JBC* **245**, 3422 (1970).
271. H. Theorell, T. Yonetani, and B. Sjöberg, *Acta Chem. Scand.* **23**, 255 (1969).
272. T.-K. Li and H. Theorell, *Acta Chem. Scand.* **23**, 892 (1969).

to the metabolic effects of these pyrazole derivatives in order to investigate their possible use to diminish the damages caused by alcohol abuse (*26,30,273–276*).

Only 4-substituted derivatives are found to increase the inhibition of pyrazole. Substitutions in position N-1, C-3, or C-5 abolish the inhibitory power (*271,277*). It has recently been shown (*277*) that by increasing the chain length in the 4-alkyl-substituted derivatives from methyl to pentyl, the inhibitory power increases, consistent with the presence of a hydrophobic active site pocket (Table XIII).

b. Amides. The stable ternary complexes of fatty acid amides, NADH, and LADH were first reported by Theorell and co-workers (*139,252,269*) and subsequently studied by Woronick (*278*). These studies have been reviewed (*1*). At that time no evidence was obtained from equilibrium fluorescence titrations for the existence of corresponding ternary complexes with the oxidized coenzyme. Sigman and Winer (*279*) have since shown that these complexes are formed but that they are much weaker than those formed with NADH (Table XIII). They concluded that there is no absolute specificity in amide binding but a relative difference in affinity, suggesting that alcohol and aldehyde share the same substrate binding site.

Substituted benzamides have recently been extensively studied in ternary complexes with LADH–NADH (*280–283*) in order to deduce and separate hydrophobic, electronic, and steric roles of the substituents for binding. The detailed analyses of these data are compatible with two different binding groups on the amide, one hydrophobic and the other electronic. An attempt has also been made to map the active site from the binding pattern of some aliphatic amides (*284*).

273. R. G. Thurman, T. Yonetani, J. R. Williamson, and B. Chance, eds., "Alcohol and Aldehyde Metabolizing Systems." Academic Press, New York, 1974.
274. R. Blomstrand, *in* "Structure and Function of Oxidation-Reduction Enzymes" (Å. Åkeson and A. Ehrenberg, eds.), p. 667. Pergamon, Oxford, 1972.
275. R. Blomstrand and G. Öhman, *Life Sci.* **13,** 107 (1973).
276. R. Blomstrand, L. Kager, and O. Lantto, *Life Sci.* **13,** 1131 (1973).
277. R. Dahlbom, B. R. Tolf, Å. Åkeson, G. Lundquist, and H. Theorell, *BBRC* **57,** 549 (1974).
278. C. L. Woronick, *Acta Chem. Scand.* **15,** 2062 (1961); **17,** 1789 and 1791 (1963).
279. D. S. Sigman and A. D. Winer, *BBA* **206,** 183 (1970).
280. R. H. Sarma and C. L. Woronick, *Res. Commun. Chem. Pathol. Pharmacol.* **2,** 177 (1971).
281. R. H. Sarma and C. L. Woronick, *Biochemistry* **11,** 170 (1972).
282. C. Hansch, J. Schaeffer, and R. Kerley, *JBC* **247,** 4703 (1972).
283. C. Hansch, K. H. Kim, and R. H. Sarma, *JACS* **95,** 6447 (1973).
284. R. H. Sarma and C. L. Woronick, *Res. Commun. Chem. Pathol. Pharmacol.* **2,** 601 (1971).

c. Auramine O. The cationic diphenylmethane dye auramine O is an unusual inhibitor since it forms strong and specific complexes with LADH, both in the absence and presence of coenzyme. The inhibitory action was first discovered by Brand and co-workers (*285*) who showed that this normally nonfluorescent dye exhibits a strong fluorescence in the presence of LADH. Subsequent investigations have shown (*286,287*) that auramine binds somewhat stronger to LADH in the presence of both NAD^+ and NADH than in the absence of coenzyme (see Tables XII and XIII). This inhibitor is furthermore displaced by substrates such as ethanol and cyclohexanol and by other inhibitors like fatty acids, amides, pyrazole, and chelating agents. Thus, it is apparent that auramine binds in the active site pocket, in the vicinity of zinc but probably not as a ligand and overlaps the substrate binding site. The deep hydrophobic pocket of LADH apparently provides a very specific binding site for this dye.

A circular dichroic study (*288*) of these complexes has shown that the CD spectrum of bound auramine is different in binary compared to ternary complexes. This finding is consistent with the view that the coenzyme-induced conformational change alters some details of the substrate binding pocket (Section II,C,3,b).

I. Kinetic Aspects

Only those kinetic studies on LADH will be reviewed here which are relevant to those aspects of the catalytic mechanism that are discussed in the next section. Other aspects of the kinetic properties of LADH can be found elsewhere (*1,7,127,289–293*).

1. *Coenzyme Binding and Dissociation*

Recent determinations (*119,294*) of the dissociation constants for binding NADH ($K = 0.33–0.37\ \mu M$ at pH 7.0 and $23.0\ \mu M$ at pH 10.0) and NAD^+ ($K = 133\ \mu M$ at pH 7.0 and $5.5\ \mu M$ at pH 10.0) to LADH agree

285. R. H. Conrad, J. R. Heitz, and L. Brand, *Biochemistry* **9**, 1540 (1970).
286. J. R. Heitz and L. Brand, *Biochemistry* **10**, 2695 (1971).
287. D. S. Sigman and A. N. Glazer, *JBC* **247**, 334 (1972).
288. J. S. Wicken and R. W. Woody, *Biochemistry* **12**, 3459 (1974).
289. W. W. Cleland, "The Enzymes," 3rd ed., Vol. 2, p. 1, 1970.
290. H. Theorell, *in* "Molecular Association in Biology" (B. Pullman, ed.), p. 89. Academic Press, New York, 1968.
291. H. Theorell, *Harvey Lect.* **61**, 17 (1967).
292. K. Dalziel, *in* "Pyridine Nucleotide-Dependent Dehydrogenases" (H. Sund, ed.), p. 3. Springer-Verlag, Berlin and New York, 1970.
293. G. A. Hamilton, *Progr. Bioorg. Chem.* **1**, 83 (1971).
294. J. D. Shore, *Biochemistry* **8**, 1588 (1969).

in magnitude with earlier values (*1*). The kinetic studies are consistent with independent binding of coenzyme to the two subunits of the molecule.

In the overall reaction one equivalent of proton is released per equivalent of alcohol that is oxidized (*1*). There is strong evidence available that this proton release is associated with NAD^+ binding. Taniguchi *et al.* (*119*) suggested, from direct measurements of NAD^+ binding, that its pH dependence can be attributed to the interaction of NAD^+ with a proton donating group of pK 8.75 in the apoenzyme. These studies confirmed earlier kinetic analyses by Dalziel (*295*). Since imidazole abolishes (*252*) the pH dependence of NAD^+ binding, it was proposed (*119*) that the proton donating group is a water molecule bound to zinc. The introduction of the positive charge of NAD^+ in the vicinity of the zinc atom shifts the pK_a of this water molecule from 8.75 to 6.85. Dunn (*296*) has shown that a displacement of bound NAD^+ by NADH at pH 8.8 is accompanied by a net uptake of one equivalence of hydrogen ions from solution for every mole of NAD^+ displaced. By transient kinetic studies (*296*) he also finds evidence that this uptake corresponds to the proton release in the overall reaction during alcohol oxidation. Furthermore, Shore *et al.* (*297*) have shown that one half equivalent of protons is liberated at pH 7.6 when saturating amounts of NAD^+ are added to the enzyme. Additional protons, resulting in a total of one proton per equivalent of enzyme, were released upon formation of a ternary complex with the substrate competitive inhibitor trifluoroethanol. This release occurs prior to and uncoupled from the catalytic hydride transfer step. They also found that NAD^+ binding results in a perturbation of the pK_a values of a group on the enzyme from 9.6 to 7.6, slightly different values from those previously calculated (*119*).

The assumption of Taniguchi *et al.* (*119*) that a water molecule is bound to the catalytic zinc atom (Section II,C,3,b) and that imidazole displaces this water molecule by forming a binary complex with LADH (Section II,H,1,b) have been conclusively confirmed by the X-ray study. Combining the X-ray results with the kinetic studies described above, Eklund *et al.* (*107*) suggested that the proton released in the overall reaction by NAD^+ binding is dissociated from the water molecule bound to zinc in the apoenzyme. They also suggested that this proton release occurs via a hydrogen bond system from the water molecule through the side chain of Ser-48 to the imidazole ring of His-51 (Fig.

295. K. Dalziel, *JBC* **238**, 2850 (1963).
296. M. F. Dunn, *Biochemistry* **13**, 1146 (1974).
297. J. D. Shore, H. Gutfreund, D. Santiago, and P. Santiago, *in* "Alcohol and Aldehyde Metabolizing Systems" (R. G. Thurman *et al.*, eds.), p. 45. Academic Press, New York, 1974.

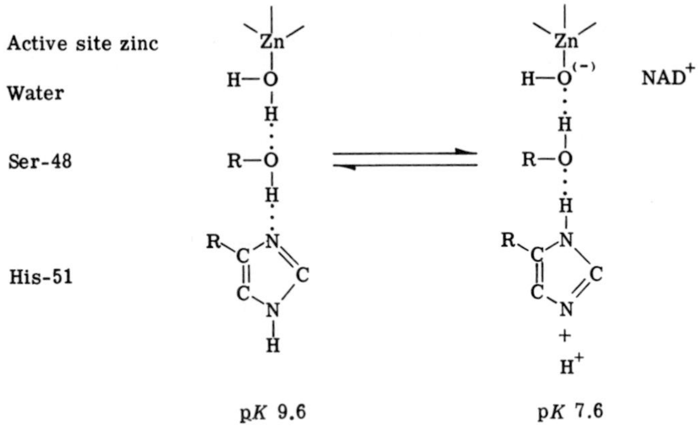

FIG. 17. Schematic diagram illustrating the hydrogen bond system of the zinc-bound water molecule, Ser-48 and His-51 and a possible mechanism for the proton release induced by NAD^+ binding.

17). The protons would thus be released at the surface of the molecule and not into the interior of the hydrophobic substrate binding pocket.

In a large number of kinetic studies (*1,294,295,298*) the dissociation of reduced coenzyme has been shown to be rate limiting for alcohol oxidation with a first-order rate constant around 3 sec^{-1} (*294*). This rate constant increases to 9 sec^{-1} in 50 m*M* chloride ions permitting the observation of an isomerization of the binary E·NADH complex (*298*) with a rate constant of 11 sec^{-1}. Shore and co-workers (*270,294,299,300*) found by using 3-acetyl pyridine AD^+ as coenzyme that the rate constant for the dissociation of the reduced form of this analog was 100 sec^{-1}, although the turnover number was only 23 sec^{-1}. Since the hydride transfer step was shown not to be rate determining (*300*), it was suggested that the rate limiting step is an isomerization with a rate constant of 23 sec^{-1} of the binary complex with the coenzyme analog. A slow intramolecular isomerization step has also been suggested (*301*) on the basis of transient kinetic data for the reduction of benzaldehyde.

Czeisler and Hollis (*302*) suggested from an NMR study that the NADH binding and dissociation is a two-step process: A rapid initial binding is followed by a slow isomerization step. They found strong evidence for a dissociation rate constant of 100 sec^{-1} for the adenine moiety of the coenzyme and, in combination with other studies (*281,303*), indica-

298. J. D. Shore and H. Gutfreund, *Biochemistry* **9,** 4655 (1970).
299. J. D. Shore and H. Theorell, *Eur. J. Biochem.* **2,** 32 (1967).
300. J. D. Shore and R. L. Brooks, *ABB* **147,** 825 (1971).
301. P. L. Luisi and R. Favilla, *Biochemistry* **11,** 2303 (1972).
302. J. L. Czeisler and D. P. Hollis, *Biochemistry* **12,** 1683 (1973).
303. D. P. Hollis, *Biochemistry* **6,** 2080 (1967).

tions of a similar rate constant for the nicotinamide moiety. They concluded that the spectroscopic changes of NADH on which all kinetic measurements have been based occur during the last slow step. The actual binding process produces no known spectroscopic changes but is visible to the NMR technique. A previous report on spectroscopic evidence for this two-step process (*304*) was later reported to be in error (*195*).

The equilibrium and rate constants for NADH binding to the three isozymes EE, ES, and SS of the horse enzyme have been determined (*305*). Differences in binding to the two types of chains were found both for the binding strength and the pH dependence. Changes in the absorption spectrum (*306,307*), the fluorescence polarization spectrum (*308*), the optical rotatory dispersion spectrum (*309*), and the effect of D_2O on the fluorescence spectrum (*310*) have been studied for the binary enzyme coenzyme complexes compared to the free molecules.

The role of coenzyme binding is thus at least twofold: to induce a conformational change, as shown for NADH binding, which might change some properties of the active site pocket and, in the case of NAD^+ binding, to induce a shift in the pK_a value of a group involved in alcohol binding, presumably the zinc-bound water molecule. These two events might be related and the effect is a mutual stabilization of NAD^+ and alcohol binding.

2. *Substrate Binding*

Recent investigations (*311–316*) confirm the broad substrate specificity of LADH (Section I,B). These studies are consistent with the presence of the wide and deep hydrophobic substrate binding pocket found in the X-ray structure (Section II,C,3,b). The kinetic properties of substituted cyclohexanones (*317–319*) and the steric analog adamantanone as sub-

304. H. Theorell, A. Ehrenberg, and C. de Zalenski, *BBRC* **27,** 309 (1967).
305. H. Theorell, Å. Åkeson, B. Liszka-Kopeć, and C. de Zalenski, *ABB* **139,** 241 (1970).
306. H. F. Fisher, A. C. Haine, A. P. Mathias, and B. R. Rabin, *BBA* **139,** 169 (1967).
307. H. F. Fisher, D. L. Adija, and D. G. Cross, *Biochemistry* **8,** 4424 (1969).
308. H. Weiner, *ABB* **126,** 692 (1968).
309. A. Rosenberg, H. Theorell, and T. Yonetani, *ABB* **110,** 413 (1965).
310. C. H. Blomquist, *JBC* **244,** 1605 (1969).
311. G. R. Waller, *Nature* (*London*) **208,** 1389 (1965).
312. J.-C. Mani and J. Dornand, *C.R. Acad. Sci.* **272,** 3210 (1971).
313. R. F. Lambe and D. C. Williams, *BJ* **97,** 475 (1965).
314. B. Zagalak, P. A. Frey, G. L. Karabatsos, and R. H. Abeles, *JBC* **241,** 3028 (1966).
315. F. M. Dickinson and K. Dalziel, *Nature* (*London*) **214,** 31 (1967).
316. J.-C. Mani, R. Pietruszko, and H. Theorell, *ABB* **140,** 52 (1970).
317. V. Prelog, *Colloq. Ges. Physiol. Chem.* **14,** 288 (1963); *Pure Appl. Chem.* **9,** 119 (1964).

strates have been studied (*320*). From the relative rates of reduction of these derivatives, detailed steric requirements were suggested for a molecule to fit into the hydrophobic active site. The presence of a hydrophobic binding site for substrates was early recognized (*1*) and is also consistent with recent inhibitor studies (Section II,H) and kinetic investigations on the rate of oxidation of various aliphatic alcohols (*321–324*). The influence of substrate structure on the hydride transfer step has been studied by transient kinetic methods for different aliphatic alcohols (*324*) and substituted aromatic aldehydes (*325*).

Kinetic and spectroscopic measurements support the hypothesis that the substrate binds directly to the catalytic zinc atom. Dunn and Hutchison (*326*) have produced evidence using a chromophoric aldehyde as substrate that the carbonyl oxygen of the reaction intermediate is coordinated to zinc. They concluded that zinc acts as a Lewis acid catalyst. Similar conclusions have been reached by McFarland and co-workers from the spectral properties of the enzyme complex with 4-(2′-imidazolylazo)benzaldehyde (*327*), from the observed small electronic substituent effect of para-substituted benzaldehydes (*325*), and from the absence of a large pH effect in the hydride transfer step (*328*). Assuming the mechanisms and the subunit structures to be essentially similar in YADH and LADH, a magnetic resonance study of coenzyme and substrate binding to YADH (*329*) also support the hypothesis of direct binding of substrate to zinc.

Gilleland and Shore have proposed (*330*) from a kinetic study of zinc chelators that the maximum binding rate of substrate to zinc in LADH is 230 sec^{-1}. This value does not contradict the notion of substrate binding to zinc.

---

318. H. Dutler, M. J. Coon, A. Kull, H. Vogel, G. Waldvogel, and V. Prelog, *Eur. J. Biochem.* **22,** 203 (1971).
319. J. M. H. Graves, A. Clark, and H. J. Ringold, *Biochemistry* **4,** 2655 (1965).
320. H. J. Ringold, T. Bellas, and A. Clark, *BBRC* **27,** 361 (1967).
321. C. S. Tsai, *Can. J. Biochem.* **46,** 381 (1968).
322. K. Dalziel and F. M. Dickinson, *BJ* **100,** 34 (1966).
323. F. M. Dickinson and K. Dalziel, *BJ* **104,** 165 (1967).
324. R. L. Brooks and J. D. Shore, *Biochemistry* **10,** 3855 (1971).
325. J. W. Jacobs, J. T. McFarland, I. Wainer, D. Jeanmaier, C. Ham, K. Hamm, M. Wnuk, and M. Lam, *Biochemistry* **13,** 60 (1974).
326. M. F. Dunn and J. S. Hutchison, *Biochemistry* **12,** 4882 (1973).
327. J. T. McFarland, Y.-H. Chu, and J. W. Jacobs, *Biochemistry* **13,** 65 (1974).
328. J. T. McFarland and Y.-H. Chu, personal communication (1974).
329. D. L. Sloan, Jr. and A. S. Mildvan, *in* "Alcohol and Aldehyde Metabolizing System" (R. G. Thurman *et al.,* eds.), p. 69. Academic Press, New York, 1974.
330. M. J. Gilleland and J. D. Shore, *BBRC* **40,** 230 (1970).

From a study of the pH dependence of the hydride transfer step (*331*), it was concluded that a group with a pK of 6.4 is involved in this step. A histidine (His-51) or lysine residue was suggested. Since the zinc-bound water molecule is hydrogen bonded to Ser-48, it is possible that the side chain of this residue is also involved in binding or polarization of the substrate. The properties of this side chain must be influenced by the ionization state of the imidazole ring of His-51 since they are hydrogen bonded to each other (Fig. 17). It is thus possible that the group with a pK of 6.4 observed in the hydride transfer step is the imidazole ring of His-51, linked by hydrogen bonds to the substrate through the side chain of Ser-48.

3. *The Ordered Mechanism*

Early evidence for the sequential mechanism proposed by Theorell and Chance (*332*) has been reviewed in a previous chapter on LADH (1). The requirements of this mechanism are satisfied under certain conditions for primary alcohols and aldehydes but not for secondary alcohols (*295,322,333–336*). Thus, the first step is the binding of the coenzyme and the last and rate limiting step dissociation of the coenzyme. Formation of the productive ternary complexes E·NAD^+·Alc and E·NADH·Ald have been demonstrated (*333,334,337–339*). The interconversion of these complexes are not kinetically important for the above-mentioned conditions as required by the mechanism. It has been suggested, however, that this step is rate limiting during different conditions (*324,336*). Substrate dissociation constants for the ternary complexes have been estimated (*340*).

The reaction is not compulsory. Dalziel and Dickinson (*322*) have proposed a more general mechanism based on the formation of abortive ternary E·NADH·alcohol complexes and binary E·alcohol complexes. Formation at high alcohol concentrations of this abortive ternary complex from which NADH can dissociate had previously been suggested (*252*). Evidence for this alternative pathway was furthermore adduced by Silverstein and Boyer (*341*) on the basis of isotope exchange experiments.

331. R. L. Brooks, J. D. Shore, and H. Gutfreund, *JBC* **247**, 2382 (1972).
332. H. Theorell and B. Chance, *Acta Chem. Scand.* **5**, 1127 (1951).
333. C. C. Wratten and W. W. Cleland, *Biochemistry* **2**, 935 (1963).
334. C. C. Wratten and W. W. Cleland, *Biochemistry* **4**, 2442 (1965).
335. F. B. Rudolph and H. J. Fromm, *Biochemistry* **9**, 4660 (1970).
336. K. Bush, V. J. Shiner, Jr., and H. R. Mahler, *Biochemistry* **12**, 4802 (1973).
337. H. Theorell and T. Yonetani, *ABB, Suppl.* **1**, 209 (1962).
338. D. Palm, T. Fiedler, and D. Ruhrseitz, *Z. Naturforsch. B* **23**, 623 (1968).
339. H. Theorell and K. Tatemoto, *Acta Chem. Scand.* **24**, 3069 (1970).
340. J. D. Shore and H. Theorell, *ABB* **116**, 255 (1966).
341. E. Silverstein and P. D. Boyer, *JBC* **239**, 3908 (1964).

Dalziel and Dickinson (*342*) found that this pathway contributed significantly to the kinetic parameters when cyclohexanol was used as a substrate. Isotope exchange studies (*343*) have confirmed the partly random mechanism for this substrate. Shore and Theorell (*344*) have studied the substrate inhibition for several aliphatic alcohols resulting from the formation of abortive E·NADH·alcohol complexes.

The rate of oxidation of ethanol can be increased by the simultaneous presence of an aldehyde (*345*). It was suggested that the rate increase was resulting from the absence of an enzyme–coenzyme dissociation step. Using a suitable aldehyde this reaction can be used as an assay method (*346*).

4. *The Dismutase Reaction*

LADH catalyzes the dismutation of different aldehydes to alcohol and acid in the presence of NAD^+ at high aldehyde concentrations (*53,347,348*). Dalziel and Dickinson (*349*) found that this reaction did not influence the kinetic parameters of normal aldehyde reduction, provided reasonable substrate concentrations were used. They also suggested a mechanism for the dismutase reaction based on a catalytic role of NAD^+. Using acetyl pyridine AD^+ instead of NAD^+ it could, however, be demonstrated (*350*) that a net reduction of coenzyme occurs during this reaction. Later studies (*351*) have shown a stoichiometry of one mole NAD^+ reduced per mole acid formed. It was also shown that with octanal as a substrate the reaction exhibits a much higher maximal velocity than with acetaldehyde. The "isomerase" activity of LADH previously reported (*260,352*) was later shown to result from impurities in the LADH preparations (*353*).

5. *Half-Site Reactivity*

In a transient kinetic study of the reduction of chromophoric aldehydes, Bernhard and co-workers (*354*) observed a biphasic optical change

342. K. Dalziel and F. M. Dickinson, *BJ* **100,** 491 (1966).
343. G. R. Ainslie, Jr. and W. W. Cleland, *JBC* **247,** 946 (1972).
344. J. D. Shore and H. Theorell, *ABB* **117,** 375 (1966).
345. N. K. Gupta and W. G. Robinson, *BBA* **118,** 431 (1966).
346. C. L. Woodley and N. K. Gupta, *ABB* **148,** 238 (1972).
347. L. Kendal and A. Ramanathan, *BJ* **52,** 430 (1952).
348. R. H. Abeles and H. A. Lee, *JBC* **235,** 1499 (1960).
349. K. Dalziel and F. M. Dickinson, *Nature* (*London*) **206,** 255 (1965).
350. N. K. Gupta, *ABB* **141,** 632 (1970).
351. J. A. Hinson and R. A. Neal, *JBC* **247,** 7106 (1972).
352. J. van Eys, *JBC* **236,** 1531 (1961).
353. R. Snyder and E. W. Lee, *ABB* **117,** 587 (1966).
354. S. A. Bernhard, M. F. Dunn, P. L. Luisi, and P. Schack, *Biochemistry* **9,** 185 (1970).

of both substrate and coenzyme in the presence of excess enzyme, a rapid burst followed by a slower conversion to products. They also found that each of the two phases of the reaction corresponded to processes in amounts equal to exactly one-half of the catalytically functional enzyme sites. The results were discussed in terms of two models, one of which is favored and based on nonequivalent but interconvertible states of the two sites. The special case of such negative cooperativity observed here has also been found for GAPDH and is generally called half-site reactivity (*355,356*). Implicit in this model is the flip-flop mechanism discussed by Lazdunski (*105*) for different enzymes including LADH. Similar evidence for the nonequivalence of the subunits was also obtained in subsequent studies using a chromophoric nitroso substrate (*357*) or a method based on the freezing of the enzyme sites at the stage where NAD^+ is formed by using excess pyrazole in the reaction mixture (*358*).

On the other hand, Shore and Gutfreund (*298*) found evidence in their transient kinetic studies of LADH using ethanol as substrate that the two subunits are equivalent. Some possible causes for the different results obtained by the two groups have been discussed (*359*).

Additional support for the nonequivalence of the two subunits has been claimed by Everse (*360*) from fluorescence quenching studies of the stoichiometry of vitamin A acid binding. The nonlinear nature of such quenching (Section II,F) might, however, require more direct binding studies for confirmation, especially since the results were obtained in the absence of coenzyme.

The preliminary X-ray data on coenzyme-containing complexes are not inconsistent with coenzyme-induced asymmetry of the two subunits, since these are not related by crystallographic symmetry in such complexes (see Section II,C,1 and Table II). Considerable nonfunctional differences in conformation between chemically identical protein molecules have previously been observed (*361*). It is theoretically possible that the coenzyme-induced conformational change in LADH produces functionally similar coenzyme binding sites but nonequivalent substrate binding sites.

355. R. A. MacQuarrie and S. A. Bernhard, *JMB* **55**, 181 (1971).
356. A. Levitzki, W. Stallcup, and D. E. Koshland, Jr., *Biochemistry* **10**, 3371 (1971).
357. M. F. Dunn and S. A. Bernhard, *Biochemistry* **10**, 4569 (1971).
358. J. T. McFarland and S. A. Bernhard, *Biochemistry* **11**, 1486 (1972).
359. J. J. Holbrook and H. Gutfreund, *FEBS (Fed. Eur. Biochem. Soc.) Lett.* **31**, 157 (1973).
360. J. Everse, *Mol. Pharmacol.* **9**, 199 (1973).
361. T. Blundell, G. Dodson, D. Hodgkin, and D. Mercola, *Advan. Protein Chem.* **26**, 280 (1972).

J. A Mechanism for Catalysis

Figure 18 shows a mechanism for the oxidation of alcohols that has been suggested (*107*) by combining the results of crystallographic and kinetic studies. Seven different steps are distinguished in this reaction

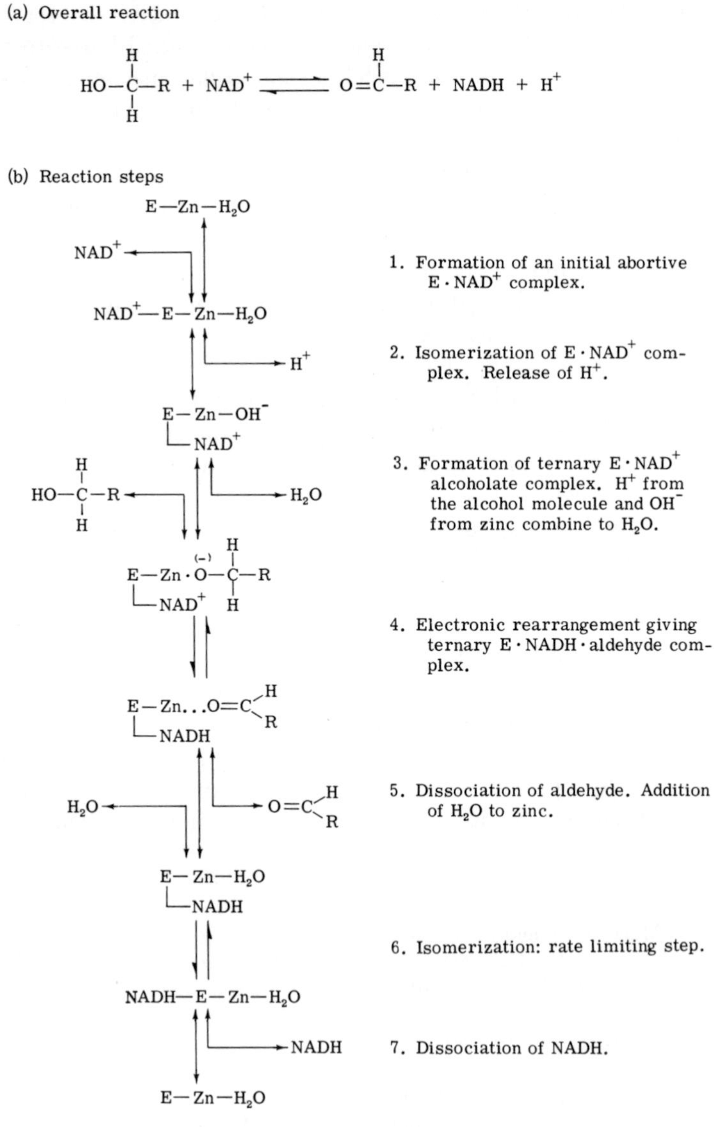

(a) Overall reaction
(b) Reaction steps
1. Formation of an initial abortive E · NAD⁺ complex.
2. Isomerization of E · NAD⁺ complex. Release of H⁺.
3. Formation of ternary E · NAD⁺ alcoholate complex. H⁺ from the alcohol molecule and OH⁻ from zinc combine to H₂O.
4. Electronic rearrangement giving ternary E · NADH · aldehyde complex.
5. Dissociation of aldehyde. Addition of H₂O to zinc.
6. Isomerization: rate limiting step.
7. Dissociation of NADH.

Fig. 18. Schematic reaction mechanism for the oxidation of alcohols by LADH.

scheme. More stages will be required as the techniques of detecting intermediates are being refined, especially in the actual hydride transfer step.

The events outlined in this scheme refer only to one of the two subunits of the LADH molecule. Possible cooperativity between the subunits has not been taken into account. Studies bearing on that question are reviewed in Section II,I,5.

The proposed mechanism is essentially electrophilic catalysis mediated by the active site zinc atom. The mechanism is valid in the pH region 6–10. The pH optimum for alcohol oxidation is around pH 8.0 (*1*). The only polar amino acid side chains which could exert any influence on the reactive part of the substrate are Ser-48 and, indirectly, His-51 (Fig. 17). No role has been assigned to these residues in the present scheme since conclusive evidence for their participation is lacking. However, their possible involvement in proton release and/or substrate binding is discussed in Sections II,I,1 and II,I,2.

The main parts of this scheme were proposed earlier by Theorell and co-workers (*119,291*) on the basis of inhibitor binding and steady-state kinetic studies. Other suggested mechanisms based on general acid-base catalysis (*297*), reduction of the enzyme (*362*), or direct participation of histidine (*363*) or cysteine (*364*) in the hydride transfer step are highly unlikely in view of the crystallographic and kinetic results reviewed in this chapter. Contrary to expectations the mechanism described here is in most details very different from that proposed for lactic dehydrogenase (*126*).

The involvement of zinc as a Lewis acid catalyst in LADH has been previously suggested by many investigators (*1,4,5,252,291,365–368*). A model compound for this reaction has been prepared and found to catalyze hydrogen transfer from the alcohol to a nicotinamide ring (*368*). Only recently, however, has strong evidence been obtained that the substrate binds directly to zinc and that the active site zinc mediates electrophilic catalysis (Section II,I,2).

It is proposed in the mechanism that alcohol binds to zinc as the negatively charged alcoholate ion and displaces the hydroxyl ion. The formation of alcoholate ion is mediated here by the hydroxyl ion bound to zinc. This base can combine with the proton of the hydroxyl group of

362. K. A. Schellenberg, *JBC* **240,** 1165 (1965).
363. H. J. Ringold, *Nature* (*London*) **210,** 535 (1966).
364. J. H. Wang, *Science* **161,** 328 (1968).
365. H. R. Mahler and J. Douglas, *JACS* **79,** 1159 (1957).
366. K. Wallenfels and H. Sund, *Biochem. Z.* **329,** 17 and 59 (1957).
367. R. H. Abeles, R. F. Hutton, and F. H. Westheimer, *JACS* **79,** 712 (1957).
368. D. J. Creighton and D. S. Sigman, *JACS* **93,** 6314 (1971).

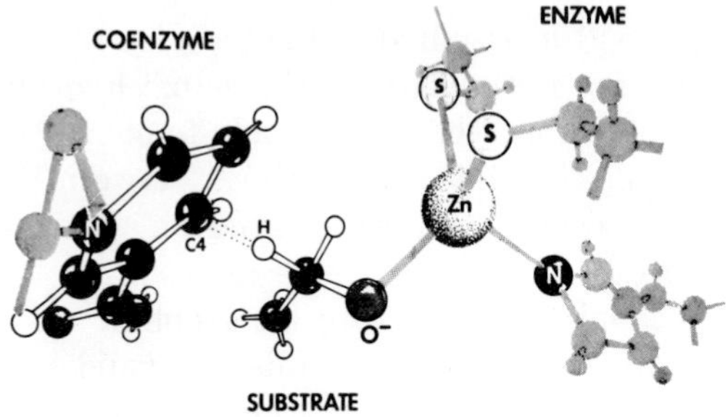

FIG. 19. Schematic diagram illustrating the assumed positions of substrate and nicotinamide in relation to the catalytic zinc atom.

the alcohol to form a water molecule. The hydrophobic nature of the substrate binding pocket might facilitate this reaction.

Base-catalyzed formation of alcoholate ion which binds to zinc in LADH has recently been suggested in a transient kinetic study (*328*) from the observed large decrease with decreasing pH of the apparent binding constant for alcohol.

Model building has shown (*107*) that the alcoholate ion can bind to zinc in the apoenzyme structure in such a way that the hydrogen atom which is to be transferred is sufficiently close to the C-4 atom of the nicotinamide ring for direct transfer to occur (see Fig. 19). The known position and conformation of bound ADP-ribose was used in the model building procedure (Section II,C,3,a). It was assumed that the remaining part of NAD^+ had the same conformation in LADH as in LDH (*126*).

Substrate binding according to this scheme is also consistent (*119*) with the observations (*1*) that negatively charged fatty acids form strong ternary complexes, presumably displacing the hydroxyl ion bound to zinc, in the presence of NAD^+. On the other hand, neutral inhibitors, like fatty acid amides, displace the water molecules bound to zinc and thus form stronger complexes with NADH than with NAD^+ (Section II,H,2).

The hydride transfer step is a direct transfer between substrate and coenzyme. The suggestion by Schellenberg (*362,369*) that tryptophan mediates this hydride transfer in the yeast enzyme has been shown chemically not to be valid for both yeast and liver enzyme (*93,370–372*). Furthermore, the tertiary structure shows that the two tryptophans of the LADH subunit are both approximately 20 Å away from the active site

369. K. A. Schellenberg, *JBC* **241,** 2446 (1966).
370. J. T. Wong and G. R. Williams, *ABB* **124,** 344 (1968).
371. J.-F. Biellmann and M. J. Jung, *Eur. J. Biochem.* **19,** 130 (1971).
372. D. Palm, *BBRC* **22,** 151 (1966).

zinc atom (Fig. 15) and cannot be directly involved in the catalytic activity.

It was early pointed out (*365,366*) that a zinc-catalyzed hydride transfer exhibited striking similarities with certain organic reactions, particularly the Meerwein–Pondorff–Oppenauer reaction (*373*). Both reactions involve hydride ion transfer to carbonyl compounds and they both require the participation of a strong Lewis acid which polarizes the reactant by binding to its carbonyl oxygen atom.

III. Yeast Alcohol Dehydrogenases

A. Purification and Molecular Properties

The most studied yeast alcohol dehydrogenase is a cytoplasmic protein, commercially available, and usually obtained from bakers' yeast. It was the first pyridine nucleotide–dependent dehydrogenase to be crystallized and several purification methods have been reported (*1*). New methods of purification include toluene plasmolysis (*374*); DEAE-cellulose chromatography (*375,376*); fractionation with protamine sulfate and calcium phosphate gel (*375*); and purification without autolysis, heat denaturation, or use of organic solvents (*377*). Two different forms have been observed (*375*) on electrophoresis in polyacrylamide gels of these preparations. Apparent homogeneous preparations have also been reported (*374*) and no evidence for heterogeneity has yet been detected by amino acid sequence analysis (*12,378,379*). Separation of YADH from simple enzyme mixtures by affinity chromatography on immobilized NAD (*380,381*) or AMP analogs (*380,382*) have also been reported. Commercially available crystalline yeast alcohol dehydrogenase has been shown to contain protein kinase activity as an impurity (*383*).

Apart from this "classic" yeast alcohol dehydrogenase which is consid-

373. J. Hine, "Physical Organic Chemistry." McGraw-Hill, New York, 1955.
374. P. J. G. Butler and G. M. T. Jones, *BJ* **118**, 375 (1970).
375. F. M. Dickinson, *BJ* **120**, 821 (1970).
376. F. M. Dickinson, *BJ* **126**, 133 (1972).
377. P. L. Coleman and H. Weiner, *Biochemistry* **12**, 3466 (1973).
378. P. J. G. Butler, Dissertation, University of Cambridge, 1967.
379. P. J. G. Butler and J. I. Harris, *Abstr. Fed. Eur. Biochem. Soc.* **5**, 186 (1968).
380. C. R. Lowe, K. Mosbach, and P. D. G. Dean, *BBRC* **48**, 1004 (1972).
381. M. K. Weibel, E. R. Doyle, A. E. Humphrey, and H. J. Bright, *Biotechnol. Bioeng. Symp.* **3**, 167 (1972).
382. R. Ohlsson, P. Brodelius, and K. Mosbach, *FEBS* (*Fed. Eur. Biochem. Soc.*) *Lett.* **25**, 234 (1972).
383. B. E. Kemp, M. Froscio, and A. W. Murray, *BJ* **131**, 271 (1973).

ered to function in fermentation (*384–386*) and to be synthesized constitutively (*384–387*), a second soluble isozyme is also present in *Saccharomyces cerevisiae* (*384–388*). This isozyme has a broader substrate specificity (*385,386,388*), is repressed by glucose (*385–387*), and is thus selectively produced by yeast grown on lactate or ethanol as a sole carbon source. The second isozyme, therefore, participates in formation of carbohydrates from ethanol during oxidative metabolism. The fermentative and oxidative isozymes are synthesized from separate gene loci. Similar isozymes appear to be present in other species, too (*389,390*) (Section IV,A).

In addition, a mitochondrial yeast alcohol dehydrogenase isozyme has been demonstrated (*387,391–395*). This third isozyme is also sensitive to glucose repression (*387*). Its synthesis may therefore be related to that of the soluble oxidative isozyme, but it may also represent a third gene locus of alcohol dehydrogenase in yeast (*387*). Finally, a multiplicity of enzyme forms has been reported for each isozyme (*394,395*), but the relationship between these multiple forms is unknown. Most studies reported below have been performed with the "classic" fermentative isozyme.

YADH is a tetramer of molecular weight 140,000–150,000 (*1,375,396–400*) as determined by ultracentrifugation and gel filtration. Chelating agents (*1*), urea (*1,398*), maleylation (*401*), sodium dodecyl

384. L. Schimpfessel, *Arch. Int. Physiol. Biochim.* **75**, 368 (1967).
385. L. Schimpfessel, *BBA* **151**, 317 (1968).
386. U. Lutstorf and R. Megnet, *ABB* **126**, 933 (1968).
387. P. W. Fowler, A. J. S. Ball, and D. E. Griffiths, *Can. J. Biochem.* **50**, 35 (1972).
388. K. Ebisuzaki and E. S. G. Barron, *ABB* **69**, 555 (1957).
389. H. M. C. Heick and M. Barrette, *BBA* **212**, 8 (1970).
390. M. J. Fernández, C. Gómez-Moreno, and M. Ruiz-Amil, *Arch. Mikrobiol.* **84**, 153 (1972).
391. L. Schimpfessel, J. Sugar, and R. Crokaert, *Arch. Int. Physiol. Biochim.* **77**, 556 (1969).
392. H. M. C. Heick, J. Willemot, and N. Bégin-Heick, *BBA* **191**, 493 (1969).
393. J. I. Wenger and C. Bernofsky, *BBA* **227**, 479 (1971).
394. J. Sugar, L. Schimpfessel, E. Rozen, and R. Crokaert, *Arch. Int. Physiol. Biochim.* **78**, 1009 (1970).
395. J. Sugar, L. Schimpfessel, E. Rozen, and R. Crokaert, *Arch. Int. Physiol. Biochim.* **79**, 849 (1971).
396. R. T. Hersh, *BBA* **58**, 353 (1962).
397. P. Andrews, *BJ* **96**, 595 (1965).
398. T. Ohta and Y. Ogura, *J. Biochem.* (*Tokyo*) **58**, 73 (1965).
399. M. Bühner and H. Sund, *Eur. J. Biochem.* **11**, 73 (1969).
400. R. Cohen and M. Mire, *Eur. J. Biochem.* **23**, 276 (1971).
401. P. J. G. Butler, H. Jörnvall, and J. I. Harris, *FEBS* (*Fed. Eur. Biochem. Soc.*) *Lett.* **2**, 239 (1969).

sulfate (*396*), sulfhydryl reagents (*1,402*), and thiol oxidation (*399*) have been used to dissociate the molecule into subunits. Most values thus observed for the molecular weight of the subunit are in the range 36,000–37,000 which is consistent with the results of structural studies (*12,86,378,379*). Association to complexes of higher molecular weight than 150,000 has also been observed (*1,329*).

A number of zinc analyses have been made with different techniques (*1,377,403–405*). The amount of zinc determined varies in the range 3.2–5.2 atoms per molecule. The generally accepted value has been four tightly bound zinc atoms per molecule or one zinc atom per subunit. However, as discussed in Section II,D, there are strong indications that the subunits of YADH and LADH have similar structures, probably including the presence of two zinc atoms.

B. Primary Structure

The amino acid sequence of a fragment containing a reactive cysteine residue was reported by Harris (*86*) and, later, that of another peptide containing a second cysteine residue by Twu *et al.* (*406,407*). The absence of a free N-terminus in the protein was demonstrated in other studies (*378,404*) and subsequently found to be resulting from blockage by an acyl group (*12*). Tryptic peptides of the enzyme were partly characterized by Butler *et al.* (*378,379*).

The primary structure of the protein was further studied by Jörnvall (*12*) who isolated peptides after proteolytic treatments with trypsin, chymotrypsin, pepsin, thermolysin, or cyanogen bromide. From this, most of the amino acid sequence of the subunit could be determined and sequence information for over half of the molecule has been reported (*12,136,137*). Segments at known positions are given in Table XIV and compared to equivalent positions in other alcohol dehydrogenases. Segments from other regions of the subunit are given in Table XV. The sequence comparisons (Table XIV) established that the yeast and mammalian alcohol dehydrogenases are distantly related. Inside a region of the N-terminal thirds of the molecules the structural identity is about 40%, which is more than at the terminal ends where no homology is obvious.

402. J. R. Heitz and B. M. Anderson, *ABB* **127,** 637 (1968).
403. B. S. Vanderheiden, J. O. Meinhart, R. G. Dodson, and E. G. Krebs, *JBC* **237,** 2095 (1962).
404. A. Arens, H. Sund, and K. Wallenfels, *Biochem. Z.* **337,** 1 (1963).
405. A. Curdel and M. Iwatsubo, *FEBS* (*Fed. Eur. Biochem. Soc.*) *Lett.* **1,** 133 (1968).
406. J.-S. Twu and F. Wold, *Biochemistry* **12,** 381 (1973).
407. J.-S. Twu, C. C. Q. Chin, and F. Wold, *Biochemistry* **12,** 2856 (1973).

TABLE XIV
Comparison of the Primary Structures of Different Types of Alcohol Dehydrogenases

---

                                        1   5   10   15
Mammalian liver[a]   Acetyl-Ser-Thr-Ala -Gly-Lys-Val-Ile -Lys-Cys -**Lys-Ala**-**Ala-Val**-Leu -Trp-**Glu**-Glu-**Lys-Lys**-
                                                                                                        Gln His
                                                                                                        Pro
Yeast[b]   Acyl-Ser -Ile -Pro-Glu-Thr-Gln -**Lys**-Gly-Val-Ile -Phe/Tyr-**Glu**-Ser -His-Gly-
Bacillus stearothermophilus[c]   Met-**Lys-Ala**-**Ala-Val**-Val -Glu-Gln-Phe-**Lys-Lys**-

20   25   30   35   40
Pro-Phe-Ser -Ile -Glu-**Glu-Val-Glu-Val**-Ala-Pro-**Pro-Lys-Ala** -His -**Glu-Val** -Arg -**Ile** -Lys-Met-Val-**Ala** -Thr-**Gly**-
            Thr   Asp Ile   Val
                                                                                                                          Ala
Lys-**Leu**-Glu-Tyr-**Lys**-Asp-Ile -Pro-**Val -Pro-Lys-Pro-Lys-Ala** -Asn-**Glu**-Leu-**Leu**/**Ile**- Asn-Val -**Lys**-Tyr/Ser -**Gly**-
Pro-Leu-Gln-Val -**Lys-Glu-Val-Glu**-Lys-**Pro-Lys**-Ile -Ser -Tyr-Gly -**Glu-Val** -**Leu** -Val-Arg-Ile -**Lys-Ala** -Cys-**Gly**-

45   50   55   60   65
Ile -**Cys**-Arg-Ser -**Asp**-Asp-**His**-Val-Val -Ser -**Gly**-Thr-**Leu**- – -Val- – -**Thr** -Pro-**Leu-Pro**-Val -Ile -Ala -**Gly-His-Glu**-Ala-
Ser   **Ala**   Ser   Phe   Ala **Val** Leu   **Gly**
Val-Cys-His-Thr-Asp-Leu-His-Ala-Trp-His-**Gly**-Asp-**Leu**-Pro-Trp-Pro-**Thr**/Lys-**Leu-Pro**-Leu-**Val**-Gly-**Gly-His-Glu-Gly**-
Val-Cys-His-Thr-Val -**Leu**-Gly-**Ala**-Ala -

70   75   80   85   90
Ala-Gly-Ile -**Val**-Glu-Ser-Ile -**Gly-Glu**-Gly-**Val**-Thr-Thr-Val -Arg-Pro-**Gly-Asp**-Lys-Val-Ile -Pro-Leu-Phe-Thr-
                                                              Cys   **Lys**   Ile
                                                                                                                             Ser
Ala-Gly-Val-**Val**-Val -Gly-Met-**Gly-Glu**-Asn-**Val**-Lys -Gly -Trp-**Lys**-Ile -**Gly-Asp**-Tyr-Ala-Gly-Ile -Lys-Trp-Leu-

95 100 105 110 115 167

Pro- – -Gln-**Cys**-Gly- Lys-**Cys**-Arg-Val -**Cys**-Lys-His -Pro-Glu-Gly-Asn-Phe-**Cys**- – -Leu-Lys-Asn-**Asp-Leu**- . . . -Glu-Lys-**Val**-

Ser Ile Ser Leu Cys Gln Thr Lys Ser Asp

Asn

Asp(Ser,Gly,**Cys**)Met-Ala -**Cys**-Glu-Tyr-**Cys**-Glu-Leu-Gly-Asn-Glu-Ser -Asn-**Cys**- – -Pro-His–Ala -**Asp-Leu**- · · · -Ala -Glu-**Val**-

170 175 180 185 190

Cys-Leu-Ile -Gly-**Cys**-Gly- Phe-Ser-**Thr**-Gly-**Tyr**-Gly-Ser-**Ala**-Val -**Lys**-Val-**Ala**-Lys-Val -Thr-

Gln

Ala -Pro -Val-Leu-**Cys**-Ala -Gly-Ile-**Thr**-Val -**Tyr**-Lys- – -**Ala**-Leu-**Lys**-Ser -**Ala**-Asn-Leu-Met-

---

[a] Top line is horse E subunit. Numbering system refers to this sequence. Known alternative residues in other mammalian alcohol dehydrogenases (Table I) are given immediately below the E-chain alternatives, but nonsequenced segments of the human enzyme deduced by homology only *(75)*, are excluded. Deletions in the horse enzyme must be assumed after residues 57, 58, and at one position between 87 and 96 (arbitrarily placed after residue 95 in the table), apart from the deletion after residue 111, which is evident from the comparisons of the mammalian enzymes (Table I). The latter deletion is also present in the yeast enzyme, which in addition has one further deletion at position 182.

[b] From Jörnvall *(12)* for first part and from Jörnvall *et al.* (137) for second part.

[c] From Bridgen *et al.* *(13)*. The bacterial protein is only included in the first part of the comparison (known segment ends at position 53).

TABLE XV
THE PRIMARY STRUCTURE OF SEGMENTS FROM YADH NOT GIVEN IN TABLE XIV

Internal segment[a]	-Val-Arg-Ala-Asn-Gly-Thr-Thr-Val-Leu-Val-Gly-Met- -Pro-Ala-Gly-Ala-Lys-
C-terminal segment[b]	-Glu-Ala-Leu-Asp-Phe-Phe-Ala-Arg-Gly-Leu-Ile- -Lys-Ser-Pro-Ile -Lys-Val-Val-Gly-Leu-Ser-Thr- -Leu-Pro-Glu-Ile-Tyr-Glu-Lys-Met-Glu-Lys-Gly- -Gln-Val-Val-Gly-Arg-Tyr-Val-Val-Asp-Thr-Ser- -Lys.

[a] From Jörnvall (*136*).
[b] From Jörnvall (*12*).

The more conserved region was suggested to have a common important function (*12*). Further aspects of the structure comparisons are discussed in Section II,D.

C. Chemical Modifications

1. *Cysteine Residues*

The yeast enzyme has 8 (*86,378,379*) to 9 (*366,399*) free SH groups per subunit and is inhibited by thiol reagents. It may also be inactivated by oxidation to form intrachain disulfide bridges (*399*). Preparations may vary in SH content (*366,399,408,409*), although the number of cysteine residues reactive to iodoacetamide is constant (*408,409*). This type of variability seems reminiscent to that of the subfractions of rat liver alcohol dehydrogenase (Section II,A,3,a).

Two of the cysteine residues are especially reactive toward chemical modification. Thus, one residue per subunit is selectively alkylated with iodoacetate (*86*) and a different one with butylisocyanate (*406,407*). In both cases the enzyme is inactivated and protected by the coenzyme against modification, suggesting that these residues are at the active sites of the enzyme. The two residues are now known (*12,137*) to be homologous to the two reactive cysteine residues in the horse enzyme, Cys-46 and Cys-174 (Section II,E,1,a), which are ligands to the active site zinc atom (Section II,C,3,b). A number of other reagents, apart from reactive coenzyme analogs, have also been shown to modify essential cysteine residues, *i.e.*, probably either of these residues. Thus, one cysteine residue

408. B. R. Rabin and E. P. Whitehead, *Nature* (*London*) **196**, 658 (1962).
409. E. P. Whitehead and B. R. Rabin, *BJ* **90**, 532 (1964).

may be labeled with iodoacetamide (*376,402,407–412*) which was the first reagent used (*408*) and which has now been found to involve both reactive sulfhydryl groups (*407*). Cysteine residues are also alkylated with α-iodopropionic acid or its amide (*413*) in which case the two enantiomers react differently, indicating highly dissymmetric surroundings of the reacting SH groups (*413*). Maleimides (*402,411,414*) as well as a dibromopyridazone (*412*), a naphthoquinone (*415*), dehydroascorbic acid or D-arabinosone (*416*) also react, and two molecules of fluoresceine mercuric acetate per active site are specifically bound to the enzyme (*402*). The effect of *p*-mercuribenzoate on activity and coenzyme binding capacity of YADH has been studied (*417*).

A reactive coenzyme analog, nicotinamide-[5-bromoacetyl-4-methylimidazole] dinucleotide, has also been found to specifically label Cys-43 in the yeast enzyme (*137*). This residue corresponds to Cys-46 in the horse enzyme (Table XIV). Selective modification of cysteine has also been reported with other NAD analogs (*418–420*), as summarized in Table XVI, but the modified cysteines have not been identified in the primary structure of the protein.

The results of cysteine modification confirm the similarities in structure and function of the active sites of mammalian and yeast alcohol dehydrogenases (Section II,D). Minor differences are, however, observed. Thus, the nicotinamide-substituted imidazole dinucleotide (*137*) selectively alkylates one of the two cysteine ligands to the catalytic zinc atom, Cys-43, in the yeast enzyme. In the horse enzyme, on the other hand, the same reagent alkylates a different ligand to the same zinc atom, Cys-174.

2. *Histidine Residues*

It has been reported (*421*) that modification with 3-(3-bromoacetylpyridinio)propyl adenosine pyrophosphate (Table XVI), followed by oxi-

410. B. R. Rabin, J. R. Cruz, D. C. Watts, and E. P. Whitehead, *BJ* **90,** 539 (1964).
411. F. Auricchio and C. B. Bruni, *BBA* **185,** 461 (1969).
412. W. Schreiber, L. Richter, and U. H. Klemens, *Hoppe-Seyler's Z. Physiol. Chem.* **349,** 1405 (1968).
413. B. Eisele and K. Wallenfels, *Eur. J. Biochem.* **6,** 29 (1968).
414. J. R. Heitz, C. D. Anderson, and B. M. Anderson, *ABB* **127,** 627 (1968).
415. N. Nakai and J.-I. Hase, *Chem. Pharm. Bull.* **19,** 460 (1971).
416. K. Uehara and S. Fujimoto, *J. Vitaminol.* (*Kyoto*) **18,** 10 (1972).
417. K. Wallenfels and B. Müller-Hill, *Biochem. Z.* **339,** 352 (1964).
418. B. V. Plapp, C. Woenckhaus, and G. Pfleiderer, *ABB* **128,** 360 (1968).
419. D. T. Browne, S. S. Hixson, and F. H. Westheimer, *JBC* **246,** 4477 (1971).
420. C. Woenckhaus, R. Jeck, E. Schättle, G. Dietz, and G. Jentsch, *FEBS* (*Fed. Eur. Biochem. Soc.*) *Lett.* **34,** 175 (1973).
421. C. Woenckhaus, M. Zoltobrocki, J. Berghäuser, and R. Jeck, *Hoppe-Seyler's Z. Physiol. Chem.* **354,** 60 (1973).

TABLE XVI

REACTIVE COENZYME ANALOGS USED FOR MODIFICATION OF RESIDUES IN YADH AND LADH

Reactive analog[a]				Ref.	Enzyme	Modified residue
I. Pyridine substitutions 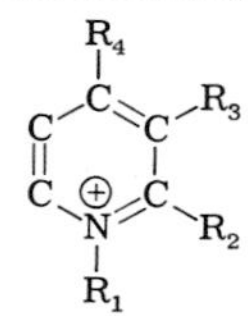						
R1	R2	R3	R4			
$(CH_2)_2NHCOCH_2Br$	H	$CONH_2$	H	*418*	YADH	Cys
ADPR	H	$CH_2OCOCHN_2$	H	*419*	YADH	Cys
ADPR	H	$N{\equiv}N$	H	*218*	YADH	Not identified
ADPR	H	$COCH_2Cl$	H	*163*	YADH,LADH	Not identified
$(CH_2)_4PPRA$	H	$COCH_2Br$	H	*148*	YADH,LADH	Not identified
$(CH_2)_3PPRA$	H	$COCH_2Br$	H	*146*	YADH	Not identified
$(CH_2)_3PPRA$	H	$COCH_2Br$	H	*421*	YADH	His
$(CH_2)_3PPRA$	H	$COCH_2Br$	H	*423*	YADH	His
$(CH_2)_3PPRA$	H	H	$COCH_2Br$	*420*	YADH	Cys
$(CH_2)_3PPRA$	H	H	$COCH_2Br$	*424*	YADH,LADH	Not identified
$(CH_2)_3PPRA$	$COCH_2Br$	H	H	*422*	YADH	Not identified
II. Adenine substituted for 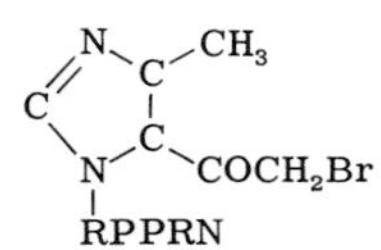				*146*	YADH,LADH	Not identified
				147	LADH	Not identified
				137	LADH	Cys-174
				137	YADH	Cys-43[b]

[a] The letters in ADPR, PPRA and RPPRN denote adenine (A), ribose (R), diphosphate or pyrophosphate (DP or PP), and nicotinamide (N).

[b] Equivalent to Cys-46 in LADH (Table XIV).

dation and total hydrolysis liberated *N*-carboxymethylhistidine. The pH dependence (*421*) of the inactivation also supported the involvement of a histidine residue close to the active site. The labeled histidine residue has, however, not been identified and the simultaneous modification of other residues, especially cysteine, may not be excluded.

3. *Arginine Residues*

Butanedione or phenylglyoxal (*162*) inactivate the enzyme and modify two arginine residues per subunit. NADH offers partial protection. It has, therefore, been suggested that, in the yeast enzyme as well as in the mammalian enzymes (Section II,E,1,e), arginine residues participate in NADH binding, perhaps through interaction with the coenzyme pyrophosphate bridge (*162*). It should, however, be noticed that Arg-47, which is involved in pyrophosphate binding in LADH (Section II,C,3,a), is exchanged for histidine in YADH (Table XIV).

4. *Uncharacterized Modifications*

Apart from those NAD analogs, which have been shown to modify cysteine or histidine residues, other reactive coenzyme analogs (*148,163,218,422–424*), also listed in Table XVI, inactivate the enzyme, but the modified residues have not been identified. Labeling of tryptophan has also been suggested but later not confirmed (Section II,J).

5. *Properties of Modified Enzymes*

Coenzyme binding capacity of YADH modified by iodoacetamide has been investigated (*376,411,425*). The modification inactivates the enzyme although coenzyme is still bound. This excluded involvement of the "essential" sulfhydryl groups of YADH in the binding of NADH (*376,411,425*). It was also found that the dissociation constant for coenzyme binding was unaffected by this modification (*376,425*).

Ternary complex formation with substrate analogs could not be observed in these studies (*376,425*) from measurements of the fluorescence enhancement of NADH. However, using NMR methods, Sloan and Mildvan detected both coenzyme and substrate binding to YADH which was modified by a spin-labeled substituted iodoacetamide (*329*). All investi-

422. R. Jeck and C. Woenckhaus, *Z. Naturforsch. C* **29**, 180 (1974).
423. M. Zoltobrocki, C. Woenckhaus, E. Schättle, R. Jeck, and J. Berghäuser, *Hoppe-Seyler's Z. Physiol. Chem.* **353**, 771 (1972).
424. C. Woenckhaus, E. Schättle, R. Jeck, and J. Berghäuser, *Hoppe-Seyler's Z. Physiol. Chem.* **353**, 559 (1972).
425. J. A. Sanderson and H. Weiner, *Fed. Proc., Fed. Amer. Soc. Exp. Biol.* **32**, 543 (1973); *Abstr. Int. Congr. Biochem., 9th, 1973* p. 108 (1973).

gated compounds were bound with the same affinity as in unmodified enzyme. It was suggested (*329*) that the modified enzyme is inactive because the substrate might be bound in a different orientation with respect to NADH, compared to the situation in the native enzyme.

6. *The Cobalt Enzyme*

Yeast synthesizes a green cobalt containing YADH (*405*) in media containing cobalt with zinc as impurity. Three times crystallized enzyme contained a total amount of bound metal (Co + Zn) of 3.5 atoms compared to a value of 4.0 for standard grown yeast. The ratio of cobalt to zinc in cobalt-grown YADH was found to be very close to 1:2. The absorption spectrum of the cobalt containing enzyme shows big differences from that of normal enzyme.

Foster *et al.* (*426*) pointed out that this spectrum was similar to that of tetrahedral cobalt(II). Moreover, the presence of a charge transfer band at 375 nm indicated that the cobalt(II) atom is bound in a roughly tetrahedral environment by at least one cysteine. They further suggested (*426*) that the metal is coordinated to three protein ligands, two of which are cysteine. A water molecule that can be displaced by a chelating ligand completes the coordination. This proposed metal binding agrees astonishingly well with the coordination of the catalytic zinc in LADH which in all probability is identical in YADH (Section II,D).

7. *The Manganese Enzyme*

A manganese containing YADH was obtained from yeast grown anaerobically in a zinc-free manganese medium (*377*). This enzyme was compared to a zinc containing YADH from the same strain grown under normal conditions. Each enzyme had approximately four atoms of metal per molecule. The manganese enzyme contained only manganese.

Thermal stability was considerably decreased in the manganese enzyme which was also much more sensitive to spontaneous oxidation than the zinc enzyme (*377*). The electrophoretic mobilities of the two enzymes are different, the pH optima differ slightly, the pH profile is more narrow for the manganese enzyme and the kinetic parameters are somewhat different (*377*).

8. *Denaturation Studies*

YADH is reversibly denatured in urea at concentrations less than 2 *M* (*398*), probably because of dissociation into subunits (Section III,A).

426. M. A. Foster, H. A. O. Hill, and R. J. P. Williams, *Biochem. Soc. Symp.* **31**, 187 (1970).

Higher concentrations give irreversible denaturation as a result of unfolding of the polypeptide chains as measured by optical rotatory methods (*1,185,427*), fluorescence quenching (*185,398*), and difference absorption spectra (*398*). Urea treated enzyme loses its capacity to bind coenzyme as a function of urea concentration (*398*): Substituted N^1-hydroxythiourea denatures YADH (*428*), probably because of oxidation of SH groups.

Heat denaturation studies show a sharp break in activity between 35° and 40° (*185*) where major changes in optical rotatory parameters also occur. These studies show that YADH is less stable than LADH. It has been reported that coenzyme destabilizes YADH (*193*).

The inactivation of YADH by X-rays has also been studied (*1,429,430*) as well as inactivation by UV light (*431*), ^{60}Co-γ-radiation (*432*) and photooxidation in different media (*433,434*).

D. Inhibitor Studies

1. *Coenzyme Competitive Inhibitors*

Studies of dinucleotides as coenzyme analogs for YADH and LADH have been discussed in Section II,G. Fragments of the coenzyme molecule are inhibitors to YADH, competitive to the coenzyme (*1,214,302,435–437*). The participation of the pyrophosphate moiety in coenzyme binding to YADH has been questioned since free pyrophosphate is not an inhibitor (*1,435*). Furthermore, a smaller inhibition constant has been found for adenosine than for AMP and ADP (*214*). On the other hand, AMP and ADP inhibit the enzyme at low pH, whereas adenosine has no effect (*437*). Evidence for possible interaction between the pyrophosphate moiety and arginine residues in YADH is discussed in Section III,C,3.

427. D. D. Ulmer and B. L. Vallee, *JBC* **236,** 730 (1961).
428. G. Clifton, S. R. Bryant, and C. G. Skinner, *J. Med. Chem.* **13,** 377 (1970).
429. F. Shimazu and A. L. Tappel, *Radiat. Res.* **23,** 210 (1964).
430. G. Gorin, M. Quintiliani, and S. K. Airee, *Radiat. Res.* **32,** 671 (1967).
431. K. Dose and G. Krause, *Photochem. Photobiol.* **7,** 503 (1968).
432. I. Ganzke, E. Tapp, and G. Siegel, *Radiat. Biol. Ther.* **11,** 329 (1970).
433. D. della Pietra and K. Dose, *Biophysik* **2,** 347 (1965).
434. R. Nilsson and D. R. Kearns, *Photochem. Photobiol.* **17,** 65 (1973).
435. B. M. Anderson and M. L. Reynolds, *ABB* **111,** 1 (1965).
436. B. M. Anderson, M. L. Reynolds, and C. D. Anderson, *BBA* **113,** 235 (1966).
437. I. Steinbrecht, H. W. Augustin, and E. Hofmann, *Acta Biol. Med. Ger.* **21,** 409 (1968).

The inhibitory power of *N*-alkyl nicotinamide derivatives has been found to increase with increasing length of the alkyl chain (*1,435,438–442*). These results indicate that there is a hydrophobic binding area close to or actually in the substrate binding site, provided the nicotinamide moiety of these derivatives and of the coenzyme molecule bind to the same site on the enzyme. Similar results were found for *N*-alkyl-substituted ammonium chlorides (*441,443*). Substitution of the amino group by a carboxyl group in *N*-benzyl-3-aminopyridinium chloride reduces the inhibition (*442*).

Flourescent dyes (*226,444,445*) and the drugs chloroquine (*446*) and propranolol (*259*) have also been found to bind to YADH in competition with coenzyme.

2. *Other Inhibitors*

The chelating agent 1,10-phenanthroline which binds to the catalytic zinc atom in LADH (Section II,C,3,b) is also an inhibitor for YADH (*1,245,427,436,440,447*). However, the inhibitory power is considerably smaller with YADH than LADH (*1*). Binding to zinc has been questioned (*436*) since nonchelating analogs such as 1,5-phenanthroline and 2,9-dimethyl-1,10-phenanthroline are better inhibitors than 1,10-phenanthroline. It has been suggested (*436*) that the hydrophobic character of these molecules, rather than their capacity to chelate zinc, is more important for binding to YADH. On the other hand, there is spectroscopic evidence that 1,10-phenanthroline binds to zinc in YADH (*1,248*). Similar evidence for binding to zinc in LADH was shown to be valid (*107*). In contrast to LADH, YADH dissociates into subunits in the presence of 1,10-phenanthroline (Section III,A).

Pyrazole, which is a very specific inhibitor for LADH (Section II,H,2,a), also inhibits YADH (*77,377,448*). The inhibition is competitive with ethanol and uncompetitive with coenzyme and acetaldehyde (*77*). 4-Subsituted pyrazole derivatives which are stronger inhibitors with

438. B. M. Anderson and C. D. Anderson, *BBRC* **16,** 258 (1964).
439. B. M. Anderson, M. L. Reynolds, and C. D. Anderson, *BBA* **99,** 46 (1965).
440. B. M. Anderson and M. L. Reynolds, *ABB* **114,** 299 (1966).
441. M. L. Fonda and B. M. Anderson, *ABB* **120,** 49 (1967).
442. J. R. Heitz and B. M. Anderson, *Mol. Pharmacol.* **4,** 44 (1968).
443. B. M. Anderson and M. L. Reynolds, *BBA* **96,** 45 (1965).
444. F. M. Dickinson, *FEBS* (*Fed. Eur. Biochem. Soc.*) *Lett.* **15,** 17 (1971).
445. A. Kotaki, M. Naoi, and K. Yagi, *BBA* **229,** 547 (1971).
446. R. Fiddick and H. Heath, *Nature* (*London*) **213,** 628 (1967).
447. F. L. Hoch and B. L. Vallee, *JBC* **221,** 491 (1956).
448. T. E. Singlevich and J. J. Barboriak, *Fed. Proc., Fed. Amer. Soc. Exp. Biol.* **29,** 275 (1970); *Biochem. Pharmacol.* **20,** 2087 (1971).

the liver enzyme than pyrazole are much weaker with the yeast enzyme (*77*). This probably reflects differences in the details of the active site pocket (Section II,D,1,b).

Amides can form fluorescent ternary complexes with reduced coenzyme (*375,407*) similar to those formed with LADH (Section II,H,2,b). The ternary complex with isobutyramide has been investigated using NMR methods (*329*), resulting in a determination of the position of the isobutyramide molecule with respect to the coenzyme. The results suggest a slight reorientation of the NADH molecule when the inhibitor binds. From measurements of the same complex with a cobalt containing YADH, a mean distance of about 7 Å has been calculated from cobalt to the CH_3 protons of the isobutyramide molecule (*449*). These results are consistent with the coordination of the inhibitor molecule to zinc.

Thiocyanate has been found to be competitive with ethanol but noncompetitive with NAD (*398*). At concentrations less than 0.3 *M* the inhibition is reversible but at higher concentrations irreversible. Low concentrations of urea (*398,450*) inhibit YADH noncompetitively with respect to NAD, NADH, alcohol, and aldehyde. Higher concentrations inhibit the enzyme irreversibly (Section III,C,8). Thiourea, guanidinium salts (*398*), phenylisopropyl hydrazine (*258*), and canavanine (*451*) are also inhibitors of YADH.

E. Kinetic Aspects

1. *Coenzyme Binding*

Since the YADH molecule is composed of four identical polypeptide chains (Section III,B), each presumably containing one coenzyme binding domain similar in structure to other dehydrogenases (Section II,D,1,b), it seems probable that the YADH molecule can bind four coenzyme molecules. Furthermore, it can be predicted that the coenzyme binds in an open conformation similar to coenzyme binding in other dehydrogenases (*2*). Direct evidence for this open conformation has been obtained by Sloan and Mildvan in an NMR investigation of coenzyme binding to a spin-labeled YADH (*329*). Indirect evidence is obtained from the similarities between LADH and YADH in binding and functional properties of various coenzyme analogs (Section II,G). There are differences

449. D. L. Sloan and J. M. Young, *Fed. Proc., Fed. Amer. Soc. Exp. Biol.* **33,** 1244 (1974).
450. K. V. Rajagopalan, I. Fridovich, and P. Handler, *JBC* **236,** 1059 (1961).
451. B. Tschiersch, *Tetrahedron Lett.* **28,** 3237 (1966).

in the details of coenzyme binding such as different pH dependences and binding strengths (*1*).

A large variety of methods have been used to study the number of coenzyme molecules bound to YADH since the publication of the previous review (*1*) producing conflicting results. Hayes and Velick (*452*) originally found, in an ultracentrifugation study, a maximum of 3.6 molecules bound NADH or NAD^+ per enzyme molecule, indicating a binding stoichiometry of one coenzyme per subunit. Subsequent studies using a gel filtration technique (*453*) have produced varied results: 5.0–5.2 (*454*), 3.3–5.1 (*411*), an average of 2.7 (*375*), and a maximum of 4 (*455*) bound coenzyme molecules per enzyme molecule. Fisher *et al.* (*307,456*) found, by using careful difference spectroscopy, a spectroscopic shift for NADH binding to YADH similar to the shift which had previously been observed for binding to LADH (*1*). A study of this spectroscopic shift in combination with CD investigations (*457*) gave 4.0 moles NADH bound per mole enzyme. On the other hand, fluorescence titration with NADH in the presence of amide gave values of 2.2–2.6 (*375*) and 3 (*425*) molecules bound per enzyme molecule. Spectroscopic studies of the binding of the coenzyme analog pyridine-3-aldehyde-AD^+ to YADH in the presence of hydroxylamine gave similar values (2.6–3.0) (*375*). All observed dissociation constants agree with those previously published (*1*).

The finding of less than four bound coenzyme molecules per YADH molecule has been interpreted in terms of negative cooperativity (*375,425,455*). Yamada and Yamato have considered a model consisting of a dimer of dimers with two different intrinsic binding constants for coenzyme binding (*455*). This model cannot, however, be reconciled with the results of Temler and Kägi who have reported (*457*) that the YADH molecule binds four coenzyme molecules and that the binding sites are equal and the binding noncooperative.

2. *Substrate Specificity*

YADH has been shown to be active toward a variety of substrates (*1*), but it differs in many ways in its substrate specificity from LADH. One of the most interesting differences is that cyclohexanol is a substrate for LADH but not for YADH (*1*). This must result from differences in the part of the substrate binding pocket that is close to

452. J. E. Hayes, Jr. and S. F. Velick, *JBC* **207,** 225 (1954).
453. J. P. Hummel and W. J. Dreyer, *BBA* **63,** 530 (1962).
454. G. Pfleiderer and F. Auricchio, *BBRC* **16,** 53 (1964).
455. T. Yamada and M. Yamato, *J. Biochem.* (*Tokyo*) **74,** 971 (1973).
456. H. F. Fisher and D. G. Cross, *Science* **153,** 414 (1966).
457. R. S. Temler and J. H. R. Kägi, *Abstr. Int. Congr. Biochem., 9th, 1973* p. 109 (1973).

zinc in LADH. Based on the comparisons described in Section II,D,1,b, the main differences in this region are the changes from Phe-93 and Ser-48 in LADH to Trp and Thr, respectively, in YADH. The effect of these changes is a smaller substrate binding pocket as judged from the LADH structure.

This same situation is also reflected in the investigations by Dickinson and Dalziel (*315,323*) on a number of secondary alcohols. They found that the yeast enzyme is completely inactive toward those secondary alcohols where both alkyl groups are larger than methyl and active with only one isomer of butan-2-ol and octan-2-ol. This is in contrast to LADH which is active toward all substrates mentioned.

The results obtained in the oxidation of different aliphatic alcohols show that, apart from methanol, alcohols are increasingly better substrates for YADH with decreasing chain length. Furthermore, primary alcohols are better substrates than secondary alcohols (*1,458,459*). Reduction of the corresponding aldehydes shows an additional maximum in rate for a chain length of 8–10 carbon atoms (*459*). A kinetic investigation of para-substituted benzaldehydes showed a significant electronic substituent effect on the rate of reduction of these substrates (*460*). This was suggested not to result from a solvent effect, although it has been shown that the real substrate of YADH is the free and not the hydrated aldehyde (*461*).

3. *The Reaction Mechanism*

The main features of the reaction mechanism for YADH are in all probability essentially the same as in LADH. The structural similarities of the catalytic domain, including the catalytic zinc and its protein ligands, are strong indications that both enzymes perform electrophilic catalysis mediated by zinc. Involvement of zinc in the catalytic action of YADH was suggested almost 20 years ago (*365,366,462*), but evidence for a direct participation has been obtained only recently (*329,449*).

Finer details of the reaction mechanism are, however, different between the two enzymes. There are differences in substrate specificity (Section III,E,2) as well as in the rate of catalyzed reaction. The rate constants for dissociation of NAD^+ and NADH are about 30-fold and 160-fold larger, respectively, for YADH compared to LADH at pH 7.0, 25° (*458*).

458. F. M. Dickinson and G. P. Monger, *BJ* **131,** 261 (1973).
459. W. Schöpp and H. Aurich, *Acta Biol. Med. Ger.* **31,** 19 (1973).
460. J. P. Klinman, *JBC* **247,** 7977 (1972); *in* "Alcohol and Aldehyde Metabolizing Systems" (R. G. Thurman *et al.*, eds.), p. 81. Academic Press, New York, 1974.
461. B. Müller-Hill and K. Wallenfels, *Biochem. Z.* **339,** 349 (1964).
462. B. L. Vallee and F. L. Hoch, *Proc. Nat. Acad. Sci. U. S.* **41,** 327 (1955).

Coenzyme dissociation is rate limiting for both enzymes (*341,463*). Furthermore, no kinetic evidence has been obtained for an isomerization of the binary YADH·NADH complex, in contrast to the corresponding LADH complex (*302*). Hydrogen–deuterium exchange studies (*464*), however, indicate a conformational change upon coenzyme binding.

There also seem to be differences in the requirements for ordered events during the catalytic mechanism. Wratten and Cleland (*333*) found from product inhibition studies that early suggestions (*365*) of a random mechanism were not compatible with their data, and they proposed an ordered mechanism. Silverstein and Boyer (*341*), using isotope exchange at equilibrium, showed that coenzyme dissociation is rate limiting at low substrate concentrations, but that a compulsory mechanism is not valid. They suggested that random dissociation of coenzyme and substrate are kinetically important at higher substrate concentrations. Alternative explanations are the formation of dead-end enzyme–substrate complexes (*465*) or formation of abortive E·NADH·alcohol complexes, from which coenzyme can dissociate (*341*). Dickinson and Monger have proposed (*458*) a compulsory mechanism for aldehyde reduction and a partly random mechanism for alcohol oxidation with compulsory order of product dissociation. They suggested that binary E·aldehyde complexes may occur but that these are not kinetically significant.

IV. Alcohol Dehydrogenases from Other Sources

A. Introduction

Alcohol dehydrogenase has a widespread occurrence in nature (*1*). This review will be limited to those enzymes for which ethanol is a good substrate, and for which pyridine nucleotides act as coenzymes. This eliminates a large number of "alcohol dehydrogenases" acting on OH groups in various steroids, sugars, polyalcohols, and other compounds. It also eliminates "alcohol dehydrogenase" of some lower organisms which requires FAD or phenazine methosulfate and other artificial electron acceptors. Neither brain nor kidney aldehyde reductases, which have different properties and little or no activity for ethanol oxidation, will be considered. Even with these limitations all reports of alcohol dehydrogenase may not be covered, since the enzyme occurs in a variety of organisms,

463. B. Müller-Hill and K. Wallenfels, *Biochem. Z.* **339**, 338 (1964).
464. Aa. Hvidt, J. H. R. Kägi, and M. Ottesen, *BBA* **75**, 290 (1963).
465. J. T.-F. Wong and C. S. Hanes, *Nature* (*London*) **203**, 492 (1964).

including different types of bacteria (Section IV,A), yeasts (*1,466*), fungi (*467*), plants (Section IV,B), insects (Section IV,C), fishes (*1,468*), amphibians (*1*), birds (*469*), and mammals (*1,62,82,470–472*). Apart from liver and yeast enzymes, those of bacterial, plant, and insect origin have been characterized further and are discussed below. Multiple molecular forms in a single species, resulting from genetic variations or secondary modified forms, are common. The variability suggests several gene duplications and other genetic reorganizations during the evolution of ADH, which is consistent with the evolutionary conclusions drawn from the structural analyses (Section II,D,2).

In higher organisms, isozymes frequently have different organal distributions (*62,66,471*). In this respect, the retinal alcohol dehydrogenase was of special interest because of its possible role in conversion of retinene (*1,473*). The retinal enzyme, however, apparently lacks this capacity and is different from the liver enzyme (*81,474*). ADH also occurs in other organs outside the liver (*62,66,475,476*). Different intracellular distributions of isozymes (*477*) and an isozyme dependency on the culture medium (*478*) are also known for other species than the *Saccharomyces* (Section III,A). Completely (Section IV,C; *479*) or partially (*480*) ADH-deficient mutants are known in some species.

B. Bacterial ADH

Alcohol dehydrogenase has been studied in several bacteria. Apart from work cited in Sund and Theorell (*1*), the following sources may be mentioned: *Clostridium acetobutylicum* (*481*), *Escherichia coli* (*482*),

466. C. J. Woscinski and D. O. McClary, *Can. J. Microbiol.* **19**, 353 (1973).
467. F. H. Gleason, *Mycologia* **63**, 906 (1971).
468. H. Hitzeroth, J. Klore, S. Ohno, and U. Wolf, *Biochem. Genet.* **1**, 287 (1968).
469. E. Castro-Sierra and S. Ohno, *Biochem. Genet.* **1**, 323 (1968).
470. J. Papenberg, J. P. von Wartburg, and H. Aebi, *Biochem. Z.* **342**, 95 (1965).
471. J. P. von Wartburg and J. Papenberg, *Psychosom. Med.* **28**, 405 (1966).
472. M. O'Keane, M. R. Moore, and A. Goldberg, *Clin. Sci.* **42**, 781 (1972).
473. A. F. Bliss, *Biol. Bull.* **97**, 221 (1949).
474. W. D. Watkins and T. R. Tephly, *J. Neurochem.* **18**, 2397 (1971).
475. W. Engel, J. Frowein, W. Krone, and U. Wolf, *Clin. Genet.* **3**, 34 (1971).
476. J. Frowein, *J. Endocrinol.* **57**, 437 (1973).
477. N. Bégin-Heick and H. M. C. Heick, *BBA* **212**, 13 (1970).
478. V. G. del Vecchio and G. Turian, *Pathol. Microbiol.* **32**, 141 (1968).
479. D. Schwartz, *J. Chromatogr.* **67**, 385 (1972).
480. R. Megnet, *Pathol. Microbiol.* **28**, 50 (1965).
481. W. M. Fogarty and J. A. Ward, *BJ* **119**, 19P (1970).
482. A. Hatanaka, O. Adachi, T. Chiyonobu, and M. Ameyama, *Agr. Biol. Chem.* **35**, 1142 (1971).

Leuconostoc mesenteroides (*483*), *Pseudomonas aeruginosa* (*484*), *Streptococcus mutans* (*485*), and *Bacillus stearothermophilus* (*486,487*). The protein from the latter species was purified by Kolb and Harris (*486*) and found to have a free N-terminus in contrast to the liver and yeast alcohol dehydrogenases. The bacterial protein was, therefore, submitted to sequence analysis in an automatic sequenator which revealed the 45 first residues (*13*). These are listed in Table XIV. It is clear that the structure of the bacterial enzyme is distantly related to those of the yeast and mammalian enzymes, but few residues are identical in all proteins at equivalent positions (Table XIV). Further aspects of this relationship are discussed in Section II,D.

C. Plant ADH

Different species of higher plants contain ADH. The peanut enzyme has been purified and shown to be a zinc metalloenzyme (*488*). From tea seeds a mixture of two isozymes of molecular weight 150,000 have been purified and found to be unstable in the absence of thiols (*489*). Partially purified pea ADH is also unstable (*490*). Other species in which the enzyme is present, apart from those previously reviewed (*1*), include oats (*491*), barley (*492*), and wheat (*493*). The occurrence of isozymes has been noticed in wheat (*494*).

Several genetic and structural studies have been performed on maize ADH. Alleles at two different genes exist for the enzyme in this species (*495–499*). The level of activity in the scutellum of maize is controlled

483. A. Hatanaka, O. Adachi, T. Chiyonobu, and M. Ameyama, *Agr. Biol. Chem.* **35,** 1304 (1971).
484. J.-P. Tassin and J.-P. Vandecasteele, *BBA* **276,** 31 (1972).
485. A. T. Brown and C. E. Patterson, *Arch. Oral Biol.* **18,** 127 (1973).
486. E. Kolb and J. I. Harris, *BJ* **124,** 76P (1971).
487. A. Atkinson, B. W. Phillips, D. S. Callow, W. R. Stones, and P. A. Bradford, *BJ* **127,** 63P (1972).
488. H. E. Pattee and H. E. Swaisgood, *J. Food Sci.* **33,** 250 (1968).
489. A. Hatanaka and T. Harada, *Agr. Biol. Chem.* **36,** 2033 (1972).
490. C. E. Eriksson, *J. Food Sci.* **33,** 525 (1968).
491. J. Berger and G. S. Avery, *Amer. J. Bot.* **30,** 290 (1943).
492. J. H. Duffus, *Phytochemistry* **7,** 1135 (1968).
493. J. Macko, G. R. Honold, and M. A. Stahmann, *Phytochemistry* **6,** 465 (1967).
494. C. R. Bhatia and J. P. Nilson, *Biochem. Genet.* **3,** 207 (1969).
495. D. Schwartz and T. Endo, *Genetics* **53,** 709 (1966).
496. D. Schwartz, *Proc. Nat. Acad. Sci. U. S.* **56,** 1431 (1966).
497. J. G. Scandalios, *Biochem. Genet.* **1,** 1 (1967).
498. D. Schwartz, *Science* **164,** 585 (1969).
499. J. G. Scandalios, *Science* **166,** 623 (1969).

by a separate gene (*500*). Two of the isozymes, corresponding to the homogeneous dimers of two alleles at one *ADH* locus, have been purified (*501*). Their molecular weights are about 60,000 and their amino acid compositions are similar, suggesting that the molecules differ by only a few amino acid residues (*501*). They have, however, been reported to differ in zinc binding capacity (*502*) and in adaptive significance for their carriers to various soils (*503*).

D. Insect ADH

1. A cockroach, *Leucophaea maderae,* was found to contain alcohol dehydrogenase in the fat bodies, the functions of which resemble the mammalian liver (*504*). A malaria mosquito, *Anopheles stephensi,* has three variants of alcohol dehydrogenase, corresponding to codominant alleles at a single, probably autosomal, locus (*505*).

2. *Drosophila ADH.* Alcohol dehydrogenase in *Drosophila* has been extensively studied. Two naturally occurring alleles at the *ADH* locus in *Drosophilia melanogaster* correspond to electrophoretically fast and slow forms of the enzyme in homozygotes and to hybrid forms in heterozygotes (*506–508*). This locus is autosomal on the second chromosome (*507*). Differences in the number of molecules produced by the two alleles have also been reported (*509*) as well as variations in the level of enzyme depending on the sex of the hybrids (*510*). Chemically or radiologically induced mutational variants, including ADH-negative strains, have also been studied (*507,511–513*).

500. Y. Efron, *Science* **170,** 751 (1970); *Mol. Gen. Genet.* **111,** 97 (1971); *Nature (London), New Biol.* **241,** 41 (1973).
501. M. R. Felder, J. G. Scandalios, and E. H. Liu, *BBA* **318,** 149 (1973).
502. M. Fischer and D. Schwartz, *Mol. Gen. Genet.* **127,** 33 (1973).
503. D. R. Marshall, P. Broué, and A. J. Pryor, *Nature (London), New Biol.* **244,** 16 (1973).
504. C. W. Despreaux, Jr. and L. L. Pierre, *Can. J. Microbiol.* **15,** 257 (1969).
505. M. P. Iqbal, R. K. Sakai, and R. H. Baker, *J. Med. Entomol.* **10,** 309 (1973).
506. F. M. Johnson and C. Denniston, *Nature (London)* **204,** 906 (1964).
507. E. H. Grell, K. B. Jacobson, and J. B. Murphy, *Science* **149,** 80 (1965); *Ann. N. Y. Acad. Sci.* **151,** 441 (1968).
508. H. Ursprung and J. Leone, *J. Exp. Zool.* **160,** 147 (1965).
509. J. B. Gibson, *Experientia* **28,** 975 (1972).
510. S. P. Pipkin and N. E. Hewitt, *J. Hered.* **63,** 267 (1972).
511. D. L. Lindsley and E. H. Grell, *in* "Genetic Variations of *Drosophila melanogaster.*" Carnegie Inst. Wash. Publ. No. 627,1968.
512. L. Gerace and W. H. Sofer, *Drosophilia Inform. Serv.* **49,** 39 (1972).
513. W. H. Sofer and M. A. Hatkoff, *Genetics* **72,** 545 (1972).

In addition to the genetic variations a multiplicity of forms has been demonstrated (*506–508,514,515*). These forms are interconvertible, and faster moving forms are obtained in the presence of NAD (*514,515*) or acetone (*516*). Conformational changes, associated with differences in heat stability (*514,516*) and protein fluorescence (*517*), have been suggested to explain the different forms which are likely to result from noncovalent binding of a heat-stable ligand probably containing nicotinamide (*514,518*). Age dependencies on the occurrencies of these forms have been reported (*519*). An analogous polymorphism has also been described for ADH from *Drosophila subobscura* (*520*).

Isozymes from *Drosophila melanogaster* have been purified (*521,522*). Somewhat different molecular weights of the protein have been deduced, but the value is likely to be close to 50,000 (*514,518,521–523*). The genetic analyses strongly indicate that the enzyme is a dimer. A subunit molecular weight close to 25,000 has been determined (*518*), although lower, unlikely values have also been reported (*523*). The number of tryptic peptides analyzed by peptide mapping (*518*) is also consistent with an enzyme molecule composed of two identical subunits. Amino acid compositions (*518,522*) show that the protein contains no methionine and only small amounts of half-cystine and tryptophan.

514. K. B. Jacobson, *Science* **159,** 324 (1968).
515. H. Ursprung and L. Carlin, *Ann. N. Y. Acad. Sci.* **151,** 456 (1968).
516. K. B. Jacobson, J. B. Murphy, J. A. Knopp, and J. R. Ortiz, *ABB* **149,** 22 (1972).
517. J. A. Knopp and K. B. Jacobson, *ABB* **149,** 36 (1972).
518. M. Schwartz, L. Gerace, J. O'Donnell, and W. Sofer, *in* "International Conference on Isozymes, 3rd" (Markert, ed.), Vol. 1, p. 725. Academic Press, New York, 1975.
519. G. R. Dunn, T. G. Wilson, and K. B. Jacobson, *J. Exp. Zool.* **171,** 185 (1969).
520. S. Lakovaara and A. Saura, *Ann. Acad. Sci. Fenn., Ser. A4* **163,** 1 (1970); *Acta Physiol. Scand.* **79,** 3A (1970).
521. W. Sofer and H. Ursprung, *JBC* **243,** 3110 (1968).
522. K. B. Jacobson, J. B. Murphy, and F. C. Hartman, *JBC* **245,** 1075 (1970).
523. K. B. Jacobson and P. Pfuderer, *JBC* **245,** 3938 (1970).

4

Lactate Dehydrogenase

J. JOHN HOLBROOK • ANDERS LILJAS • STEVEN J. STEINDEL • MICHAEL G. ROSSMANN

I. Introduction

A. Reviews

The enzyme L-lactate dehydrogenase (EC 1.1.1.27, LDH) catalyzes the reaction:

$$\underset{\text{Lactate}}{\begin{matrix} CH_3 \\ | \\ HCOH \\ | \\ COOH \end{matrix}} + NAD^+ \rightleftharpoons \underset{\text{Pyruvate}}{\begin{matrix} CH_3 \\ | \\ C{=}O \\ | \\ COOH \end{matrix}} + NADH + H^+$$

Everse and Kaplan (*1*) have recently reviewed the properties of LDH, while those of the coenzyme are discussed in Popják's (*2*) contribution to "The Enzymes." A review of dehydrogenases in general has been given by Colowick *et al.* (*3*). More recent work is summarized in the article of the symposium edited by Sund (*4*). Although some lower organisms contain D-specific LDH (EC 1.1.1.28), we shall confine this review primarily to the L-specific enzyme.

B. Historical

That lactic acid is the end product of anaerobic glycolysis in muscle tissue has been known for all of this century (Fig. 1). Cell-free extracts able to catalyze the oxidation of lactate to pyruvate were first obtained in 1932 (*5*). Warburg (*6*) and von Euler (*7*) and their colleagues discovered the above reaction [Eq. (1)] and associated it with the chemical properties of a coenzyme. Racker (*8*) demonstrated in 1950 that the forward reaction also involved the release of a proton. The first purified enzyme was reported by Straub (*9*) in 1940, while the first micrographs of LDH crystals were shown by Kubowitz and Ott (*10*).

The question as to whether the molecule was a trimer or tetramer (*9*) was eventually settled in the classic paper by Markert and Møller (*11*). Different cellular species of LDH were shown to exist by several groups (*12–17*). Combining this evidence with their own work, Markert and Møller were able to demonstrate that LDH was a tetrameric molecule. The five different permutations of two different polypeptide chains readily explained the electrophoretic patterns. The distribution of these

1. J. Everse and N. O. Kaplan, *Advan. Enzymol.* **28,** 61 (1973).
2. G. Popják, "The Enzymes," 3rd ed., Vol. 2, p. 115, 1970.
3. S. P. Colowick, J. van Eys, and J. H. Park, *Comp. Biochem.* **14,** 1 (1966).
4. H. Sund, ed., "Pyridine Nucleotide-Dependent Dehydrogenases." Springer-Verlag, Berlin and New York, 1970.
5. I. Banga, A. Szent-Györgyi, and L. Vargha, *Hoppe-Seyler's Z. Physiol. Chem.* **210,** 228 (1932).
6. O. Warburg and W. Christian, *Biochem. Z.* **287,** 291 (1936).
7. H. von Euler, H. Albers, and F. Schlenk, *Hoppe-Seyler's Z. Physiol. Chem.* **240,** 113 (1936).
8. E. Racker, *JBC* **184,** 313 (1950).
9. F. B. Straub, *BJ* **34,** 483 (1940).
10. F. Kubowitz and P. Ott, *Biochem. Z.* **314,** 94 (1943).
11. C. L. Markert and F. Møller, *Proc. Nat. Acad. Sci. U.S.* **45,** 753 (1959).
12. A. Meister, *JBC* **184,** 117 (1950).
13. J. B. Neilands. *Science* **115,** 143 (1952).
14. T. Wieland and G. Pfleiderer, *Biochem. Z.* **329,** 112 (1957).
15. I. Haupt and H. Giersberg, *Naturwissenschaften* **45,** 268 (1958).
16. T. Wieland, G. Pfleiderer, I. Haupt, and W. Wörner, *Biochem. Z.* **332,** 1 (1959).
17. G. Pfleiderer and D. Jeckel, *Biochem. Z.* **329,** 370 (1957).

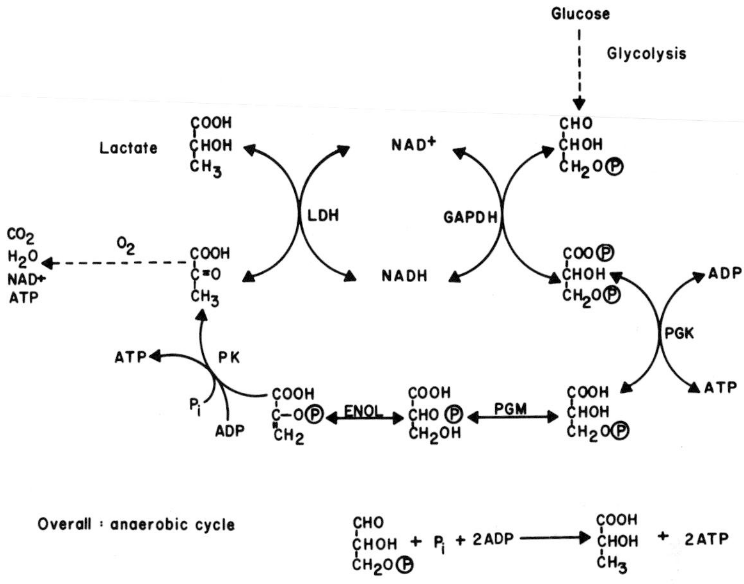

FIG. 1. Metabolic roles of lactate dehydrogenase. Other enzymes referred to are GAPDH (glyceraldehyde-3-phosphate dehydrogenase), PGK (phosphoglycerate kinase), PGM (phosphoglycerate mutase), PK (pyruvate kinase), and ENOL (enolase).

two polypeptide chains was dependent on whether the extract originated in aerobic tissue such as heart (where the H_4 isozyme predominates) or in anaerobic tissue as in skeletal muscle (where the M_4 isozyme predominates) (*18*). In Table I are listed cross-references for the nomenclature of LDH isozymes.

The first X-ray diffraction results were reported in 1964 (*19*) for pig M_4 enzyme and confirmed the 140,000 MW for the tetramer. Good LDH crystals were, however, not obtained until the dogfish, *Squalus acanthias*, M_4 enzyme (*20*) was used. These crystals established the 222 symmetry of the molecule (*21*), and a complete three-dimensional structure was reported in 1970 (*22*).

18. G. Pfleiderer and E. D. Wachsmuth, *Klin. Wochenschr.* **39**, 352 (1961). See also refs. *61* and *171*.
19. B. Pickles, B. A. Jeffery, and M. G. Rossmann, *JMB* **9**, 598 (1964).
20. A. Pesce, T. P. Fondy, F. E. Stolzenbach, D. Castillo, and N. O. Kaplan, *JBC* **242**, 2151 (1967).
21. M. G. Rossmann, B. A. Jeffery, P. Main, and S. Warren, *Proc. Nat. Acad. Sci. U. S.* **57**, 515 (1967).
22. M. J. Adams, G. C. Ford, R. Koekoek, P. J. Lentz, Jr., A. McPherson, Jr., M. G. Rossmann, I. E. Smiley, R. W. Schevitz, and A. J. Wonacott, *Nature (London)* **227**, 1098 (1970).

TABLE I
NOMENCLATURE OF KNOWN LDH ISOZYMES[a]

Chain identifications		Electrophoretic identication of homotetramer	Primary occurrence
M	A	LDH-5	Anaerobic skeletal muscle, tumors
M′	A′		Salmonoid fish muscle
H	B	LDH-1	Aerobic cardiac muscle
H′	B′		Salmonoid fish cardiac muscle
E			Photoreceptors of teleost fish
X	C		Testes of mammals and birds
F			Liver of gadoid fish

[a] The electrophoretic labels LDH-1, 2, 3, 4, and 5 refer to the H_4, H_3M, H_2M_2, HM_3, and M_4 tetramers.

The enzyme, when isolated from a wide variety of sources, is specific for the A side of the nicotinamide ring (*2,23,24*). Organisms examined cover the range from mammals and plants (*25*) to bacteria (*26,27*). The A side specificity of the dogfish enzyme has been demonstrated by Biellmann and Rosenheimer (*28*). The β-glycosidic bond is essential in both parts of the dinucleotide for activity (*29*). The redox potential of the pyridine nucleotide reaction is —0.320 V (*30,31*). Enzyme from a given species is specific either for D- or L-lactate. In some lower organisms such as horseshoe crab (*32*) and *E. coli* (*33*) the enzyme is specific for D-lactate, and there are indications that in these cases the active molecule is only a dimer (*34*).

23. F. H. Westheimer, H. F. Fisher, E. E. Conn, and B. Vennesland, *JACS* **73**, 2403 (1951).
24. H. F. Fisher, E. E. Conn, B. Vennesland, and F. H. Westheimer, *JBC* **202**, 687 (1953).
25. F. A. Loewus and H. A. Stafford, *JBC* **235**, 3317 (1960).
26. D. Dennis and N. O. Kaplan, *JBC* **235**, 810 (1960).
27. F. A. Loewus, H. R. Levy, and B. Vennesland, *JBC* **223**, 589 (1959).
28. J.-F. Biellmann and N. Rosenheimer, *FEBS* (*Fed. Eur. Biochem. Soc.*) *Lett.* **34**, 143 (1973).
29. S. Shifrin and N. O. Kaplan, *Nature* (*London*) **183**, 1529 (1959).
30. B. M. Anderson and N. O. Kaplan, *JBC* **234**, 1226 (1959).
31. K. Burton and T. H. Wilson, *BJ* **54**, 86 (1953).
32. G. L. Long and N. O. Kaplan, *Science* **162**, 685 (1968).
33. M. I. Siegel, P. W. Sciulli, and B. Rosenberg-Bassin, *Comp. Biochem. Physiol. B* **45**, 613 (1973).
34. R. K. Selander and S. Y. Yang, *Science* **169**, 179 (1970).

The NAD^+ binding structure found in LDH occurs frequently in other dehydrogenases and other proteins (*35*). Thus a careful study of LDH can provide a basis for the understanding of the structures and mechanisms of related enzymes. In LDH the problem of catalysis is presented in stark simplicity. The complications of metal ions (as in liver alcohol dehydrogenase), linked substrate phosphorylation (as in glyceraldehyde-3-phosphate dehydrogenase), or of ammonia uptake (as in glutamate dehydrogenase) are absent. Indeed, LDH is the only simpler dehydrogenase where both structure and sequence are known at this time.

C. Isozymes

1. *General*

The concept of multiple molecular forms (isozymes) of LDH by Markert and Møller (*11*) has stimulated many investigations into the nature, function, and control of isozymes (*36,37*). Although there are only two major structural genes (corresponding to the M and H chains), there is a complex variety of other LDH genes which can be expressed in some tissues at certain stages of development. These isozymes are distinguishable by their chemical composition and their kinetic and immunological properties (*38–40*). The difference in isozyme composition in various tissues has been correlated to different metabolic requirements (*18,41–43*). The known genes have been related to a phylogenetic tree (*44*).

2. *Known Genes*

a. The M and H Loci: Regulation and Properties of Isozymes. Most vertebrates are known to possess M and H type of isozymes although their relative proportion varies among different tissues; for instance, in rat heart tissue 78% of the polypeptide chains are of the H variety as

35. M. G. Rossmann, A. Liljas, C.-I. Brändén, and L. J. Banaszak, Chapter 2. this volume.
36. R. D. Cahn, N. O. Kaplan, L. Levine, and E. Zwilling, *Science* **136**, 962 (1962).
37. C. L. Markert and G. S. Whitt, *Experientia* **24**, 977 (1968).
38. J. S. Nisselbaum and O. Bodansky, *JBC* **238**, 969 (1963).
39. T. P. Fondy, A. Pesce, I. Freedberg, F. E. Stolzenbach, and N. O. Kaplan, *Biochemistry* **3**, 522 (1964).
40. N. O. Kaplan, *in* "Evolving Genes and Proteins" (V. Bryson and H. J. Vogel, eds.), p. 190. Academic Press, New York, 1968.
41. A. C. Wilson, R. D. Cahn, and N. O. Kaplan, *Nature* (*London*) **197**, 331 (1963).
42. G. Hathaway and R. S. Criddle, *Proc. Nat. Acad. Sci. U. S.* **56**, 680 (1966).
43. E. Goldberg and C. Hawtrey, *J. Exp. Zool.* **165**, 309 (1967).
44. R. S. Holmes, *FEBS* (*Fed. Eur. Biochem. Soc.*) *Lett.* **28**, 51 (1972).

opposed to only 11% in rat leg muscle tissue (*45*). There is a preference for dimer formation in the lower organisms such as tapeworm and crab although the monomer weight remains roughly constant at around 35,000. In tapeworm a single LDH, akin to the H type, has been found (*46*). Both types are present in various crabs (*47,48*). Only one type of LDH has been found in *E. coli* (*49,50*).

Hybridization studies between combinations of H and M subunits *in vitro* produce a binomial distribution of the five possible isozymes; thus, for equal quantities of H and M subunits isozymes are formed in the proportion 1:4:6:4:1 (*51,52*). Subunit interchange can be induced by repeated freezing and thawing, dilution, or by mild denaturation. With increasing evolutionary distance the hybridization process becomes more difficult and the results more variable. It has been concluded (*52*) that the M chains from different species are more like each other than they are to the H chains of the same species. These conclusions are supported by chemical and immunological results (*20,39,40,53,54*). An antiserum against the M_4 isozyme of one species will cross react with M_4 isozyme from several species but not with its own H_4 isozyme (*36,54*). The H_4 isozyme is a much poorer immunogen than the M_4 isozyme (*55*), and in some cases anti-H_4 serum will cross react with M_4 isozymes of other species (*53,56,57*).

Identification of equivalent subunits in different species is not always straightforward. Thus, while in most cases the H_4 isozyme migrates fastest to the anode during electrophoresis, there can be a reversal of the direction of migration as in pigeon (*58*) and newt (*59*) tissues. Identification with respect to other species must then refer to the isozyme domi-

45. I. H. Fine, N. O. Kaplan, and D. Kuftinec, *Biochemistry* **2,** 116 (1963).
46. W. F. Burke, R. W. Gracy, and B. G. Harris, *Comp. Biochem. Physiol. B.* **43,** 345 (1972).
47. G. L. Long and N. O. Kaplan, *ABB* **154,** 696 (1973).
48. G. L. Long and N. O. Kaplan, *ABB* **154,** 711 (1973).
49. E. M. Tarmy and N. O. Kaplan, *JBC* **243,** 2579 (1968).
50. E. M. Tarmy and N. O. Kaplan, *JBC* **243,** 2587 (1968).
51. S. N. Salthe, O. P. Chilson, and N. O. Kaplan, *Nature* (*London*) **207,** 723 (1965).
52. O. P. Chilson, L. A. Costello, and N. O. Kaplan, *Biochemistry* **4,** 271 (1965).
53. A. Pesce, R. H. McKay, F. E. Stolzenbach, R. D. Cahn, and N. O. Kaplan, *JBC* **239,** 1753 (1964).
54. C. L. Markert and E. Appella, *Ann. N. Y. Acad. Sci.* **103,** 915 (1963).
55. J. F. Burd and M. Ustegui-Gomez, *BBA* **310,** 238 (1973).
56. S. Avrameas and K. Rajewski, *Nature* (*London*) **201,** 405 (1964).
57. K. Rajewski, S. Avrameas, P. Grabar, G. Pfleiderer, and E. D. Wachsmuth, *BBA* **92,** 248 (1964).
58. R. G. Rose and A. C. Wilson, *Science* **153,** 1411 (1966).
59. D. A. Wright and F. H. Moyer, *Comp. Biochem. Physiol. B* **44,** 1011 (1973).

nance in the specific tissue and its properties in relation to those of other species.

The expression of different genes during ontogeny has been documented (*60–66*). In early stages of the embryo the *H* and *M* genes are expressed at the same rate but, as organs differentiate, their isozyme distribution tends toward that of the adult tissue. The distribution of LDH isozymes can also be altered by temperature (*33*) or by variation of available oxygen (*67–69*). The LDH in serum responds in quantity and isozyme distribution to the presence of anaerobic tumors (*46*) and organ transplants (*70*).

Various reasons for the advantages gained in the specialization of major LDH isozymes have been proposed (*68,69,71–74*). One possibility may be in the greater degree of pyruvate inhibition of the H_4 isozyme (*1,36,69*) as a result of the formation of the abortive ternary LDH:NAD–pyruvate complex (see Section III,C,3).

b. The C Locus. LDH-X has been found in the testes and spermatozoa of mammals (*75–77*) and birds (*78,79*). Its presence can only be detected in postpubertal testes (*43,80,81*) and is under hormonal control (*82*). When injected into female rabbits it can cause termination of pregnancy

60. B. Prochazka and E. D. Wachsmuth, *J. Exp. Zool.* **182,** 201 (1972).
61. G. Pfleiderer and E. D. Wachsmuth, *Biochem. Z.* **334,** 185 (1961).
62. E. D. Wachsmuth, G. Pfleiderer, and T. Wieland, *Biochem. Z.* **340,** 80 (1964).
62a. E. D. Wachsmuth, *Klin. Wochensch.* **43,** 69 (1965).
63. B. Karlsson and L. Palmer, *Comp. Biochem. Physiol. B* **38,** 299 (1971).
64. E. Appella and C. L. Markert, *BBRC* **6,** 171 (1961).
65. C. L. Markert and H. Ursprung, *Develop. Biol.* **5,** 363 (1962).
66. D. T. Lindsay, *J. Exp. Zool.* **152,** 75 (1963).
67. S. Lindy and M. Rajasalmi, *Science* **153,** 1401 (1966).
68. D. Dawson, T. Goodfriend, and N. O. Kaplan, *Science* **143,** 929 (1964).
69. N. O. Kaplan, J. Everse, and J. Admiraal, *Ann. N. Y. Acad. Sci.* **151,** 400 (1968).
70. J. J. Nora, D. A. Cooley, B. L. Johnson, S. C. Watson, and J. D. Milam, *Science* **164,** 1079 (1969).
71. E. S. Vesell and P. E. Pool, *Proc. Nat. Acad. Sci. U. S.* **55,** 756 (1966).
72. T. Wuntch, R. F. Chen, and E. S. Vesell, *Science* **167,** 63 (1970).
73. T. Wuntch, E. S. Vesell, and R. F. Chen, *JBC* **244,** 6100 (1969).
74. T. Wuntch, R. F. Chen, and E. S. Vesell, *Science* **169,** 480 (1970).
75. A. Blanco and W. H. Zinkham, *Science* **139,** 601 (1963).
76. E. Goldberg, *Science* **139,** 602 (1963).
77. R. Stambaugh and J. Buckley, *JBC* **242,** 4053 (1967).
78. W. H. Zinkham, *Science* **164,** 185 (1969).
79. W. H. Zinkham, A. Blanco, and L. Kupchyk, *Science* **144,** 1353 (1964).
80. C. O. Hawtrey and E. Goldberg, *Ann. N. Y. Acad. Sci.* **151,** 611 (1968).
81. R. Lindahl and K. Mayeda, *Comp. Biochem. Physiol. B* **45,** 265 (1973).
82. A. D. Winer and B. Nikitovitch-Winer, *FEBS (Fed. Eur. Biochem. Soc.) Lett.* **16,** 21 (1971).

(*83*). This isozyme is chemically and immunologically distinct from the more widely distributed forms of LDH (*84,85*); however, its subunits (designated C) can form enzymically active hybrids with H *in vitro* and with M *in vivo* and *in vitro* (*86*). It has been isolated from mouse testes in a crystalline form (*85*) that is suitable for X-ray diffraction studies. (*87*).

c. The E Locus. The distinctive E isozyme (*88,89*) is synthesized in cells of the teleost (bony fish) nervous tissue, particularly in the photoreceptor cells. The LDH *E* gene appears to have arisen from a duplication of the LDH *H* gene near the time of the adaptive radiation of the teleosts (*90*). The kinetic properties of the E_4 isozyme coupled with its developmental and cellular specificity suggest that it may play an important role in visual metabolism (*88,91*). This locus is referred to as *C* by Shaklee *et al.* (*92*).

d. The F Locus. Studies on LDH from gadoid fish (e.g., haddock and cod) have shown the presence of an LDH isozyme specific to liver tissue. Genetic and evolutionary variation of the LDH isozyme in these fish indicate that it is encoded at a separate *F* gene (*93*).

e. The M′ and H′ Loci. Cytological and biochemical studies have verified the existence of duplicate *M* (*M* and *M′*) and *H* (*H* and *H′*) loci in salmonoid fish (trout) (*94*). A diferential synthesis of M and M′ subunits takes place during embryonic development of trout. During early stages of embryogenesis, M′ subunits predominate, but as development progresses, increasing amounts of M_4 LDH can be identified (*95*).

3. *Evolution of LDH Genes*

Similarity of physicochemical properties, ability to hybridize, and immunological data, as well as the occurrence of certain genes only in re-

83. E. Goldberg, *Science* **181,** 458 (1973).
84. E. Goldberg, *Proc. Nat. Acad. Sci. U. S.* **68,** 349 (1971).
85. E. Goldberg, *JBC* **247,** 2044 (1972).
86. E. Goldberg, *ABB* **109,** 134 (1965); T. E. Wheat and E. Goldberg *in* "Isozymes" (C. L. Markert, ed.), Vol. 3, pp. 325–346. Academic Press, 1975 (in press).
87. A. D. Adams, M. J. Adams, M. G. Rossmann, and E. Goldberg, *JMB* **78,** 721 (1973).
88. G. S. Whitt, *J. Exp. Zool.* **175,** 1 (1970).
89. G. S. Whitt, W. F. Childers, and T. E. Wheat, *Biochem. Genet.* **5,** 257 (1971).
90. J. J. Horowitz and G. S. Whitt, *J. Exp. Zool.* **180,** 13 (1972).
91. G. S. Whitt and G. M. Booth, *J. Exp. Zool.* **174,** 215 (1970).
92. J. B. Shaklee, K. L. Kepes, and G. S. Whitt, *J. Exp. Zool.* **185,** 217 (1973).
93. G. F. Sensabaugh, Jr. and N. O. Kaplan, *JBC* **247,** 585 (1972).
94. T. Wuntch and E. Goldberg, *J. Exp. Zool.* **174,** 233 (1970).
95. E. Goldberg, J. P. Cuerrier, and J. L. Ward, *Biochem. Genet.* **2,** 335 (1969).

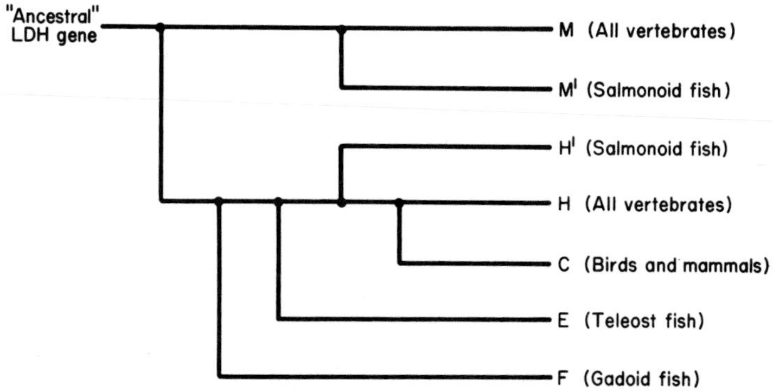

FIG. 2. Evolution of vertebrate lactate dehydrogenase isozymes.

lated species, have made it possible to suggest an evolutionary tree relating all known vertebrate LDH isozymes (*44*) (Fig. 2). Whitt and his colleagues (*90*) were able to show that the E_4 isozyme appears to be absent in all orders of fish except the teleosts.

Gene duplication clearly plays an important role in the evolution of LDH as it does in the evolution of dehydrogenases in general. It may be caused either by unequal crossover of chromosomes or by chromosomal duplication.

4. *D-Lactate-Specific Enzymes*

Some lower organisms contain D-lactate-specific LDH. In a survey of Anthropoidea, Mollusca, Annelida, and Coelenterata, Long and Kaplan (*32*) showed that while all species have LDH, they never have both D- and L-specific forms.

If the structural interpretation of L-specificity in dogfish LDH is correct (see Section II,A,3,f,i), namely, that specificity is provided by the ligands arginine-171, histidine-195, and the C-4 atom of the nicotinamide ring, then some change in structure or mode of substrate binding might be anticipated in D-specific enzymes. In the horseshoe crab, arginine-171 is probably replaced by lysine, and this might give rise to other modes of interaction with the substrate. Possibly this may be reflected in the preference for dimers in LDH from horseshoe crab (*34,47,48*) and fungi (*96*). However, LDH from *E. coli* is probably tetrameric (*49,50*).

D. ASSAY

The enzyme is usually assayed by following the rate of decrease of absorbance at 340 nm as NADH is oxidized by pyruvate (*10,97*). Al-

96. H. B. LéJohn, *JBC* **246,** 2116 (1971).
97. A. Kornberg, "Methods in Enzymology," Vol. 1, p. 441, 1955.

though the NADH concentration is not critical, the pyruvate concentration must be chosen carefully. Inhibition at high pyruvate concentration (*10*) leads to an optimal value which varies with pH (*98*), isozyme, temperature (*99*), and species. At pH 7.2 and 25°, 0.3 mM pyruvate is normally suitable for the H_4 isozyme assay and 2 mM pyruvate for the M_4 isozyme. Such an assay medium can also be used to specifically reveal the enzyme after electrophoresis (*14,100*).

The assay based on the observation of NADH production from lactate and NAD^+ has also been used (*101*). The NADH can be detected from its ultraviolet spectrum, but may also be coupled with phenazine methosulfate and a tetrazolium salt to give a blue color suitable for automated analysis and histological stains (*102*). The equilibrium is unfavorable at neutral pH and the reaction is best carried out at pH 9–10. The reaction can also be followed by observing the proton uptake in a pH-stat (*103*).

The enzyme from lobster tail muscle, when assayed by lactate reduction of NAD^+, is exceptional in requiring the presence of NADH for full activity (*104*). The enzyme from *Bacillus subtilis* has a specific enzymic activity which depends upon protein concentration (*105*). Enzymes from *E. coli* and other microorganisms can be activated by pyruvate (*50*) or nucleotide triphosphates (*96*).

E. Isolation

Lactate dehydrogenase is found in the cytoplasm and can be liberated into solution when the cells are broken with a homogenizer or by osmotic shock. The procedure introduced by Straub (*9*) consisted of extraction of the enzyme, adsorbing onto calcium phosphate gel, and fractionating in a series of acetone and ammonium sulfate steps. The method of Jecsai (*106*) gives crystalline M_4 enzyme from pig skeletal muscle simply by repeated ammonium sulfate fractionation. Although much refinement was achieved with these methods, and multienzyme isolation procedures were

98. A. D. Winer and G. W. Schwert, *JBC* **231,** 1065 (1958).
99. P. G. W. Plagemann, K. F. Gregory, and F. Wroblewski, *Biochem. Z.* **334,** 37 (1961).
100. T. Wieland and G. Pfleiderer, *Angew. Chem.* **69,** 199 (1957).
101. J. B. Neilands, "Methods in Enzymology," Vol. 1, p. 449, 1955.
102. M. M. Nachlas, D. G. Walker, and A. M. Seligman, *JBBC* **4,** 29 (1958).
103. J. J. Holbrook and R. A. Stinson, *BJ* **131,** 739 (1973).
104. H. Kaloustian, F. E. Stolzenbach, J. Everse, and N. O. Kaplan, *JBC* **244,** 2891 (1969).
105. A. Yoshida and E. Freese, *BBA* **99,** 56 (1965).
106. G. Jecsai, *Acta Physiol. Hung.* **20,** 341 (1962).

developed (*107*), the most usual methods of isolation are now based upon ion-exchange chromatography on cellulose or other supports. Since the isozymes differ greatly in their charge, a preliminary investigation by starch gel electrophoresis is desirable (*108*). These techniques were the basis of the isolation of sixteen H_4 or M_4 enzymes reported by Pesce *et al.* (*20*).

Affinity chromatography promises to provide methods for specifically isolating LDH. The behavior of pig H_4 and M_4 enzyme has been examined on chemically well-defined columns in which either AMP or NAD^+ is linked via a 5–6 Å-long spacer arm of, for example, 6-aminohexanoic acid to Sepharose (*109–117*). Such columns would be expected to absorb either adenine nucleotide binding enzymes or dehydrogenases as a class. However, separation can be achieved by choosing the correct nucleotide to displace a specific enzyme; for example, a mixture of glyceraldehyde-3-phosphate dehydrogenase and LDH can be fractionated in a column of Sepharose-6-aminohexanoyl-NAD^+ by eluting the former with NAD^+ and the latter with NADH (*110*). Many of the properties of the enzymes bound to the column suggest that the adsorption involves binding of the nucleotide at the enzyme's active site; for example, pig H_4 and pig M_4 are both bound at a pH where they have opposite surface charge. Furthermore, the interaction of the bound NAD^+ is strengthened by sulfite (*112*). The order of elution of the five pig isozymes with a gradient of NADH (*114*) correlates with the relative affinity of the isozymes for NADH in solution (*118*).

A yet more specific isolation procedure is based on an affinity column

107. G. Beisenherz, H. J. Boltze, T. Bücher, R. Czok, K. H. Garbade, E. Meyer-Arendt, and G. Pfleiderer, *Z. Naturforsch. B* **8**, 555 (1953).
108. I. H. Fine and L. A. Costello, "Methods in Enzymology," Vol. 6, p. 958, 1963.
109. K. Mosbach, H. Guilford, P.-O. Larsson, R. Ohlsson, and M. Scott, *BJ* **125**, 20P (1971).
110. K. Mosbach, H. Guilford, R. Ohlsson, and M. Scott, *BJ* **127**, 625 (1972).
111. R. Ohlsson, P. Brodelius, and K. Mosbach, *FEBS* (*Fed. Eur. Biochem. Soc.*) *Lett.* **25**, 234 (1972).
112. C. R. Lowe and P. D. G. Dean, *BJ* **133**, 515 (1973).
113. C. R. Lowe, M. J. Harvey, D. B. Craven, and P. D. G. Dean, *BJ* **133**, 499 (1973).
114. P. Brodelius and K. Mosbach, *FEBS* (*Fed. Eur. Biochem. Soc.*) *Lett.* **35**, 223 (1973).
115. N. O. Kaplan, J. Everse, J. E. Dixon, F. E. Stolzenbach, C. Lee, C. T. Lee, S. S. Taylor, and F. Mosbach, *Proc. Nat. Acad. Sci. U. S.* **71**, 3450 (1974).
116. F. Widmer, J. E. Dixon, and N. O. Kaplan, *Anal. Biochem.* **55**, 282 (1973).
117. J. E. Dixon, F. E. Stolzenbach, J. T. Berenson, and N. O. Kaplan, *BBRC* **52**, 905 (1973).
118. R. A. Stinson and J. J. Holbrook, *BJ* **131**, 719 (1973).

carrying bound substrate analog. Substantial amounts of pig H_4 enzyme have been purified on a column in which oxamate is bound via a 6-aminohexanoly Sepharose and eluted with a decreasing gradient of NADH (*119,120*). In contrast to the usual preparations of pig H_4 and beef H_4 enzymes, material prepared in this way consistently gives large single crystals. In this respect it should be mentioned that pig H_4 enzyme contains tightly bound nucleotide which can only be removed by charcoal treatment or a yet tighter binding analog such as NAD-sulfite (*121*).

By far the most stable preparations of the enzyme are those linked to an insoluble matrix. LDH bound to amino glass with glutaraldehyde can be stored active for 35 days at pH 3.2 (*117*).

II. Molecular Properties

A. Structure

1. *Composition*

The amino acid compositions of a number of lactate dehydrogenases (H_4, M_4, and X_4) are given in Table II. Chemical modification of one cysteine per subunit (*122*) results in loss of activity (see Section II,B). Hence, cysteine in LDH has received particular attention. No disulfide bridges have been reported for LDH (*123*). The enzyme does not contain any metals (*53,124*).

2. *Amino Acid Sequence*

The complete sequence of dogfish M_4 LDH, consisting of **329** amino acid residues, has been determined (*125,126*) and is shown in Fig. **3**. Numbering in Fig. **3** was introduced by Rossmann *et al.* (*127*) and will be used throughout this article even though a few discontinuities occur.

119. W. Eventoff, K. W. Olsen, and M. L. Hackert, *BBA* **341,** 327 (1974).
120. P. O'Carra and S. Barry, *FEBS* (*Fed. Eur. Biochem. Soc.*) *Lett.* **21,** 281 (1972).
121. T. Wieland, P. Duesberg, G. Pfleiderer, A. Stock, and E. Sann, *ABB, Suppl.* **1,** 260 (1962).
121a. G. R. Jago, L. W. Nichol, K. O'Dea, and W. H. Sawyer, *BBA* **250,** 271 (1971).
122. S. F. Velick, *JBC* **233,** 1455 (1958).
123. G. DiSabato, A. Pesce, and N. O. Kaplan, *BBA* **77,** 135 (1963).
124. G. Pfleiderer, D. Jeckel, and T. Wieland, *Biochem. Z.* **330,** 296 (1959).
125. S. S. Taylor, S. S. Oxley, W. S. Allison, and N. O. Kaplan, *Proc. Nat. Acad. Sci U. S.* **70,** 1790 (1973).
126. S. S. Taylor, private communication (1975) and *JBC* (1975) (submitted for publication).
127. M. G. Rossmann, M. J. Adams, M. Buehner, G. O. Ford, M. L. Hackert, P. J. Lentz, Jr., A. McPherson, Jr., R. W. Schevitz, and I. E. Smiley, *Cold Spring Harbor Symp. Quant. Biol.* **36,** 179 (1971).

TABLE II

AMINO ACID COMPOSITION OF SOME LACTATE DEHYDROGENASES

Residue	H			M					LDH-X	Bacterial LDH			D-Lactate LDH
										L-Lactate		D-Lactate	
	Beef (Ref. *20*)	Chicken (Ref. *20*)	Pig (Ref. *62*)	Beef (Ref. *20*)	Chicken (Ref. *20*)	Pig (Ref. *62*)	Dogfish (Ref. *20*)[b]	Dogfish (Refs. *125,126*)[c]	Mouse (Ref. *85*)	*B. subtilis* (Ref. *105*)	*S. cremoris* (Ref. *121a*)	*E. coli* (Ref. *49*)	Horseshoe crab (Ref. *47*)
Lysine	26	25	23	26	28	25	30	29	25	23	22	17	22
Histidine	7	8	7	8	16	12	11	11	7	9	7	8	7
Arginine	9	9	9	11	11	11	7	9	11	6	10	18	11
Aspartic acid	33	32	34	32	31	31	31	24	36	39	36	30	37
(asparagine)								14					
Threonine	14	19	15	12	13	12	14	12	18	17	16	16	17
Serine	23	27	24	22	28	20	25	26	23	18	18	16	18
Glutamic acid	33	31	31	30	26	26	24	14	25	33	36	34	38
(glutamine)								9					
Proline	12	10	12	13	11	13	10	10	12	11	9	12	20
Glycine	24	24	24	25	26	25	24	24	34	31	24	22	23
Alanine	19	22	19	20	20	19	19	19	20	32	37	26	27
Half-cystine (Ref. *131*)	4	6	4	6	7	5	10	7	4	4	1	3	5
Valine	32	31	37	29	30	35	31	34	36	29	33	22	27
Methionine	9	6	9	8	8	8	10	11	6	6	6	8	16
Isoleucine	23	17	21	23	21	23	20	20	21	20	19	17	19
Leucine	35	37	34	34	30	35	33	35	39	25	25	32	34
Tyrosine	7	8	7	7	5	7	6	7	6	13	10	12	9
Phenylalanine	5	5	5	7	7	7	6	7	8	13	13	12	14
Tryptophan	?	?	?	?	?	?	6	7	?	2	2	2–3	5
Sum	315	317	315	313	318	314	313	329	331	331	324	328	349

[a] The values are calculated on the basis of a subunit molecular weight of 35,000.

[b] Refers to experimental amino acid composition.

[c] Refers to the composition based on sequence.

ACETYL

THR – ALA – LEU – LYS – ASP – LYS – LEU – ILE – GLY – HIS – 10 LEU – ALA – THR – SER – GLN – GLU – PRO – ARG – SER – 20 TYR – – ASN – LYS – ILE – THR – VAL – VAL – GLY – CYS –

30 ASX – ALA – VAL – GLY – MET – ALA – ASP – ALA – ILE – SER – 40 VAL – LEU – MET – LYS – ASP – LEU – ALA – ASP – GLU – VAL – 50 ALA – LEU – VAL – ASP – VAL – MET – GLU – ASP – LYS – LEU –

60 LYS – GLY – GLU – MET – MET – ASP – LEU – GLN – HIS – GLY – 70 SER – LEU – PHE – LEU – HIS – THR – ALA – LYS – ILE – VAL – 80 SER – GLY – – LYS – ASP – TYR – SER – VAL – SER – ALA –

90 GLY – SER – LYS – LEU – VAL – VAL – ILE – THR – ALA – GLY – 100 ALA – ARG – GLN – GLN – – GLU – GLY – GLU – SER – ARG – 110 LEU – ASN – LEU – VAL – GLN – ARG – ASN – VAL – ASN – ILE –

120 PHE – LYS – PHE – ILE – ILE – PRO – ASP – ILE – VAL – LYS – 130 HIS – SER – 132A PRO – 132B ASP – CYS – (ILE – LEU – LYS – GLU – LEU – HIS – PRO – 140 GLU – LEU – GLY – THR – ASN – LYS) – ASP – LYS – ASN – ASP –

150 TRP – LYS – LEU – SER – GLY – LEU – PRO – MET – HIS – ARG – 160 ILE – ILE – GLY – SER – GLY – CYS – ASN – LEU – ASP – SER – 170 ALA – ARG – PHE – ARG – TYR – LEU – MET – GLY – GLU – ARG –

180 LEU – GLY – VAL – HIS – SER – CYS – (SER, GLY, VAL, LEU) – 190 TRP – VAL – ILE – GLY – GLU – HIS – GLY – ASP – SER – VAL – 200 PRO – SER – VAL – TRP – SER – GLY – MET – TRP – ASN – ALA –

210 LYS – LEU – HIS – LYS – ASP – VAL – VAL – ASP – SER – ALA – 220 TYR – GLU – VAL – ILE – LYS – LEU – LYS – GLY – TYR – THR – 230 SER – TRP – ALA – ILE – GLY – LEU – VAL – VAL – SER – ASN –

240 PRO – VAL – ASP – VAL – LEU – THR – TYR – VAL – ALA – TRP – 250 LYS – GLY – CYS – SER – VAL – ALA – ASP – LEU – ALA – GLU – 260 THR – ILE – MET – LYS – ASN – LEU – CYS – ARG – VAL – HIS –

270 PRO – VAL – SER – THR – MET – VAL – LYS – ASP – PHE – TYR – 280 GLY – ILE – LYS – ASP – ASN – VAL – PHE – LEU – SER – LEU – 290 PRO – CYS – VAL – LEU – ASN – ASP – HIS – GLY – ILE – SER –

300 – ASN – ILE – VAL – LYS – MET – LYS – LEU – LYS – PRO – 310 ASN – GLU – GLU – GLN – GLN – LEU – GLN – LYS – SER – ALA – 320 THR – THR – LEU – TRP – ASP – ILE – GLN – LYS – ASP – LEU – 330 LYS – PHE

FIG. 3. Amino acid sequence of dogfish M_4 LDH (*126*). Some doubt remains in the two sections enclosed by parentheses. Until these are removed the amino acid numbering will be that introduced by Rossmann *et al.* (*127*). There is no crystallographic evidence for residue Ile between Cys 133 and Leu 134.

Various N- and C-terminal sequences (*128–130*) have been determined (Fig. 4), and show good homology. However, residues from 298 to 315 in pig H_4 LDH show little homology with the dogfish M_4 LDH sequence and are difficult to accommodate in the dogfish LDH structure. Heart enzymes may all have one extra residue at the C-terminus compared to the muscle enzyme. Arginine peptides isolated from pig M_4 and H_4 as well as from chicken M_4 (*130*) are in complete agreement with the sequence from arginine-101 to arginine-115 in the dogfish M_4 enzyme. Sequences of essential cysteine peptides (*47,131–134*) are given in Fig. 5. Possible homologies, where only composition is known, are also indicated.

Many types of reagents have been used to identify residues in the active site of the enzyme (see Section II,B). Woenckhaus *et al.* (*135*) have modified an essential histidine residue in the pig H_4 enzyme and isolated the labeled peptide. The sequence of this peptide is in very close agreement with residues 191–203 in the dogfish M_4 enzyme (Fig. 6).

3. *Three-Dimensional Structure*

a. Crystals and Electron Density Maps. Crystallization of this enzyme has usually been from ammonium sulfate. Pesce *et al.* (*20*) summarize the crystallization of lactate dehydrogenases from a number of species. Crystals of three kinds of trout M_4 LDH were obtained by Wuntch and Goldberg (*94*). Goldberg (*85*) also crystallized mouse LDH-X. Some properties of LDH crystals are listed in Table III (*136*; also see

128. L. D. Stegink, B. M. Sanborn, M. C. Brummel, and C. S. Vestling, *BBA* **251,** 31 (1971).
129. K. Mella, J. J. Torff, E. T. J. Folsche, and G. Pfleiderer, *Hoppe-Seyler's Z. Physiol. Chem.* **350,** 23 (1969).
130. G. Pfleiderer, C. Woenckhaus, D. Jeckel, and K. Mella, *in* "Pyridine Nucleotide-Dependent Dehydrogenases" (H. Sund, ed.), p. 145. Springer-Verlag, Berlin and New York, 1970.
131. J. J. Holbrook, G. Pfleiderer, K. Mella, M. Valz, W. Lescowac, and R. Jeckel, *Eur. J. Biochem.* **1,** 476 (1967).
132. T. P. Fondy, J. Everse, G. A. Driscoll, F. Castillo, F. E. Stolzenbach, and N. O. Kaplan, *JBC* **240,** 4219 (1965).
133. J. J. Holbrook and G. Pfleiderer, *Biochem. Z.* **342,** 111 (1965).
134. W. S. Allison, J. Admiraal, and N. O. Kaplan, *JBC* **244,** 4743 (1969).
135. C. Woenckhaus, J. Berghäuser, and G. Pfleiderer, *Hoppe-Seyler's Z. Physiol. Chem.* **350,** 473 (1969).
136. M. J. Adams, M. Buehner, K. Chandrasekhar, G. C. Ford, M. L. Hackert, A. Liljas, P. J. Lentz, Jr., S. T. Rao, M. G. Rossmann, I. E. Smiley, and J. L. White, *in* "Protein-Protein Interactions" (R. Jaenicke and E. Helmreich, eds.), p. 139. Springer-Verlag, Berlin and New York, 1972.

DOGFISH	M_4	N-ACETYL – THR – ALA – LEU
PIG	H_4	
PIG	M_4	
RAT	M_4	N-ACETYL – ALA – ALA
RABBIT	M_4	N-ACETYL – ALA – ALA
CHICKEN	H_4	N-ACETYL – ALA – THR
BEEF	H_4	N-ACETYL – ALA – THR

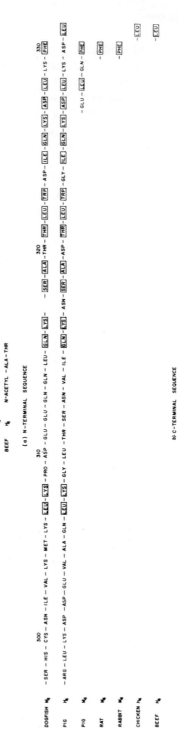

(a) N-TERMINAL SEQUENCE

(300, 310, 320, 330)

DOGFISH	M_4	– SER – HIS – CYS[300] – ASN – ILE – VAL – LYS – MET – LYS – [LEU] – [LYS] – PRO – ASP[310] – GLU – GLU – GLN – GLN – LEU – [GLN] – [LYS] – – [SER] – [ALA] – THR[320] – [THR] – [LEU] – [TRP] – ASP – [ILE] – [GLN] – [LYS] – [ASP] – [LEU] – LYS – [PHE][330]
PIG	H_4	– ARG – LEU – LYS – ASP – ASP – GLU – VAL – ALA – GLN – [LEU] – [LYS] – GLY – LEU – THR – SER – ASN – VAL – ILE – [GLN] – [LYS] – ASN – [SER] – [ALA] – ASP – [THR] – [LEU] – [TRP] – GLY – [ILE] – [GLN] – [LYS] – [ASP] – [LEU] – LYS – ASP – [LEU]
PIG	M_4	– GLU – [LEU] – GLN – [PHE]
RAT	M_4	– [PHE]
RABBIT	M_4	– [PHE]
CHICKEN	H_4	– [LEU]
BEEF	H_4	– [LEU]

b) C-TERMINAL SEQUENCE

FIG. 4. Various N- and C-terminal sequences. Amino acids which remain unchanged in similar isozymes have been enclosed in a box.

TABLE III
CRYSTALLINE FORMS OF LDH

Species	Isozyme	Ref.	Complex	Space group	Cell dimensions (Å)	Subunits/asymmetric unit	Directions of molecular 2-fold axis
Dogfish	M_4	*21*	Apo	F422	$a = 146.8$ $c = 155.35$	1	Molecular center at $\frac{1}{4}, \frac{1}{4}, \frac{1}{4}$, $P \parallel x$, $Q \parallel z$, $R \parallel y$
Dogfish	M_4	*136a*	With NAD^+ diffused	$C4_22_12$	$a = 146.9$ $c = 155.8$	2	Molecular center at $\frac{1}{4}, \frac{1}{4}, -\frac{1}{4}$, $Q \parallel z$
Dogfish	M_4	*136b*	With NAD^+ co-crystallized	I2	$a = 149.0$ $b = 149.0$ $c = 146.0$ $\beta = 98°$	4	Packing similar to M_4 apo dogfish
Dogfish	M_4	*136c*	With NAD-pyruvate or NAD:oxalate or NADH:oxamate	$C42_12$	$a = 134.5$ $c = 85.9$	1	Molecular center at $\frac{1}{2}, \frac{1}{2}, 0$, $P \parallel z$, $Q \parallel x$, $R \parallel y$
Pig	M_4	*136d*	With NAD-pyruvate or NADH:oxamate	$P22_12_1$	$a = 86.8$ $b = 60.8$ $c = 136.9$	2	$P \parallel x$, Q in yz plane 13° from z (noncrystallographic) molecular center at $x = 0.08$, $y = 0$, $z = 0$
Pig	H_4	*119,136d*	With NAD:oxalate or NADH:oxamate	C2	$a = 162.0$ $b = 60.7$ $c = 138.5$ $\beta = 93.2$	4	$P \parallel a^*$, Q in yz plane 12.5° from $-z$ (noncrystallographic) molecular center at $x = 0.29$, $y = 0.5$, $z = 0.25$
Pig	M_4	*19*	With NAD^+ co-crystallized	C2	$a = 143.6$ $b = 148.0$ $c = 84.3$ $\beta = 97°$	4	—
Chicken	M_4	*21*	Apo	F222	$a = 190.0$ $b = 108.0$ $c = 620.0$	8	—
Mouse	X	*87,136e*	Apo or holo	P1	$a = 84.8$ $b = 76.7$ $c = 63.9$ $\alpha = 109.7$ $\beta = 89.5°$ $\gamma = 96.5°$	4	P approximately $\parallel$ to x, Q in yz plane approximately 53° from z (both noncrystallographic). Choice of molecular center is arbitrary.

		165
Dogfish	M_4	- ILE - ILE - GLY - SER - GLY - CYS - ASN - LEU - ASP - SER - ALA - ARG -
Pig	H_4	- VAL - ILE - GLY - SER - GLY - CYS - ASN - LEU - ASP - SER - ALA - ARG -
Chicken	H_4	- VAL - ILE - GLY - SER - GLY - CYS - ASN - LEU - ASP - THR - ALA - ARG -
Bullfrog	M_4	(VAL, ILE, GLY, SER, GLY, CYS, ASX, LEU, ASX, SER, ALA, ARG)
Chicken	M_4	(VAL, ILE, GLY, SER, GLY, CYS, ASX, LEU, ASX, SER, ALA, ARG)
Beef	H_4	(VAL, ILE, GLY, SER, GLY, CYS, ASX, LEU, ASX, SER, ALA, ARG)
Horseshoe crab		(PHE)+(VAL, ILE, GLY, SER) (GLY, CYS, ASX, LEU, ASX, THR, ALA, LYS)
Halibut	M_4	(VAL, VAL, GLU, CYS, ASX, LEU, ASX, GLU, ALA LYS)

FIG. 5. Essential cysteine peptides. Sequences in parentheses are based only on composition and homology.

		195
Dogfish	M_4	- VAL - ILE - GLY - GLN - HIS - GLY - ASP - SER - VAL - PRO - SER - VAL - TRP -
Pig	H_4	- ILE - LEU GLY - GLU - HIS - GLY - ASP - SER - VAL - PRO - SER - VAL - TRP -

FIG. 6. Essential histidine peptides.

19,21,87,119,136a–136e). Table IV gives a summary of the existing and anticipated electron density maps from different lactate dehydrogenases (*22,126,136a,d,e,137–139d*). The molecule itself is best referred to by

136a. M. J. Adams, D. J. Haas, B. A. Jeffrey, A. McPherson, Jr., H. L. Mermall, M. G. Rossmann, R. W. Schevitz, and A. J. Wonacott, *JMB* **41**, 159 (1969).
136b. A. Liljas, private communication (1973).
136c. I. E. Smiley, R. Koekoek, M. J. Adams, and M. G. Rossmann, *JMB* **55**, 467 (1971).
136d. M. L. Hackert, G. C. Ford, and M. G. Rossmann, *JMB* **78**, 665 (1973).
136e. A. D. Adams and M. G. Rossmann, private communication (1974).
137. M. J. Adams, G. C. Ford, A. Liljas, and M. G. Rossmann, *BBRC* **53**, 46 (1973).
137a. P. M. Wassarman and P. J. Lentz, Jr., *JMB* **60**, 509 (1971).
138. K. Chandrasekhar, A. McPherson, Jr., M. J. Adams, and M. G. Rossmann, *JMB* **76**, 503 (1973).
139. M. J. Adams, A. Liljas, and M. G. Rossmann, *JMB* **76**, 519 (1973).
139a. M. J. Adams, M. Buehner, K. Chandrasekhar, G. C. Ford, M. L. Hackert, A. Liljas, M. G. Rossmann, I. E. Smiley, W. S. Allison, J. Everse, N. O. Kaplan, and S. S. Taylor, *Proc. Nat. Acad. Sci. U. S.* **70**, 1968 (1973).

TABLE IV
ANTICIPATED AND EXISTING ELECTRON DENSITY MAPS OF LDH

Independent structure determinations	Species	Ref.	Complex with	Nominal resolution[a] (Å)
A	Dogfish M_4	*136a,130*	Apo	2.0
	Dogfish M_4	*137a*	Tetraiodofluorescein	4.0
	Dogfish M_4	*138*	Adenosine	2.8
	Dogfish M_4	*138*	AMP	2.8
	Dogfish M_4	*138*	ADP	2.8
	Dogfish M_4	*139*	Citrate (pH 6.0 and 7.8)	2.8
	Dogfish M_4	*22,138*	NAD^+	5.0
B	Dogfish M_4	*136c,138,139a*	NAD-pyruvate	3.0
	Dogfish M_4	*139b*	NAD:oxalate	3.0
	Dogfish M_4	*139b*	NADH:oxamate	3.0 (2.0)
C	Pig M_4	*136d*	NADH:oxamate	6.0
	Pig M_4	*139c*	NAD-pyruvate	(3.0)
D	Pig H_4	*139d*	NADH:oxamate	6.0
		139d		(2.5)
E	Mouse X_4	*136e*	Apo or holo	7.0 (3.0)

[a] Values given in parentheses indicate that work is in progress.

means of an orthogonal (P,Q,R) system of axes associated with the structure rather than any arbitrary crystal cell. These axes [defined by Rossmann *et al.* (*140*) and restated in a previous chapter (*35*)] coincide with the molecular 2-fold axes. This nomenclature should not be confused with that of Levitzki (*141*).

The first high resolution electron density map of LDH was of the dogfish M_4 apoenzyme at 2.8 Å resolution (*22*) and was interpreted in terms of a continuous polypeptide chain. The peptides at the two termini were known. The dodecapeptide containing the essential thiol group was located with the cysteine in position 165. Later a peptide containing the

139b. J. L. White, M. Buehner, and M. G. Rossmann, private communication (1974).
139c. S. J. Steindel and M. G. Rossmann, private communication (1974).
139d. W. Eventoff, M. L. Hackert, and M. G. Rossmann, private communication (1974).
140. M. G. Rossmann, M. J. Adams, M. Buehner, G. C. Ford, M. L. Hackert, A. Liljas, S. T. Rao, L. J. Banaszak, E. Hill, D. Tsernoglou, and L. Webb, *JMB* **76**, 533 (1973).
141. A. Levitzki, *FEBS (Fed. Eur. Biochem. Soc.) Lett.* **24**, 301 (1972).

essential histidine was placed in the electron density with the histidine at position 195 (*127*). The amino acid sequence and its relationship to the three-dimensional structure at a nominal resolution of 2.0 Å was determined by continuous exchange of chemical (*125,126*) and crystallographic (*142*) information.

The three-dimensional structure of the ternary complex of the dogfish enzyme with NAD^+ and pyruvate is a separate and independent structure determination (*136c*). The two maps can be interpreted with the same overall fold of polypeptide chain and amino acid sequence. A list of α-carbon coordinates and dihedral angles for both the apo- and ternary complex structures is given in Table V.

b. Secondary Structure. The nomenclature used in describing the helices (αA, αB, . . . , αH) and the β structure (βA, βB, . . . , βH) is defined in Fig. 7 (*136*). A diagram of the main chain hydrogen bonding pattern of dogfish M_4 apo-LDH and also of its ternary complex with

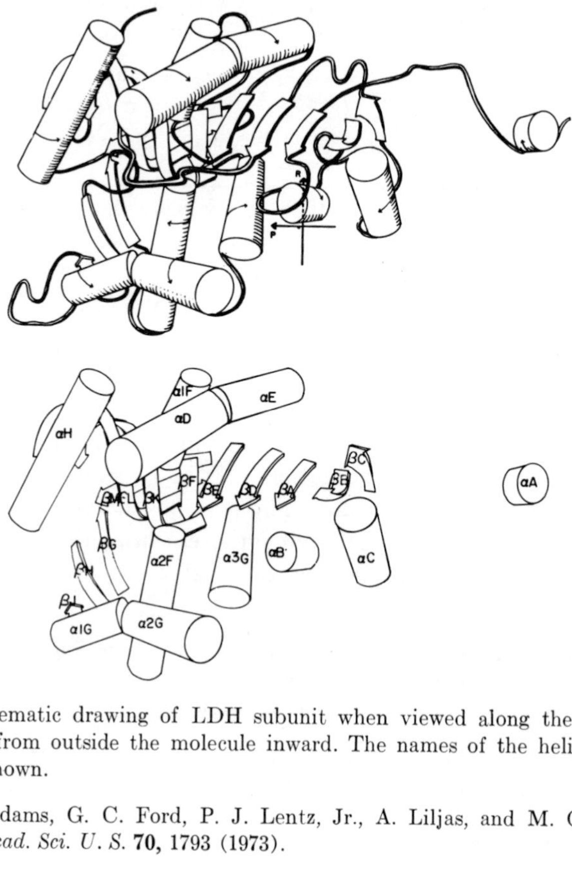

FIG. 7. Schematic drawing of LDH subunit when viewed along the molecular *Q* axis looking from outside the molecule inward. The names of the helical and sheet regions are shown.

142. M. J. Adams, G. C. Ford, P. J. Lentz, Jr., A. Liljas, and M. G. Rossmann, *Proc. Nat. Acad. Sci. U. S.* **70,** 1793 (1973).

NAD^+-pyruvate is shown in Fig. 8. The enzyme has an extensive amount of secondary structure. Ramachandran diagrams are shown in Fig. 9. Estimates of the proportion of amino acid residues located in the helices depend on the precise definition of secondary structure and vary around 40%. Three different β structures are found in LDH and account for around 23% of all residues. The amino terminal half of the molecule contains a six-stranded parallel sheet whereas the carboxy terminal half of the subunit has two three-stranded antiparallel β structures. The parallel sheet which is mainly in the interior of the molecule, exposes two edges to the solvent and one to a subunit interface. Thus, most of the residues in this structure are hydrophobic. The high concentration of β-branched residues such as valine, isoleucine, and threonine as well as the twist present in all observed β structures (*143*) can be seen in Fig. 10. The two small antiparallel β structures are fairly irregular but do have the same twist as the parallel sheet. Each β structure consists of two long and one short strand coiled around an imaginary axis to form a helix where the sense is left-handed on following the path of the polypeptide chain.

LDH also has a number of reverse turns (β bends, 3_{10}-bends). Modifications of the criteria given by Crawford *et al.* (*144*) and Lewis *et al.* (*145*) have been used to recognize the turns listed in Table VI. In some cases turns of type II have other residues than glycine in position three, contradicting the prediction by Venkatachalam (*146*) but in agreement with the observations of Crawford *et al.* (*144*).

c. *Folding of the Polypeptide Chain.* The domain structure (*147*) of the polypeptide chain of one subunit of dogfish M_4 LDH is illustrated in Fig. 11. A view from the same direction and another perpendicular to it are given as stereoscopic pairs in Fig. 12.

Subunit structure will be described in terms of (a) the amino terminal arm, (b) that part which *binds* the coenzyme, and (c) the part containing *catalytic* residues. The dinucleotide binding part is subdivided into two mononucleotide binding domains (B_1 and B_2) and these have been observed in other dehydrogenases (*35*). The catalytic part is also subdivided into two domains. The domain concept is illustrated by a distance plot (*147,148*) for LDH (Fig. 13) and discussed in a preceding review (*35*).

143. C. Chothia, *JMB* **75,** 295 (1973).
144. J. L. Crawford, W. N. Lipscomb, and C. G. Schellman, *Proc. Nat. Acad. Sci. U. S.* **70,** 538 (1973).
145. P. N. Lewis, F. A. Momany, and H. A. Scheraga, *BBA* **303,** 211 (1973).
146. C. M. Venkatachalam, *Biopolymers* **6,** 1425 (1968).
147. M. G. Rossmann and A. Liljas, *JMB* **85,** 177 (1974).
148. D. C. Phillips, *Biochem. Soc. Symp.* **30,** 11 (1970).

TABLE V

C-Alpha Coordinates and Dihedral Angles for Apo-Dogfish M_4 LDH and Dogfish M_4 LDH:NAD-Pyruvate

APO

		P	Q	R	PHI	TAU	PSI	CHI1	CHI2	CHI3	CHI4	CHI5
ACE		0.0	0.0	0.0	0	-57	0					
THR	1	-37.0	6.3	18.7	0	110	-49	-32				
ALA	2	-33.9	7.7	17.1	149	110	133					
LEU	3	-31.9	7.3	13.9	-58	101	-70	-179	14			
LYS	4	-28.8	6.6	15.9	-32	112	-55	-115	-160	89	156	
ASP	5	-30.7	4.1	18.0	-81	137	-17	-50	-60			
LYS	6	-33.2	1.9	16.0	-85	99	-16	-18	-135	-58	-115	
LEU	7	-30.0	1.1	14.1	-106	117	-56	-156	123			
ILE	8	-26.9	1.3	16.3	-88	113	71	-89	-30			
GLY	9	-28.0	-1.0	19.1	-76	101	-158					
HIS	10	-25.3	-2.8	21.1	-174	120	159	30	-80			
LEU	11	-24.0	-5.7	23.1	-91	105	-180	-33	-166			
ALA	12	-22.9	-5.6	26.7	-92	105	70					
THR	13	-19.4	-4.6	25.9	-124	113	-42	93				
SER	14	-19.0	-.9	26.2	9	112	96	-129				
GLN	15	-15.3	-.6	26.5	-131	99	133	158	163	-10		
GLU	16	-13.7	2.9	26.1	-65	118	140	-71	-3	24		
PRO	17	-11.8	3.8	23.0	0	0	143					
ARG	18	-8.0	3.6	23.4	-145	96	133	51	-127	123	98	4
SER	19	-5.8	4.6	20.5	-102	99	175	-120				
TYR	20	-2.3	3.2	20.1	-153	125	-29	-67	-74			
ASN	22	-.1	4.0	17.1	-124	82	87	-56	-47			
LYS	23	-1.2	7.6	17.6	-109	96	113	163	-168	-155	127	
ILE	24	1.0	9.6	15.3	-103	101	127	-116	119			
THR	25	.9	13.2	16.2	-115	112	148	-33				
VAL	26	2.0	16.3	14.2	-140	165	110					
VAL	27	2.4	19.9	15.4	128	145	104					
GLY	28	2.5	22.4	12.6	119	98	126					
CYS	29	-.1	22.4	9.9	-136	121	77	-66				
ASP	30	.7	23.8	6.4	179	95	-149	176	-39			
ALA	31	4.0	23.4	4.6	-8	124	-118					
VAL	32	6.6	21.1	6.0	-62	-161	108					
GLY	33	4.0	20.5	8.7	-20	115	-47					
MET	34	.7	20.1	6.9	-74	111	-18	163	61	62		
ALA	35	2.5	17.9	4.3	-71	107	-64					
ASP	36	3.7	15.2	6.7	-39	109	-42	-53	151			
ALA	37	.1	15.1	7.8	-59	113	-46					
ILE	38	-1.2	14.4	4.3	-57	102	-65	-174	173			
SER	39	1.5	11.7	4.0	-40	127	-69	-101				
VAL	40	1.4	9.5	7.1	-42	158	117					
LEU	41	-2.3	8.7	7.0	-49	106	-38	-53	124			
MET	42	-1.9	8.3	3.2	-57	105	-38	-152	160	-165		
LYS	43	.6	5.6	4.2	-76	108	-40	-90	-90	-137	-103	
ASP	44	-1.7	4.4	7.0	94	116	41	-130	2			
LEU	45	.5	4.3	10.1	-81	113	-70	-89	-28			
ALA	46	-1.5	5.9	12.9	-64	117	153					
ASP	47	-4.8	4.6	14.2	-112	103	27	72	38			
GLU	48	-5.7	8.1	15.3	170	94	133	139	130	-148		
VAL	49	-3.9	11.2	13.9	-127	-154	98					
ALA	50	-4.0	14.3	16.0	136	98	141					
LEU	51	-2.6	17.6	14.6	-129	108	124	90	82			
VAL	52	-1.6	20.5	16.8	-112	-151	115					
ASP	53	-.9	24.1	16.1	150	115	158	-170	80			
VAL	54	-.7	27.7	17.5	-92	107	125					
MET	55	-3.4	29.5	15.5	35	114	148	58	164	83		

TERN

		P	Q	R	PHI	TAU	PSI	CHI1	CHI2	CHI3	CHI4	CHI5
ACE		0.0	0.0	0.0	0	125	0					
THR	1	-35.8	8.7	19.8	0	110	-37	-73				
ALA	2	-33.1	6.1	19.0	-153	110	39					
LEU	3	-32.8	6.2	15.2	-49	111	-71	5	76			
LYS	4	-29.0	5.8	15.0	-50	101	-49	148	-154	125	171	
ASP	5	-29.7	3.6	18.0	-50	124	-50	-82	-117			
LYS	6	-32.2	1.1	16.8	-68	114	-24	-100	-109	74	-126	
LEU	7	-30.2	.4	13.7	-82	104	-41	-171	161			
ILE	8	-27.1	-.0	15.8	-120	119	96	24	146			
GLY	9	-27.4	-1.7	19.2	-91	110	172					
HIS	10	-24.6	-2.7	21.6	-111	110	-173	-2	-69			
LEU	11	-23.7	-5.9	23.5	-117	107	173	-23	129			
ALA	12	-23.6	-6.7	27.2	-76	112	-11					
THR	13	-19.8	-6.6	27.1	-67	117	-75	170				
SER	14	-18.7	-3.0	26.8	72	126	70	171				
GLN	15	-15.0	-2.4	26.4	-93	94	105	-172	131	6		
GLU	16	-15.3	1.4	26.5	-51	117	162	-68	-104	-84		
PRO	17	-13.9	3.8	23.8	0	0	160					
ARG	18	-10.2	4.5	23.8	-175	106	-124	7	127	-160	-160	-12
SER	19	-7.6	4.1	21.0	163	108	-161	-146				
TYR	20	-3.9	3.2	20.6	-159	115	-75	-85	-64			
ASN	22	-1.6	4.7	18.0	-85	118	114	-56	-90			
LYS	23	-1.2	8.5	18.2	-108	127	139	-85	-87	-104	-159	
ILE	24	.1	11.3	15.9	-129	111	130	-113	160			
THR	25	.4	15.0	16.8	-125	113	133	-58				
VAL	26	1.2	17.9	14.5	-120	174	117					
VAL	27	2.3	21.4	15.3	134	175	110					
GLY	28	1.9	24.1	12.6	117	114	73					
CYS	29	-1.1	23.4	10.3	-65	106	111	-56				
ASP	30	.3	24.5	6.9	173	114	-144	48	-138			
ALA	31	3.0	23.7	4.4	-54	107	-99					
VAL	32	5.5	21.5	6.2	-29	166	110					
GLY	33	2.6	20.4	8.4	-53	110	-66					
MET	34	-.1	20.1	5.7	-45	110	-49	-72	-17	-59		
ALA	35	2.5	18.3	3.6	-58	109	-49					
ASP	36	3.4	16.0	6.5	-55	102	-68	-82	71			
ALA	37	-.2	15.4	7.2	-37	114	-48					
ILE	38	-.9	14.7	3.6	-59	107	-45	-141	94			
SER	39	2.0	12.2	3.7	-66	114	-61	-54				
VAL	40	1.4	10.5	7.0	-64	143	115					
LEU	41	-2.4	10.6	6.7	-34	118	-27	-110	96			
MET	42	-2.7	9.0	3.3	-97	107	-26	106	153	12		
LYS	43	.2	6.7	3.9	-69	106	-46	157	175	-72	-105	
ASP	44	-1.6	5.3	7.0	99	114	108	-89	35			
LEU	45	-.1	5.2	10.5	-149	114	-78	-68	166			
ALA	46	-1.8	6.7	13.5	-69	119	142					
ASP	47	-5.1	5.5	15.0	-63	103	-47	-89	79			
GLU	48	-5.9	9.2	15.5	-139	109	143	125	-166	-111		
VAL	49	-4.0	12.4	14.5	-112	-10	115					
ALA	50	-3.6	15.7	16.4	145	117	130					
LEU	51	-3.2	19.2	14.9	-123	104	138	-94	166			
VAL	52	-2.7	22.3	17.1	-152	-43	120					
ASP	53	-2.2	26.0	16.5	115	119	176	-98	-67			
VAL	54	-2.8	29.6	17.4	-106	-57	103					
MET	55	-4.5	30.1	14.1	42	104	64	-108	171	161		

GLU	56	-7.1	29.5	14.7	-54	123	-77	-106	-155	3		
ASP	57	-7.7	29.4	10.9	-38	121	-66	-46	84			
LYS	58	-5.1	26.9	9.7	-53	120	-31	-120	151	-112	-175	
LEU	59	-6.0	24.2	12.3	-70	114	-47	-158	50			
LYS	60	-9.7	23.9	11.5	-72	111	-42	-141	164	-172	-172	
GLY	61	-9.2	24.3	7.7	-51	109	-69					
GLU	62	-6.9	21.3	7.4	-54	109	-59	-37	-53	152		
MET	63	-8.5	19.4	10.2	-35	101	-81	-163	-116	-139		
MET	64	-11.8	19.6	8.4	-48	133	-24	-152	157	146		
ASP	65	-10.9	19.2	4.7	-78	105	-54	157	-163			
LEU	66	-9.0	15.9	5.2	-50	109	-43	-62	154			
GLN	67	-12.1	14.8	7.2	-43	113	-46	-130	177	-109		
HIS	68	-14.1	14.6	3.9	-52	100	-77	-67	53			
GLY	69	-11.4	12.4	2.4	-34	115	-29					
SER	70	-11.6	10.4	5.5	-20	116	-30	-99				
LEU	71	-13.0	7.7	3.3	-81	124	-7	142	81			
PHE	72	-10.1	6.9	1.0	-115	109	29	-37	101			
LEU	73	-7.7	6.2	3.8	-153	99	174	18	76			
HIS	74	-7.9	4.0	6.9	-119	113	50	-35	129			
THR	75	-7.7	6.5	9.7	-110	102	144	-51				
ALA	76	-10.5	6.8	12.3	-93	110	-72					
LYS	77	-10.1	10.0	14.3	-102	115	144	-47	152	-110	178	
ILE	78	-8.3	13.1	13.3	-145	99	132	-20	-174			
VAL	79	-8.3	16.2	15.5	-143	50	113					
SER	80	-6.8	19.7	15.6	152	105	141	53				
GLY	81	-5.6	21.1	18.9	-126	117	138					
LYS	83	-5.0	24.6	20.2	-104	92	-56	149	159	43	154	
ASP	84	-4.7	23.2	23.7	-63	108	-161	-160	56			
TYR	85	-2.4	20.2	24.4	-141	112	22	-100	131			
SER	86	-4.8	17.7	26.0	-82	100	-27	-81				
VAL	87	-5.0	16.4	22.5	-64	-167	123					
SER	88	-1.5	14.9	22.2	-19	99	-28	-30				
ALA	89	-3.0	12.7	24.8	-3	120	88					
GLY	90	-1.0	9.5	24.5	75	122	74					
SER	91	1.4	9.4	21.5	-115	119	174	-44				
LYS	92	4.3	7.5	20.1	-112	96	-58	-58	-116	-133	-145	
LEU	93	5.4	10.3	17.7	-121	124	125	-102	-166			
VAL	94	5.2	14.1	17.9	-103	155	101					
VAL	95	6.2	15.3	14.4	112	161	93					
ILE	96	7.2	18.8	15.6	111	124	102	150	110			
THR	97	7.5	21.7	13.2	-122	107	140	99				
ALA	98	7.1	25.4	14.2	161	110	164					
GLY	99	9.0	28.6	15.0	129	122	-157					
ALA	100	10.7	31.6	13.6	-84	122	-180					
ARG	101	12.0	32.8	10.2	-122	136	-33	-170	177	174	94	20
GLN	102	14.9	35.1	9.5	98	109	87	-85	-156	-87		
GLN	103	13.9	38.6	8.6	-81	121	-11	-87	15	56		
GLU	105	17.2	39.9	7.3	150	111	-42	-131	45	94		
GLY	106	19.5	40.3	10.3	-87	87	177					
GLU	107	20.3	36.7	11.4	-169	112	71	109	155	155		
SER	108	21.9	36.3	14.8	-77	103	107	-100				
ARG	109	22.4	32.6	14.8	-40	122	-69	-107	-84	-142	-5	-20
LEU	110	22.0	31.4	18.4	-57	101	-45	5	-175			
ASN	111	19.7	34.2	18.2	-57	110	-51	-141	-6			
LEU	112	17.1	32.6	15.9	-63	107	-35	-166	170			
VAL	113	17.8	29.3	17.7	-57	170	97					
GLN	114	17.0	31.3	20.8	-64	113	-46	-103	-91	-119		
ARG	115	13.7	32.4	19.2	-63	109	-45	-127	178	-35	149	-9
ASN	116	12.9	28.8	18.3	-53	106	-39	-84	-151			
VAL	117	13.3	28.2	22.0	-53	152	118					
ASN	118	10.6	30.4	23.5	-68	106	-64	-40	130			
ILE	119	8.1	28.6	21.4	-54	111	-50	-103	135			
PHE	120	9.8	25.2	22.0	-68	111	-48	-80	76			
LYS	121	10.5	26.0	25.7	-42	100	-49	162	-56	138	124	
PHE	122	6.7	26.1	25.6	-65	122	-72	-146	178			
ILE	123	5.2	23.4	23.4	-52	124	-46	-43	126			

GLU	56	-8.0	29.1	14.9	-31	112	-64	-145	139	-112		
ASP	57	-9.2	29.8	11.4	-42	118	-72	-66	60			
LYS	58	-6.8	27.8	9.3	-40	111	-65	-134	133	122	-155	
LEU	59	-7.0	24.8	11.5	-41	116	-43	-94	106			
LYS	60	-10.8	24.7	11.1	-56	99	-75	-167	-100	164	-158	
GLY	61	-10.4	24.7	7.3	-43	115	-42					
GLU	62	-7.4	22.5	7.5	-65	108	-78	-100	35	99		
MET	63	-8.6	19.8	10.0	-43	114	-44	105	127	157		
MET	64	-12.1	20.0	8.5	-65	117	-58	-143	-143	36		
ASP	65	-11.4	19.8	4.8	-61	113	-54	-170	-121			
LEU	66	-8.5	17.3	4.9	-52	101	-56	-167	72			
GLN	67	-10.8	15.4	7.3	-39	103	-44	-123	167	56		
HIS	68	-13.0	15.7	4.2	-37	108	-70	-165	-102			
GLY	69	-10.9	13.2	2.3	-48	104	-38					
SER	70	-10.9	11.5	5.7	-1	106	-52	177				
LEU	71	-12.9	9.0	3.7	-72	125	-12	95	89			
PHE	72	-10.6	8.4	.8	-79	111	-10	-62	147			
LEU	73	-8.3	6.4	3.1	-146	99	-155	34	63			
HIS	74	-8.8	5.3	6.8	-152	119	113	-7	69			
THR	75	-7.5	6.7	10.1	-122	114	-56	-113				
ALA	76	-10.3	7.2	12.6	133	110	-80					
LYS	77	-9.9	10.6	14.3	-134	119	142	177	141	-157	144	
ILE	78	-8.2	13.9	13.6	-127	104	123	99	126			
VAL	79	-8.2	16.5	16.4	-112	-167	117					
SER	80	-7.5	20.2	16.6	150	113	157	47				
GLY	81	-7.2	23.0	19.2	166	113	-173					
LYS	83	-4.9	25.4	21.0	-142	111	52	-33	143	-55	-154	
ASP	84	-4.3	24.0	24.5	-152	113	19	-75	-78			
TYR	85	-2.1	21.0	24.0	26	109	56	-43	87			
SER	86	-4.7	19.1	26.1	-120	101	-45	-107				
VAL	87	-6.1	17.0	23.2	-43	-126	121					
SER	88	-3.1	14.7	22.5	-53	118	-41	-157				
ALA	89	-2.3	13.8	26.1	14	114	85					
GLY	90	-.2	10.7	25.4	104	123	47					
SER	91	1.3	10.6	21.9	-102	122	145	72				
LYS	92	4.1	8.6	20.3	-83	114	-57	-141	156	172	-113	
LEU	93	4.9	10.4	17.1	-97	116	91	-136	-57			
VAL	94	5.1	14.2	17.8	-109	-153	113					
VAL	95	6.5	16.7	15.4	132	44	114					
ILE	96	6.9	20.5	15.4	143	114	156	126	92			
THR	97	7.7	23.2	12.8	-131	108	-86	18				
ALA	98	6.7	26.8	13.5	99	112	166					
GLY	99	6.9	30.0	11.5	-93	115	125					
ALA	100	10.3	31.3	10.4	158	107	73					
ARG	101	11.3	30.4	6.8	-106	115	150	5	118	106	116	8
GLN	102	14.1	31.5	4.6	-54	117	142	-57	-95	-36		
GLN	103	13.8	34.6	2.4	-108	117	59	-96	-28	43		
GLU	105	15.7	33.8	-.7	67	101	-79	130	138	28		
GLY	106	18.6	35.7	.8	-53	102	162					
GLU	107	20.2	33.7	3.6	-78	122	-141	-50	-153	-164		
SER	108	20.5	34.0	7.4	82	112	34	-101				
ARG	109	18.8	30.6	8.1	-58	103	-34	172	171	-87	160	-1[illegible]
LEU	110	20.9	30.9	11.3	-69	111	-34	-86	156			
ASN	111	19.5	34.4	12.0	-50	99	-51	-114	-119			
LEU	112	16.4	32.3	12.4	-34	116	-8	-38	174			
VAL	113	17.8	31.5	15.8	-73	68	113					
GLN	114	17.0	34.8	17.6	-49	110	-69	-154	41	-89		
ARG	115	13.5	35.3	16.2	-53	118	-14	50	141	99	108	-111
ASN	116	12.8	31.6	16.8	-88	124	-48	-114	-13			
VAL	117	13.3	30.6	20.4	-55	133	110					
ASN	118	10.7	33.3	21.3	-29	108	-71	-81	77			
ILE	119	7.8	31.2	20.2	-69	115	-49	121	-128			
PHE	120	9.4	27.7	20.2	-50	115	-45	-180	106			
LYS	121	10.4	28.0	23.9	-47	111	-63	-177	-141	-140	109	
PHE	122	6.9	28.2	25.3	-46	114	-40	-78	117			
ILE	123	5.8	25.4	22.9	-49	105	-43	-48	146			

TABLE V (*Continued*)

APO	P	Q	R	PHI	TAU	PSI	CHI1	CHI2	CHI3	CHI4	CHI5
ILE 124	7.5	20.5	24.1	-51	119	-52	-83	-46			
PRO 125	6.9	20.0	27.9	0	0	-52					
ASN 126	3.1	20.2	27.4	-55	104	-59	-156	6			
ILE 127	3.6	17.8	24.5	-42	98	-58	-98	-164			
VAL 128	5.2	15.8	27.3	-51	173	119					
LYS 129	2.5	16.3	29.9	-47	112	-38	-134	-161	103	170	
HIS 130	.1	14.5	27.6	-96	124	1	-67	111			
SER 131	2.1	11.7	26.1	-163	99	76	11				
PRO 132	4.9	11.0	28.6	0	0	-29					
ASP 132	5.9	8.1	26.3	-89	124	27	-66	-82			
CYS 133	6.9	10.0	23.1	-87	132	147	39				
LEU 134	9.5	9.9	20.4	-100	92	133	-62	-46			
LYS 135	9.8	13.5	19.2	-113	114	124	-53	-122	-133	80	
GLU 136	10.6	14.4	15.6	-106	117	134	-64	-96	157		
LEU 137	12.0	17.7	14.3	-112	123	129	-124	89			
HIS 138	11.6	19.4	10.9	-116	111	-31	-52	170			
PRO 139	11.8	23.1	12.0	0	0	133					
GLU 140	15.2	24.5	11.0	-51	133	164	-110	17	132		
LEU 141	18.7	25.0	12.5	67	81	62	-120	176			
GLY 142	17.6	21.3	13.0	-77	101	-67					
THR 143	21.1	20.2	14.1	-21	99	-72	-5				
ASP 144	20.5	22.7	16.8	-52	128	-72	-150	-129			
LYS 145	16.8	23.1	17.8	-49	117	-10	-53	162	126	-112	
ASN 146	16.8	19.3	18.4	-64	99	-70	-59	103			
LYS 147	18.7	19.8	21.7	-43	114	-46	-175	-111	-167	-171	
GLN 148	16.6	22.9	22.4	-48	109	-62	-30	-131	71		
ASP 149	13.4	20.9	22.5	-53	106	-51	137	8			
TRP 150	15.3	18.2	24.3	-48	118	-49	-176	62			
LYS 151	16.6	20.3	27.2	-65	107	-34	-94	-99	119	174	
LEU 152	13.0	21.6	27.5	-62	104	-58	-99	-36			
SER 153	11.7	18.0	27.7	-59	139	-47	85				
GLY 154	13.7	15.5	29.9	87	114	2					
LEU 155	13.0	12.6	27.5	-71	103	129	-25	176			
PRO 156	16.1	10.4	27.7	0	0	147					
MET 157	18.4	10.5	24.7	-60	118	113	78	179	78		
HIS 158	17.2	8.2	22.0	137	103	8	64	-102			
ARG 159	13.7	9.7	22.0	-130	102	7	-81	171	36	144	-26
ILE 160	15.0	13.0	20.5	-123	105	88	-11	37			
ILE 161	15.6	12.3	16.8	-98	113	130	2	-168			
GLY 162	16.8	15.1	14.5	-65	113	-115					
SER 163	16.0	14.9	10.8	170	88	-49	-166				
GLY 164	19.4	16.6	10.1	-8	114	96					
CYS 165	20.2	15.0	6.8	104	134	-40	-106				
ASN 166	17.8	12.2	5.6	-42	117	-41	-154	64			
LEU 167	16.5	14.3	2.8	-81	104	-54	-134	44			
ASP 168	19.8	16.1	2.1	-31	109	-75	-62	-142			
SER 169	21.5	12.9	1.1	-66	117	-33	46				
ALA 170	18.4	11.0	.1	-54	104	-52					
ARG 171	18.0	13.5	-2.7	-62	109	-52	-17	174	144	178	4
PHE 172	21.7	13.6	-3.4	-42	109	-23	172	75			
ARG 173	21.0	10.0	-4.4	-72	124	-41	-52	-133	-65	-152	50
TYR 174	18.7	10.3	-7.4	-88	117	-15	106	-85			
LEU 175	20.4	13.2	-9.1	-78	88	-54	50	146			
MET 176	23.3	10.8	-8.6	-40	102	-33	-179	-168	131		
GLY 177	20.7	8.8	-10.5	-75	128	-31					
GLU 178	19.8	10.8	-13.7	-79	99	-17	-38	-86	64		

TERN	P	Q	R	PHI	TAU	PSI	CHI1	CHI2	CHI3	CHI4	CHI5
ILE 124	8.1	23.3	25.1	-41	124	-59	-50	-9			
PRO 125	6.3	23.1	28.4	0	0	-42					
ASN 126	3.1	22.3	26.6	-62	114	-65	-58	-28			
ILE 127	4.1	19.6	24.2	-56	119	-27	-55	108			
VAL 128	6.2	17.7	26.8	-85	117	112					
LYS 129	3.8	17.9	29.8	-45	114	-44	-91	179	-84	177	
HIS 130	1.1	15.9	27.9	-92	119	-18	-42	-162			
SER 131	3.3	13.8	25.6	-111	107	60	63				
PRO 132	5.3	12.1	28.4	0	0	-44					
ASP 132	5.6	9.0	26.2	-98	114	35	-167	-87			
CYS 133	6.2	10.9	23.0	-113	115	154	64				
LEU 134	9.1	11.3	20.7	-112	105	137	22	-170			
LYS 135	9.8	14.7	19.1	-105	125	119	-43	62	175	175	
GLU 136	10.6	15.4	15.4	-108	110	123	-33	-164	-32		
LEU 137	11.4	19.0	14.4	-102	120	98	-63	99			
HIS 138	10.8	20.0	10.8	-78	106	-11	173	154			
PRO 139	11.2	23.5	12.1	0	0	153					
GLU 140	14.2	25.5	10.7	-50	114	153	-127	19	-133		
LEU 141	17.7	24.7	12.0	67	104	60	-49	108			
GLY 142	17.0	21.0	12.0	-37	104	-77					
THR 143	20.0	20.4	14.2	-44	117	-67	95				
ASP 144	20.1	23.6	16.3	-43	121	-71	-170	-130			
LYS 145	16.5	24.0	17.5	-48	111	-43	-91	-168	75	131	
ASN 146	16.4	20.3	18.3	-39	108	-70	-160	16			
LYS 147	18.8	20.8	21.2	-39	116	-56	-140	-111	-133	-137	
GLN 148	17.0	23.8	22.7	-38	112	-66	-79	14	-81		
ASP 149	13.8	21.8	23.4	-57	111	-52	166	-170			
TRP 150	15.7	18.6	24.3	-46	114	-67	-178	73			
LYS 151	17.9	20.1	27.0	-55	101	-45	-179	-126	-155	167	
LEU 152	14.8	22.2	27.7	-49	108	-64	-134	-30			
SER 153	12.6	19.2	28.1	-73	109	-56	48				
GLY 154	15.4	16.7	28.7	85	114	40					
LEU 155	13.8	13.5	27.4	-82	96	145	-77	-100			
PRO 156	16.5	10.9	27.1	0	0	121					
MET 157	18.1	11.4	23.7	-10	119	-84	-87	87	131		
HIS 158	16.8	8.3	21.8	-25	122	-20	62	-59			
ARG 159	13.2	9.7	22.0	-101	112	-8	-84	149	47	98	-44
ILE 160	14.2	13.2	20.6	-110	111	125	82	64			
ILE 161	14.8	13.6	16.9	-131	107	135	-43	-6			
GLY 162	15.4	17.0	15.2	-78	112	-74					
SER 163	14.8	16.4	11.5	145	107	-35	-115				
GLY 164	17.9	17.9	10.0	-1	132	-102					
CYS 165	19.8	16.0	7.3	-80	124	17	-114				
ASN 166	17.2	13.6	6.2	-54	111	-68	86	77			
LEU 167	16.4	15.1	2.8	-59	113	-32	78	153			
ASP 168	20.0	16.2	2.4	-37	103	-72	-105	-15			
SER 169	20.8	12.6	1.3	-41	117	-66	-46				
ALA 170	17.7	12.0	-.9	-38	118	-52					
ARG 171	18.3	15.0	-3.1	-64	104	-45	-106	174	-43	-147	42
PHE 172	21.9	14.1	-3.1	-62	107	-57	-163	46			
ARG 173	21.1	10.4	-3.6	-54	106	-51	-164	127	91	119	-74
TYR 174	18.5	11.5	-6.2	-58	122	-54	150	65			
LEU 175	20.3	14.0	-8.3	-52	99	-43	-82	148			
MET 176	22.7	11.1	-8.3	-43	122	-52	-112	157	-96		
GLY 177	20.7	8.6	-10.3	-61	116	-25					
GLU 178	20.2	11.0	-13.2	-56	111	-52	-111	-39	-94		

ARG 179	23.6	10.8	-13.8	-76	107	-29	-112	170	108	-157	-8
LEU 180	23.4	7.0	-14.1	-116	122	-27	-43	-170			
GLY 181	19.9	5.8	-15.0	85	106	63					
VAL 182	19.1	3.9	-11.8	-152	-133	100					
HIS 183	16.9	4.8	-8.8	-169	123	118	149	50			
SER 184	17.6	6.5	-5.4	2	113	-58	-92				
CYS 185	17.0	3.1	-3.8	-40	116	-74	-32				
LEU 186	19.9	1.1	-5.2	-65	111	-55	-13	174			
VAL 187	22.1	4.1	-6.1	-49	-63	96					
SER 188	23.3	4.1	-2.5	105	119	111	178				
GLY 189	25.3	6.9	-.9	-138	114	-178					
TRP 190	25.7	9.1	2.1	-150	128	139	-69	77			
VAL 191	25.9	12.6	3.6	-132	176	99					
ILE 192	25.9	12.8	7.4	127	110	-161	72	-41			
GLY 193	26.7	15.4	10.1	65	126	-112					
GLN 194	25.6	19.0	9.6	-62	133	-52	-106	-112	-145		
HIS 195	22.7	19.4	7.2	99	102	75	-168	108			
GLY 196	24.1	21.7	4.5	112	115	-151					
ASP 197	27.4	23.7	4.3	-51	100	115	-63	-64			
SER 198	29.3	21.4	6.6	106	108	11	-46				
VAL 199	27.4	18.3	5.4	-99	62	109					
PRO 200	30.3	16.0	4.6	79	0	143					
SER 201	29.4	13.2	2.2	-119	102	139	57				
VAL 202	31.3	9.9	2.4	-93	145	114					
TRP 203	32.8	8.7	-.8	114	117	4	-56	119			
SER 204	33.6	5.2	.5	-99	105	3	100				
GLY 205	29.9	4.5	.6	-103	96	-56					
MET 206	29.5	5.8	-3.0	-19	112	92	-85	-25	126		
TRP 207	28.4	2.5	-4.5	-161	108	159	70	-84			
ASP 208	25.5	1.0	-6.4	-149	87	139	131	-171			
ALA 209	25.7	-2.8	-6.6	95	90	-101					
LYS 210	28.1	-2.8	-3.6	-179	89	38	42	107	166	171	
LEU 211	30.8	-1.4	-5.9	-104	134	100	-60	-109			
HIS 212	33.8	.8	-4.9	173	-144	-142	61	-32			
LYS 213	33.4	4.3	-6.2	-48	105	112	139	130	-163	29	
ASP 214	36.8	5.7	-5.1	132	88	-44	-138	-73			
VAL 215	37.9	2.2	-6.2	-71	-165	106					
VAL 216	36.8	2.2	-9.9	-81	159	109					
ASP 217	37.7	5.9	-9.7	-33	103	119	-138	96			
SER 218	40.2	7.0	-7.0	-86	121	-12	172				
ALA 219	39.8	10.8	-7.3	-139	122	-179					
TYR 220	37.8	13.9	-8.0	-175	111	-109	19	54			
GLU 221	38.0	17.1	-6.1	-57	110	85	-157	-32	39		
VAL 222	39.0	19.2	-9.2	91	165	116					
ILE 223	40.8	16.2	-10.9	-24	123	-6	80	131			
LYS 224	37.9	14.6	-12.8	-154	108	-177	-111	-71	168	148	
LEU 225	34.3	15.2	-13.8	-50	119	75	-91	-39			
LYS 226	33.2	17.2	-10.9	4	124	75	-10	-76	-118	179	
GLY 227	31.5	14.7	-8.7	-84	110	-26					
TYR 228	31.9	16.8	-5.6	-63	98	-23	-99	33			
THR 229	29.5	18.7	-7.8	-57	102	-62	-132				
SER 230	27.0	16.0	-7.2	-34	109	-56	168				
TRP 231	27.5	16.5	-3.5	-50	109	-68	145	-129			
ALA 232	27.1	20.3	-3.6	-39	112	-55					
ILE 233	24.0	19.9	-5.8	-36	107	-58	-37	-62			
GLY 234	22.4	18.3	-2.7	-30	113	-67					
LEU 235	22.5	21.4	-.6	-69	114	-39	95	151			
VAL 236	22.3	23.8	-3.5	-72	73	111					
VAL 237	19.4	21.9	-5.0	-33	100	106					
SER 238	17.2	22.7	-2.1	-65	122	-75	-157				
ASN 239	16.9	26.5	-2.2	-42	127	-38	-28	121			
PRO 240	15.8	27.0	-5.8	0	0	-25					
VAL 241	12.5	25.5	-4.7	-46	172	101					
ASP 242	11.8	28.9	-3.2	-61	113	-30	-49	179			
VAL 243	12.5	30.2	-6.7	-21	47	97					
LEU 244	8.7	30.1	-6.3	-52	109	-54	61	67			

ARG 179	23.8	10.7	-14.3	-70	112	-25	-79	-165	164	175	5
LEU 180	23.7	7.0	-13.6	-80	120	-52	-96	-53			
GLY 181	20.6	5.6	-15.3	97	101	52					
VAL 182	19.4	3.8	-12.2	-169	178	86					
HIS 183	17.7	5.7	-9.4	176	122	166	-162	-95			
SER 184	17.7	7.0	-5.8	-24	113	-38	65				
CYS 185	16.7	3.4	-5.0	-44	111	-96	-55				
LEU 186	19.9	1.3	-5.4	-62	116	-33	-109	-64			
VAL 187	22.1	4.4	-5.9	-67	123	106					
SER 188	22.9	4.7	-2.3	89	109	108	-10				
GLY 189	24.8	8.0	-1.8	-126	111	128					
TRP 190	25.6	9.7	1.5	-106	127	134	-69	89			
VAL 191	25.8	13.3	2.7	-137	42	102					
ILE 192	25.8	13.9	6.5	117	114	128	127	-60			
GLY 193	26.1	17.1	8.5	-164	110	-109					
GLN 194	23.4	19.4	9.8	-143	120	-6	88	-142	-115		
HIS 195	21.2	20.8	7.1	104	126	21	-151	73			
GLY 196	22.6	24.0	5.7	-152	112	147					
ASP 197	25.1	25.1	3.1	-29	112	113	32	166			
SER 198	28.4	23.6	4.2	130	110	-16	76				
VAL 199	27.1	20.2	5.1	-48	20	111					
PRO 200	30.1	17.8	4.7	115	0	151					
SER 201	29.8	15.1	2.2	-124	105	142	35				
VAL 202	31.3	11.6	2.6	-113	20	108					
TRP 203	31.7	10.7	-1.1	64	114	-20	-2	59			
SER 204	34.1	8.0	.0	-56	107	-18	-26				
GLY 205	30.8	6.1	.5	-106	104	69					
MET 206	29.9	6.2	-3.2	-129	99	125	-64	-151	144		
TRP 207	29.0	2.8	-4.6	-168	108	178	76	-106			
ASP 208	26.3	.8	-6.4	-151	99	148	-103	-115			
ALA 209	25.4	-2.8	-5.8	57	116	-100					
LYS 210	28.3	-4.4	-4.0	-136	106	11	76	-149	156	158	
LEU 211	31.0	-2.3	-5.7	-114	113	130	109	169			
HIS 212	32.5	.9	-4.3	-76	112	131	-153	85			
LYS 213	32.4	3.9	-6.6	-60	123	-67	-175	143	24	-89	
ASP 214	35.5	6.1	-6.0	-31	102	-57	-94	-84			
VAL 215	37.2	2.8	-6.8	-39	-171	113					
VAL 216	35.6	2.6	-10.3	-59	88	115					
ASP 217	36.6	6.2	-11.1	-38	107	-36	-33	-74			
SER 218	39.5	6.3	-8.7	115	110	-99	173				
ALA 219	39.8	9.6	-6.9	-111	118	-161					
TYR 220	38.2	13.1	-6.6	-131	151	-130	-76	-152			
GLU 221	37.0	16.6	-6.1	-79	115	130	121	-39	37		
VAL 222	39.5	19.5	-6.2	161	110	91					
ILE 223	39.3	19.1	-10.0	-19	103	-106	-15	-161			
LYS 224	37.3	16.4	-11.7	44	113	-158	-144	54	129	78	
LEU 225	34.0	17.3	-13.3	-100	109	37	-73	-84			
LYS 226	32.7	19.1	-10.2	69	123	41	-72	-67	-175	62	
GLY 227	31.4	16.3	-8.0	-51	96	-33					
TYR 228	31.9	19.4	-5.8	-44	111	-56	-143	-95			
THR 229	28.9	21.1	-7.4	-50	113	-54	-49				
SER 230	26.6	18.2	-6.8	-31	107	-67	-137				
TRP 231	26.8	18.8	-3.1	-51	119	-65	177	-135			
ALA 232	26.7	22.6	-2.9	-36	121	-46					
ILE 233	23.3	23.1	-4.6	-88	106	23	141	107			
GLY 234	22.2	20.3	-2.3	-104	84	-45					
LEU 235	23.0	23.1	.1	-68	104	-52	-32	-92			
VAL 236	21.8	25.6	-2.5	-63	153	112					
VAL 237	18.7	23.4	-3.1	-36	144	111					
SER 238	17.0	25.5	-.4	-55	107	-76	-113				
ASN 239	17.6	28.8	-2.1	-19	133	-74	-167	-77			
PRO 240	16.1	29.0	-5.6	0	0	-74					
VAL 241	13.0	26.9	-5.0	-13	157	118					
ASP 242	11.5	29.7	-2.8	-49	112	22	-56	127			
VAL 243	11.1	31.6	-6.1	-91	142	85					
LEU 244	8.1	29.5	-5.0	2	105	-4	-168	100			

TABLE V (*Continued*)

APO

	P	Q	R	PHI	TAU	PSI	CHI1	CHI2	CHI3	CHI4	CHI5
THR 245	8.3	29.3	-2.6	-162	136	-36	-152				
TYR 246	9.5	26.5	-.4	171	107	-162	67	-119			
VAL 247	9.4	22.7	-.3	-136	18	102					
ALA 248	6.5	20.6	1.1	140	117	-40					
TRP 249	6.5	17.1	-.2	-71	111	34	-43	109			
LYS 250	9.4	16.5	2.2	-85	96	-16	125	104	145	-114	
GLY 251	6.5	14.8	4.0	-44	106	-33					
CYS 252	7.9	11.9	2.0	-54	104	-76	-23				
SER 253	11.2	11.9	3.9	-30	116	-60	-155				
VAL 254	9.6	12.1	7.3	-55	-168	113					
ALA 255	7.1	9.4	6.6	-48	117	-59					
ASP 256	9.5	6.5	6.2	-56	117	-32	-104	40			
LEU 257	11.3	7.3	9.4	-69	110	-43	-103	55			
ALA 258	8.2	7.2	11.5	-65	110	-47					
GLU 259	7.0	3.9	9.9	-54	111	-44	159	160	69		
THR 260	10.3	2.2	10.8	-59	107	-59	-31				
ILE 261	10.1	3.4	14.4	-53	107	-60	-93	-177			
MET 262	6.4	2.7	14.5	-52	119	-50	-77	-77	-50		
LYS 263	6.4	-.7	12.7	-95	107	28	-42	-176	154	174	
ASN 264	9.7	-1.4	14.5	31	122	62	-118	30			
LEU 265	12.0	-2.0	11.6	-88	122	-26	-67	176			
CYS 266	15.4	-1.0	12.8	59	107	62	-52				
ARG 267	16.4	.5	9.5	-102	96	165	-101	-152	-48	-141	12
VAL 268	19.4	2.8	9.5	-113	180	111					
HIS 269	18.9	6.5	8.6	118	120	151	-38	124			
PRO 270	21.0	9.6	8.1	0	0	45					
VAL 271	19.7	10.6	11.5	-68	147	120					
SER 272	21.0	13.4	13.7	147	112	112	109				
THR 273	21.8	12.5	17.2	-118	113	-174	-14				
MET 274	23.8	13.8	20.2	-88	94	124	107	115	52		
VAL 275	27.3	12.6	19.1	-94	-139	106					
LYS 276	28.2	12.8	22.8	34	126	-42	-175	80	173	163	
ASP 277	31.3	10.6	22.7	-152	124	41	-74	23			
PHE 278	32.4	9.2	19.3	-108	111	-167	-58	-81			
TYR 279	35.1	10.1	16.8	14	92	-139	-52	-63			
GLY 280	36.3	13.1	18.9	-102	135	-4					
ILE 281	33.7	15.6	19.9	-61	120	145	-70	143			
LYS 282	33.0	16.6	23.5	-134	123	63	150	115	-173	102	
ASP 283	29.8	18.5	24.3	-171	135	37	-114	-1			
ASN 284	26.2	17.4	24.1	59	112	52	-46	-24			
VAL 285	25.8	18.3	20.4	-118	-93	114					
PHE 286	23.6	16.8	17.7	145	112	110	-52	57			
LEU 287	25.2	16.3	14.3	-125	95	166	-81	-158			
SER 288	24.0	13.6	11.9	-81	112	147	-51				
LEU 289	25.4	10.1	12.0	-154	103	157	-120	102			
PRO 290	24.3	6.8	10.4	0	0	77					
CYS 291	21.4	6.5	12.9	-86	126	127	-124				
VAL 292	19.3	3.4	13.5	-94	-64	117					
LEU 293	15.8	2.9	14.9	163	133	169	20	-158			
ASN 294	12.4	1.6	16.1	-172	103	-156	61	23			

TERN

	P	Q	R	PHI	TAU	PSI	CHI1	CHI2	CHI3	CHI4	CHI5
THR 245	8.2	30.9	-1.5	99	121	-30	-50				
TYR 246	9.2	27.9	.5	179	113	-153	58	-104			
VAL 247	9.1	24.1	.8	-115	53	100					
ALA 248	5.7	22.6	.5	135	113	1					
TRP 249	6.5	18.9	.3	-165	126	35	-49	81			
LYS 250	9.9	17.5	1.4	-82	91	-3	112	-95	162	18	
GLY 251	7.1	16.2	3.7	-33	120	-20					
CYS 252	8.1	12.7	2.8	-70	109	-52	-29				
SER 253	11.4	12.9	4.5	-43	112	-53	-80				
VAL 254	9.8	13.7	7.8	-60	-167	108					
ALA 255	7.2	11.0	7.1	-44	109	-46					
ASP 256	10.0	8.5	6.5	-62	108	-54	-170	-151			
LEU 257	11.9	9.6	9.5	-41	114	-56	-43	-179			
ALA 258	9.1	8.9	11.9	-70	106	-27					
GLU 259	8.4	5.9	9.7	-38	108	-67	-153	169	-152		
THR 260	11.0	3.8	11.4	-48	110	-51	58				
ILE 261	9.9	5.2	14.8	-50	114	-84	-58	-152			
MET 262	6.1	4.6	14.9	-30	113	-57	-112	-99	70		
LYS 263	6.6	1.1	13.5	-71	113	-35	34	-133	-43	-168	
ASN 264	9.9	.5	15.4	78	118	48	-160	41			
LEU 265	12.1	-1.2	12.8	-92	128	-11	-34	117			
CYS 266	15.7	-.2	13.4	64	118	75	-60				
ARG 267	17.1	1.3	10.2	-121	104	166	-120	-168	-56	133	5
VAL 268	20.0	3.7	9.7	-100	159	114					
HIS 269	19.1	7.3	8.9	112	117	154	-88	-93			
PRO 270	20.3	10.9	9.4	0	0	177					
VAL 271	18.9	13.1	12.1	156	84	102					
SER 272	20.3	15.2	14.9	-170	110	123	94				
THR 273	21.7	13.8	18.2	-128	110	-146	15				
MET 274	24.5	14.6	20.6	-117	113	120	-130	-162	113		
VAL 275	28.2	14.6	19.5	-107	-89	102					
LYS 276	30.2	14.9	22.8	107	116	-32	-174	142	156	-141	
ASP 277	32.9	12.2	22.8	-135	110	51	-59	-58			
PHE 278	33.5	11.5	19.1	-116	119	-81	-74	-100			
TYR 279	35.7	13.7	16.9	-106	99	-137	-53	166			
GLY 280	36.7	17.2	17.9	-66	110	-53					
ILE 281	33.2	18.3	18.9	-39	113	120	-58	80			
LYS 282	33.0	18.0	22.7	-151	107	79	164	-141	58	-145	
ASP 283	30.2	20.2	24.1	171	111	51	-153	178			
ASN 284	26.8	18.9	23.1	43	113	85	67	123			
VAL 285	26.5	19.8	19.4	-119	-95	106					
PHE 286	23.2	18.4	18.3	81	111	101	-102	-119			
LEU 287	23.6	18.1	14.6	-131	115	-165	69	84			
SER 288	23.1	15.9	11.5	-79	115	111	-121				
LEU 289	24.2	12.3	12.1	-159	115	148	-126	173			
PRO 290	23.2	8.7	11.7	0	0	135					
CYS 291	21.4	7.4	14.8	-150	87	160	164				
VAL 292	19.1	4.4	14.3	-116	155	105					
LEU 293	15.4	4.7	15.2	130	125	150	17	116			
ASN 294	12.6	2.7	16.7	-161	116	-176	70	-56			

ASX 295	10.0	2.2	19.0	-28	121	90	-84	-30		
GLY 296	11.0	5.7	20.3	35	110	50				
ILE 297	14.5	4.3	20.4	-167	112	149	161	132		
SER 298	17.7	4.1	18.4	-97	98	69	-13			
HIS 299	17.7	.4	19.0	-17	126	97	-163	116		
CYS 300	20.8	-.7	20.8	46	122	-7	-140			
ASN 301	23.0	.8	18.2	-157	123	161	-15	-124		
ILE 302	24.1	3.8	16.0	-147	93	138	165	-169		
VAL 303	26.6	3.1	13.2	-72	130	125				
LYS 304	30.3	3.8	14.1	104	111	113	-32	131	63	102
MET 305	32.1	4.7	10.9	-111	118	171	36	-152	-49	
LYS 306	35.6	5.6	9.7	-113	110	93	-26	-65	-135	174
LEU 307	35.6	9.1	8.2	-88	98	150	-23	146		
LYS 308	39.0	10.1	7.0	-66	93	164	-91	134	-174	-135
PRO 309	40.3	12.5	9.7	0	0	-35				
ASP 310	39.3	15.3	7.3	-58	111	-36	123	119		
GLU 311	35.7	14.3	7.7	-86	113	-46	-70	155	-38	
GLU 312	35.6	13.0	11.3	-53	106	-47	139	160	90	
GLN 313	37.0	16.5	12.2	-72	114	-13	-138	-131	-23	
GLN 314	34.4	18.1	9.9	-71	95	-53	-134	-65	-91	
LEU 315	32.0	16.6	12.4	-43	105	-59	-91	-131		
GLN 316	34.1	18.3	15.1	-40	121	-75	-50	-117	-92	
LYS 317	34.3	22.0	14.1	-51	112	-42	-33	-126	167	175
SER 318	30.7	21.8	13.0	-55	99	-54	-95			
ALA 319	30.1	20.5	16.5	-57	113	-48				
THR 320	32.5	23.0	18.1	-47	113	-54	-119			
THR 321	30.5	26.0	16.9	-73	112	-37	95			
LEU 322	27.2	24.2	17.2	-47	105	-68	-94	130		
TRP 323	27.6	24.0	21.0	-55	121	-46	156	-123		
ASP 324	29.4	27.3	21.7	-47	110	-58	-105	-86		
ILE 325	26.4	29.3	20.5	-32	119	-69	-58	-99		
GLN 326	23.8	28.2	23.0	-12	105	77	147	-77	15	
LYS 327	23.9	31.9	24.1	-125	103	37	-167	68	60	161
ASP 328	20.1	32.0	24.7	-112	117	155	-12	119		
LEU 329	17.9	32.6	27.7	-68	118	-138	41	116		
LYS 330	15.1	30.5	29.1	-4	116	109	-118	-49	-172	-143
PHE 331	16.5	27.0	29.5	132	114	-96	142	-87		

ASX 295	9.5	2.4	18.9	-33	128	81	-107	-48		
GLY 296	9.3	5.5	20.9	77	110	28				
ILE 297	13.1	5.9	20.8	-175	109	162	136	160		
SER 298	16.0	6.3	18.4	-97	103	70	77			
HIS 299	17.1	2.7	18.9	3	117	51	-136	101		
CYS 300	20.1	3.9	20.9	9	125	60	-35			
ASN 301	22.8	3.4	18.2	-153	106	126	-71	-97		
ILE 302	24.7	6.1	16.4	-117	94	128	-53	167		
VAL 303	26.7	4.5	13.6	-84	172	118				
LYS 304	30.4	5.0	13.5	117	116	131	-58	166	-119	168
MET 305	32.3	5.2	10.2	-139	109	159	60	-140	173	
LYS 306	35.8	6.4	9.1	-88	114	95	22	154	-136	-77
LEU 307	35.4	9.8	7.5	-75	112	124	-45	149		
LYS 308	38.2	10.7	5.1	-50	102	154	-76	0	178	116
PRO 309	40.2	13.5	6.6	0	0	-52				
ASP 310	39.0	15.9	3.8	-45	102	-59	-94	-173		
GLU 311	35.6	14.8	4.9	-88	104	-50	-148	29	57	
GLU 312	36.7	13.9	8.4	-31	113	-70	175	79	23	
GLN 313	37.3	17.4	9.6	-55	111	-45	-13	-170	170	
GLN 314	34.3	18.7	7.6	-39	113	-47	-128	-5	-98	
LEU 315	32.0	17.0	10.0	-69	108	-48	148	51		
GLN 316	34.1	18.1	13.0	-55	99	-62	172	-157	84	
LYS 317	34.1	21.5	11.3	-30	113	-67	-133	-10	178	-175
SER 318	30.3	21.8	11.2	-37	111	-61	-56			
ALA 319	30.1	21.3	14.9	-43	109	-50				
THR 320	32.6	24.1	15.3	-55	112	-62	-120			
THR 321	30.8	26.6	13.1	-58	110	-41	100			
LEU 322	27.4	25.4	14.4	-56	117	-55	-135	-101		
TRP 323	27.9	26.0	18.1	-51	117	-29	-119	112		
ASP 324	28.7	29.7	17.5	-39	107	-66	170	-112		
ILE 325	24.9	30.4	17.5	-45	124	-39	-15	-78		
GLN 326	23.9	28.6	20.7	-70	104	-68	-147	-121	-94	
LYS 327	26.7	30.1	22.8	-9	115	-54	-104	-53	-165	-174
ASP 328	24.8	33.4	22.8	-52	107	-36	-99	-141		
LEU 329	22.1	31.4	24.5	77	120	-70	-176	-83		
LYS 330	19.3	33.9	24.8	36	106	59	-80	73	97	120
PHE 331	17.4	31.3	26.8	-92	112	21	-124	85		

TABLE VI

β BENDS—APO AND TERNARY LDH

β bends—Apo LDH

Residues	Sequence				φ_{i+1}	ψ_{i+1}	φ_{i+2}	ψ_{i+2}	$C_{\alpha i}$–$C_{\alpha i+3}$ (Å)	H bond: O–N (Å) distance	H bond: Angle (°)	Comment: Crawford	Comment: Lewis
11–14	Leu	Ala	Thr	Ser	−92	70	−124	−42	7.6	4.5	12	0	IV
30–33	Asp	Ala	Val	Gly	0	−120	−73	−3	5.7	3.1	11	NR	IV
46–49	Ala	Asp	Glu	Val	−112	27	170	133	5.9	5.7	12	NR	IV
70–73	Ser	Leu	Phe	Leu	−81	−8	−114	30	6.0	3.9	16	RT	I
84–87	Asp	Tyr	Ser	Val	−140	22	−82	−27	6.9	5.2	14	0	IV
88–91	Ser	Ala	Gly	Ser	−3	88	75	74	6.3	3.7	7	0	IV
102–106	Gln	Gln	Glu	Gly	−81	−11	150	−42	6.9	6.1	11	0	IV
124–127	Ile	Pro	Asn	Ile	0	−52	−55	−50	4.7	3.0	20	NR	IV
128–131	Val	Lys	His	Ser	−51	−38	−96	1	5.3	3.2	17	RT	VI
130–132B	His	Ser	Pro	Asp	−163	76	0	−29	8.7	6.6	11	0	VI
137–140	Leu	His	Pro	Glu	−116	−31	0	133	8.2	6.8	10	0	VI
139–142	Pro	Glu	Leu	Gly	−51	164	67	62	6.1	4.1	12	0	IV
142–145	Gly	Thr	Asp	Lys	−21	−72	−52	−72	5.2	3.1	14	NR	IV
156–159	Pro	Met	His	Arg	−59	113	137	8	6.3	3.4	8	0	IV
163–166	Ser	Gly	Cys	Asn	−8	96	104	−40	6.1	3.2	4	0	IV
178–181	Glu	Arg	Leu	Gly	−76	−29	−116	−26	5.2	4.4	16	0	I
183–186	His	Ser	Cys	Leu	2	−58	−40	−74	6.0	2.9	5	0	IV
196–199	Gly	Asp	Ser	Val	−51	115	106	11	4.9	2.4	6	RT	II
202–205	Val	Trp	Ser	Gly	−80	4	−99	3	5.8	3.3	16	NR	IV
207–210	Trp	Asp	Ala	Lys	−150	139	91	−103	5.3	5.8	10	NR	IV
212–215	His	Lys	Asp	Val	−56	−32	−44	−60	5.1	2.5	18	NR	IV
215–218	Val	Val	Asp	Ser	−42	−30	−116	−34	6.1	4.0	14	0	IV
220–223	Tyr	Glu	Val	Ile	9	75	89	−26	4.8	2.3	7	NR	IV
237–240	Val	Ser	Asn	Pro	−24	−75	−42	−38	6.4	3.4	9	0	IV
242–245	Asp	Val	Leu	Thr	−21	−52	−95	−54	3.6	2.6	23	NR	IV
244–247	Leu	Thr	Tyr	Val	−162	−36	171	−162	9.6	8.7	7	0	VIII
261–264	Ile	Met	Lys	Asn	−52	−50	−95	28	4.8	2.9	18	RT	I
275–278	Val	Lys	Asp	Phe	−24	−42	−152	41	6.2	3.6	12	0	IV
278–281	Phe	Tyr	Gly	Ile	14	−139	−102	−4	6.5	3.6	8	0	IV
294–297	Asn	Asx	Gly	Ile	−26	87	36	49	5.3	2.8	13	NR	IV
298–301	Ser	His	Cys	Asn	−17	97	46	−7	6.2	3.3	11	0	IV
308–311	Lys	Pro	Asp	Glu	0	−35	−58	−36	5.4	2.9	17	0	IV

0–3	Ace	Thr	Ala	Leu	0	−37	−153	38	5.4	5.6	11	0	IV
10–13	His	Leu	Ala	Thr	−117	173	−76	−11	8.3	6.8	10	0	VII
31–34	Ala	Val	Gly	Met	−29	−53	−62	−66	4.9	2.7	14	NR	IV
86–89	Ser	Val	Ser	Ala	−43	−53	−64	−41	5.8	3.4	14	NR	IV
88–91	Ser	Ala	Gly	Ser	14	86	104	47	6.1	4.1	14	0	IV
91–94	Ser	Lys	Leu	Val	−83	−57	−97	91	6.6	5.2	13	0	IV
103–107	Gln	Glu	Gly	Glu	67	−79	−53	162	6.6	5.2	13	0	IV
128–131	Val	Lys	His	Ser	−50	−44	−93	−18	5.1	3.3	18	NR	IV
130–132B	His	Ser	Pro	Asp	−112	60	0	−44	8.4	5.6	12	0	VI
139–142	Pro	Glu	Leu	Gly	−50	153	67	60	6.3	3.8	3	0	VI
157–160	Met	His	Arg	Ile	−25	−20	−101	−8	5.3	2.4	3	NR	IV
163–166	Ser	Gly	Cys	Asn	0	−102	−80	17	6.4	3.5	8	0	IV
182–185	Val	His	Ser	Cys	−134	166	−24	−38	7.7	6.7	10	0	VII
184–187	Ser	Cys	Leu	Val	−44	−96	−62	−33	5.1	3.7	5	NR	VI
196–199	Gly	Asp	Ser	Val	−29	114	130	−16	5.9	3.0	8	NR	IV
203–206	Trp	Ser	Gly	Met	−56	−18	−106	69	5.3	3.0	14	NR	IV
208–211	Asp	Ala	Lys	Leu	57	−100	−136	11	5.6	2.8	7	NR	IV
214–217	Asp	Val	Val	Asp	−39	−59	−52	−38	5.2	2.8	15	NR	IV
215–218	Val	Val	Asp	Ser	−52	−38	−96	−36	4.5	3.1	20	NR	IV
220–223	Tyr	Glu	Val	Ile	−79	130	161	−19	7.0	3.9	11	0	IV
241–244	Val	Asp	Val	Leu	−64	22	−91	2	5.5	2.4	16	NR	IV
248–251	Ala	Trp	Lys	Gly	−165	35	−82	−3	7.2	5.4	13	0	IV
261–264	Ile	Met	Lys	Asn	−30	−57	−71	−35	4.8	2.5	15	NR	IV
294–297	Asn	Asx	Gly	Ile	−33	81	76	27	5.3	2.8	11	NR	IV
308–311	Lys	Pro	Asp	Glu	0	−52	−45	−59	5.0	3.0	19	NR	IV
326–329	Gln	Lys	Asp	Leu	−9	−54	−52	−36	5.0	2.1	4	NR	IV

[a] All reverse turns where found by visual inspection of skeletal models. A computer search was then made to find other sets of four residues in which φ_{i+1}, ψ_{i+1}, φ_{i+2}, and ψ_{i+2} were within ± 30 of the values for the turn. In most cases this lead to either no new turns or to turns in a helix. If the criteria selected a turn in a helix, the search limit for the appropriate angle was decreased until no turns in helices were located. In general, this lead to almost as many types as reverse turns. The turns are classified by the Crawford (*144*) system and Lewis (*145*) system.

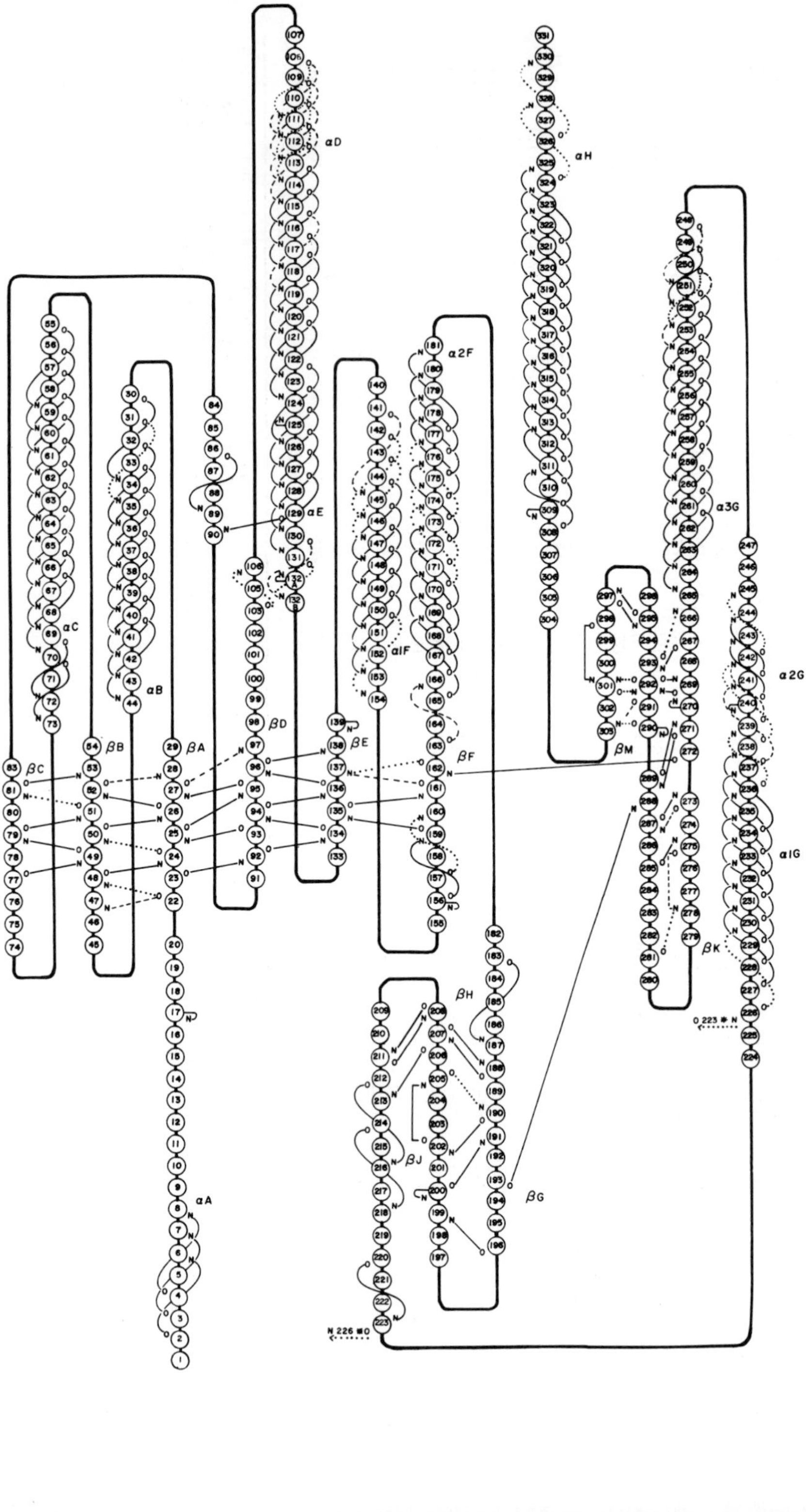

αA
αB
βA
βB
αC
βC
αD
αE
βD
βE
α1F
α2F
βF
βG
βH
βJ
α1G
α2G
α3G
βK
βM
αH
O 223 ≡ N
N 226 ≡ O

An unusual feature of the subunit is the N-terminal 20 residues that extend from the main "body" of the subunit. This "arm" is important in the interaction between subunits (see Section II,A,4,a).

The dinucleotide binding part consists of two mononucleotide binding domains each of similar structure as shown in the exploded view of Fig. 11. Domain B_1 is associated with AMP binding and B_2 with NMN binding. It is suggested that the two domains are the result of gene duplication (*35*). Domains B_1 and B_2 are related by an approximate twofold rotation axis with a small translational element (*149*). This axis is located between βA and βD. The loop and helix αD which are between βD and αE (residues 98–118) are only found in the second domain and are intimately involved with the binding of the pyrophosphate moiety of the coenzyme (see Section II,A,3,e).

The third and the fourth domains also show some degree of similarity as indicated by the distance plot (see above). Each of these two domains could be described as a three-stranded antiparallel β structure with immediate returns from one strand into the next. This piece of polypeptide chain is preceded and followed by a helix. There are an approximate 30° rotation and 25 Å translation which relate these two domains. An important active site residue, histidine-195, occurs between the two first strands of structure in domain C_1. There is no corresponding residue in domain C_2.

d. Location of Side Chains. The location of the amino acid side chains as determined chemically (*125,126*) is mostly consistent with both available electron density maps and agrees with the generalizations arrived at from studies of other proteins (*150,151*). Stretches of inconsistencies occur between residues 134 and 145, and at residue 300.

The ion pairs within the dogfish M_4 subunit are listed in Table VII. Aspartate-168 and histidine-195 are close enough to form an ion pair in the apoenzyme, and this may be important in the catalytic function (see Section III,G). However, these residues are in the wrong orientation to form a hydrogen bond.

149. S. T. Rao and M. G. Rossmann, *JMB* **76**, 241 (1973).
150. B. Lee and F. M. Richards, *JMB* **55**, 379 (1971).
151. A. Shrake and J. A. Rupley, *JMB* **79**, 351 (1973).

FIG. 8. Main chain hydrogen bonding scheme for dogfish M_4 LDH. Continuous lines show hydrogen bonds where they occur both in the apoenzyme and in the LDH:NAD-pyruvate ternary complex. Dotted bonds occur only in the ternary complex while dashed lines are bonds in the apo-structure alone. The nomenclature for the secondary structure is also indicated.

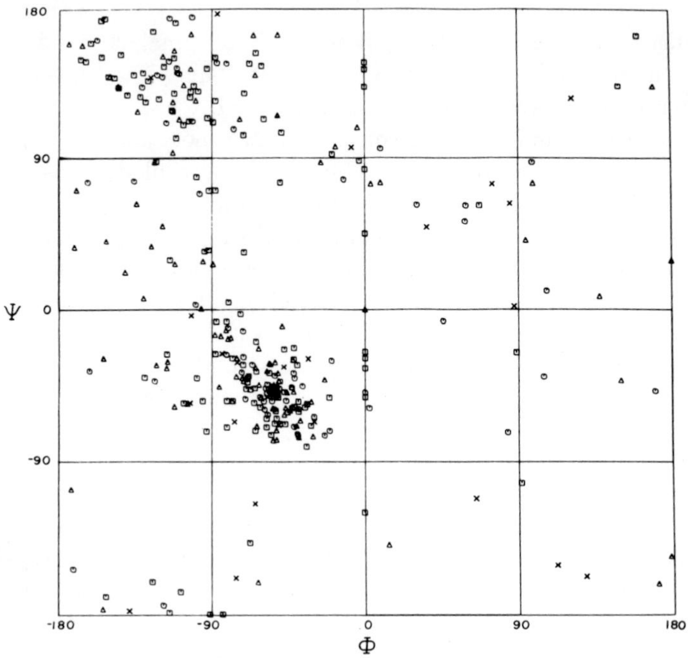

(a)

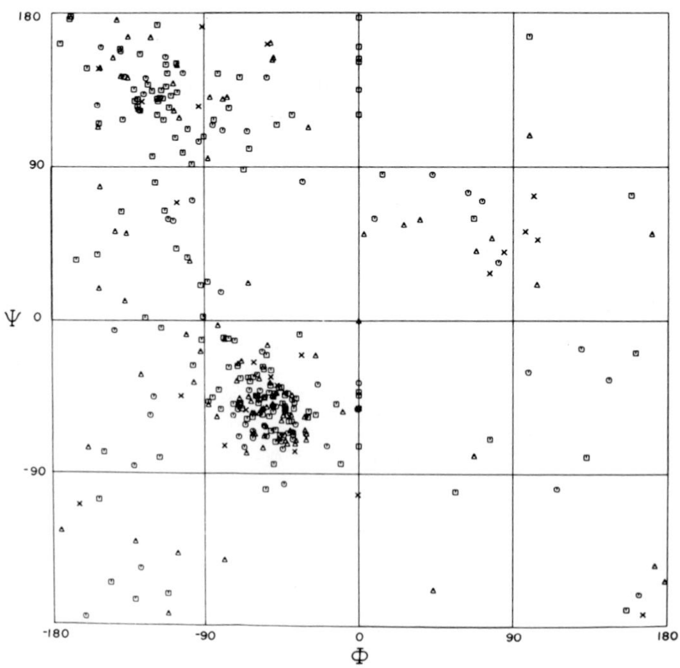

(b)

TABLE VII
POSSIBLE INTERNAL ION PAIRS[a]

Negative charge			Positive charge			Distance (Å)	Comment
Residue	No.	Atom	Residue	No.	Atom		
Asp	30	$C\delta$	Lys	58	$N\zeta$	3.5	A
Asp	36	$C\delta$	His	138	$N\delta_1$	3.7	A
Asp	36	$C\delta$	His	138	$N\delta_1$	4.2	T
Asp	44	$C\delta$	His	74	$N\delta_1$	3.6	A
Asp	44	$C\delta$	His	74	$N\delta_1$	4.7	T
Glu	48	$C\epsilon$	Lys	23	$N\zeta$	3.4	T
Glu	107	$C\epsilon$	Arg	109	$N\eta_1$	8.3	T
Glu	140	$C\epsilon$	Arg	109	$N\epsilon$	3.4	T
Asp	149	$C\delta$	Lys	135	$N\zeta$	3.4	A
Asp	168	$C\delta$	His	195	$N\epsilon_2$	4.3	A
Asp	197	$C\delta$	Arg	109	$N\eta_1$	6.2	T
Glu	259	$C\epsilon$	Lys	263	$N\zeta$	3.7	A
Asp	328	$C\delta$	Lys	330	$N\zeta$	3.0	T

[a] Atom symbols are designated by IUPAC-IUB (1970) nomenclature (*151a*). "A" designates that the paired interaction is found in the apoenzyme subunit only, while "T" designates the ternary complex only. Arginine-109 could form an ion pair to either glutamate-107 or aspartate-197. Both possible pairs are indicated.

Table VIII lists the side chain hydrogen bonds within one subunit. Table IX gives the distances of all tyrosines and tryptophans within the molecule to the center of the nicotinamide in the red subunit and these are depicted in Fig. 14. Table IX also describes the environment of these residues. Amino acid residues that have been chemically modified will be further discussed in Section II,B. Tryptophan has not been modified in LDH, but Shallenberg had implicated it in the catalytic mechanism of many dehydrogenases (*152*). His data could not be confirmed (*153,154*). As can be inferred from Table IX, there is no tryptophan in the neighborhood of the active site.

151a. IUPAC-IUB, Commission on Biochemical Nomenclature, *JBC* **245,** 648 (1970).
152. K. A. Schellenberg, *JBC* **242,** 1815 (1967).
153. J. J. Holbrook, P. A. Roberts, B. Robson, and R. A. Stinson, *Abstr. Int. Congr. Biochem., 8th, 1969* p. 83 (1970).
154. W. S. Allison, H. B. White, and M. J. Connors, *Biochemistry* **10,** 2290 (1971).

FIG. 9. Ramachandran diagrams for the two states of dogfish M_4 LDH: (a) apo and (b) ternary. (□) Hydrophobic residues Ala, Phe, Val, Pro, Met, Leu, and Ile; (○) uncharged hydrophilic residues Asn, Gln, Ser, Thr, and Cys; (△) charged hydrophilic residues Asp, Glu, Lys, Arg, His, Trp, and Tyr; and (×) glycine. Note the dominance of hydrophobic residues in the β-structure area of the diagram ($0 < \psi < 180$, $-180 < \varphi < 0$).

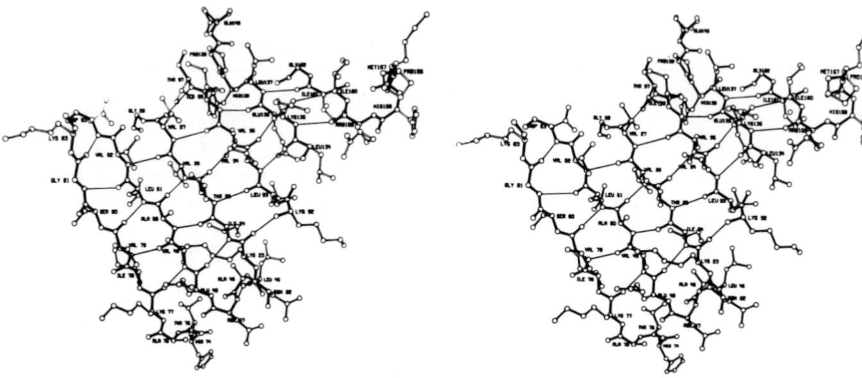

FIG. 10. Parallel pleated sheet area in dogfish M_4 LDH ternary complex. There is a twist of about 100° between the extreme edges of the strand.

The environments of other residues can be constructed from the coordinate lists (*137*) and the information given in Table V.

e. The Coenzyme Binding Site. (*i*) *Crystallographic studies.* The coenzyme is in an open conformation. Dihedral angles of the coenzyme and of coenzyme fragments were presented by Chandrasekhar *et al.* (*138*). Comparisons of these angles for NAD^+ in LDH and in soluble malate dehydrogenase show striking similarities (*155*), as they also do for bound NAD^+ in other dehydrogenases (*35*). Figure 15 gives a summary of the conformations for adenosine, AMP, ADP, and NAD^+ in binary complexes with the enzyme. These are compared to the conformation of NAD^+ in the ternary complex. There are minor but significant differences in the location on the protein and the conformation of these compounds. There is a significant movement of helix α2G toward the coenzyme site when changing from either the apo or a binary to a ternary complex.

The interactions of the coenzyme with the protein that produce the correct conformation needed for catalysis have been partially discussed by Rossmann *et al.* (*35*). Figure 16 shows two views of bound NAD and the surrounding protein. All observed interactions are listed in Table X.

The adenine binds in a hydrophobic pocket. Tyrosine-85 provides an interaction during the initial stages of coenzyme binding. Aspartate-53 also gives some specificity to the adenine orientation. The same interactions occur in glyceraldehyde-3-phosphate dehydrogenase. The adenine ribose is positioned by two hydrogen bonds to its hydroxyls. One of the two hydrogen bond acceptors, aspartate-53, together with its neighboring residues moves 2 Å to provide sufficient room for the coenzyme. This movement has been observed in all binary and ternary complexes where

155. L. E. Webb, E. J. Hill, and L. J. Banaszak, *Biochemistry* **12**, 5101 (1973).

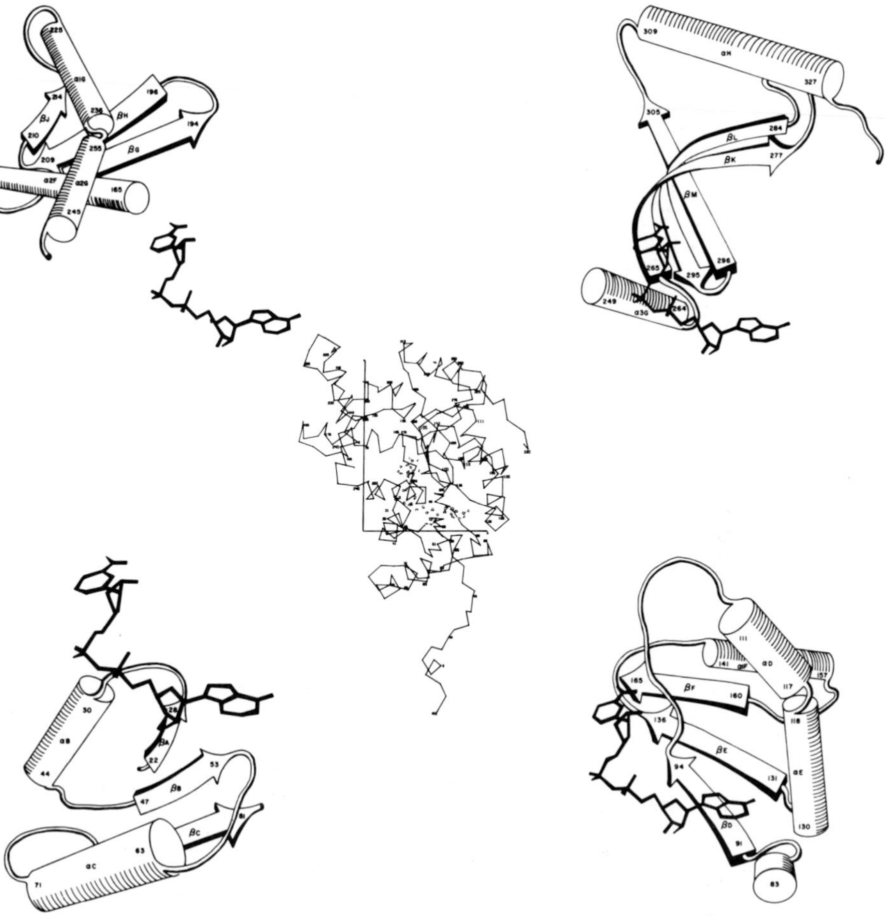

FIG. 11. Diagrammatic representation of domain structure with "blow-up" of each of the four principal parts.

the ribose is present (*138*). In the ternary complex, one of the negative charges of the pyrophosphate is balanced by interaction with arginine-101, while the other is probably solvated. The adenine and nicotinamide phosphates could form hydrogen bonds between the main chain amino groups of residues 31 and 32, respectively. Homologous bonds have been found in glyceraldehyde-3-phosphate dehydrogenase. Tyrosine-246 helps to position arginine-101. Glycines-28 and -99 are located close to the adenine ribose and phosphate, respectively. Any other larger residue would have disrupted the coenzyme binding site severely. These glycines are highly conserved in all dehydrogenases of known structure (*35*).

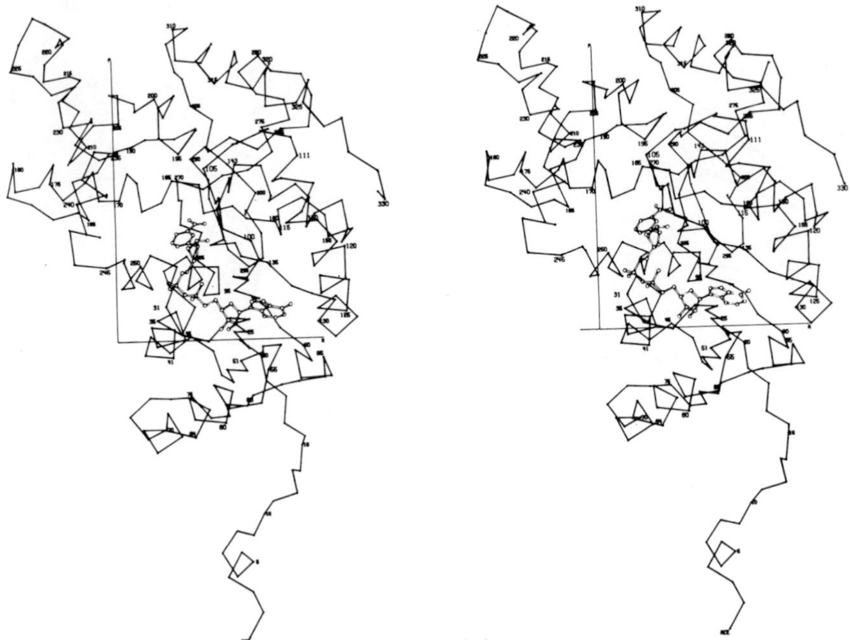

(a)

(b)

Hydrogen bonds to the nicotinamide ribose hydroxyls orients the NMN half of the coenzyme. The nicotinamide ring itself is supported by hydrophobic residues valine-32 and -247 and by a hydrogen bond between its carboxyamide and the ε-amino group of lysine-250. When in the oxidized form, it could make a charge interaction with glutamate-140. Figure 17 gives a simplified representation of the polar interactions between the protein and NAD-pyruvate. Figure 18 shows a stereo view of the substrate environment.

There are two permitted orientations for the nicotinamide ring which can be obtained by rotation about the glycosidic bond (*156,157*). These correspond to exposing the A and B side of the ring to the substrate site. Assuming, therefore, the known A side specificity, the position of the carboxyamide group relative to the substrate site and protein follows (Fig. 18). The carboxyamide group may be assumed to be planar with the nicotinamide ring and its carbonyl oxygen pointed backward toward the N-1 atom by comparison with other known structures (*158*). The present

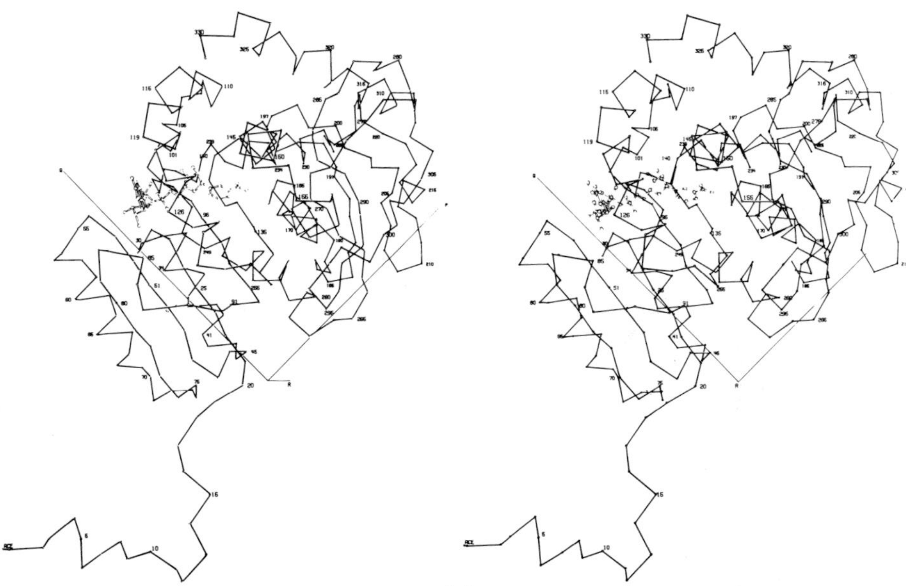

(c)

FIG. 12. Views of single polypeptide chain seen (a) along *Q* as in Fig. 11 and (b) and (c) along *R*. Conformation is that of the apoenzyme in (a) and (b) and that of the ternary complex in (c). The NAD⁺ has been included in the apo-structure for orientation purposes.

156. S. Arnott and D. W. L. Hukins, *BJ* **130,** 453 (1972).
157. M. Sundaralingam, *Biopolymers* **7,** 821 (1969).
158. J. K. Frank, N. N. Thayer, and I. C. Paul, *JACS* **95,** 5386 (1973).

TABLE VIII
SIDE CHAIN HYDROGEN BONDS WITHIN ONE SUBUNIT[a]

Proton donor			Proton acceptor			
Residue	No.	Atom	Residue	No.	Atom	Comment
His	10	$N_{\delta 1}$	Ile	8	O	A
His	10	$N_{\delta 1}$	Asn	301	$O_{\delta 1}$	T
Arg	18	N_{ϵ}	Glu	16	O	AT
Arg	18	$N_{\eta 1}$	Tyr	20	O_{η}	AT
Lys	23	N_{ζ}	Ala	89	O	A
Thr	25	$O_{\gamma 1}$	Asp	91	$O_{\delta 1}$	AT
Lys	43	N_{ζ}	Asn	259	$O_{\delta 1}$	T
Ser	19	O_{γ}	Asp	47	$O_{\delta 2}$	A
Thr	75	$O_{\gamma 1}$	Ala	46	O	T
Ser	86	O_{γ}	Asp	84	$O_{\delta 2}$	A
Ser	88	O	Asp	91	$O_{\delta 2}$	A
Lys	92	N_{ζ}	Tyr	20	O	A
Lys	92	N_{ζ}	Asn	295	$O_{\delta 1}$	T
Gln	102	$O_{\epsilon 1}$	Glu	107	$O_{\epsilon 2}$	A
Ser	108	O_{γ}	Asn	111	$O_{\delta 1}$	AT
Arg	109	N_{ϵ}	Ser	238	O_{γ}	T
Gln	114	$N_{\epsilon 1}$	Leu	329	O	A
Thr	143	$O_{\gamma 1}$	Gln	194	O	AT
Lys	145	N_{ζ}	Asn	116	$O_{\delta 1}$	AT
Lys	147	N_{ζ}	Asn	284	$O_{\delta 1}$	T
Ser	153	O_{γ}	Asp	149	$O_{\delta 2}$	T
Trp	150	$N_{\epsilon 1}$	Leu	155	O	AT
Ser	153	O_{γ}	Trp	150	O	A
His	158	$N_{\delta 1}$	Met	157	O	AT
Arg	159	$N_{\eta 1}$	Pro	132A	O	A
Arg	159	$N_{\eta 1}$	Cys	133	O	T
Ser	169	O_{γ}	Val	191	O	AT
Arg	171	$N_{\eta 1}$	Ser	238	O_{γ}	A
Arg	173	$N_{\eta 1}$	Val	187	O	A
Arg	173	$N_{\eta 2}$	Cys	185	O	AT
Arg	179	$N_{\eta 1}$	Ser	230	O_{γ}	A
Ser	318	O_{γ}	Glu	194	$O_{\epsilon 2}$	AT
His	195	$N_{\delta 1}$	Gly	164	O	T
Gly	196	N	Asp	168	$O_{\delta 1}$	A
Ser	198	O_{γ}	Asp	197	O	AT
His	212	$N_{\epsilon 2}$	Ser	204	O	T
Tyr	220	O_{η}	Asp	214	$O_{\delta 1}$	AT
Ser	253	O_{γ}	Asn	166	$O_{\delta 1}$	AT
His	269		Asp	256		*
Arg	267					*

TABLE VIII (*Continued*)

Proton donor			Proton acceptor			
Residue	No.	Atom	Residue	No.	Atom	Comment
Lys	43		Glu	259		*
Lys	263					*
Arg	267					*
Arg	267	$N_{\eta 1}$	Thr	260	$O_{\gamma 1}$	AT
Arg	267	$N_{\eta 1}$	His	269	$N_{\epsilon 2}$	AT
Ser	272	O_{γ}	Asn	146	$O_{\delta 1}$	A
Lys	276	N_{ζ}	Asn	284	$O_{\delta 1}$	A
Ser	288	O_{γ}	Gly	164	O	A
Asp	295	N	Asn	264	$O_{\delta 1}$	AT
His	299	$N_{\delta 1}$	Asn	294	$O_{\delta 1}$	AT
Asn	301	$N_{\delta 2}$	His	299	O	AT
Gln	313	$N_{\epsilon 2}$	Gln	316	$O_{\epsilon 1}$	A
Trp	323	$N_{\epsilon 1}$	Asp	283	O	T
Lys	330	N_{ζ}	Gln	114	$O_{\epsilon 1}$	T

[a] See Table VII for description of symbols (O and N used to designate main chain atoms). An asterisk indicates the residues involved in a complex hydrophilic region found near the molecular center of the ternary complex.

maps do not distinguish between a flat and a puckered dihydronicotinamide ring.

The 400-fold better binding of NADH over NAD^+ is not well understood in terms of structure at this time. An alteration of interaction between lysine-250 and the nicotinamide moiety might contribute to this change. The nicotinamide ring is positioned between the possibly negatively charged glutamate-140 and the positively charged histidine-138 in the ternary complex. Aspartate-36 polarizes the imidazole ring in both the apoenzyme and the ternary complex, although histidine-138 possibly undergoes a translation (Fig. 19). While the electrostatic field between histidine-138 and glutamate-140 might provide an explanation for the coenzyme binding constants, residue 138 requires sequence confirmation (see Section II,A,3,d).

(*ii*) *Coenzyme analog studies*. A large number of NAD analogs have been synthesized in an attempt to determine those portions of the coenzyme that are necessary for productive binding. A list has been given by Colowick *et al.* (*3*) covering the literature through 1964, which has been extended by Everse and Kaplan (*1*). Some difficulty exists in comparing all these results because of the often scanty and incomplete data of the published enzymic properties. Where comparative studies between different enzymes have been made (mostly LDH, liver, and yeast alcohol

TABLE IX

DISTANCES IN Å FROM NICOTINAMIDE RING IN THE RED SUBUNIT TO ALL TYROSINES AND TRYPTOPHANS IN THE MOLECULE[a]

		Subunit, distances in Å				
Residue	No.	Red	Blue	Yellow	Green	Environment
Tyr	20	31.8	39.8	37.2	27.3	Arg-18; residues in green subunit: Lys-92, Asp-132B, Asn-295
Tyr	85	19.6	55.8	32.1	49.9	Val-52, Asn-126, adenine ring
Tyr	174	17.7	36.4	29.7	48.5	Ile-175, Glu-178, His-183; residue in yellow subunit: His-68
Tyr	220	28.2	40.7	48.6	59.2	Trp-203, Asp-214, Glu-221, Lys-224
Tyr	228	24.3	49.2	46.2	64.6	Trp-203, Glu-221
Tyr	246	10.4	50.9	20.1	53.1	Arg-101, Thr-245, Ala-248
Tyr	279	25.0	45.5	51.6	57.6	Phe-278, Met-305, Leu-307
Trp	150	19.9	50.1	44.8	52.3	Lys-147, Met-157, Ile-160, Met-274, Phe 286
Trp	190	19.7	33.4	41.0	46.6	Ser-169, Ile-188, Val-268, Pro-290, Met-305; residues in blue subunit: Leu-186, Ala-209, Lys-210
Trp	203	23.5	42.6	45.6	58.7	Ser-201, Tyr-220, Glu-221, Tyr-228
Trp	207	27.5	27.7	44.9	45.9	Val-186, Ile-188; residues in blue subunit: Ile-188, Tyr-207, His-212
Trp	231	16.5	44.6	40.0	57.3	Phe-172, Val-199, Ser-201, Val-235
Trp	249	13.1	42.6	16.2	44.6	Ile-38, Ala-248; residues in yellow subunit: Ile-38, Met-34
Trp	323	21.7	56.9	49.3	63.1	Leu-147, Asp-283, Asn-284, Val-285, Asn-326, Lys-327

[a] Environment of each residue in the red subunit is also given.

dehydrogenase and, less frequently, malate and glyceraldehyde-3-phosphate dehydrogenase), the general pattern is the same. In view of the exceedingly similar NAD binding sites in all these enzymes (*35*) this is not surprising.

Where in the past it has been necessary to deduce the structures from the peroperties, it is the purpose of this section to discuss the analog properties in terms of the known LDH structure.

The conformation of the NAD^+ cofactor is probably "closed" a large part of the time while in solution (*159–161*). McPherson (*162*) showed

159. R. H. Sarma and N. O. Kaplan, *Biochemistry* **9**, 539 (1970).
160. R. H. Sarma, M. Moore, and N. O. Kaplan, *Biochemistry* **9**, 549 (1970).
161. R. H. Sarma and N. O. Kaplan, *Biochemistry* **9**, 557 (1970).
162. A. McPherson, Jr., *JMB* **51**, 39 (1970).

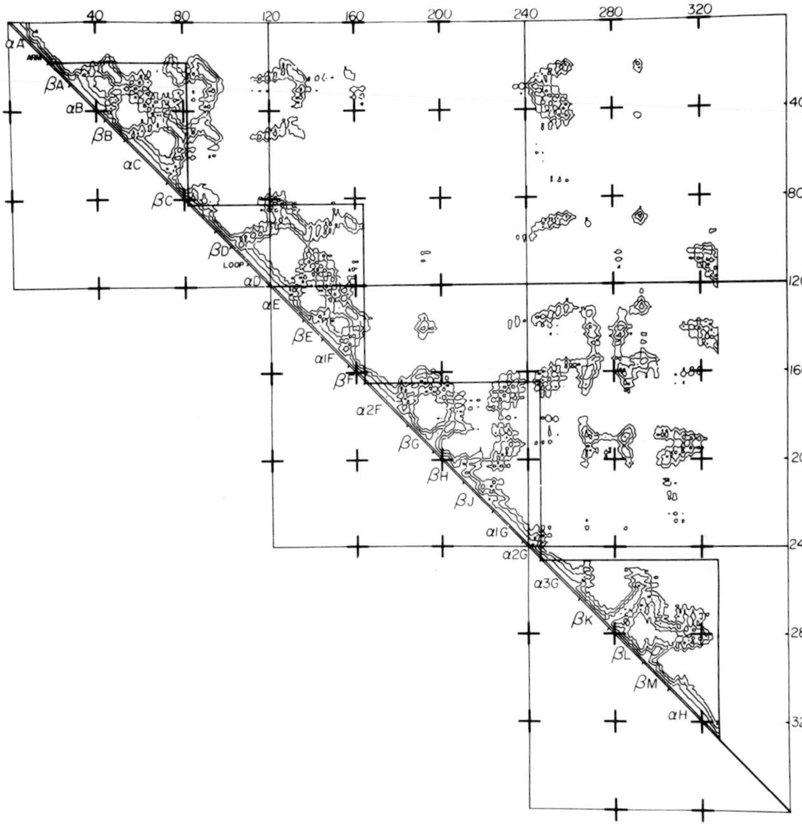

FIG. 13. The distance plot for LDH with contours at 0, 4, 8, and 16 Å between C_α atoms. Different domains have been identified along the diagonal. The first two are each a mononucleotide binding structure which together give the complete NAD^+ binding protein found in all currently known dehydrogenases.

that NMN^+ alone does not inhibit LDH, but it does in the presence of AMP. Hence, the binding of the AMP portion of the coenzyme must generate a site for the nicotinamide moiety. In view of the probable order of these events, studies relating to the adenine moiety will be discussed first, followed progressively by the adenine ribose, the phosphates, the nicotinamide ribose and finally the nicotinamide itself. Chandrasekhar *et al.* (*138*) have indicated the corresponding progressive conformational changes of the NAD^+ molecule as it binds to the protein (Fig. 15).

The adenine moiety binds into a nonspecific hydrophobic pocket with possible hydrogen bonds to the N-1 and N-3 atoms. These observations

TABLE X
INTERACTIONS WITH COENZYME

Coenzyme moiety	Residue	Type
Adenine	Val-27	Hydrophobic
	Val-52	Hydrophobic
	Asp-53	H bond to N-3[a]
	Val-54	Hydrophobic
	Tyr-85	H bond to N-1
	Ile-96	Hydrophobic
	Ala-98	Hydrophobic
	Ile-119	Hydrophobic
	Ile-123	Hydrophobic
Adenine ribose	Gly-28	Hydrophobic
	Asp-30	H bond to O-3′[b]
	Asp-53	H bond to O-2′[a]
	Met-55	Hydrophobic
	Lys-58	H bond to O-3′
Pyrophosphate	Ala-31	Hydrophobic
	Lys-58	Charge interaction
	Gly-99	Hydrophobic
	Arg-101	Charge interaction
	Tyr-246	H bond to Arg-101
Nicotinamide ribose	Val-32	Hydrophobic
	Thr-97	Hydrophobic
	Carbonyl on 98	H bond to O-3′[c]
	Arg-109	H bond to Glu-140
	Pro-139	Hydrophobic
	Amino on 140	H bond to O-2′
Nicotinamide	Val-32	Hydrophobic
	His-138	Polarized by Asp-36
	Glu-140	Charge interaction (?)
	Val-247	Hydrophobic
	Lys-250	H bond to O-7
Substrate	Arg-109	
	Arg-171	H bond to carboxy
	His-195	
	Ser-238	

[a] Aspartate-53 can potentially hydrogen bond to either N-3 of adenine or O-3′ of ribose.

[b] In Adams *et al.* (*139a*) the main chain carbonyl of residue 29 is implicated. However, aspartate-30 is probably a better candidate.

[c] In Adams *et al.* (*139a*) the main chain carbonyl of residue 100 is implicated. However, the main chain carbonyl of residue 98 is better oriented.

COMPOUNDS FOR TABLES XI AND XII

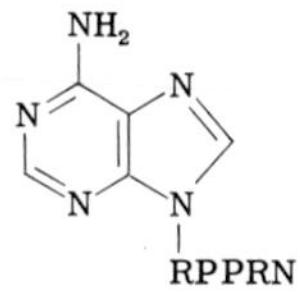

NAD
(Nicotinamide adenine dinucleotide)
(I)

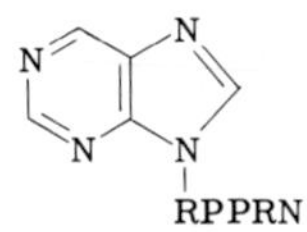

NPD
(*N*-Purine-D)
(II)

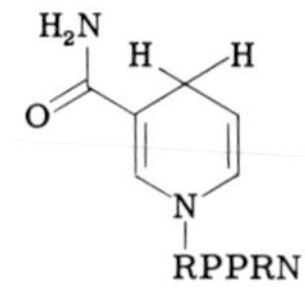

NNDH
(*N*-Nicotinamide-D)
(III)

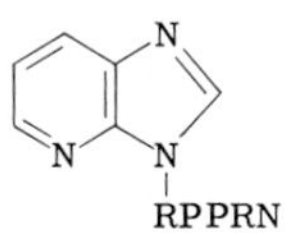

N1DD
(*N*-1-Deazapurine-D)
(IV)

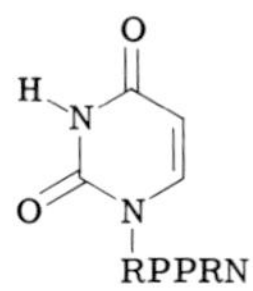

NUD
(*N*-Uracyl-D)
(V)

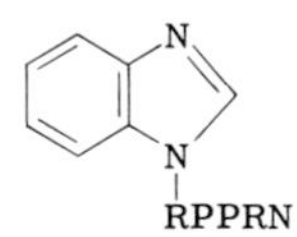

NBD
(*N*-Benz-D)
(VI)

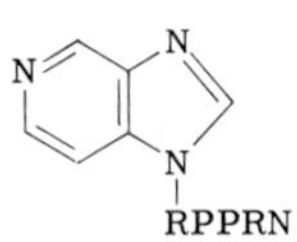

N3DD
(*N*-3-Deazapurine-D)
(VII)

NPhD
(*N*-Phenyl-D)
(VIII)

NMNPR
(*N*-Mononucleotide PR)
(IX)

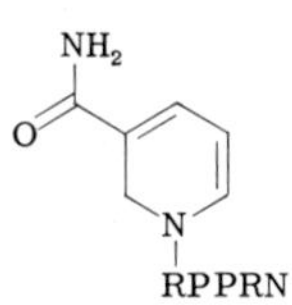

NND+
(*N*-Nicotinamide-D)
(X)

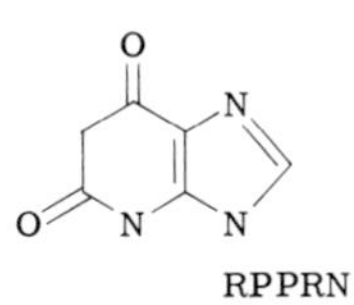

NHD
(*N*-Hypoxanthine-D)
(XI)

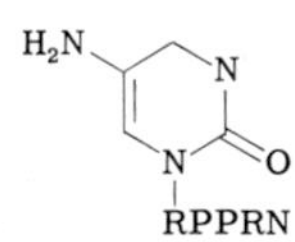

NCD
(*N*-Cytosine-D)
(XII)

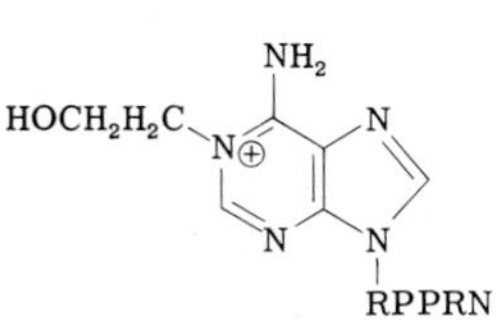

(XIII)

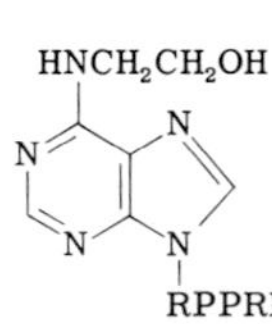

(XIV)

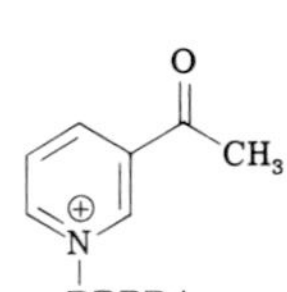

3AcPyAD
(3-Acetylpyridine-AD)
(XV)

TABLE XI
MODIFICATIONS TO ADENINE MOIETY OF NAD—KINETIC CONSTANTS[a]

	Oxidized analog				Reduced analog			
	Compound	Ref.	$K_m \times 10^4$ (molar)	V_{max}[b]	Compound	Ref.	$K_m \times 10^4$ (molar)	V_{max}[b]
I	NAD⊕	*162a*	0.75	17,000	NADH		0.1	80,000
II	NPD⊕	*162b*	1.8	18,000				
III	NNDH⊕	*162a*	2.0	14,000	$NNDH_2$	*162a*	0.6	55,000
IV	N1DD⊕	*162b, 162c*	2.8	15,000				
V	NUD⊕	*162d*	5.6	15,000				
VI	NBD⊕	*162a–162c*	6.0	15,000	NBDH	*162a*	0.3	66,000
VII	N3DD⊕	*162c*	6.2	20,000				
VIII	NPhD⊕	*162c, 162e*	8.7	15,000				
IX	NMNPR⊕	*162a, 162c, 162e*	60	500	NMNPRH	*162a*	1	9,600
X	NND²⊕	*162a*	100	300	NNDH⊕	*162a*	10	3,500

[a] All results for pig H_4 LDH at pH 9.5 and 25°C.
[b] V_{max} is defined as moles NADH/moles LDH/min.

TABLE XII
MODIFICATIONS TO ADENINE MOIETY OF NAD—RELATIVE ACTIVITIES

Compound	Ref.	Beef H_4 LDH pH 7.1 (%)	Beef H_4 LDH pH 9.5 (%)	Pig H_4 LDH pH 7.1 (%)	Pig H_4 LDH pH 9.5 (%)	Rabbit M_4 LDH pH 7.1 (%)	Rabbit M_4 LDH pH 9.5 (%)
I NAD⊕	*163*	100	100		100		100
XIV	*163a*	52	55			104	81
XI NHD⊕	*163*				50		50
XII NCD⊕	*163*				55		17
X NND[2]⊕	*163*				0.12		0.015
IX NMNPR⊕	*163*				0.13		0.025
NMN	*163*				0		0
NR	*163*				0		0
XIII	*163a*	2.0	11.0			0.5	4.0
XV 3AcPyAD	*163*				83		14
3AcPyMNPR	*163*				0.87		0.28
3AcPyMN	*163*				0		0

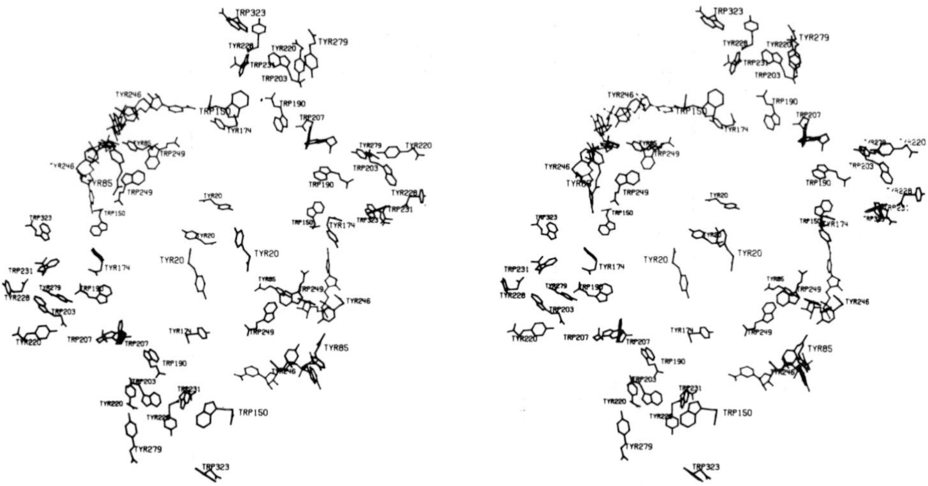

FIG. 14. Stereo view of tyrosines, trytophans, and bound NAD in an LDH molecule.

are fairly consistent with modifications of the adenine moiety in the coenzyme [Tables XI (*162a–162e*) and XII (*163,163a*)].

The best analogs are those which possess both a large aromatic group and have the ability to form hydrogen bonds as in the true coenzyme. The absence of any kind of aromatic anchor, as in compound IX, results in exceedingly poor kinetic properties. Extension of the amino group in position 6, which points outward into solution, produces little effect (e.g., compound XIV). The presence of positive charge (compound X) is clearly objectionable, possibly because such a property may repel arginine-101 and hence hinder the collapse of the loop. This demonstrates that the presence of positive charge on nicotinamide is a distinguishing feature by which the enzyme recognizes the two differing ends of the oxidized coenzyme. It is less clear how the enzyme distinguishes between the bases of the reduced coenzyme.

The adenine ribose ring is bound to the protein by both of its hydroxy groups. The conservation of aspartate-53, which binds to the O-2′ atom in other dehydrogenases, suggests that this is a most important association. It is thus not surprising that NADP, with its extra phosphate

162a. C. Woenckhaus and D. Scherr, *Hoppe-Seyler's Z. Physiol. Chem.* **354**, 53 (1973).
162b. C. Woenckhaus and G. Pfleiderer, *Biochem. Z.* **341**, 495 (1965).
162c. C. Woenckhaus and P. Zumpe, *Z. Naturforsch. B* **23**, 484 (1968).
162d. C. Woenckhaus and D. Scherr, *Z. Naturforsch. B* **26**, 106 (1971).
162e. C. Woenckhaus and M. H. Volz, *Chem. Ber.* **99**, 1712 (1966).
163. G. Pfleiderer, E. Sann, and F. Ortanderl, *BBA* **73**, 39 (1963).
163a. H. G. Windmueller and N. O. Kaplan, *JBC* **236**, 2716 (1961).

on the O-2′ position, is virtually inactive (*163*). Both rat liver LDH (*164*) and beef heart LDH (*165*) show a significant increase of activity with NADPH as pH decreases (Table XIII), possibly involving an altered form of binding in which the charge repulsion between aspartate-53 and the O-2′ phosphate is eliminated.

Substitution of the cyclic sugar by an aliphatic chain does not cause complete inactivity (Table XIV). When the length of the chain is such as to roughly span the space covered by the ribose ($n = 3$ or 4 in the compound A-CH_2—$(CH_2)_n$—PPRN), then the activity is optimized (*166*).

The primary function of the phosphates is probably to attract the guanadinium group of arginine-101 and thereby aid the conformational change of the loop. When the phosphates are blocked (*167*) there is complete inactivity.

In the ternary enzyme–coenzyme–substrate complex the nicotinamide

TABLE XIII
NADPH ACTIVITY[a]

Enzyme	pH	Relative activity with NADH (%)	Relative activity with NADPH (%)
Beef H_4 LDH	7.5	100	0.15
Beef H_4 LDH	6.0	90	2.0

[a] Results from Fawcett and Kaplan (*165*).

TABLE XIV
EFFECT OF INCREASING SPACER BETWEEN BASES[a]

A—CH_2—$(CH_2)_n$—PPRN

No. of methylene groups (n)	$K_m \times 10^4$ (molar)	$V_{max} \times$ min	Compound	$K_m \times 10^4$ (molar)	V_{max} (moles/moles/min)
			NAD (dihydro)	0.1	140,000
1	50	30	$n = 1$ (dihydro)	30	4,000
2	100	150	$n = 2$ (dihydro)	20	5,000
3	60	200	$n = 3$ (dihydro)	10	12,000
4	100	100	$n = 4$ (dihydro)	20	12,000

[a] From Jeck and Wilhelm (*166*). Species and isozyme unknown.

164. F. Navazio, B. B. Ernster, and L. Ernster, *BBA* **26,** 416 (1957).
165. C. P. Fawcett and N. O. Kaplan, *JBC* **237,** 1709 (1962).
166. R. Jeck and G. Wilhelm, *Justus Liebigs Ann. Chem.* p. 531 (1973).
167. G. Pfleiderer, E. Sann, and A. Stock, *Chem. Ber.* **93,** 3083 (1960).

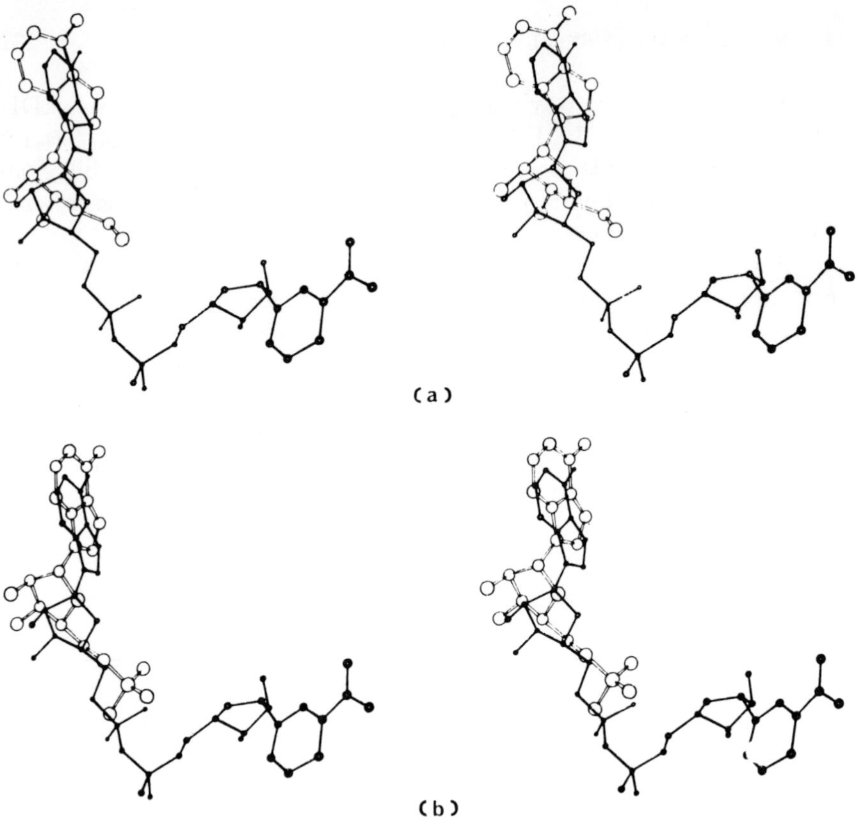

FIG. 15. Stereo drawing depicting the difference in the conformation of the coenzyme and its fragments when bound to LDH. (See facing page.)

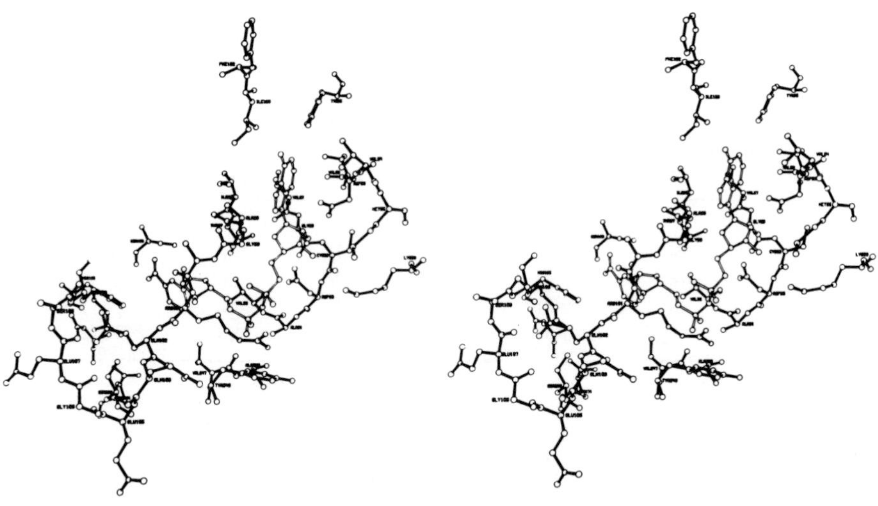

(a)

FIG. 16. Views roughly parallel to (a) the Q and (b) the R axis showing bound NAD in the ternary complex.

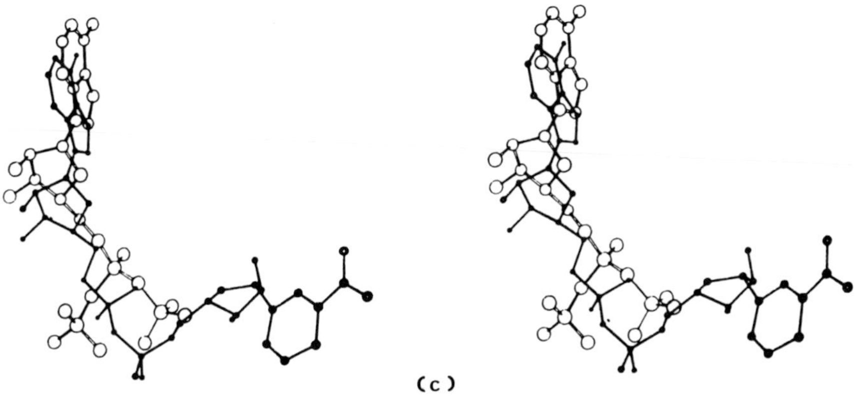

(c)

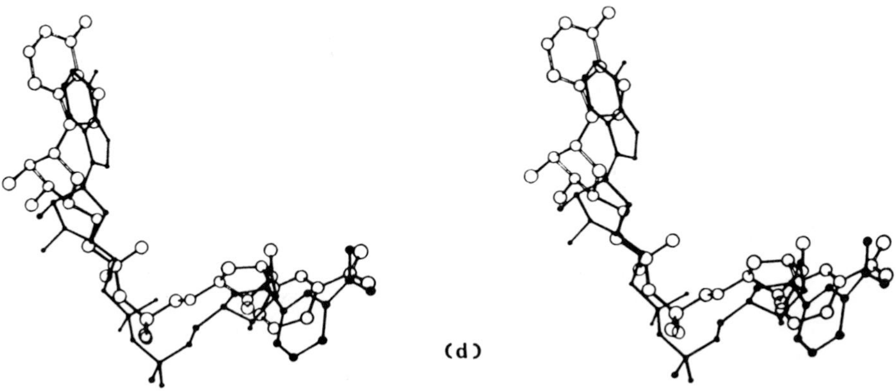

(d)

FIG. 15. (Cont.). This is demonstrated with reference to the NAD⁺ structure in the ternary complex (in solid black). Shown are (a) adenosine, (b) AMP, (c) ADP, and (d) NAD⁺ in the binary complex (taken from Chandrasekhar *et al. 138*).

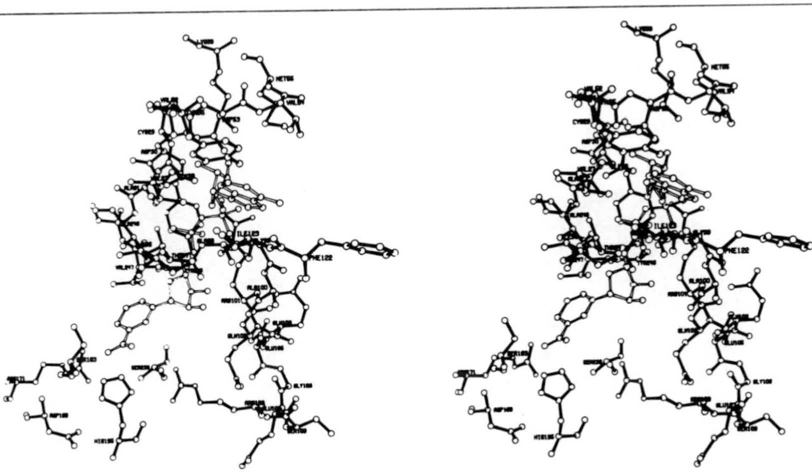

(b)

FIG. 16. (Cont.).

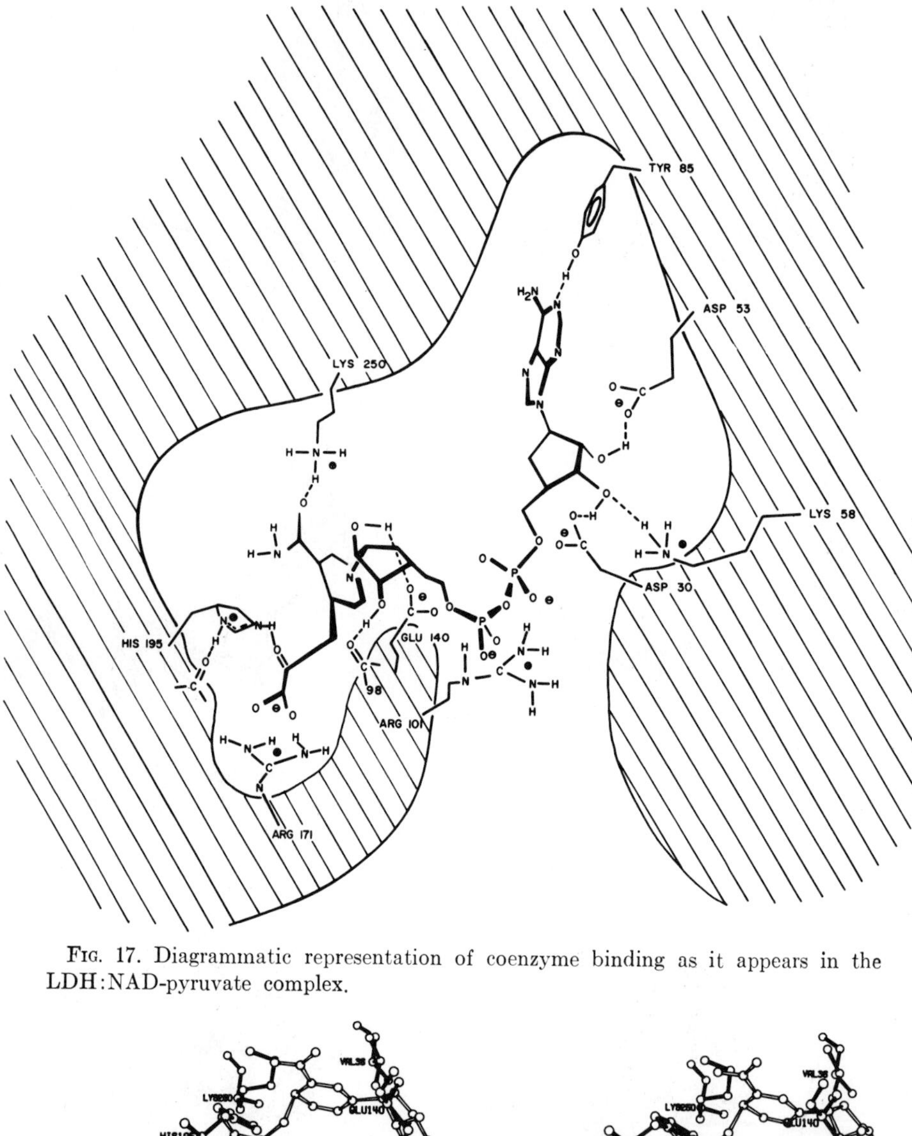

FIG. 17. Diagrammatic representation of coenzyme binding as it appears in the LDH:NAD-pyruvate complex.

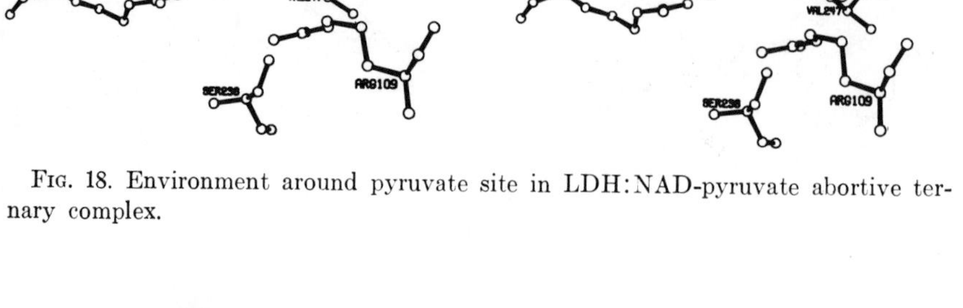

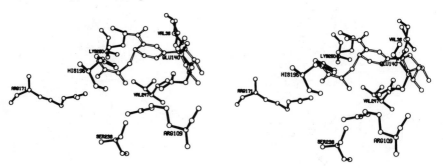

FIG. 18. Environment around pyruvate site in LDH:NAD-pyruvate abortive ternary complex.

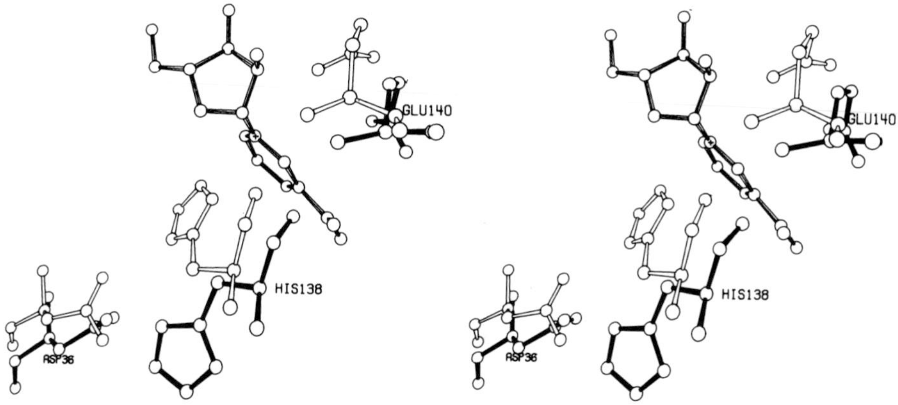

FIG. 19. Plausible conformational change in histidine-138 between the apoenzyme (darkened bonds) and its abortive ternary complex with NAD^+ and pyruvate (open bonds). Note also residue glutamate-140 and the positive charge on the N-1 atom of the oxidized coenzyme.

ribose moiety is bound by residues found on the flexible loop. It thus gives a final adjustment to the position of the nicotinamide ring prior to catalysis. The analog compound with the nicotinamide ribose replaced by the aliphatic chain —$(CH_2)_5$— is a good inhibitor (*168*) which is consistent with structural results.

The nicotinamide moiety must be oriented in such a way as to present the hydrogen on the C-4 atom in the correct position with respect to the substrate. It is free of immediate protein environment, although the carboxyamide group is probably associated with lysine-250.

A substitution on the 4 position of the nicotinamide must and does (*169,170*) produce an inactive analog, but substitution on the 6 position leaves the full activity of the coenzyme (*170*). The role of the carboxyamide group in the 3 position (Table XV) has been extensively reviewed (*30,163*). Only substituents in the 3 position containing a carbon-oxygen, -sulfur, or -nitrogen double bond are active. The amide group may serve primarily to provide the pyridine ring with the appropriate redox potential by means of a conjugated double bond. The ratio of activity between analogs with different substitutions at the 3 position has been used to investigate dissimilarities among lactate dehydrogenases (*171*).

f. Substrate Site. (*i*) *Crystallographic studies.* The position of the substrate binding site has been determined by the pyruvate in the

168. K. H. Göbbler and C. Woenckhaus, *Justus Liebigs Ann. Chem.* **700,** 180 (1966).
169. C. Woenckhaus, E. Schättle, R. Jeck, and J. Berghäuser, *Hoppe-Seyler's Z. Physiol. Chem.* **353,** 559 (1972).
170. G. Pfleiderer, J. J. Holbrook, and T. Wieland, *Biochem. Z.* **338,** 52 (1963).
171. N. O. Kaplan and M. M. Ciotti, *Ann. N. Y. Acad. Sci.* **94,** 701 (1961).
171a. M. R. Banner and S. B. Rosalki, *Nature* (*London*) **213,** 726 (1967).

TABLE XV

Modifications at the 3 Position of the Nicotinamide Group

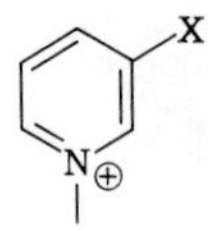

X	Relative rates: Rabbit M_4 LDH	Relative rates: Beef H_4 LDH	Ref.
$-C(=O)NH_2$	1.0	1.0	*30*
$-C(=O)CH_3$	0.14		*163*
$-C(=O)NHOH$	0.2	0.1	*30*
$-C(=O)NHNH_2$	0.33	0.09	*30*
$-CH=N-OH$	0.07	0.01	*30*
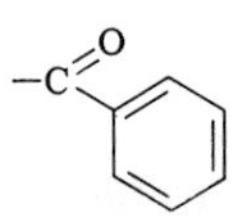	0	0	*30*
$-NH_2$	0	0	*30*
$-NH-C(=O)CH_3$	0	0	*30*
$-CH=CH-C(=O)NH_2$	0	0	*30*
$-C(=O)-CH(CH_3)_2$	1.25	0.37	*30*
$-C(=S)NH_2$	0.03	0.41	*30*

LDH:NAD-pyruvate ternary complex. This complex is isomorphous with respect to the LDH:NAD:oxalate and LDH:NADH:oxamate ternary complexes. All three compounds show clearly the position of the substrate site between the nicotinamide and histidine-195. The same position is occupied by a sulfate ion in the apo-structure. It can be replaced by various anions (*172*) but not oxamate. Although the position of the substrate site (close to the subunit center) is thus well characterized, the orientation of the studied inhibitors in the electron density can only be deduced indirectly by considering the interactions with the protein. The covalent nature of the adduct NAD-pyruvate adds another constraint to the orientation of the substratelike part of this inhibitor. Difference electron density maps among the ternary complexes show peaks in this region which have yet to be interpreted.

In the ternary complex the position of the loop excludes bulk water from the active site. The negative charge of the substrate needs to be neutralized. Two arginine residues 109 and 171 as well as histidine-195 are available to provide a partner for an ion pair. Arginine-171 seems to be best suited. If the inhibitor or substrate analog molecules used in the crystallographic studies are replaced in the model by the real substrates, then histidine-195 would be conveniently located to form a hydrogen bond with the carbonyl of pyruvate or the hydroxyl of lactate (Figs. 18 and 20). This hydrogen bond is not present in the NAD-pyruvate

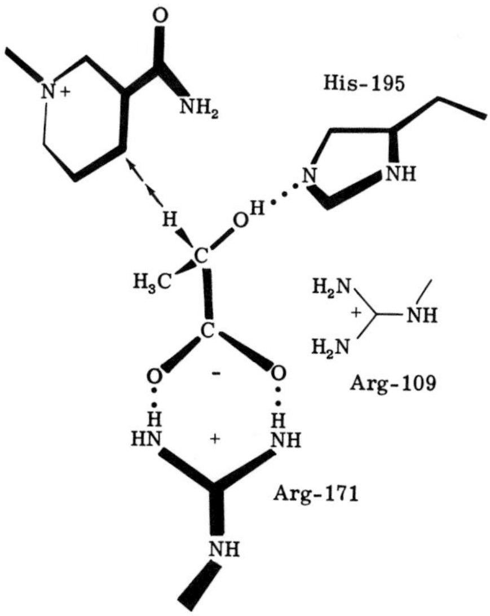

FIG. 20. Diagrammatic representation of anticipated substrate binding in the active ternary intermediate.

172. A. Meister, *JBC* **197**, 309 (1952).

complex. The charged residues enclosed in the active site when the loop is down include arginine-109 and -171, histidine-195, glutamate-107 and -140, and aspartate-168. Other changes arise from the pyridinium cation in NAD^+ and the carboxyl group of the substrate. The distances between these are given in Table XVI. Undoubtedly, these groups form a basis for complicated charge interactions of importance for the catalytic function. This will be further discussed in Section III,G.

TABLE XVI
DISTANCES BETWEEN CHARGED GROUPS IN ACTIVE SITE IN LDH : NAD-PYRUVATE COMPLEX[a]

Residue	No.	Atom	Residue	No.	Atom	Distance (Å)
Lys	250	N_ϵ	Glu	107	C_δ	18.8
			Arg	109	C_ξ	10.4
			Glu	140	C_δ	9.5
			Asp	168	C_γ	9.7
			Arg	171	C_ξ	8.1
			His	195	$N_{\epsilon 2}$	6.1
			His	195	$N_{\delta 1}$	6.6
			Nic		N_1	7.2
			Pyr		C_3	6.1
Glu	107	C_δ	Arg	109	C_ζ	8.7
			Glu	140	C_δ	11.8
			Asp	168	C_γ	14.4
			Arg	171	C_ζ	14.4
			His	195	$N_{\epsilon 2}$	13.2
			His	195	$N_{\delta 1}$	13.9
			Nic		N_1	15.7
			Pyr		C_3	14.4
Arg	109	C_ζ	Glu	140	C_δ	4.1
			Asp	168	C_γ	9.2
			Arg	171	C_ζ	8.4
			His	195	$N_{\epsilon 2}$	5.0
			His	195	$N_{\delta 1}$	6.6
			Nic		N_1	7.2
			Pyr		C_3	6.4
Glu	140	C_δ	Asp	168	C_γ	11.6
			Arg	171	C_ζ	10.8
			His	195	$N_{\epsilon 2}$	5.7
			His	195	$N_{\delta 1}$	7.6
			Nic		N_1	4.4
			Pyr		C_3	7.4

TABLE XVI (*Continued*)

Residue	No.	Atom	Residue	No.	Atom	Distance (A)
Asp	168	C_γ	Arg	171	C_ζ	6.3
			His	195	$N_{\epsilon 2}$	6.2
			His	195	$N_{\delta 1}$	4.6
			Nic		N_1	11.9
			Pyr		C_3	6.5
Arg	171	C_ζ	His	195	$N_{\epsilon 2}$	6.6
			His	195	$N_{\delta 1}$	6.9
			Nic		N_1	10.3
			Pyr		C_3	4.4
His	195	$N_{\epsilon 2}$	Nic		N_1	5.8
			Pyr		C_3	3.0
His	195	$N_{\delta 1}$	Nic		N_1	7.6
			Pyr		C_3	4.0
Nic		N_1	Pyr		C_3	6.2

[a] See Table VII for description of nomenclature. "Nic" refers to the nicotinamide ring of the coenzyme and "Pyr" refers to the pyruvate substrate in the abortive ternary complex.

(*ii*) *Substrate analog studies.* L-Lactate dehydrogenase shows a broad specificity for α-keto acids (XVI) and α-hydroxy acids (XVII).

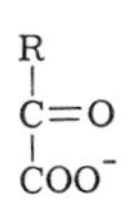

(XVI)

(XVII)

Substrates have been examined with various isozymes from various species. Superimposed upon the variation due to the *R* group (Table XVII) are the relative specificities of different isozymes ($X_4 < H_4 < M_4$). Meister (*12,172*) examined a series of straight chain α-keto acids from C_2 (glyoxalate) to C_9 (α-ketononanoate). The K_m values showed a minimum at pyruvate (C_3) and increased up to C_5. After C_5 increasing chain length to C_9 had little effect on K_m. Inspection of the model suggests that these extra atoms should lie in the sizable gap between arginine-101 and glycine-196.

The maximum velocity with C_2, C_3, and C_4 keto acids is similar, but then rapidly diminishes from C_5 to C_9. An interaction of atoms C-5 to C-9 of the bound substrate with the enzyme is apparently unfavorable

TABLE XVII

KINETIC CONSTANTS FOR SUBSTRATE ANALOGS

Reactant	pH	Parameter[a]	H_4	Isozyme M_4	X_4	Ref.
Pyruvate						
CH_3COCOO^-	7.4	V_{max}[b]	256		666	*251*
	7.4	K_m	5.0×10^{-5}	2.0×10^{-4}	5.2×10^{-5}	*251*
Lactate						
$CH_3CHOHCOO^-$	9.0	K_m	3.7×10^{-3}	5.5×10^{-3}	2.1×10^{-3}	*251*
α-Ketobutyrate						
$CH_3CH_2COCOO^-$	7.4	V_{max}[c]	1.03	0.28		*171a*
	7.4	K_m	1×10^{-3}	3.5×10^{-3}	0.048×10^{-3}	*251*
α-Ketovalerate						
$CH_3CH_2CH_2COCOO^-$	7.4	V_{max}	0.016[c]		684[b]	*174,175*
	7.4	K_m			8×10^{-6}	*174,251*
α-Hydroxyvalerate						
$CH_3CH_2CH_2CHOHCOO^-$	9.0	K_m			2×10^{-3}	*251*

[a] K_m in moles/liter.

[b] V_{max} in μmoles/mg/min.

[c] Relative to pyruvate = 1.

for catalysis. The importance of the interactions beyond C_4 are also seen in the behavior of a series of α,γ-diketo acids (*12*). From C_4 to C_9 these acids all have about the same K_m value as α-ketobutyrate and a maximum velocity of about one-tenth of that for pyruvate. This indicates that an interaction of the C_4 keto group, probably with glutamate-140, orients the substrate to prevent C_5–C_9 making unfavorable interactions.

The effects of introducing a carboxyl group into the side chain of the α-keto acid vary with isoenzymes. With H_4 and M_4 the rates observed with oxaloacetate and α-ketoglutarate are so small that they could result from trace contamination of LDH with malate and glutamate dehydrogenases. However, the X_4 enzyme is almost as active with α-ketoglutarate as with pyruvate (*173*). Since the X_4 enzyme is also very active with C_5 α-keto acids (*174*) and has a low K_m value this suggests that the region around C_5 will be more open in the X_4 enzyme than in the H_4 or M_4 isozymes. The reaction with 2-ketobutyrate has been used to analyze the proportion of H and M subunits in tissue extracts (*175*).

Compounds in which the methyl group of pyruvate is substituted with a benzene ring have about the same K_m value as C_4 keto acids, but they have very low maximum velocities (*12,103*) and are useful in mechanistic studies.

g. Conformational Changes. (*i*) *Apoenzyme to ternary complex.* Conformational differences with respect to the apoenzyme will be described starting from the adenine portion of the cofactor.

A minor movement of aspartate-53 and its associated polypeptide chain occurs on binding the adenosine part of the coenzyme (Section II,A,3,e) and is illustrated in Fig. 21a. There is a larger movement of the loop connecting βD and αE, including the helix αD (residues 98–120) involving main chain displacement of up to 11 Å (Fig. 21). In crystals of ternary complexes this loop is down and encloses the coenzyme and substrate. In the apoenzyme it extends into the solvent. Arginine-101 forms an ion pair with the pyrophosphate group of the coenzyme in the ternary complex. Its formation may be a trigger for the conformational change of this loop and subsequent rearrangements in the subunit. The guanidinium group of arginine-109 moves 14 Å and changes from being completely exposed to the solvent to having a close interaction with groups around the substrate site. The two connected helices αD and αE are more angled to each other in the ternary complex.

The movement of the N-terminal part of helix αD is associated with a movement of the neighboring C-terminal part of helix αH. Thus, αH

173. L. Schatz and H. L. Segal, *JBC* **244,** 4393 (1969).
174. C. O. Hawtrey and E. Goldberg, *J. Exp. Zool.* **174,** 451 (1973).
175. S. B. Rosalki and J. H. Wilkinson, *Nature* (*London*) **188,** 1110 (1960).

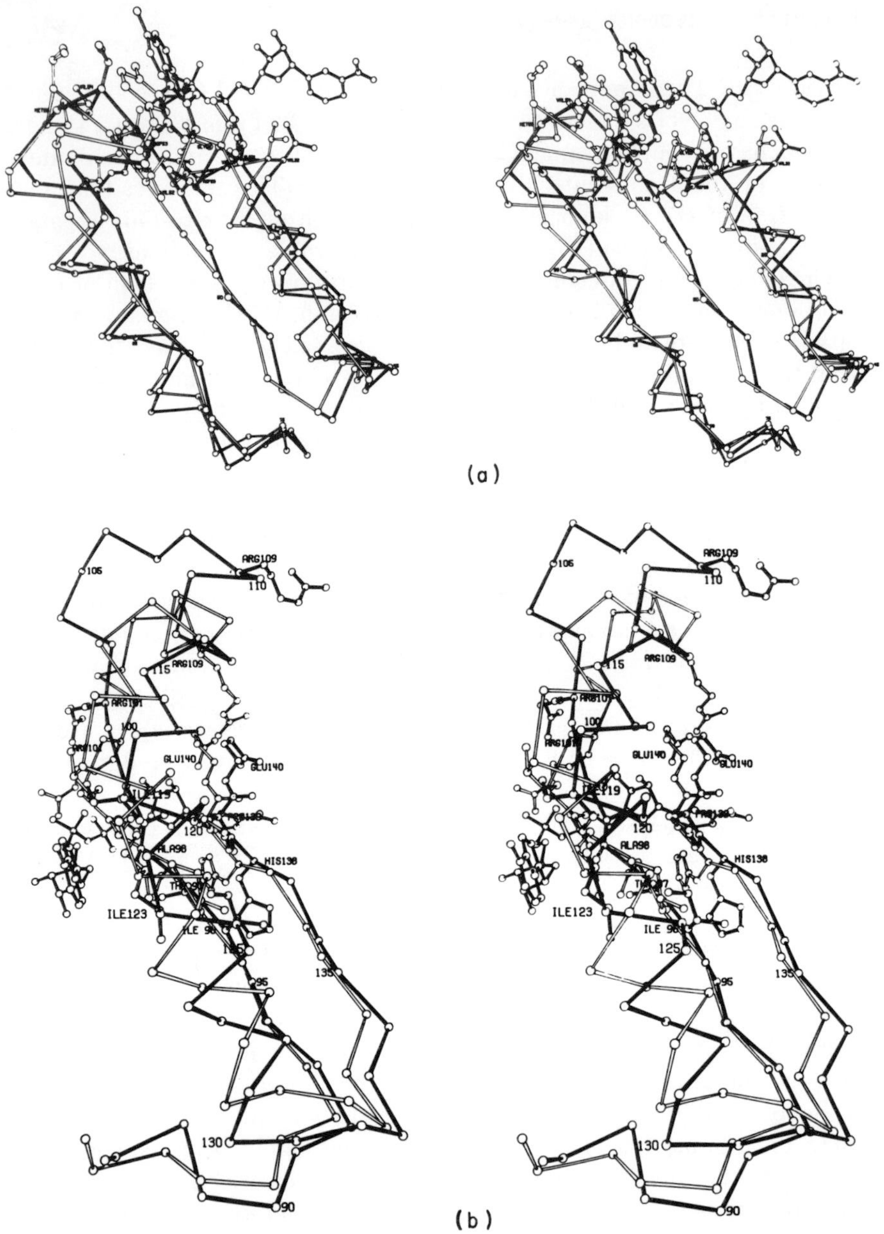

FIG. 21. Superposition of the LDH apoenzyme (darkened lines) and LDH:NAD-pyruvate ternary complex (open lines) showing conformational changes. The coenzyme is added to help orient the viewer. The residues shown in Table X involved in conenzyme interactions have been included. Elsewhere only C_α atoms are shown. All views are roughly along the R axis from the outside of the molecule. (a) Residues 22–85, (b) residues 86–140, (c) residues 141–150 and 260–330, and (d) residues 150–260.

pulls the tip of the loop βK–βL along by about 3 Å. These concerted movements caused by the formation of the ternary complex are illustrated in Fig. 21c. The helix α2G and to some extent helices α1G and α3G move small but significant amounts in order to approach the coenzyme. A small but very important conformational change is related to the loop βG–βH. This piece of polypeptide chain stretches out more in the ternary complex than in the apoenzyme and results in histidine-195 pointing deeper into the active site and into contact with the substrate (Fig. 21d).

A summary of the conformational changes observed on binding coenzyme and substrate is given in Fig. 22, where the movements of the C_α atoms with regard to the substrate site and each other are shown. The differences are mostly toward the substrate in the ternary complex, and

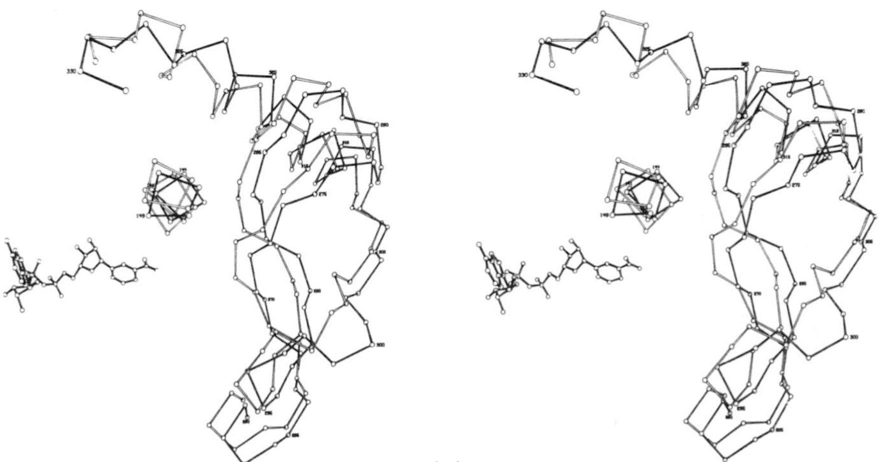

(c)

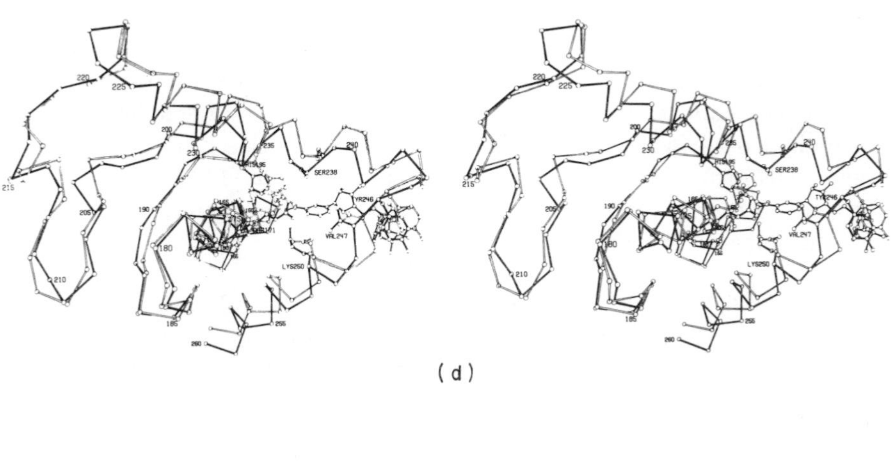

(d)

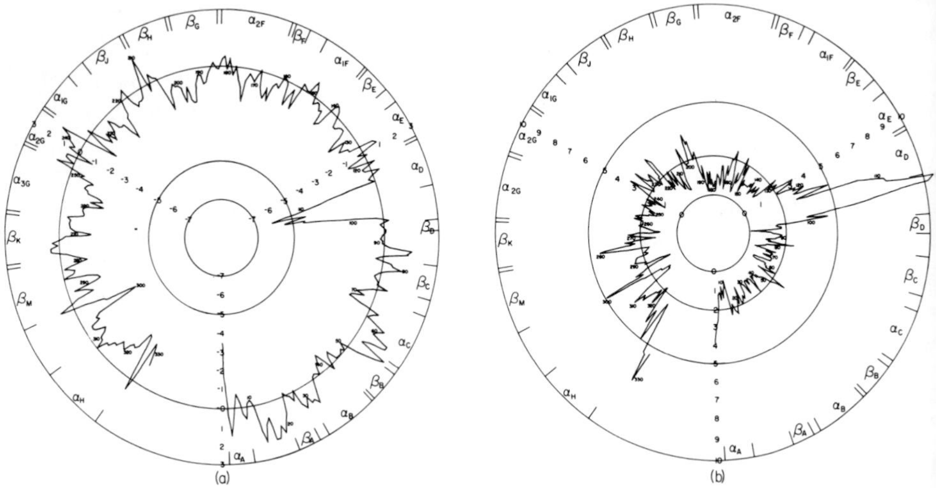

FIG. 22. Relative movements of C_α atoms between the apoenzyme and the LDH: NAD-pyruvate complex. The distance for each residue is plotted along the radius of a polar graph where the angle is proportional to sequence number. (a) Change in radial component of C_α atoms from substrate. Atoms which move toward the substrate in the ternary complex are plotted "negatively." (b) Magnitude of complete movement between equivalent C_α atoms.

hence the subunit on the whole undergoes a contraction when the coenzyme and the substrate bind.

(*ii*) *Binary complex intermediate.* The conformation of the enzyme in the intermediate binary complex has been more difficult to study. Soaking apoenzyme crystals with coenzyme causes a loss of crystal symmetry where only the molecular *Q* axis is retained. Under more severe conditions the crystals break up. The better binding NADH form requires a lower concentration than NAD^+ to produce these effects. Presumably the crystal shattering is brought about by some initial conformational changes in the loop. In this connection it is noteworthy that the loop is involved in intermolecular crystal contacts which would be broken if the loop folds or moves. Conformation of the loops in both subunits of soluble malate dehydrogenase (where there is bound NAD^+) more closely resembles the apo-LDH structure than that of the ternary complex (*149*). Thus, binding of coenzyme alone probably induces a relatively minor conformational change of the loop.

4. *The Oligomer*

a. Structure. In the case of the apoenzyme as well as the ternary complexes the tetrameric LDH molecule has perfect 222 symmetry. A view down each twofold axis is given in Fig. 23. Each subunit in the tetramer has been color coded [see Section VI,C in Rossmann *et al.* (*35*) and Rossmann *et al.* (*140*)]. We shall consider the contact of the red subunit with its neighbors. These are summarized in Fig. 24. Figure 25 gives a symbolic representation of these interactions.

The red–yellow subunit contacts are the most extensive. They are generated by helices αC, αB, α3G, and to some extent by α2F with the corresponding Q-axis related set, α3G, αB, α2F, and αC, respectively. Contacts of the red with the blue subunit are primarily between side chains of β structures. The red–green contacts are almost entirely related to the N-terminal arm. The absence of the N-terminal residues, and the existence of the dominant Q-axes interactions, would thus cause a breakdown of the tetramer into Q-axis dimers (see Figs. 24 and 25).

The active sites of the four subunits are relatively far from each other. The shortest distance (20 Å) between two of them is between the adenine riboses across the Q axis. The closest approach to an active site from a side chain belonging to a different subunit is also across the Q axis. Histidine-68 in the yellow subunit is situated 14 Å from the nicotinamide of the coenzyme in the red subunit. These distances are unlikely to facilitate interaction between different active centers.

b. Denaturation and Renaturation. The general processes of folding of a polypeptide chain into an active protein have been reviewed by Wetlaufer and Ristow (*176*). The active fold is both energetically and kinetically controlled.

Denaturation and renaturation of LDH have been studied by Kaplan and co-workers (*1,177*). The polypeptide chain has been unfolded in concentrated solutions of urea (*178,179*), guanidinium hydrochloride (*180,181*), or lithium chloride (*177*). Inactivation resulting from denaturation is inhibited by coenzyme (*178*). The denatured enzyme can be reactivated by dilution or dialysis in the presence of reducing agents such as β-mercaptoethanol (*141*). The rate of reactivation is accelerated by coenzyme. No intermediates between unfolded polypeptide chain and active tetramers have been positively identified (*181*).

c. Hybridization. The ability to form hybrids from a mixture of homotetramers provides an opportunity to observe the assembly of LDH molecules (*51,52,141*). The catalytic properties of hybrids are usually in pro-

176. D. B. Wetlaufer and S. Ristow, *Annu. Rev. Biochem.* **42,** 135 (1973).
177. A. S. Levi and N. O. Kaplan, *JBC* **246,** 6409 (1971).
178. G. DiSabato and N. O. Kaplan, *JBC* **240,** 1072 (1965).
179. D. Jeckel and G. Pfleiderer, *Hoppe-Seyler's Z. Physiol. Chem.* **350,** 903 (1969).
180. O. P. Chilson, G. B. Kitto, J. Pudles, and N. O. Kaplan, *JBC* **241,** 2431 (1966).
181. P. M. Wassarman and J. W. Burgner, *JMB* **67,** 537 (1972).

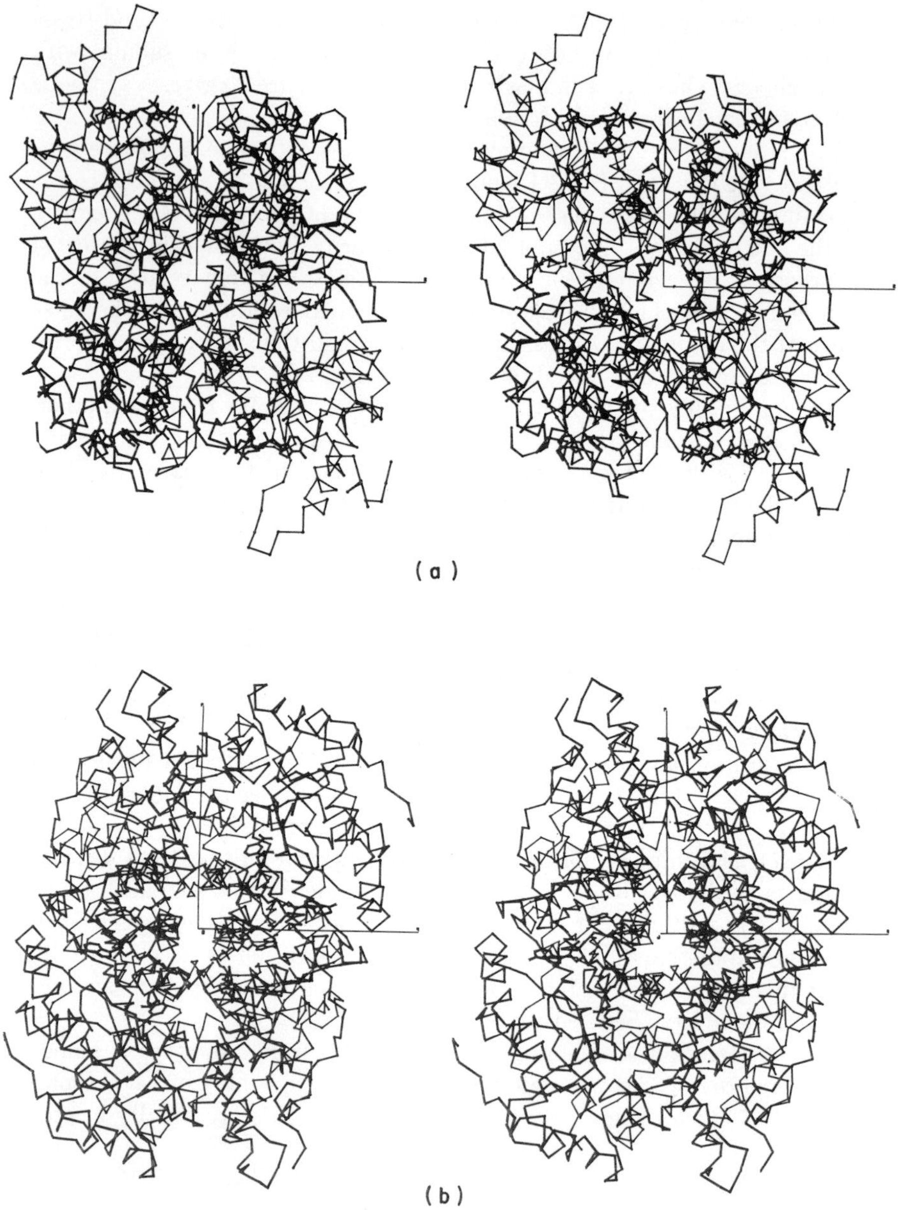

FIG. 23. Three different views of the C_α backbone in the tetrameric apoenzyme. The coenzyme has been added for orientation purposes. (a) Viewed parallel to P, (b) viewed parallel to Q, and (c) viewed parallel to R.

portion to the ratio of H to M subunits, suggesting that each subunit acts independently (*182*). It has been proposed that the association of different subunits may, however, precede the establishment of full catalytic and normal physical properties (*141*). This process is dependent on a variety of variables such as pH, temperature, pressure (*183*), the presence of halides, other ions and coenzyme (*34,184*). The pH dependence of association varies with species and isozymes (*185*). The rate of hybridization is in part dependent upon the ability to dissociate the tetramers into monomers, and the effect of salts on this process follows the Hofmeister series (*185,186*). The presence of coenzyme also inhibits dissociation and the consequent hybridization (*177,187,188*). Adams *et al.* (*139*) showed that there is an anion binding site close to the *P* axis in the red–blue interface.

The subunit–subunit interaction must be important in the assembly process. It is worth noting that in the case of glyceraldehyde-3-phosphate dehydrogenase the amino acids in the subunit interface are far more conserved than other amino acids elsewhere in that enzyme (*189*). If this

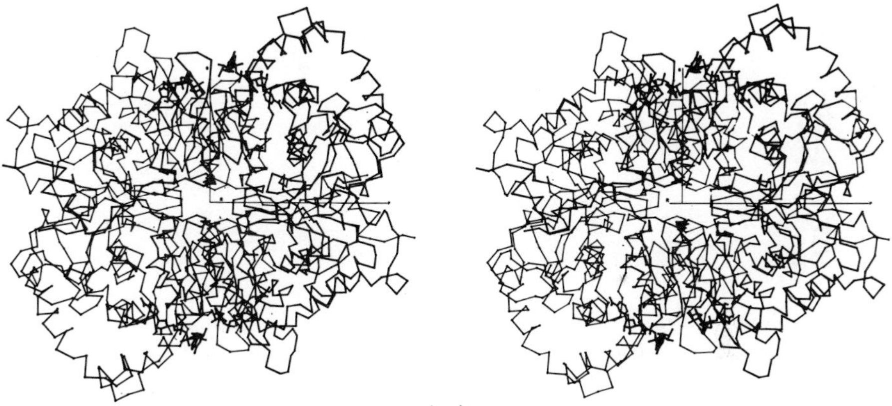

(c)

FIG. 23. (Cont.).

182. M. J. Bishop, J. Everse, and N. O. Kaplan, *Proc. Nat. Acad. Sci. U. S.* **69,** 1761 (1972).
183. R. Jaenicke, R. Koberstein, and B. Teuscher, *Eur. J. Biochem.* **23,** 150 (1971).
184. J. Südi and M. G. Khan, *FEBS* (*Fed. Eur. Biochem. Soc.*) *Lett.* **6,** 253 (1970).
185. O. P. Chilson, G. B. Kitto, and N. O. Kaplan, *Proc. Nat. Acad. Sci. U. S.* **53,** 1006 (1965).
186. F. Hofmeister, *Arch. Exp. Pathol. Pharmakol.* **24,** 247 (1888).
187. C. L. Markert and E. J. Massaro, *Science* **162,** 695 (1968).
188. J. Clausen and R. Hustrulid, *BBA* **167,** 221 (1968).
189. M. Buehner, G. C. Ford, D. Moras, K. W. Olsen, and M. G. Rossmann, *JMB* **90,** 25 (1974).

(a)

	RED	P AXIS BLUE	Q AXIS YELLOW	R AXIS GREEN
—	ALA 2		VAL 216, ASP 217,	
	LEU 3		ARG 179, LEU 180, LYS 213, ASP 217	
	LYS 4		ARG 179, LEU 180	
	ASP 5			LYS 304
	LYS 6		VAL 215, VAL 216, ASP 217	VAL 303, LYS 304, LYS 306
	LEU 7		LEU 180, ASP 208, LEU 211, VAL 216	ILE 302, VAL 304, LYS 304
	ILE 8		LEU 180, ASP 208	CYS 300, ASN 301, ILE 302, VAL 303
				LYS 304
	GLY 9			PHE 278, CYS 300, ASN 301, ILE 302
				LYS 304
	HIS 10			PHE 278, CYS 300, ASN 301, ILE 302
	LEU 11			MET 157, MET 274, THR 273, PHE 278
ARM				SER 298, ILE 302
	ALA 12			MET 157, HIS 158
	THR 13			MET 157, HIS 158, ILE 297, SER 298
	SER 14			ILE 297
	GLN 15			HIS 158, ARG 159, ASP 295, GLY 296
				ILE 297
	PRO 17			ASP 295, HIS 299
	SER 19			ASP 295
	TYR 20			ASN 22, LYS 92, ASP 295
	ASN 22			TYR 20, ASN 22, ASP 47, LYS 92
				MET 262
—	MET 34		TRP 249	
	ASP 35		TRP 249	
	ILE 38		ILE 38, MET 42, TRP 249	
	SER 39		MET 42,	
α_B	MET 42		ILE 38, SER 39, MET 42	
	LYS 43		PHE 72	
—	ASP 44			ARG 259, LYS 263
—	ASP 47			ASN 22, MET 262, LYS 263, ASP 264
	GLU 56		ASP 243	
	ASP 57		ASP 243, VAL 244	
	LYS 58		VAL 244	
	LEU 59		ASP 243	
	LYS 60		ASP 243, VAL 244	
	GLY 61		ASP 243, VAL 244, LEU 245	
α_C	GLU 62		LEU 245, TRP 249	
	MET 64		TYR 174, PRO 240, VAL 237	
	ASP 65		VAL 241, LEU 245, TYR 247, ALA 248	
			TRP 249, LYS 250,	
	LEU 66		TRP 249, CYS 252	
	GLN 67		TYR 174, GLU 179	
	HIS 68		TYR 174, ARG 171, SER 184, VAL 237	
			SER 238, CYS 252	
	GLY 69		CYS 252	
—	SER 70		TYR 174, HIS 183, SER 184	
	LEU 71		ALA 170, ARG 173 HIS 183, SER 184	
			CYS 185	
	PHE 72		LYS 43, ASN 166, LEU 167, ALA 170	
			CYS 252, SER 253, ASP 256	
β_C	LEU 73		HIS 183, CYS 252	LEU 265
	HIS 74			GLU 259, LYS 263, LEU 265, ARG 267
	THR 75			LYS 263, ASN 264, LEU 265
—	ALA 76			LYS 263, ASN 264, LEU 265
—	LYS 77			ASN 264
	LYS 92			TYR 20, ASN 22
β_F	MET 157			LEU 11, ALA 12, THR 13
	HIS 158			ALA 12, THR 15, GLN 15
—	ARG 159			GLN 15
	ASN 166		PHE 72	
	LEU 167		PHE 72	
	ALA 170		LEU 71, PHE 72	
α_{2F}	ARG 171		HIS 68	
	ARG 173		LEU 71	
	TYR 174		MET 64, GLN 67, HIS 68, SER 70	
	GLU 178		GLN 67	

were to be generally true it would explain why hybridization is useful in measuring evolutionary distances. Since only one LDH sequence is available, the degree of conservation of residues in subunit interfaces is as yet unknown.

There are theoretically three ways in which the subunits can be combined to make an M_2H_2 hybrid. Levitzki (*141*) has been able to separate the M_2H_2 electrophoretic band into two parts which might show that more than one of the M_2H_2 hybrids can be formed.

Whether or not the subunit is active on its own has been widely discussed (*178,183,190*) but is somewhat of an academic question since an isolated subunit might undergo conformational changes because the normally internal and hydrophobic surfaces would be exposed to solvent. The subunit interactions in dogfish M_4 LDH are indeed more hydro-

190. R. Jaenicke and G. Pfleiderer, *BBA* **60**, 615 (1962).

	RED	P AXIS BLUE	Q AXIS YELLOW	R AXIS GREEN
	LEU 180	ASN 301	LEU 3, LYS 4, ILE 8	
α_{2F}	GLY 181	CYS 266, VAL 292, HIS 299, ASN 301		
—	VAL 182	LEU 265, CYS 266, ARG 267, VAL 268 VAL 29[illegible]		
	HIS 183	LEU 265, CYS 266, ARG 267	SER 70, LEU 71, LEU 73	
	SER 184		HIS 68, SER 70, LEU 71	
β_G	CYS 185	ARG 267	LEU 71	
	LEU 186	CYS 266, ARG 267, VAL 268, HIS 269		
	SER 188	TRP 207, ALA 209		
	GLY 189	LYS 210		
	TRP 190	TRP 207, ASP 208, ALA 209, LYS 210		
—	SER 201	LYS 210		
	VAL 202	LYS 210, LEU 211		
	TRP 203	LYS 210		
	SER 204	LYS 210, HIS 212		
β_H	GLY 205	TRP 207, LYS 210		
	TRP 207	SER 188, TRP 190, GLY 205, TRP 207 LYS 210, HIS 212		
	ASP 208	TRP 190, VAL 268, VAL 303	LEU 7, ILE 8	
—	ALA 209	SER 188, TRP 190, VAL 268, PRO 290 VAL 303		
	LYS 210	GLY 189, TRP 190, SER 201, VAL 202 TRP 203, SER 204, GLY 205, TRP 207 HIS 212, MET 305		
	LEU 211	TYR 279, VAL 303, LYS 304, MET 305 LYS 306	LEU 7	
β_J	HIS 212	SER 204, TRP 207, LYS 210, HIS 212		
	LYS 213		LEU 3	
	VAL 215	LYS 306	LYS 6	
	VAL 216	LYS 304, LYS 306	ALA 2, LEU 3, LYS 6	
—	ASP 217		ALA 2, LEU 3 LYS 6	
	VAL 237		MET 64, HIS 68	
	SER 238		HIS 68	
	PRO 240		MET 64	
	VAL 241		GLY 61, ASP 65	
α_{2G}	LEU 244		GLU 56, ASP 57, LEU 59, LYS 60 GLY 61	
	THR 245		ASP 57, LYS 58, LYS 60, GLY 61	
—	VAL 247		GLU 62, ASP 65	
	ALA 248		ASP 65	
	TRP 249		LEU 34, CYS 35, ASP 38, GLU 62 ASP 65, LEU 66	
	LYS 250		ASP 65	
	CYS 252		LEU 66, HIS 68, GLY 69, PHE 72 LEU 73	
	SER 253		PHE 72	
α_{3G}	ASP 256		PHE 72	
	GLU 259			ASP 44, HIS 74
	MET 262			ASN 22, ASP 47
	LYS 263			ASP 44, ASP 47, HIS 74, THR 75 ALA 76
	ASN 264			ARG 18, ASP 47, THR 75, ALA 76 LYS 77
—	LEU 265	VAL 182, HIS 183		LEU 73, HIS 74, THR 75, ALA 76
	CYS 266	GLY 181, VAL 182, HIS 183, LEU 186		
	ARG 267	VAL 182, HIS 183, CYS 185, LEU 186		HIS 74
	VAL 268	VAL 182, LEU 186, ALA 209, ASP 208		
	HIS 269	LEU 186		
β_K	THR 273			LEU 11
	MET 274			LEU 11
	PHE 278			GLY 9, HIS 10, LEU 11
—	TYR 279	LEU 211		
	PRO 290	ALA 209		
	VAL 292	GLY 181, VAL 182		
β_M	ASP 295			GLN 15, PRO 17, ARG 18, SER 19 TYR 20
—	GLY 296			GLN 15, ARG 18
	ILE 297			THR 13, SER 14, GLN 15, ARG 18
	SER 298			LEU 11, THR 13
	HIS 299	GLY 181		PRO 17
	CYS 300			ILE 8, GLY 9, HIS 10
	ASN 301	LEU 180, GLY 181		ILE 8, GLY 9, HIS 10
	ILE 302			LEU 7, ILE 8, GLY 9, HIS 10, LEU 11
	VAL 303	ASP 208, ALA 209, LEU 211		LYS 6, LEU 7, ILE 8
	LYS 304	LEU 211, VAL 216		ASP 5, LYS 6, LEU 7, ILE 8, GLY 9
	MET 305	LYS 210, LEU 211		
	LYS 306	VAL 215, VAL 216, LEU 211		LYS 6

FIG. 24. All residues which are involved in a subunit–subunit contact for the apoenzyme are shown in (a). A residue is assumed to be involved in a contact if it has an atom within 5.0 Å of any atom in the red subunit. In (b), (c), and (d) are shown the hydrophilic interactions (within 3.8 Å) for the red–blue, red–yellow, red–green interfaces, respectively. Only half of all the interactions are shown in (b), (c), and (d), the others being related by molecular 2-fold axes. Any interactions listed in (a) and not shown in (b), (c), or (d) are hydrophobic or weaker hydrophilic interactions. (See following page.)

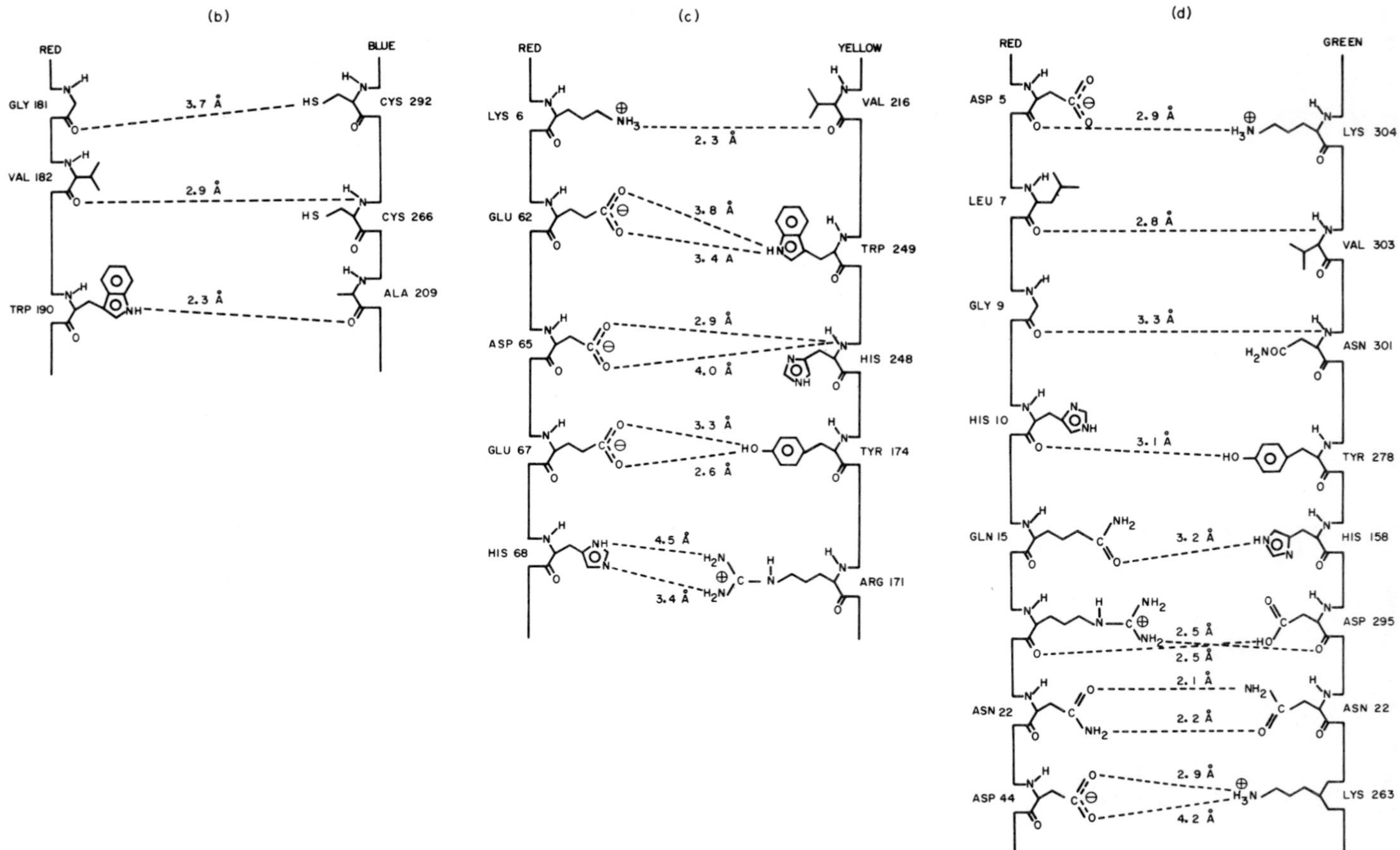

(b)
RED
BLUE
GLY 181
VAL 182
TRP 190
CYS 292
CYS 266
ALA 209
3.7 Å
2.9 Å
2.3 Å
(c)
RED
YELLOW
LYS 6
GLU 62
ASP 65
GLU 67
HIS 68
VAL 216
TRP 249
HIS 248
TYR 174
ARG 171
2.3 Å
3.8 Å
3.4 Å
2.9 Å
4.0 Å
3.3 Å
2.6 Å
4.5 Å
3.4 Å
(d)
RED
GREEN
ASP 5
LEU 7
GLY 9
HIS 10
GLN 15
ASN 22
ASP 44
LYS 304
VAL 303
ASN 301
TYR 278
HIS 158
ASP 295
ASN 22
LYS 263
2.9 Å
2.8 Å
3.3 Å
3.1 Å
3.2 Å
2.5 Å
2.5 Å
2.1 Å
2.2 Å
2.9 Å
4.2 Å

FIG. 24 b–d.

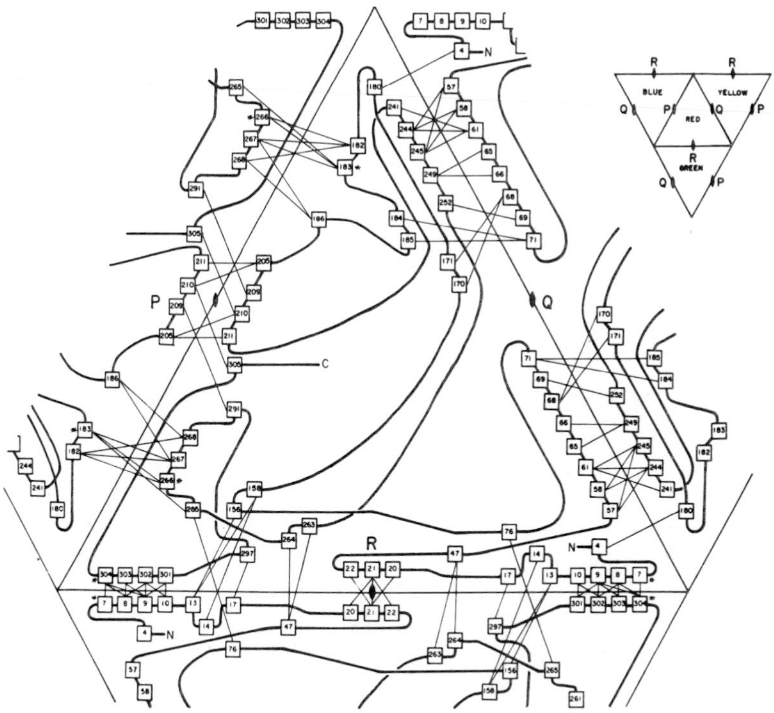

FIG. 25. Diagrammatic representation of contacts between residues in neighboring subunits. Each subunit is givn a color code: red, yellow, green, or blue. Thus, the red subunit contacts the other three by virtue of the *Q*, *R*, and *P* axes, respectively.

phobic than the molecular surface. Jaenicke (*191*) showed that the molecule can dissociate into inactive dimers at very low concentration.

B. MODIFICATIONS OF THE COVALENT STRUCTURE

1. *Mild Proteolysis*

Attempts have been made to devise conditions of mild proteolysis to discover whether fragments of the whole molecule are active. Apo-pig M_4-LDH is more susceptible to proteolysis than apo-H_4-LDH, although no active cleavage products were found (*192*). If the enzyme–NAD–sulfite complex is treated with trypsin, then 90% enzymic activity is retained

191. R. Jaenicke, *in* "Pyridine Nucleotide-Dependent Dehydrogenases" (H. Sund, ed.), p. 71. Springer-Verlag, Berlin and New York, 1970.
192. L. Selmeci and E. Pósch, *Experientia* **27**, 888 (1971).

after some peptide bonds are cleaved (*193*). However, if the nucleotide is subsequently removed, the enzymic activity is lost. The cleaved enzyme has unchanged ORD, but shows increased exposure of thiols. Some of these results can be rationalized from the molecular structure. First, the ternary complex is a more compact structure than the apoenzyme. Second, in the apoenzyme, the arginine-rich loop (residues 98–120) is exposed to the solvent, and thus exposed to proteolysis by trypsin. In the ternary complex the susceptible bonds are shielded by the protein. If the loop is removed by cleavage at the arginines, the enzyme will have lost some essential catalytic residues.

2. *Modifications of Side Chains*

A group has been described as essential if its modification leads to an irreversible loss of activity, a definition that will be followed here.

a. Cysteines. Considerable interest has been devoted to the possible involvement of cysteine residues in the catalytic activity of dehydrogenases (*131,194*). Kubowitz and Ott (*10*) found that LDH could be crystallized as an inactive mercuric salt and reactivated by thiols. Later, Neilands (*195*) and also Velick (*122*) reported that *p*-mercuribenzoate inhibited the enzyme but did not seriously affect the binding of coenzyme. However, in contrast, Schwert (*196*) found that the presence of coenzyme prevented this inactivation. Organic mercurials could not be used to identify the thiol since they migrated from essential to nonessential cysteines (*197*).

Of the many cysteines in LDH, it is unusual for the essential thiol to be the first to react with a covalent modifying reagent (*132,133*). In the dogfish enzyme the nonessential cysteine-266 reacts first with mercurials. In Pig H_4 the essential thiol reacts with maleimide (*198*) in the native state. In other species and isozymes the protein has to be first dissociated (*132*) or denatured (*199*) before the essential thiol can be covalently labeled.

The modification of the essential thiol with small reagents alters neither the binding of NADH nor its fluorescence (*198,200*). However,

193. D. Jeckel, R. Anders, and G. Pfleiderer, *Hoppe-Seyler's Z. Physiol. Chem.* **354,** 737 (1973).
194. G. Pfleiderer and D. Jeckel, *Eur. J. Biochem.* **2,** 171 (1967).
195. J. B. Neilands, *JBC* **208,** 225 (1954).
196. G. W. Schwert, *Ann. N. Y. Acad. Sci.* **75,** 311 (1958).
197. W. Gruber, K. Warzecha, G. Pfleiderer, and T. Wieland, *Biochem. Z.* **336,** 107 (1962).
198. J. J. Holbrook, *Biochem. Z.* **344,** 141 (1966).
199. A. H. Gold and H. L. Segal, *Biochemistry* **4,** 1506 (1965).
200. W. S. Allison, *Ann. N. Y. Acad. Sci.* **151,** 180 (1968).

substrates or substrate analogs can no longer bind to the enzyme–NADH complex (*201*).

The essential thiol has been recognized as residue 165 in the crystal structure of dogfish M_4 LDH. It is located at a distance of 10 Å from the substrate binding site, accessible only from the active center but nevertheless separated from it by a main chain. Any group modifying this thiol will be situated between aspartate-168 and histidine-195 preventing movement of the histidine toward the substrate binding site (Fig. 21d). Arginine-109 would also be hindered in adopting its position in the ternary complex by this modification. Thus it would not be possible to form the necessary hydrogen bond from the imidazole to pyruvate or lactate. Conversely, in the ternary complex, where the loop is down, cysteine-165 is completely inaccessible to modifying agents. This illustrates that a group can be essential in that its modification prevents conformational changes necessary in the formation of a catalytic complex. Whatever the role of this thiol and its surrounding sequence, it is conserved in all analyzed lactate dehydrogenases including L- as well as D-lactate-specific enzymes (Fig. 5).

b. Histidines. It has long been speculated that the group with an apparent p*K* of 7 involved in binding substrate to LDH is histidine. It could act as a source and sink of the proton in the dehydrogenase reaction (*202–204*). The first specific covalent labeling of an essential histidine was by Woenckhaus and co-workers using 3-bromoacetyl pyridine (*135*). They isolated and sequenced the corresponding peptide (see Section II,A,2). The sequence fitted the electron density with the histidine at position 195 (*127*; also see Section II,A,3,f).

Berghäuser *et al.* (*205*) used bromopyruvate to modify the same histidine. In the presence of NAD^+ this reagent is a substrate site directed inhibitor. In the apoenzyme both the essential cysteine and the histidine were modified; while only the histidine reacted in the presence of NAD^+. The histidine is apparently at the substrate binding site because the modified enzyme binds NADH normally but cannot form a ternary complex.

Analogs of NAD^+ containing alkylating functions (*169*) are also known to inhibit the enzyme and these might react with histidine-195.

201. J. J. Holbrook and R. A. Stinson, *BJ* **120,** 289 (1970).
202. A. D. Winer, G. W. Schwert, and D. B. S. Miller, *JBC* **234,** 1149 (1959).
203. W. B. Novoa, A. D. Winer, A. J. Glaid, and G. W. Schwert, *JBC* **234,** 1143 (1959).
204. D. B. Millar and G. W. Schwert, *JBC* **238,** 3249 (1963).
205. J. Berghäuser, I. Falderbaum, and C. Woenckhaus, *Hoppe-Seyler's Z. Physiol. Chem.* **352,** 52 (1971).

c. Tyrosines. The early experiments on modifying tyrosine were difficult to interpret because the reagents used often resulted in undetected modification of other residues, especially cysteine (*195,206–208*). DiSabato (*209*) obtained specific modification of tyrosine, and inhibition after formation of 1 to 2 moles of diiodotyrosine. It was also observed that 10 out of 30 tyrosines could be titrated in the native enzyme between pH 8 and 11. Binding of coenzyme raised the average value of pK_{app} above its normal value of 10.3, indicating that at least one tyrosine was close to the coenzyme binding site.

A more selective modification, with concomitant loss of activity, has been achieved using diazotized [^{35}S]sulfanilic acid (*130,194*). The free thiol groups were protected by mercuric chloride before the tyrosine modification was performed. It was found that an incorporation of 1.5 radioactive sulfur atoms per subunit inhibited the activity. The labeled peptide was isolated and partially sequenced. By comparison with the dogfish LDH sequence the labeled tyrosine corresponds to residue 220. However, this tyrosine is nowhere near the active center of the subunit. It is most improbable that it can be associated with catalysis.

It is also possible to attenuate the enzymic activity by reacting tyrosine with tetranitromethane (*210*), but no peptide has been isolated. Possibly tyrosine-85 or -246 which are involved in binding of the coenzyme may be relevant to some of these results.

d. Arginines. Kinetic measurements show coenzyme binding is relatively independent of hydrogen ion concentration below pH 9 (see Section III,D). Schwert and co-workers have suggested that the coenzyme may interact with the guanidinium group of an arginine (*98,211*). Yang and Schwert (*212*) modified the enzyme unspecifically with butanedione. They concluded that three arginyl residues might be essential to enzymic activity.

Berghäuser and Falderbaum (*213*) used phenylglyoxalate to inactivate LDH. They found that one arginyl residue per subunit was modified. In both these cases ternary complexes protected modification of arginines. Since the modified enzyme could bind coenzyme but could not form a

206. A. P. Nygaard, *Acta Chem. Scand.* **10**, 397 (1956).
207. S. K. Dube, O. Roholt, and D. Pressman, *JBC* **238**, 613 (1963).
208. M. C. Shen and P. M. Wassarman, *BBA* **221**, 405 (1970).
209. G. DiSabato, *Biochemistry* **4**, 2288 (1965).
210. D. Jeckel, R. Anders, and G. Pfleiderer, *Hoppe-Seyler's Z. Physiol. Chem.* **352**, 769 (1971).
211. G. W. Schwert, B. R. Miller, and R. J. Peanasky, *JBC* **242**, 3245 (1967).
212. P. C. Yang and G. W. Schwert, *Biochemistry* **11**, 2218 (1972).
213. J. Berghäuser and I. Falderbaum, *Hoppe-Seyler's Z. Physiol. Chem.* **352**, 1189 (1971).

ternary complex, arginines were implicated in substrate binding. The chemical modifications might be explained in terms of arginine-109 and -171 involved in substrate binding and arginine-101 involved in the triggering of the loop movement during ternary complex formation.

e. Lysines. Lysine-250 is available near the substrate binding site, yet no chemical modification study has implicated any lysines in the mechanism (*214,215*). However, it has not been possible to modify more than half of the lysines. Lysine modification can cause instability of the oligomer (*214*), and can change the immunological characteristics (*56,57*).

f. Methionines. Dickerson *et al.* (*216*) have reviewed the binding of $PtCl_4^{2-}$ to proteins studied by X-ray diffraction. It appears that this anion almost exclusively binds to methionines. In dogfish M_4 LDH methionine-55, -63, -64, and -157 react with $PtCl_6^{2-}$ in decreasing order (Table XVIII).

3. *Modifications in Crystals*

In Table XVIII are shown the relative substitutions of various sites in the dogfish M_4 apoenzyme and ternary complex crystals. The shielding of the essential cysteine-165 at site B in the ternary complex is quite clear. A partial shielding of cysteine-165 was used in the formation of the methylmercurynitrate derivative of the apoenzyme. Here the deleterious effect of substitutions at the B site was avoided by temporarily blocking it with NAD^+ (*217*). The shielding of groups by molecular packing can be observed with respect to the surface cysteine-133 (site E). This residue reacts in the ternary complex crystals but is completely inactive in crystals of the apoenzyme where it is hidden by an intermolecular contact.

C. Physical Properties

Some representative physicochemical properties of LDH are given in Table XIX.

Attempts have been made to detect conformational changes by measur-

214. G. Pfleiderer, J. J. Holbrook, L. Zaki, and D. Jeckel, *FEBS* (*Fed. Eur. Biochem. Soc.*) *Lett.* **1**, 129 (1968).
215. P. C. Yang and G. W. Schwert, *JBC* **245**, 4886 (1970).
216. R. E. Dickerson, D. Eisenberg, J. Varnum, and M. L. Kopka, *JMB* **45**, 77 (1969).
217. M. J. Adams, A. McPherson, Jr., M. G. Rossmann, R. W. Schevitz, and A. J. Wonacott, *JMB* **51**, 31 (1970).
217a. R. Jaenicke and S. Knopf, *Eur. J. Biochem.* **4**, 157 (1968).
218. A. Bolking, D. S. Markovich, M. V. Volkenstein, and P. Zavodzky, *BBA* **132**, 271 (1967).

TABLE XVIII

RELATIVE OCCUPANCIES (IN TERMS OF ELECTRONS) AT DIFFERENT SITES FOR VARIOUS HEAVY ATOM REAGENTS IN DOGFISH LDH[a]

Site	LDH:NAD-pyruvate complex					Apoenzyme		
	$HgCl_2$	HMB	$PtCl_6^{2-}$	MMN(1)	MMN(2)	$HgCl_2$	HMB	MMN
A Cys-266	130	90	—	60	81	95	84	45
B Cys-165	—	—	—	16	—	109	—	36
C Met-55	—	—	82	—	—	—	—	—
D Met-176	—	—	—	50	10	—	—	67
E Cys-133	50	54	—	36	11	—	—	—
F Met-63 and Met-64	—	—	66	—	—	—	—	—
G Met-157 and Met-274	—	—	8	—	—	—	—	—

[a] HMB is hydroxymercuribenzoate. MMN (1 or 2) is a derivative of methylmercurynitrate. The B site was partially blocked by the temporary presence of NAD^+ in the apoenzyme.

TABLE XIX
SOME REPRESENTATIVE PROPERTIES OF APO-LACTATE DEHYDROGENASES

Property	Beef H_4 (Ref. *53*)	Beef M_4 (Ref. *53*)	Chicken H_4 (Refs. *20,53*)	Chicken M_4 (Refs. *20,53*)	Pig H_4 (Refs. *179, 190,217a*)	Pig M_4 (Refs. *217a,218*)	Dogfish M_4 (Ref. *20*)
Sedimentation coefficient ($s^0_{20,w}/10^{13}$ cm sec^{-1})	7.45	7.32	7.31	7.33	7.38	7.6	7.54
Diffusion coefficient ($D^0_{20,w}/10^7$ cm^2 sec^{-1})	5.47	4.47	4.53	4.90	5.05	5.28	
Partial specific volume ($\bar{v}$/ml g^{-1})	0.75	0.747	0.74	0.74	0.74	0.74	0.74
Molecular weight (M/10^3)	131	153	151	140	142	142	141 144[a]
$E_{1\,cm}^{1\,mg/ml}$ 280 nm	1.5	1.29	1.36	1.56	1.4	1.45	1.48
Specific activity[b]	350	525	300	680	360	700	770
$[m']_{233}$ (degs)			−5880	−6630	−5170		−5570
λ_C (nm) (Drude plot)			262	254	254	272	250
% Helix (from ORD)			41	46	32	40	39

[a] Calculated from the nearly complete sequence.
[b] μmole NADH min^{-1} mg^{-1} at pH 7.5, 25°.

ing differences in ORD parameters of the enzyme on combining with coenzyme. No changes were detected during combination of NADH with beef H_4 or rabbit M_4 (*219*); or of NADH, NAD-SO_3^- or of NADH and oxamate with pig H_4 (*179,220*). Combination of NADH with pig M_4 has been reported to give ORD changes (*218*) similar to those on the formation of the ternary complex with NADH and oxamate.

The helix content, estimated for a number of species, varied from 32% to 46% based upon $[m']_{233}$ and from 38 to 51% based upon b_0. For dogfish M_4 LDH it was estimated to be 39% from $[m']_{233}$ and 55% from b_0 (*20*). These values bracket the range (40–45%) measured from the crystal structure determination. In the more recent work by Chen *et al.* (*221*) the values for dogfish LDH secondary structure have been used to construct calibration curves for the interpretation or ORD and CD spectra. Krigbaum and Knutton (*222*) predicted 33% helix, 26% sheet, and 46% "turn" in LDH by a procedure dependent upon amino acid composition (compare Section II,A,3,b).

D. Fluorescence and Spectroscopy

The optical properties (see Table XX) of the protein and coenzyme have been important tools for investigating the mechanism of LDH.

1. *Protein Fluorescence*

Proteins which contain tryptophan fluoresce at about 350 nm when excited at 270–305 nm. The fluorescence of tyrosine at 305 nm is not visible in LDH; this is normal for proteins which also contain tryptophan (*223*). The fluorescence of tryptophan residues in LDH is much higher in the native apoprotein than in the unfolded protein or its alkaline hydrolysate

TABLE XX
Fluorescence and Absorption Properties

Compound	Ultraviolet absorbance peaks (nm)	Fluorescence activation wavelengths (nm)	Fluorescence emissions wavelengths (nm)
Protein	280	270–305	350
NAD^+	260	None	None
NADH	260, 340	260, 340	470

219. I. Listowsky, C. S. Furfine, J. J. Bethell, and S. Englard, *JBC* **240,** 4253 (1965).
220. H. d'A. Heck, *JBC* **244,** 4375 (1969).
221. Y. H. Chen, J. T. Yang, and H. M. Martinez, *Biochemistry* **11,** 4120 (1972).
222. W. R. Krigbaum and S. P. Knutton, *Proc. Nat. Acad. Sci. U. S.* **70,** 2809 (1973).
223. F. W. J. Teale and G. Weber, *BJ* **65,** 476 (1957).

(*122*), and this is consistent with the structural finding that the environment of the tryptophan residues is generally hydrophobic (Table IX). The tryptophans most likely to become exposed to water on dissociation of the tetramer to monomers are residues 190, 207, and 249.

Changes in the protein fluorescence on binding coenzyme and substrate could occur by (a) their direct interaction with a tryptophan at the binding site, (b) radiationless transfer of electronic excitation energy from tryptophan to an acceptor ligand according to the Förster mechanism (*224*), and (c) a ligand-induced conformational change resulting in a changed protein environment of tryptophan residues.

Direct interaction with tryptophan residues [see (a) above] close to the active site has been considered an explanation for the 87% quenching of the fluorescence of LDH from many species on binding NADH (*225*). The lack of quenching of protein fluorescence by NAD^+ at the same site makes direct interaction an unlikely explanation (*226*). The absence of tryptophan residues from the active site pocket confirms this (Table IX and Fig. 14).

Radiationless energy transfer [see (b) above] from tryptophan to the dihydronicotinamide ring of NADH is calculated to occur with 50% efficiency at 25 Å. Thus, the fluorescence of some tryptophan residues in the molecule will be quenched when only one NADH is bound to the tetramer. Binding of each successive NADH molecule to the tetramer will reduce the protein fluorescence by the same factor. The total observed fluorescence for thermodynamically independent binding sites obeys the relationship

$$F = [1 - \alpha(1 - x)]^n$$

where F is the relative intensity of fluorescence with respect to the apoprotein, n is the number of NADH binding sites per molecule, α is the fractional occupancy of the NADH binding sites, and x is a constant termed the "geometric quenching factor" which depends upon the protein and the acceptor. This nonlinear phenomenon is also observed when NADH binds to other dehydrogenases (*227*) and when potential acceptors of tryptophan fluorescence bind to other proteins (*228*). Only compounds that form a complex with the enzyme and which have an absorption spectrum between 310 and 400 nm result in major quenching of LDH protein fluorescence (see Table XXI).

224. T. Förster, *Discuss. Faraday Soc.* **27,** 7 (1959).
225. R. H. MacKay and N. O. Kaplan, *BBA* **79,** 273 (1964).
226. J. J. Holbrook, *BJ* **128,** 921 (1972).
227. J. J. Holbrook, D. W. Yates, S. J. Reynolds, R. W. Evans, C. Greenwood, and M. G. Gore, *BJ* **128,** 933 (1972).
228. J. J. Holbrook, *Biochem. Soc. Trans.* **1,** 615 (1973).
228a. J. J. Holbrook, *BJ* **133,** 847 (1973).

TABLE XXI
LACTATE DEHYDROGENASE FLUORESCENCE CHARACTERISTICS

Compound	Longwave ultraviolet absorption		Nucleotide fluorescence			Fluorescence of protein ~ 335 nm (apoprotein = 1)	State of protonation of His-195 at pH 6–8	Ref.
	ϵ_{mM}	λ_{max} (nm)	$\lambda_{ex} \pm 10$	$\lambda_{em} \pm 10$	Intensity (free NADH = 1)			
Apoenzyme	0	—	—	—	0	1	pK = 6.7 ± 0.2	*289*
E(NAD⁺)—B:	0	—	—	—	0	1 ± 0.1	pK = 6.7 ± 0.2	*118,289*
E(NAD⁺)(oxalate)—B:	0	—	—	—	0	0.8 ± 0.1		*229*
E(NAD⁺)(lactate)—B:	0	—	—	—	0	1 ± 0.1	Deprotonated[a] (pK < 5)	*269*
E:NAD-pyruvate	6 ± 0.5	327	—	—	0	0.12		*262,275*
E(3APAD⁺)(enolpyruvate)	Large	337				0.12		*260*
E(thioNAD⁺)(hydroxylamine)	Large	334				0.39		*260*

E (3APAD, hydroxylamine)	Large	323				0.55		*260*
E—reduced naphthyridone	Large	340	340	450	>1	≪1		*265*
E (NADH)	6.2	337	340	440	2.5	0.13 − 0.17	p*K* = 6.7 ± 0.2	*118*
E—BH⁺ (NADH, lactate)	6.2 ± 1	340 ± 10	340	440	~5	0.13	Protonated[b]	*249,262*
E—BH⁺ (NADH, pyruvate) (all forms)	6.2 ± 1	340 ± 10	340	440	0.25	0.13	Protonated	*269*
E—BH⁺ (NADH, nitrophenylpyruvate)	6.2 ± 1	340 ± 10	340	440	0.25	0.12	Protonated (p*K* > 9)	*118,228a*
E—BH⁺ (NADH, oxamate)	6.2 ± 1	340 ± 10	340	440	0.25	0.12	Protonated (p*K* > 11)	*118*

[a] See pH dependence of K_m for lactate (Fig. 27).
[b] See Fig. 29.

Small changes in protein fluorescence resulting from ligand-induced changes [see (c) above] in the environment of tryptophans could be evidenced by the quenching in protein fluorescence of the ternary

NAD
/
E complex where the spectral properties for quenching are not
\\
oxalate

consistent for an energy transfer mechanism (*229*).

2. *Coenzyme Fluorescence*

The dihydronicotinamide ring of NADH fluoresces weakly at 470 nm when excited within its absorption band. The intensity of fluorescence in water is low and can be increased by solution in organic solvents (*230*) and shifted to shorter wavelengths (330 nm) by cleaving the pyrophosphate bond and thus removing the interaction with adenine (*231*). The fluorescence is increased on combination with the enzyme (*122,225,232*) but can be either enhanced or quenched when a ternary complex forms.

III. Catalytic Properties

A. Reactions Catalyzed

The equilibrium constant for the reduction of NAD^+ depends on the substrate. Some values are given in Table XXII (*233,233a*).

1. *Ketopyruvate Reduction*

Ketopyruvate undergoes both enolization and hydration. An aqueous solution contains:

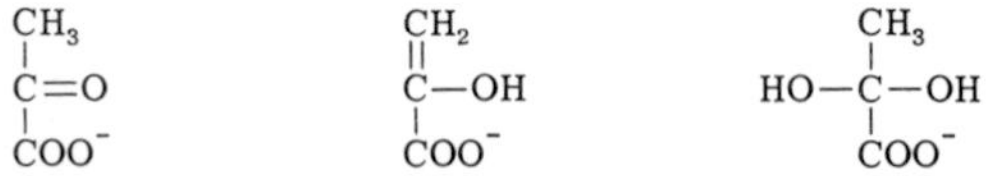

(XVIII) (XIX) (XX)

229. J. J. Holbrook, unpublished observations (1974).
230. C. H. Blomquist, *Proc. Werner-Gren Cent. Int. Symp. Ser.* **18**, 647 (1972).
231. G. Weber, *Nature* (*London*) **180**, 1409 (1957).
232. A. D. Winer, W. B. Novoa, and G. W. Schwert, *JACS* **79**, 6571 (1957).
233. M. T. Hakala, A. J. Glaid, and G. W. Schwert, *JBC* **221**, 191 (1956).
233a. R. S. Lane and E. E. Dekkar, *Biochemistry* **8**, 2958 (1969).

TABLE XXII
EQUILIBRIUM CONSTANT FOR VARIOUS SUBSTRATES

$$K = \frac{[\text{ketosubstrate}][\text{NADH}][\text{H}^+]}{[\text{hydroxysubstrate}][\text{NAD}^+]}$$

Substrate	K	Ref.
Pyruvate	2.8×10^{-12}	*233*
Glyoxalate[a]	1.0×10^{-9}	*233a*
2-Ketobutyrate	3.0×10^{-11}	*233a*
2-Keto-4-hydroxybutyrate	1.2×10^{-10}	*233a*

[a] No allowance is made for the proportion of the molecules which are hydrated.

The equilibrium mixture contains 94% keto and 6% diol (*234*) and only traces of enol. Only the keto form (XVIII) is a substrate (*234–236*). The rate at which the diol (XX) dehydrates to the ketone (XVIII) correlates well with the slow appearance of active substrate when a solution of the diol (XX) is brought to pH 6.9 (*235*). Pyruvate enolizes too slowly for the enol form to be the true substrate (*236*). Pyruvate transiently freed of enolpyruvate is a better substrate than the equilibrium mixture of forms XVIII, XIX, and XX (*237*) since the formation of the enzyme:NAD-pyruvate adduct is prevented.

Tienhaara and Meany (*235*) have emphasized the K_m values should be interpreted only when the proportion of unhydrated or unenolized α-keto acid is known and have suggested that fluoropyruvate may be a far better substrate than first thought (*238*).

2. *Glyoxalate Oxidation and Reduction*

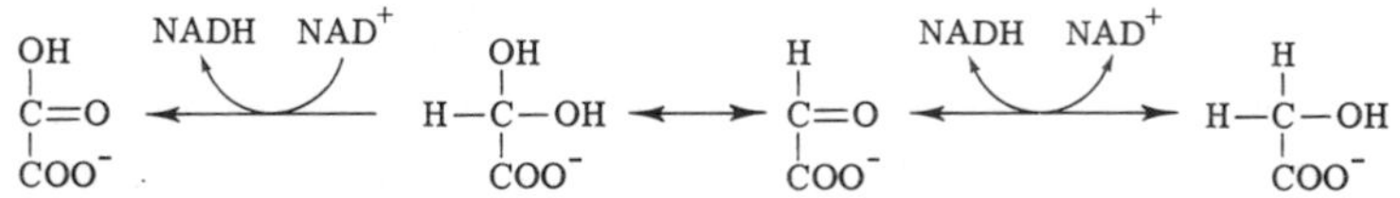

(XXI) (XXII) (XXIII) (XXIV)

Lactate dehydrogenase reduces an α-keto acid to a hydroxy acid with NADH. The enzyme will, however, also oxidize glyoxalate (XXIII) to

234. Y. Pocker, J. E. Meany, B. J. Nist, and C. Zadorojny, *J. Phys. Chem.* **73,** 2879 (1969).
235. R. Tienhaara and J. E. Meany, *Biochemistry* **12,** 2067 (1973).
236. M. Hegazi and J. E. Meany, *J. Phys. Chem.* **76,** 3121 (1972).
237. C. J. Coulson and B. R. Rabin, *FEBS (Fed. Eur. Biochem. Soc.) Lett.* **3,** 333 (1969).
238. E. H. Eisman, H. A. Lee, and A. D. Winer, *Biochemistry* **4,** 606 (1965).

oxalate (XXI) with NAD^+ or reduce glyoxalate to glycolate (XXIV) with NADH (*172,239–241*). The turnover numbers of the pig muscle enzyme (*240*) and the pig heart enzyme (*241*) for these substrates are similar to those with pyruvate and lactate.

At first sight the oxidation of an aldehyde (XXIII) to an acid (XXI) appears to be an atypical reaction. However, Duncan and Tipton (*240*) pointed out that glyoxalate exists principally as the hydrate (XXII), an isoelectronic structural analog of lactate, and suggested that the reaction is still the NAD^+ oxidation of a hydroxyl group (XXII) to a carbonyl group (XXI). Warren (*241*) showed that the steady-state inhibition patterns by oxamate and oxalate with the glyoxalate substrates were similar to those observed with pyruvate and lactate (*203*) and suggested that the same active site functions for both.

3. *Cyanide Addition*

The enzyme will also catalyze the addition of CN^- to the 4 position of the nicotinamide ring to give NAD-CN. The nonenzymic reaction is sufficiently slow at low pH that it is possible to observe a 30-fold enhancement with pig heart LDH (*242*).

B. Steady-State Kinetics

1. *Noninhibiting Substrate Concentrations*

Lactate dehydrogenase kinetics (*98,211,233,240,241,243–245*) at noninhibitory levels of pyruvate and lactate fit Eqs. (2) and (3).

$$V_f/v = 1 + K_O/\text{O} + K_L/\text{L} + K_{OL}/\text{OL} \tag{2}$$
$$V_r/v = 1 + K_R/\text{R} + K_P/\text{P} + K_{RP}/\text{RP} \tag{3}$$

where V_f is the maximum forward (lactate oxidation) rate; V_r the maximum reverse rate; K_O, K_L, K_R, and K_P are Michaelis constants for NAD^+, lactate, NADH, and pyruvate, respectively; and K_{OL} and K_{PR} are complex kinetic constants. Plots of $1/v$ against 1/[substrate] at various fixed coenzyme concentrations yield a pattern of intersecting straight lines which, using Cleland's (*246*) terminology, are characteristic of a

239. S. Sawaki, H. Hattori, and K. Yamada, *J. Biochem.* (*Tokyo*) **62,** 263 (1963).
240. R. J. S. Duncan and K. F. Tipton, *Eur. J. Biochem.* **11,** 58 (1969).
241. W. A. Warren, *JBC* **145,** 1675 (1970).
242. D. Gerlach, G. Pfleiderer, and J. J. Holbrook, *Biochem. Z.* **343,** 354 (1965).
243. V. Zewe and H. J. Fromm, *JBC* **237,** 1668 (1962).
244. V. Zewe and H. J. Fromm, *Biochemistry* **4,** 782 (1965).
245. G. W. Schwert, *JBC* **244,** 1285 (1969).
246. W. W. Cleland, *BBA* **67,** 104 (1963).

bi-bi sequential mechanism. Studies of product inhibition patterns (*247,248*) could theoretically decide between a compulsory, random, or preferred order of addition of substrate and coenzyme to give a ternary complex. However, in the case of LDH the compulsory order of binding, that is, coenzyme first and substrate second, was mainly established as a result of equilibrium binding experiments (Sections III,D and III,E).

Schwert's group (*211,245*) obtained the values of individual kinetic constants over a wide range of pH. Some values at pH 8 are shown in Fig. 26 for beef H_4 LDH. Meaningful results are only obtained using highly purified coenzymes and lactate. The slowest step in the forward reaction, at saturating concentrations of NAD^+ and lactate all at pH 8, is the rate of dissociation of NADH from the binary complex (50 sec^{-1}). This basic conclusion is supported by transient kinetic experiments (Section III,E,5) although the actual process may be an isomerization prior to the step in which NADH is liberated.

In the reverse direction the rate limiting step is either associated with the redox reaction (whose rate was not determined) or is the rate of dissociation of the E–NAD^+ complex. Transient kinetic experiments establish that this step actually is an isomerization of the ternary complex with NADH (Section III,E,5).

At higher pH values the steady-state kinetics predicted that the binding of NADH would become weaker depending on an apparent pK =

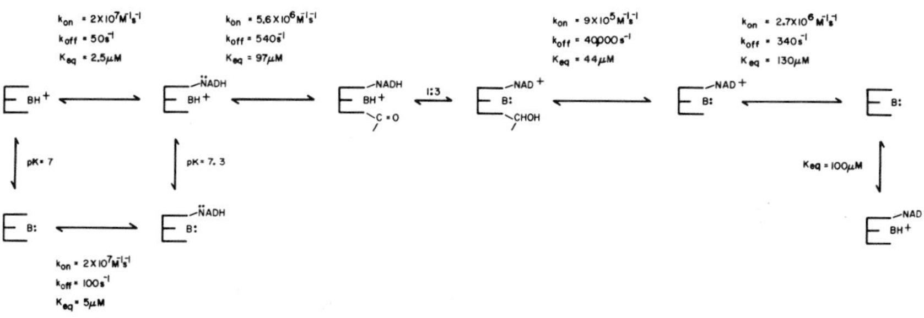

FIG. 26. Steady-state kinetic mechanism for beef H_4 LDH at pH 8.0 (adapted from Schwert *et al.* (*211*)). Pyruvate and lactate are indicated by >C=O and >CHOH, respectively. B is a base which is the proton sink in the reaction. All equilibria involving proton exchange with the solvent are fast compared to other steps.

247. H. J. Fromm, *BBA* **81**, 413 (1964).
248. S. R. Anderson, J. R. Florini, and C. S. Vestling, *JBC* **239**, 2991 (1964).

9.15. This is borne out by equilibrium binding experiments (Section III,D,1).

At lower pH values it was necessary to postulate a "dead-end" complex

$$\text{E}\begin{matrix} \nearrow \text{NAD}^+ \\ \searrow \text{Pyr} \end{matrix}$$

(*211*); otherwise, the rate of dissociation of NAD^+ in the reverse reaction would have been slower than V_r. The problem was not encountered in transient kinetic studies, perhaps because these are at such high protein concentrations as to be completed before any enzyme:NAD-pyruvate can form.

One of the most fruitful results of the steady-state kinetic studies has been the analysis of the pH dependencies (*211*). The main conclusions (Fig. 27) are

(a) V_r is pH independent,
(b) V_f is constant down to pH 7 and thereafter decreases only slightly (see Section III,C),
(c) below pH 9 the Michaelis constants for NAD^+ and NADH are pH independent,

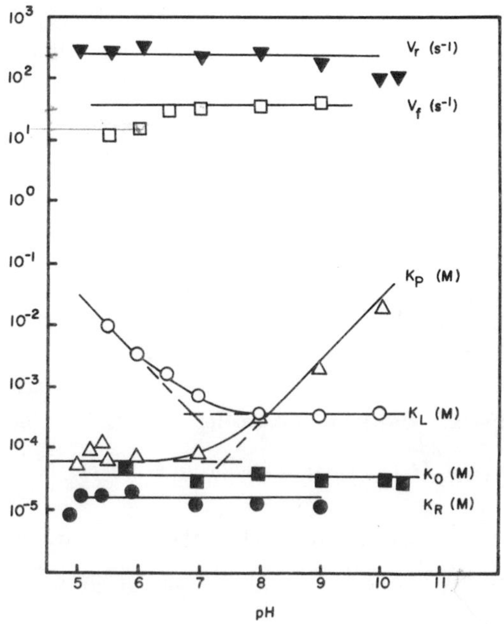

FIG. 27. pH dependence of kinetic constants of H_4 LDH. Reverse reaction: beef H_4; forward reaction: pig H_4.

(d) kinetic constants involving lactate are constant down to about pH 7 while below pH 7 they increase consistent with $pK_{app} = 7$ (*249*), and
(e) kinetic constants involving pyruvate are low at pH 5.5 and increase depending upon $pK_{app} \simeq 7.2$.

The mechanism given in Section III,G fulfills these steady-state requirements.

Silverstein and Boyer (*250*) used the rate of isotope exchange at equilibrium in the presence of high concentrations of substrates to support a compulsory order mechanism at neutral pH. However, at high pH they suggested the mechanism may only be a preferred order of addition of substrate and coenzyme.

2. *Steady-State Inhibitors*

Pyruvate	Oxamate	Lactate	Oxalate
(XXV)	(XXVI)	(XXVII)	(XXVIII)

a. Substrate Analogs. Inhibition by excess substrate will be discussed in Section III,C. Much of our knowledge of the mechanism of the enzyme has come from inhibitor studies with carboxylic acids (*203*), particularly oxamate and oxalate. Oxamate (XXVI) is a competitive inhibitor of pyruvate (XXV). It also can form a 10-times weaker complex with

  NAD$^+$
 /
E

or free enzyme. Conversely, oxalate (XXVIII) competes with lactate (XXVII) for NAD$^+$, but also forms a weaker complex with

  NADH
 /
E

. The chemical structure of the inhibitory ternary complexes will be considered in Section III,E. Oxamate (XXVI) is isoelectronic and isosteric with pyruvate (XXV) and is thought to be a perfect structural analog. The structural analogy between oxalate (XXVIII) and lactate (XXVII) is less obvious. The kinetically determined dissociation con-

       NADH
      /
stant of oxamate from E-BH$^+$ is small at pH 5.5 and increases with
      \
       oxamate

249. J. J. Holbrook and A. Lodola, in preparation.
250. E. Silverstein and P. D. Boyer, *JBC* **239**, 3901 (1964).

pH consistent with no binding when a group with p*K* of 7 is deprotonated. The inhibition by other anions, for example, acetate, implies that Tris-carboxylate buffers should not be used for mechanistic studies of this enzyme. LDH-X is much more sensitive to citrate, succinate, malate, aspartate, and glutamate as inhibitors than LDH H_4 or M_4 (*251*).

b. Miscellaneous Inhibitors. The enzyme is inhibited by *o*-phenanthroline, but does not give the color change with this reagent that would occur if there were bound zinc (*124*). Many compounds containing sulfite, cyanide, hydroxylamine, or sulfide can form complexes with enzyme-bound NAD^+ and thus appear to compete with lactate (*252–255*). Mercapto acids may act as inhibitors in the same way (*256,257*). These compounds are considered as coenzyme derivatives in Section III,D,2,d.

C. Inhibition by Substrates

1. *Pyruvate*

The susceptibility of lactate dehydrogenase to inhibition by pyruvate was noted by Kubowitz and Ott (*10*). All enzymes isolated are inhibited by pyruvate, although in general the effect is more pronounced with the heart than the muscle isozymes (*20,99*). The pyruvate concentration required for inhibition is less at low pH than at high pH (*98*).

Molecular investigation of the inhibition was initiated by Fromm's (*258*) discovery that the enzyme forms a complex with NAD^+ and pyruvate which has an absorption band at 325 nm and a protein fluorescence of 0.12 relative to the apoprotein. A similar compound can be formed if NAD^+ is replaced by 3-acetylpyridine-AD^+ which has an absorption band at 355 nm (*259,260*). Subsequent structural investigations have been summarized by Kaplan and co-workers (*260,261*). Two observations show that formation of the LDH:NAD-pyruvate complex is the cause of inhibition by high concentrations of pyruvate. First, the dissociation

251. L. J. Battellino and A. Blanco, *J. Exp. Zool.* **174,** 173 (1970).
252. H. Tanayama and C. S. Vestling, *JBC* **20,** 586 (1956).
253. G. Pfleiderer, D. Jeckel, and T. Wieland, *Biochem. Z.* **329,** 104 (1957).
254. O. Meyerhoff, P. Ohlmeyer, and W. Möhle, *Biochem. Z.* **297,** 113 (1938).
255. G. Pfleiderer, D. Jeckel, and T. Wieland, *Biochem. Z.* **329,** 370 (1957).
256. R. R. Chaffee and W. L. Barlett, *BBA* **39,** 370 (1960).
257. J. van Eys, F. E. Stolzenbach, L. Sherwood, and N. O. Kaplan, *BBA* **27,** 63 (1958).
258. H. J. Fromm, *BBA* **52,** 199 (1961).
259. C. S. Vestling and U. Künsch, *ABB* **127,** 568 (1968).
260. J. Everse, R. E. Barnett, C. J. R. Thorne, and N. O. Kaplan, *ABB* **143,** 444 (1971).
261. L. J. Arnold, Jr. and N. O. Kaplan, *JBC* **249,** 652 (1974).

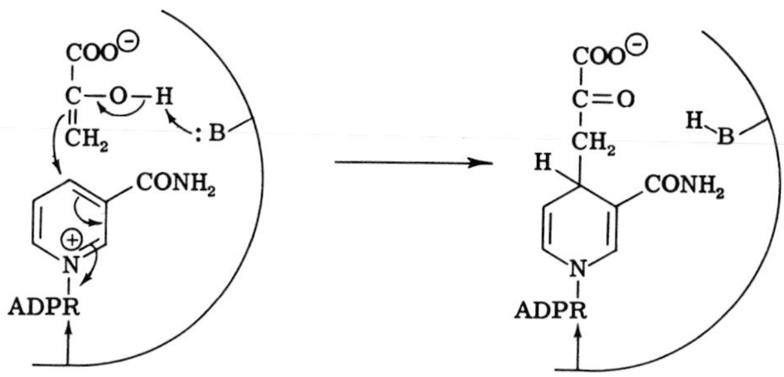

FIG. 28. Enzyme-catalyzed formation of the inhibitory LDH:NAD-pyruvate ternary complex.

constant of pyruvate from this complex is similar to the inhibition constant for pyruvate (*262*). Second, the steady-state inhibition by pyruvate appears slowly and correlates well with the formation of the 325-nm absorbing compound (*263*). Coulson and Rabin (*237*) showed that enolpyruvate was the effective inhibitor and suggested that it forms a compound with enzyme-bound NAD^+ (Fig. 28). The enzyme-bound complex is extremely stable (*264*). Even up to 0.1 *M* pyruvate there does not appear to be a site on the enzyme–NAD^+ complex which reversibly binds enolpyruvate (*263*).

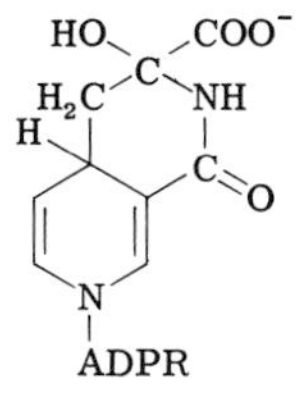

(XXIX)

The compound (XXIX) formed between NAD^+ and pyruvate at high pH in the absence of enzyme is an inhibitor (*265*) but is not the compound responsible for inhibition of LDH by excess pyruvate (*266*).

2. *Lactate*

LDH is inhibited by very high concentrations of lactate. As with most inhibitors, the H_4 isozyme (K_i = 26 m*M*) is more sensitive than the M_4

262. H. J. Fromm, *JBC* **238**, 2938 (1963).
263. H. Gutfreund, R. Cantwell, C. H. McMurray, R. S. Criddle, and G. Hathaway, *BJ* **106**, 683 (1968).
264. G. DiSabato, *BBRC* **33**, 688 (1968).
265. J. Everse, C. E. Zoll, L. Kahn, and N. O. Kaplan, *Bioorg. Chem.* **1**, 207 (1971).
266. R. F. Ozols and G. V. Marinetti, *BBRC* **34**, 712 (1969).

isozyme (K_i = 130 mM) (*267*). The variation of this inhibition with pH has not been well studied, and at low pH the situation is unclear since pyruvate is a common impurity in lactate. Taking typical assay conditions (10 mM lactate and 10 mM NAD^+ at pH 6) and using $K_{eq} = 3 \times 10^{-12}$ M, then only 15 μM pyruvate and NADH are produced at equilibrium. Hence, even slight impurities of pyruvate at this pH will give reduced initial rates.

Inhibition may result from the formation of the abortive ternary complex $\mathrm{E}{<}^{\mathrm{NADH}}_{\mathrm{lactate}}$. NADH in this complex is twice as fluorescent as in $\mathrm{E}{\diagup}^{\mathrm{NADH}}$ (*232,249,262,268,269*). The pH dependence of the formation of the complex is typical of complexes $\mathrm{E}{<}^{\mathrm{NADH}}_{\mathrm{carboxylate}}$ with an apparent pK of about 7 (see Section III,E,3).

3. *Biological Significance of the Abortive Complex, Enzyme:NAD-Pyruvate*

It is advantageous for aerobic tissues (e.g., heart and brain) to retain a significant amount of pyruvate which can be oxidized in the citric acid cycle and result in the production of ATP. On the other hand, in anaerobic tissue such as in skeletal muscle, where there are occasional high demands for energy and where no ATP production is possible via the citric acid cycle, it is essential for there to be an effective glycolytic metabolism with lactic acid as the end product (Fig. 1). The latter can then be recirculated in the bloodstream.

It has been proposed that the molecular basis for the evolution of H and M isozymes is that the M form is less susceptible to inhibition by pyruvate (*36*). This has been questioned since pyruvate levels in anaerobic muscle at 37° do not approach those required for inhibition (Table XXIII). Lactate levels do increase dramatically, and resistance to lactate inhibition has been suggested as an alternative explanation (*267*).

The recent recognition that the LDH:NAD-pyruvate complex is very

267. R. Stambaugh and D. Post, *JBC* **241,** 1462 (1966).
268. A. D. Winer and G. W. Schwert, *JBC* **234,** 1155 (1959).
269. J. R. Whittaker, D. W. Yates, N. G. Bennett, J. J. Holbrook, and H. Gutfreund, *BJ* (in press).

TABLE XXIII
In Vivo METABOLITE CONCENTRATIONS

Muscle	Pyruvate (m*M*)	Lactate (m*M*)	Reference
Resting	0.1–0.3	1.5	*71,267,269a*
Anaerobic	0.7	20–48	*71,269a,269b*

stable at physiological levels of enzyme has given renewed support to the proposition that LDH is under metabolic control by pyruvate (*1*). However, LDH:NAD-pyruvate formation is caused by enolpyruvate and a final decision must await the demonstration of this compound in tissues. Enolpyruvate bound to protein has been considered as an intermediate in the pyruvate kinase (*270*) and the phosphoenolpyruvate carboxytransphosphorylase (*271*) reactions although its presence free in solution was not demonstrated.

D. COENZYME BINDING

Binary complexes have been examined to obtain information about
(a) the number of coenzyme binding sites per subunit,
(b) the interaction or independence of the sites,
(c) the environment of the coenzyme and electronic structure of the nicotinamide ring, and
(d) any possible sequential changes of conformation in the binding.

1. *LDH:NADH Complexes*

a. Number of Binding Sites. The 340-nm absorption peak of NADH is shifted to shorter wavelength and thus gives a negative difference spectrum at 365 nm when NADH binds to LDH. Observation of this difference (*272*) showed that each subunit binds one NADH. Many other methods have been used to substantiate this finding which is also quite apparent from the crystallographic results.

b. Independence of NADH Sites. Observation of the equilibrium of NADH binding requires sensitive fluorimetric techniques because the dissociation constant

$$K_{E,NADH} = \frac{[E][NADH]}{[E^{NADH}]}$$

269a. J. Sacks and J. H. Morton, *Amer. J. Physiol.* **186,** 221 (1956).
269b. S. Karpatki, E. Helmreich, and C. F. Cori, *JBC* **239,** 3139 (1964).
270. I. A. Rose, *JBC* **235,** 1170 (1960).
271. J. M. Willard and I. A. Rose, *Biochemistry* **12,** 5241 (1973).
272. B. Chance and J. B. Neilands, *JBC* **199,** 383 (1952).

is very low (Table XXIV). The equilibrium has been examined by following the enhancement of NADH fluorescence (*118,122,202,262,273*), polarization of NADH fluorescence (*122*), quenching of protein fluorescence (*122,225*), and calorimetry (*274*). When due allowance for the nonlinearity in protein fluorescence (*226*) was made (*118*) all the results led to the conclusion that LDH H_4 and M_4 isozymes have four independent noninteracting binding sites. The atypical result with the beef M_4 enzyme (*273*) was not repeated with continuous titration (*118*).

Binding is insensitive to protein concentration with pure NADH (*118*). The dissociation constants are decreased at high temperature, in high salt and in high organic solvent concentration (*118*). In general, M_4 binds NADH weaker than H_4. The binding is not very sensitive to pH from pH 5.5 to pH 9, but thereafter becomes much weaker (*118,275*).

c. Environment of Bound NADH. The blue shift and increase in fluorescence intensity when NADH is bound to LDH are reminiscent of the changes brought about by preventing the interaction of the nicotinamide with the adenine ring (*231*) and by dissolving the NADH in a nonpolar solvent, respectively. The fluorescence of the dihydronicotinamide ring in solution can be excited by 260 nm radiation absorbed by the adenine ring. Lack of a 260-nm excitation peak for NADH bound to LDH (Table XXI) is consistent with the open structure of coenzyme.

d. Kinetics of NADH Binding. The kinetics of the combination of NADH with LDH were first examined by relaxation methods (*220,276*). Although an initial study (*276*) suggested that the combination might be a two-step process, the later work (*220*) with improved instrumentation showed that the process was the single step bimolecular reaction

$$\mathrm{E} + \mathrm{NADH} \underset{k_{\mathrm{off}}}{\overset{k_{\mathrm{on}}}{\rightleftharpoons}} \mathrm{E}^{-\mathrm{NADH}}$$

The formation of the complex can also be examined by rapid mixing techniques, which show that all four sites of the tetramer combine with NADH at the same rate. Similarly, NADH dissociates from all four sites at the same rate (*269*). The good agreement of $k_{\mathrm{off}}/k_{\mathrm{on}}$ with the equilibrium dissociation constant (measured under the same conditions) implies that the conformational changes observed crystallographically on binding NADH to the four noninteracting sites must occur simultaneously with the binding process. Misleading conclusions can be reached

273. S. R. Anderson and G. Weber, *Biochemistry* 4, 1948 (1965).
274. H. H. Hinz and R. Jaenicke, *BBRC* **54,** 1432 (1973).
275. A. D. Winer, *Acta Chem. Scand.* 17, 203 (1963).
276. G. Czerlinski and F. Hommes, *BBA* **79,** 46 (1964).

TABLE XXIV
EQUILIBRIUM AND KINETIC VALUES OF DISSOCIATION CONSTANTS

Enzyme	NADH at pH 8.0			NAD⁺ at pH 8.0			Oxamate from E(NADH)(oxamate) at pH 6.0			Ref.
	K_d (μM)	k_{on} (M^{-1} sec^{-1})	k_{off} (sec^{-1})	K_d (mM)	k_{on} (M^{-1} sec^{-1})	k_{off} (sec^{-1})	K_d (μM)	k_{on} (M^{-1} sec^{-1})	k_{off} (sec^{-1})	
Pig H_4		3.3×10^7	32	0.35	5.8×10^6 (for thio-NAD⁺)	2.4×10^6	18	8×10^6	140	*220*
Pig H_4		4.0×10^7	50							*269*
	0.5			0.4						*118*
Pig M_4		6.3×10^7	450							*279*
	3.7			0.3						*118*
Beef H_4	2.0									*275*
				0.3						*118*
Beef M_4	1.7									*118*
Dogfish M_4	3.6									*118*
Rabbit M_4	3.5									*118*
Lobster M_4	7.0									*118*

(*277*) when nonlinearity of protein fluorescence has been neglected and when the on reaction is examined under conditions of less than 80% complex formation (*269,278*).

During the steady-state production of NADH from lactate and NAD^+, the enzyme has quenched protein fluorescence, and enhanced NADH fluorescence, which suggests that it is present as an $\mathrm{E}{}^{\diagup \mathrm{NADH}}$ complex. However, the rate of dissociation of NADH from $\mathrm{E}{}^{\diagup \mathrm{NADH}}$ is 200–400 sec^{-1} for pig M_4 and 50 sec^{-1} for pig H_4. These values are greater than the steady-state rate for both isozymes, especially at low pH. It is possible that a slow isomerization occurs after pyruvate has dissociated, although the formation of the abortive complex $\mathrm{E\text{-}BH^+}{}^{\diagup \mathrm{NADH}}_{\diagdown \mathrm{lactate}}$ would also explain the results (*269,279,280*).

2. *LDH:NAD⁺ Complexes*

a. Number of Binding Sites. Takenaka and Schwert (*281*) found one NAD^+ binding site per subunit by ultracentrifugation and this is confirmed by gel filtration (*118*). NAD^+ and NADH compete for the same binding sites (*268*). This is also clear from the crystallographic investigations.

b. Independence of NAD⁺ Sites. The NAD^+ sites are approximately identical and independent although, because of the weak binding, none of the results is very precise (*118,281*). K_{E,NAD^+} is not dependent upon pH (*118*). The optical properties of NAD^+ are unsuitable for binding studies, although thio-NAD^+, with a difference spectrum at 360 nm on binding, can be studied and binds at equivalent sites (*220*).

277. J. Everse, R. L. Berger, and N. O. Kaplan, *Proc. Werner-Gren Cent. Int. Symp. Ser.* **18**, 691 (1972).
278. H. Gutfreund, *in* "Enzymes, Physical Principles," p. 122. Wiley (Interscience), New York, 1972.
279. H. Gutfreund and R. A. Stinson, *BJ* **121**, 235 (1971).
280. J. J. Holbrook and H. Gutfreund, *FEBS (Fed. Eur. Biochem. Soc.) Lett.* **31**, 157 (1973).
281. Y. Takenaka and G. W. Schwert, *JBC* **223**, 157 (1956).

c. Environment of Bound NAD$^+$. Both NAD^+ and pyridine model compounds will form adducts at the 4 position of the nicotinamide ring:

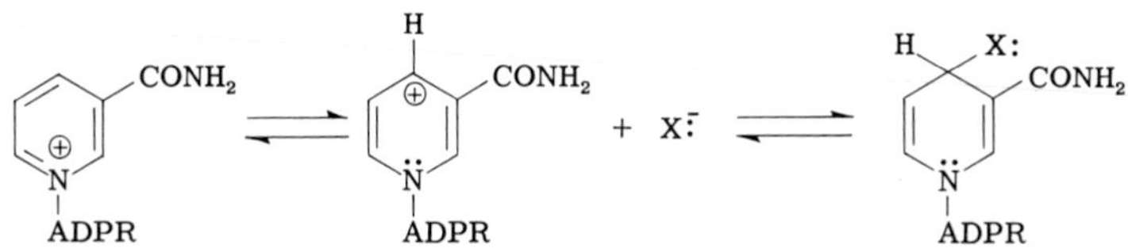

(XXX) (XXXI) (XXXII)

Reactions studied were with anions such as cyanide (*254,257*), sulfide (*252*), sulfite (*255*), and hydroxylamine (*254*). The adducts (XXXII) have an absorption band at wavelengths above 300 nm similar to NADH. The stability of the adducts increases if substituents at the N-1 position of the nicotinamide ring can donate electrons to the ring. The concentration of X^- needed to form the adducts (XXXII) is a probe for the degree to which oxidized nicotinamide is stabilized with a positive charge at the 4 position (XXXI). When bound to LDH the concentration of sulfite required to form the adduct with NAD^+ is 10^4 to 10^5 times less than when NAD^+ is in free solution (*255,282*). Thus, one function of the enzyme is to bind the NAD^+ in a manner such as to stabilize a fractional positive charge on the C-4 atom of nicotinamide. This would facilitate acceptance of a formal hydride ion, H^-. Similar conclusions follow from consideration of the NAD^+-sulfide complex (*252*).

d. Kinetics of NAD$^+$ *Binding.* NAD^+ binding kinetics have not been measured. Thio-NAD^+ binds in a simple bimolecular process. The ratio of k_{off} to k_{on} compares well with the equilibrium dissociation constant and implies that the conformational changes observed crystallographically must also occur simultaneously with NAD^+ binding.

E. Substrate Binding

1. *Absence of Enzyme–Substrate Compounds*

The acceptance of a compulsory order of substrate binding implies that pyruvate and lactate, or their analogs, cannot bind to the enzyme in the absence of nucleotides. Tests of this prediction have been made at neutral pH.

Pyruvate does not bind to the H_4 apoenzyme with $K_d < 15$ mM (*201,281*). Since the K_m is at least 100 times less than this limit, any interaction with the apoenzyme is not significant for catalysis.

282. G. Pfleiderer, *Colloq. Ges. Physiol. Chem.* **14,** 300 (1963).

Lactate (*281*) does not bind to the apoenzyme with $K_d < 1$ mM. This limiting value is of the same order of magnitude as the K_m value. However, indirect evidence from affinity chromatography (*110*) and heat denaturation experiments (*283*) suggests that the dissociation constant must be greater than 0.1 M.

There have been persistent reports that addition of NAD^+ (*259,275,284*) or NAD^+ analog (*285*) to the enzyme in the absence of added substrate gives rise to a weak absorption band between 300 and 400 nm. The early reports certainly resulted from the presence of lactate in the system. By using gloves to exclude lactate from the skin it has been possible to mix the pig H_4 enzyme with ultrapure NAD^+ and observe only the production of 0.2 mole NADH or NAD-pyruvate per subunit (*269*). In another recent extensive investigation of the reaction, the amount of tightly bound "lactate" was not reduced below 0.25 mole per subunit (*284*).

2. *Enzyme–Anion Complexes*

Steady-state kinetics studies of the inhibition of the enzyme by oxamate and oxalate (*203*) hinted at the existence of enzyme–oxamate complexes. No binding of 0.15 mM oxamate was detected in the ultracentrifuge. However, Südi (*283*) reported that oxalate (15 mM) and oxamate (50 mM) protect M_4 enzyme from heat denaturation. Oxaloacetate and fructose 1,6-diphosphate are effective at protecting against thermal denaturation at 1 mM concentration (*286*). It is possible that these anions bind at sites similar to those detected in the dogfish M_4 LDH molecule (*139*).

3. *LDH:Coenzyme:Inhibitor Complexes*

a. Complexes with NADH. The $\overset{\text{NADH}}{\text{E-BH}^+}$ binary complex binds carboxylic acids (*268*). These complexes are a 1:1:1 adduct of subunit:NADH:carboxylate (*103,287*) formed at equivalent and noninteracting sites (*103*) on the tetramer. The kinetics of formation of the

283. J. Südi, *BBA* **212,** 213 (1970).
284. H. A. White, C. J. Coulson, C. M. Kemp, and B. R. Rabin, *FEBS* (*Fed. Eur. Biochem. Soc.*) *Lett.* **34,** 155 (1973).
285. H. Holzer and H. D. Soling, *Biochem. Z.* **336,** 201 (1962).
286. E. S. Vesell and K. L. Yielding, *Ann. N. Y. Acad. Sci.* **151,** 678 (1968).
287. W. B. Novoa and G. W. Schwert, *JBC* **236,** 2150 (1961).

E<NADH, oxamate> complex have been examined by temperature jump relaxation. The binding of oxamate to E–NADH is a simple bimolecular process, and the ratio of k_{off} to k_{on} (Table XXIV) compares well to a thermodynamic equilibrium constant (*219*). Oxamate cannot dissociate before NADH from this complex.

The apparent dissociation constant of oxamate is pH dependent and increases with pH in a way (Fig. 29) which indicates that oxamate can only bind to E-BH$^+$–NADH if a base of pK 6.7 is protonated (*103,268*). The fluorescence of NADH is normally enhanced by forming a ternary complex with a carboxylic acid. Only with the pyruvate analog, oxamate, is it quenched, and this observation prompted Winer and Schwert (*268*) to suggest that in this complex the NADH is stabilized in a resonance configuration resembling NAD$^+$ (Fig. 30).

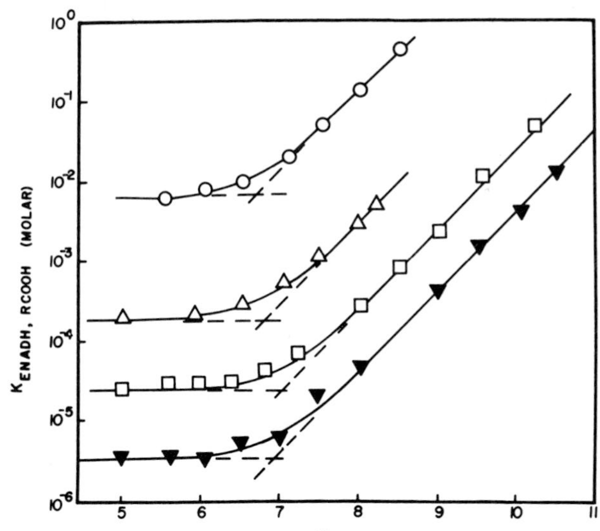

FIG. 29. The apparent dissociation constants of nitrophenylpyruvate (△), oxamate (▲), and lactate (○) from the E–NADH complex of H_4 LDH; and of oxamate from the E–NADH complex of M_4 LDH (□), as a function of pH.

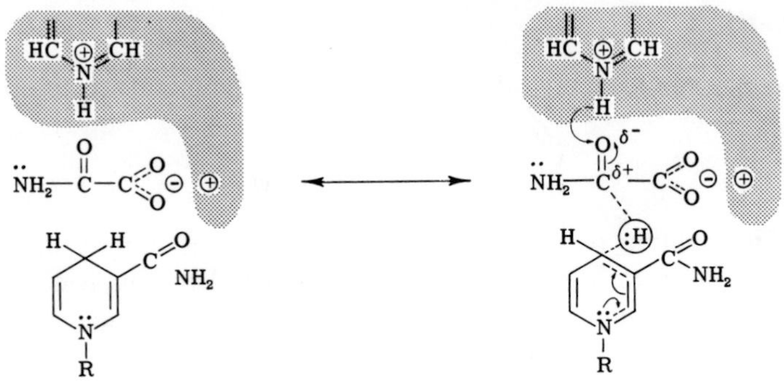

FIG. 30. Stabilization of NADH on binding oxamate (*268*).

b. Complexes with NAD^+. The complex $\mathrm{E}<^{\mathrm{NAD^+}}_{\mathrm{oxalate}}$ was predicted from steady-state studies and detected in ultracentrifuge binding experiments (*287*). Steady-state inhibition results suggest that formation of this complex is favored below pH 8 (*203*). Oxamate also forms a complex with $\mathrm{E}{}^{\diagup \mathrm{NAD^+}}$ (*287*) but weaker than with $\mathrm{E}{}^{\diagup \mathrm{NADH}}$. Transient kinetic studies (*269*) show that there are two distinct binding constants of oxamate to $\mathrm{E}{}^{\diagup \mathrm{NAD^+}}$. This phenomenon merits further study. Trifluorolactate may prove to be a more suitable analog of lactate since it binds to $\mathrm{E}{}^{\diagup \mathrm{NAD^+}}$ with a dissociation constant similar in magnitude to the K_m for lactate (*288*). Ternary complexes with NAD^+ are not fluorescent at 400–500 nm and show no difference spectra above 320 nm.

4. *Active Ternary Complexes*

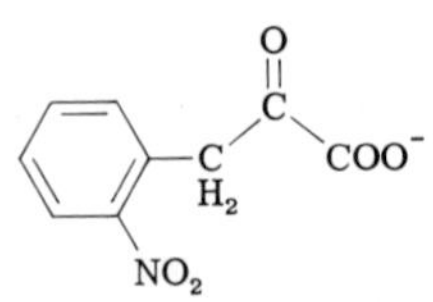

(XXXIII)

288. A. Pogolotti and J. A. Rupley, *BBRC* **55,** 1214 (1973).

Poor substrates (low V_{max}) of a suitable kind (low K_m) can provide useful information about the mechanism since they allow "central" complexes to build up to levels which are unattainable when the catalytic step is rapid. Nitrophenylpyruvate (NPP; XXXIII) is reduced by NADH at 1/2000th of the rate of pyruvate, and is independent of pH. The substrate rapidly forms a ternary complex with E-BH⁺ (NADH) which has quenched NADH fluorescence. This process can be observed to occur before the spectrum of NADH is destroyed in the catalytic step. The apparent binding constant of the substrate to E (NADH) shows that a stable complex only forms when a group of pK 6.7 in the binary complex is protonated (Figs. 29 and 31). The rate of the catalytic step is unchanged if NADD replaces NADH. This indicates that the rate limiting process is an isomerization rather than a step involving C—H bond fission (*103*). The uptake of a proton by a group of pK 6.7 when nitrophenylpyruvate binds at pH 8 has been directly measured with a glass electrode (*228*). Binding of nitrophenylpyruvate is blocked by carbethoxylation of the essential histidine (*289*).

The quenched fluorescence of the active ternary complex E-BH⁺ (NADH, NPP) is similar to the quenched fluorescence of NADH in the active ternary complex of glutamate dehydrogenase with NADH and oxoglutarate reported by di Franco and Iwatsubo (*290*). This further reinforces the

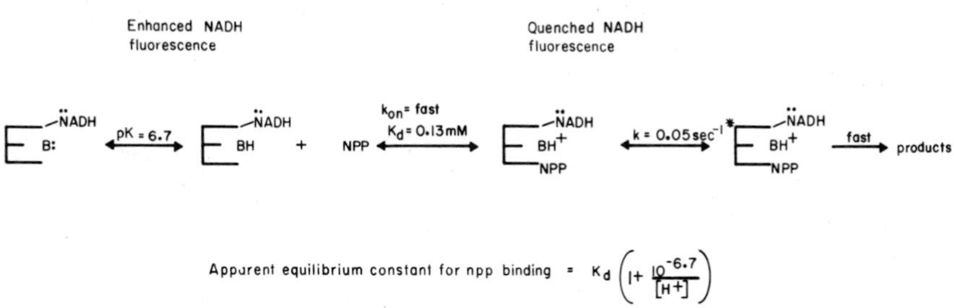

FIG. 31. Binding and reduction of *o*-nitrophenylpyruvate (NPP) by LDH.

289. J. J. Holbrook and V. A. Ingram, *BJ* **131,** 729 (1973).
290. A. di Franco and M. Iwatsubo, *Eur. J. Biochem.* **30,** 517 (1972).

view that E-BH⁺ (with NADH and oxamate bound) (which also has quenched NADH fluorescence) is a good model for the active ternary complex.

5. *Ternary Complexes Observed by Transient Kinetics*

Early transient kinetic studies of pig H_4 and M_4 LDH are summarized by Gutfreund (*291*). Resolution of the intermediates in the LDH reaction has depended upon measurement of protein fluorescence, NADH fluorescence, NADH absorbance at 340 nm, and proton uptake with dyes (*269*). These characteristics are listed in Table XXI.

a. Forward Reaction. When LDH is rapidly saturated with NAD^+ and lactate three phases are observed in the subsequent formation of NADH (Fig. 32).

Phase 1. An "instantaneous" (<1 msec) formation of 0.1–0.3 mole NADH per subunit. This instantaneously formed compound has absorbance at 340 nm, little NADH fluorescence, and quenched protein fluorescence. No proton is liberated.

Phase 2. A first-order process in which all the remaining sites become saturated with NADH. During this process NADH fluorescence is enhanced but protein fluorescence remains quenched. Protons are liberated at the same rate as NADH is produced.

Phase 3. The steady-state production of NADH. NADH fluorescence remains enhanced and protein fluorescence quenched. Protons are liberated at the same rate as NADH is produced.

Südi (*291a*) has also examined the transient changes in NADH fluorescence during the forward reaction and has observed these three phases.

Using the properties listed in Table XXI the three phases are identified with the three steps shown in Eq. (4).

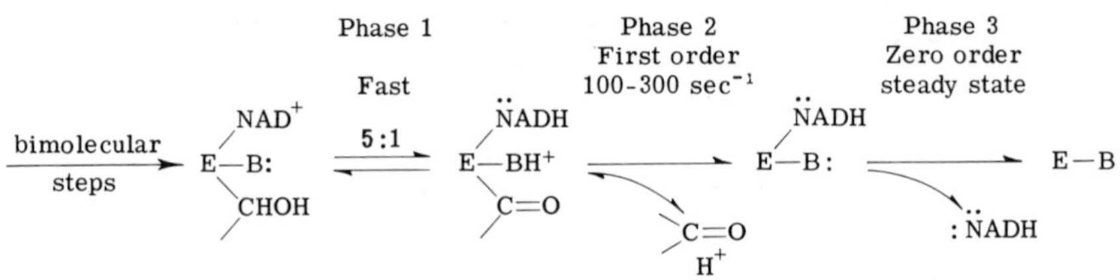

(4)

291. H. Gutfreund, *in* "Probes of Structure and Function of Macromolecules and Membranes" (B. Chance, T. Yonetani, and A. S. Mildvan, eds.), Vol. 2, p. 119. Academic Press, New York, 1971.

291a. J. Südi, *Biochem. J.* **139**, 261 (1974).

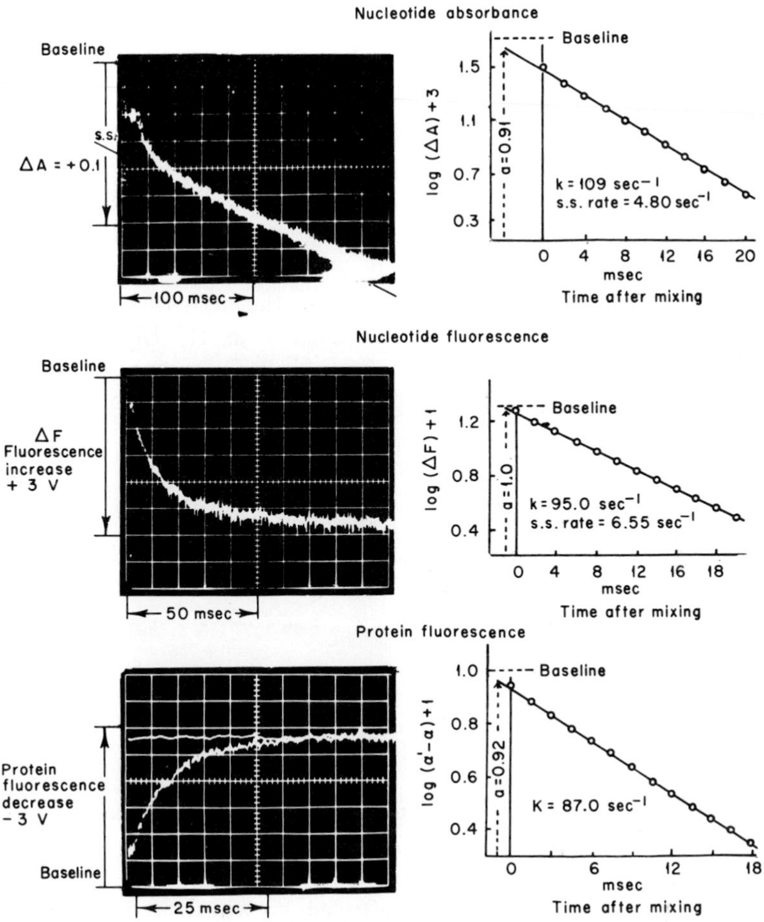

FIG. 32. Reaction of lactate dehydrogenase with NAD^+ and lactate at pH 6 monitored by nucleotide absorbance, nucleotide fluorescence, and protein fluorescence. The derivative plots on the right represent the initial transient approach to a steady-state rate of NADH production for nucleotide absorbance and fluorescence. To obtain the derivative plot for protein fluorescence, the fluorescence was measured as a fraction of the fluorescence of unliganded enzyme at each time, and the fractional fluorescence was then converted into fractional saturation of the enzyme with ligand by using the equation $F^{1/n} = 1 - \alpha(1 - x)$ (see Section II,D,1). Steady-state rate (s.s.) was 4.8 sec^{-1}.

When oxamate at high concentration is included in the reaction mixture then phase 2 is completed without the liberation of a proton to the solvent. Thus charge must be conserved on the protein during the redox step and during isomerizations of ternary complexes.

The equilibrium constant for the rapid interconversion of ternary complexes is 0.2 [Eq. (5)], while for the same reaction in solution [Eq. (6)] it is 3×10^{-5} at pH 7.

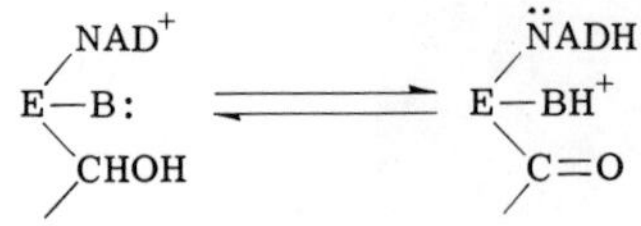

(5)

$$NAD^{+} + \rangle CHOH \rightleftharpoons \ddot{N}ADH + \rangle C{=}O + H^{+} \qquad (6)$$

One role of the protein is thus to stabilize the activated forms of the reactants. The above results are obtained with H_4 LDH. With M_4 LDH (*292*) the mechanism is qualitatively the same except that the equilibrium for Eq. (5) is completely to the right and only phases 2 and 3 are observed.

b. Reverse Reaction. The E—NADH complex is rapidly mixed with a saturating concentration of pyruvate. Absorbance of NADH at 340 nm disappears and two phases, shown in Eq. (7) can be resolved (*269,279,293*)

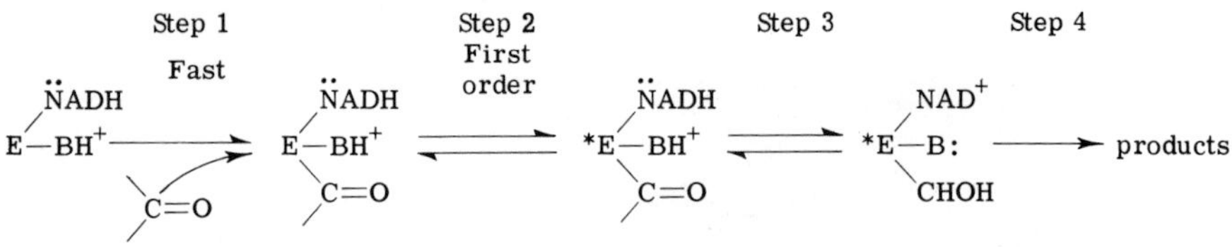

(7)

Phase 1 [identified as step 1 in Eq. (7)] is complete within the mixing time and gives a compound retaining absorbance at 340 nm, but with quenched NADH and protein fluorescence. Phase 2 is a first-order process in which absorbance at 340 nm is destroyed, protein fluorescence appears, and NADH fluorescence remains quenched. If this first-order process were after the redox step [step 4 in Eq. (7)], then the first turnover of NADH would be very fast. This is not observed. If the first-order process were the redox step itself [step 3 in Eq. (7)] then it would be slower with NADD than with NADH by a factor of 6 to 7, as with alcohol dehydrogenase (*293*). No appreciable isotope effect is measured. Thus the first-order phase must be identified with an isomerization of the ternary complex with NADH [step 2 in Eq. (7)] before the redox step (*269,279*). Südi (*293a*) has also observed two phases in the reverse reaction and has deduced the "on" rate for pyruvate. The kinetics do not indicate the

292. N. G. Bennett and H. Gutfreund, *BJ* **135,** 81 (1973).
293. J. D. Shore and H. Gutfreund, *Biochemistry* **9,** 4665 (1970).
293a. J. Südi, *Biochem. J.* **139,** 257 (1974).

nature of the isomerizaton. The changes in thermal stability (*177*), sedimentation coefficient (*294*), and three-dimensional structure still need to be correlated with a given step.

The rate limiting isomerization is the same whether, on average, only the first, or all four, sites on the tetramer are allowed to turnover. This confirms that the four subunits are kinetically independent of one another.

F. State of Protonation of Histidine–195

Identification by Woenckhaus and his colleagues (*135*) of an essential histidine opened the way for a study of its pK and an identification of it as residue 195 crystallographically (see Sections II,A,3,a and II,B,2,b). Fortunately, this histidine in pig H_4 LDH reacts so rapidly with diethylpyrocarbonate (*289*) that changes in reactivity can be directly related to the pK of the group. The reaction (8) can be followed from the ultra-

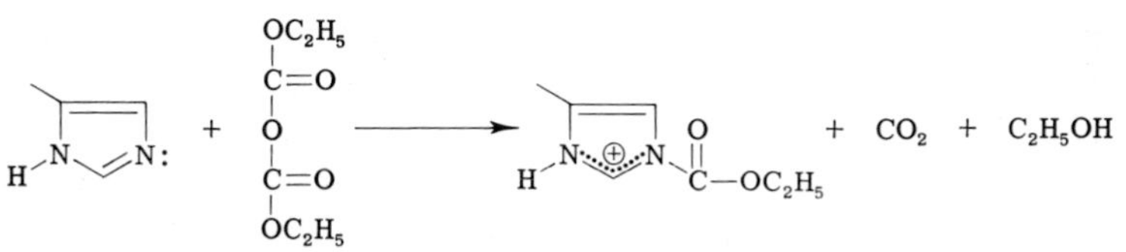

(8)

violet absorption of the product. In model reactions only unprotonated imidazoles react. The essential histidine in LDH reacts 10 times more rapidly than a free imidazole of the same pK in water; thus, the changes in reactivity with pH cannot reflect changes in groups which control access to histidine-195 and the pK can be assigned as 6.7. When the enzyme is saturated with NADH, the reactivity of the histidine only changes by a factor of 3 and the pK remains at 6.7. In protonated ternary complexes of NADH with oxamate or nitrophenylpyruvate the histidine does not react. Since protonated imidazoles do not react with diethylpyrocarbonate, and a group with pK 6.7 is protonated in the ternary complex, it is possible to identify the apparent pK of substrate binding with the pK of histidine-195 (*103*).

G. A Mechanism for Catalysis

Figure 33 shows a mechanism that has been proposed for the enzyme (*103,280*). More steps may be required as the resolution of the techniques for detecting intermediates improves and conformational changes become

294. R. S. Criddle, C. H. McMurray, and H. Gutfreund, *Nature* (*London*) **220**, 109 (1968).

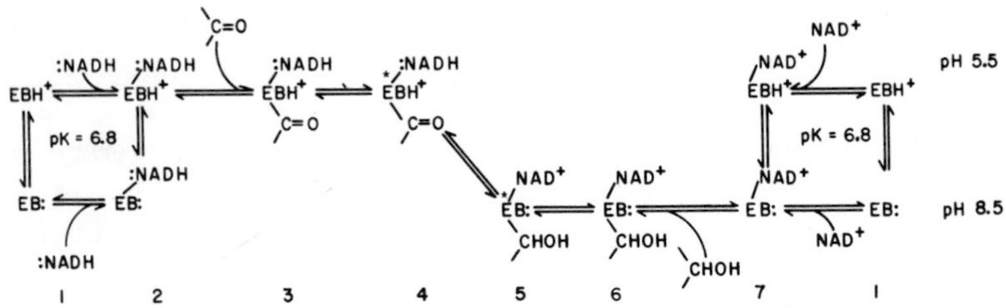

FIG. 33. A mechanism for LDH. The base B: is the imidazole of histidine-195. Pyruvate is >C=O and lactate is >CHOH. Forms shown as E* are isomers different from those shown as E.

more precisely described. Since all four active centers in the LDH molecule are equivalent they can be represented by the processes occurring at a single active site. The reaction is described as proceeding through seven distinct states not counting those which only differ by their state of protonation. The mechanism relates to the pH range of 5.5–8.5. It is consistent with the ordered binding of coenzyme and substrate. Let us now consider the reaction mechanism from left to right.

The affinity of enzyme for NADH (step $1 \rightarrow 2$) is fairly insensitive to hydrogen ion concentration over the range pH 5.5–8.5. Binding of the coenzyme becomes weak above pH 8.5 although there is as yet no satisfactory structural explanation for this. Crystallographic experiments show that there is a conformational change on binding NADH; however, the kinetics show that this change must be very fast. For the sake of simplicity we have not included these fast changes in Fig. 33. By analogy with s-malate dehydrogenase, the loop may have moved slightly at this stage.

The binding of pyruvate to the binary E–NADH complex (step $2 \rightarrow 3$) compulsorily protonates histidine-195 (see Fig. 34). The pK for this group is 6.8 in the apoenzyme and also in the E–NADH complex. Stabilization of the structure shown in Fig. 34b is explained by $N_{\delta 1}$ of histidine-195 acting as a hydrogen bond donor to the main chain carbonyl of glycine-164 in the LDH:NAD-pyruvate complex (Table VIII).

In stage 3 (Fig. 33) the carboxyl group of pyruvate probably forms an ion pair with arginine-171 and the keto group of pyruvate forms a hydrogen bond with $N_{\varepsilon 2}$ of histidine-195. The formation of the inactive

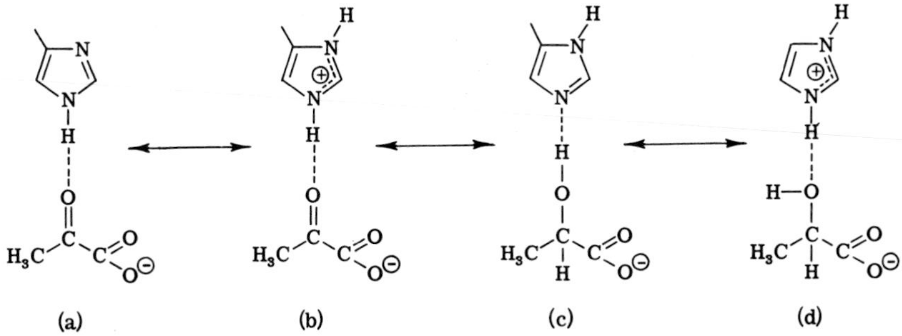

FIG. 34. The relationship between the state of protonation of histidine-195 and the ability to bind substrates. Structures (a) and (d) would be inactive and may not exist.

ternary complex is fast. Because of similarities in fluorescence behavior Holbrook and Stinson (*103*) suggested that the structure of the active ternary complex in stage 4 is the same as that of the inhibitor complex with oxamate and thus identical with one of the complexes studied crystallographically. Indirect evidence indicates that once the inactive ternary complex is formed there is a slow process ("isomerization," step $3 \rightarrow 4$) yielding the active complex. It could be argued that the fast formation of complex 3 is the binding of the pyruvate to the binary complex and that the slow isomerization reflects the extensive conformational changes in the protein and coenzyme released by the movement of the loop.

The chemical interconversion between the active ternary complex with NADH and pyruvate and the one with NAD⁺ and lactate (steps $4 \rightarrow 5$) is a rapid equilibrium and needs only a transfer of electrons. Even though the steps after the initial reduction of NAD⁺ bound to the enzyme are less well understood, it seems plausible that the enzyme must undergo a reverse isomerization and conformational changes (step $5 \rightarrow 6$) equivalent to steps $4 \rightarrow 5$.

The formation of abortive ternary complexes of enolpyruvate with

$$E\diagup^{NAD^+}$$ and lactate with $$E\text{-}BH^+\diagup^{NADH}$$ have been discussed earlier and could be added to the scheme.

The mechanism emphasizes that the biologically significant processes are those involved in binding, orienting, and activating the reactants. The catalytic steps are very fast in comparison. The phrase "induced fit" (*295*) could be applied to the mechanism since the binding of the substrate triggers an essential conformational change in the formation of

295. D. E. Koshland, Jr., *Proc. Nat. Acad. Sci. U. S.* **44**, 98 (1958).

the active ternary complex. The work on LDH illustrates how the fusion of many different experimental approaches can elucidate how a protein catalyzes a normally unfavorable reaction.

Acknowledgments

The authors would like to thank Dr. Susan S. Taylor for sequence information prior to publication, Connie Braun for the preparation of some of the diagrams, and William Boyle for photographic services. They are deeply indebted to Sandy Hurto and Sharon Wilder for help in the organization and preparation of the manuscript. The crystallographic investigation on LDH was supported by N.I.H. (grant No. GM10704) and N.S.F. (grant No. GB29596x) and many of the mechanistic studies were supported by SRC grant (BSR6099) and travel grants to J.J.H. and M.G.R. from the Welcome Foundation and Purdue Research Foundation, respecively. One of the authors (S.J.S.) was supported by an American Cancer Society Postdoctoral Fellowship (PF870).

5

Glutamate Dehydrogenases

EMIL L. SMITH • BRIAN M. AUSTEN • KENNETH M. BLUMENTHAL • JOSEPH F. NYC

I. Introduction

A. Function and Equilibrium

L-Glutamate dehydrogenases (EC 1.4.1.2–4) catalyze the interconversion of α-ketoglutarate and L-glutamic acid:

$$\alpha\text{-Ketoglutarate} + \text{NAD(P)H} + \text{NH}_4^+ + \text{H}^+ \rightleftharpoons \text{L-glutamate} + \text{NAD(P)}^+ + \text{H}_2\text{O}$$

The enzymes are recognized to be important because of the pivotal positions in metabolism occupied by glutamate and α-ketoglutarate and the ability of these compounds to enter into many types of pathways. Glutamate dehydrogenases provide a route for incorporation of nitrogen into organic compounds, and thus a link between carbohydrate and amino acid metabolism. Although the forward reaction as written above is undoubtedly the primary role of the enzyme in many organisms, the reverse reaction, i.e., oxidation of glutamate, is of importance in some organisms or in specialized tissues or cell components.

The equilibrium for the reaction is greatly in favor of glutamate formation. From the equation above

$$K = \frac{[\text{NAD(P)H}][\alpha\text{-ketoglutarate}^{2-}][\text{NH}_4^+][\text{H}^+]}{[\text{NAD(P)}^+][\text{glutamate}^{+2-}][\text{H}_2\text{O}]}$$

The most recent data are those of Engel and Dalziel (*1*). At 38° and 0.25 ionic strength, which approximates conditions in the liver, $K = 6.97 \times 10^{-15}$ *M* with NAD^+ and 4.48×10^{-15} *M* with $NADP^+$. At 25° and 0.1 ionic strength, $K = 1.54 \times 10^{-15}$ *M* and 0.86×10^{-15} *M*, respectively. *K* is strongly dependent on ionic strength. $\Delta H = 16.9$ kcal/mole for the NAD^+ reaction and 17.5 kcal/mole for $NADP^+$; $\Delta G^\circ = 18.45$ kcal/mole for NAD^+ and 19.16 kcal/mole for $NADP^+$.

There are at least three types of glutamate dehydrogenases which differ in coenzyme specificity: those specific for either NAD or NADP and those that can function with both. For brevity, the first two will be abbre-

1. P. C. Engel and K. Dalziel, *BJ* **105,** 691 (1967).

viated as NAD-GDH or NADP-GDH, particularly since many species contain both types of coenzyme-specific GDH's.

Molecular studies on various GDH's are now proceeding rapidly since elucidation of the primary structures of some of these enzymes. It is essential to note, however, that these enzymes vary not only in coenzyme specificity but also in other properties, e.g., induction and repression of synthesis by metabolites, regulation of activity by purine nucleoside di- and triphosphates, e.g., ADP, GDP, ATP, GTP, and other ligands, and in molecular properties. The successful isolation from some species of modified forms of the enzyme produced by mutant strains, particularly of *Neurospora*, now permits identification of residues important for maintenance of normal activity.

Although the GDH's are, in many organisms, the catalysts responsible for formation of glutamate, it is well to note that some organisms possess *glutamate synthases* which utilize glutamine rather than ammonia as the source of amino groups in the analogous reductive reaction:

$$\alpha\text{-Ketoglutarate} + \text{NADPH} + \text{H}^+ + \text{glutamine} \rightleftharpoons 2\ \text{glutamate} + \text{NADP}^+$$

B. Historical Background

Neubauer (*2*) and Knoop (*2a*) demonstrated in the early part of this century the metabolic interconversion of α-amino and α-keto acids. In 1920, Thunberg (*3*) observed that glutamate was oxidized in the presence of frog muscle, and, in 1936, Weil-Malherbe (*4*) prepared an extract from brain tissue which catalyzed glutamate oxidation and identified the product, α-ketoglutarate, as its 2,4-dinitrophenylhydrazone.

von Euler and co-workers (*5*) and Dewan (*6*) recognized that dinucleotide coenzymes were essential for the reversible reaction. Adler *et al.* (*7*) noted that extracts of some plants were active only in the presence of NAD, whereas NADP was required for extracts of *E. coli* and yeast.

In 1951, the enzyme was crystallized from bovine liver independently

2. O. Neubauer, *Deut. Arch. Klin. Med.* **95,** 211 (1909).
2a. F. Knoop, *Hoppe-Seyler's Z. Physiol. Chem.* **67,** 489 (1910); F. Knoop and E. Kertess, *ibid.* **71,** 153 (1911).
3. T. Thunberg, *Skand. Arch. Physiol.* **40,** 1 (1920).
4. H. Weil-Malherbe, *BJ* **30,** 665 (1936).
5. H. von Euler, E. Adler, G. Günther, and N. B. Das, *Hoppe-Seyler's Z. Physiol. Chem.* **254,** 61 (1938).
6. J. G. Dewan, *BJ* **32,** 1378 (1938).
7. E. Adler, V. Hellström, G. Günther, and H. von Euler, *Hoppe-Seyler's Z. Physiol. Chem.* **255,** 14 and 27 (1938).

by Strecker (*8*) and by Olson and Anfinsen (*9*), while Fincham (*10*) made the important observation that single gene mutants of *Neurospora crassa* lacked the GDH activity of the wild-type strain, but would grow well if any one of a number of L-amino acids was supplied in the medium.

Observations that bovine liver GDH polymerizes in concentrated solution (*11*) and may be regulated, not only by purine nucleotides (*12*) but also by a whole host of ligands, have led to many studies, as recent reviews will attest (*13–19*). This article will summarize the above types of investigations with major emphasis on recent studies that include elucidation of complete or partial amino acid sequences of the enzymes from several sources.

C. Distribution and Coenzyme Specificity

Glutamate dehydrogenases are widely distributed; only a few organisms lack these enzymes. As indicated above, it was recognized early that, whereas GDH's from animals utilize NAD or NADP, the enzymes from other sources are specific for one of these coenzymes. In addition, studies of glutamate synthesis in fungi led to the discovery of two distinct enzymes in bakers' yeast, one requiring NAD and the other NADP (*20*), and later in *Fusarium oxysporum* (*21*), in *Neurospora crassa* (*22*), and other organisms.

From evidence discussed in Section III,A, it is generally held that the enzyme linked to NAD is involved primarily in oxidation of glutamate, whereas the one linked with NADP is responsible for biosynthesis. The

8. H. J. Strecker, *ABB* **32,** 448 (1951).
9. J. A. Olson and C. B. Anfinsen, *Fed. Proc., Fed. Amer. Soc. Exp. Biol.* **10,** 230 (1951).
10. J. R. S. Fincham, *JBC* **182,** 61 (1950).
11. J. A. Olson and C. B. Anfinsen, *JBC* **197,** 67 (1952).
12. C. Frieden, *JBC* **234,** 815 (1959).
13. C. Frieden, "The Enzymes," 2nd ed., Vol. **7,** p. 3, 1963.
14. C. Frieden, *Brookhaven Symp. Biol.* **17,** 98 (1964).
15. C. Frieden, *in* "The Mechanism and Action of Dehydrogenases" (G. W. Schwert and A. D. Winer, eds.), p. 197. University Press, Lexington, Kentucky, 1969.
16. H. F. Fisher, *in* "The Mechanism and Action of Dehydrogenases" (G. W. Schwert and A. D. Winer, eds.), p. 221. University Press, Lexington, Kentucky, 1969.
17. C. Frieden, *in* "The Role of Nucleotides for the Function and Conformation of Enzymes" (H. M. Kalckar *et al.,* eds.), p. 194. Munksgaard, Copenhagen, 1969.
18. B. R. Goldin and C. Frieden, *Curr. Top. Cell. Regul.* **4,** 77 (1971).
19. H. F. Fisher, *Advan. Enzymol.* **39,** 369 (1973).
20. H. Holzer and S. Schneider, *Biochem. Z.* **329,** 361 (1957).
21. B. D. Sanwal, *ABB* **93,** 377 (1961).
22. B. D. Sanwal and M. Lata, *Can. J. Microbiol.* **7,** 319 (1961).

type of regulation that some microorganisms exert over the levels of these enzymes (see Section III) is in contrast to the intense allosteric control by purine nucleoside di- and triphosphates (*17*) that is exhibited by animal GDH's at the level of preformed enzyme.

1. *Fungi*

Distribution and properties of GDH's in over 40 species of fungi have been reviewed by LéJohn (*23*) who has concluded that, whereas higher fungi of the classes Deuteromycetes, Ascomycetes, and Basidiomycetes possess two distinct enzymes, lower fungi, members of the Phycomycetes and Myxomycetes (slime molds), have only one enzyme active with NAD. However, vegetative cell homogenates of *Dictyostelium discoideum* possess both an enzyme that is active with NAD and NADP, and one active only with NAD (*24*).

Some of the enzymes from lower fungi, investigated by LéJohn, are modulated by purine nucleotides, but generally in a different way from the regulation of GDH's from animal sources; for example, whereas GDH's from *Achlya* and *Saprolegnia* are only weakly affected by purine nucleotides at concentrations of about 1 m*M*, that from *Pythium* is completely inhibited, in the direction of oxidative deamination, by 1 m*M* AMP (*25*).

2. *Bacteria*

Types of GDH's found in bacteria are listed in Table I (*7,26–38*). *Escherichia coli* (*7,27*) and *S. typhimurium* (*26*) possess only biosynthetic enzymes, specific for NADP, and may metabolize glutamate for supply of energy by other routes. Studies of mutants of *E. coli* (Section

23. H. B. LéJohn, *Nature (London)* **231**, 164 (1971).
24. F. J. DeToma and W. H. R. Langridge, *Fed. Proc., Fed. Amer. Soc. Exp. Biol.* **33**, 1427 (1974).
25. H. B. LéJohn, R. M. Stevenson, and R. Meuser, *JBC* **245**, 5569 (1970).
26. J. W. Coulton and M. Kapoor, *Can. J. Microbiol.* **19**, 427 (1973).
27. J. L. Meers and L. K. Pedersen, *J. Gen. Microbiol.* **70**, 277 (1972).
28. P. V. Phibbs and R. W. Bernlohr, *J. Bacteriol.* **106**, 375 (1971).
29. M. M. Hong, S. C. Shen, and A. E. Braunstein, *BBA* **36**, 288 (1959).
30. I. Shiio and H. Ozaki, *J. Biochem. (Tokyo)* **68**, 633 (1970).
31. W. M. Johnson and D. W. S. Westlake, *Can. J. Microbiol.* **18**, 881 (1972).
32. O. M. Kew and C. A. Woolfolk, *BBRC* **39**, 1126 (1970).
33. H. B. LéJohn, I. Suzuki, and J. A. Wright, *JBC* **243**, 118 (1968).
34. J. Krämer, *Arch. Microbiol.* **71**, 226 (1970).
35. A. B. Hooper, J. Hansen, and R. Bell, *JBC* **242**, 288 (1967).
36. E. L. Winnacker and H. A. Barker, *BBA* **212**, 225 (1970).
37. R. Bachofen and H. Neeracher, *Arch. Mikrobiol.* **60**, 235 (1968).
38. G. Yarrison, D. W. Young, and G. L. Choules, *J. Bacteriol.* **110**, 494 (1972).

TABLE I
DISTRIBUTION OF GLUTAMATE DEHYDROGENASES IN BACTERIA

Organism	Coenzyme specificity		Ref.
	NAD	NADP	
Escherichia coli	−	+	*7*
Salmonella typhimurium	−	+	*26*
Aerobacter aerogenes	−	+	*27*
Bacillus licheniformis	−	+	*28*
Bacillus cereus	−	+	*28*
Bacillus subtilis 168T	−	−	*28*
Bacillus subtilis IRC 1 and 7	−	+	*29*
Bacillus megaterium	−	−	*28*
Bacillus mycoides	−	−	*28*
Bacillus anthracoides	−	+	*29*
Brevibacterium flavum	?	+	*30*
Micrococcus (Peptococcus) aerogenes	+	+	*31,32*
Thiobacillus novellus	+	+	*33*
Hydrogenomonas H16	+	+	*34*
Nitrosomonas europaea	?	+	*35*
Clostridium SB_4	+	−	*36*
Rhodospirillum rubrum	+	−	*37*
	Active with NAD + NADP		
Mycoplasma laidawii	+		*38*

III,B,1), grown in a medium containing glutamate as sole carbon source, suggest that glutamate is degraded mainly by transamination with oxaloacetate and the aspartate produced forms fumarate by the action of aspartase (*39*).

GDH from the anaerobe *Clostridium* SB_4 is NAD specific; the enzyme is probably responsible for the production of α-ketoglutarate, required as a substrate for transamination reactions and for the citric acid cycle (*36*). No NADP-specific activity has been detected in this organism nor in *Rhodospirillum rubrum* (*37*), but *Clostridium kluyveri* may possess both NAD and NADP linked GDH's (*40*).

Several species of *Bacillus* (see Table I) are devoid of GDH activity. The alanine or other amino acid dehydrogenases of these species are probably responsible for synthesis of amino acids from ammonia and intermediary metabolites (*41*). Tempest *et al.* (*42*) identified a previously

39. J. Vendor, K. Jayaraman, and H. Rickenberg, *J. Bacteriol.* **90,** 1304 (1965).
40. J. R. Stern and G. Bambers, *Biochemistry* **5,** 1113 (1966).
41. A. Meister, *in* "Biochemistry of the Amino Acids," 2nd ed., Vol. 1, p. 313. Academic Press, New York, 1965.
42. D. W. Tempest, J. L. Meers, and C. M. Brown, *BJ* **117,** 405 (1970).

unknown pathway for glutamate biosynthesis in *A. aerogenes*. One of the reactions of this pathway involves the transfer of the amide group of glutamine to α-ketoglutarate to form two molecules of glutamate by NADP-dependent glutamate synthase. When coupled with glutamine synthetase, the net results are biosynthesis of glutamate and, with the additional activity of various transaminases, assimilation of ammonia into amino acids. Because of the higher affinity for ammonia of glutamine synthetase than glutamate dehydrogenase, this pathway may function at much lower levels of free ammonia and probably occurs in many other microorganisms (*27,43*). In *B. megaterium*, which lacks GDH, glutamate may be synthesized exclusively by this route (*44*).

It is noteworthy that glutamate synthase occurs in a complex with glutamate dehydrogenase in *E. coli* (*45*). The two activities can be copurified 30-fold. An important mechanism of regulation may be provided by association of the two enzymes *in vivo*.

Some aerobic cocci grown in media containing glutamate are rich in NAD-GDH, which comprises about 10% of the total protein of the organisms (*31,32*). A distinct NADPH-GDH may also exist in *Micrococcus* (*32*) and, in this respect, *Micrococcus* may be similar to certain facultative chemoautotrophic bacteria (see Table I) which make two separate enzymes with functions similar to those of *Neurospora*.

An interesting GDH is found in *Mycoplasma laidawii*, a bacterium that has a very small genome and lacks cell walls (*38*). This is the only known microorganism possessing a single enzyme that is active with both NAD and NADP. Activity observed in the presence of NADPH and α-ketoglutarate is about 30% of that with NADH; the percentage remains constant through six stages of purification. Isolated enzyme exhibits a single band (active with reduced or oxidized forms of NAD or NADP) after electrophoresis on polyacrylamide. In addition, either NAD or NADP stabilizes the enzyme against inactivation by heat. Molecular weight and subunit size are similar to those of bovine liver GDH, but purine nucleotides have little effect on activity.

Of all bacterial GDH's, only the one from *Thiobacillus novellus* is strongly affected by a purine nucleotide, AMP, but in a fashion which is distinct from that in which animal enzymes are regulated (*33*). Since few regulatory features have been established for bacterial enzymes, it is unlikely that GDH constitutes a major point of control of nitrogen metabolism in these organisms.

43. J. L. Meers, D. W. Tempest, and C. M. Brown, *J. Gen. Microbiol.* **64,** 187 (1970).
44. C. Elmerich and J.-P. Aubert, *BBRC* **42,** 371 (1971).
45. M. A. Savageau, A. M. Kotre, and N. Sakamoto, *BBRC* **48,** 41 (1972).

3. *Plants*

Relatively few detailed studies have been performed on the GDH's of plants. An enzyme active with NAD but not NADP has been purified 1250-fold from pea roots (*46*). When assayed by reductive amination, however, the rate of reaction is 1.8 times faster with NADPH than NADH. An NAD-linked enzyme has been isolated from soybean cotyledons (*47*), and a similar enzyme from corn leaves (*48*) has been shown not to be affected by purine nucleotides (*17*).

Although plant NAD-linked activity is probably mitochondrial in origin, it is possible that an NADP-linked enzyme is located in chloroplasts where it may provide a mode of incorporation of ammonia into carbon skeletons that are produced by photosynthesis. Leech and Kirk (*49*), by separating chloroplasts from mitochondria of broad beans, obtained fractions that were enriched in activity associated with NADP, and the ratio of NAD to NADP activity in chloroplasts from lettuce leaves was found to be 1:1, whereas the ratio in mitochondria was 7:1 (*50*).

GDH's of higher plants may bear some relationship to those isolated from *Chlorella pyrenoidosa* (*51*). A constitutive enzyme, active with both NADH and NADPH, is separable by ion-exchange chromatography from an NADPH-specific enzyme that is induced by ammonia. However, glutamate synthase of higher plants may be the primary catalyst involved in glutamate formation (*52*).

4. *Animals*

The GDH's that have been purified from animal tissues are listed in Table II (*11,26,28,30–33,35,36,38,46–48,51,53–73*). These enzymes func-

46. E. Pahlich and K. W. Joy, *Can. J. Biochem.* **49**, 127 (1971).
47. J. King and W. Y.-F. Wu, *Phytochemistry* **10**, 915 (1971).
48. W. A. Bulen, *ABB* **62**, 173 (1956).
49. R. M. Leech and P. R. Kirk, *BBRC* **32**, 685 (1968).
50. P. J. Lea and D. A. Thurman, *J. Exp. Bot.* **23**, 440 (1972).
51. V. R. Shatilov, V. G. Ambartsamyan, and V. L. Kretovich, *Dokl. Akad. Nauk SSSR* **207**, 1229 (1972).
52. D. K. Dougall, *BBRC* **58**, 639 (1974).
53. F. G. Lehmann and G. Pfleiderer, *Hoppe-Seyler's Z. Physiol. Chem.* **350**, 609 (1969).
54. S. Seyama, T. Saeki, and N. Katunuma, *J. Biochem.* (*Tokyo*) **73**, 39 (1973).
55. H. Arnold and K. P. Maier, *BBA* **251**, 133 (1971).
56. S. Grisolia, C. L. Quijada, and M. Fernandez, *BBA* **81**, 61 (1964).
57. A. Younes, R. Durand, and D. Gautheron, *C.R. Acad. Sci., Ser. D* **273**, 907 (1971).

tion with either NAD or NADP, but there are marked differences among species in their capacities for the two coenzymes, e.g., the chicken and many mammalian enzymes have almost equal activities with either coenzyme, but the frog liver (*60*) and dogfish liver enzymes (*63*) are much more active in the presence of NAD than NADP. Most animal GDH's are inhibited by GTP and activated by ADP. There are, however, quantitative differences among these enzymes; some of these properties are described in more detail in Section VII.

Levels of the enzyme vary greatly in different mammalian tissues. Normally, liver is the richest source (54 IU/g), followed by kidney cortex (11 IU/g), with lower amounts in brain, gastric mucosa, lymph nodes, and lung (*74*). GDH's from various organs of the same species are similar, if not identical. Complete cross-reaction occurs between antibodies induced by bovine liver GDH and extracts of bovine spleen, brain, and heart (*75*); the enzymes from liver, brain, kidney, heart, and skeletal muscle are affected identically by purine nucleotides (*17*). One difference has been noted concerning the effect of phosphate, which stimulates ADP activation of the enzyme of porcine heart (*76*).

Tetrahymena pyriformis possesses a single mitochondrial GDH which is active with both NAD and NADP. Only NAD-related activity is stim-

58. H. Kubo, M. Iwatsubo, H. Watari, and T. Soyama, *J. Biochem.* (*Tokyo*) **46,** 1171 (1959).
59. J. E. Snoke, *JBC* **223,** 271 (1956).
60. L. A. Fahien, B. O. Wiggert, and P. P. Cohen, *JBC* **240,** 1083 (1965).
61. L. A. Fahien and P. P. Cohen, "Methods in Enzymology," Vol. 17A, p. 839, 1970.
62. F. M. Veronese, private communication.
63. L. Corman, L. M. Prescott, and N. O. Kaplan, *JBC* **242,** 1383 (1967).
64. F. M. Veronese, J. F. Nyc, Y. Degani, D. M. Brown, and E. L. Smith, *JBC* **249,** 7922 (1974).
65. K. M. Blumenthal and E. L. Smith, *JBC* **248,** 6002 (1973).
66. J. C. Wootton, *Biochem. Soc. Trans.* **1,** 1250 (1973).
66a. K. M. Blumenthal and E. L. Smith, in preparation.
67. D. Doherty, "Methods in Enzymology," Vol. 17A, p. 850, 1970.
68. A. Fourcade, *Bull. Soc. Chim. Biol.* **50,** 1671 (1968).
69. R. Venard and A. Fourcade, *Biochimie* **54,** 1381 (1972).
70. J. S. Price and F. H. Gleason, *Plant Physiol.* **49,** 87 (1972).
71. H. B. LéJohn and R. M. Stevenson, *JBC* **245,** 3890 (1970).
72. H. B. LéJohn and S. Jackson, *JBC* **243,** 3447 (1968).
73. F. M. Veronese, E. Boccù, and L. Conventi, *BBA* (in press).
74. E. Schmidt, *in* "Methods of Enzymatic Analysis" (H. U. Bergmeyer, ed.), p. 752. Academic Press, New York, 1965.
75. N. Talal and G. M. Tomkins, *Science* **146,** 1309 (1964).
76. C. Godinot and D. Gautheron, *FEBS* (*Fed. Eur. Biochem. Soc.*) *Lett.* **13,** 235 (1971).

TABLE II

SOURCES FROM WHICH GLUTAMATE DEHYDROGENASE HAS BEEN PURIFIED

Source	Coenzyme specificity	Degree of purification[a]	Specific activity[b]	Crystalline form	Ref.
Human liver	NAD + NADP	×276 Homogeneous	87[c]	Needles	*53*
Rat kidney	NAD + NADP	×221 Homogeneous	22.8[c]	Needles	*54*
Rat liver	NAD + NADP	×236 Homogeneous	185[c,d]	Needles	*55*
Bovine brain	NAD + NADP	×35 Partially purified	1.2[e]		*56*
Bovine intestinal mucosa	NAD + NADP	×33 Partially purified	0.48[e]		*56*
Bovine liver	NAD + NADP	×83 Homogeneous	30[e]	Needles	*11*
Porcine heart	NAD + NADP	×47 Homogeneous	10–12[c]		*57*
Porcine liver	NAD + NADP	Homogeneous	3.6[f]	Needles	*58*
Chicken liver	NAD + NADP	×55 Homogeneous	43[c]	Thin hexagonal plates	*59*
Rana catesbeiana (frog liver)	NAD + NADP[g]	×36 Homogeneous	25[c]	Stubby needles	*60*
Rana catesbeiana (tadpole liver)	NAD + NADP[h]	×360	25[c]		*61*
Tuna liver	NAD + NADP	Homogeneous			*62*
Squalus acanthias (dogfish liver)	NAD + NADP[i,j]	×440 Homogeneous	4[f]	Needles	*63*
Pisum sativum (roots)	NAD[k]	×1250	49[c]		*46*
Zea mays (leaves)	NAD	×45	3.5[c]		*48*
Glycine max. (cotelydons)	NAD	×250	1.9[c]		*47*

Neurospora crassa	NAD	×81	390[c]		*64*
		Homogeneous			
	NADP	×142	140[e]	"Trigonal" (*66*)	*65*
		Homogeneous			
Saccharomyces cerevisiae	NAD	×258			*67*
		Minor contamination			
	NADP	×1750			*68*
		Homogeneous			*69*
Apodachlya brachynema	NAD	×50	20[f]		*70*
Achlya sp. (1969)	NAD	×118			*71*
		Homogeneous by sedimentation in sucrose density gradients			
Pythium debaryanum	NAD	×127			*71*
Blastocladiella emersonii	NAD	× >100			*72*
Chlorella pyrenoidosa	NAD + NADP	×37	6.9[c]		*51*
	NADP	×63	1.8[e]		*51*
Salmonella typhimurium	NADP	×190	54[e]		*26*
		Minor contamination			
Escherichia coli	NADP	×860	250[e]		*73*
		Homogeneous			
Peptococcus aerogenes	NAD	×13	64.5[f]		*31*
		Contained additional minor component with NAD-GDH activity			
Micrococcus aerogenes	NAD[l]	×20	268[c]		*32*
		Homogeneous			
Brevibacterium flavum	NADP	×65	71[e]		*30*
		Partially purified			
Bacillus licheniformis	NADP	×55	26.2[e]		*28*
		Partially purified			
Clostridium SB_4	NAD	×250	485[e]	Rhombohedric	*36*
		Homogeneous			

TABLE II (*Continued*)

Source	Coenzyme specificity	Degree of purification[a]	Specific activity[b]	Crystalline form	Ref.
Thiobacillus novellus	NAD	×250			*33*
	NADP	×250			*33*
Nitrosomonas europaea	NADP	×160	10.3[m]		*35*
Mycoplasma laidawii	NAD + NADP	×570 ~70% pure	79.5[c]		*38*

[a] Preparations are described as homogeneous where sufficient evidence is available to show that this is so.

[b] Specific activity is expressed in international units per milligram of protein. An international unit is the amount of enzyme required to convert 1 μmole of substrate per minute. No attempt has been made to standardize conditions. Hence, strict comparisons are not possible.

[c] Measured by following oxidation of NADH.

[d] Measured in the presence of 2 m*M* ADP.

[e] Measured by following oxidation of NADPH.

[f] Measured by following reduction of NAD.

[g] Eight times more reactive with NAD than NADP.

[h] Tadpole liver GDH is about as reactive with NAD as frog enzyme, but much less reactive with NADP.

[i] Fifteen times more reactive with NAD than NADP.

[j] Six times more reactive with NADH than NADPH.

[k] Also reactive with NADPH, but not with NADP.

[l] At pH 7, the enzyme has 3% of the activity with NADPH as it has with NADH. No activity is observed with NADP.

[m] Measured by following reduction of NADP.

ulated by ADP and inhibited by GTP; activity with NADPH is not affected (*77*). It has been suggested that this enzyme is intermediate in type between animal and fungal glutamate dehydrogenases, of which only the NAD-related type is affected by purine nucleotides.

D. Cellular Location

The GDH's of animal cells are in mitochondria (*78*), and the ease with which the enzyme is solubilized suggests that it is in the matrix and not attached to mitochondrial membranes. Submitochondrial fractionation techniques have confirmed this location.

As with most mitochondrial proteins, biosynthesis occurs on microsomes and GDH is then transported into mitochondria (*79*). The observed binding of GDH to phosphatidylserine, which *in vivo* is found in microsomal membranes, and to cardiolipin, which is found in mitochondrial membranes, has led to the suggestion, not as yet demonstrated experimentally, that these phospholipids are involved in transport (*79,80*).

Some studies suggest that the principal pathway of glutamate utilization in liver mitochondria is by transamination (*81*). GDH decreases the distribution coefficient of glutamate-oxaloacetate aminotransferase on Sephadex G-200, possibly by forming a complex with that enzyme (*82*). In addition, in the presence of NADPH and NH_4^+, GDH appears to catalyze the conversion of the pyridoxal phosphate form of the aminotransferase to the pyridoxamine form, which catalyzes the formation of α-amino acids from α-keto acids (*83*). This reaction is interesting in view of the inhibition of GDH by pyridoxal phosphate (*84*) (See Section V,A). If the complex exists in mitochondria, it may provide an efficient mode of dehydrogenation of amino acids that are not normally good substrates of GDH (*82*).

There are conflicting reports about the presence of GDH in nuclei of rat liver cells and Chang's liver cells (*85–87*). Damage to mitochondria,

77. A. Hooper, K. R. Terry, and K. Kemp, *Fed. Proc., Fed. Amer. Soc. Exp. Biol.* 33, 696 (1974).
78. C. de Duve, R. Wattiaux, and P. Baudhuin, *Advan. Enzymol.* **24**, 291 (1962).
79. C. Godinot and H. A. Lardy, *Biochemistry* **12**, 2051 (1973).
80. C. Godinot, *Biochemistry* **12**, 4029 (1973).
81. P. Borst, *BBA* **57**, 256 (1962).
82. L. A. Fahien and S. E. Smith, *JBC* **249**, 2696 (1974).
83. L. A. Fahien and S. E. Smith, *ABB* **135**, 136 (1969).
84. D. Piszkiewicz and E. L. Smith, *Biochemistry* **10**, 4538 (1971).
85. M. Banay-Schwartz and H. J. Strecker, *Int. J. Biochem.* **1**, 371 (1970).
86. G. di Prisco and H. J. Strecker, *Eur. J. Biochem.* **12**, 483 (1970).
87. K. S. King and C. Frieden, *JBC* **245**, 4391 (1970).

which results in redistribution of soluble enzymes and their appearance in nuclear fractions, may occur during subcellular fractionation. Claims (*85,86*) that there are kinetic differences between GDH's released from nuclei and the mitochondrial enzyme have not been substantiated (*87*). More recently, the properties of the enzymes from mitochondrial and nuclear fractions have been further studied; they appear to be similar in most respects (*88*).

GDH's of fungal cells are in the soluble phase of the cytosol. In 1963, it was noted that an NAD-GDH of *Puccinia helianthi* was separated from mitochondria and microsomes by fractional centrifugation (*89*). Both NAD- and NADP-GDH's of *S. carlsbergensis* and *S. cerevisiae* are cytoplasmic (*90,91*). Two possible explanations were offered (*90*). Location outside mitochondria may represent an adaptation to strong repression exerted by glucose on all mitochondrial enzymes in yeast. Alternatively, evolution of a mitochondrial GDH is a late acquisition and coincides with expression of an enzyme that is not specific in its requirements for NAD or NADP. Existence in *Dictyostelium discoideum* of both a mitochondrial enzyme, which is active with both NAD and NADP, and an extramitochondrial enzyme, which is active only with NAD, tends to support the latter view (*24*).

E. Metamorphosis

There is a substantial increase in specific activity of GDH of tadpole liver during metamorphosis occurring naturally or induced by thyroxine. Studies on the purified enzymes from tadpole and frog have shown differences in kinetic constants for NAD and NADP, and in the effects of temperature on initial rates observed with NAD as coenzyme. The molecular weight of tadpole GDH is somewhat higher than that of the frog enzyme. Thyroxine not only increases the rate of glutamate dehydrogenase biosynthesis but also increases the rate of conversion of some, as yet unknown, precursor of the enzyme. Incidental to these studies was a measure of the turnover time of GDH, which appears to have a half-life of 24 hr in frog liver (*92,93*).

88. G. di Prisco and F. Garofano, *BBRC* **58,** 683 (1974).
89. J. E. Smith, *Can. J. Microbiol.* **9,** 345 (1963).
90. C. P. Hollenberg, W. F. Riley, and P. Borst, *BBA* **201,** 13 (1970).
91. P. S. Perlman and H. R. Mahler, *ABB* **136,** 245 (1970).
92. B. O. Wiggert and P. P. Cohen, *JBC* **241,** 210 (1966).
93. J. B. Balinsky, G. E. Shambaugh, and P. P. Cohen, *JBC* **245,** 128 (1970).

II. Molecular Properties

A. Purification

Several procedures have been reported for crystallization of GDH since the original descriptions in 1951 (*8,9*). A general method involving initial extraction of the mitochondrial fraction of liver homogenates with cetyltrimethylammonium bromide yields enzyme of reasonably high specific activity (*94*). GDH's have also been purified to varying degrees of homogeneity from some microorganisms and a few plants (see Table II).

An affinity column formed by the reaction of the amino group of L-glutamic acid with 1-*N*-bromoacetyl-6-*N*-Sepharose hexyldiamine has been employed as a purification tool (*65*). It may be possible to develop the use of columns containing dinucleotides attached to polymers for this purpose. Substitution on the adenine rather than the pyridine ring would yield a more effective ligand. Bovine liver GDH could not be recovered from a column obtained by attaching NAD or NADP to ε-aminocaproyl Sepharose by treatment with a carbodiimide. This enzyme could be eluted, however, from a derivative formed by linking AMP through its amino group and a six-carbon spacer molecule to Sepharose (*95*).

B. Polymerization

1. *Significance*

Olson and Anfinsen (*11*) observed that a non-Gaussian gradient curve is produced by sedimenting bovine liver enzyme, and suggested that disaggregation or unfolding of the protein occurs in dilute solutions. The variation of light scattering obtained by increasing the concentration of the bovine or porcine liver enzymes demonstrates that polymerization occurs at protein concentrations above 0.1 mg/ml (*58*).

A number of ligands, which also influence the rates of the catalytic reactions, affect the polymerization reaction. Substances such as GTP (*96*), thyroxine (*97*), and diethylstilbestrol (*98*), which inhibit GDH, favor disaggregation in the presence of reduced coenzyme. On the other

94. L. A. Fahien, M. Strmecki, and S. Smith, *ABB* **130**, 449 (1969).
95. C. R. Lowe, K. Mosbach, and P. D. G. Dean, *BBRC* **48**, 1004 (1972).
96. J. Wolff, *JBC* **237**, 236 (1962).
97. J. Wolff, *BBA* **26**, 387 (1957).
98. G. M. Tomkins, K. L. Yielding, and J. Curran, *Proc. Nat. Acad. Sci. U. S.* **47**, 270 (1961).

hand, ADP, which enhances enzymic activity, favors aggregation in the presence of NADH (*12*). Concentrations of GDH in many organs are at levels at which polymerized forms may be present; hence, it has been suggested that the effects of modifiers upon polymerization may be physiologically significant.

It is now known that GDH's from different sources vary greatly in their ability to polymerize. Liver GDH's from pig (*99*) and man (*58*) polymerize in concentrated solutions to the same extent as the bovine enzyme, whereas chicken (*100*) and bullfrog (*60*) enzymes aggregate to a lesser extent. The enzymes purified from rat liver (*87*), rat kidney (*54*), dogfish liver (*63*), and tuna liver (*62*) do not polymerize, and no polymerization of GDH's from nonanimal sources has been reported (see Table III) (*101–105*).

Studies discussed below have provided much insight into factors that determine the quaternary structure of a protein. However, in view of the fact that rat GDH, which is kinetically and structurally similar to the bovine enzyme, does not polymerize, it seems unlikely that the phenomenon of polymerization plays any major physiological role.

Below concentrations of 0.1 mg/ml, bovine liver GDH is predominantly in a form of 330,000 molecular weight which contains six identical polypeptide chains (subunits). Most evidence suggests that this form, to which the term oligomer (or hexamer) will be applied, is the smallest enzymically active unit. In the ensuing discussion, the terms polymerization and aggregation will refer to formation of species of molecular weight greater than 330,000; association and dissociation will refer to interactions among subunits.

2. *Mechanism*

Angular dependence of light scattered from concentrated solutions of the bovine enzyme is characteristic of scattering from a rod-shaped polymer (*58,102*). Studies of the small-angle X-ray diffraction (*106*) and viscosity (*107*) of enzyme solutions as a function of protein concentration have also shown that the enzyme aggregates to form linear polymers.

99. P. Dessen and D. Pantaloni, *Eur. J. Biochem.* **8,** 292 (1969).
100. C. Frieden, *BBA* **62,** 423 (1962).
101. D. Pantaloni and P. Dessen, *Eur. J. Biochem.* **11,** 510 (1969).
102. H. Eisenberg and G. M. Tomkins, *JMB* **31,** 37 (1968).
103. K. Moon and E. L. Smith, *JBC* **248,** 3082 (1973).
104. M. Cassman and H. K. Schachman, *Biochemistry* **10,** 1015 (1971).
105. P. J. Anderson and P. Johnson, *BBA* **181,** 45 (1969).
106. H. Sund, I. Pilz, and M. Herbst, *Eur. J. Biochem.* **7,** 517 (1969).
107. E. Reisler and H. Eisenberg, *Biopolymers* **9,** 877 (1970).

TABLE III
MOLECULAR WEIGHTS

Source	Coenzyme specificity	Oligomeric MW ($\times 10^{-5}$)	Method*	Ref.	Association	Subunit MW ($\times 10^{-4}$)	Method	Ref.
Human liver	NAD(P)			*53*	Yes			
Porcine liver	NAD(P)	3.10	*a*	*101*	Yes	5.2	*b*	*101*
Porcine heart	NAD(P)	2.50	*c*	*57*				
Bovine liver	NAD(P)	3.01–3.25	*a*	*102*	Yes	5.5393	*d*	*103*
		3.00–3.40	*e*	*104*				
Rat liver	NAD(P)	3.30–3.70	*e*	*87*	No	4.3–5.3	*f*	*87*
Rat kidney	NAD(P)			*54*	No			
Chicken liver	NAD(P)	3.06–3.46	*c*	*105*	Yes[g]	5.5658	*d*	*103*
Frog liver	NAD(P)	2.5	*a*	*60*	Yes[g]			
Dogfish liver	NAD(P)	3.1–3.5	*e*	*63*	No			
Tuna liver	NAD(P)	3.12–3.50	*e*	*62*	No	5.1–5.6	*f*	*62*
Pisum sativum (roots)	NAD	2.1	*h*	*46*				
Neurospora crassa	NAD	4.53–5.07	*e*	*64*	No	11.1–12.1	*f*	*64*
	NADP	2.80–2.96	*e*	*66*	No	4.8438	*d*	*66*
Saccharomyces cerevisiae	NADP	2.70–2.90	*a*	*69*	No	4.8	*i*	*69*
Achlya sp. (1969)	NAD	2.25	*j*	*71*				
Pythium debaryanum	NAD	2.25	*j*	*71*				
Blastocladiella emersonii	NAD	2.1–2.5	*j*	*72*				
Salmonella typhimurium	NADP	2.8	*h*	*26*				
Escherichia coli	NADP	2.40–2.54	*h, j*	*73*		4.6	*i*	*73*
Peptococcus aerogenes	NAD	2.7	*h*	*31*				
Clostridium SB_4	NAD	3.5–3.9	*h*	*36*	No			
		2.68–2.82	*e*	*36*				
Thiobacillus novellus	NAD	1.1–1.3	*j*	*33*				
	NADP	1.2–1.4	*j*	*33*				
Mycoplasma laidawii	NAD(P)	2.4–2.6	*j*	*38*		4.8	*i*	*38*

* Methods used:
- *a*, Light scattering.
- *b*, Light scattering under denaturing conditions.
- *c*, Sedimentation velocity.
- *d*, Sequence determination.
- *e*, Sedimentation equilibrium.
- *f*, Sedimentation equilibrium under denaturing and reducing conditions.
- *g*, Associates to a lesser extent than the beef enzyme.
- *h*, Gel chromatography.
- *i*, SDS-polyacrylamide electrophoresis.
- *j*, Sedimentation in sucrose density gradients.

Mass per unit length (2340 daltons/Å) and the cross-sectional radius of gyration (30 Å) of the polymers are independent of size, as expected for a linear mode of polymerization (*106*).

Colman and Frieden (*108*) proposed that the polymerization could be described as a closed equilibrium between oligomer and a polymer containing four oligomers without intermediate steps, but if this type of equilibrium were to occur, two peaks would probably be formed in the ultracentrifuge whereas Olson and Anfinsen (*11*) had seen only one.

Apparent molecular weights, (M_{App}) as obtained, for example, by light scattering techniques, are related to the true, weight average, molecular weights (M_w) by an expression

$$M_{App}^{-1} = M_w^{-1} + 2A_2c$$

where A_2 is the second virial coefficient, and higher terms of the expansion are not included. Ignoring the nonideality term, Sund and Burchard (*109*) suggested that their data were consistent with a stepwise association to yield a limited polymer consisting of eight oligomers. Eisenberg and Tomkins (*102*) calculated a value of 313,000 for the molecular weight of the oligomer by extrapolation to zero protein concentration of the intensities of light scattered by solutions of bovine enzyme in the presence, and the absence, of the disaggregating ligands NADH and GTP. Their data were later shown to be consistent with a model in which polymerization proceeds in a stepwise manner to yield polymers of indefinite length (*110*). Over a limited range of protein concentration (3–8 mg/ml) the polymerization can be described by a unique equilibrium constant and a second virial coefficient that is independent of polymer size. This is not the case over the complete concentration range, however (*111*), but a value of 2.0 ± 0.1 ml/mg is obtained for the equilibrium constant at 20° at concentrations below 0.4 mg/ml. Higher molecular weights at lower protein concentrations are obtained in the presence of toluene or benzene, and a linear relationship can be established between radius of gyration of the polymeric length and size, up to molecular weights of 3.5 million, when toluene is added to the enzymic solution (*112*).

Analyses of high-speed equilibrium measurements in the ultracentrifuge suggest, however, that equilibrium constants for each step are not identical, but the oligomer–dimer equilibrium is favored, while lower equilibrium constants are found for larger aggregates (*104*).

108. R. F. Colman and C. Frieden, *JBC* **241**, 3661 (1966).
109. H. Sund and W. Burchard, *Eur. J. Biochem.* **6**, 202 (1968).
110. P. W. Chun and S. J. Kim, *Biochemistry* **8**, 1633 (1969).
111. E. Reisler and H. Eisenberg, *Biochemistry* **10**, 2659 (1971).
112. E. Reisler, J. Pouyet, and H. Eisenberg, *Biochemistry* **9**, 3095 (1970).

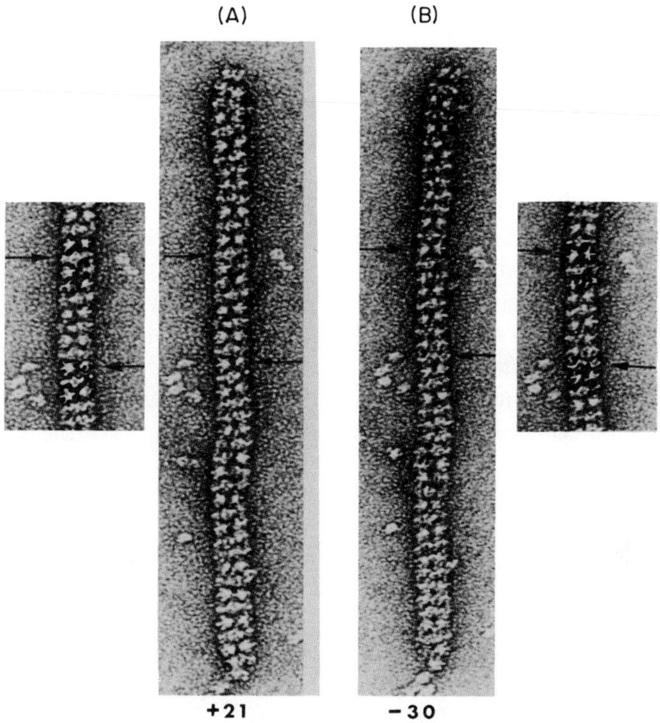

FIG. 1. An electron micrograph of a tubular structure arising from association of four polymer chains of glutamate dehydrogenase. The chains are arranged in helical array and are related by a 4-fold rotation axis coincident with the helix axis. This arrangement gives rise to the characteristic rings or annuli, which change in appearance as the tube is tilted about the helix axis from +21° (A) to −30° (B). The tube diameter is 190–200 Å. Courtesy of Josephs and Borisy (*113*).

Linear polymers have been observed by electron microscopy (*113,114*). The linear tubes depicted in Fig. 1 consist of four polymer chains arranged in helical array. The two-dimensional network visible in the electron micrographs of Munn (*114*) probably represents an initial stage in formation of crystals.

The forces involved in the polymerization reaction have not been well characterized. The influences of pH and ionic strength upon the degree of aggregation suggest that electrostatic interactions are involved (*104,115*). Assistance of polymerization by D_2O may result from in-

113. R. Josephs and G. Borisy, *JMB* **65,** 127 (1972).
114. E. A. Munn, *BBA* **285,** 301 (1972).
115. C. Frieden, *JBC* **237,** 2396 (1962).

creased strength of deuterium bonds as opposed to hydrogen bonds (*116*). The effect of dioxane, which depolymerizes the polymeric forms (*58*), may indicate participation of hydrophobic forces, but no change in the environment of aromatic residues occurs upon polymerization (*117*). The influence of temperature depends upon the ionic strength and buffer composition of the solution (*104,111,115*) and is thus not readily interpreted. Sequence determination (see Section IV) of the rat liver enzyme which, as discussed above, does not polymerize may aid in identifying the amino acid residues involved.

Rates of disaggregation induced by dilution have been measured by stopped-flow experiments (*118*). Disaggregation is a rapid process, with a half-time of about 50 msec at 25°. Activation energies measured at different protein concentrations are the same, providing more evidence that the bimolecular steps that give linear polymers have the same equilibrium constants.

3. *Nucleotide Effects*

It was originally thought that the polymeric form was active GDH, whereas the oligomeric form was relatively inactive (*119*). To account for the wide specificity of the enzyme, this idea was extended by assuming that purine nucleotides shifted the equilibrium between an oligomeric form that possessed little GDH activity but more alanine dehydrogenase activity, and a polymeric form that was more active with glutamate (*98*). However, these notions were based on observations made under assay conditions in which the enzyme was in dilute solution; furthermore, when both alanine and glutamate dehydrogenase activities are measured directly, at high enzyme concentrations, but with the inclusion of inhibitors, there is no correlation of specific activities with degree of polymerization (*120*). In addition, other investigators demonstrated that dehydrogenations of alanine and glutamate have different rate-limiting steps (*121*).

Studies of the binding of GTP to GDH in the presence of NADH yielded Scatchard plots that deviate from linearity at higher protein concentrations, indicating that GTP has greater affinity for the disaggregated enzyme than the polymeric form (*122*). In the presence of NADH and

116. R. F. Henderson, T. R. Henderson, and B. M. Woodfin, *JBC* **245,** 3733 (1970).
117. D. G. Cross and H. F. Fisher, *ABB* **110,** 222 (1965).
118. H. F. Fisher and J. R. Bard, *BBA* **188,** 168 (1969).
119. C. Frieden, *JBC* **234,** 809 (1959).
120. H. F. Fisher, D. G. Cross, and L. L. McGregor, *BBA* **99,** 165 (1965).
121. M. Iwatsubo and D. Pantaloni, *Bull. Soc. Chim. Biol.* **49,** 1563 (1967).
122. C. Frieden and R. F. Colman, *JBC* **242,** 1705 (1967).

protein concentrations of 2 mg/ml, the dissociation constant for ADP and acetylated enzyme, a form which retains activity but cannot associate, is 2.5-fold higher than for native enzyme, and consequently it was deduced that ADP binds more tightly to the polymerized form. However, in the absence of coenzyme, the dissociation constant of ADP is independent of enzyme concentration, and ADP has no effect on the state of aggregation in the absence of GTP and NADH (*123*).

Fisher (*19*) has suggested that GTP influences polymerization by competing, as its subsite on the protein surface, with the binding of an aggregating oligomer. Effective binding of GTP is ensured by binding of NAD(P)H, while ADP may interfere with binding of GTP, but binds in such a manner that an additional protein molecule is now free to approach.

When NADH and GTP are added to solutions of GDH, multiphasic changes occur that may be followed by stopped-flow techniques (*124*). A slower phase ($k = 1.5\ sec^{-1}$) is characterized by a change in the extinction coefficient at 365 nm of the bound nicotinamide ring, while a faster step ($k = 60\ sec^{-1}$) is associated with decrease in turbidity, and thus depolymerization of the enzyme. Addition of these substances produces disaggregation much faster than dilution of the enzyme, and the changes show less variation with temperature. Thus, different rate-limiting steps may be involved. The actions of NADH at its two binding sites per subunit (see Section VII) have been identified. If the enzyme is incubated with NADH (but not NADPH) before GTP is added, depolymerization is notably slower, and, whereas binding at the active site is associated with depolymerization, binding at the regulatory site is responsible for protection against depolymerization. Since the processes that result in inhibition by GTP of the oxidation of NAD(P)H with α-ketoglutarate occur much faster (in about 2 msec) than depolymerization, it is difficult to see how regulation can be achieved by control of aggregation.

There are marked variations in the rates of depolymerization depending upon which coenzyme and which purine nucleotide are used (*125*). It has been suggested that the slow rates mentioned above control the rate of conversion of a form of enzyme acting preferentially with one coenzyme to one acting preferentially with another (*18*), but no direct test of this hypothesis has been performed.

Kinetic parameters, including binding constants of GTP and ADP, are independent of the molecular size of polymers of GDH that have been

123. H. F. Fisher, J. M. Culver, and R. A. Prough, *BBRC* **46,** 1462 (1972).
124. J. M. Jallon, A. di Franco, and M. Iwatsubo, *Eur. J. Biochem.* **13,** 428 (1970).
125. C. Y. Huang and C. Frieden, *JBC* **247,** 3638 (1972).

"fixed" by glutaraldehyde (*126*). In this type of experiment, it is difficult to isolate effects of molecular size from those resulting from modification of functional groups. Most cross-linkages were formed between subunits rather than between oligomers.

C. Structure of the Oligomer

1. *Molecular Weight*

Differences between reported values for the minimal molecular weight of the bovine enzyme, which have varied between 280,000 (*109*) and 400,000 (*108*), emphasize the difficulty of measuring molecular weights of proteins that aggregate since it is now recognized that the true molecular weight is 332,000 as judged from the amino acid sequence of the six identical subunits (*103*). Molecular weights of the oligomeric, enzymically active forms of various GDH's and of their substituent polypeptide chains are recorded in Table III.

The bovine enzyme was demonstrated to be composed of polypeptide chains that are identical and not covalently linked from the following studies. Alanine is the sole NH_2-terminal amino acid and threonine is the sole COOH-terminal amino acid of the enzyme (*127,128*). Physicochemical studies performed on enzyme completely denatured by high concentrations of guanidine hydrochloride (*102,129,130*) indicated subunits of identical size.

The subunit molecular weight of 55,390 calculated from the sequence (*103*) agrees with values obtained by sedimentation-equilibrium (*104,130*) and by light scattering measurements (*102*). A value of 57,000 is obtained by spectrophotometric titration with NAD(P)H in the presence of glutamate (*131*), and a lower value (45,000–50,000) is obtained by gel chromatography on G-200 suspended in a solution of SDS (*132*). Values of 313,000 (*102*) and 320,000 (*104*) have been obtained for the molecular weight of the smallest, enzymically active unit, yielding a value of 5.7, close to 6, for the number of subunits in the oligomer.

Chicken liver GDH has three additional residues in its polypeptide

126. R. Josephs, H. Eisenberg, and E. Reisler, *Biochemistry* **12**, 4060 (1973).
127. B. Jirgensons, *JACS* **83**, 3161 (1961).
128. E. Appella and G. M. Tomkins, *JMB* **18**, 77 (1966).
129. E. Marler and C. Tanford, *JBC* **239**, 4217 (1964).
130. M. Landon, M. D. Melamed, and E. L. Smith, *JBC* **246**, 2360 (1971).
131. R. R. Egan and K. Dalziel, *BBA* **250**, 47 (1971).
132. M. Pagé and C. Godin, *Can. J. Biochem.* **47**, 401 (1969).

chain (*133*), whereas the NADP-dependent GDH from *Neurospora* possesses 50 residues less than the bovine enzyme (see Section IV). Subunits of the enzymes from porcine and tuna liver are similar in size to that of bovine liver. The chain of the NADP-specific *Neurospora* enzyme appears similar in size to those of the NADP-linked enzymes from bakers' yeast and *E. coli*, the enzyme of dual specificity from *Mycoplasma laidawii*, and possibly also the rat liver enzyme (see Table III).

The subunit size of bovine GDH is larger than that of most other dehydrogenases, but is close to that reported for glucose-6-phosphate dehydrogenase of human erythrocytes (*134*).

The size of the NAD-GDH from *Neurospora* (480,000) is very different from others that have been reported. The active enzyme possesses four identical subunits, each almost twice the size of those discussed above (*64*) ; its properties are discussed in Section IV.

2. *The Hexameric Model*

Eisenberg and Reisler (*135*) proposed a model for the oligomeric structure of bovine liver GDH by considering the number of ways of packing six subunits in a symmetrical array. The model has dihedral symmetry, and the threefold axis and one of the twofold axes are shown in Fig. 2. Polymerization occurs in the direction of the threefold axis.

The reduced specific viscosities of GDH in solution with 10^{-3} *M* GTP and 10^{-3} *M* NADH have been measured, and a value of intrinsic viscosity of $[\eta] = 3.2$ ml/g obtained by extrapolation to zero protein concentration (*107*). The transverse and axial rotary frictional coefficients of macroscopic models, similar to the physical model depicted in Fig. 2, were measured, and the viscosity calculated from these coefficients agrees with the measured value; however, it is not possible to define whether the subunits adopt a staggered or eclipsed conformation, as viewed down the threefold axis, and other models may give the same result.

Low-angle X-ray scattering measurements have yielded a value for the linear mass of glutamate dehydrogenase polymers of 2340 daltons/Å, and a diameter of approximately 86 Å (*106*). From the value of 332,000 for the molecular weight, i.e., six times the subunit molecular weight, the length of the oligomer should be 142 Å. From these values and the model proposed, it may be seen that the subunits, if in eclipsed conformations, may approximate to elongated prolate ellipsoids with axial ratios of 1:1.65; however, the X-ray scattering curve shows a high subsidiary

133. K. Moon, D. Piszkiewicz, and E. L. Smith, *JBC* **248,** 3093 (1973).
134. M. C. Ratazzi, *BBRC* **31,** 16 (1968).
135. H. Eisenberg and E. Reisler, *Biopolymers* **9,** 113 (1970).

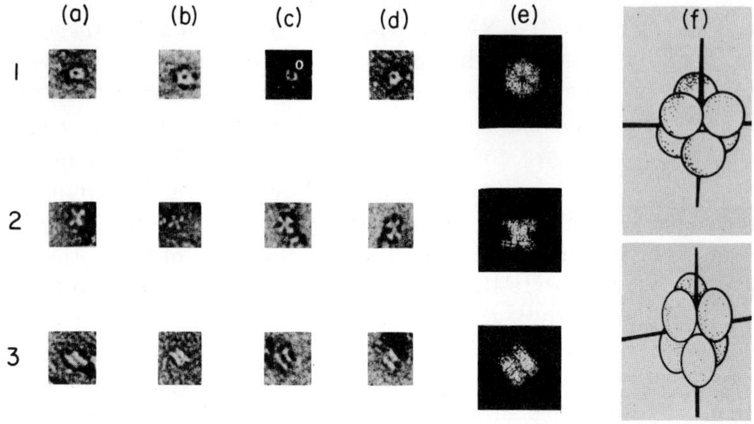

FIG. 2. Columns (a) to (d), electron micrographs of isolated molecules of glutamate dehydrogenase in various orientations. Column (e), computed projections of a model having spherical subunits. Column (f), two models of gluatamate dehydrogenase. Both models yield similar shadowgraph projections. The vertical axis is the 3-fold axis of each molecule, and the horizontal axis represents one of the three identical 2-fold axes. Courtesy of Josephs (*139*).

maximum suggesting that the oligomer is far more loosely packed than expected from the model (*106*).

A variety of profiles has been reported in electron micrographs of individual molecules of bovine GDH (*114,136–139*); however, the images obtained may be influenced by the staining conditions used. Images obtained (*139*) by staining the enzyme negatively with 1% uranyl acetate over holes are shown in Fig. 2. The profiles of the particles have been identified with various projections of the model, the forms of which are shown in the final column of Fig. 2. The top form has spherical subunits, and the bottom elliptical, but both have subunits in staggered configurations. The doughnut shapes of row 1 are compatible with a projection of either model viewed down the threefold axis. The cross-shaped patterns of row 2 may arise when the direction of view of the model is 40° from the threefold axis. The two parallel separated blobs of density of row 3 may be anticipated by a view down the plane of the twofold axes of the model. In electron micrographs of polymers, individual oligomers appear delineated along the length at spaces of 100–120 Å, somewhat shorter

136. G. D. Greville and R. W. Horne, *JMB* **6**, 506 (1963).
137. R. Valentine, *in* "PreCongress Abstracts of the Fourth European Regional Congress" (D. S. Bocciarelli, ed.), p. 3. Rome, 1968.
138. A. M. Fiskin, E. F. J. van Bruggen, and H. F. Fisher, *Biochemistry* **10**, 2396 (1971).
139. R. Josephs, *JMB* **55**, 147 (1971).

than the length obtained by X-ray scattering, and the width is about 80 Å. These dimensions are better accommodated by elliptical rather than spherical subunits.

Electron micrographs obtained by Fisher and his colleagues (*138*) have been interpreted on the basis of a previous proposal that dehydrogenases are composed of morphopoietic units, each of 20,000 molecular weight. One of the several models compatible with images observed is composed by positioning a dimeric structure of two of the morphopoietic units at each vertex of an octahedron.

3. *Formation of Trimers*

The various intersubunit binding domains of the oligomer would be expected to cleave under different conditions to yield trimers or dimers. Formation of dimers and trimers in the presence of SDS has been reported, but details of conditions were not described (*140*). In addition, the porcine enzyme in 33% dioxane retains 50% of its activity for several days and exhibits a molecular weight of 140,000 (*58*). Propellor shaped trimers have been seen in electron micrographs (*136,137,139*) and micrographs of tilted fields containing these particles reveal that they consist of three subunits, presumably formed by cleavage between layers of trimers (*139*).

Areas close to binding domains that hold together layers of three subunits react preferentially with the cross-linking reagent dimethyladipimidate to form trimers, which are recognized by their electrophoretic mobility on polyacrylamide in SDS (*141*). It has not yet been possible to separate active trimers.

Treatment of the tetrameric NAD-GDH of *Neurospora* with tetrathionate yields an active enzyme in which only two of the subunits are cross-linked (*64*).

4. *Immunochemistry*

Antigen–antibody reactions of human and bovine liver GDH's exhibit atypical Heidelberger curves (*142,143*). Equivalence is reached with equimolar amounts of antigen and antibody, but as antigen concentration is increased with fixed amount of antibody, more precipitate forms in which the ratio of antigen to antibody is increased. This situation prob-

140. J. Krause, K. Markau, M. Minssen, and H. Sund, *in* "Pyridine Nucleotide-Dependent Dehydrogenases" (H. Sund, ed.), p. 279. Springer-Verlag, Berlin and New York, 1970.
141. F. Hucho and M. Janda, *BBRC* **57**, 1080 (1974).
142. F. G. Lehmann, *BBA* **235**, 259 (1971).
143. L. A. Fahien, H. G. Steinman, and R. McCann, *JBC* **241**, 4700 (1966).

TABLE IV
CHIROPTICAL PARAMETERS OF GLUTAMATE DEHYDROGENASE

Enzyme source	λ_0 (nm)	a_0	b_0	$A^*_{(eq)193}$	$A^*_{(eq)225}$	$[m']_{233}$ deg cm^2 dmole^{-1}	Ref.
Bovine liver	212	60	−220	962	−654	−4570	*105*
Chicken liver	212	25	−190	815	−577	−4230	*105*
Saccharomyces cerevisiae (NADP)	212	−45	−150	465	−390	−2820	*69*

ably arises from formation of complexes with polymerized enzyme, in which not all antigenic sites are available (*142,143*), and probably gives rise to the multiple precipitin bands observed when concentrated solutions of the enzyme are placed in wells of a double-diffusion plate (*75*).

Complexes of GDH and antibody retain activity but antibody interferes with the actions of the allosteric modifiers ADP and GTP. There is no evidence of competition at modifier sites, but antibodies may cause conformational changes which interfere with regulation (*143*).

Complete cross-reaction occurs between antibodies to bovine enzyme and the enzymes of sheep, pig (*75*), and frog (*143*); partial cross-reaction with the enzymes of pigeon and rat; and no cross-reaction with GDH's from microorganisms (*75*).

D. TERTIARY STRUCTURE

Preliminary X-ray diffraction data for crystalline chicken liver GDH and *Neurospora* NADP-GDH are characteristic of unit cells too complex for straightforward determination of tertiary structure.

Animal GDH's are unusual proteins in that they dextrorotate plane-polarized visible light (*127*); the positive rotations do not result from polymerization (*144*). In contrast, the NADP-GDH from bakers' yeast is levorotatory (*69*). The reported lack of reproducibility of rotations recorded in the ultraviolet, which may have been caused by denaturation at low ionic strength, can be prevented by including glycerol in the solution (*105*). Parameters calculated by the procedures of Moffitt (*145*), and of Shechter and Blout (*146*), and recorded in Table IV indicate a high α-helix content.

For bovine GDH it has been calculated that 58% of the tyrosyl residues, 22% of the tryptophanyl, and 75% of the phenylalanyl residues

144. J. E. Churchich and F. Wold, *Biochemistry* **2**, 781 (1963).
145. W. Moffitt, *J. Chem. Phys.* **25**, 467 (1956).
146. E. Shechter and E. R. Blout, *Proc. Nat. Acad. Sci. U. S.* **51**, 695 and 794 (1964).

are buried in the interior of the molecules and inaccessible to small molecules such as glucose and D_2O (*147*). Of the 36 thiols in the oligomer, 26 react readily with silver ions, while the remaining 10 react at a reduced rate (*148*). Binding of NADH increases the availability of the last 10.

E. Stability, Denaturation, and Dissociation into Subunits

Many animal GDH's are more stable in buffers containing a polyvalent anion such as phosphate than in those containing a monovalent cation such as Tris (*149*). Loss of activity in Tris is accompanied by a considerable conformational change which may be monitored by measuring the increase of reactivity of sulfhydryl groups and the loss of intrinsic fluorescence. Denaturation is prevented by NaCl (*150*), suggesting that electrostatic forces aid in retaining the integrity of the protein molecule.

Reduced coenzymes exhibit biphasic effects on denaturation. At concentrations below 10^{-5} *M*, NADH and NADPH increase the rate of inactivation, whereas at higher concentrations they stabilize the protein (*149*). In the presence of NADH, diethylstilbestrol, zinc ions, and GTP enhance the rate of denaturation (*150*) whereas ADP protects the enzyme under most mild inactivating conditions (*149*). Indeed, it was found necessary to add ADP to α-ketoglutarate and NADH in Tris buffer to obtain reproducible conditions under which rat liver GDH would not inactivate during the assay procedure (*87*). Activity could also be retained by lowering pH from 8 to 7, or by adding glycerol.

In 0.2 *M*-phosphate buffer (pH 7.6), bovine liver GDH is stable for an hour at 41°, but loses activity rapidly and irreversibly at 50° (*151*) unless sodium sulfate (0.5 *M*) is added to the buffer (*58*).

Stabilities of GDH's from microorganisms vary greatly. Whereas the NADP-GDH from *Neurospora* retains activity for several days at 50° and pH 7.2 (*152*), the NAD-GDH inactivates rapidly at raised temperatures and low ionic strengths, but is more stable in the presence of NAD (1 m*M*) and the competitive inhibitor isophthalate (*64*). NAD-GDH's from lower fungi are also reported to be unstable (*71*), but those isolated from *Clostridium* SB_4 (*36*) and *Peptococcus aerogenes* (*31*) are stable at 50°, and so are the NADP-dependent enzymes of *Salmonella* (*26*) and *E. coli* (*73*).

147. D. G. Cross and H. F. Fisher, *Biochemistry* **5,** 880 (1966).
148. K. S. Rogers, *BBA* **263,** 309 (1972).
149. C. Frieden, *JBC* **238,** 146 (1963).
150. M. W. Bitensky, K. L. Yielding, and G. M. Tomkins, *JBC* **240,** 1077 (1965).
151. J. A. Olson and C. B. Anfinsen, *JBC* **202,** 841 (1953).
152. R. W. Barratt and W. N. Strickland, *ABB* **102,** 66 (1963).

Dissociation of bovine liver GDH into subunits is always accompanied by loss of activity, and sometimes by unfolding of the polypeptide chains. When the pH is raised from 10 to 12, or lowered from 5 to 2, dissociation is accompanied by changes in the ultraviolet spectra which are characteristic of perturbation of tyrosyl chromophores (*153,154*). Changes in molecular weight show the same dependence on pH as spectral changes, and may be fitted to theoretical titration curves for ionizing groups with pK_a values of 3.7 and 10.8. One interpretation of these results is that some of the forces holding subunits together involve hydrogen bonds between phenolic hydroxyls and side chain carboxyl groups (*153,154*). In contrast, disaggregation of polymerized forms of the protein to the active hexamer is not accompanied by any spectral change (*117*).

Guanidine hydrochloride at a concentration of at least 2 *M* dissociates the bovine enzyme, although a protease, which may contaminate some commercial preparations of the enzyme, is active under these conditions (*104*). A cooperative transition occurs as the concentration of urea is increased through 4 *M*, resulting in dissociation into subunits (*58*). The NADP-GDH's from *Salmonella* (*26*) and *E. coli* (*73*) are more resistant to these denaturing agents; activity is retained in the presence of 4 *M* urea or 2.5 *M* guanidine hydrochloride.

Dissociation of the bovine enzyme into subunits occurs in the presence of SDS, but prolonged incubation (*132*) or relatively high concentrations of SDS (*115*) are necessary to achieve complete dissociation. Although loss of intrinsic fluorescence occurs in 0.4 m*M* SDS, the polypeptide chains retain considerable order as exhibited by their ORD spectra (*155*). Similar ORD parameters are given by solutions of both chicken and bovine liver GDH's in 0.1 *M* SDS (*105*).

F. Composition

Amino acid compositions (see Table V) of various GDH's are strikingly similar and distinct from the compositions of most other dehydrogenases; for example, the ratio of arginine to lysine in GDH's is relatively high.

Appella and Tomkins (*128*) detected four tryptophan residues per subunit in the bovine liver enzyme by spectral analysis and titration with *N*-bromosuccinimide. A value of 3.4 was obtained by colorimetric titration with *p*-dimethylaminobenzaldehyde (*130*), while three tryptophan

153. H. F. Fisher, L. L. McGregor, and U. Power, *BBRC* **8,** 402 (1962).
154. H. F. Fisher, L. L. McGregor, and D. G. Cross, *BBA* **65,** 175 (1962).
155. K. S. Rogers and S. C. Yusko, *JBC* **247,** 3671 (1972).

TABLE V
AMINO ACID ANALYSES[a]

Amino acid residue	Human liver Ref. (*53*) Residues per 50,000 g protein	Bovine liver Ref. (*103*) Residues/subunit[b] (MW 55,393)	Rat liver Ref. (*87*) Residues per 50,000 g protein	Chicken liver Ref. (*133*) Residues/subunit[b] (MW 55,658)	Dogfish liver Ref. (*63*) Residues per 50,000 g protein[c]	*Neurospora crassa* NAD-linked Ref. (*64*) Residues/subunit (MW 116,000)	*Neurospora crassa* NADP-linked Ref. (*66a*) Residues/subunit[b] (MW 48,438)	*E. coli* NADP-linked Ref. (*73*) Residues/subunit (MW 46,000)	*Clostridium* SB_4 NAD-linked Ref. (*36*) Residues per 50,000 g protein
Lysine	32	33	26	35	28	65	29	22	29
Histidine	14	14	12	16	9	25	9	9	11
Arginine	26	30	22	28	20	63	17	18	17
Aspartic acid	45	29	39	48	43	114	38	41	45
Asparagine		21							
Threonine	23[d]	28	21[d]	27	23	48	16	22	13
Serine	30[d]	30	23[d]	28	21	74	29	29	15
Glutamic acid	49	32[e]	33	48	45	103	51	54	45
Glutamine		13							
Proline	15	21	19	20	21	49	13	16	18
Glycine	43	47	38	48	44	66	54	51	50
Alanine	37	37	28	40	32	71	52	45	32
Cysteine	5[f]	6	n.d.[g]	7	n.d.[g]	9	6	6	3
Valine	25[h]	34	26[h]	33	22	69	33	34	30
Methionine	9	13	10	15	14	20	9	10	11
Isoleucine	30[h]	37	28[h]	35	27	67	18	15	26
Leucine	28	31	26	32	26	92	38	31	26
Tyrosine	16	18	14	17	15	41	15	12	17[i]
Phenylalanine	19	23	15	23	19	47	18	15	22
Tryptophan	n.d.[g]	3	4[i]	3	n.d.[g]	7[j]	7	4	4[i]

[a] Where the subunit molecular weight is unknown, results are expressed as number of residues per 50,000 g of protein.
[b] Obtained by sequence determination.
[c] Average of values obtained by acid hydrolysis for 24 and 48 hr.
[d] Not corrected for destruction by acid hydrolysis.
[e] This figure includes one residue that may be glutamine.
[f] Estimated by titration with DTNB.
[g] Not determined.
[h] Not corrected for incomplete release by acid hydrolysis.
[i] Determined by spectral analysis.
[j] By autoanalysis after hydrolysis in *p*-toluenesulfonic acid.

residues were located during determination of the sequence (*103,156*). Isolation of an additional chymotryptic peptide, Glu-Trp, has been reported, but there is no indication of its location in the polypeptide chain (*157*).

No disulfide bridges have been detected in any GDH. Estimates of the sulfhydryl content of the bovine enzyme have given values of 7.2 by titration in guanidine hydrochloride with *o*-iodosobenzoate (*158*) and 6.0 by titration with DTNB in guanidine hydrochloride (*128,150*) or 0.2% SDS (*108*). Performic acid oxidation yields eight cysteic acid residues, and eight residues of carboxamidomethylcysteine are obtained by alkylation with iodoacetamide (*128*). Amperometric titration with silver ions results in formation of 6.7 mercaptide ions (*148*). These variations may have been caused, in part, by adoption of different methods to determine protein concentration. The complete sequences gave six cysteine residues (*103,156*) in the bovine and seven in the chicken enzyme (*133*).

GDH was originally thought to be a zinc metalloprotein. Emission spectroscopy yielded an average value of 1 g-atom of zinc per mole of protein (MW 332,000), but no adequate explanation of the large coefficient of variation (30%) of this value was offered (*159*). It was also noted that the enzyme is inhibited by metal chelators, including 1,10-phenanthroline, but later observations that nonchelating analogs of 1,10-phenanthroline are even better inhibitors threw doubt on the belief that zinc is essential for enzymic activity (*160*). Zinc content may be reduced to 0.015 g-atom/mole of oligomer by gel chromatography in the presence of EDTA without losing activity, and the zinc-free enzyme is inhibited to the same extent by 1,10-phenanthroline. However, the enzyme has a high affinity for zinc, and a value of 4.26×10^{-6} *M* has been measured for the dissociation constant of the complex (*161*).

G. Electrophoretic and Spectrophotometric Properties

Electrophoretic mobilities of bovine liver GDH in phosphate buffer at pH 6 to 9.5 were measured by Olson and Anfinsen (*11*). Insolubility and instability precluded measurements at lower pH, but they judged the isoelectric point to be between pH 4 and 5. An isoelectric point above

156. E. L. Smith, M. Landon, D. Piszkiewicz, W. J. Brattin, T. J. Langley, and M. D. Melamed, *Proc. Nat. Acad. Sci. U. S.* **67**, 724 (1970).
157. V. Witzemann, R. Koberstein, and H. Sund, *Eur. J. Biochem.* (in press).
158. L. Hellerman, K. A. Schellenberg, and O. K. Reiss, *JBC* **233**, 1468 (1958).
159. S. J. Adelstein and B. L. Vallee, *JBC* **233**, 589 (1958).
160. K. L. Yielding and G. M. Tomkins, *BBA* **62**, 327 (1962).
161. R. F. Colman and D. S. Foster, *JBC* **245**, 6190 (1970).

neutrality may be calculated from the composition (Table V) which indicates an excess of cationic residues; the substantial difference may indicate that many basic residues are involved in hydrogen bonding or buried in the interior of the molecule. Relative electrophoretic mobilities of GDH's from a number of sources undoubtedly depend, at least in part, on the abilities of proteins to polymerize (*55,63,105*).

The extinction coefficient of bovine GDH has been subject to some debate. A value of $E_{1\,cm}^{0.1\,\%} = 0.97\ cm^2\ mg^{-1}$ at 279 nm was obtained by Olson and Anfinsen (*11*). More recently, it was noted that light scattering from a solution of about 0.5 mg/ml contributes about 20% to extinction at 279 nm, and a corrected value of $E_{1\,cm}^{0.1\,\%} = 0.83\ cm^2\ mg^{-2}$, equivalent to a molar extinction coefficient of $46.5 \times 10^6\ cm^2\ mole^{-1}$, was obtained; this gave closer agreement to the extinction coefficient calculated from the tyrosine and tryptophan content found by sequence determination (*162*). Other workers found a corrected value $E_{1\,cm}^{0.1\,\%} = 0.89\ cm^2\ mg^{-1}$ (*131*). Since glutamate dehydrogenase polymerizes, a correction for light scattering should always be made.

III. Genetics and Regulation of Enzyme Synthesis

A. Fungi

As already noted (Section I,C,1), the higher fungi Deuteromycetes, Ascomycetes, and Basidiomycetes can produce both an NAD-GDH and an NADP-GDH (*23,25,163,164*). Synthesis of the NAD-enzymes of Oomycetes and Hypochytridiomycetes is repressed when the organisms are grown in the presence of either glucose or sucrose and limited amounts of amino acids (*71*). Under these conditions, the levels of NAD-enzymes in the Chytridiales, Blastocladiales, and Mucorales are not affected significantly. Glutamate and other amino acids that can serve as precursors of glutamate can act as inducers of the repressed NAD-enzyme in Oomycetes and Hypochytriomycetes. More detailed information concerning this type of regulation is known in a number if species, particularly *Neurospora crassa, Aspergillus nidulans,* and *Saccharomyces cerevisiae.*

1. *Neurospora crassa*

a. Regulation of Synthesis. Neurospora crassa contains distinct GDH's one specific for NADP, and the other for NAD (*22,165*), with

162. A. D. B. Malcolm, *Hoppe-Seyler's Z. Physiol. Chem.* **352,** 883 (1971).
163. P. J. Casselton, *Sci. Prog.* (*London*) **57,** 207 (1969).
164. W. M. Dowler, P. D. Shaw, and D. Gottlieb, *J. Bacteriol.* **86,** 9 (1963).
165. B. D. Sanwal and M. Lata, *Nature* (*London*) **190,** 286 (1961).

the former being biosynthetic, and the latter degradative since its formation is induced by glutamate (*22,166,167*). The structures of the two enzymes are determined by two unlinked genes (*168*), and the two proteins are antigenically unrelated (*169*). The levels of the two enzymes are concurrently regulated by a repression–derepression type of mechanism. Thus, in the presence of glutamate or a nitrogenous precursor (ammonia, urea, alanine, aspartate, etc.) formation of NAD-GDH is induced while that of NADP-GDH is repressed (*166,169,170*), resulting in an inverse relationship between the relative amounts of the two enzymes (*166,170*).

It has been suggested (*171*) that the actual agent *in vivo* for the regulation of the two enzymes is NH_4^+. Induction of the NAD-GDH and repression of NADP-GDH are proportional to $[NH_4^+]$ in the medium, while the concentration of glutamate in the cells increases to a maximum at an external $[NH_4^+]$ of 0.1 g/100 ml. Increases of $[NH_4^+]$ beyond this level give further induction of the NAD-GDH, as would be expected if ammonia rather than glutamate serves as a regulator of the two enzymes. In addition, it was reported that NH_4^+ at high levels as well as glutamate can repress the NADP-specific enzyme (*172*), and it was deduced that the presence of both substances is not essential for regulation (*166*). However, since NH_4^+ disproportionately represses the NADP-GDH only at high concentrations, this decrease could result from a general ammonia toxicity (*167*).

Synthesis of NAD-GDH is repressed and that of NADP-GDH is induced by increasing concentrations of sucrose or glucose in the medium (*173*). Citrate, pyruvate, and succinate do not exert a significant effect on NAD-GDH although pyruvate and citrate do induce NADP-GDH. Additional studies have indicated that the regulation of the two GDH's is controlled by some balance between the internal amino acids and glucose metabolites (*174*).

The decrease in the NADP-GDH activity with exogenous supplements like urea reflects repression of gene action rather than enzyme inhibition (*166,169*). The *am* designation is used for the locus specifying NADP-GDH, and the synthesis of the corresponding inactive proteins (CRM) by am (amination-deficient) mutants can be regulated by urea in the

166. B. D. Sanwal and M. Lata, *BBRC* **6,** 404 (1962).
167. W. N. Strickland, *Aust. J. Biol. Sci.* **22,** 425 (1969).
168. S. I. Ahmed and B. D. Sanwal, *Genetics* **55,** 359 (1967).
169. B. D. Sanwal and M. Lata, *ABB* **98,** 420 (1962).
170. B. D. Sanwal and M. Lata, *ABB* **97,** 582 (1962).
171. C. S. Stachow and B. D. Sanwal, *BBA* **139,** 294 (1967).
172. R. W. Barratt, *J. Gen. Microbiol.* **33,** 33 (1963).
173. M. Kapoor and A. K. Grover, *Can. J. Microbiol.* **16,** 33 (1970).
174. W. N. Strickland, *Aust. J. Biol. Sci.* **24,** 905 (1971).

same way as the active enzyme (*169*). Mutants deficient in the NADP-GDH locus (*am*) but possessing an antigenically related inactive protein (CRM) fall into two groups with regard to the regulation of NAD-GDH by urea during growth: in one group, induction by urea of NAD-GDH is accompanied by a decrease in the CRM related to the NADP-GDH as in wild-type; in other am mutants, however, urea has no effect on either the CRM or the NAD-GDH.

b. Mutant Strains. The only report (*168*) on mutations in *N. crassa* that are concerned with the NAD-GDH indicates that it is not linked with the *am* gene responsible for synthesis of the NADP-dependent enzyme.

The *am* locus in *N. crassa* specifies the structure of NADP-GDH (*175,176*). Allelic mutants of the am series, mapping within a short segment on the right arm of linkage group V, produce aberrant forms of the enzyme (*177,178*) and have no effect on the NAD-enzyme. These am mutants require an exogenous supply of L-amino acids for normal growth without a prolonged lag phase (*10*). Glutamate, aspartate, alanine, and ornithine are the most active in promoting growth; the corresponding α-keto acids are inactive. The increased growth of am mutants on amino acid supplements is accompanied by increased specific activity of the NAD-GDH, which may adopt a biosynthetic role under these conditions and negate to some extent the deficiency in the endogenous glutamate caused by the loss of NADP-GDH (*170*). The presence of this second GDH probably accounts for the fact that am mutants grow on unsupplemented medium after a growth lag. The am mutants are subnormal in their capacity to assimilate ammonium nitrogen (*10*).

Nineteen mutations at the *am* locus in *N. crassa* have been described (*175,178,179*) and are specified by numerical superscripts forming the series am^1 to am^{19}. Some of the properties of these mutant strains are given in Table VI. Crosses between different am mutants produce low frequencies of wild-type progeny (*180,181*) as expected from mutations at different sites within the same gene. The present status of the linear map of the *am* gene is shown in Fig. 3 (*182,183*). There is no obvious

175. J. R. S. Fincham, *J. Gen. Microbiol.* **11**, 236 (1954).
176. J. A. Pateman, *J. Genet.* **55**, 444 (1957).
177. J. R. S. Fincham, *J. Gen. Microbiol.* **5**, 793 (1951).
178. J. R. S. Fincham, *JMB* **4**, 257 (1962).
179. J. R. S. Fincham and D. R. Stadler, *Genet. Res.* **6**, 121 (1965).
180. J. A. Pateman, *Nature* (*London*) **181**, 1605 (1958).
181. J. A. Pateman, *Genetics* **45**, 839 (1960).
182. J. R. S. Fincham, *Genet. Res.* **9**, 49 (1967).
183. D. R. Smyth, *Aust. J. Biol. Sci.* **26**, 1355 (1973).

TABLE VI
VARIOUS PROPERTIES ASSOCIATED WITH *Neurospora* MUTANT STRAINS

Mutant strains	CRM[a]	Complementation[b]		Reversions at *am* locus	
		Strains	Ref.	Occurrence	Ref.
am^{1}	+	am^{2}, am^{3}, am^{14}, am^{19}	*175,178,192–195, 198–200*	+	*176,201*
am^{2}	+	am^{1}, am^{14}	*194*	+	*176,178,202,203*
am^{3}, am^{12c}	+	am^{1}, am^{14}	*194*	+	*176,178,204,205*
am^{4}	+			−	*206*
am^{5}	−			+	*206*
am^{6}	−			+	*206*
am^{7}, am^{13c}	+	am^{14}		+	*206*
am^{8}	−			−	*206*
am^{9}	−			−	*206*
am^{10}				−	*206*
am^{11}	−			−	*206*
am^{14}	−	am^{1}, am^{2}, am^{3}, am^{7}, am^{19}	*179, 197, 207*	+	*201*
am^{15}	−			−	*201*
am^{16}	−			−	*201*
am^{17}	−			+	*201*
am^{18}	−			+	*201*
am^{19}	+	am^{1}, am^{14}	*179,196,197*	+	*185,201*

[a] The presence of cross-reacting material to antiserum prepared against normal NADP-GDH (*189*).
[b] The allele with which complementation yields active NADP-GDH.
[c] Recurrences of the same mutational event (*178,189*).

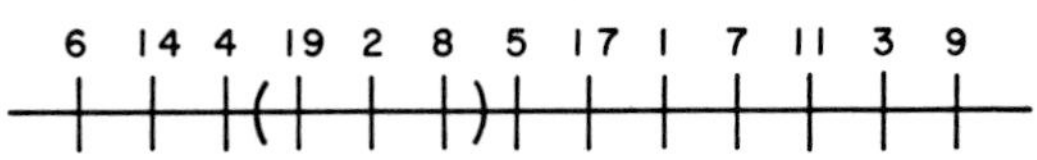

FIG. 3. Map of mutant sites based on frequencies of am^{+} recombinants in *N. crassa*. Amino acid replacements of am^{19} (residue 141) and am^{2} (residue 142) cannot be aligned by recombination frequency (see text).

relationship between the map position of a given am mutant and the properties of the protein product of the mutant gene.

Sequence studies have shown a replacement of lysine-141 with methionine in am^{19} and of histidine-142 with glutamine in am^{2}. In two revertants of am^{19}, leucine replaced isoleucine-191 in strain R1 and arginine replaced glutamine-391 in strain R24 (*184*) (also see Section IV).

184. J. R. S. Fincham, personal communication.

Mutations involving the *am* locus produce aberrant forms of NADP-GDH (*10,165,175,177*) which can be distinguished from each other and from the normal enzyme by various criteria (Table VI), particularly in their enzymic capacities since most are inactive (*178,179,185,186*). The enzyme variants of am mutants often show "conditional" activity (*187*), which implies that although they are inactive under conditions where the normal enzyme is functional, they can show some activity under selective conditions or after special treatment.

Each of the proteins from am^1, am^2, am^3, am^4, am^7, and am^{19} is found in about the same amount per unit weight of mycelium as is wild-type NADP-GDH, while am^{14} does not produce a major protein component resembling the normal enzyme (*179*). Of the altered forms of NADP-GDH that have been isolated, only the proteins from am^4 and am^{19} differ electrophoretically from the normal (*179,188*). With antiserum prepared against normal enzyme, Roberts and Pateman (*189*) looked for CRM in all the am mutants by using both immunoelectrophoresis and the technique of double diffusion in agar gel (Table VI). Antibodies against the proteins produced by the mutants am^1 and am^2 behave in antienzyme studies in the same way as antibodies produced against the normal enzyme (*190*). Although the mutations affected the catalytic sites of these two proteins, antibodies were produced that are similar to those prepared against wild-type protein.

c. Complementation of Mutants. When different mutations produce a deficiency of the same enzyme and fall within the same short chromosome segment, they may be considered as part of the same locus (*175*). Usually, two such allelic mutations will not complement each other to form a nonmutant phenotype when present together in the same cell. With the am mutants of *N. crassa* certain pairs of alleles can promote enzyme formation jointly, though unable to do so individually (Table VI). This type of interallelic or intracistronic complementation is a general phenomenon (*191*) involving the formation of hetero-oligomeric proteins.

In the am series of *N. crassa* a complementation enzyme would be a hybrid molecule composed of six polypeptide chains derived from different mutant sources, the defect in one type of chain being compensated

185. T. K. Sundaram and J. R. S. Fincham, *JMB* **10**, 423 (1964).
186. D. B. Roberts, *J. Bacteriol.* **91**, 1888 (1966).
187. J. R. S. Fincham, *Brit. Med. Bull.* **18**, 14 (1962).
188. R. W. Barratt, *Rec. Genet. Soc. Amer.* **31**, 72 (1962); *Genetics* **47**, 941 (1962).
189. D. B. Roberts and J. A. Pateman, *J. Gen. Microbiol.* **34**, 295 (1964).
190. D. B. Roberts, *J. Bacteriol.* **94**, 958 (1967).
191. M. J. Schlesinger and C. Levinthal, *JMB* **7**, 1 (1963).

to some extent by the presence of the corresponding normal portion of the other chain (*178*). The complementing pairs of am mutants, when existing in the common cytoplasm of a heterokaryon, are often able to form mycelia which will grow almost normally on unsupplemented medium. The hybrid GDH that is formed under these conditions apparently can express its enzyme function *in vivo*.

Fincham (*178*) described a procedure for isolating normal NADP-GDH and its mutant analogs and used these materials for studies on the *in vitro* complementation of these proteins (*192–197*). During complementation two or more types of hybrids can be formed whose proportions are determined by the input ratios of the two proteins (*198–207*). Since the sedimentation characteristics of the hybrid enzymes showed that they are the same size as the wild-type enzyme (*192,195*), the hybrid variations result from the association of different ratios of the two kinds of participating subunits (*196*). The hybrids from a complementing pair can often be distinguished by their enzymic and physical properties (*192–194,196*). Those formed from wild-type protein interacting with am^4, am^{14}, and am^{19} monomers produced a less stable form of the enzyme (*197,207*), and similar studies with am^2 and am^3 subunits suggested that these can also impose their faulty conformations on the wild-type monomers with which they complement (*194*). Although these hybrids often show lower enzymic activity, the hybridization mixture from am^1 and wild type retained the properties of normal GDH (*194*).

d. Reversion of the am Locus. Reversions in am^1, am^2, and am^3 were obtained by irradiation with ultraviolet light (*176*). These reversions result from further mutations in the *am* region, and suppressor mutations affecting NADP-GDH were not found. All of the secondary mutants have

192. J. R. S. Fincham and A. Coddington, *JMB* **6**, 361 (1963).
193. J. R. S. Fincham and A. Coddington, *Cold Spring Harbor Symp. Quant. Biol.* **28**, 517 (1963).
194. A. Coddington, *BJ* **99**, 9 (1966).
195. A. Coddington and J. R. S. Fincham, *JMB* **12**, 152 (1965).
196. A. Coddington, J. R. S. Fincham, and T. K. Sundaram, *JMB* **17**, 503 (1966).
197. T. K. Sundaram and J. R. S. Fincham, *JMB* **29**, 433 (1967).
198. J. R. S. Fincham, *J. Gen. Microbiol.* **21**, 600 (1959).
199. J. R. S. Fincham and J. A. Pateman, *Nature (London)* **179**, 741 (1957).
200. J. A. Pateman and J. R. S. Fincham, *Heredity* **12**, 317 (1958).
201. D. R. Stadler, *Genet. Res.* **7**, 18 (1966).
202. J. R. S. Fincham, *BJ* **65**, 721 (1957).
203. J. R. S. Fincham and H. R. Garner, *BJ* **103**, 705 (1967).
204. J. R. S. Fincham and P. A. Bond, *BJ* **77**, 96 (1960).
205. J. A. Pateman and J. R. S. Fincham, *Genet. Res.* **6**, 419 (1965).
206. J. A. Pateman, *J. Gen. Microbiol.* **23**, 393 (1960).
207. T. K. Sundaram and J. R. S. Fincham, *J. Bacteriol.* **95**, 787 (1968).

some enzymic activity. Revertants from *am*[1] give an enzyme indistinguishable from the wild type, suggesting frequent reversal of the original mutational change. Partial revertants from *am*[2] and *am*[3] produce enzymes that differ from the wild-type dehydrogenase. These organisms grow more or less like the wild type despite their aberrant enzymes. At least eight types of revertants are obtained from the am[3] strain; these differed from each other in their variable forms of the enzyme (*205*). Only three mutants in the am[4] to am[11] series give revertants as a result of ultraviolet irradiation (Table VI). Since the other strains in this series do not give revertants at a detectable frequency, it became evident that reversion might be used as one of the criteria for differentiating among these strains (*206*). Reversions by ultraviolet irradiation were also investigated in the remaining mutants (am[14] to am[19]), and genetic analysis confirms the previous observation (*176*) that these secondary mutations result from events in the *am* region (*201*). Reversion frequencies of *am*[14], *am*[17], and *am*[19] are an order of magnitude greater than those of *am*[1] and *am*[18], and there are differences in the heat stabilities at pH 6.5 of the revertant enzymes (*201*), e.g., nearly all of those from *am*[14] and *am*[19] are more labile than wild-type enzyme, whereas all revertants of *am*[1] and *am*[18] and nearly all of those from *am*[17] produce enzymes with similar stability to that of wild-type enzyme.

With the availability of the sequence of the normal enzyme (Section IV), it has become possible to identify the amino acid changes in the revertants. Thus far, only two revertants of *am*[19] have been identified (Section III,A,1,b).

2. *Miscellaneous Molds*

In *Aspergillus nidulans* the activity of NADP-GDH is greatly decreased when the mold is grown on L-glutamate or on high levels of either NH_4^+ or urea, and it has been suggested that glutamate alone determines the rate of synthesis of NADP-GDH; other nitrogen sources might express their effects only on glutamate levels in the mycelium (*208*). The regulation of several enzymes involved in the turnover of nitrogen is abnormal in gdh-A strains lacking NADP-GDH resulting from mutations thought to involve the structural gene for this enzyme (*209*). A role has been postulated for NADP-GDH in ammonium regulation by which enzyme present at regulatory sites in the cell could complex NH_4^+ and thereby affect nitrogen turnover by other systems (*210*).

208. J. A. Pateman, *BJ* **115,** 769 (1969).
209. J. R. Kinghorn and J. A. Pateman, *Biochem. Soc. Trans.* **1,** 675 (1973).
210. J. A. Pateman, J. R. Kinghorn, E. Dunn, and E. Forbes, *Biochem. Soc. Trans.* **1,** 674 (1973).

The NAD-GDH in *A. nidulans* is repressed by glucose or some metabolite derived from it and derepressed in media where glutamate, aspartate, or alanine are in each case the sole source of both carbon and nitrogen (*209*). Mutant strains deficient in this enzyme (designated gdh-B) cannot utilize glutamate as a carbon source because the glutamate-repressed level of NADP-GDH is inadequate for the deamination if this substrate. These strains do not show the altered regulation of other enzymes observed in the gdh-A mutations. A mutation at a third locus *gdh*-C results in a strain showing less catabolite repression; that is, NAD-GDH is significantly derepressed in the presence of glucose (*209*). This work suggests that repression of NAD-GDH by glucose may be a positive control system in which glucose mediates some function of the *gdh*-C gene product which is required in the synthesis of NAD-GDH.

Regulation of GDH synthesis in higher fungi containing both the NAD-specific and the NADP-specific enzymes has also been studied in *Aspergillus niger* (*211–213*), *Fusarium oxysporum* (*21*), *Coprinus lagopus* (*214*), and *Schizophyllum commune* (*215*). In *A. niger* (*213*) and *F. oxysporum* (*21*) the levels of the two dehydrogenases are related to the age of the cultures, and organisms grown on media containing NH_4^+ have high levels of NADP-GDH and low amounts of NAD-GDH during early stages of growth. Maximal specific activity of the NAD-GDH is obtained after two days of growth, at which time the NADP-GDH level is much lower than during early growth. In *C. lagopus* the two GDH's do not appear to be under the direct regulation by either glutamate or NH_4^+ (*214*). Some of the results support the view that a product of glucose metabolism represses the synthesis of the NAD-GDH and derepresses or induces that of the NADP-GDH, and evidence was obtained that this regulator is α-ketoglutarate. It was concluded that more than one molecule is involved in the complete system of regulation.

The NADP-GDH in *S. commune* is depressed during vegetative growth of mycelium on glucose-containing media with NH_4^+ as sole nitrogen source and increased when glutamate is the nitrogen source (*215*). Intracellular NH_4^+ and NADP-GDH levels are inversely related during spore germination and homokaryotic mycelium growth (*216*). The NAD-GDH in *S. commune* is increased by NH_4^+ and unaffected by glutamate. The

211. V. K. Shah and C. V. Ramakrishnan, *Enzymologia* **26,** 3 (1963).
212. V. L. Kretovich, A. S. Demina, and V. I. Yakovleva, *Dokl. Akad. Nauk SSSR* **159,** 1169 (1964).
213. J. C. Galbraith and J. E. Smith, *J. Gen. Microbiol.* **59,** 31 (1969).
214. M. O. Fawole and P. J. Casselton, *J. Exp. Bot.* **23,** 530 (1972).
215. D. W. Dennen and D. J. Niederpruem, *Life Sci.* **4,** 93 (1964).
216. D. W. Dennen and D. J. Niederpruem, *J. Bacteriol.* **93,** 904 (1967).

regulation of the enzymes in this basidiomycete appears to differ from that observed in other species where it is the NAD-GDH that is usually derepressed by glutamate.

Limited information is available concerning regulation of the NAD-GDH in lower fungi. In *Colletotrichum gloesporioides* this enzyme is believed to be induced when the mold is grown on 2-deoxy-D-glucose (*217*). In *Apodachlya brachynema* it is repressed by glucose as a sole carbon source, whereas glutamate, proline, alanine, or ornithine plus aspartate as sole carbon sources induced synthesis of the enzyme (*70*). The NAD-GDH in *Dictyostelium discoidem* does not change during development of the slime mold, and the rate of glutamate oxidation *in vivo* is believed to be controlled by substrate throughout differentiation (*218–220*). Studies with the slime mold, *Physarum polycephalum,* showed that the induction of spherulation by growth on a medium containing 0.5 *M* mannitol results in about a 50% increase in the specific activity of NAD-GDH as a result of *de novo* synthesis of the enzyme (*221,222*).

The NAD-GDH is believed to be repressed by NH_4^+ in *Saccharomyces cerevisiae* (*223,224*), and this apparent difference in the GDH behavior of yeasts has attracted some interest (*163*). The specific activity of NAD-GDH is low in cultures containing NH_4^+, both in the presence and absence of glutamate, and NADP-GDH activity is similar in cells cultured in either NH_4^+ or glutamate but reduced in cultures containing both nitrogen sources. The effects of variations in carbon sources (*225–229*) indicate that NAD-GDH is repressed in normal yeast cells grown aerobically on media with high concentrations of sugars (*225,226,228*) while the NADP-GDH content is increased (*225–228*). This sugar-stimulated increase of NADP-GDH does not occur when excess ammonia in the medium allows a large amino acid pool to accumulate in the yeast cells (*227*), indicating that the endogenous level of amino acids is also an important factor in this regulatory mechanism.

217. G. L. Greene, *Mycologia* **61,** 902 (1969).
218. B. E. Wright and M. L. Anderson, *BBA* **31,** 310 (1959).
219. M. Brühmüller and B. E. Wright, *BBA* **71,** 50 (1963).
220. B. E. Wright and S. Bard, *BBA* **71,** 45 (1963).
221. A. Huettermann, S. M. Elsevier, and W. Eschrich, *Arch. Mikrobiol.* **77,** 74 (1971).
222. A. Huettermann, and I. Chet, *Arch. Mikrobiol.* **78,** 189 (1971).
223. H. Holzer and G. Hierholzer, *BBA* **77,** 329 (1963).
224. G. Hierholzer and H. Holzer, *Biochem. Z.* **339,** 175 (1963).
225. E. S. Polakis and W. Bartley, *BJ* **97,** 284 (1965).
226. C. Beck and H. K. von Meyenburg, *J. Bacteriol.* **96,** 479 (1968).
227. C. M. Brown and B. Johnson, *J. Gen. Microbiol.* **64,** 279 (1970).
228. I. Nunez de Castro, M. Ugarte, A. Cano, and F. Mayor, *Eur. J. Biochem.* **16,** 567 (1970).
229. K. W. Thomulka and A. G. Moat, *J. Bacteriol.* **109,** 25 (1972).

A mutant strain 4324c of *S. cerevisiae* lacks NADP-GDH (gdh-A mutation) but does have NAD-GDH (*230*). The sensitivity of the general amino acid permease to inhibition by ammonia is strongly depressed by this mutation, and the inhibition of amino acid uptake observed in normal strains after growth in the presence of NH_4^+ is reduced considerably in strain 4324c. Although intracellular ammonia and α-ketoglutarate were eliminated as possible effectors involved in this regulation, the exact nature of the altered regulation in the gdh-A mutation is not known.

Candida utilis has high levels of NAD-GDH when grown on either glutamate or some other amino acids. Addition of either ammonia or glutamine to yeast so adapted can result in rapid and extensive inactivation of NAD-GDH at a faster rate than can be accounted for by the immediate cessation of enzyme synthesis (*231*). The NADP-GDH is not subject to rapid changes in activity under these conditions. In *Saccharomyces carlsbergensis* synthesis of NAD-GDH is also repressed by NH_4^+ and derepressed by glutamate; the reverse is true for NADP-GDH (*90,232*).

B. Bacteria

1. *Bacilli*

The NADP-GDH is induced by NH_4^+ in a number of bacilli (*27,28,233–235*), and glutamate is a repressor of synthesis of the enzyme under some growth conditions (*27,28*). The NAD-GDH is found in *B. brevis* only when grown on media containing both glutamate and NH_4^+ (*235*).

Normal strains of *B. subtilis* possess an NAD-dependent L-alanine dehydrogenase and are devoid of either NAD-GDH or NADP-GDH (*29,236*). am^+ mutants of this species are known which contain a fairly active NADP-GDH and lack demonstrable alanine dehydrogenase activity (*29,236*), and it has been suggested that there is conversion of the alanine dehydrogenase to an enzyme with glutamate dehydrogenase activity by mutations induced by nitrous acid (*237*).

2. *Escherichia coli*

Escherichia coli contains only NADP-GDH and cannot grow on glutamate as sole source of carbon. A mutation leading to the loss of glutamate

230. M. Grenson and C. Hou, *BBRC* **48**, 749 (1972).
231. A. R. Ferguson and A. P. Sims, *J. Gen. Microbiol.* **69**, 423 (1971).
232. H. Westphal and H. Holzer, *BBA* **89**, 42 (1964).
233. S. V. Shestakov, V. V. Groshev, and V. D. Filippov, *Biol. Nauki* **11**, 104 (1969).
234. S. V. Shestakov and V. D. Filippov, *Dokl. Akad. Nauk SSSR* **169**, 705 (1966).
235. V. D. Filippov, *Tr. Mosk. Obshchest. Ispyt. Prir.* **28**, 126 (1968).
236. J. M. Wiame and A. Pierard, *Nature (London)* **176**, 1073 (1955).
237. S. C. Shen, M. M. Hung, W. C. Chen, and T. T. Kun, *Sci. Sinica* **12**, 545 (1963).

decarboxylase endows *E. coli* with the ability to utilize glutamate as sole source of carbon by the transamination with oxaloacetate (*39,238–240*). Glutamate represses GDH in these mutants and also to a lesser extent in wild type (*39,239,241*). A mutant strain of *E. coli,* designated $W1317^G$, lacks NADP-GDH and glutamate synthase (Section I,C,2). Other studies have indicated that the loci for the two enzymes are unlinked (*242*).

3. *Miscellaneous Bacteria*

The regulation of NADP-GDH by a repression–derepression type of mechanism has been studied in some of the bacteria possessing this enzyme (Table I). The NADP-GDH is repressed by glutamate in *B. anitratum* (*243*), *R. japonicum* (*244,245*), *and A. aerogenes* (*43*). In the last case the repression is most effective when nitrogen-limiting levels of glutamate are used in the growth medium. However, the level of this enzyme does not appear to be regulated by glutamate in *S. typhimurium* (*26*). The level of NADP-GDH is also decreased by nitrate in *B. anitratum* (*243*), by nitrogen-limiting levels of NH_4^+ in *A. aerogenes* and *P. fluorescens* (*43*), and by aspartate in *R. japonicum* (*244,245*). Examination of *A. aerogenes* grown on limiting nitrogen shows that this organism possesses high levels of glutamine synthetase and glutamate synthase, and low levels of GDH (*42,43*).

The presence of both NAD-GDH and NADP-GDH has been reported in several bacteria (Table I). NADP-GDH is induced by NH_4^+ and repressed by nitrogen starvation in *Hydrogenomonas* H16 (*34*). The level of this enzyme in *Hydrogenomonas eutropha* is unaffected by several nitrogen sources (*246,247*), and the same is true in *Neisseria meningitidis* where neither glutamate nor NH_4^+ has any observed regulatory effect (*248*). The NADP-GDH in *Thiobacillus novellus* is repressed by arginine, histidine, and aspartate (*249*). The NAD-GDH is induced by glutamate in *T. novellus, N. meningitidis,* and *H. eutropha* (*246–250*), but not in *Hydrogenomonas* H16 (*34*) where the enzyme is induced by nitro-

238. Y. S. Halpern and H. E. Umbarger, *Bacteriol. Proc.* p. 124 (1959).
239. Y. S. Halpern and H. E. Umbarger, *J. Bacteriol.* **80,** 285 (1960).
240. Y. S. Halpern and H. E. Umbarger, *J. Gen. Microbiol.* **26,** 175 (1961).
241. F. Varricchio, *BBA* **177,** 560 (1969).
242. M. A. Berberich, *BBRC* **47,** 1498 (1972).
243. K. Jyssum and P. E. Jones, *Acta Pathol. Microbiol. Scand.* **64,** 387 (1965).
244. P. Mooney and P. F. Fottrell, *BJ* **110,** 17P (1968).
245. P. F. Fottrell and P. Mooney, *J. Gen. Microbiol.* **59,** 211 (1969).
246. A. A. Joseph and R. L. Wixom, *BBA* **201,** 295 (1970).
247. H. G. Truper, *BBA* **111,** 565 (1965).
248. K. Jyssum and B. Borchgrevink, *Acta Pathol. Microbiol. Scand.* **48,** 361 (1960).
249. H. B. LéJohn, *BBRC* **28,** 96 (1967).
250. H. B. LéJohn and B.-E. McCrea, *J. Bacteriol.* **95,** 87 (1968).

gen starvation. The NAD-GDH is also induced in *N. meningitidis* grown on medium with NH_4^+ as the only nitrogen source (*248*). The NAD-GDH appears to be the predominant form of the GDH in *Clostridium* SB_4, *Azobacter chroococcum,* and *Micrococcus aerogenes* (Table I) and is one of the most abundant proteins in crude extracts of *M. aerogenes* grown on media containing glutamate or histidine (*31,32,251*).

C. Plants

Regulation of GDH synthesis has been investigated in only a few higher plants. Both glutamine and glutamate induce the NAD-GDH in the roots of pea seedlings (*252*). Infiltration of ammonium phosphate into these roots caused induction of NAD-GDH and NADP-GDH after a 5-hr exposure, and the NADP-GDH was shown to be repressed by γ-aminobutyric acid (*253*).

NAD-GDH is rapidly induced in rice plant roots by NH_4^+, whereas in shoots the enzyme is less affected. In roots NAD-GDH is found in both the soluble fractions and in mitochondria. Induction by NH_4^+ causes an increase of enzyme mainly in the mitochondria. A new band of GDH activity was detected on the zymograms of polyacrylamide gel electrophoresis, and this inducible enzyme is active with both NAD and NADP (*254*).

The NAD-GDH concentration increases six- to eightfold in detached oat leaves floated on a medium containing NH_4^+ (*255*). Glutamate, glutamine, urea, or α-ketoglutarate does not affect the specific activity. Induction is prevented by cycloheximide. Gel electrophoresis indicates that the induction process correlates with the synthesis of a new slower moving isozyme which appears to be regulated by the intracellular ammonia concentration. It has also been shown that glucose stimulates both NAD-GDH and NADP-GDH activity in grape leaves (*256*).

Chlorella pyrenoidosa has an NADP-GDH that is induced by ammonium ion and also a constitutive GDH that utilizes either NAD or NADP (Table II). Glutamate, alanine, and α-ketoglutarate do not affect the specific activity of the NADP-GDH. In *Chlorella vulgaris* the

251. W. M. Johnson and D. W. S. Westlake, *Can. J. Microbiol.* **18,** 875 (1972).
252. V. L. Kretovich, G. Sh. Tkemaladze, T. I. Karyakina, E. A. Romanova, and L. I. Sidel'nikova, *Dokl. Akad. Nauk SSSR* **190,** 222 (1970).
253. V. L. Kretovich, T. I. Karyakina, and G. Sh. Tkemaladze, *Izv. Akad. Nauk SSSR, Ser. Biol.* **5,** 759 (1969).
254. T. Kanamori, S. Konishi, and E. Takahashi, *Physiol. Plant.* **26,** 1 (1972).
255. I. Barash, T. Sadon, and H. Mor, *Nature* (*London*) *New Biol.* **244,** 150 (1973).
256. M. Sh. Gordeziani, D. F. Kintsurashvili, and M. K. Gogoberidze, *Soobshch. Akad. Nauk. Gruz. SSR* **66,** 441 (1972).

NADP-GDH increases after nitrogen starvation (*257*), as is also found in *Neurospora* (*172*).

IV. Primary Structures

The amino acid sequences of three GDH's have been determined; these are presented in Figs. 4 and 5, with B representing the bovine (*258*), C the chicken (*258*), and N the *Neurospora* NADP-dependent enzyme (*259*). Some peptides obtained from the rat liver enzyme (R) are also shown (*260*); in this instance the tryptic peptides have been placed only by homology.

A tentative structure for bovine GDH was first proposed in 1970 (*156,261–263*) based on studies of peptides from tryptic (*130,264*) and peptic (*265*) digests, from tryptic digests of the maleylated protein (*266*), and of those obtained by cleavage with cyanogen bromide (*267,268*). Subsequently, investigation of the sequence of chicken GDH (*133,258*) indicated that the proposed tentative sequence of the bovine enzyme, which lacked definitive overlaps at two segments of the chain, was in error, mainly because of transposition of one segment of the chain (*258*). Definitive overlaps were subsequently established in the bovine polypeptide chain by chemical cleavage with hydroxylamine and with the bromine adduct of 2-(2-nitrophenylsulfenyl)-3 methylindole (*103*). The final ordering of peptides in bovine GDH was facilitated both by these studies and by the fact that although the bovine and chicken enzymes display very extensive homology there exist enough points of difference, particu-

257. I. Morris and P. J. Syrett, *J. Gen. Microbiol.* **38,** 21 (1965).
258. K. Moon, D. Piszkiewicz, and E. L. Smith, *Proc. Nat. Acad. Sci. U. S.* **69,** 1380 (1972).
259. J. C. Wootton, G. K. Chambers, A. A. Holder, A. J. Baron, J. G. Taylor, J. R. S. Fincham, K. M. Blumenthal, K. E. Moon, and E. L. Smith, *Proc. Nat. Acad. Sci. U. S.* **71,** 4361 (1974).
260. C. J. Coffee and C. Frieden, unpublished personal communication from Dr. C. Frieden.
261. M. Landon, W. J. Brattin, T. J. Langley, M. D. Melamed, D. Piszkiewicz, and E. L. Smith, *Fed. Proc., Fed. Amer. Soc. Exp. Biol.* **29,** 319 (abstr.) (1970).
262. E. L. Smith, M. Landon, and D. Piszkiewicz, *Abstr., Int. Congr. Biochem., 8th, 1969* p. 89 (1970).
263. M. Landon, D. Piszkiewicz, and E. L. Smith, *in* "Oxidation Reduction Enzymes" (A. Åkeson and A. Ehrenberg, eds.), p. 53. Pergamon, Oxford, 1972.
264. M. Landon, D. Piszkiewicz, and E. L. Smith, *JBC* **246,** 2374 (1971).
265. D. Piszkiewicz, M. Landon, and E. L. Smith, *JBC* **248,** 3067 (1973).
266. W. J. Brattin, Jr. and E. L. Smith, *JBC* **246,** 2400 (1971).
267. T. J. Langley and E. L. Smith, *JBC* **246,** 3789 (1971).
268. M. Landon, T. J. Langley, and E. L. Smith, *JBC* **246,** 3802 and 3807 (1971).

larly of methionine residues, to render the more intractable regions of the bovine enzyme amenable to cleavage by cyanogen bromide in the chicken GDH.

Examination of the bovine and chicken sequences (Fig. 4) indicates that the amino terminal end of the chain is fairly polar whereas the carboxy terminal portion is relatively hydrophobic. There are no unusual features in composition or sequence of either of the vertebrate enzymes except that the amino terminal residue of chicken GDH is consistently found as cysteic acid (*133,258*) the first time this residue has been encountered in a protein. It is unknown whether or not it exists in this form *in vivo,* but isolation of the enzyme under reducing conditions still gave NH_2-terminal cysteic acid, whereas six other residues were always found as cysteine.

The sequence of the chicken GDH is remarkably similar to that of the bovine, with only 27 differences in the 500 amino acid residues which both enzymes possess in common, plus the three additional residues at the NH_2-terminus of the chicken enzyme. The incomplete sequence of the rat liver enzyme (Fig. 4) also indicates great similarity to those of the other two vertebrates and as yet offers no clue as to why the rat enzyme does not polymerize.

Engel (*269*) has suggested that a region of internal homology may exist within bovine GDH, relating residues 114–163 with residues 269–318. It was suggested that portions of the latter region may include parts of the regulatory sites of the enzyme.

Comparison of the sequence of the NADP-enzyme of *Neurospora* (*259*) with those of the vertebrate enzymes reveals that although the two types of enzymes possess very different regulatory properties, significant sequence homology exists, particularly within the first 200 residues of the chain which includes the region containing the two lysine residues that have been implicated in substrate binding (Section V,A). The homology is less striking in the remainder of the sequence although short regions of similarity do exist. As will be discussed in a later section, residues that have been shown by chemical modification to influence binding of allosteric regulators by bovine GDH have all been found, thus far, in the COOH-terminal region. Since the *Neurospora* enzyme does not display the allosteric responses to GTP and ADP exhibited by vertebrate GDH's, it might be anticipated that sequence differences would be more pronounced in this region. The separation into polar and nonpolar regions, mentioned above in relation to the chicken and bovine enzymes, also exists in the *Neurospora* sequence although to a less striking degree.

269. P. C. Engel, *Nature* (*London*) **241,** 118 (1973).

Although the NADP-GDH of *Neurospora* is shorter by 48 residues than the bovine enzyme, there is considerable homology between the proteins, dependent to some extent on the presence of a number of small gaps (Fig. 5). All of these proteins contain a uniquely reactive lysine residue, number 126 of the vertebrate chains, of unusually low pK_a, and this residue is located in the part of the sequence exhibiting the greatest homology. In the immediate vicinity of this lysine residue, all the proteins are almost identical (see Section, V,A). Within a 113 residue segment (residues 55–167 of the *Neurospora* sequence) 40 residues are identical to the bovine and an additional 44, mostly conservative substitutions, are related by single base changes in the respective codons.

Examination of the nature of the residues conserved in the *Neurospora* NADP-GDH and the vertebrate enzymes indicates that the bulk of the identical residues must be critical for maintenance of conformation rather than for catalysis (*259,270*). In this regard, Rossmann *et al.* (*271*) have noted that conservation of a three-dimensional domain responsible for coenzyme binding can be demonstrated for a number of other dehydrogenases and kinases. Among the dehyrogenases investigated, the striking feature is conservation of a three-dimensional structure involved in coenzyme binding rather than a strict retention of linear sequence. This coenzyme domain consists of a number of short alternating stretches of β-pleated sheet and α-helix. Although the three-dimensional structure of any glutamate dehydrogenase has yet to be determined, sufficient linear sequence homology exists among residues 245–280 of bovine and chicken GDH's and the residues demonstrated by X-ray analysis to be involved in coenzyme binding in lactate, ethanol, and glyceraldehyde-3-phosphate dehydrogenases. Although the proposed coenzyme binding site of the GDH's is not situated within the region of maximal homology between the vertebrate and *Neurospora* enzymes, comparison of residues 245–280 of bovine and chicken GDH's with residues 219–254 of the *Neurospora* NADP-GDH reveals the presence of significant sequence homology and the conservation of most of the residues to which a functional significance has been ascribed by Rossmann *et al.* This is discussed below.

In addition to the similarities discussed above, bovine and chicken GDH's and *Neurospora* GDH (NADP) exhibit short stretches of homology relating *Neurospora* residues 179–197 and 330–348 to bovine residues 198–216 and 342–360, respectively. The precise roles of the homologous residues in these regions is still not apparent.

270. K. M. Blumenthal, K. E. Moon, and E. L. Smith, *JBC* **250** (in press) (1975).
271. M. G. Rossmann, A. Liljas, C.-I. Brändén, and L. J. Banaszek, Chapter 2, this volume.

```
              1                        10                                    20                 25
B           Ala-Asp-Arg-Glu-Asp-Arg-Pro-Asn-Phe-Phe-Lys-Met-Val-Glu-Gly-Phe-Phe-Asp-Arg-Gly-Ala-Ser-Ile-Val-Glu-
    -3
C   CyS-Glu-Ala-Ala-Asp(Lys-Glu-Asp)Asp-Pro-Asn-Phe-Phe-Lys-Met-Val-Glu-Gly-Phe-Phe-Asp(Arg-Gly-Ala-Ser-Ile-Val-Glu-
R           Thr-Asp-Lys-Glx-Asx-Asx-Pro-Asx-Phe-Phe-Arg Met-Val-Glx-Gly-Phe-Phe-Asp-Arg Gly-Ala-Ser-Ile(Val-Glx-

                       30                  35                  40                  45                  50
B           Asp-Lys-Leu-Val-Glu-Asp-Leu-Lys-Thr-Arg-Gln-Thr-Gln-Glu-Gln-Lys-Arg-Asn-Arg-Val-Arg-Gly-Ile-Leu-Arg-
C           Asp-Lys-Leu-Val-Glu)Gly-Leu-Arg-Thr-Arg(Glu-Ser)Met-Glu-Gln-Arg-Arg-His-Arg-Val-Arg-Gly(Ile-Leu)Arg-
R           Asx-Lys)Leu-Val-Glx-Asx-Leu-Lys Thr-Arg(Asx-Thr-Glx-Glx-Glx-Lys-Arg)Asx-Arg

                       55                  60                  65                  70                  75
B           Ile-Ile-Lys-Pro-Cys-Asn-His-Val-Leu-Ser-Leu-Ser-Phe-Pro-Ile-Arg-Arg-Asp-Asp-Gly-Ser-Trp-Glu-Val-Ile-
C          (Ile-Ile-Lys-Pro-Cys-Asn)His-Val-Leu-Ser(Val-Ser-Phe-Pro)Ile-Lys-Arg-Asp(Asp-Gly-Glx-Trp-Glu-Val-Ile-
R           Ile-Ile-Lys-Pro-Cys-Asx(His)Val(Leu-Ser-Leu-Ser-Phe-Pro-Pro)Arg Arg Asx-Asx-Gly(Ser-Trp-Glx-Val-Ile-

                       80                  85                  90                  95                 100
B           Glu-Gly-Tyr-Arg-Ala-Gln-His-Ser-His-Gln-Arg-Thr-Pro-Cys-Lys-Gly-Gly-Ile-Arg-Tyr-Ser-Thr-Asp-Val-Ser-
C           Glu-Gly)Tyr-Arg-Ala-Gln-His(Ser-His-Gln)Arg(Thr-Pro)Cys-Lys-Gly(Gly-Ile)Arg-Tyr-Ser(Leu-Asp-Val-Ser-
R           Glx-Gly)Tyr-Arg                         Thr-Pro-Cys-Lys                 Tyr-Ser-Thr-Asx-Val-Ser-

            101        105                 110                 115                 120                125
B           Val-Asp-Glu-Val-Lys-Ala-Leu-Ala-Ser-Leu-Met-Thr-Tyr-Lys-Cys-Ala-Val-Val-Asp-Val-Pro-Phe-Gly-Gly-Ala-
C           Val-Asp-Glu-Val)Lys Ala-Leu-Ala(Ser-Leu)Met-Thr-Tyr-Lys(Cys-Ala-Val-Val-Asp-Val-Pro-Phe-Gly-Gly-Ala-
R           Val-Asx-Glx-Val-Lys Ala-Leu-Ala-Ser-Leu-Met-Thr-Tyr-Lys Cys-Ala-Val-Val-Asp-Val-Pro-Phe-Gly-Gly-Ala-

                       130                 135                 140                 145                150
B           Lys-Ala-Gly-Val-Lys-Ile-Asn-Pro-Lys-Asn-Tyr-Thr-Asp-Glu-Asp-Leu-Glu-Lys-Ile-Thr-Arg-Arg-Phe-Thr-Met-
C           Lys(Ala-Gly-Val)Lys-Ile(Asn-Pro)Lys(Asn-Tyr-Thr-Asp-Glu-Asp-Leu-Glu)Lys(Ile-Thr)Arg(Arg)Phe-Thr-Met-
R           Lys Ala-Gly(Val-Lys)Ile-Asx-Pro-Lys Asx-Tyr-Thr-Asx-Glx-Asx-Leu-Glu-Lys             Arg Phe-Thr-Met-

                       155                 160                 165                 170                175
B           Glu-Leu-Ala-Lys-Lys-Gly-Phe-Ile-Gly-Pro-Gly-Val-Asp-Val-Pro-Ala-Pro-Asn-Met-Ser-Thr-Gly-Glu-Arg-Glu-
C           Glu-Leu-Ala-Lys-Lys-Gly-Phe-Ile-Gly-Pro-Gly(Val-Asp-Val-Pro-Ala-Pro-Asn)Met(Ser-Thr-Gly-Glu)Arg-Glu-
R           Glx(Leu-Ala-Lys)Lys Gly-Phe-Ile-Gly-Pro-Gly-Ile-Asp-Val-Pro-Ala     Asx-Met-Ser(Thr-Gly-Glu)Arg Glu-

                       180                 185                 190                 195                200
B           Met-Ser-Trp-Ile-Ala-Asp-Thr-Thr-Ala-Ser-Tyr-Ile-Gly-His-Tyr-Asp-Ile-Asn-Ala-His-Ala-Cys-Val-Thr-Lys-
C           Met-Ser-Trp-Ile(Ala-Asp-Thr)Thr-Ala(Ser-Tyr-Ile)Gly-His-Tyr-Asp(Ile-Asn)Ala-His(Ala-Cys-Val-Thr-Lys-
R           Met-Ser(Trp- X -Ala-Asx-Thr-Thr-Ala-Ser-Tyr- X -Gly-His-Tyr-Asx- X -Asx-Ala-His-Ala-Cys-Val-Thr-

                       205                 210                 215                 220                225
B           Pro-Gly-Ile-Ser-Gln-Gly-Gly-Ile-His-Gly-Arg-Ile-Ser-Ala-Thr-Gly-Arg-Gly-Val-Phe-Gly-His-Ile-Glu-Asn-
C           Pro-Gly-Ile-Ser-Gln-Gly-Gly-Ile)His-Gly-Arg-Ile-Ser(Ala-Thr-Gly)Arg-Gly(Leu-Phe)Gly-His-Ile-Glu-Asn-
R               Gly- X -Ser-Glx-Gly-Lys             (Ile-Ser-Ala-Glx-Gly-Lys)

                       230                 235                 240                 245                250
B           Phe-Ile-Glu-Asn-Ala-Ser-Tyr-Met-Ser-Ile-Leu-Gly-Met-Thr-Pro-Gly-Phe-Gly-Asp-Lys-Thr-Phe-Ala-Val-Gln-
C           Phe-Ile(Glu-Asn-Ala-Ser)Tyr-Met-Ser-Ile-Leu-Gly-Met(Thr-Pro-Gly-Phe-Gly-Asp)Lys(Thr-Phe-Ala-Val-Gln-
R           Phe-Ile-Glx-Asx                         Met-Thr-Pro-Gly
```

FIG. 4. Amino acid sequences of bovine (B), chicken (C), and rat (R) liver glutamate dehydrogenases. The chicken enzyme possesses three additional residues at the NH_2-terminal end with the NH_2-terminal residue indicated as cysteine although it was found as cysteic acid. The unpublished data (*260*) for the rat enzyme were

```
                    255                 260                 265                 270                 275
Gly-Phe-Gly-Asn-Val-Gly-Leu-His-Ser-Met-Arg-Tyr-Leu-His-Arg-Phe-Gly-Ala-Lys-Cys-Val-Ala-Val-Gly-Glu-
Gly-Phe-Gly-Asn-Val)Gly(Leu-His-Ser)Met-Arg-Tyr-Leu-His-Arg-Phe-Gly-Ala-Lys(Cys-Val-Ala-Val-Gly-Glu)
    Phe-Gly-Asx                             (Glx-Leu-His-Arg)Phe-Gly-Ala-Lys(Cys-Val-Ala-Val-Gly-Glx-

                    280                 285                 290                 295                 300
Ser-Asp-Gly-Ser-Ile-Trp-Asn-Pro-Asp-Gly-Ile-Asp-Pro-Lys-Glu-Leu-Glu-Asp-Phe-Lys-Leu-Gln-His-Gly-Thr-
Phe(Asp-Gly-Ser-Ile)Trp(Asn-Pro-Asp-Gly-Ile-Asp-Pro)Lys-Glu(Leu-Glu-Asp-Tyr)Lys-Leu(Gln-His)Gly-Thr-
Ser-Asx-Gly-Ser-Ile-Trp-Asx-Pro-Asx-Gly-Ile-Asx-Pro-Lys)(Glu-Leu)Glu-Asp-Phe-Lys Leu-Glx-His-Gly(Ser-

                    305                 310                 315                 320                 325
Ile-Leu-Gly-Phe-Pro-Lys-Ala-Lys-Ile-Tyr-Glu-Gly-Ser-Ile-Leu-Glu-Val-Asp-Cys-Asp-Ile-Leu-Ile-Pro-Ala-
Ile-Met-Gly-Phe-Pro-Lys-Ala-Gln-Lys-Leu-Glu(Gly-Ser-Ile)Leu(Glu-Thr-Asp-Cys-Asp-Ile-Leu-Ile-Pro-Ala-
Ile-Leu-Gly-Phe-Pro-Lys)Ala-Lys Val-Tyr-Glu(Gly-Ser-Ile-Leu-Glx-Ala-Asx-Cys-Asx-Ile-Leu-Ile-Pro-Ala-

                    330                 335                 340                 345                 350
Ala-Ser-Glu-Lys-Gln-Leu-Thr-Lys-Ser-Asn-Ala-Pro-Arg-Val-Lys-Ala-Lys-Ile-Ile-Ala-Glu-Gly-Ala-Asn-Gly-
Ala-Ser-Glu)Lys-Gln-Leu-Thr-Lys-Ala-Asn-Ala-His-Lys-Val-Lys-Ala-Lys(Ile-Ile-Ala-Glu-Gly-Ala-Asn-Gly-
Ala-Ser-Glx)Lys(Gln-Leu)Thr-Lys                   Ala-Lys Ile-Ile-Ala-Glx-Gly-Ala-Asx-Gly-

                    355                 360                 365                 370                 375
Pro-Thr-Thr-Pro-Glx-Ala-Asp-Lys-Ile-Phe-Leu-Glu-Arg-Asn-Ile-Met-Val-Ile-Pro-Asp-Leu-Tyr-Leu-Asn-Ala-
Pro-Thr-Thr-Pro-Gln-Ala-Asp)Lys-Ile-Phe-Leu-Glu-Arg-Asn-Ile-Met(Val-Ile-Pro-Asp-Leu-Tyr)Leu(Asn-Ala-
Pro-Thr(Thr-Pro-Glx-Ala-Asx-Lys-Ile-Phe-Leu-Glx)Arg                           Leu-Tyr

                    380                 385                 390                 395                 400
Gly-Gly-Val-Thr-Val-Ser-Tyr-Phe-Glx-Leu-Lys-Asn-Leu-Asn-His-Val-Ser-Tyr-Gly-Arg-Leu-Thr-Phe-Lys-Tyr-
Gly-Gly-Val-Thr-Val-Ser-Ala-Phe-Glx-Glx)Lys(Asn-Leu-Asn-His-Val-Ser-Tyr-Gly-Arg-Leu-Thr-Phe)Lys-Tyr-
                                            Asx-Leu-Asx-His(Val-Ser-Tyr-Gly-Arg)Leu-Thr-Phe-Lys Tyr-

                    405                 410                 415                 420                 425
Glu-Arg-Asp-Ser-Asn-Tyr-His-Leu-Leu-Met-Ser-Val-Gln-Glu-Ser-Leu-Glu-Arg-Lys-Phe-Gly-Lys-His-Gly-Gly-
Glu-Arg-Asp(Ser-Asn)Tyr(His-Leu-Leu)Met-Ser(Val-Gln-Glu-Ser)Leu-Glu-Arg-Lys-Phe-Gly-Lys(His-Gly-Gly-
Glu-Arg Asn)(Ser-Asx-   His-Leu-Leu-Met-Ser-Val-Glx-Glx-Ser-Leu-Glx-Arg)Lys Phe-Gly-Lys His-Gly-Gly-

                    430                 435                 440                 445                 450
Thr-Ile-Pro-Ile-Val-Pro-Thr-Ala-Glu-Phe-Gln-Asp-Arg-Ile-Ser-Gly-Ala-Ser-Glu-Lys-Asp-Ile-Val-His-Ser-
Thr-Ile-Pro)Val(Val-Pro-Thr-Ala-Glu)Phe(Gln-Asp)Arg-Ile-Ser-Gly(Ala-Ser-Glu)Lys-Asp-Ile-Val-His(Ser-
Thr-Ile-Pro-Val-Val-Pro-Thr-Ala-Glu-Phe(Glx-Asp)Arg Ile-Ser-Gly-Ala-Ser-Glx-Lys Asx-Ile-Val(His-Ser-

                    455                 460                 465                 470                 475
Gly-Leu-Ala-Tyr-Thr-Met-Glu-Arg-Ser-Ala-Arg-Gln-Ile-Met-Arg-Thr-Ala-Met-Lys-Tyr-Asn-Leu-Gly-Leu-Asp-
Gly-Leu-Ala)Tyr-Thr-Met-Glu-Arg-Ser-Ala-Arg-Gln-Ile-Met-Arg-Thr-Ala-Met-Lys-Tyr-Asn(Leu-Gly-Leu-Asp-
Gly-Leu-Ala-Tyr)Thr-Met(Glx-Arg)Ser-Ala-Arg(Gln-Ile)Met-Arg Thr-Ala(Met-Lys)Tyr-Asx-Leu-Gly-Leu-Asx-

                    480                 485                 490                 495                 500
Leu-Arg-Thr-Ala-Ala-Tyr-Val-Asn-Ala-Ile-Glu-Lys-Val-Phe-Arg-Val-Tyr-Asn-Glu-Ala-Gly-Val-Thr-Phe-Thr-COOH
Leu)Arg(Thr-Ala-Ala-Tyr-Val-Asn-Ala-Ile-Glu)Lys-Val-Phe-Lys-Val(Tyr-Asn-Glu-Ala-Gly)Leu-Thr(Phe-Thr-COOH)
Leu-Arg Thr-Ala-Ala-Tyr(Val-Asx-Ala-Ile-Glx)Lys(Val-Phe-Lys)Val-Tyr(Asx-Glx-Ala-Gly-Val)Thr-Phe-Thr
```

kindly furnished by Dr. Carl Frieden and represent alignment of tryptic peptides by homology only; missing regions have not been studied. Sequences containing identical residues have been enclosed in boxes. X indicates unidentified residues.

1 10 14 20

(B) Ala-Asp-Arg-Glu-Asp-Asp-Pro-Asn-Phe-Phe-Lys-Met-Val-Glu-Gly-Phe-Phe-Asp-Arg-Gly-Ala-Ser-Ile-Val-Glu-Asp-Lys-
(N) Acetyl-Ser-Asp-Leu-Pro-Ser-Glu-Pro-Glu-Phe-Glu-Gln-Ala-Tyr-Lys-

10

40

Leu-Val-Glu-Asp-Leu-Lys-Thr-Arg-Gln-Thr-Gln-Glu-Gln-Lys-Arg-Asn-Arg-Val-Arg-Gly-Ile-Leu-Arg-Ile-Ile-Lys-Pro-
Glu-Leu-Ala-Tyr-Thr-Leu-Glu-Asn-Ser-Ser-Leu-Phe-Gln-Lys-His-Pro-Glu-Tyr-Arg-Thr-Ala-Leu-Thr-Val-Ala-Ser-Ile-

20 30 40

60 80

Cys-Asn-His-Val-Leu-Ser-Leu-Ser-Phe-Pro-Ile-Arg-Arg-Asp-Asp-Gly-Ser-Trp-Glu-Val-Ile-Glu-Gly-Tyr-Arg-Ala-Gln-
Pro-Glu-Arg-Val-Ile-Gln-Phe-Arg-Val-Val-Trp-Glu-Asp-Asp-Asp-Gly-Asn-Val-Gln-Val-Asn-Arg-Gly-Tyr-Arg-Val-Gln-

50 60

100

His-Ser-His-Gln-Arg-Thr-Pro-Cys-Lys-Gly-Gly-Ile-Arg-Tyr-Ser-Thr-Asp-Val-Ser-Val-Asp-Glu-Val-Lys-Ala-Leu-Ala-
Phe-Asn-Ser-Ala-Leu-Gly-Pro-Tyr-Lys-Gly-Gly-Leu-Arg-Leu-His-Pro-Ser-Val-Asn-Leu-Ser-Ile-Leu-Lys-Phe-Leu-Gly-

70 80 90

120 126

Ser-Leu-Met-Thr-Tyr-Lys-Cys-Ala-Val-Val-Asp-Val-Pro-Phe-Gly-Gly-Ala-Lys-Ala-Gly-Val-Lys-Ile-Asn-Pro-Lys-Asn-
Phe-Glu-Gln-Ile-Phe-Lys-Asn-Ala-Leu-Thr-Gly-Leu-Ser-Met-Gly-Gly-Gly-Lys-Gly-Gly-Ala-Asp-Phe-Asp-Pro-Lys-Gly-

100 110 113 120

140 160

Tyr-Thr-Asp-Glu-Asp-Leu-Glu-Lys-Ile-Thr-Arg-Arg-Phe-Thr-Met-Glu-Leu-Ala-Lys-Lys-Gly-Phe-Ile-Gly-Pro-Gly-Val-
Lys-Ser-Asp-Ala-Glu-Ile-Arg-Arg-Phe-Cys-Cys-Ala-Phe-Met-Ala-Glu-Leu-His-Lys-His- Ile-Gly-Ala-Asp-Thr-

130 140 142 143

180

Asp-Val-Pro-Ala-Pro-Asn-Met-Ser-Thr-Gly-Glu-Arg-Glu-Met-Ser-Trp-Ile-Ala-Asp-Thr-Tyr-Ala-Ser-Thr-Ile-Gly-His-
Asp-Val-Pro-Ala-Gly-Asp-Ile-Gly-Val-Gly-Gly-Arg-Glu-Ile-Gly-Tyr-Met-Phe-Gly-Ala-Tyr-Arg-Lys-Ala-Ala-Asn-Arg-

150 160 170

200

Tyr-Asp-Ile-Asn-Ala-His-Ala-Cys-Val-Thr-Lys-Pro-Gly-Ile-Ser-Gln-Gly-Gly-Ile-His-Gly-Arg-Ile-Ser-Ala-Thr-Gly-
Phe-Glu-Gly-Val- Leu-Thr-Gly-Lys-Gly-Leu-Ser-Trp-Gly-Gly-Ser-Leu-Ile-Arg-Pro-Glu-Ala-Thr-Gly-

178 179 180 190

220 240

Arg-Gly-Val-Phe-Gly-His-Ile-Glu-Asn-Phe-Ile-Glu-Asn-Ala-Ser-Tyr-Met-Ser-Ile-Leu-Gly-Met-Thr-Pro-Gly-Phe-Gly-
Tyr-Gly-Leu-Val-Tyr-Tyr-Val-Gly-His-Met-Leu-Glu Tyr-Ser-Gly-Ala-Gly-Ser-Tyr-Ala-

200 209 210

260

Asp-Lys-Thr-Phe-Ala-Val-Gln-Gly-Phe-Gly-Asn-Val-Gly-Leu-His-Ser-Met-Arg-Tyr-Leu-His-Arg-Phe-Gly-Ala-Lys-Cys-
Gly-Lys-Arg-Val-Ala-Leu-Ser-Gly-Ser-Gly-Asn-Val-Ala-Gln-Tyr-Ala-Ala-Leu-Lys-Leu-Ile-Glu-Leu-Gly-Ala-Thr-Val-

220 230 240

280

Val-Ala-Val-Gly-Glu-Ser-Asp-Gly-Ser-Ile-Trp-Asn-Pro-Asp-Gly-Ile-Asp-Pro-Lys-Glu-Leu-Glu-Asp-Phe-Lys-Leu-Gln-
Val-Ser-Leu-Ser-Asp-Ser-Lys-Gly-Ala-Leu-Val-Ala-Thr-Gly-Glu-Ser-Gly-Ile-Thr-Val-Glu-Asx-Ile-Asx-Ala-Val-Met-

250 260 270

300 320

His-Gly-Thr-Ile-Leu-Gly-Phe-Pro-Lys-Ala-Lys-Ile-Tyr-Glu-Gly-Ser-Ile-Leu-Glu-Val-Asp-Cys-Asp-Ile-Leu-Ile-Pro-
Ala-Ile-Lys-Glu-Ala-Arg-Gln-Ser-Leu-Thr-Ser-Phe-Gln-His-Ala-Gly-His-Leu-Lys-Trp-Ile-Glu-Gly-Ala-Arg-Pro-Trp-

280 290

340

Ala-Ala-Ser-Glu-Lys-Gln-Leu-Thr-Lys-Ser-Asn-Ala-Pro-Arg-Val-Lys-
Leu-His-Val-Gly-Lys-Val-Asp-Ile-Ala-Leu-Pro-Cys-Ala-Thr-Glu-Asp-Glu-Val-Ser-Lys-Glu-Glu-Ala-Glu-Gly-Leu-Leu-

300 310 320

341 360

Ala-Lys-Ile-Ile-Ala-Glu-Gly-Ala-Asn-Gly-Pro-Thr-Thr-Pro-Gln-Ala-Asp-Lys-Ile-Phe-Leu-Glu-Arg-Asn-
Ala-Ala-Gly-Cys-Lys-Phe-Val-Ala-Glu-Gly-Ser-Asn-Met-Gly-Cys-Thr-Leu-Glu-Ala-Ile-Glu-Val-Phe-Glu-Asn-Asn-Arg-

330 340 350

380

Ile-Met-Val-Ile-Pro-Asp-Leu-Tyr-Leu-Asn-Ala-Gly-Gly-Val-Thr-Val-Ser-Tyr-Phe-Glx-Leu-Lys-Asn-Leu-Asn-His-Val-
Lys-Glu-Lys-Lys-Gly-Glu-Ala-Val-Trp-Tyr-Ala-Pro-Gly-Lys-Ala-Ala-Asn-Cys-Gly-Gly-Val-Ala-Val-Ser-Gly-Leu-Glu-

360 370

400

Ser-Tyr-Gly-Arg-Leu-Thr-Phe-Lys-Tyr-Glu-Arg-Asp-Ser-Asn-Tyr-His-Leu-Leu-Met-Ser-Val-Gln-Glu-Ser-Leu-Glu-Arg-
Met-Ala-Gln-Asn-Ser-Gln-Arg-Leu-Asn-Trp-Thr-Gln-Ala-Glu-Val-Asp-Glu-Lys-Leu-Lys-Asp-Ile-Met-Lys-Asn-Ala-Phe-

380 390 400

420 440

Lys-Phe-Gly-Lys-His-Gly-Gly-Thr-Ile-Pro-Ile-Val-Pro-Thr-Ala-Glu-Phe-Gln-Asp-Arg-Ile-Ser-Gly-Ala-Ser-Glu-Lys-
Phe-Asn-Gly-Leu-Asn-Thr-Ala-Lys-Thr-Tyr-Val-Glu-Ala-Ala-Glu-Gly-Gln-Leu-Pro-Ser-Leu-Val-Ala-Gly-Ser-Asn-Ile-

410 420 430

450 460

Asp-Ile-Val-His-Ser-Gly-Leu-Ala-Tyr-Thr-Met-Glu-Arg-Ser-Ala-Arg-Gln-Ile-Met-Arg-Thr-Ala-Met-Lys-Tyr-Asn-Leu-
Ala-Gly-Phe-Val-Lys-Val-Ala-Gln-Ala-Met-His-Asp-Gln-Gly-Asp-Trp-Ser-Lys-Asn-COOH

440 450

480

Gly-Leu-Asp-Leu-Arg-Thr-Ala-Ala-Tyr-Val-Asn-Ala-Ile-Glu-Lys-Val-Phe-Arg-Val-Tyr-Asn-Glu-Ala-Gly-Val-Thr-Phe-

500

Thr-COOH

	190	205
LDH	Trp-Val-Ile-[Gly]-Gln-[His-Gly]-Asp-Ser-Val-[Pro]-Ser-[Val]-Trp-Met-Asx	
	418	432
GDH	Arg-Lys-Phe-[Gly]-Lys-[His-Gly]-Gly-Thr-Ile-[Pro]-Ile-[Val]-Pro-Thr-Ala	

	209	220
G3PDH	[Gly]-Ala-[Ala-Lys-Ala]-Val-Gly-[Lys]-Val-Ile-[Pro]-Glu	
	123	134
GDH	[Gly]-Gly-[Ala-Lys-Ala]-Gly-Val-[Lys]-Ile-Asn-[Pro]-Lys	

FIG. 6. Apparent homologous sequences in lactate dehydrogenase (LDH) and bovine GDH, and in glyceraldehyde-3-phosphate dehydrogenase (G3PDH) and bovine GDH.

Two additional examples of homology between bovine and chicken GDH's and other pyridine nucleotide–dependent dehydrogenases can be cited although their possible significance is unclear. Comparison of the amino acid sequence in the vicinity of the active site histidine (residue 195) (*272*) of dogfish M_4 lactate dehydrogenase (residues 190–205) with residues 418–432 of bovine and chicken GDH's (Fig. 6) reveals that among these 15 residues five are identical, including the histidine, and eight are related by single base changes in the respective codons. This is noteworthy in view of reports, to be discussed later, that histidine may be involved in the function and regulation of bovine GDH. Another instance of homology involves a region of glyceraldehyde-3-phosphate dehydrogenase (*156*) and the portion of the sequence of the GDH's including Lys-126 (see below) which is reactive with various amino group reagents producing inactivation of the enzymes. The homology, shown in Fig. 6, is also striking.

Studies of am mutants of *Neurospora* NADP-GDH (*184*) (Section III,A,1,b) have shown that in am^{19} Lys-141 is replaced by methionine and results in an inactive enzyme; however, since revertants of am^{19} involve substitution of leucine for Ile-191 in strain R1 and arginine for Gln-391 in strain R24, it is evident that the change in am^{19} involves,

272. S. S. Taylor, S. S. Oxley, W. S. Allison, and N. O. Kaplan, *Proc. Nat. Acad. Sci. U. S.* **70**, 1790 (1973).

FIG. 5. Amino acid sequences of bovine (B) (*258*) and *Neurospora* (N) (*259*) NADP glutamate dehydrogenases. The alignments have been made to obtain maximal homology of the much shorter *Neurospora* sequence. In addition to the difference in lengths reflected at the NH_2- and COOH-terminal ends, two gaps have been introduced in the *Neurospora* sequence and one in the bovine. Boxes indicate identical residues; underlines represent residues that differ by a single base change in the amino acid codons.

TABLE VII
IDENTICAL RESIDUES IN BOVINE AND *Neurospora* GDH'S

Amino acid	No. of identities[a]	Amino acid	No. of identities[a]
Lysine	10	Cysteine	0
Histidine	0	Glycine	21
Arginine	5	Alanine	8
Aspartic acid	4	Valine	6
Asparagine	2	Methionine	0
Threonine	3	Isoleucine	1
Serine	2	Leucine	5
Glutamic acid	4	Tyrosine	2
Glutamine	2	Phenylalanine	2
Proline	3	Tryptophan	0

[a] The number of identities is derived from the alignment of the two proteins as shown in Fig. 5.

undoubtedly, a residue important for conformation. Thus, Lys-141 cannot be involved in the active site. Further studies of the amino acid replacements in the am mutants and their revertants should aid in providing clearer definition of the roles of certain residues.

Table VII gives the number of identical residues of various kinds occupying the same positions in the bovine and *Neurospora* enzymes. Clearly, most of these residues must be involved in conformation rather than mechanism in view of the large number of glycine residues occupying identical positions in the sequences.

It is surprising that in comparing the sequences of the bovine and chicken GDH's and *Neurospora* NADP-GDH there is no coincidence of residues of cysteine, methionine, histidine, and tryptophan. This does not, by any means, exclude such residues from location in the active site or participation in catalytic action in these homologous enzymes. Since the polypeptide chains are of different lengths, residues may assume the same position in the three-dimensional structure despite differences in sequence. Nevertheless, ascription of special roles to such residues must be treated cautiously. The possible involvement of these four types of residues in the activity of the GDH's will be discussed in subsequent sections.

Sequence studies are presently in progress on the NAD-GDH of *Neurospora*. Unlike the enzymes previously discussed, the NAD-specific GDH is a tetramer with four identical subunits, each with a molecular weight near 120,000. Peptides that have thus far been sequenced, including two long sequences involving up to 100 residues each, display no obvious homology with any of the GDH's for which the primary structures have been elucidated (*273–275*).

V. Chemical Modification Studies—Binding of Substrates and Coenzymes

The reactive residues of glutamate dehydrogenase have been probed with a variety of reagents and methods. Unless noted otherwise, the studies have been conducted with the bovine enzyme.

A. Lysine Residues

Colman and Frieden (*108*) demonstrated in 1966 that acetylation of one amino group per subunit with acetic anhydride produces 80% inactivation. More extensive acetylation alters the degree of polymerization and certain kinetic parameters (*276*). Almost simultaneously, Anderson *et al.* (*277*) reported the reversible inhibition of GDH by pyridoxal 5′-phosphate and certain other aromatic aldehydes. The inhibition was attributed to formation of a Schiff base since reduction with $NaBH_4$ results in irreversible inactivation. It was estimated that approximately one residue of ε-pyridoxyllysine had been formed per subunit. In 1969, Holbrook and Jeckel (*278*) inactivated the enzyme by reaction with a substituted maleimide and, subsequently, obtained the partial sequence of a tryptic peptide containing a modified lysine residue (Fig. 7).

Piszkiewicz *et al.* (*279*) isolated a peptide containing ε-pyridoxyllysine from a tryptic digest of GDH which had been inactivated by the procedure of Anderson *et al.* (*277*). The sequence of the peptide demonstrates that the substituted residue is Lys-126 both in the bovine (*279*) and, later, in the chicken enzymes (*133,280*).

273. F. M. Veronese, Y. Degani, J. F. Nyc, and E. L. Smith, *JBC* **249,** 7936 (1974).
274. B. M. Austen and E. L. Smith, unpublished data.
275. B. M. Austen, Y. Degani, R. G. Duggleby, J. F. Nyc, E. L. Smith, and F. M. Veronese, unpublished data.
276. R. F. Colman and C. Frieden, *JBC* **241,** 3652 (1966).
277. B. M. Anderson, C. D. Anderson, and J. E. Churchich, *Biochemistry* **5,** 2893 (1966).
278. J. J. Holbrook and R. Jeckel, *BJ* **111,** 689 (1969).
279. D. Piszkiewicz, M. Landon, and E. L. Smith, *JBC* **245,** 2622 (1970).
280. In the original publication, the reactive lysine was identified as residue 97. Because of the transposition already mentioned, the correct number of this residue is 126.

		Reagent	Reference
Bovine liver	Cys-Ala-Val-Val-Asp-Val-Pro-Phe-Gly-Gly-Ala-Lys*-Ala-Gly-Val-Lys	PLP	279
Chicken liver	Cys-Ala-Val-Val-Asp-Val-Pro-Phe-Gly-Gly-Ala-Lys*-Ala-Gly-Val-Lys	PLP	133
Bovine liver	Cys-Ala-Val-Val-Asp-Val-Pro-Phe-Gly-Gly-Ala-Lys*-Ala-Gly-Val-Lys	Cyanate	282
Bovine liver	(Asx,Ala,Val,Val,Asx,Val,)Phe-Gly-Ala(Gly,Lys*)Ala-Gly-Val-Lys	ASPM	278
Bovine liver	(Asp,Val,Pro,Phe,Gly,Gly,Ala,Lys*,Ala,Gly,Val,Lys)	IASA	284
Chicken liver[a]	(Asp,Val,Pro,Phe,Gly,Gly,Ala,Lys*,Ala, ,Val,Lys)	IASA	284
Porcine liver[b]	(Asp,Val,Pro,Phe,Gly,Gly,Ala,Lys*,Ala, ,Val,Lys)	IASA	284
N. crassa (NADP)	Met-Gly-Gly-Gly-Lys*-Gly-Gly-Ala-Asp	PLP	66, 259
N. crassa (NADP)	Asn-Ala-Leu-Thr-Gly-Leu-Ser-Met-Gly-Gly-Gly-Lys -Gly-Gly-Ala-Asp	Unmodified	285

FIG. 7. Sequences obtained by treating GDH with various inhibitory reagents. The homologous lysine residue (*), number 126 in the bovine sequence, is found in each instance. [a] The peptide was not homogeneous by analysis and contained only 1.9 residues of glycine as compared to the 3 residues known from the sequence (*285*). [b] The peptide was not homogeneous by analysis, containing only 2.0 residues of glycine and, in addition, 0.7 residues of glutamate, which is absent in this peptide in all of the vertebrate GDH's which have been sequenced.

The ϵ-amino group of the reactive lysine residue has an abnormally low pK_a of 7.7–8.0 at 30° as measured by the rate of inactivation of the enzyme as a function of pH with pyridoxal (*84,280*), pyridoxal 5′-phosphate (*280,281*), or cyanate (*282*). In the last instance, the reactive lysine was demonstrated to be the same as that found with pyridoxal phosphate. It should be noted that the rate of oxidation of glutamate has been shown to be dependent upon an ionizable group in the enzyme complex having a pK_a of 7.7–7.8 (*283*).

Pyridoxal phosphate forms a noncovalent complex ($K = 2.5 \times 10^{-3}$ M) with the enzyme before imine formation (*281*), while no such evidence was observed for reaction with pyridoxal (*84*) or cyanate (*282*). The phosphate group must, therefore, be involved in rapid formation of the noncovalent complex, and it seems likely that the existence of this complex explains why the phosphate derivative is a more effective inhibitor than pyridoxal. Apparent equilibrium constants for maximal imine formation have been determined as 3.2×10^{-4} M for pyridoxal at pH 8.5 (*84*) and 4.4×10^{-5} M for PLP at pH 7.7 (*281*).

Other reagents such as *N*(*N′*-acetyl-4-sulfamoylphenyl)maleimide (*278*) and 4-iodoacetamidosalicylic acid (*284*) also inactivate bovine GDH at slightly alkaline pH; these reagents also react with Lys-126. With the latter reagent, reaction was also demonstrated with the homolo-

281. D. Piszkiewicz and E. L. Smith, *Biochemistry* **10,** 4544 (1971).
282. F. M. Veronese, D. Piszkiewicz, and E. L. Smith, *JBC* **247,** 754 (1972).
283. K. S. Rogers, *JBC* **246,** 2004 (1971).
284. J. J. Holbrook, P. A. Roberts, and R. B. Wallis, *BJ* **133,** 165 (1973).

gous lysine residue in the porcine (*284*) and chicken enzymes (*284*) (Fig. 7) (*285*).

The NADP-specific enzyme of *Neurospora* contains a lysine residue with $pK_a = 7.6$ at 34°, which reacts with pyridoxal phosphate or *N*-ethylmaleimide to yield an inactive enzyme (*65*). The heat of ionization of the ϵ-amino group as judged by its reaction with *N*-ethylmaleimide is 12.6 ± 2.0 kcal/mole. The labeled residue, Lys-113, is in a homologous sequence with that of the vertebrate enzymes (Fig. 5).

Although Piszkiewicz and Smith (*281*) reported that GDH is partially protected by NADH or NADPH from pyridoxal phosphate inactivation, other evidence has suggested that Lys-126 may be involved in the substrate binding site. In this respect, Brown *et al.* (*286*) have shown that, although PLP-GDH forms apparently normal complexes with NADPH, it is unable to bind substrate.

Deppert *et al.* (*287*) have reported that reaction with 5-diazo-1*H*-tetrazole at pH 5.5 results in the inactivation of bovine GDH as well as disaggregation of the polymer. Reaction with the tetrazole modifies 1.2 amino groups per chain with no loss of cysteine, histidine, or tyrosine. The modified protein could still react with pyridoxal phosphate and this was ascribed to Lys-126, but it was not demonstrated. Modification by tetrazole affects both substrate and coenzyme binding, although the effect on coenzyme binding might have been indirect. In the presence of both α-ketoglutarate and NADH, only the substrate site amino group is protected. The same authors (*288*) later reported that, at pH 8.8, GDH could be reversibly inactivated by glyoxal, a competitive inhibitor relative to α-ketoglutarate ($K_i = 2$ m*M*). Treatment of the enzyme–glyoxal complex with $NaBH_4$ results in irreversible inactivation and loss of 0.8 amino group per subunit. The amino group was reported to have a $pK_a = 8.2$. By reducing with tritiated borohydride, they identified the site of reaction with glyoxal as Lys-27 (*288*). It should be noted that this residue is constant in all GDH's of known sequence.

The NAD-GDH of *Neurospora* is also inactivated by reaction with pyridoxal 5′-phosphate or cyanate (*289*); however, the reactions could not be performed at the low pH values used with the aforementioned enzymes. At higher pH values, as many as 4–5 lysine residues per subunit

285. J. C. Wootton, G. K. Chambers, J. G. Taylor, and J. R. S. Fincham, *Nature (London), New Biol.* **241**, 42 (1973).
286. A. Brown, J. M. Culver, and H. F. Fisher, *Biochemistry* **12**, 4367 (1973).
287. W. Deppert, F. Hucho, and H. Sund, *Eur. J. Biochem.* **32**, 76 (1973).
288. I. Rasched, H. Jörnvall, and H. Sund, *Eur. J. Biochem.* (in press).
289. Y. Degani, F. M. Veronese, and E. L. Smith, *JBC* **249**, 7929 (1974).

are reactive with pyridoxal phosphate and eight with cyanate, thus precluding unequivocal identification of the residue (or residues) responsible for inactivation of the enzyme (*273*).

B. Histidine Residues

Tudball and Thomas (*290*) have suggested that an imidazole might be involved in proton transfer from glutamate to NAD on the basis of experiments involving photooxidation in the presence of Rose Bengal. Inactivation is correlated with destruction of eight of the 14 histidine residues in the subunit with no loss of tryptophan. Subsequently, it was reported (*291*) that oxidation of four histidines per subunit results in a 90% loss of enzymic activity. Reaction of GDH with diethylpyrocarbonate (DEP) results in a 50% loss of catalytic activity with acylation of one residue of histidine per subunit. DEP does not react with cysteine, tyrosine, tryptophan, or lysine residues under their experimental conditions, but the modification of histidine has rather complex effects on the catalytic and regulatory properties of the enzyme. At pH 6.1, three residues of histidine can be acylated in a biphasic reaction, the fast portion of which results in modification of a single histidine and 50% inactivation of the enzyme. Acylation of the other two histidines is slower and appears to have little effect on the residual activity; however, reaction at these sites abolishes the GTP effect but not the response to ADP. Neither glutamate nor NAD protects at pH 6.1; however, at pH 7.5 glutamate does protect against both inactivation and modification. Acylation of a single histidine at pH 6.1, followed by further reaction at pH 7.5, results in reaction of a second histidine at the higher pH with complete loss of residual activity; this second residue is protected by L-glutamate.

The effects of DEP on GDH were further investigated by Wallis and Holbrook (*292*) with results in conflict with those described above. At pH 6.0, rapid modification of a single residue of histidine per subunit results in a 1.4–2.0-fold activation of the enzyme while further modification gives slow inactivation. The activation is reversible by hydroxylamine, known to regenerate histidine from the ethoxyformyl derivative, and is prevented by inclusion of 1 m*M* NADH in the reaction mixture, suggesting that the activation might result from loss of the known inhibition produced by excess NADH. The activation phase of the reaction is observed when DEP and subunit concentrations were equimolar. DEP-activated GDH could be further activated by ADP, and the level of ADP

290. N. Tudball and P. Thomas, *BJ* **123**, 421 (1971).
291. N. Tudball, R. Bailey-Wood, and P. Thomas, *BJ* **129**, 419 (1972).
292. R. B. Wallis and J. J. Holbrook, *BJ* **133**, 183 (1973).

necessary for half-maximal activation is decreased fivefold by the modification. The effects of GTP on native and DEP-activated GDH are qualitatively and quantitatively identical.

The role of histidine has also been investigated by photooxidation at pH 7.6 with PLP as sensitizer (*293*). In these studies, destruction of a single histidine per subunit results in approximately 30% inactivation, loss of the ability of the protein to polymerize, and abolition of the GTP effect, while the ADP response is unchanged. Oxidation of a second histidine results in complete inactivation of the enzyme.

The above results do not afford as yet any clear picture of the precise roles played by histidine residues in bovine GDH, in view of the lack of stoichiometry of inactivation and the inability to identify specific residues in the sequence. Assignment of precise functions to individual residues must await use of more specific reagents than those employed in the experiments summarized above. It perhaps should be emphasized that amino acid analysis as the sole method of identifying possible loss of residues in a peptide chain as large as that of bovine GDH is somewhat uncertain.

C. Cysteine Residues

The role of cysteine in bovine GDH has been the subject of many investigations, but unequivocal evidence for a catalytic role for sulfhydryl groups has not been presented. Malcolm and Radda (*294*) reported that acid hydrolysates of enzyme which had been 90% inactivated with 4-iodoacetamidosalicylic acid (IASA) contained 0.93 residue of *S*-carboxymethylcysteine (recalculated value) per 55,200 molecular weight and that no other modified residues were observed; however, it was acknowledged that the primary site of reaction may have been the ε-amino group of a lysine residue and that the carboxymethyl group may have been transferred to cysteine during acid hydrolysis. Subsequently, Holbrook *et al.* (*284*) showed that the site of reaction of IASA with GDH, at pH 7.5, is Lys-126 and have discussed the reasons for this having been overlooked in the earlier studies.

Piszkiewicz and Smith (*295*) attempted to determine whether any of the cysteine residues were specifically reactive with iodoacetate. With the ^{14}C-labeled reagent, the reaction was stopped at various times of alkylation. At various degrees of reaction, tryptic peptide patterns were examined by two-dimensional electrophoresis and chromatography. At each

293. F. Hucho, U. Markau, and H. Sund, *Eur. J. Biochem.* **32**, 69 (1973).
294. A. D. B. Malcolm and G. K. Radda, *Eur. J. Biochem.* **15**, 555 (1970).
295. D. Piszkiewicz and E. L. Smith, unpublished data.

level of reaction, all cysteine-containing peptides were labeled in equivalent amount. It appears that reaction with [^{14}C]iodoacetate produces an unfolding of the protein resulting in the equivalent labeling of cysteine residues in the peptide chain.

The GDH's of *Neurospora* present a different case with respect to the role of sulfhydryl groups although here, again, no direct role in catalysis has been observed. Degani *et al.* (*289*) have found a single reactive thiol group per subunit in the native NAD-GDH. The enzyme is inhibited to varying degree depending on which thiol reagent is employed and is not inactivated by tetrathionate even though mixed disulfides are formed. Iodoacetate and iodoacetamide do not react with the thiol group. Detailed kinetic analysis has shown that modification with *N*-ethylmaleimide, producing apparently 72% inactivation under standard assay conditions, is without effect on V_{max} but increases K_m for both NAD and glutamate approximately fivefold. Thus, the modified thiol group cannot be involved directly in catalysis, but its reaction with a bulky or charged substituent results in a conformational change altering the K_m values but not V_{max}.

Gore and Greenwood (*296*) have reported that the NADP-GDH of *Neurospora* is inhibited by *p*-mercuribenzoate and that this inhibition is prevented by inclusion of coenzyme in the reaction mixture. In their experiments, prior treatment with the mercurial precluded binding of reduced coenzyme. However, Blumenthal and Smith (*65*) have shown that all thiol groups of the native enzyme are reactive with iodoacetate and that this treatment is without effect on catalytic activity, thus excluding any possible direct role for thiol groups in the activity of this enzyme.

D. Other Modifications

Modification of bovine GDH with 2-hydroxy-5-nitrobenzylbromide strongly reduces the degree of polymerization but decreases V_{max} by only 20% and is without effect on any of the K_m values or regulatory properties (*297*). These results were interpreted to indicate that tryptophan is involved only in polymerization; this is in conflict with results of certain spectrophotometric studies discussed below.

Rosen *et al.* (*298*) have presented evidence for involvement of methionine in the activity and regulation of bovine GDH. At pH 5.6–6.0 the enzyme is inactivated by IASA in a pseudo-first-order process. At these

296. M. G. Gore and C. Greenwood, *BJ* **127**, 31P (1972).
297. V. Witzemann, R. Koberstein, H. Sund, I. Rasched, H. Jörnvall, and K. Noack, *Eur. J. Biochem.* (in press).
298. N. L. Rosen, L. Bishop, J. B. Burnett, M. Bishop, and R. F. Colman, *JBC* **248**, 7359 (1973).

pH values there was stated to be no reaction with Lys-126 which does react with this reagent at higher pH values (*284*). Modification of four residues per subunit results in 90% inactivation and the loss of two residues each of cysteine and methionine. The activity is protected by a combination of substrate, ADP, and NAD. Inactivation was attributed to modification of methionine because conditions which protect the SH groups do not prevent either inactivation or destruction of methionine. Furthermore, regeneration of methionine by reducing agents causes partial recovery of activity. In addition to inactivation, alkylation of methionine apparently also results in decreased sensitivity to inhibition by GTP. The above study does not clearly differentiate whether the modified methionine residues affect the conformation of the protein or play a more active role in catalysis. This will be discussed further in a subsequent section.

E. Spectrophotometric Studies

Spectrophotometric studies have suggested possible involvement of tryptophan in the activity of bovine GDH. Fisher and Cross (*299*) noted that binding of reduced coenzyme alters the tryptophan spectrum of the enzyme. Summers and Yielding (*300*) reported that binding of NADH quenches protein fluorescence at 350 nm and that upon excitation of the E-NADH complex at 295 nm, a new emission band, resulting from bound NADH, appears at 465 nm. These data were interpreted as indicative of energy transfer between a protein tryptophan residue and the nicotinamide moiety of bound NADH, thus implying that the two groups are close. A similar effect has been reported in the NAD-GDH of *Neurospora* (*171*). Subsequently, Cross *et al.* (*301*) demonstrated that formation of the E-NADPH binary complex, and the abortive ternary complexes E-NADPH-L-glutamate and E-NADP-α-ketoglutarate are all characterized by a red shift in the tryptophan absorption spectrum. It appears likely, therefore, that a tryptophan residue is located in or near the coenzyme binding site.

F. Substrate Site—Competitive Inhibitors

The earliest experiments mapping the active site of bovine GDH by Caughey *et al.* (*302*) established that certain dicarboxylic acids, such

299. H. F. Fisher and D. G. Cross, *Science* **153**, 414 (1966).
300. M. R. Summers and K. L. Yielding, *BBA* **223**, 374 (1970).
301. D. G. Cross, L. L. McGregor, and H. F. Fisher, *BBA* **289**, 28 (1972).
302. W. S. Caughey, J. D. Smiley, and L. Hellerman, *JBC* **224**, 591 (1957).

as glutarate and isophthalate, and a group of *m*-halobenzoates are potent competitive inhibitors. A common feature of these compounds is the presence of either two carboxylate groups or a single carboxylate and a highly electronegative element. Examination of three-dimensional models showed that in the case of the more effective cyclic inhibitors the two electron withdrawing substituents are separated by about 7.5 Å, and that the straight chain dicarboxylates could achieve a similar conformation. Isophthalate, an extremely potent competitive inhibitor ($K_i = 5.6 \times 10^{-4}$ *M*) for the bovine enzyme (*302*) has its two carboxyl groups held rigidly at this 7.5 Å spacing, thus accounting for its efficiency of binding. It was postulated, therefore, that the substrate binding site must contain two positively charged groups having this spacing. Fisher (*16*) proposed that binding of glutamate with maximal efficiency required not only the α- and γ-carboxyl groups, but also the α-amino group, and the α- and both γ-hydrogens. However, γ-monofluoroglutamate as a substrate has a K_m similar to that of glutamate (see below), suggesting that both γ-hydrogens are not essential.

Isophthalate is also a potent competitive inhibitor for the *Neurospora* enzymes: with NAD-GDH, $K_i = 6.1 \times 10^{-5}$ *M* (*64*) and with NADP-GDH, $K_i = 3.6 \times 10^{-4}$ *M* (*303*), suggesting similar requirements for these enzymes. Among other competitive inhibitors, α-ketoglutarate oxime is one of the more potent for the bovine enzyme (*304*). Table VIII lists a number of competitive inhibitors for bovine GDH (*283,287,302,304,305*).

Bovine GDH can catalyze oxidative deamination of a number of monocarboxylic L-amino acids, e.g., leucine, alanine, norvaline, and α-aminobutyrate (*306*); this indicates that the active site has a more stringent requirement for the α-carboxyl group. Replacement of the α-carboxyl group of glutamate with tetrazole diminishes binding to a greater degree than replacement of the γ-carboxyl (*305*). Other α-amino and α-keto acids are also substrates for GDH (*306*). As indicated in Table IX (*306–308*), all are utilized at a lower rate than the glutamate-α-ketoglutarate system. In addition, reactions involving monocarboxylate substrates have much higher pH optima than those observed for deamination of glutamate. At physiological pH most substrates other than glutamate

303. K. M. Blumenthal and E. L. Smith, *BBRC* **62,** 78 (1975).
304. E. Kun and B. Achmatowicz, *JBC* **240,** 2619 (1965).
305. J. K. Elwood, R. M. Herbst, and G. L. Kilgour, *JBC* **240,** 2073 (1965).
306. J. Struck and I. W. Sizer, *ABB* **86,** 260 (1960).
307. B. Jollès-Bergeret, *BBA* **146,** 45 (1967).
308. K. H. Bässler and C. H. Hammar, *Biochem. Z.* **330,** 446 (1958).

TABLE VIII
COMPETITIVE INHIBITORS OF BOVINE GDH

Reaction	Inhibitor	K_i	Ref.
Glutamate + NAD	Glutarate	5.8×10^{-4}	*302*
	α-Ketoglutarate	7.3×10^{-4}	*302*
	Fumarate	6.8×10^{-3}	*302*
	Isophthalate	5.6×10^{-4}	*302*
	m-Iodobenzoate	4.6×10^{-4}	*302*
	m-Bromobenzoate	5.4×10^{-4}	*302*
	m-Chlorobenzoate	1.02×10^{-3}	*302*
	m-Nitrobenzoate	3.40×10^{-3}	*302*
	5-Bromofuroate	5.9×10^{-5}	*302*
	5-Chlorofuroate	6.3×10^{-5}	*302*
	5-Nitrofuroate	1.7×10^{-4}	*302*
	D-Glutamate	2.0×10^{-3}	*302*
α-Ketoglutarate + NADH + NH_4^+	α-Monofluoroglutarate	6.3×10^{-3}	*304*
Glutamate + NADP	α-Monofluoroglutarate	7.1×10^{-4}	*304*
	α-Ketoglutarate oxime	7.3×10^{-5}	*304*
Glutamate + NAD	α-Monofluoroglutarate	3.3×10^{-4}	*304*
	α-Ketoglutarate oxime	9.7×10^{-6}	*304*
Glutamate + NAD	α-Tetrazole	2.30×10^{-2}	*305*
	γ-Tetrazole	5.3×10^{-3}	*305*
	Ditetrazole	1.4×10^{-3}	*305*
α-Ketoglutarate + NADH + NH_4^+	Glyoxal	2.0×10^{-3}	*287*
Glutamate + NAD	Thiodiglycolic acid	3.3×10^{-4}	*283*
+ NADP		3.1×10^{-4}	*283*
Glutamate + NAD	Oxydiglycolic acid	2.2×10^{-4}	*283*
+ NADP		2.1×10^{-4}	*283*
Glutamate + NAD	Iminodiacetic acid	1.5×10^{-4}	*283*
+ NADP		1.3×10^{-4}	*283*

or α-ketoglutarate are very poorly bound; thus, the *in vivo* significance of these other substrates appears to be doubtful.

As in the case of many enzymes that can utilize either the amide group of glutamine or ammonia, it has been found (*303*) that glutamine can serve as a nitrogen donor with both bovine GDH and the *Neurospora* NADP-GDH at a rate about 40% that of ammonia. Asparagine can donate its amide group only with the *Neurospora* NADP-GDH at about 10% of the rate with ammonia. The optimal rate for utilization of the amides is at pH 8.4 whereas it is 7.6–7.8 with ammonia. ADP increases the rate of reductive amination with glutamine about threefold with bovine GDH, an effect similar to that found with ammonia. The NAD-GDH of *Neurospora* does not utilize glutamine or asparagine as a nitrogen donor.

TABLE IX
Substrate Specificity of Bovine GDH

Substrate[a]	Relative rate of oxidation[b]	pH optimum for oxidation	Ref.
Glutamate	100	8.3–8.6	*306*
Homocysteinesulfinate	65	9.0	*307*
Norvaline	17	9.5	*308*
Leucine	1.7	9.7	*306*
α-Aminobutyrate	2.3	9.5–10	*306*
Valine	1.6	9.5–10	*306*
Norleucine	1.6	9.5–10	*306*
Isoleucine	0.95	9.5–10	*306*
Methionine	0.82	9.5–10	*306*
Alanine	0.27	9.5–10	*306*

[a] All amino acids were of the L configuration except norleucine which was used as a racemic mixture.

[b] Reaction rates were determined at pH 9.0.

G. Coenzyme Site and Specificity

In this section we shall consider only the coenzyme active site; the regulatory sites for purine nucleotides will be discussed later. Cross and Fisher (*309*) have divided the coenzyme site into two subsites, one specific for the amide portion of the nicotinamide and the other for the adenosine diphosphate moiety. Binding of coenzyme at the amide subsite causes perturbations in the tyrosine and tryptophan absorption regions. Presumably, the tryptophan involved is the one found by Cross *et al.* (*301*) to have a red-shifted spectrum upon coenzyme binding.

Kaplan *et al.* (*310*) demonstrated that the 3-acetylpyridine and pyridine-3-aldehyde analogs of NAD are utilized by bovine GDH with a greater efficiency than NAD itself. This has been attributed (*309*) to the fact that the 3-acetylpyridine analog is incapable of binding at the amide subsite; thus, since dissociation of reduced coenzyme may be the rate-limiting step in glutamate oxidation, the velocity with the modified coenzyme is enhanced. Replacement of the 5′-AMP moiety of the coenzyme by formycin, 2-aminopurine ribonucleoside, or 7-deazapurine ribonucleoside diminishes, but does not abolish, catalytic activity (*311*); thus, the

309. D. G. Cross and H. F. Fisher, *JBC* **245,** 2612 (1970).
310. N. O. Kaplan, M. M. Ciotti, and F. E. Stolzenbach, *JBC* **221,** 833 (1956).
311. D. C. Ward, T. Horn, and E. Reich, *JBC* **247,** 4014 (1972).

adenosine diphosphate subsite (*301*) does not display absolute specificity.

Nicotinamide mononucleotide is not a cofactor for bovine GDH (*311a*). However, replacement of the adenine moiety of the coenzymes with either hypoxanthine, cytidine, or nicotinamide yields a derivative which will function as coenzyme, although with diminished efficiencies of 91, 53, and 3% (relative to the activity with NAD), respectively. Nicotinamide mononucleotide phosphoriboside is 0.4% as effective as NAD. These data, which give further indication of the lack of specificity in the adenosine diphosphate site, show that although the adenosine moiety of the coenzyme may be altered in a number of ways, it must be present in some form in order for the analog to function as a cofactor.

As discussed previously (Section IV), residues 245–280 of bovine GDH display sequence homology with the residues comprising the coenzyme binding domain of the dehydrogenases of known three-dimensional structure and sequence. Moreover, the three residues to which a functional role has been attributed are all conserved in GDH and those residues, believed to lie facing the β-pleated sheet regions, are either identical or functionally conserved in GDH when compared to other dehydrogenases. It seems likely, therefore, that residues 245–280 of bovine GDH comprise a portion of the adenylate subsite within the active site for coenzyme [subsite II in the Cross and Fisher model (*309*)].

Gore and Greenwood (*296*) have reported that the NADP-GDH of *Neurospora* utilizes both the 3-acetylpyridine and deamino analogs of NADP. In both cases, V_{max} is lower than with NADP and, in addition, the deamino analog is less efficiently bound.

Neurospora NADP-GDH is rapidly inactivated upon reaction with tetranitromethane (*311b*). This inactivation is completely prevented by the presence of coenzyme (NADP) or nicotinamide mononucleotide (NMN) but not by substrate, NADH, or 2′-monophosphoadenosine-5′-diphosphoribose. Analysis indicates that the primary effect of modification is nitration of a single residue of tyrosine per polypeptide chain. The reactive tyrosine was identified by isolation of a single, uniquely labeled peptide after hydrolysis with trypsin followed by cleavage with cyanogen bromide as tyrosine-168. This residue is not present in the part of the sequence that had been previously implicated as involved in binding the adenylate portion of the coenzyme but is present in the sequences of bovine and chicken GDH's. Both NMN and 2′-monophosphoadenosine-5′-diphosphoribose act as competitive inhibitors of NADP in the oxidation of glutamate with K_i values of 4.65×10^{-4} M and

311a. G. Pfleiderer, E. Sann, and F. Ortanderl, *BBA* **73,** 39 (1963).
311b. K. M. Blumenthal and E. L. Smith, *JBC* **250** (in press) (1975).

4.30×10^{-4} *M*, respectively. Thus, the specific protection afforded by NADP and NMN, but not by 2′-monophosphoadenosine-5′-diphosphoribose, indicates that tyrosine-168 is involved in binding the nicotinamide portion of the coenzyme.

Reaction of NADP-GDH of *Neurospora* with 1,2-cyclohexanedione results in a biphasic loss of enzyme activity (*311c*). At the end of the rapid phase of the reaction ($t_{1/2}$ = 1.5 min) the enzyme activity is diminished by approximately 60% with the simultaneous loss of one residue of arginine per subunit. After 60 min, the enzyme activity is competely lost with the modification of a total of two arginine residues per subunit. Reaction of bovine liver GDH with cyclohexanedione causes a rapid loss of about 45% of the enzyme activity and modification of about 1.5 residues of arginine per subunit. More prolonged treatment results in reaction of an additional four residues of arginine per subunit, but is without further effect on the residual activity. The activity of the *Neurospora* enzyme is not protected by substrate, coenzyme, or a combination of both; however, the activity of the bovine enzyme is partially protected by high levels of NAD or NADP. Although the K_m for α-ketoglutarate is unchanged by limited modification of either enzyme with cyclohexanedione, the K_m for coenzyme is increased about twofold for the *Neurospora* enzyme and about 1.5-fold for the bovine enzyme. The K_i of the *Neurospora* dehydrogenase for the competitive inhibitor 2′-monophosphoadenosine-5′-diphosphoribose is unchanged by the enzyme modification, but nicotinamide mononucleotide, a competitive inhibitor for the native *Neurospora* enzyme, does not inhibit the GDH with one modified arginine residue. This finding implies that the modified arginine is at or near the nicotinamide binding site of the enzyme.

VI. The Kinetic Mechanism

A. Kinetic Studies

Study of the kinetics of bovine GDH has been complicated both by the polymerization–depolymerization phenomena mentioned earlier (Section II,B) and by the allosteric effects of the coenzymes. An additional difficulty is that the purine nucleotide allosteric effectors influence the degree of polymerization of the protein. Although these phenomena complicate the interpretation of data, they may not be significant in a discussion of enzymic properties since the specific activity is independent of the degree of polymerization (*18*). Moreover, as already noted, the rat

311c. K. M. Blumenthal and E. L. Smith, *JBC* **250** (in press) (1975).

liver enzyme, which does not polymerize, is subject to allosteric control by GTP and ADP (*18,87*), further indicating that allosteric control and polymerization are independent phenomena.

Kinetic studies of the bovine enzyme have been complex and often conflicting. The subject has been thoroughly reviewed in the past, and recently by Dalziel (*312*). Hence, the present discussion will attempt only to afford a picture of some aspects of the mechanism of the reaction.

As discussed above, the 3-acetylpyridine and pyridine-3-aldehyde analogs of NAD^+ are utilized more efficiently than the natural coenzyme. Olson and Anfinsen (*151*) showed that NAD is used more effectively than NADP. Levy and Vennesland (*313*) established that the β-4-hydrogen of the coenzyme is utilized by bovine GDH. More recently, it was shown that both yeast and pea GDH's also display B specificity (*314*).

Various nucleotides, including NAD or NADH, exert different allosteric effects upon the GDH reaction, making it rather difficult to interpret kinetic experiments. Most of our knowledge concerning the mechanism has been derived from studies utilizing NADP(H).

For many years, it was felt that the GDH mechanism involved a compulsory order of binding with E-coenzyme being the first complex formed. This was based upon the unpublished observations of Frieden (*315*) that free enzyme could bind NADPH and that incubation of free enzyme with NH_4^+ and α-ketoglutarate in $H_2{}^{18}O$ failed to incorporate heavy oxygen into α-ketoglutarate. It should be noted that ^{18}O exchange would be anticipated if, as discussed below, α-iminoglutarate is an intermediate in the reaction. Kinetic studies (*315*) suggested that the data were consistent with the compulsory order of binding reduced coenzyme, NH_4^+, α-ketoglutarate. Later studies, however, employing more sensitive methods, have indicated that this model is incorrect. Engel and Dalziel (*316*), studying the oxidative deamination of glutamate and norvaline, reported that although their data did not unequivocally indicate random binding, they were able to rule out certain compulsory order mechanisms. Initial rate studies on the reductive amination of α-ketoglutarate were later shown to be inconsistent with any of the six possible compulsory order mechanisms (*317*); it was concluded that the mechanism must involve at least partially random order binding. The authors noted that earlier conclusions (*315*) regarding compulsory order were based on insufficient data

312. K. Dalziel, Chapter 1, this volume.
313. H. R. Levy and B. Vennesland, *JBC* **228,** 85 (1957).
314. D. D. Davies, A. Teixeira, and P. Kenworthy, *BJ* **127,** 335 (1972).
315. C. Frieden, *JBC* **234,** 2891 (1959).
316. P. C. Engel and K. Dalziel, *BJ* **115,** 621 (1969).
317. P. C. Engel and K. Dalziel, *BJ* **118,** 409 (1970).

and suggested that a fully random order mechanism with rapid equilibrium between free and bound substrate and coenzyme provided the best fit of the experimental data (*318*).

Cross *et al.* (*301*) reported that GDH could bind α-ketoglutarate, as judged by changes in the absorption spectrum. Since Frieden (*315*) had shown that coenzyme is bound by free enzyme, the data of Cross *et al.* can be interpreted as supportive of random order binding. The results of stopped-flow measurements on the NADP-linked deamination of L-glutamate led Colen *et al.* (*319*) to postulate a mechanism involving random order binding of substrate and cofactor, rapid equilibrium between the rather weak binary complexes, and cooperativity in the formation of the ternary complex E-NADP-glutamate. Bates and Frieden (*320*) have presented results of stopped-flow and computer simulation studies on oxidation of coenzyme which tend to support the random order model.

A study of the rates of exchange of the glutamate-α-ketoglutarate, NAD-NADH, and NADP-NADPH pairs (*321*) has supported the concept of partially random, or alternative, order binding of substrates and coenzymes. Substrates and coenzymes are free to bind in any order but dissociate more freely from binary than from ternary complexes, in agreement with results cited earlier (*317,319*). However, the order cannot be fully random since the rates of exchange for the various coenzyme and substrate pairs are not identical.

Iwatsubo and Pantaloni (*121*) demonstrated that the rate-limiting step in oxidation of glutamate is release of the reduced coenzyme, confirming an earlier conclusion based on steady-state measurements (*315*). Subsequently, Fisher *et al.* (*322*) and DiFranco and Iwatsubo (*323*) indicated that reduction of NADP can be resolved into a three-part process, a rapid, first-order burst followed by a slower, zero-order, second phase, then an approach to equilibrium. Fisher *et al.* (*322*) demonstrated that replacement of the α-hydrogen of the substrate by deuterium results in a significant slowing of the burst phase but has little effect on subsequent steps, thus indicating that hydride ion abstraction is involved in the burst.

DiFranco and Iwatsubo (*323*) observed that two distinct E-NADPH complexes, one weakly and the other strongly fluorescent, are observed

318. P. C. Engel and K. Dalziel, *in* "Pyridine Nucleotide-Dependent Dehydrogenases" (H. Sund, ed.), p. 245. Springer-Verlag, Berlin and New York, 1970.
319. A. H. Colen, R. A. Prough, and H. F. Fisher, *JBC* **247,** 7905 (1972).
320. D. J. Bates and C. Frieden, *JBC* **248,** 7885 (1973).
321. E. Silverstein and G. Sulebele, *Biochemistry* **12,** 2164 (1973).
322. H. F. Fisher, J. R. Bard, and R. A. Prough, *BBRC* **41,** 601 (1970).
323. A. DiFranco and M. Iwatsubo, *Biochimie* **53,** 153 (1971).

during glutamate oxidation: Complex 1 is tentatively identified as E-NADPH-α-ketoglutarate. After liberation of α-ketoglutarate complex 2, an abortive ternary complex E-NADPH-glutamate, may be formed in the presence of high levels of substrate followed by dissociation to free enzyme, glutamate, and reduced coenzyme. Supportive of this proposed scheme is the fact that complex 1 exhibits a blue-shifted nicotinamide absorbance peak at 330 nm, a feature which is also shown by E-NADPH-α-ketoglutarate but not by E-NADPH, E-NADPH-glutamate, or E-NADPH-NH_4^+. Furthermore, complex 2 shows a red-shifted spectrum, a feature characteristic of either E-NADPH or E-NADPH-glutamate. In addition, no deuterium isotope effect can be observed in the second portion of the reaction. That the third phase represents release of free NADPH is indicated by the appearance of a normal NADPH absorption spectrum.

B. Reaction Mechanism

Hochreiter and Schellenberg (*324,325*) have proposed that α-iminoglutarate is an intermediate in the GDH reaction. Incubation of GDH with α-ketoglutarate and ammonia leads to formation of α-iminoglutarate, which is trapped by borohydride reduction to yield L-glutamate. That the formation of iminoglutarate is enzyme-mediated is indicated by the finding that competitive inhibitors prevent its formation. Furthermore, reduction of the nonenzymic reaction product of α-ketoglutarate and NH_3 yields DL-glutamate. It is possible that the complex E-NADPH-iminoglutarate is responsible for the blue-shifted complex 1 spectrum discussed above. As previously noted (*18*), formation of α-iminoglutarate as an intermediate strengthens the case for random order binding.

It is interesting to consider the possible role of the essential amino group of Lys-126 in a reaction scheme involving α-iminoglutarate. From the work of Brown *et al.* (*286*) Lys-126 is somehow involved in substrate binding; however, since the apparent pK is in the range 7.6–8.0, it is unlikely that it represents one of the two cationic groups required for binding of the carboxylates since it would be approximately 50% unprotonated at the pH optimum for α-ketoglutarate reduction and almost completely unprotonated at the optimum for glutamate oxidation.

A more attractive hypothesis would involve condensation of the uncharged ϵ-amino group with α-ketoglutarate to form a Schiff base with release of a molecule of water. Attack of ammonia on such a Schiff base

324. M. C. Hochreiter and K. A. Schellenberg, *JACS* **91,** 6531 (1969).
325. M. C. Hochreiter, D. R. Patek, and K. A. Schellenberg, *JBC* **247,** 6271 (1972).

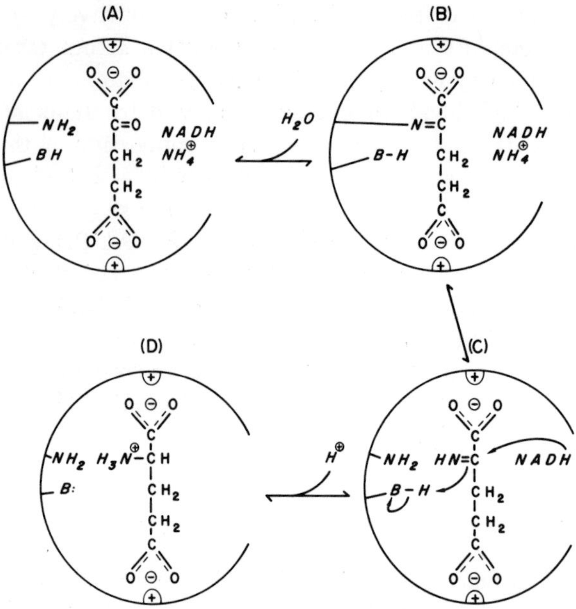

FIG. 8. Postulated pathway involving formation of a Schiff base intermediate (B) of the enzyme lysine-126 with α-ketoglutarate. This leads to formation of enzyme-bound α-iminoglutarate (C). Reduction of this intermediate yields glutamate (D).

intermediate would yield bound α-iminoglutarate which could then be reduced by NAD(P)H to yield glutamate. Thus, the complex reaction would, in effect, involve discrete steps for dehydration and reduction. In the reverse direction, the initial step would be hydride ion abstraction from glutamate, presumably mediated by an enzyme base, and transferral to coenzyme, to form iminoglutarate, followed by reaction of this intermediate with the ε-amino group of Lys-126 to yield the Schiff base with loss of ammonia. Hydrolysis of the Schiff base, which may be catalyzed by a protonated group such as an ε-ammonium group, would then generate α-ketoglutarate (see Fig. 8).

The above hypothetical mechanism is, of course, based on the view that α-iminoglutarate is indeed an intermediate in the reaction. It has been shown that α-imino acids are intermediates in the reactions catalyzed by the flavin-containing L- and D-amino acid oxidases (*326,327*). In addition, such a mechanism would be consistent with the observed difference in pH optima for the forward and reverse reactions (Table

326. L. Hellerman and D. S. Coffey, *JBC* **242,** 582 (1967).
327. E. W. Hafner and D. Wellner, *Proc. Nat. Acad. Sci. U. S.* **68,** 987 (1971).

TABLE X
pH Optima for Various Glutamate Dehydrogenases

Source	pH optimum		Ref.
	Glutamate oxidation	Reductive amination	
Bovine liver	8.5–9.0	7.8	*88,328,329*
Chicken liver	8.5–9.0	7.8	*59,328*
Human liver[a]	8.0	7.8	*53*
Apodachyla	9.0	8.2	*70*
Neurospora (NADP)	8.6–9.0	7.6	*330*
Yeast (NADP)[b]	7.6	8.0	*56*
Blastocladiella emersonii	8.8	7.9	*331*
Clostridium SB_4	9.4	7.8	*36*
Nitrosomonas europaea	8.4–10	7.7	*35*
E. coli	9.0	7.9	*73*
Mycoplasma laidlawii			
with NAD(H)	9.6	9.1	
with NADP(H)	8.9	8.9	*38*
Pea root	8.2	7.9	*46*
Soybean cotyledon	9.3	8.0	*47*

[a] The pH optimum for glutamate oxidation may in fact be higher than the value given, pH 8.0. The concentration of glutamate was 0.0144 *M*, or 6.25-fold higher than K_m, which may not be sufficient to saturate the enzyme over the entire pH range. In contrast, Smith and Piszkiewicz (*328*) used 0.05 *M* glutamate in determining the pH activity profile for bovine and chicken GDH's.

[b] This is the only reported instance of a pH optimum higher for reductive amination than for glutamate oxidation; however, the concentration of glutamate that was used was equivalent to the K_m for glutamate at pH 7.6 and may well have been too low to saturate the enzyme over the pH range tested.

X) (*328–331*). The amination reaction for most GDH's is optimal near pH 7.6–8.0 whereas for the deamination reaction it is in the region of 8.5–9.0. The difference is seen most clearly when ADP is used as the activator of the bovine enzyme: pH optima are then near pH 8.0 for reduction and pH 9.0 for oxidation.

The data in Table X for various GDH's vary considerably in the conditions used for the determinations. In some instances the substrate concentrations were not sufficiently high to reflect V_{max} values. The differences in pH optima appear to be between 1.0 and 1.5 pH units for the

328. E. L. Smith and D. Piszkiewicz, *JBC* **248,** 3089 (1973).
329. H. J. Strecker, *ABB* **46,** 128 (1953).
330. K. M. Blumenthal and E. L. Smith, unpublished data.
331. T. Sanner, *BBA* **250,** 297 (1971).

most reliable data. This would suggest that the activities of these diverse GDH's reflect similar mechanisms. The higher optimum for glutamate oxidation could result from a requirement for deprotonation of the ϵ-amino group of Lys-126 and partial deprotonation of the amino group of glutamate since there would be electrostatic repulsion between the substrate and the enzyme amino groups if both were positively charged. A schematic representation of this proposed mechanism is shown in Fig. 8.

The behavior of invertebrate and plant GDH's has been less extensively studied than that of the bovine enzyme. The question of compulsory order as opposed to random order binding, which has been resolved only with great difficulty for bovine GDH, has been investigated with only a few other GDH's. In each case, for Phycomycetes GDH (NAD) (*332*), GDH (NADP) of *Brevibacterium flavum* (*30*), soybean GDH (NAD) (*47*), and both the NAD- and NADP-dependent GDH's of *Thiobacillus novellus* (*33*), compulsory order binding has been reported in which coenzyme binds first and NH_4^+ last. However, since the more refined methods employed for investigation of the mechanism of bovine GDH have not been applied to any of these systems, the question of random vs. ordered mechanism cannot be said to have been resolved, particularly since the methods thus far employed did not give decisive results with bovine GDH.

The kinetic constants for a number of GDH's are summarized in Table XI. The K_m values for α-ketoglutarate, NAD^+, $NADP^+$, and NADPH are all rather similar, regardless of the source of the enzyme, whereas K_m values for glutamate, ammonia, and NADH vary rather widely. Enzymes having nearly identical primary structures, such as bovine and chicken GDH, display great similarity in K_m values, but the *Neurospora* GDH (NADP) is significantly different from the bovine although there is considerable sequence homology for these two enzymes.

VII. Regulatory Behavior

The GDH's that have been studied extensively are regulated either by induction-repression effects (Section III) or by allosteric effects. The effectors can be divided into three general classes: purine nucleoside di- and triphosphates, substrates and coenzymes, and certain steroid hormones and lipids.

Since the observation by Frieden in 1963 (*333*) that GTP inhibits and

332. R. M. Stevenson and H. B. LéJohn, *JBC* **246**, 2127 (1971).
332a. L. Corman and N. O. Kaplan, *JBC* **242**, 2840 (1967).
333. C. Frieden, *JBC* **238**, 3286 (1963).

TABLE XI

K_m VALUES FOR VARIOUS GLUTAMATE DEHYDROGENASES

Source	Ref.	L-Glutamate	α-Keto-glutarate	NH_4^+	NAD	NADH	NADP	NADPH
Bovine liver	*315*	1.8×10^{-3}	7.0×10^{-4}	3.2×10^{-3}	7.0×10^{-4}	2.4×10^{-5}	4.7×10^{-5}	2.5×10^{-5}
Chicken liver	*59*	2.0×10^{-3}			6.1×10^{-4}			
Frog liver	*60*	1.8×10^{-3}	5.0×10^{-3}	5.0×10^{-4}		2.0×10^{-4}	5.0×10^{-4}	2.0×10^{-4}
Dogfish liver	*332a*	2.5×10^{-3}	4.5×10^{-3}	8.0×10^{-2}			8.0×10^{-5}	4.0×10^{-4}
Achlya	*71*	3.1×10^{-3}	3.3×10^{-3}	2.6×10^{-2}	6.1×10^{-4}	5.5×10^{-5}		
Pythium	*71*	1.67×10^{-3}	3.85×10^{-3}	1.1×10^{-2}	1.4×10^{-4}	6.7×10^{-4}		
Fusarium (NAD)	*21*	1.1×10^{-2}	2.1×10^{-3}	1.7×10^{-2}	2.5×10^{-4}	8.0×10^{-5}		
Fusarium (NADP)	*21*	5.0×10^{-2}	1.7×10^{-3}	1.1×10^{-2}			2.5×10^{-5}	3.0×10^{-5}
N. crassa (NAD)	*22*	5.5×10^{-3}	4.6×10^{-3}	1.7×10^{-2}	3.3×10^{-4}	5.5×10^{-4}		
N. crassa (NADP)	*22*	4.5×10^{-2}	5.3×10^{-3}	1.0×10^{-2}			5.0×10^{-5}	1.25×10^{-4}
S. cerevisiae (NADP)	*56*	1.0×10^{-2}	9.5×10^{-4}	1.1×10^{-2}			7.1×10^{-5}	8.5×10^{-5}
N. europaea	*35*	6.7×10^{-3}	4.3×10^{-3}	1.6×10^{-2}			7.9×10^{-6}	4.9×10^{-5}
T. novellus (NAD)	*33*	1.3×10^{-3}	6.6×10^{-4}	5.0×10^{-4}	1.9×10^{-4}	4.0×10^{-6}		
T. novellus (NADP)	*33*	3.6×10^{-2}	7.4×10^{-3}	7.5×10^{-3}			6.1×10^{-5}	7.7×10^{-5}
Clostridium SB_4	*36*	1.8×10^{-3}	6.5×10^{-4}	3.2×10^{-4}	1.0×10^{-5}	1.0×10^{-5}		
B. emersonii	*72*	1.4×10^{-3}	5.6×10^{-4}	4.0×10^{-2}	4.0×10^{-4}	3.3×10^{-5}		
Pea root	*46*	7.3×10^{-3}	3.3×10^{-3}	3.8×10^{-2}	6.5×10^{-4}	8.6×10^{-5}		
Glycine max.	*47*	8×10^{-3}	1.2×10^{-3}	9.4×10^{-3}	2.1×10^{-4}	1.5×10^{-5}		
B. flavum	*30*	10×10^{-2}	5.7×10^{-3}	3.1×10^{-3}			3.3×10^{-5}	2.7×10^{-5}

ADP activates the enzyme, attempts have been made both to clarify the mechanism and to determine the sites of these effects. Frieden (*334*) noted that the enzyme appears to possess a site capable of interaction with either ADP or GTP, but that interaction at this site is dependent upon prior binding of coenzyme. It was further reported that there is a single GTP site per subunit and that the 2′-OH group of the nucleotide is essential since the deoxy analog of GTP is only about 10% as inhibitory as GTP itself. It was noted that GTP could be classified as an uncompetitive inhibitor since the K_m for glutamate is increased and the V_{max} diminished. All of the vertebrate enzymes tested display GTP inhibition and all but frog liver GDH are activated by ADP. It appears that only those GDH's that lack coenzyme specificity are subject to regulation by purine nucleotide regulation although the NAD-GDH of *Neurospora* is inhibited by GMP. In this case, however, inhibition is competitive with coenzyme, and the inhibition constant of 50 μM is 500 times larger than that for the effect of GTP on bovine GDH.

The problem of identifying the residues involved at the sites of interaction with purine derivatives has been studied by various methods. Colman and Frieden (*276*) reported that acetylation of bovine GDH increases the dissociation constant for GTP and decreases that for ADP. Subsequently, it was found (*108*) that amino groups are the primary site of acetylation and that two types of amino groups could be modified, one involved in catalysis and the other in the polymerization site. Goldin and Frieden (*335*) later observed that reaction of the enzyme with trinitrobenzenesulfonate (TNBS) alters the allosteric properties at a level of modification causing only a small degree of inactivation. The precise nature of the effects depends upon which coenzyme is used: K_{GTP} is increased with NAD but is unaffected with NADP; however, with NADP, K_{ADP} is decreased by trinitrophenylation as is the maximal possible extent of activation by ADP. More extensive modification produces nondissociable aggregates. High concentrations of NAD(H) or NADH + GTP protect against these effects while other combinations of substrates and cofactors do not. It seems likely that these complex effects result from conformational changes produced by the bulky trinitrophenyl group. Coffee *et al.* (*336*) reported later that the sites of TNBS modification are Lys-425 and Lys-428 of the bovine enzyme with the latter reacting more rapidly than the former; reaction at one of these residues apparently precludes modification at the other within the same polypeptide

334. C. Frieden, *JBC* **240,** 2028 (1965).
335. B. R. Goldin and C. Frieden, *Biochemistry* **10,** 3527 (1971).
336. C. J. Coffee, R. A. Bradshaw, B. R. Goldin, and C. Frieden, *Biochemistry* **10,** 3516 (1971).

chain. Alternatively, three subunits of the oligomer are modified at Lys-425 and three at Lys-428. No reaction of Lys-126 was detected; indeed, after modification of one group per chain, activity in the absence of allosteric effectors is largely unaffected. Clark and Yielding (*337*) have noted that limited reaction with TNBS results in diminished sensitivity to GTP but is without effect on activation by ADP.

Price and Radda (*338*) found that *N*-acetylimidazole could acetylate up to six tyrosine residues without loss of activity or alteration of K_m for substrate; however, reaction of about one tyrosine per subunit results in desensitization toward GTP, but the response to ADP is not abolished even by extensive *O*-acetylation. Essentially the same results are observed upon nitration with tetranitromethane (TNM). Acetylation does not grossly alter the molecular weight, as measured by sedimentation velocity, or the conformation, as determined by ORD. The GTP site is not protected by NADH alone, but is partially protected (25–50%) by GTP and is at least 75% protected by inclusion of both GTP and NADH in the reaction mixture. Piszkiewicz *et al.* (*339*) confirmed these findings by modification with TNM. The reaction is biphasic with initial rapid formation of one residue of 3-nitrotyrosine per subunit. The primary site of reaction is tyrosine-406 in the linear sequence (*340*). Later (*328*) the same effect was obtained with chicken GDH; with both enzymes there is no influence on activation by ADP. Further, the pH optima of the enzymes are not influenced by the degree of nitration or the inhibition by GTP or activation by ADP (*328*).

DiPrisco (*341*) reported that under certain conditions reaction of GDH with 1-fluoro-2,4-dinitrobenzene led to desensitization to both GTP inhibition and ADP activation by modification of three residues of tyrosine per hexamer. Attempts to confirm these results in this laboratory (*342*) have been unsuccessful.

Malcolm and Radda (*294*) have reported that reaction of GDH with IASA results in desensitization toward GTP concurrent with alkylation of a single cysteine residue per subunit. In agreement with this, Nishida and Yielding (*343*) have found that reaction of the enzyme with methylmercuric iodide also causes allosteric desensitization. LéJohn and Jackson

337. C. E. Clark and K. L. Yielding, *ABB* **143,** 158 (1971).
338. N. C. Price and G. K. Radda, *BJ* **114,** 419 (1969).
339. D. Piszkiewicz, M. Landon, and E. L. Smith, *JBC* **246,** 1324 (1971).
340. Based on the initial tentative sequence, the reactive residue was identified as Tyr-412. In the present sequence it is Tyr-406.
341. G. DiPrisco, *Biochemistry* **10,** 585 (1971).
342. F. M. Veronese and E. L. Smith, unpublished data.
343. M. Nishida and K. L. Yielding, *ABB* **141,** 409 (1970).

(*344*) have shown that treatment of *B. emersonii* GDH with $HgCl_2$ desensitizes this enzyme toward its allosteric effectors as well. Although the effects of ADP and GTP are mutually exclusive, they are not bound at the same site. Fisher *et al.* (*123*) showed that prior binding of ADP prevents binding of GTP, suggesting that if distinct sites are involved they must be proximal.

Oxidation of glutamate can be stimulated by high levels of L-leucine (*345*); binding of leucine or ADP excludes binding the other, thus suggesting a common site for the two effectors. It was later observed (*346*) that the effects of ADP and leucine appear to be at least partially additive, thus suggestive of separate sites. It is now known that the effects of ADP are mimicked by a number of monocarboxylic amino acids, all of which appear to interact at the same site (*347*). Prough *et al.* (*348*) have reported that the ADP and leucine sites are distinct but probably lie close to one another on the enzyme surface since binding of one influences that of the other. It was also shown that, like ADP, leucine activates the enzyme by enhancing the dissociation of reduced coenzyme.

It appears that the primary effect of both GTP and ADP is upon binding of coenzyme. Yielding and Holt (*349*) reported that binding of NADH is weakened in the presence of ADP and slightly strengthened with GTP. At higher levels of GTP (3×10^{-4} *M*), the dissociation constant for NADPH is decreased approximately 20-fold (*350*) with both bovine and dogfish GDH. Dalziel and Egan (*351*) have shown that the effects of GTP and ADP on binding of the oxidized coenzymes are similar to those described above. Markau *et al.* (*352*) have suggested that the effect of ADP on coenzyme dissociation may result from competition for a portion of the coenzyme site.

Related to the effects of the purine nucleotides are the allosteric effects of NAD(H). Frog (*60*), bovine (*151*), and dogfish (*332a*) GDH's all display nonlinear kinetics with respect to NAD and NADH, and Frieden (*119*) has pointed out that the data indicate an activating, albeit nonactive site for NAD, while binding of reduced coenzyme at the second site is inhibitory. An exception is the GDH of *Drosophila,* which appears to

344. H. B. LéJohn and S. Jackson, *BBRC* **33,** 613 (1968).
345. K. L. Yielding and G. M. Tomkins, *Proc. Nat. Acad. Sci. U. S.* **47,** 983 (1961).
346. A. Hershko and S. H. Kindler, *BJ* **101,** 661 (1966).
347. R. A. Prough and H. F. Fisher, *Biochemistry* **11,** 2479 (1972).
348. R. A. Prough, J. M. Culver, and H. F. Fisher, *JBC* **248,** 8528 (1973).
349. K. L. Yielding and B. B. Holt, *JBC* **242,** 1079 (1967).
350. D. A. Malencik and S. R. Anderson, *Biochemistry* **11,** 2766 (1972).
351. K. Dalziel and R. R. Egan, *BJ* **126,** 975 (1972).
352. K. Markau, J. Schneider, and H. Sund, *FEBS (Fed. Eur. Biochem. Soc.) Lett.* **24,** 32 (1972).

be activated by NADH (*353*). Pantaloni and Dessen (*101*) have shown that NADP(H) is not bound at the regulatory site, presumably because of the presence of the 2′-phosphate. These workers suggested that ADP and GTP are also bound to the coenzyme regulatory site, and this has been confirmed by Jallon and Iwatsubo (*354*). Koberstein *et al.* (*355*) have shown that α-NADH is bound only at the coenzyme regulatory site and can be displaced from this site by ADP. Recently, it has been reported (*356*) that NADPH is also bound to the regulatory site of bovine GDH although its binding is also about 10-fold weaker than that of NADH. Coffee *et al.* (*336*) have noted that trinitrophenylation of Lys-425 or Lys-428 abolishes the inhibitory effects of NADH in addition to its effects upon GTP binding.

The interrelationships among the various sites described above have been demonstrated by the studies of Cross and Fisher (*309*) and recently reviewed by Fisher (*19*). In agreement with earlier conclusions, ADP and GTP are bound at distinct but mutually exclusive sites; the ADP and leucine sites are also different although the effects of these ligands are similar. Prough *et al.* (*348*) have suggested that these two sites are probably close to one another and near the amide subsite portion of the coenzyme site on the basis of the observation that ADP has no effect on the rate of oxidative deamination when the 3-acetylpyridine analog of NAD is used as cofactor. Since ADP interferes with binding of coenzymes at the nonactive coenzyme sites, presumably the binding of coenzyme at this site involves the adenosine phosphate moiety of the NAD(H).

Since each of the binding sites discussed above probably consists of some sort of "pocket" on the enzyme surface, it is hardly surprising that so many different types of residues have been found to be involved at the sites of ADP and GTP binding, particularly since the effects undoubtedly result from conformational changes in the enzyme. The above studies all clearly demonstrate that separate sites must exist at least for the GTP and ADP effects. The exact locations will probably not be known until the three-dimensional structure of at least one of the vertebrate enzymes has been solved.

A number of other substances has been reported to influence the activity of bovine GDH. Wolff (*96*) observed that the enzyme is markedly inhibited by thyroxine and concluded that the site of interaction is the allosteric site for coenzyme. Such a mode of inhibition would be similar

353. P. A. Bond and J. H. Sang, *J. Insect Physiol.* **14**, 341 (1968).
354. J.-M. Jallon and M. Iwatsubo, *BBRC* **45**, 964 (1971).
355. R. Koberstein, J. Krause, and H. Sund, *Eur. J. Biochem.* **40**, 543 (1973).
356. J. Krause, M. Bühner, and H. Sund, *Eur. J. Biochem.* (in press).

to that described earlier for certain estrogenic steroids (*357*). Most effective of this class of inhibitors is diethylstilbestrol (DES). Earlier, it was suggested that binding of DES may involve enzymic sulfhydryl groups (*358*), and more recently, a carboxyl group (*359*). It should be noted that the DES effects are observed at levels of DES much higher than the levels of this substance normally effective *in vivo* as a hormone analog.

Acetylcholine, serotonin, norepinephrine, and tryptamine are noncompetitive inhibitors whereas epinephrine and dihydroxyphenylalanine are activators of GDH (*360*). Mitochondrial lipids have also been reported to be inhibitory, among them phosphatidylcholine, phosphatidylethanolamine, cardiolipin, and phosphatidylserine (*361*).

The physiological roles of ADP and GTP as regulators of the activity of vertebrate GDH's seem obvious, the enzyme being inhibited when the energy supply is high and activated when it is low. This would suggest that the primary role of the enzyme in animal tissues is to supply energy by oxidation of glutamate thus furnishing α-ketoglutarate to the tricarboxylic acid cycle despite the fact that the equilibrium for the reaction greatly favors synthesis of glutamate (Section I,A). Thus, one would suppose that, at high energy charge, i.e., high levels of ATP and GTP, little glutamate would be supplied by this route for aminotransferase reactions.

The reversible inhibition by pyridoxal 5′-phosphate could provide another possible type of physiological control since high levels of this cofactor would be available when aminotransferases are inactive. Whether PLP actually plays such a role is unknown.

VIII. Concluding Remarks

Regardless of the precise roles that ADP and GTP play in the regulation of glutamate dehydrogenases of animal origin, it is evident that the regulatory behavior of these enzymes is of considerable interest. The hexameric structure of these enzymes is itself not too common and the fact that each polypeptide chain has specific and distinct binding sites for coenzymes, substrates, ADP, and GTP represents an interesting challenge

357. K. L. Yielding and G. M. Tomkins, *Proc. Nat. Acad. Sci. U. S.* **46,** 1483 (1960).
358. M. Bitensky, K. L. Yielding, and G. M. Tomkins, *JBC* **240,** 663, 668 (1965).
359. J. Kallos and K. P. Shaw, *Proc. Nat. Acad. Sci. U. S.* **68,** 916 (1971).
360. G. Kaur and M. S. Kanungo, *Indian J. Biochem.* **7,** 170 (1970).
361. J. H. Juillard and D. C. Gautheron, *FEBS (Fed. Eur. Biochem. Soc.) Lett.* **25,** 343 (1972).

in solving the interrelated problems of structure, function, and regulatory behavior. The large number of other types of compounds which can bind to GDH and influence its enzymic activity also pose interesting problems, although the physiological role of many of these substances in regulation is obscure at best.

Acknowledgments

Most of the research reported from the authors' laboratory has been supported by grants GM 11061 and GM 10935 from the National Institute of General Medical Sciences.

6

Malate Dehydrogenases

LEONARD J. BANASZAK • RALPH A. BRADSHAW

I. Introduction

The oxidation of L-malate in most living organisms is catalyzed by two distinct types of pyridine nucleotide–dependent enzymes. In one case the principal product is oxaloacetate, while in the other it is pyruvate and CO_2. The enzymes of the malate-oxaloacetate class, which utilize NAD^+, have been referred to as "simple" dehydrogenases, while enzymes of the malate-pyruvate type, which, in contrast, use $NADP^+$, have been designated "decarboxylating" dehydrogenases and are commonly known as "malic" enzymes (*1*).

The "simple" malate dehydrogenases occur in virtually all eukaryotic cells in at least two unique forms identified as mitochondrial (m-MDH) and soluble or cytoplasmic (s-MDH) according to their cellular location (*2*). In certain plant tissues, a third form of the enzyme has been identi-

1. E. Kun, "The Enzymes," 2nd ed., Vol. 7, p. 149, 1963.
2. A. Delbrück, H. Schimassek, K. Bartsch, and T. Bücher, *Biochem. Z.* **331**, 297 (1959).

fied in the microbody or glyoxysome fraction (*3,4*). A number of physical and kinetic parameters indicate that the microbody form is apparently unique, although definitive structural data to establish its identity are lacking. In addition, there have been numerous reports suggesting that both the cytoplasmic and mitochondrial enzymes can occur in multiple subforms. These observations have generally been based on electrophoretic criteria and appear to result either from genetic differences or stuctural alterations arising either intracellularly or during preparative manipulations. However, unlike the isozymes defined by cellular or organelle origin, the subforms do not usually show distinguishing kinetic or mechanistic characteristics and their physiological importance remains uncertain.

The soluble isozyme is generally considered to take part in the cytoplasmic side of the "malate shuttle," providing a means of transporting NADH equivalents, in the form of malate, across the mitochondrial membrane. The mitochondrial enzyme, in addition to its role in the other half of the malate shuttle, is also a necessary component of the tricarboxylic acid cycle. The microbody malate dehydrogenase found in some plants appears to function in the glyoxylate cycle (*5*) or possibly in photorespiration (*6*).

This review will deal with the NAD^+-dependent malate dehydrogenases of animal tissue for which the most structural, mechanistic, and regulatory data are available. Particular attention will be directed toward the similarities and differences characterizing the cytoplasmic and mitochondrial enzymes.

II. NAD^+-Dependent Malate Dehydrogenases

A. Distribution and Preparation

Malate dehydrogenase has been identified in a wide variety of sources and purified to homogeneity from a number of them. The majority of studies have been carried out with enzyme isolated from either pig or beef heart. The first apparently pure preparation was obtained by Wolfe and Neilands (*7*). It differs from previously reported procedures (*8,9*)

3. R. A. Curry and I. P. Ting, *ABB* **158,** 213 (1973).
4. V. Rocha and I. P. Ting, *ABB* **147,** 114 (1971).
5. R. W. Breidenback, *Ann. N. Y. Acad. Sci.* **168,** 342 (1969).
6. R. K. Yamozaki and N. E. Tolbert, *BBA* **178,** 11 (1969).
7. R. G. Wolfe and J. B. Neilands, *JBC* **221,** 61 (1956).
8. S. Ochoa, "Methods in Enzymology," Vol. 1, p. 735, 1955.
9. F. B. Straub, *Hoppe-Seyler's Z. Physiol. Chem.* **275,** 63 (1942).

which used ammonium sulfate and ethanol fractionations of acetone powders, primarily in the addition of a zinc-ethanol precipitation step. From subsequent studies (*10–12*) that defined the cellular isozymes, the product of this procedure was identified as the mitochondrial form. The first homogeneous preparation of pig heart cytoplasmic malate dehydrogenase was obtained by Thorne and Cooper (*12*) who utilized ammonium sulfate precipitation, DEAE-cellulose chromatography, and starch column electrophoresis. The material obtained readily crystallized from concentrated solutions of ammonium sulfate. Gerding and Wolfe (*13*) have reported a large-scale purification of the cytoplasmic enzyme which uses negative batch adsorption with both CM-cellulose (to remove the mitochondrial isozyme) and DEAE-cellulose followed by positive adsorption on DEAE-cellulose, ammonium sulfate precipitation, and hydroxyapatite chromatography. A somewhat different preparative-scale method has been reported recently by Glatthaar *et al.* (*14*); it has the advantage that both malate dehydrogenase isozymes as well as L-3-hydroxyacyl-CoA dehydrogenase and the isozymes of aspartate aminotransferase can be isolated from the same starting material. This procedure uses an initial fractionation of the ammonium sulfate precipitate by CM-cellulose chromatography. Cytoplasmic malate dehydrogenase is further purified on columns of DEAE-Sephadex and Sephadex G-100. The mitochondrial enzyme is isolated from a different portion of the initial CM-cellulose fractionation by gel filtration and another CM-cellulose chromatography step. Somewhat different methods have been reported by Leskovac (*15*), who utilizes calcium phosphate gel adsorption as well as ammonium sulfate precipitation, gel filtration, and DEAE-Sephadex chromatography and can obtain both malate dehydrogenase isozymes and lactate dehydrogenase, and by Gregory *et al.* (*16*), who obtain mitochondrial enzyme from acetone powders by ammonium sulfate precipitation and chromatography on Bio-Rex 70 and DEAE-cellulose.

Purification of the malate dehydrogenase isozymes from beef heart have utilized similar procedures (*17–22*) to those already described for

10. A. Delbrück, E. Zebe, and T. Bücher, *Biochem. Z.* **331,** 273 (1959).
11. G. S. Christie and J. D. Judah, *Proc. Roy. Soc., Ser. B* **141,** 420 (1953).
12. C. J. R. Thorne and P. M. Cooper, *BBA* **81,** 397 (1963).
13. R. K. Gerding and R. G. Wolfe, *JBC* **244,** 1164 (1969).
14. B. E. Glatthaar, G. R. Barbarash, B. E. Noyes, L. J. Banaszak, and R. A. Bradshaw, *Anal. Biochem.* **57,** 432 (1974).
15. V. Leskovac, *Glas. Hem. Drus., Beograd* **34,** 509 (1969).
16. E. M. Gregory, F. J. Yost, Jr., M. S. Rohrbach, and J. H. Harrison, *JBC* **246,** 5491 (1971).
17. S. Englard, L. Siegel, and H. H. Breiger, *BBRC* **3,** 323 (1960).
18. L. Siegel and S. Englard, *BBRC* **3,** 253 (1960).

the pig heart enzymes. As with the pig heart cytoplasmic form, crystalline s-MDH is readily obtained from the final step by treatment with ammonium sulfate. A somewhat simplified procedure for the preparation of beef heart s-MDH, which uses a tandem arrangement of CM- and DEAE-Sephadex columns, has been reported by Guha *et al.* (*23*). Thus the separation of the two isozymes of MDH can be achieved either by careful separation of the intact mitochondria in the initial extraction step or by anion or cation exchange chromatography of total homogenates. The latter methods seem to be preferable and this is most clearly demonstrated by the work of Thorne (*24*) in the fractionation of the rat liver isozymes. The ion exchange methods and electrophoretic separations (*25*) depend on the fact that from most sources s-MDH has an acidic pI and m-MDH has a basic pI value. However, it should be noted that in tuna heart, the charge properties for the isozymes are reversed with the mitochondrial enzyme being more acidic (*26*). In this case, the isozymes were separated by a salt gradient on a column of DEAE-cellulose.

Homogeneous preparations of malate dehydrogenase have also been isolated from chicken heart (*27*), horse heart (*28*), *Drosophila virilis* (*29*), and *Neurospora crassa* (*30,31*). In the last case, it was reported by Munkres and Richards (*30*) that this eukaryote contained only a single isozyme. However, subsequent studies by Kitto *et al.* (*31*) established the presence of the two forms and identified the form isolated by Munkres and Richards as that associated with the mitochondria. Both malate dehydrogenase isozymes have also been found in the fungus *Phycomyces blakesleeanus* (*32*).

Malate dehydrogenase has also been isolated from several prokaryotes. Yoshida (*33*) purified the enzyme from *Bacillus subtilis* utilizing ammo-

19. L. Siegel and S. Englard, *BBA* **54,** 67 (1961).
20. L. Siegel and S. Englard, *BBA* **64,** 101 (1962).
21. S. Englard and H. H. Breiger, *BBA* **56,** 57 (1962).
22. F. C. Grimm and D. G. Doherty, *JBC* **236,** 1980 (1961).
23. A. Guha, S. Englard, and I. Listowsky, *JBC* **243,** 609 (1968).
24. C. J. R. Thorne, *BBA* **42, 175** (1960).
25. C. J. R. Thorne, L. I. Grossman, and N. O. Kaplan, *BBA* **73,** 193 (1963).
26. G. B. Kitto and R. G. Lewis, *BBA* **139,** 1 (1967).
27. G. B. Kitto and N. O. Kaplan, *Biochemistry* **5,** 3966 (1966).
28. C. J. R. Thorne, *BBA* **59,** 624 (1962).
29. M. S. McReynolds and G. B. Kitto, *BBA* **198,** 165 (1970).
30. K. D. Munkres and F. M. Richards, *ABB* **109,** 466 (1965).
31. G. B. Kitto, M. E. Kottke, L. H. Bertland, W. H. Murphey, and N. O. Kaplan, *ABB* **121,** 224 (1967).
32. G. Sulebele and E. Silverstein, *ABB* **133,** 425 (1969).
33. A. Yoshida, *JBC* **240,** 1113 (1965).

nium sulfate precipitation, and calcium phosphate gel, DEAE-cellulose and ECTEOLA-cellulose chromatography. Murphey *et al.* (*34*) also isolated the enzyme from this source as well as from *Bacillus stearothermophilus* and *Escherichia coli* using combinations of gel filtration and ion-exchange chromatography. In each case, only a single form of the enzyme was found. This observation was confirmed in *E. coli* by means of genetic mutants (*35*). Based on these observations, it is probable that all eukaryotes contain at least two malate dehydrogenase isozymes, one associated with the mitochondria and one found in the cytoplasm, while prokaryotes contain only a single form.

B. Molecular Properties

1. *Molecular Weight*

The reported values for the molecular weight of the malate dehydrogenase isozymes from mammalian sources have varied over the range of 15,000–74,000 including many intermediate values. However, more recent reports have placed the values for both enzymes consistently in the 60,000–70,000 range. Murphey *et al.* (*36*), using calibrated columns of Sephadex G-100, examined a large number of malate dehydrogenase preparations, both as homogeneous and crude or partially purified preparations, including pig s- and m-, chicken s- and m-, tuna s- and m-, and *N. crassa* m-, and *E. coli*. In each case, identical elution positions were observed corresponding to a molecular weight of 67,000. This study eliminates the possibility that the mitochondrial and cytoplasmic isozymes vary markedly in size. In addition, X-ray diffraction studies (*37*) and amino acid sequence analysis (*38*) have clearly established that the cytoplasmic form of pig heart malate dehydrogenase to be about 72,000. Sequence (*39*) and compositional analyses (*40*) of the mitochondrial en-

34. W. H. Murphey, C. Barnaby, F. J. Lin, and N. O. Kaplan, *JBC* **242,** 1548 (1967).
35. J. B. Courtright and U. Henning, *J. Bacteriol.* **102,** 722 (1970).
36. W. H. Murphey, G. B. Kitto, J. Everse, and N. O. Kaplan, *Biochemistry* **6,** 603 (1967).
37. L. J. Banaszak, D. Tsernoglou, and M. Wade, *in* "Probes of Structure and Function of Macromolecules and Membranes" (B. Chance, T. Yonetani, and A. S. Mildvan, eds.), Vol. 2, p. 71. Academic Press, New York, 1971.
38. B. E. Glatthaar, M. Wade, G. R. Barbarash, and R. A. Bradshaw, unpublished experiments.
39. M. Sutton, J. S. Garavelli, B. E. Glatthaar, and R. A. Bradshaw, unpublished experiments.
40. B. E. Noyes, B. E. Glatthaar, J. S. Garavelli, and R. A. Bradshaw, *Proc. Nat. Acad. Sci. U. S.* **71,** 1334 (1974).

zyme from the same source suggest a slightly lower value of about 68,000. Several independent investigations with this and other sources support these values.

Murphey *et al.* (*36*) also observed that the enzyme isolated from certain *Eubacteriales* possessed significantly higher molecular weights of about 117,000. However, under denaturing conditions, these enzymes were reduced in size to units identical to that obtained from the dissociation of malate dehydrogenase obtained from eukaryotic sources. Thus, they concluded that these higher molecular weight enzymes contained more of the same size of subunits found in the malate dehydrogenases possessing molecular weights of about 67,000. The value for *B. subtilis* malate dehydrogenase obtained by Yoshida (*33*) was consistent with that observed by Murphey *et al.* (*36*).

Consiglio *et al.* (*41*) have reported that pig heart mitochondria also contain a larger malate dehydrogenase with a sedimentation coefficient of 9 S, as compared to the 67,000 molecular weight form which has a sedimentation coefficient of about 5 S. This protein has a molecular weight of 138,000 and can be dissociated to two proteins of about equal molecular weight (5 S) by the action of thyroxine (*42*). When resolved on polyacrylamide gel electrophoresis, only one band possesses malate dehydrogenase activity. Amino acid analysis indicates that the 9 S form is not a homogeneous dimer of 67,000 molecular weight units. The simplest interpretation of these data is that the normal 5 S protein combines with another protein of about the same size of undefined function either *in vivo* or *in vitro* to produce the 9 S complex. The physiological importance of this entity remains obscure.

2. *Subunit Structure*

Several experimental observations suggest that malate dehydrogenases are composed of identical subunits of molecular weight 33,000–35,000. The cytoplasmic enzymes from both pig and beef heart have been examined by sedimentation analyses in the ultracentrifuge under denaturing conditions (*43,44*) and by tryptic peptide mapping (*13,44*). In addition, only carboxyl terminal alanine was found in the beef heart enzyme (*43*). The amino terminus of the supernatant enzyme is apparently blocked (*13,44*). By analogy with several other cytoplasmic enzymes, it is probable that the polypeptide chain has an N-acetyl group in this position. Crystallo-

41. E. Consiglio, S. Varrone, and I. Covelli, *Eur. J. Biochem.* **17,** 408 (1970).
42. I. Covelli, E. Consiglio, and S. Varrone, *BBA* **184,** 678 (1969).
43. C. Wolfenstein, S. Englard, and I. Listowsky, *JBC* **244,** 6415 (1969).
44. M. J. Wade, Ph.D. Thesis, Washington University, St. Louis, Missouri, 1971.

graphic analysis of the cytoplasmic enzyme from pig heart, completed to a resolution of 3.0 Å (*45*), has confirmed the chemical analysis and in combination with partial sequence data has established that only minor sequence differences at best could distinguish the two polypeptide chains.

The mitochondrial enzyme from pig heart has been shown by similar analyses to possess the same type of structure as the cytoplasmic enzyme (*40,46*). However, in contrast to the cytoplasmic form, the amino terminus is not blocked, and the sequence of the first 28 residues has been determined to be (*39*):

1 10
Ala-Lys-Val-Ala-Val-Leu-Gly-Ala-Ser-Gly-Gly-Ile-Gly-Gln-Pro-Leu-Ser-Leu-Leu-
20 28
Leu-Lys-Asn-Ser-Pro-Leu-Val-Ser-Arg-

These data also provide a clear demonstration of the presence of only a single polypeptide sequence in m-MDH.

Only one report has suggested that malate dehydrogenase contains nonidentical subunits. Mann and Vestling (*47*), by studying rat liver m-malate dehydrogenase, isolated two forms of this enzyme, designated A and C, and obtained apparent differences in the tryptic peptide maps. However, in light of the data on the subforms of m-malate dehydrogenase from other species (described below), and the low yield and impurity of the forms isolated in this study, these results are open to some question. Thus, in the cases examined for malate dehydrogenase from animal sources, with the exception noted, the data are consistent only with a dimeric structure composed of two identical or very similar polypeptide chains. In the absence of disulfide bonds, these subunits must be associated by noncovalent bonds. The chemical evidence confirms this conclusion.

3. *Amino Acid Composition*

The amino acid composition of the homogeneous preparations of both the cytoplasmic and mitochondrial isozymes from a number of sources has been determined. For the most part, little significant variation in composition is found in the independent determinations of enzyme isolated from the same source. A summary of a representative group of

45. E. Hill, D. Tsernoglou, L. Webb, and L. J. Banaszak, *JMB* **72,** 577 (1972).
46. T. Devenyi, S. J. Rogers, and R. G. Wolfe, *Nature* (*London*) **210,** 489 (1966).
47. K. G. Mann and C. S. Vestling, *Biochemistry* **9,** 3020 (1970).

amino acid compositions is listed in Table I (*21,26–28,33,39,40,44,48,49*). In each case the composition is given for the subunit, the minimum unique unit in each enzyme. In addition to the compositions listed in Table I, other analyses for pig heart mitochondrial (*50*) and supernatant (*13*), beef heart mitochondrial (*20*) and *B. subtilis* (*33*) malate dehydrogenase have been reported.

Not unexpectedly, there is a marked similarity in the composition of the mitochondrial and supernatant isozymes when compared with the same isozyme from other species, although in the latter case, the two salmon enzymes, which appear to be genetic variants, differ somewhat more than do the malate dehydrogenases from pig, beef, and chicken sources. As noted previously (*40*), there is also quite a pronounced similarity between the profiles presented by the two cellular isozymes. The most notable difference in the amino acid composition is the absence of tryptophan in mitochondrial malate dehydrogenase.

The amino acid composition of *B. subtilis* and *E. coli* enzymes, calculated for a minimum subunit of 33,500, show a greater divergence. Whereas the *E. coli* enzyme is not unsimilar to both the mitochondrial and cytoplasmic isozymes from eukaryotic sources, the *B. subtilis* protein is decidedly different. Interestingly, the *E. coli* enzyme possesses a molecular weight of about 67,000 as compared to the 117,000 of the *B. subtilis* protein (*37*). Primary structure analyses will be required to ascertain the extent to which either of these prokaryotic enzymes are sequentially related to either of the eukaryotic forms of the enzyme.

4. *Nature of the Subforms*

Both of the principal cellular isozymes of animal malate dehydrogenases have been shown to exist in multiple subforms by electrophoretic techniques. In the past, these forms—the origin of which has been attributed to conformational differences (*51*), combination of similar, but nonidentical polypeptide chains (*52*), and proteolytic degradation (*53*)—have also been called isozymes. However, this terminology is inappropriate since recent studies (*14,54*) have indicated that the probable origin,

48. E. Heyde and S. Ainsworth, *BJ* **109,** 663 (1968).
49. G. S. Bailey, A. C. Wilson, J. E. Halver, and C. L. Johnson, *JBC* **245,** 5927 (1970).
50. B. H. Anderton, *Eur. J. Biochem.* **15,** 562 (1970).
51. G. B. Kitto, P. M. Wassarman, and N. O. Kaplan, *Proc. Nat. Acad. Sci. U. S.* **56,** 578 (1966).
52. R. J. Kulick and F. W. Barnes, *BBA* **167,** 1 (1968).
53. M. Cassman and R. King, *BBA* **257,** 143 (1972).
54. B. E. Glatthaar, G. R. Barbarash, J. S. Garavelli, L. J. Banaszak, and R. A. Bradshaw, in preparation.

TABLE I

Amino Acid Composition of the Cytoplasmic and Mitochondrial Malate Dehydrogenases from Animal and Bacterial Sources

Amino acid	Pig heart m-	Beef heart m-	Horse heart m-	Chicken heart m-	Tuna heart m-	Pig heart s-	Beef heart s-	Chicken heart s-	Salmon AA	s BB	*B. subtilis*	*E. coli*
Lysine	26	24	26	26	20	31	35	28	24	24	19	20
Histidine	5	5	6	6	6	4	4	6	8	7	1	3
Arginine	8	8	8	10	9	10	10	10	14	14	12	9
Aspartic acid	25	23	21	26	30	39	36	34	39	35	27	23
Threonine	21	19	21	22	20	16	12	15	10	14	21	17
Serine	18	16	18	21	20	22	16	18	25	24	17	16
Glutamic acid	25	24	25	30	25	27	29	30	26	27	32	33
Proline	23	20	22	20	21	12	11	14	12	14	13	14
Glycine	29	25	27	30	29	23	22	30	24	23	28	33
Alanine	33	30	33	33	36	32	29	30	36	42	23	32
$\frac{1}{2}$-Cystine	7	8	7	7	7	5	5	4	6	8	0	3
Valine	27	26	27	24	32	26	25	26	28	29	31	30
Methionine	6	6	7	6	6	8	8	7	10	6	4	4
Isoleucine	21	21	18	20	16	19	18	21	19	18	27	16
Leucine	28	26	30	28	30	32	29	30	32	28	34	35
Tyrosine	5	6	5	4	4	8	6	8	7	4	11	5
Phenylalanine	11	11	11	15	10	11	10	12	10	11	8	12
Tryptophan	0	0	0	0	0	5	6	6	4	3	0	2
Total	318	303	312	328	321	330	311	329	334	331	308	307
Reference	*40*	*48*	*28*	*27*	*26*	*44*	*21*	*27*	*49*		*33*	*39*

at least in pig heart m-malate dehydrogenase, is because of deamidation of asparaginyl and/or glutaminyl residues. Consequently, they have been designated as subforms in this review.

Initial characterization of the subforms of pig heart m-malate dehydrogenase by Thorne *et al.* (*25*) and Kitto *et al.* (55) indicated that the isolated enzymes possessed similar catalytic properties, resistance to thermal inactivation, and indistinguishable amino acid compositions. It was also established that they did not result from aggregation (*55*). However, differences in reactivity toward iodine and *p*-hydroxymercuribenzoate and in their optical rotatory dispersion spectra were noted (*51*). Consequently, it was proposed that the subforms were distinguished by conformational differences rather than differences in the covalent structure. This conclusion was not confirmed by Schechter and Epstein (*56*), who were unable to demonstrate interconversion of the subforms following reversible denaturation with guanidine-HCl. However, using chicken heart m-MDH, Kitto *et al.* (*57*) were able to effect changes in electrophoretic mobility and, hence, presumably interconversion between the subforms.

As a result of improved isolation procedures, Glatthaar *et al.* (*14*) have isolated preparative amounts of two of the subforms corresponding to the enzymes designated A and C (*51*). Under appropriate conditions, most of the enzyme can be recovered as subform A, the most basic band as judged electrophoretically. It was established that conversion to the more acidic bands can accompany the purification, particularly the formation of subform C from B (*54*). Hybridization experiments have established that form A and form C are homogeneous dimers of the XX and YY type and that band B is an XY hybrid. Conversion of band A into bands B and C can also be effected by an as-yet unidentified protein component of human or rabbit serum. That this conversion is not proteolytic in nature has been established by amino and carboxyl terminal analyses (*54*). These experiments established that the subforms B and C as well as the even more acidic ones (designated D and E) are formed at the expense of the most basic one, subform A, most probably by deamidation although other forms of chemical modification have not been eliminated. However, B and C cannot be converted to bands B or A which precludes the "conformer" explanation. In light of these experiments and in the absence of any evidence that these subforms are physiologically significant, they are best considered as preparation artifacts at present. Although definitive data are lacking, it is probable that a similar explanation applies to the subforms observed for the cytoplasmic enzyme.

55. G. B. Kitto, P. M. Wassarman, J. Michjeda, and N. O. Kaplan, *BBRC* **22,** 75 (1966).

56. A. N. Schechter and C. J. Epstein, *Science* **159,** 997 (1968).

57. G. B. Kitto, F. E. Stolzenbach, and N. O. Kaplan, *BBRC* **38,** 31 (1970).

C. Structure of the NAD⁺-Dependent Cytoplasmic Malate Dehydrogenase

Of the two forms of the NAD⁺-dependent malate dehydrogenases, detailed molecular structural data are available for only one of them, s-MDH. Single crystal X-ray diffraction studies on s-MDH from porcine heart have described most of the molecular structure of this enzyme as well as the conformation of the bound coenzyme (*45,58*). Because of the close structural homology of s-MDH to dogfish LDH (*45*), a description of the structure will be limited to important features which relate mainly to the malate dehydrogenases. Chapter 2, this volume, describes in detail the structural homology among other dehydrogenases and related enzymes.

1. *Crystal Structure of Pig Heart s-MDH*

The first characterization of crystals of s-MDH included a survey of the crystalline enzyme as obtained from heart muscle from several different mammals (*59*). The X-ray reflections to about 5.0 Å resolution were recorded from three different reciprocal lattice zones. The similarities in the X-ray data from s-MDH crystals obtained from pig, horse, beef, and lamb indicate that at least to low resolution the structure of these enzymes are basically the same. Thus the structural results described for the porcine enzyme, at least to the first approximation, should apply to s-MDH from other sources.

The basic structure of porcine s-MDH is shown schematically in Fig. 1. The reader is referred to the original papers for detailed stereo drawings of the polypeptide chain folding (*45,58*). The structure was originally determined at 3.0 Å resolution and then at 2.5 Å resolution. Although the electron density map at 2.5 Å resolution has been interpreted in terms of a Kendrew–Watson model, the final detailed structure has not yet been correlated with the amino acid sequence. Model building studies on just the polypeptide conformation have not produced any significant changes from the original 3.0 Å model (*45*).

The nomenclature used to describe the structure of s-MDH, as shown in Fig. 1, applies as well to dogfish LDH (*45,60*) and is the same as that used in the homology chapter (Chapter 2, this volume). Extended polypeptide chain is given the prefix β and helical regions are

58. L. E. Webb, E. Hill, and L. J. Banaszak, *Biochemistry* **12,** 5101 (1973).
59. L. J. Banaszak, *JMB* **22,** 389 (1966).
60. M. J. Adams, G. C. Ford, R. Koekoek, P. J. Levitz, A. McPherson, Jr., M. G. Rossmann, I. E. Smiley, R. W. Schevitz, and A. J. Wonacott, *Nature* (*London*) **227,** 1098 (1970).

labeled α. No distinction is made between alpha and other forms of protein helices. Each segment of secondary structure is then labeled with a letter depending on where it occurs in the chain, starting at the amino terminal end. Turns connecting defined segments of secondary structure are given a combined notation such as βAαB.

The two identical subunits of s-MDH are related by approximate twofold symmetry. The right hand side of the subunit shown in Fig. 1 interacts with the other subunit in the dimer. This can be seen in Fig. 1 if one imagines a rotation of 180° around the molecular dyad. Therefore, the contact region between the two subunits is comprised of three pairs of well-defined helices, αB, αC, and α3G. Although it is not apparent in

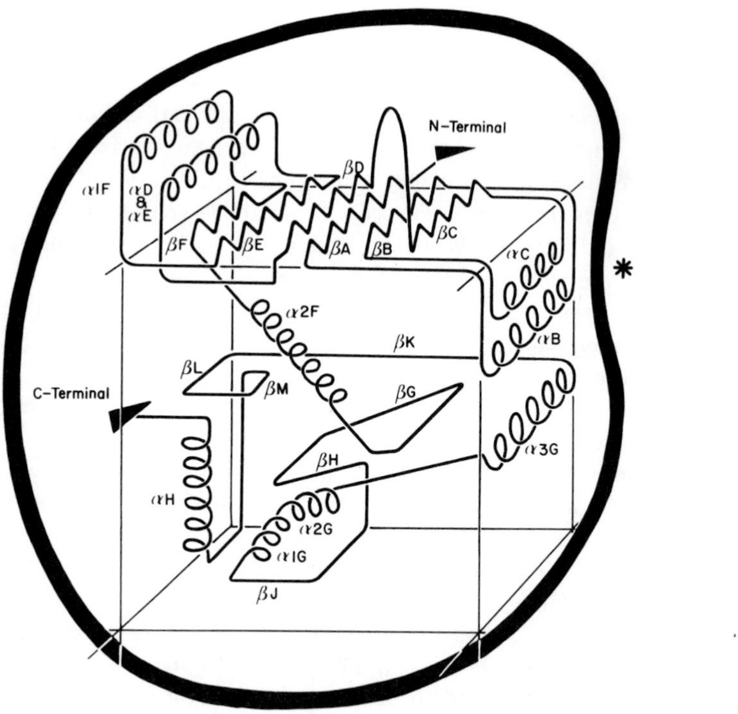

FIG. 1. Schematic drawing of the polypeptide conformation of one subunit of pig heart s-MDH. The polypeptide conformation of one subunit of s-MDH from pig heart is shown in a highly schematic way. Segments of secondary structure are divided into two types. Extended polypeptide chain is labeled with a prefix β and helical regions with an α. The asterisk marks the approximate position of the molecular twofold symmetry axis (dyad). The dyad would be a line perpendicular to the plane of the drawing passing through the asterisk. The other subunit can be pictured by an imaginary rotation (180°) of the schematic drawing around the dyad (*).

Fig. 1, the only other part of the polypeptide chain involved in the subunit:subunit contact is the α2Gα3G turn which comes very close to the αC helix of the paired subunit. As has been the case with other polymeric enzymes, the region closest to the symmetry axis is devoid of atoms giving the appearance of a small hole through the center of the dimer (*61*).

The question of whether or not there is a structural basis for a stable (and active) monomer cannot be answered with the molecular model. There are no residues of one subunit which contribute to what is currently thought of as the active site of the other subunit. Furthermore, most of the segments of chain comprising the subunit interface have many more nonbonding interactions with atoms in their own polypeptide chain than with the paired subunit. The exception is the helix αC. To a first approximation, this segment appears to have at least as many contacts with the paired subunit as it has with portions of its own subunit. Thus, upon dissociation, one could speculate that it might take up an altered orientation with respect to the remainder of its polypeptide chain. Should this occur, a conformational difference between a polypeptide chain in the dimeric form and one in the monomeric form would probably exist.

An equally difficult question to answer at this stage of the structural study is whether or not cooperative changes occur simultaneously with binding of the coenzyme. In hemoglobin, for example, the binding of oxygen triggers small conformational changes in one polypeptide chain and this results in a subunit rearrangement (*61,62*). Evidence for a slight asymmetry in the binding of NAD^+ by s-MDH has not yet been confirmed by the demonstration of the conformational differences between the two polypeptide chains of the crystalline dimer (*63,64*), but some subtle differences are thought to exist (*58*).

When s-MDH is viewed down the dyad, as shown in Fig. 1, the first half of the polypeptide chain (βA to βF) is folded into a reasonably defined region. The second half of the molecule (α2F to αH) comprises the remaining part of the molecule. It should be emphasized that the distribution of atoms is continuous; there is no visible bilobal structure as has been seen in other enzymes (*65*). The first six segments of the extended chain (βA to βF) form a twisted region of parallel β sheet structure. The schematic drawing in Fig. 1 does not show this twist. The amino terminal part of the sheet (βA, βB, and βC) is on the surface of the

61. M. F. Perutz, *JMB* **13,** 646 (1965).
62. M. F. Perutz, *Nature* (*London*) **228,** 726 (1970).
63. B. E. Glatthaar, L. J. Banaszak, and R. A. Bradshaw, *BBRC* **46,** 757 (1972).
64. D. Tsernoglou, E. Hill, and L. J. Banaszak, *Cold Spring Harbor Symp. Quant. Biol.* **36,** 171 (1971).
65. C. C. F. Blake, P. R. Evans, and R. K. Scopes, *Nature* (*London*) **235,** 195 (1972).

molecule. The remaining three strands (βD, βE, and βF) are almost totally buried within the subunit.

The second part of the structure from α2F to αH consists mainly of helical regions and antiparallel segments of extended chain. These segments with their interconnecting turns are often twisted or bent giving the chain the appearance of a deformed hairpin. Residues from this part of the molecule are near the substrate binding region and the nicotinamide end of the bound coenzyme. The C-terminal end segment, αH, is the longest alpha helix and contains 24 residues. There is a total of about 138 amino acid residues (approximately 40%) in helical conformations.

2. *Interaction with Coenzyme*

a. Equilibrium Binding of NAD^+ and NADH. Evidence presented in the early part of this chapter has shown that both s-MDH and m-MDH are dimers of identical or nearly identical subunits. Therefore, there should be two equivalent binding sites per molecule of enzyme for both the reduced and oxidized forms of the coenzyme. Recent studies by Holbrook and Wolfe indicate that two equivalent sites do exist for both m-MDH and s-MDH from porcine heart (*66*). They determined both the number of NADH sites per enzyme molecule and the dissociation constant by following the enhancement of NADH fluorescence which occurs upon binding. The same data were obtained for NAD^+ by studying its competition for NADH sites. For both s-MDH and m-MDH, two equivalent binding sites were found for both NADH and NAD^+ (*66*). At pH 8.0 in Tris buffer, the NADH dissociation constant was 2.16 and 1.4 μM for m-MDH and s-MDH, respectively (*66*). At the same ionic conditions, the dissociation constant for NAD^+ was 585 μM for s-MDH and 480 μM for m-MDH (*66*), indicating little significant differences in this property of the two enzymes.

Cassman and King have reported cooperative binding of NADH to beef heart s-MDH (*67*). They also found evidence for the existence of s-MDH monomer from sedimentation equilibrium analysis at low protein concentrations (*67*). The fluorescence studies of the binding of NADH to beef heart s-MDH were interpreted in terms of a model in which both monomer and dimer bind NADH with cooperative binding to the dimer (*67*). The dissociation constant for the first site was about 25 μM and for the second site about 0.18 μM.

Cassman later showed that two forms of s-MDH may be present in

66. J. J. Holbrook and R. G. Wolfe, *Biochemistry* **11**, 2499 (1972).
67. M. Cassman and R. C. King, *Biochemistry* **11**, 4937 (1973).

the enzyme preparations, and these results are described in Section II,D,1.

Holbrook and Wolfe also reviewed the results of earlier investigations on the number of binding sites for both forms of the coenzyme to s-MDH and m-MDH (*66*). These results are in general agreement with two binding sites per dimer. Some variation in the values for the dissociation constants is apparent, but this could result from the variety of buffers and pH values used in the individual studies. The conflict between normal and cooperative binding of NADH to s-MDH cannot be resolved at the present time.

b. Conformation of NAD⁺ Bound to s-MDH. The crystals used in the X-ray analysis of pig heart s-MDH were grown in the presence of NAD^+ using ammonium sulfate solutions at about pH 5 (*37*). Chemical analyses of the crystals showed that they contain at least one mole of bound NAD^+ per mole of enzyme (*63*). When they were soaked in high concentrations of NAD^+ and a difference electron density map between the soaked and native crystals was determined, some additional binding of the coenzyme (*64*) was found. However, this occurred on only one subunit, suggesting some asymmetry in the binding sites. Later reports demonstrated the presence of some bound coenzyme in both subunits (*58*). The additional binding which occurs on soaking the crystals in high concentrations of NAD^+ overlays coenzyme present in the native crystals and it is likely that only the relative occupancies differ in the two subunits. The observation that only one of the chemically identical subunits will bind additional coenzyme can be accounted for in two ways (*64*). The simplest explanation is that only one of the two sites is accessible to the NAD^+ in the crystalline state. This is clearly feasible because of the asymmetric environment of the two subunits in the crystalline lattice (*68*). An alternative explanation is that the binding of coenzyme may involve a conformational change which increases the affinity of the subunit for the coenzyme. If crystal packing forces do not allow an equivalent change in both subunits, a weaker site is present in one subunit (*64*). The data of Holbrook and Wolfe (*66*) indicate that the asymmetry found in the crystalline state does not occur in solution.

Webb *et al.* (*58*) have studied the conformation of NAD^+ bound to s-MDH using the electron density maps of the native protein. While small conformational differences may exist in the conformation of the polypeptide chain around the binding sites of the two subunits, the structure of the bound dinucleotide may be interpreted as being in basically the same configuration. The conformation of NAD^+ bound to s-MDH

68. D. Tsernoglou, E. Hill, and L. J. Banaszak, *JMB* **69**, 75 (1972).

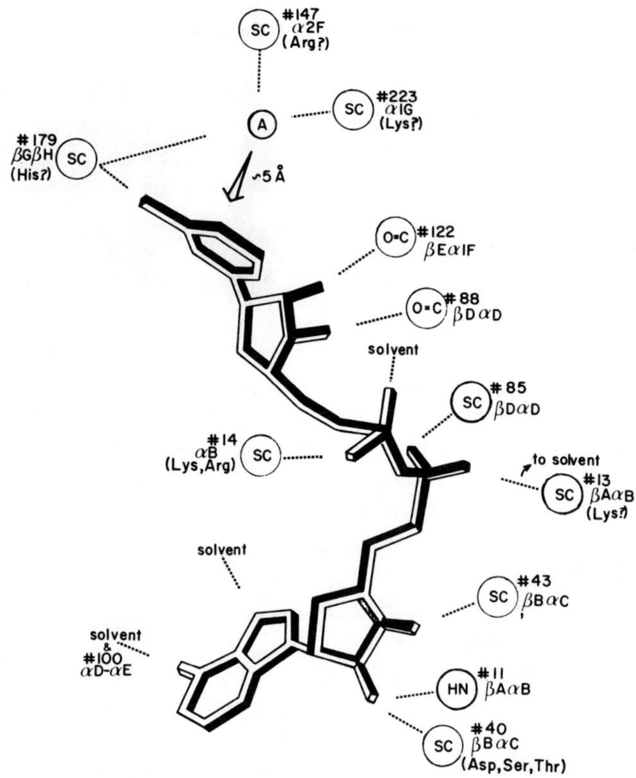

FIG. 2. Schematic drawing of the conformation of NAD⁺ when bound to pig heart s-MDH. The reader is referred to accurate stereo drawings or the crystallographic coordinates for details on bond torsional angles or interatomic distances (*58,69*). The adenine end of NAD⁺ is at the bottom of the drawing and the pyridine end of the dinucleotide near the top. Also shown on the drawing are contacts made with regions of the enzyme molecule. These are tentative assignments and are described in detail in the original reference (*69*). The position A is the anion binding site referred to in the text. Segments of secondary structure are labeled as in Fig. 1. The letters SC indicate that atoms from an amino acid side chain are involved in the interaction as opposed to groups such as C=O which arise from the polypeptide chain. The numbering scheme for the residues in the polypeptide chain scheme is tentative and may change upon completion of the amino acid sequence and model refinement.

is shown schematically in Fig. 2. The reader is again advised to see the original stereoviews of the NAD⁺ structure for details of the conformation (*58,69*). This is basically the same structure as when the coenzyme is

69. L. J. Banaszak and L. Webb, *in* "Structure and Conformation of Nucleic Acids and Protein-Nucleic Acid Interactions" (M. Sundralingam, ed.), 4th Annu. Steinbock Symp., Univ. Park Press, Baltimore, Maryland, 1974 (in press).

bound to dogfish LDH and probably other dehydrogenases (*58,70;* see also Chapter 2).

As can be seen in Fig. 2, the bound dinucleotide is in an open or extended conformation. Both the pyridine and adenine bases have an *anti* position about their respective glycosyl bonds. A detailed description of the various torsional angles which define the conformation of nucleotides has been given by Sundaralingam (*71*). The puckering of the ribose rings, although not shown in Fig. 2, was C-3′-endo for the adenine ribose and C-2′-endo for the nicotinamide ribose. These configurations produced a better fit of the dinucleotide to the electron density. However, because of the resolution of the X-ray data, the puckering of the ribose rings cannot be obtained directly from the electron density (*58*).

The open conformation of NAD^+ when bound to s-MDH is probably significantly different from the conformation in solution (*72*). Evidence from a number of NMR studies tend to indicate that in dilute aqueous solutions both NAD^+ and NADH tend to have a folded conformation probably involving stacking of the adenine and nicotinamide bases (*72–74*). The folded form of the dinucleotide only represents a thermodynamic preference and intermediate unfolded configurations are also present (*72*). The presence of methanol or urea increases the preference of the dinucleotide for a more unfolded conformation (*72*). Differences in this conformational preference between NADH and NAD^+ in the presence of urea and methanol have also been observed (*72*). The presence of methanol or urea may in fact mimic the environmental conditions in the active site of s-MDH.

D. Catalytic Properties

1. *Kinetic Analyses*

When an enzyme contains multiple subunits and the reaction which it catalyzes involves multiple substrates, detailed steps in the catalytic processes may become very complicated. The malate dehydrogenases from mammals contain two subunits, and the substrates include both a form of the dicarboxylic acid and the dinucleotide coenzyme. The third

70. M. J. Adams, M. Buehner, K. Chandrasekhar, G. C. Ford, M. L. Hackert, A. Liljas, M. G. Rossmann, I. E. Smiley, W. S. Allison, J. Everse, N. O. Kaplan, and S. S. Taylor, *Proc. Nat. Acad. Sci. U. S.* **70,** 1968 (1973).
71. M. Sundaralingam, *Jerusalem Symp. Chem. Biochem.* **5,** 417 (1973).
72. G. McDonald, B. Brown, D. Hollis, and C. Walter, *Biochemistry* **11** 1920 (1972).
73. A. Jardetzky and N. G. Wade-Jardetzky, *JBC* **241,** 85 (1966).
74. R. H. Sarma and N. O. Kaplan, *Biochemistry* **9,** 539 (1970).

reactant is a proton from the solvent. To attempt a description of the catalytic mechanism one would first like to know the stoichiometry, affinity, and order of substrate(s) binding. Dehydrogenases in general are thought to operate via either a compulsory ordered addition of substrates or a partially ordered pathway although this may not be true for every dehydrogenase (*75*). Next, one would like to know the various steps in the catalytic pathway and which of these steps is or can be rate-limiting. Various forms of substrate inhibition also become more involved since with multiple substrates, abortive ternary complexes may exist. Typically, with the malate dehydrogenases under appropriate conditions one could form an abortive complex such as MDH:L-malate:NADH.

The section which follows attempts to describe some of the more recent observations which have been made on the NAD^+-dependent malate dehydrogenases. Considerably more data on the mitochondrial form are available, but for reasons of simplicity, both forms are discussed in the same section.

The basic scheme of an ordered catalytic pathway for the malate dehydrogenases is shown schematically in Fig. 3. Included in Fig. 3 is one form of an abortive ternary complex. Silverstein and Sulebele have

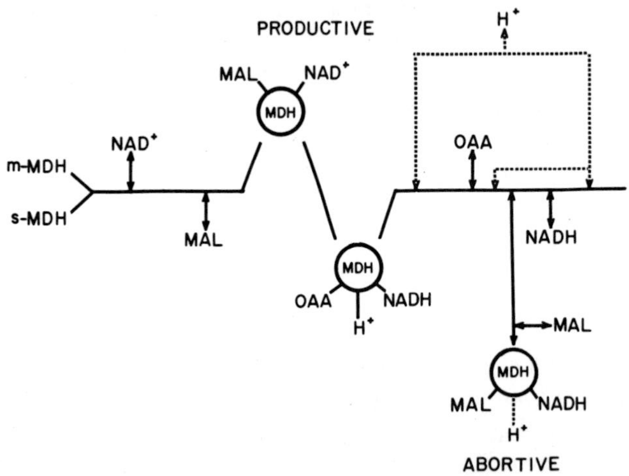

FIG. 3. Catalytic pathway for an ordered reaction mechanism for malate dehydrogenase. The steps in the s-MDH catalyzed oxidation for L-malate or reduction of oxaloacetate are represented by a schematic set of ordered binding reactions. The substrates L-malate and oxaloacetate are abbreviated MAL and OAA, respectively. The addition of a proton, a necessary step in the catalytic reaction, is indicated in dotted lines at three different positions in the pathway. This is to show that there is no data on where it participates in an ordered scheme. An example of what has been called an abortive ternary complex is also shown.

75. E. Silverstein and G. Sulebele, *Biochemistry* **8,** 2543 (1969).

shown that a compulsory ordered mechanism would explain data they obtained on bovine heart s-MDH and m-MDH and pig heart m-MDH (*75,76*). They studied isotopic exchange rates of reactants at equilibrium. If the only binary complexes which can be formed are MDH–NAD^+ or MDH–NADH then increasing levels of oxaloacetate and malate should depress the $NAD^+ \leftrightarrows$ NADH isotopic exchange but not the isotopic exchange rate of malate $\rightleftharpoons$ oxaloacetate. This was observed with pig heart m-MDH at pH 8.0 (*75*). By increasing the concentration of nonreactive pairs of substrates such as NADH and malate but maintaining equilibrium, evidence for the existence of the abortive ternary complex MDH–NADH–malate was obtained. However, no evidence for an abortive MDH–NAD^+–oxaloacetate was found (*75*). With m-MDH at pH 8.0, marked differences in the exchange rates of the substrates are taken as evidence that the rate-limiting step in the net reaction is not the interconversion of productive ternary complexes shown in Fig. 3 (*75*). These isotopic exchange data did not seem to be compatible with a reciprocating compulsory mechanism which will be discussed in a later section.

Identical experiments carried out with both s-MDH and m-MDH from beef heart resulted in similar changes in the isotopic exchange rates (*76*). For both enzymes, the data were compatible with a compulsory ordered mechanism involving binary complexes of the MDH's with the coenzyme but not with the dicarboxylic acid substrates (*76*). The isotopic exchange rates were very similar for m-MDH from pig or beef heart, and the results were qualitatively the same for s-MDH from beef heart (*75,76*). Despite some of the striking differences in the physical–chemical properties of m-MDH and s-MDH, the two forms appear to have closely related catalytic mechanisms.

The compulsory ordered mechanism arrived at through isotopic exchange rates is basically in agreement with mechanisms postulated on the basis of initial velocity studies. Heyde and Ainsworth showed that for beef heart m-MDH the initial velocity pattern in the absence of products and the product inhibition pattern are consistent with an ordered mechanism (*77*). Raval and Wolfe obtained similar results with pig heart m-MDH; data obtained by initial velocity studies in both directions are in agreement with a compulsory ordered mechanism (*78,79*). Substrate inhibition by oxaloacetate also occurs with pig heart m-MDH (*80*). Similar initial velocity studies on beef heart s-MDH did

76. E. Silverstein and G. Sulebele, *BBA* **185**, 297 (1969).
77. E. Heyde and S. Ainsworth, *JBC* **243**, 2413 (1968).
78. D. N. Raval and R. G. Wolfe, *Biochemistry* **1**, 1112 (1962).
79. D. N. Raval and R. G. Wolfe, *Biochemistry* **1**, 263 (1962).
80. D. N. Raval and R. G. Wolfe, *Biochemistry* **2**, 220 (1963).

not appear to be totally compatible with a compulsory ordered mechanism. Data obtained at pH 5 indicated that any catalytic mechanism for s-MDH should include the presence of some kinetically active binary complexes such as s-MDH–OAA (*81*).

To summarize, the data so far obtained for both s-MDH and m-MDH are in agreement with a compulsory ordered mechanism as shown in Fig. 3. Many of these data have been obtained around pH 8.0. The presence of an anion in the substrate site as is suggested by the crystallographic studies on s-MDH has not yet been included in the reaction pathway (*58*). The anion binding site will be discussed in more detail in the description of the active site. In addition, the ordered mechanism does not yet include the path for the proton which is necessary for the formation of the productive ternary complex which includes MDH, NADH, and oxaloacetate. The rate-limiting step for both directions of the catalytic reaction appears to be dissociation of the coenzyme and not the interconversion of productive ternary complexes (*82*). Last of all and only with respect to this ordered substrate addition, s-MDH and m-MDH appear to behave in a qualitatively similar manner.

In the case of pig heart m-MDH, Harada and Wolfe have suggested that the compulsory ordered mechanism must be modified to account for the inhibitory effect of hydroxymalonate (*83,84*). Hydroxymalonate is the three carbon dicarboxylic acid which is an analog of malate. Ketomalonate is reduced by m-MDH with half of the turnover number observed with oxaloacetate, but no appreciable oxidation of hydroxymalonate was observed (*83*). When malate is oxidized by m-MDH, hydroxymalonate inhibits this reaction uncompetitively. However, hydroxymalonate is a competitive inhibitor of the enzymic reaction which reduces oxaloacetate (*83*). To explain these effects, Harada and Wolfe have suggested a working hypothesis which consists of a reciprocating compulsory ordered mechanism (*84*). Productive ternary complexes of dimeric m-MDH such as shown in Fig. 3 would include the reciprocal intermediate. Such intermediate complexes are shown in Fig. 4. The presence of an obligatory pentenary (reciprocal) intermediate does not seem to agree with the results obtained by measuring isotopic exchange rates at equilibrium (*75,76*). Note in Fig. 4 one form of an abortive complex contains productive binding on one subunit (NAD^+ and malate) but is still a dead-end complex (*84*). Harada and Wolfe pointed out that a reciprocating compulsory ordered mechanism would be conclusively disproved if an

81. M. Cassman and S. Englard, *JBC* **241,** 787 (1966).
82. C. Frieden, personal communication.
83. K. Harada and R. G. Wolfe, *JBC* **243,** 4123 (1968).
84. K. Harada and R. G. Wolfe, *JBC* **243,** 4131 (1968).

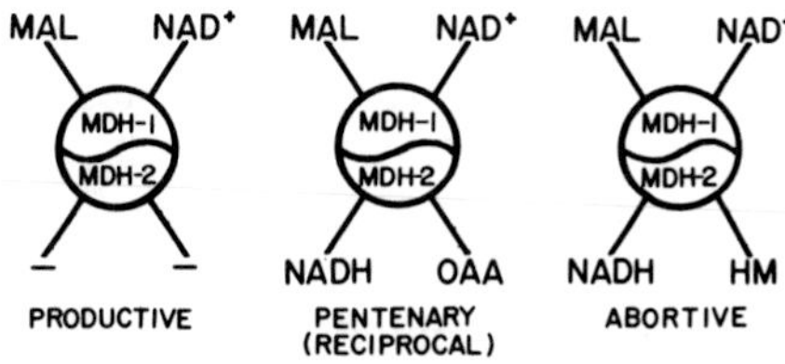

FIG. 4. Schematic representation of intermediates for a reciprocating m-MDH mechanism. The two identical subunits of m-MDH are illustrated within the circles by MDH-1 and MDH-2. L-Malate and oxaloacetate are abbreviated MAL and OAA, respectively. The first two complexes which are shown are productive forms of enzyme–substrate complexes. The last form would represent an abortive complex in the reciprocating mechanism (*84*).

enzymically active monomeric form of m-MDH could be unequivocally demonstrated (*84*). No such data has yet been obtained. Therefore it is still possible that the simple compulsory ordered mechanism at least for m-MDH might include more complex intermediates of the sort described by a reciprocating system.

Because of the large number of potential functional groups on NAD^+ which may be involved in any binding process, fragments of this coenzyme should also bind to dehydrogenases. This has been shown to be the case for a number of dehydrogenases. For pig heart m-MDH, Oza and Shore have shown that ATP, ADP, and AMP are competitive inhibitors for NADH (*85*). However, with increasing concentrations of the adenine nucleotides, complete inhibition was not obtainable (*85*). When ADP-ribose was used, total displacement of NADH from the binary complex was obtainable (*85*). ADP-ribose is quite similar to NAD^+ and NADH in both the charge distribution around the pyrophosphate bridge and the number of potential hydrogen binding groups.

ATP, ADP, and AMP inhibit the malate dehydrogenase from *E. coli* but in an allosteric manner (*86*). Sanwal has suggested that in the case of MDH from *E. coli* there is an additional (allosteric) site for NADH and the adenine nucleotides affect activity by binding at this site(s) (*86*).

For beef heart s-MDH, there is evidence that D-fructose 1,6-diphosphate inhibits the binding of NADH (*87*). Other compounds such as D-malate, citrate, and alloisocitrate enhance the binding of NADH to s-MDH (*87*). In the presence of D-fructose 1,6-diphosphate, the binding of NADH becomes clearly cooperative (*87*). Cassman and Vetterlein have obtained evidence that beef heart s-MDH may be obtained in two

85. N. B. Oza and J. D. Shore, *ABB* **154**, 360 (1973).
86. B. D. Sanwal, *JBC* **244**, 1831 (1961).
87. M. Cassman, *BBRC* **53**, 666 (1973).

forms: s-MDH_a and s-MDH_b. D-Fructose 1,6-diphosphate has no effect on the binding of NADH to s-MDH_b but cooperatively inhibits the binding of s-MDH_a (*88*). Evidence has been obtained that s-MDH_a contains 1.3–1.8 moles/mole of enzyme of covalently bound phosphate while s-MDH_b contains only 0.3–0.6 mole of phosphate per mole of enzyme (*88*). Sodium dodecyl sulfate polyacrylamide gel electrophoresis indicates both enzymes have the same subunit molecular weight, and no significant differences between their amino acid compositions were evident (*88*). It is possible that both phosphorylation of s-MDH and the effect of D-fructose 1,6-diphosphate have some regulatory significance *in vivo* (*88*). Since no electrophoretic microheterogeneity was tested for, further evidence that s-MDH is indeed phosphorylated needs to be obtained.

The malate activation of m-MDH has been explained in terms of malate binding at some effector site, a site other than the active site (*89*). It is suggested that the malate activation can be accounted for primarily by a tighter binding of NAD^+ (*89*). These data and product inhibition patterns are in general agreement with a compulsory ordered mechanism described in an earlier section.

There are a number of nonphysiological compounds which affect the activity of malate dehydrogenases. Metal compounds such as platinum (II) complexes inhibit pig heart m-MDH at low metal concentrations (1–4 moles of ions per mole of enzyme) (*90*). Similar platinum compounds inhibit pig heart s-MDH (*91*). In the case of s-MDH, the inhibition is essentially irreversible but the rate of inhibition is reduced in the presence of coenzyme (*91*). The steric relationships of these metal binding sites relative to the coenzyme binding sites were determined by X-ray crystallographic studies (*91*). Phenols and substituted phenols are inhibitors of pig heart m-MDH and appear to be competitive with NAD^+ (*92*).

2. *The Structure of the Active Site*

a. Chemical Modification. A number of amino acid residue types have been examined for their role as potential active site residues using various reagents in both s- and m-malate dehydrogenases. Only histidine has been clearly implicated as a catalytic active residue, although the modification of other types of amino acids can result in inactivation.

88. M. Cassman and D. Vetterlein, *Biochemistry* **13**, 684 (1974).
89. M. Telegdi, D. V. Wolfe, and R. G. Wolfe, *JBC* **248**, 6484 (1973).
90. M. E. Friedman, B. Musgrove, K. Lee, and J. E. Teggins, *BBA* **250**, 286 (1971).
91. M. Wade, D. Tsernoglou, E. Hill, L. Webb, and L. J. Banaszak, *BBA* **322**, 124 (1973).
92. R. T. Wedding, C. Hansch, and T. R. Fukuto, *ABB* **121**, 9 (1967).

The reaction of pig s-MDH with sulfhydryl reagents clearly indicates that these residues are not involved in the catalytic mechanism of the enzyme. Leskovac and Pfleiderer (*93*) and Banaszak *et al.* (*37*) have shown that two residues of cysteine react with Ellman's reagent, 5,5′-dithiobis-2-nitrobenzoic acid (DTNB), with no loss in activity. Skilleter *et al.* (*94*) and Guha *et al.* (*23*), working with the cytoplasmic enzyme from ox kidney and beef heart, found 1.5 to 2.0 of 6 (ox kidney) and 3 of 6 (beef heart) reactive cysteine residues using DTNB and *p*-mercuribenzoate (pMB), respectively. Neither reaction resulted in the loss of enzymic activity. Humphries *et al.* (*95*) have reported that 4,4′-bisdimethylaminodiphenylcarbinol (BDC-OH), a reagent specific for thiol groups, is without effect on pig heart s-MDH.

In contrast, m-MDH combines with mercurials in low concentrations to produce activation and, in high concentrations, causes inactivation (*96–98*). Silverstein and Sulebele (*99,100*) examined this phenomenon with pMB and found that maximum inactivation occurred after modification of 3–4 cysteinyl residues. Reaction of 7–8 residues produced greater but transient activation followed by a proportional inactivation. They also found no dependence of the activation on a NADH-X compound as proposed by Devenyi *et al.* (*98*). Stimulation by Hg^{2+} ions was reported by Kuramitsu (*96*) for pig heart m-MDH, but the activation was dependent on the presence of the substrate L-malate.

The inactivation by excess pMB was protected by NAD^+ or NADH but not by L-malate or oxaloacetate (*99*). Pfleiderer *et al.* (*97*) showed that pig heart m-MDH lost its ability to bind NADH after modification of 5 cysteinyl residues by pMB. In contrast, Siegel and Ellison (*101*) found that beef heart m-MDH still retained 25% of its enzymic activity after all thiol groups were modified with pMB.

The reaction of pig heart m-MDH with DTNB did not produce the stimulation observed with mercurials. Seguin and Kosicki (*102*) reported that native pig heart m-MDH did not react with this reagent, whereas Skilleter *et al.* (*94*) found modification of 3 to 4 out of 10 to 11 thiol groups in a time-dependent reaction. There was no loss of activity.

93. V. Leskovac and G. Pfleiderer, *Hoppe-Seyler's Z. Physiol. Chem.* **350,** 484 (1969).
94. D. N. Skilleter, N. M. Lee, and E. Kun, *Eur. J. Biochem.* **12,** 533 (1970).
95. B. A. Humphries, M. S. Rohrbach, and J. H. Harrison, *BBRC* **50,** 493 (1973).
96. H. K. Kuramitsu, *JBC* **243,** 1016 (1968).
97. G. Pfleiderer, E. Hohnholz-Merz, and D. Gerlach, *Biochem. Z.* **336,** 371 (1962).
98. T. Devenyi, S. J. Rogers, and R. G. Wolfe, *BBRC* **23,** 496 (1966).
99. E. Silverstein and G. Sulebele, *Biochemistry* **9,** 274 (1970).
100. G. Sulebele and E. Silverstein, *Biochemistry* **9,** 283 (1970).
101. L. Siegel and J. S. Ellison, *Biochemistry* **10,** 2856 (1971).
102. R. J. Seguin and G. W. Kosicki, *Can. J. Biochem.* **45,** 659 (1967).

Gregory *et al.* (*16*) and Humphries *et al.* (*95*) have examined the effect of *N*-ethylmaleimide (NEM) and BDC-OH on pig heart m-MDH and have obtained evidence in each case for the modification of two "essential" thiol groups per enzyme molecule. Both modifications were inhibited by coenzyme. Sensitivity of m-MDH to thiol modification has also been demonstrated by means of thyroxine, iodine cyanide, and molecular iodine, all of which cause oxidation of thiol groups accompanied by inactivation in this enzyme (*103*).

Although some discrepancies exist, it appears that s-MDH is essentially insensitive to thiol modification whereas m-MDH possesses a thiol group on each subunit that is located near the coenzyme binding site. Its integrity appears to be essential for enzymic activity.

The involvement of a specific histidyl residue in the catalytic mechanism of each enzyme appears to be well established. Holbrook *et al.* (*104*), utilizing diethylpyrocarbonate at pH 6.5, have shown that 0.58 ± 0.1 residue of carbethoxyhistidine is formed per NADH binding site. This modification of pig heart s-MDH completely inhibited the enzyme. The modification reaction is inhibited by oxaloacetate and hydroxymalonate but not by coenzyme. NADH binding is not prevented by the modification, but the carbethoxylated enzyme cannot form a ternary complex with NADH and hydroxymalonate.

The modification of a single histidine residue in m-MDH with iodoacetamide has been demonstrated by Anderton (*50*) and by Gregory and Harrison (*105*). In both cases, the product of the reaction was shown to be 3-carboxamidohistidine. Interestingly, in contrast to the studies of Holbrook *et al.* (*104*) with s-MDH, the modification of m-MDH by iodoacetamide can be totally inhibited by NADH and the substrates L-malate and oxaloacetate reduced the rate of inactivation (*106*). Gregory *et al.* (*106*) have also reported the isolation of a labeled tripeptide, carboxyamidomethyl-His-Gly-Gly, from a pronase digest of enzyme labeled with [^{14}C]iodoacetamide. Recently, Foster and Harrison (*107*) reported the amino acid sequence of a tryptic peptide containing this sequence to be Val-Ser-Val-Pro-Ile-His-Gly-Gly-Val-Ala-Gly-Lys. However, from sequence studies of the enzyme by Sutton *et al.* (*39*), a similar peptide was isolated which was found to have a somewhat different sequence: Val-Ser-Val-Pro-Ile-Val-Gly-Gly-His-Ala-Gly-Lys. When this latter sequence is compared to the region of lactate dehy-

103. S. Varrone, E. Consiglio, and I. Covelli, *Eur. J. Biochem.* **13**, 305 (1970).
104. J. J. Holbrook, A. Lodola, and N. P. Illsley, *BJ* **139**, 797 (1974).
105. E. M. Gregory and J. H. Harrison, *BBRC* **40**, 995 (1970).
106. E. M. Gregory, M. S. Rohrbach, and J. H. Harrison, *BBA* **243**, 489 (1971).
107. M. Foster and J. H. Harrison, *BBA* **351**, 295 (1974).

drogenase containing the active histidine at position 195 (*108*), 6 of the 12 residues are found to be identical.

m-MDH:	Val-Ser- Val -Pro- Val-Ile-Gly -Gly- His -Ala-Gly-Lys
LDH:	Val-Ser- Gly -Trp- Val-Ile-Gly -Gln- His -Gly-Asp-Ser

This homology suggests that the modification observed by Anderton (*50*) and Gregory and Harrison (*105*) is, indeed, a catalytically active residue and that the extensive homology among dehydrogenases (Rossmann *et al.*, Chapter 2, this volume) includes m-MDH as well.

Leskovac (*109*) has reported that a histidyl residue is also modified in m-MDH by iodoacetate. However, the labeled peptide isolated from this modification is completely distinct in composition from that found by Foster and Harrison (*107*) and by Sutton *et al.* (*39*) and does not appear to be related to any of the histidyl residues in the sequence of this enzyme (*39*). In the absence of further sequence data this discrepancy cannot be resolved at present.

Leskovac and Pfleiderer (*93*) have reported that iodoacetate produces a specific inactivation of s-MDH as well. However, in this case, the site of modification is a methionyl residue. The reaction is inhibited by NADH and NAD^+ but not by substrates. Leskovac (*110*) has reported the composition of the tryptic-chymotryptic peptide labeled by [^{14}C]iodoacetate as

(Ser, Pro, Gly, Val, Met)–Arg

Confirmation of these observations is provided by the observations of Friedman *et al.* (*90*), Melius *et al.* (*111*), and Wade *et al.* (*91*) that platinum (II) complexes inhibit the binding of coenzyme. In view of the known preference for platinum heavy-atom derivatives to bind at methionyl residues, it may be postulated that a methionyl residue is located at or near the binding site of coenzyme in s-MDH.

Experiments suggesting that tryptophan was a catalytically active residue in dehydrogenases were reported by Schellenberg (*112–114*). Although earlier reports showed that m-MDH was devoid of this residue,

108. S. S. Taylor, S. S. Oxley, W. S. Allison, and N. O. Kaplan, *Proc. Nat. Acad. Sci. U. S.* **70**, 1790 (1973).
109. V. Leskovac, *Bull. Soc. Chim. Beograd* **38**, 307 (1973).
110. V. Leskovac, *Croat. Chem. Acta* **43**, 183 (1971).
111. P. Melius, J. E. Teggins, M. E. Friedman, and R. W. Guthrie, *BBA* **268**, 194 (1972).
112. K. A. Schellenberg, *JBC* **240**, 1165 (1965).
113. K. A. Schellenberg, *JBC* **241**, 2446 (1966).
114. K. A. Schellenberg, *JBC* **242**, 1815 (1967).

Chan and Schellenberg (*115*) found evidence for its presence and reported its role in the enzymic reaction. Sequence analysis (*39*) has established that the enzyme is indeed devoid of tryptophan. Furthermore, Allen and Wolfe (*116*) have shown that s-MDH does not contain significant amounts of tritium after incubation with NAD-$\alpha\beta^3$H or DL-[2-^{3}H]malate. Because of these findings, the report (*117*) that pig heart m-MDH is inactivated by 2-hydroxy-5-nitrobenzyl bromide, a reagent reported to be highly selective for tryptophan, must be attributed to modification at another site.

Yost and Harrison (*118*) have reported that m-MDH is irreversibly inactivated by pyridoxal 5′-phosphate (PLP). In view of the usual reversibility of this reagent when reacted with the ϵ-amino group of lysine, and the unusual spectral properties of the derivative formed, it was concluded that the reaction of m-MDH with PLP proceeds via modification of lysine followed by the formation of a secondary stable complex. Such a complex might be a thiazolidine-like compound formed by the reaction of the Schiff base with a neighboring cysteinyl residue. The role of the lysine residue in the active site of m-MDH remains to be clarified.

b. Crystal Data. Since the determination of the amino acid residues in the active site of s-MDH has not yet been completed, it can only be presented in a schematic way. As seen in Fig. 5, the binding site for the coenzyme, as determined by the X-ray studies, occupies a deep cleft in the otherwise globular structure. Whether or not this space exists in the absence of bound NAD$^+$ or NADH has not yet been determined, but a crystalline form of pig heart s-MDH without bound NAD$^+$ has been reported (*37,64*).

The adenine end of the bound coenzyme is close to solvent but still is located within a groove in the surface. In contrast, the pyridine ring is completely buried in the coenzyme binding cavity. Residues from the polypeptide chain which are close to atoms in the bound dinucleotide were shown schematically in Fig. 2. The counterionic side chain(s) for the double negative charge on the pyrophosphate linkage of the bound dinucleotide is still unknown (*60*). In dogfish LDH there appears to be a single arginine residue (*70*) near this part of the coenzyme. The position of the substrate binding site is probably related to the location of an ion near the C-4 atom of the nicotinamide ring (*58*). This is also shown schematically in Fig. 2. The closest approach of the electron density of this ion to the C-4 atom of the nicotinamide ring is about 5 Å, and the

115. T-L. Chan and K. A. Schellenberg, *JBC* **243,** 6284 (1968).
116. L. M. Allen and R. G. Wolfe, *BBRC* **41,** 1518 (1970).
117. K. A. Schellenberg and E. W. McLean, *BBA* **191,** 727 (1969).
118. F. J. Yost, Jr. and J. H. Harrison, *BBRC* **42,** 516 (1971).

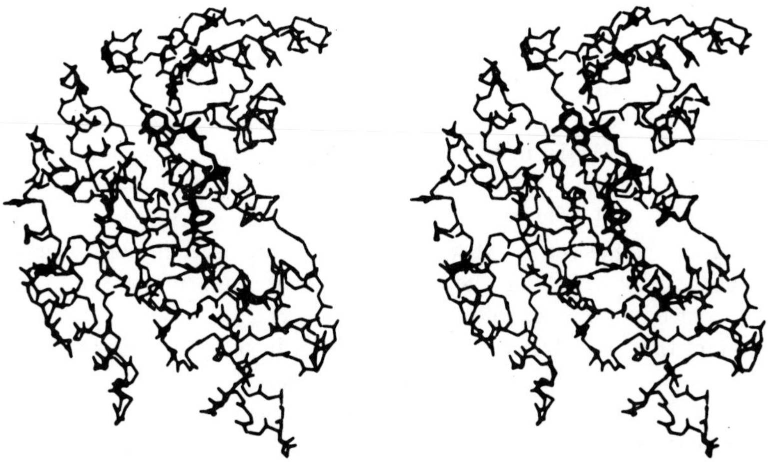

FIG. 5. Stereophotograph of a model of one subunit of pig heart s-MDH with bound coenzyme. The bound NAD^+ is indicated by the heavy black portion. The polypeptide chain of one subunit of s-MDH is also shown as a continuous line with the single branch points representing the carbonyl oxygen of an amino acid unit. The beta carbons have been omitted to reduce the complexity of the illustration. The adenine end of the dinucleotide is near the surface of the coenzyme subunit while the nicotinamide end is toward the inner part of the molecule.

ion is on the A side of the nicotinamide ring (*58*). If this ion site does in fact represent the would-be position of L-malate in a ternary complex, a direct transfer of a hydride ion to the coenzyme would seem possible. Certainly, the present interpretation of this part of the active site region rules out transfer via an intervening side chain (*58,60*).

The presence of an ion at what is the probable position of the substrate in s-MDH means that the binary complex (s-MDH-NAD^+) found in the crystals used for the X-ray analysis is in fact a dead-end ternary complex. If these crystal studies are related to the solution properties of s-MDH, any catalytic mechanism for the reaction should however include a reaction step involving dissociation of this ion. In this respect, it is worth noting that the ion site is also occupied in the crystalline apoenzyme form of dogfish LDH (*70*).

E. Environment of m-MDH

At the present time there is not much data which can be used to describe special properties of m-MDH related to its presence in the mitochondrion. Studies on the distribution of various enzymes inside the organelle indicate that m-MDH is one example of an intermediate class of mitochondrial enzymes. It is clearly not found in only the inner or outer membranes (*119*). It appears to be associated with most components and this probably means that it is in the inner membrane matrix. However, there are some data which indicate that lipids do affect the properties of m-MDH in very dilute solutions (*120*). Lysolecithin has been shown to reduce the rate of thermal inactivation of m-MDH (*121*). The mitochondrial MDH, as opposed to s-MDH, is rapidly inactivated at elevated temperatures. However, lysolecithin is a minor component of mitochondrial lipids. On the other hand, lecithin, a major lipid component, produced no measurable stabilization (*121*).

The fact that m-MDH seems to be located in both the inner membrane fraction and the inner membrane matrix may be due to ionic effects (*122,123*). Increasing levels of succinate cause the release of m-MDH and mitochondrial aspartate aminotransferase from the inner membrane fraction of mitochondria (*122*). Increasing levels of monovalent and bivalent cations produce the same effect as succinate (*123*). Cycling of the mitochondrial subfractions from sucrose to a sucrose medium containing millimolar concentrations of cation, and then back to sucrose, indicated that the release of these enzymes from the inner membrane fraction was reversible (*123*).

There is evidence, therefore, that the mitochondrial form of m-MDH does have structural properties which can result in its association with the inner membrane of mitochondria. Furthermore, this property is not unique to m-MDH since it is a property of other enzymes such as the aspartate aminotransferase. The molecular basis of this property is unknown at this time.

Acknowledgments

The authors gratefully acknowledge the helpful comments of Dr. C. Frieden and the technical assistance of Mr. G. Barbarash while preparing this manuscript. The perspective drawings of s-MDH and NAD^+ were done by Mr. J. Ahearn. Dr. L. J. Banaszak is U.S.P.H.S. Research Career Development Awardee, GM 14357, and Dr. R. A. Bradshaw is U.S.P.H.S. Research Career Development Awardee, AM 23968.

119. L. Ernster and B. Kuylenstierna, *in* "Membranes of Mitochondria and Chloroplasts" (E. Racker ed.), p. 172. Van Nostrand-Reinhold, Princeton, New Jersey, 1970.
120. J. W. Callahan and G. W. Kosicki, *Can. J. Biochem.* **45,** 839 (1967).
121. G. H. Dodd, *Eur. J. Biochem.* **33,** 418 (1973).
122. A. Waksman and A. Rendon, *BBRC* **42,** 745 (1971).
123. A. Rendon and A. Waksman, *BBRC* **42,** 1214 (1971).

7

Cytochromes c

RICHARD E. DICKERSON • RUSSELL TIMKOVICH

I. Introduction

The term cytochrome *c* is both a spectral and a structural classification, related more to the heme and its attachment to the polypeptide chain than to the protein which surrounds it. The most familiar member of the class is cytochrome *c* from the mitochondrial respiratory chain, which has a single heme group per chain of 103–113 amino acids, and a reduction potential of +260 mV. However, other members of this class have as many as four hemes, and reduction potentials covering the entire range between −290 and +400 mV. All are hemoproteins, with a heme *c* (Fig. 1) covalently attached to the polypeptide chain by thioether bridges to two (or in exceptional cases, one) cysteine residues. In contrast, cytochromes *b*, myoglobin, and hemoglobin all have a protoheme instead (Fig. 2), with the cysteine bridges replaced by vinyl groups, and with the heme not attached by covalent bonds to the protein. Cyto-

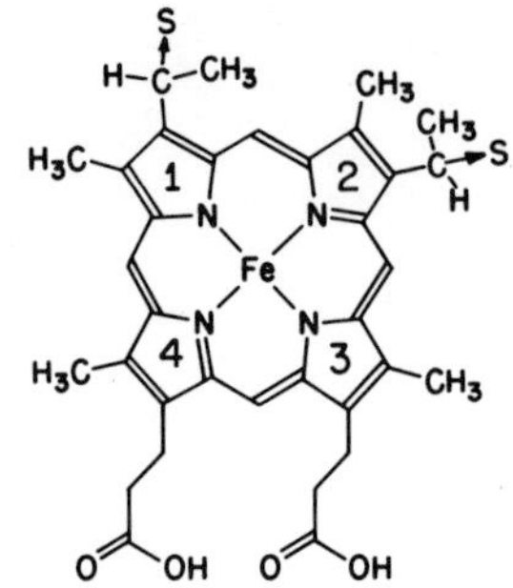

FIG. 1. Heme *c*, with the conventional numbering of pyrrole rings. The sulfur atoms are from cysteine side chains on the protein.

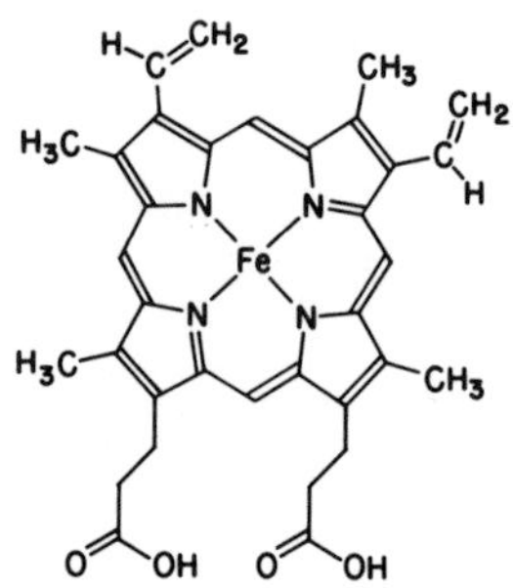

FIG. 2. Protoheme or heme *b*, also found in myoglobin and hemoglobin.

chromes *a* differ from *b* in the attachment of other substituents around the heme ring.

The iron atom in all cytochromes alternates between the oxidized Fe(III) state in which the iron has a single unpaired electron and heme has a formal charge of +1, and the reduced Fe(II) state with no unpaired electrons and no formal charge. Cytochromes all have a characteristic three-banded absorption spectrum in the reduced state, shown in Fig. 3 (data from ref. *1*), with an α-absorption band around 550–604 nm in the yellow, a β band around 520–546 nm in the green, and a γ or Soret band around 400–450 nm in the far violet. The remaining unabsorbed wavelengths give the heme its characteristic red color.

One of the most striking features of the heme group is the delocalization of electrons among the π orbitals of the porphyrin ring. The Soret band in the absorption spectrum represents the excitation of delocalized π electrons to unoccupied levels of the porphyrin ring of similar angular momentum (*2*). Absorption in the α and β region arises from a compar-

1. E. Margoliash and N. Frohwirt, *BJ* **71**, 570–572 (1959).
2. J. E. Falk, "Porphyrins and Metalloporphyrins," Chapter 6. Elsevier, Amsterdam, 1964.

TABLE I
ABSORPTION WAVELENGTHS OF THE PRINCIPAL CLASSES OF CYTOCHROMES

Heme protein	Absorption band α (nm)	β (nm)	γ (Soret) (nm)
Cytochrome *c*	550–558	521–527	415–423
Cytochrome *b*	555–567	526–546	408–449
Cytochrome *a*	592–604	Absent	439–443
Oxyhemoglobin	577	542	415

able excitation to levels of quite different angular momentum. This absorption is weaker because such transitions are more nearly forbidden.

The positions of the α, β, and Soret bands differ among the various classes of cytochromes as shown in Table I and at the bottom of Fig. 3. For comparison, the positions of the same three bands in oxyhemoglobin are also indicated. Replacing the two thioether bridges of cytochromes *c* by vinyl groups in cytochromes *b*, and replacing a methyl by a formyl group in cytochromes *a*, both enlarge the delocalization of electrons in the porphyrin ring, lower the electronic energy level spacings slightly, and shift the three absorption bands to somewhat longer wave-

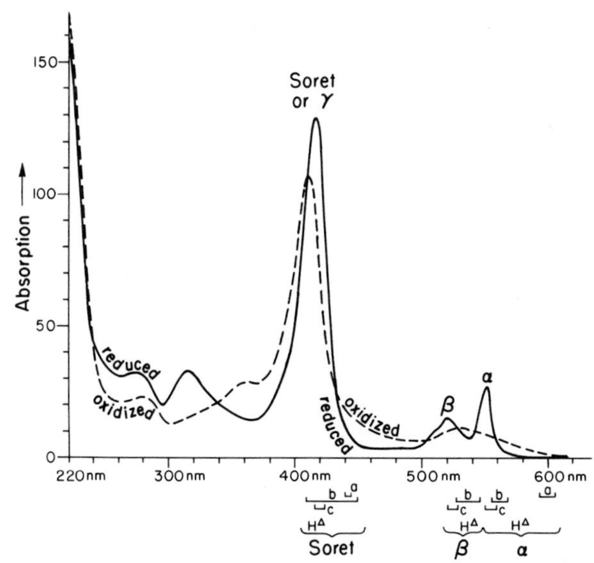

FIG. 3. Absorption spectra of oxidized and reduced horse heart cytochrome *c*. Wavelength ranges for the α, β, and Soret bands in cytochrome *a*, *b*, and *c*, and hemoglobin are indicated at the bottom.

lengths. It is interesting that in the two known examples of *single* covalent heme attachment, in cytochromes *c* from *Euglena* and *Crithidia*, where cysteine-14 is replaced by alanine and the thioether bridge by a vinyl, the α-band absorption is shifted toward cytochrome *b* values, from 550 to 558 nm. The absorption spectrum of cytochrome *c*, which is the basis for classification, thus depends on the side groups around the porphyrin ring and the way that they are connected to the protein chain.

All cytochromes *c* are oxidation–reduction proteins involved in either respiration or photosynthesis. In eukaryotic organisms—cells with both nuclei and mitochondria—there are only three *c*-type cyctochromes: *c* and c_1 from the mitochondrial respiratory chain, and c_6 or *f* from green plant photosynthesis. These are compared in Table II and discussed in Section II. The field in prokaryotes—bacteria and blue-green algae—is much more diverse, and one is left with the impression that *c*, c_1, and *f* in eukaryotes are only the most successful survivors of a very stiff biochemical competition. Prokaryotic cytochromes *c* are the subject of Section III.

Cytochromes *c* can be organized with only minor violence to the data into four principal classes as shown in Table III: single heme cytochromes with a large positive reduction potential, high-spin cytochromes *c′* with just five ligands to the iron and a myoglobinlike spectrum, cytochromes with a flavin group attached, and low potential and often multiheme cytochromes involved in sulfate respiration. Only the first of these categories is encountered in eukaryotes; the low potential end of the photosynthetic and respiratory machinery in higher organisms has been taken over by other kinds of redox proteins.

II. Eukaryotic Cytochromes c

A. Respiratory Cytochrome *c*

The respiratory cytochrome *c* found in the mitochondria of all eukaryotes is a protein with one heme *c* (Fig. 1) and 103–113 amino acids in a single polypeptide chain with no disulfide crosslinks. It is small and easily extracted from ruptured mitochondria, in contrast to the other protein components of the respiratory chain, which are larger and generally membrane-bound. For this reason, cytochrome *c* was the first cytochrome to be studied extensively, and has received the most attention from biochemists. The story of its discovery by MacMunn in 1884, its violent rejection by chemist and editor Hoppe-Seyler, subsequent neglect until rediscovered by David Keilin in 1925, and extensive characterization by Hugo Theorell and his school, is a familiar story which need not be repeated here. Good accounts of this history are to be found in reviews

TABLE II
COMPARISON OF EUKARYOTIC CYTOCHROMES *c*

Type	Source and function	Absorption spectrum			E_0'	Hemes	Molecular weight	Amino acids
		α	β	Soret				
$f = c_6$	Green plant photosynthesis	554.5	524	422	+365	1	*ca.* 38,000	324
c_{553}	Algal photosynthesis	550–555	521–523	425–427	+340 to +400	1	10,000–13,000	*ca.* 100
c	Mitochondrial respiration	550	521	415	+260	1	12,200	103–113
c_1	Mitochondrial respiration	552.5	522.5	417	+225	1	*ca.* 40,000	389

TABLE III
PRINCIPAL CLASSES OF CYTOCHROME *c*

Type and examples	E_0'	Hemes	MW/chain
I. High potential: eukaryotic and prokaryotic photosynthesis and respiration—c, c_1, c_2, c_4, c_5, c_6 (=f), c_{551}, etc.	+200 to +390	1	9,000–81,000
II. High-spin iron: c' from some pseudomonads, *Alcaligenes*, and several purple photosynthetic bacteria	0 to +130	1	13,000–16,000
III. Flavocytochromes			
c_{553} from *Chlorobium*	+98	1	50,000
c_{552} from *Chromatium*	+10	2	72,000
IV. Low potential			
c_{553} from *Desulfovibrio*	−100 to 0	1	9,000
c_3 from *Desulfovibrio*, c_7	−200	3 or 4	12,000
c_{551} from purple nonsulfurs, others	−150 to −250	1 or 2	16,000–21,000

by Margoliash and Schejter (*3*), Lemberg and Barrett (*4*), and especially in the recounting by Keilin himself (*5*).

Because of its universality and ease of extraction and purification (*3,6*), cytochrome *c* has been the subject of amino acid sequence determinations in more different species than any other protein. At the time of writing, cytochromes *c* from 67 species of plants, animals, and microorganisms have been completely sequenced, and the comparisons have been the raw material for an extensive study of eukaryotic evolution, considered later in the chapter. If the 103 amino acids of tuna cytochrome are taken as standard, then many other species have one or two extra amino acids at the carboxy terminus, and most invertebrates and plants have four to nine extra residues at the amino terminus. The amino termini in all vertebrates and high plants are blocked by acylation, $H_3C—CO—NH—$, but most lower organisms have free initial $H_2N—$ groups. The only known example of a deletion or insertion of amino acids within the chain is a single deletion in the cytochrome *c* of *Euglena*, and the integrity of the main body of polypeptide chain has been maintained during eukaryotic evolution.

3. E. Margoliash and A. Schejter, *Advan. Protein Chem.* **21**, 113–286 (1966).
4. R. Lemberg and J. Barrett, "The Cytochromes." Academic Press, New York, 1973.
5. D. Keilin, "The History of Cell Respiration and Cytochrome." Cambridge Univ. Press, London and New York, 1966.
6. E. Margoliash and O. F. Walasek, "Methods in Enzymology," Vol. **10**, pp. 339–348, 1967.

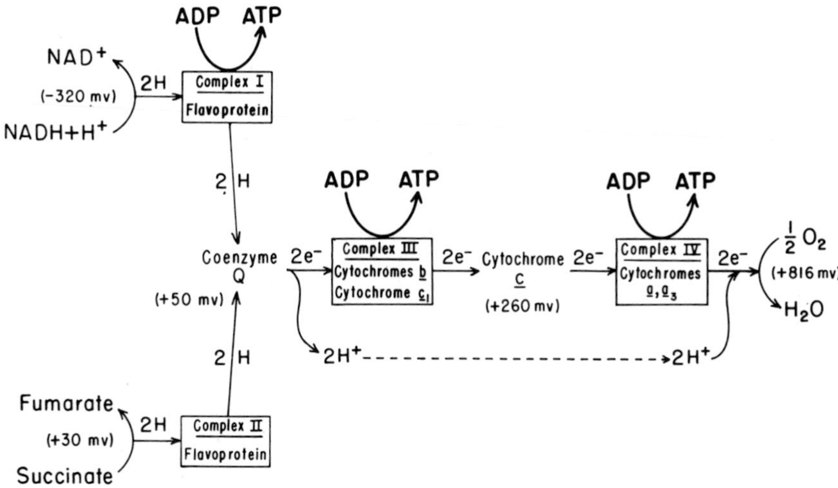

FIG. 4. The structure of the mitochondrial respiratory electron transport chain. Values in millivolts are standard (pH 7) reduction potentials for the components indicated.

The biological role of cytochrome *c*, diagramed in Fig. 4, is that of an electron-carrying shuttle between two macromolecular complexes in the inner membrane of the mitochondrion. At the beginning of the respiratory chain, NADH and succinate in the interior of the mitochondrion are oxidized by the flavoproteins of Complexes I and II, respectively. These complexes each pass their reducing equivalents to a common carrier, coenzyme Q, which in turn reduces the *b*-type cyctochromes of Complex III. These cytochromes reduce cytochrome c_1, also a part of the same complex, and the reducing electrons are passed to a more mobile small protein molecule, cytochrome *c*. This molecule reduces cytochrome *a* of Complex IV. Electrons then are passed from *a* to copper atoms, to cytochrome a_3, and finally are used to reduce O_2 to water, completing the respiratory chain. Three large drops in free energy or increases in reduction potential occur at Complexes I, III, and IV, and the free energy released is utilized to synthesize ATP. According to Wilson and coworkers (*7*), the coenzyme Q and cytochrome *c* shuttles are each roughly equipotential with their respective reductants and oxidants, and almost all of the free energy drop occurs within Complexes I, III, and IV, where the energy released can be coupled to ATP synthesis. Reduction potentials for key components in the respiratory chain are indicated in Fig. 4.

The probable locations of the components of the respiratory chain are

7. D. F. Wilson, P. L. Dutton, M. Erecinska, J. G. Lindsay, and N. Sato, *Accounts Chem. Res.* **5**, 234–241 (1972).

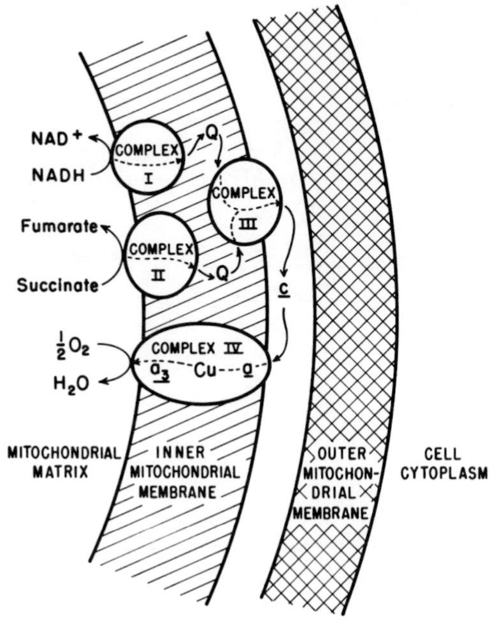

FIG. 5. Possible spatial arrangement of components of the respiratory chain in the inner mitochondrial membrane. Adapted from reference *9*.

shown in Fig. 5. The mitochondrion is surrounded by two membranes. Reoxidation of NADH and succinate, and reduction of O_2, take place at the inner surface of the inner membrane, and cytochrome *c* is found on the outer surface of this membrane (*8,9*). In spite of the deceptive concreteness of Fig. 5, it is not known whether cytochrome *c* actually migrates from Complex III to IV, rotates in place to turn the same face to III and IV during reduction and reoxidation, or merely sits passively between the two complexes as a channel or "wire" for electron flow (*10*). Even though the three-dimensional structure of the molecule has been worked out by X-ray diffraction methods (see below), the actual pathway of the electron in to the heme and out again is a source of continued controversy. Current ideas on this topic will be discussed later.

8. E. Racker, A. Loyter, and R. O. Christiansen, *in* "Probes of Structure and Function of Macromolecules and Membranes; Johnson Research Foundation Colloquium" (B. Chance, T. Yonetani, and A. S. Mildvan, eds.), pp. 407–410. Academic Press, New York, 1971.

9. L. E. Bennett, *in* "Current Research Topics in Bioinorganic Chemistry" (S. J. Lippard, ed.), pp. 1–176. Wiley, New York, 1973.

10. B. Chance, C.-P. Lee, L. Mela, and D. DeVault, *in* "Structure and Function of Cytochromes" (K. Okunuki, M. D. Kamen, and I. Sekuzu, eds.), pp. 475–484. Univ. of Tokyo Press, Tokyo, 1968.

1. *The Structure of the Cytochrome c Molecule*

Eukaryotic cytochromes *c* from several species have been studied by X-ray crystallographic methods as listed in Table IV (*11–31*). The initial analysis of ferricytochrome *c* from horse heart was begun by Dickerson and co-workers at the California Institute of Technology in 1963, but was halted at the 2.8 Å stage because of the poor quality of the horse crystals (*11–14*). Difference map methods comparing horse cytochrome with bonito (which crystallized in the same form) showed no major changes in the overall molecular folding of the two vertebrate cytochromes (*12*), confirming what had been expected from amino acid sequence comparisons. The inference was strong that the basic folding pat-

11. R. E. Dickerson, M. L. Kopka, J. Weinzierl, J. Varnum, D. Eisenberg, and E. Margoliash, *JBC* **242,** 3015–3017 (1967).

12. R. E. Dickerson, T. Takano, D. Eisenberg, O. B. Kallai, L. Samson, A. Cooper, and E. Margoliash, *JBC* **246,** 1511–1535 (1971).

13. R. E. Dickerson, T. Takano, O. B. Kallai, and L. Samson, *in* "Structure and Function of Oxidation Reduction Enzymes" (Å. Åkeson and A. Ehrenberg, eds.), pp. 69–83. Pergamon, Oxford, 1972.

14. R. E. Dickerson, *Sci. Amer.* April, 58–72 (1972).

15. R. Swanson, B. Trus, O. B. Kallai, and R. E. Dickerson, in preparation.

16. T. Takano, R. Swanson, O. B. Kallai, and R. E. Dickerson, *Cold Spring Harbor Symp. Quant. Biol.* **36,** 397–403 (1971).

17. T. Takano, O. B. Kallai, R. Swanson, and R. E. Dickerson, *JBC* **248,** 5234–5255 (1973).

18. R. E. Dickerson, T. Takano, and O. B. Kallai, *in* "Conformation of Biological Molecules and Polymers" (E. D. Bergmann and B. Pullman, eds.), Vol. V, pp. 695–707. Academic Press, 1973.

19. T. Takano, O. B. Kallai, and R. E. Dickerson, in preparation.

20. T. Ashida, T. Ueki, T. Tsukihara, A. Sugihara, T. Takano, and M. Kakudo, *J. Biochem.* (*Tokyo*) **70,** 913–924 (1971).

21. T. Ashida, N. Tanaka, T. Yamane, T. Tsukihara, and M. Kakudo, *J. Biochem.* (*Tokyo*) **73,** 463–465 (1973).

22. T. Tsukihara, T. Yamane, N. Tanaka, T. Ashida, and M. Kakudo, *J. Biochem.* (*Tokyo*) **73,** 1163–1167 (1973).

23. N. Tanaka, T. Yamane, T. Tsukihara, T. Ashida, and M. Kakudo, *J. Biochem.* (*Tokyo*) (in press).

24. Y. Morita and S. Ida, *JMB* **71,** 807–808 (1972).

25. Y. Morita, F. Yagi, S. Ida, K. Asada, and M. Takahashi, *FEBS* (*Fed. Eur. Biochem. Soc.*) *Lett.* **31,** 186–188 (1973).

26. R. Timkovich and R. E. Dickerson, *JMB* **72,** 199–203 (1972).

27. R. Timkovich and R. E. Dickerson, *JMB* **79,** 39–56 (1973).

28. R. Timkovich and R. E. Dickerson, in preparation.

29. R. Almassy and R. E. Dickerson, in progress.

30. F. R. Salemme, S. T. Freer, Ng. H. Xuong, R. A. Alden, and J. Kraut, *JBC* **248,** 3910–3921 (1973).

31. F. R. Salemme, J. Kraut, and M. D. Kamen, *JBC* **248,** 7701–7716 (1973).

TABLE IV
X-RAY CRYSTAL STRUCTURE ANALYSES OF CYTOCHROMES *c*

Oxidation state	Species, type	Resolution of analysis (Å)	No. of derivatives[a]	Crystal form	Where studied	Date	Ref.
		A. Eukaryotic respiratory cytochrome *c*					
III	Horse	4	2[b]	$P4_3$	Caltech	1967	*11*
III	Horse, bonito	2.8	2[b]	$P4_3$	Caltech	1970	*12–14*
III	Tuna	2	4[c]	$P4_3$[j]	Caltech	1974	*15*
II	Tuna	2.45	3[d]	$P2_12_12$	Caltech	1971	*16–18*
II	Tuna	2	3[d]	$P2_12_12$	Caltech	1974	*19*
II	Bonito	4	2[e]	$P2_12_12_1$	Osaka U.	1971	*20*
II	Bonito	2.3	3[f]	$P2_12_12_1$	Osaka U.	1972	*21–23*
III	Rice	Preliminary	—	$P6_1$	Kyoto U.	1972	*24*
III	Spinach	Preliminary	—	$P2_12_12$	Kyoto U.	1973	*25*
		B. Bacterial respiratory cytochrome *c*					
III	*M. denitrificans* c_{550}	4	3[g]	$P2_12_12_1$	Caltech	1973	*26,27*
III	*M. denitrificans* c_{550}	2.5	3[g]	$P2_12_12_1$	Caltech	1974	*28*
III	*P. aeruginosa* c_{551}	4	3[h]	$P2_12_12_1$	Caltech	1974	*29*
		C. Bacterial photosynthetic cytochrome c_2					
III	*R. rubrum* c_2	2	3[i]	$P2_12_12_1$	U.C. San Diego	1972	*30,31*

[a] Derivatives used:
[b] K_2PtCl_4, mersalyl.
[c] K_2PtCl_4, $(NH_4)_2Pt(CN)_4$, $K_3Ir(NO_2)_6$, $NaAu(CN)_2$.
[d] K_2PtCl_4, $(NH_4)_2Pt(CN)_4$, K_2HgI_4.
[e] K_2PtCl_4, $K_3UO_2F_5$.
[f] K_3IrCl_6, $K_3UO_2F_5$, $(CH_3)_2SnCl_2$.
[g] K_2PtCl_4, $(NH_4)_2Pt(CN)_4$, $UO_2(NO_3)_2$.
[h] K_2PtCl_4, $UO_2(NO_3)_2$, $NaAu(CN)_2$.
[i] $[Os(NH_3)_6]I_3$, $UO_2(NO_3)_2$, $NaAuCl_4$.
[j] Two independent molecules per crystallographic asymmetric unit.

tern seen in Fig. 6 extended throughout all of the eukaryotic cytochromes *c* and reflected a common evolutionary heritage.

The analysis of reduced tuna cytochrome in a different crystal form

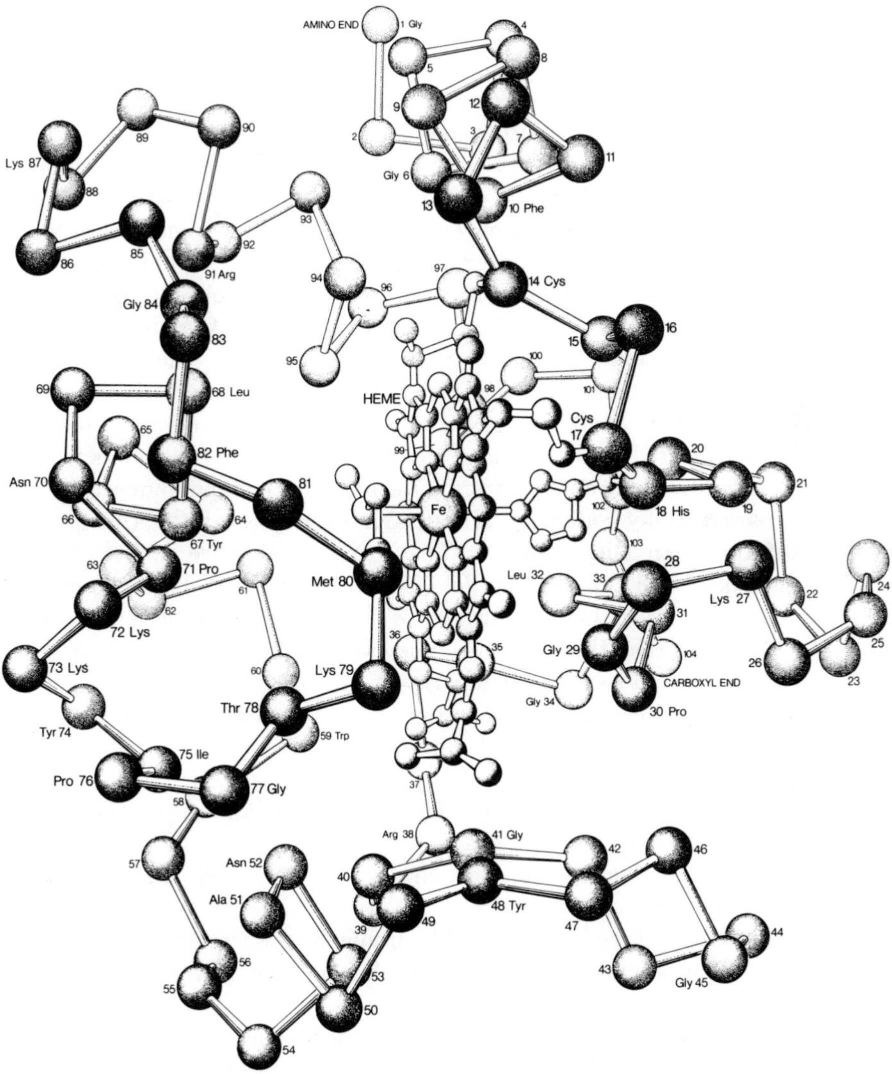

FIG. 6. Alpha-carbon skeleton of horse heart oxidized cytochrome *c*. Only α-carbon atoms are shown, with —CONH—amide groups represented by straight bonds. Only those side chains are shown which are bonded to the heme group, which is seen here nearly in an edge view. Adapted from reference *14*. Copyright by R. E. Dickerson and I. Geis.

was begun by Takano and Dickerson at Caltech in 1969. The crystals were far superior, and the analysis was completed to 2.45 Å in 1970 and 2 Å by 1973 (*16–19*). In parallel with this work, Kakudo and co-workers at the Protein Institute in Osaka carried out a study of reduced bonito cytochrome *c*, producing a 4 Å map in 1971 and extending this to 2.3 Å the following year (*20–23*).

After reexamining the problem of the horse crystals, Swanson at Caltech discovered a new crystal form of tuna oxidized cytochrome, and she and Trus carried out an entirely new structure analysis which led to a 2 Å map by mid-1974 (*15*). The coming months will be devoted to refinement, and it can be said at last that the high resolution analysis of both forms of cytochrome *c* will soon be completed. At the time of writing, good high resolution electron density maps are available for reduced cytochrome *c* from tuna and bonito, and for the oxidized molecule there exist a good 2 Å map for tuna and a less dependable 2.8 Å map for horse.

Table IV also lists the structure analyses of three bacterial cytochromes. Logically these should be delayed to the second half of the chapter, but the results have had such an important influence on thinking about oxidation–reduction mechanisms that their discussion here is mandatory. A key structure study from an evolutionary standpoint was that of *R. rubrum* c_2 by Salemme, Kraut, and colleagues at the University of California, San Diego (*30,31*). That study established in one stroke that the eukaryotic "cytochrome fold" also extended to a bacterial cytochrome and included the cytochromes of photosynthesis as well as respiration. The fact that this structural homology had been predicted on the basis of amino acid sequence comparisons (*32*) did not lessen the excitement of seeing direct confirmation from the molecular model. Only one of three possible conclusions can be drawn:

1. The electron transport chains of photosynthesis and respiration have a common evolutionary origin, or
2. One of these electron transport chains has "borrowed" a key component from the other, or
3. Nature has played an incredibly ironic joke on scientists by reinventing the same molecule for two similar but not identical purposes.

Although we are not yet able to state with assurance which portions of the cytochrome molecule are essential for its proper operation and which are secondary, cytochromes *c* and c_2 appear to resemble one another far more than could be demanded by any reasonable functional argument.

32. K. Dus, K. Sletten, and M. D. Kamen, *JBC* **243**, 5507–5518 (1968).

The more one compares their structures, the more impossible the third option becomes.

Reduced tuna cytochrome *c* (Fig. 7) may be taken as the archetype of the cytochrome fold. The heme group sits buried in the hydrophobic interior of the molecule and is exposed to the external world only along one edge: pyrrole ring 2 and the ring 2-to-ring 3 edge of the heme. (See Fig. 1 for the standard ring numbering.) The relatively short polypeptide chain can do little more than surround the heme group and give it a hydrophobic environment. It builds up what is conventionally called the right side of the molecule with an initial α-helix (residues 1–12), a short bend to attach the heme covalently at cysteine-14 and -17 to supply histidine-18 as the first of the two axial iron ligands, a 20's loop to define the right side of the molecule, a large 30's loop to the bottom rear and forward again, and residues in the 40's to build up the bottom of the molecule. By the halfway point in the polypeptide chain, the heme has been attached to the protein chain and the right half of the molecule has been constructed. The left half continues with a one-turn 50's helix and chain leading toward the bottom rear, a 60's helix and 70's loop defining the structure of the left side of the molecule, the second axial ligand to the heme iron at methionine-80, and a straight run of chain up to the top rear of the molecule, where the polypeptide concludes with the carboxyl terminal α-helix from residues 88 to 101, capped with two or three more amino acids.

The heme is held in its hydrophobic crevice by covalent bonds to cys-

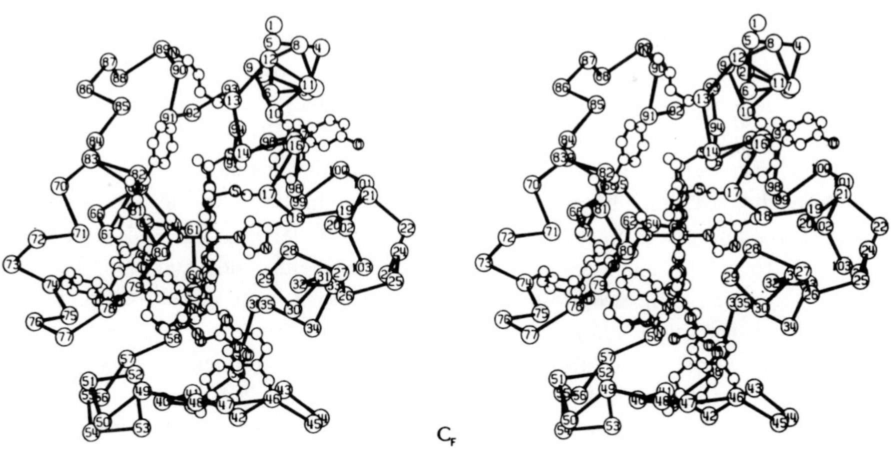

FIG. 7. Stereo pair α-carbon drawing of reduced tuna cytochrome *c*. Only those side chains which bond to the heme, and a few key lysines and aromatic groups, are shown. The heme is seen in vertical edge view at the center. Adapted from reference 7.

teine-14 and -17, and axial ligand bonds to the ϵ-nitrogen of histidine-18 and the sulfur of methionine-80. The more deeply buried of the two propionic acid side chains of the heme, on tetrapyrrole ring 4, is held in place by hydrogen bonds to tyrosine-48 and tryptophan-59, and possibly to asparagine-52 as well. The half-exposed propionic acid on ring 3 appears to be hydrogen bonded to threonine-49 and -78. All of the amino acid positions mentioned so far in this paragraph are totally unchanged through 67 species of eukaryotes with two exceptions: position 49, which in half the species is the equally good hydrogen bonding serine residue, and cysteine-14, which in *Euglena* and *Crithidia* is replaced by alanine. These latter two species lead to the remarkable conclusion that both of the covalent thioether bridges from heme to polypeptide chain are not essential for molecular stability, although two such links are the almost universal pattern. Bacterial cytochromes c_2 *and* c_{550} also show the equivalent of heme attachment by cysteine-14 and -17, histidine-18, and methionine-80, and propionic acid hydrogen bonding by tyrosine-48, tryptophan-59, and threonine–serine-49. All of these correspondences, and others, reinforce the impression that eukaryotic and bacterial cytochromes are not merely analogous, as are trypsin and subtilisin, but are evolutionarily homologous in the manner of trypsin, chymotrypsin, and elastase.

The wrapping of the polypeptide chain around the heme group can be summarized by the diagram in Fig. 8. Each amino acid side chain is represented by a circle: thin if the side chain is external, thick if the group is buried in the interior, and half and half if the chain is half-buried in the surface. The 28 totally invariant residues among the 67 species whose sequences have been examined to date are indicated by underlined capitals. A black dot extending from the amino acid toward the heme signifies that the side chain is packed against the heme group. Two of the hydrogen bonds to the heme propionic acids are indicated, as are the covalent heme connections. This diagram, while schematic, illustrates one principle of construction of the molecule: The polypeptide chain alternately loops away from the central heme and then folds back again to bring a hydrophobic residue in contact with it as though the heme was the hydrophobic "core" around which the protein chain was folded. As we shall see later, there is experimental evidence that the heme is necessary for proper folding of the molecule.

The 67 species of eukaryotes which have been examined so far show 60 different amino acid sequences. The tuna sequence is given in Table V. For each position in the table, the entry in line N tells how many of the 60 sequences have the same residue as tuna, and line O tells how many other amino acids are observed at this position. The most highly

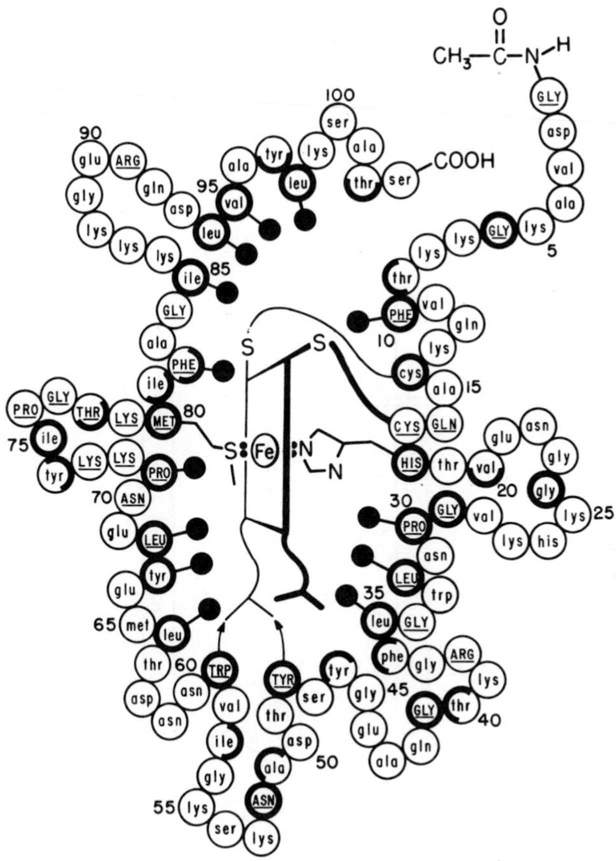

FIG. 8. Chain-packing diagram of tuna reduced cytochrome *c*, showing evolutionarily invariant amino acids (underlined capitals), buried side chains (heavy circles), and side chains packed against the heme (black dots).

conserved groups are those which attach to the heme, the hydrophobic residues packed around the heme, certain key glycines occupying tight regions in the molecule, aromatic rings, lysines maintaining surface charge distributions, and two mysteriously invariant arginines. Although individual cytochromes from distantly related species can differ in as many as 63 amino acids, the particular chemical character of each part of the molecule is strongly conserved.

Aromatic amino acids seem to be specially conserved during evolution. The positions at which such side chains occur in any of the 60 known sequences are given in Table VI, along with the number of sequences carrying each type of aromatic group. Some positions—33, 35, 61, 64, and

TABLE V

AMINO ACID SEQUENCE OF TUNA HEART CYTOCHROME c[a]

P	1	2	3	4	5	6	7	8	9	10	11	12	13	14	15	16	17	18
A	Gly	Asp	Val	Ala	Lys	Gly	Lys	Lys	Thr	Phe	Val	Gln	Lys	Cys	Ala	Gln	Cys	His
N	60	36	24	5	34	60	30	54	2	60	25	26	48	58	47	60	60	60
O	—	3	8	5	6	—	4	3	3	—	4	6	1	1	3	—	—	—
P	19	20	21	22	23	24	25	26	27	28	29	30	31	32	33	34	35	36
A	Thr	Val	Glu	Asn	Gly	Gly	Lys	His	Lys	Val	Lys	Pro	Asn	Leu	Trp	Gly	Leu	Phe
N	59	54	44	3	59	35	29	55	58	13	60	60	58	60	4	60	51	53
O	1	5	4	6	1	3	4	1	2	3	—	—	2	—	5	—	3	2
P	37	38	39	40	41	42	43	44	45	46	47	48	49	50	51	52	53	54
A	Gly	Arg	Lys	Thr	Gly	Gln	Ala	Glu	Gly	Tyr	Ser	Tyr	Thr	Asp	Ala	Asn	Lys	Ser
N	58	60	32	30	60	32	34	13	59	33	52	60	33	26	59	60	58	5
O	2	—	3	2	—	2	2	5	1	1	2	—	1	5	1	—	2	7
P	55	56	57	58	59	60	61	62	63	64	65	66	67	68	69	70	71	72
A	Lys	Gly	Ile	Val	Trp	Asn	Asn	Asp	Thr	Leu	Met	Glu	Tyr	Leu	Glu	Asn	Pro	Lys
N	42	35	32	8	60	4	2	19	56	57	22	33	59	60	36	60	60	60
O	4	2	2	8	—	8	5	7	4	2	7	5	1	—	2	—	—	—
P	73	74	75	76	77	78	79	80	81	82	83	84	85	86	87	88	89	90
A	Lys	Tyr	Ile	Pro	Gly	Thr	Lys	Met	Ile	Phe	Ala	Gly	Ile	Lys	Lys	Lys	Gly	Glu
N	60	58	58	60	60	60	60	60	25	60	28	60	24	59	58	21	10	38
O	—	1	2	—	—	—	—	—	3	—	4	—	1	1	2	7	8	2
P	91	92	93	94	95	96	97	98	99	100	101	102	103	104	105			
A	Arg	Gln	Asp	Leu	Val	Ala	Tyr	Leu	Lys	Ser	Ala	Thr	Ser	—	—			
N	60	5	58	58	4	53	59	57	55	6	39	54	11	35	59			
O	—	7	2	1	2	2	1	1	4	9	5	4	7	5	1			

[a] Abbreviations: P, amino acid position number along chain from N-terminus; A, amino acid at this position in tuna cytochrome c; N, number of species out of 60 with this amino acid; and O, number of *other* amino acids found among 60 sequences.

TABLE VI
OCCURRENCE OF AROMATIC RESIDUES IN CYTOCHROME *c*[a]

Amino acid	Position[b]													
	$\underline{10}$	$\underline{33}$	35	$\underline{36}$	$\underline{46}$	$\underline{48}$	$\underline{59}$	61	64	65	$\underline{67}$	$\underline{74}$	$\underline{82}$	$\underline{97}$
Tyrosine	—	1	—	2	33	60	—	1	—	23	59	58	—	59
Phenylalanine	60	1	2	53	27	—	—	—	1	8	1	2	60	1
Tryptophan	—	4	—	—	—	—	60	—	—	—	—	—	—	—
Other	—	54	58	5	—	—	—	59	59	29	—	—	—	—

[a] Number observed out of a set of 60 sequences.
[b] Underlined positions are aromatic in tuna cytochrome *c*.

65—are only occasionally occupied by aromatic rings. Position 36 is isolated on the back of the molecule, and its phenylalanine, tyrosine, valine, or isoleucine appears to play only a structural role as a hydrophobic residue. The other eight are invariably aromatic: always tryptophan (59), always tyrosine (48), always phenylalanine (10 and 82), or a choice between tyrosine and phenylalanine (46, 67, 74, 97). Except for tyrosine-48, hydrogen-bonded to the buried heme propionic group, aromaticity seems to be more important than the presence of the —OH group.

The eight "critical" aromatic groups are not distributed at random over the molecule but tend to occur in pairs. Tyrosine-48 and phenylalanine–tyrosine-46 are found at the bottom of the heme crevice, where 48 helps to tie down the heme and 46 provides a continuation of the hydrophobic patch formed by the edge of the heme. These residues are also present in identical positions in bacterial cytochromes c_2 and c_{550}. Phenylalanine-10 and tyrosine–phenylalanine-97 occur at the right of the molecule, in the midst of a region of packed hydrophobic side chains which has been called the "right channel" (*12*). It is important to remember that this is not an open channel giving solvent access to the heme but a conduit of hydrophobicity from the heme to the molecular surface. Phenylalanine and tyrosine are found in identical positions in c_{550} from the respiratory bacterium *Micrococcus denitrificans* (*27,28*), but in c_2 from the photosynthetic bacterium *Rhodospirillum rubrum* they are replaced by two other aromatic groups slightly lower in the right channel (*30*). The positions corresponding to 10 and 97 in the tuna *c* are occupied by serine and alanine in c_2, and the aromatic pair are supplied by a phenylalanine and tyrosine at positions equivalent to tuna 20 and 98. (One important *caveat* should be noted because of what it tells us about structural readjustments between homologous proteins. Although the tyrosine in c_2 occupies a position in space corresponding to residue 98 in tuna, its amino acid sequence homology clearly is with residue 97 as the neighboring amino acids in c and c_2 make clear. The c_2 molecule has not achieved a tyrosine lower down in the right channel by eliminating the ring at one position and adding it at the next; it has rotated the entire α-helix by 110° along its axis, bringing the still-homologous tyrosine residue to a new position in space. The structural and sequence homologies hence are one position out of phase.)

On the opposite side of the molecule, tyrosine–phenylalanine-74, tryptophan-59, and tyrosine–phenylalanine-67 form an invariantly aromatic triplet of neighboring rings leading from the heme to the surface, in what has been called the "left channel" (*12*), subject to the same qualifications as the right channel. Both bacterial cytochromes also have tyrosines and tryptophans at positions corresponding to 67 and 59, and c_2 has the equivalent of phenylalanine-74. These aromatic rings have been suggested as a possible electron pathway for reduction of cytochrome (*13,17,18*), but such ideas have been clouded by the recent observation that the position analogous to residue 74 in c_{550} is occupied by leucine (*33*). If the residues are not involved in electron transfer, then the absolute invariance of this hydrophobic channel from the surface to the back left of the heme, from respiratory eukaryotes to photosynthetic bacteria, becomes quite puzzling.

The last of the "critical" eight aromatic rings is phenylalanine-82, totally invariant over all 60 eukaryotic sequences and the two bacterial cytochromes whose three-dimensional structures are known. It is found nested against the heme, closing the upper left of the heme crevice as shown in Fig. 6. Although this residue was formerly believed from the two-derivative, 2.8 Å resolution electron density map of horse oxidized cytochrome to be swung out and away from the heme in ferricytochrome (*12–14*), the four-derivative, 2.0 Å map of tuna ferricytochrome has revealed this to be incorrect (*15*). In both oxidation states, phenylalanine-82 appears to lie next to the heme, and the crevice remains "closed" in the sense observed in Fig. 7.

The structures of ferrocytochrome c from tuna (*17*) and bonito (*23*) are in essential agreement, and Fig. 9 of ref. *23* shows a stereo comparison of the heme and left channel aromatic ring orientations in these two molecules and in horse ferricytochrome. Because of the poor quality of the horse map, these latter orientations must be considered provisional, and in a sense this chapter is one year premature. A detailed comparison of accurate ring orientations in oxidized and reduced cytochrome will soon be possible, but cannot yet be made. Although the crystal forms obtained

33. R. Timkovich, R. E. Dickerson, and E. Margoliash, *JBC* (1975) (in press).

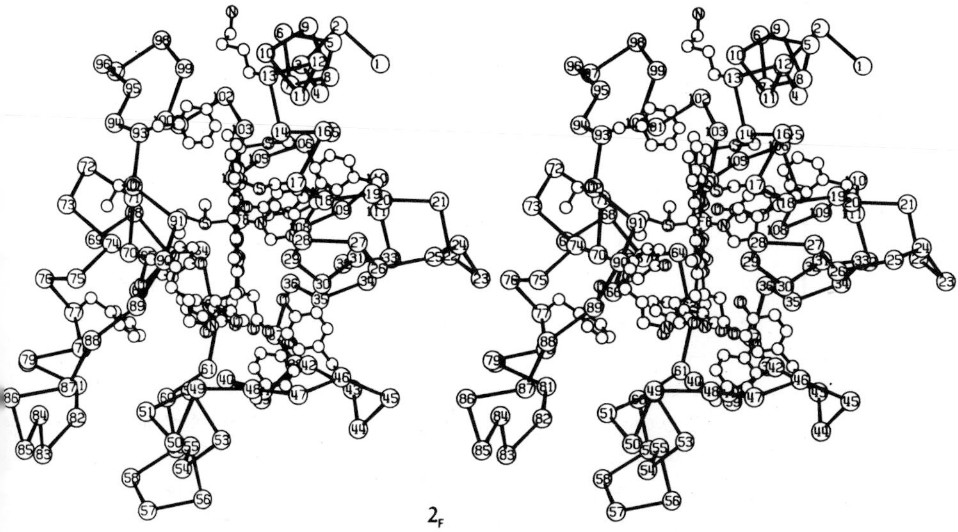

FIG. 9. Stereo pair α-carbon drawing of oxidized cytochrome c_2 from *R. rubrum*, with the same orientation and side chains shown as Fig. 7. Coordinates courtesy of F. R. Salemme and J. Kraut.

from the two oxidation states are different, Kakudo and co-workers reported that reduced cytochrome crystals from bonito can be reoxidized in the crystalline state and that difference maps using data from those crystals in both oxidation states indicate no gross conformational change of the type that would be expected with an opening up of the heme crevice at phenylalanine-82 in the ferric state (*22*). Takano and Dickerson observed exactly the same behavior by difference map methods with tuna cytochrome crystals (*34*), which was one of the main factors leading to a reexamination of the phenylalanine-82 problem and the search for a better crystal form.

The three-dimensional folding of bacterial cytochromes c_2 and c_{550} as described in refs. *30–31* and *26–28* is shown in Figs. 9 and 10, and the comparison of amino acid sequences with that of tuna *c* appears in Table VII. The sequence of c_{550} was unknown prior to the X-ray analysis, but an alignment of *c* and c_2 sequences substantially identical to that of Table VII had been proposed by Dus *et al.* (*32*) when they reported the sequence of c_2 from *Rhodospirillum rubrum*. Recently, the c_2 sequences from three rhodopseudomonads have been completed by Ambler and co-workers, and these have been included in Table VII for comparison.

The principal structural differences from the eukaryotic *c* molecule are insertions in the 50's helix and 70's loop which are shared by both

34. T. Takano and R. E. Dickerson, unpublished (1972).

TABLE VII

Comparison of Bacterial and Eukaryotic Cytochromes c, c_2, and c_{550}[a]

	1				5					10						15					20					25				
M. denitrificans c_{550}[b]	@[c]N	E	G	D	A	A	K	G	E	K	E	F	—	N	K	C	K	A	C	H	M	I	Q	A	P	D	G	T	D	I
R. capsulata c_2[d]		h	G	D	A	A	K	G	E	K	E	F	—	N	K	C	K	T	C	H	S	I	I	A	P	D	G	T	E	I
R. rubrum c_2[e]	h[f]	E	G	D	A	A	A	G	E	K	·V	S	—	K	K	C	L	A	C	H	T	F	—	—	—	D	Q	G	G	A
R. palustris c_2[d]		p	E	D	A	A	K	G	E	A	V	F	—	K	Q	C	M	T	C	H	—	—	—	—	—	R	A	D	K	N
R. spheroides c_2[d]	pE[g]	E	G	D	P	E	A	G	A	K	A	F	—	N	Q	C	Q	T	C	H	V	I	V	D	D	S	G	T	T	I
Tuna c		@[c]	G	D	V	A	K	G	K	K	T	F	V	Q	K	C	A	Q	C	H	T	V	E	—	—	N	G	G	—	—
			1				5					10					15					20								
Invariant eukaryote[h]			*					*				*						*	*	*										

			30					35					40					45					50						55		
M. denitrificans c_{550}[b]	—	—	K	G	G	K	T	G	P	N	L	Y	G	V	V	G	R	K	I	A	S	E	E	G	F	K	—	Y	G	E	G
R. capsulata c_2[d]	V	—	K	G	A	K	T	G	P	N	L	Y	G	V	V	G	R	T	A	G	T	Y	P	E	F	K	—	Y	K	D	S
R. rubrum c_2[e]	N	—	—	—	—	K	V	G	P	N	L	F	G	V	F	E	N	T	A	A	H	K	D	N	Y	A	—	Y	S	E	S
R palustris c_2[d]	M	—	—	—	—	—	V	G	P	A	L	G	G	V	V	G	R	K	A	G	T	A	A	G	F	T	—	Y	S	P	L
R. spheroides c_2[d]	A	G	R	N	A	K	T	G	P	N	L	Y	G	V	V	G	R	T	A	G	T	Q	A	D	F	K	G	Y	G	E	G
Tuna c	—	—	K	—	H	K	V	G	P	N	L	W	G	L	F	G	R	K	T	G	Q	A	E	G	Y	S	—	Y	T	D	A
			25						30					35					40					45						50	
Invariant eukaryote[h]								*	*		*		*				*			*								*			

			60					65					70					75					80					85		
M. denitrificans c_{550}[b]	I	L	E	V	A	E	K	N	P	D	L	T	W	T	E	A	N	L	I	E	Y	V	T	D	P	K	P	L	V	K
R. capsulata c_2[d]	I	—	—	V	A	L	G	A	S	G	F	A	W	T	E	E	D	I	A	T	Y	V	K	D	P	G	A	F	L	K
R. rubrum c_2[e]	Y	—	—	T	E	M	K	A	K	G	L	T	W	T	E	A	N	L	A	A	Y	V	K	D	P	K	A	F	V	L
R. palustris c_2[d]	N	—	—	H	N	S	G	E	A	G	L	V	W	T	Q	E	D	I	I	A	Y	L	P	D	P	N	A	Y	L	K
R. spheroides c_2[d]	M	—	—	K	E	A	G	A	K	G	L	A	W	D	E	E	H	F	V	Q	Y	W	Q	D	P	T	K	F	L	K
Tuna c	N	K	S	—	—	—	K	G	—	—	I	V	W	N	N	D	T	L	M	E	Y	L	E	N	P	K	K	Y	I	—
							55							60					65					70					75	
Invariant eukaryote[h]	*												*									*		*	*	*	*			

			90											95					100						105					
M. denitrificans c_{550}[b]	K	M	T	D	D	K	G	—	—	—	—	—	—	A	K	T	K	M	T	F	K	—	M	G	K	—	N	Q	A	—	D
R. capsulata c_2[d]	E	K	L	D	D	K	K	—	—	—	—	—	—	A	K	T	G	M	A	F	K	—	L	A	K	—	G	G	E	—	D
R. rubrum c_2[e]	E	K	S	E	D	P	K	—	—	—	—	—	—	A	K	S	K	M	T	F	K	—	L	T	K	D	D	Q	I	E	N
R. palustris c_2[d]	K	F	L	T	D	K	G	Q	A	D	K	A	T	G	S	T	K	M	T	F	K	—	L	A	N	D	Q	Q	R	K	D
R. spheroides c_2[d]	E	Y	T	G	D	A	K	—	—	—	—	—	—	A	K	G	K	M	I	F	K	—	L	K	K	E	A	D	A	H	N
Tuna *c*	—	—	—	—	—	P	G	—	—		—	—	—	—	—	—	T	K	M	I	F	A	G	I	K	K	K	G	E	R	Q	D
																	80					85					90			
Invariant eukaryote[h]						*	*									*	*	*		*		*							*		

	110					115					120					125					130				
M. denitrificans c_{550}[b]	V	V	A	F	L	A	Q	B	B	P	B	A	G	Z	G	Z	A	A	G	A	G	S	B	S	Z.
R. capsulata c_2[d]	V	A	A	Y	L	A	S	V	V	K.															
R. rubrum c_2[e]	V	I	A	Y	L	K	T	L	K.																
R. palustris c_2[d]	V	V	A	Y	L	A	T	L	K.																
R. spheroides c_2[d]	I	W	A	Y	L	Q	Q	V	A	V	R	P.													
Tuna *c*	L	V	A	Y	L	K	S	A	T	S.															
		95					100																		

[a] Abbreviations: A, alanine; B, Asp or Asn; C, cysteine; D, aspartic acid; E, glutamic acid; F, phenylalanine; G, glycine; H, histidine; I, isoleucine; K, lysine; L, leucine; M, methionine; N, asparagine; P, proline; Q, glutamine; R, arginine; S, serine; T, threonine; V, valine; W, tryptophan; X, trimethylysine; Y, tyrosine; Z, Glu or Gln. Asparagine at position 81/70 in *R. rubrum* (*32*) has been observed by Ambler as aspartic acid instead.

[b] From Timkovich *et al.* (*33*).

[c] Acylated amino terminus.

[d] From R. Ambler, T. Meyer, and S. Kennel, private communication.

[e] From Dus *et al.* (*32*).

[f] Free amino terminus.

[g] Pyroglutamic acid at amino terminus.

[h] Invariant residue among 60 eukaryotic sequences.

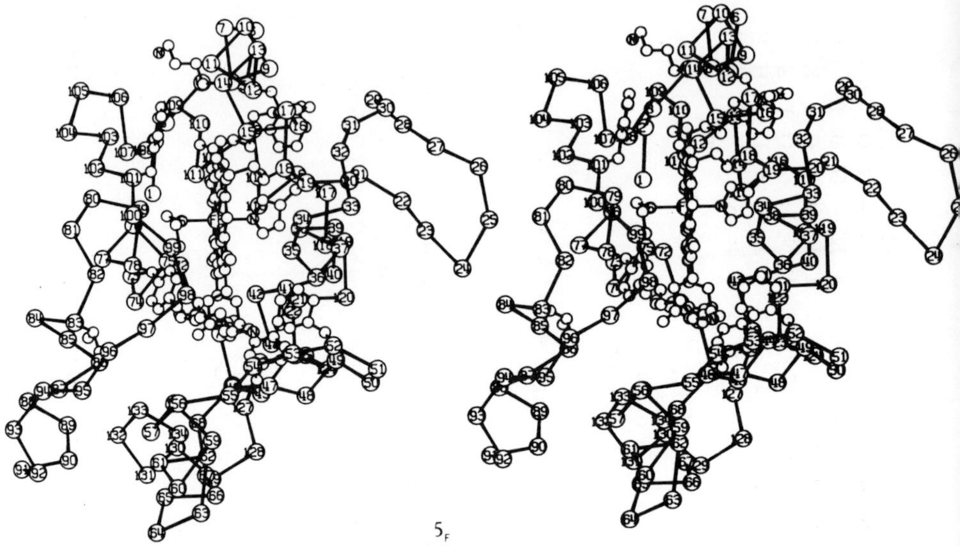

FIG. 10. Stereo pair α–carbon drawing of oxidized cytochrome c_{550} from *M. denitrificans,* with same orientation and side chains as Figs. 7 and 9. Observe the increasing amount of extra chain in Figs. 9 and 10 in comparison with Fig. 7.

c_2 and c_{550}, and an insertion in the 20's loop and elongation of the carboxyl terminus by 15 residues which are peculiar to c_{550}. The only other changes in chain path are the deletion of c residue 11 and a slight shortening of region 83–91 in both bacterial proteins. Among the three *Rhodopseudomonas* cytochromes c_2 whose sequences have just become known, *R. capsulata* and *R. spheroides* share the 20's loop insertion with c_{550}, and *R. palustris* has an even more extended 80's loop. *Micrococcus denitrificans* c_{550} and *R. capsulata* c_2 are remarkably similar molecules, differing by no more than the cytochromes c of spinach and *Euglena.* In spite of the difference in metabolic roles, respiration vs. photosynthesis, c_{550} seems like a perfectly normal member of the c_2 class.

The equivalence in essential side chains among these three cytochromes now can be seen: heme attachment groups 14, 17, 18, and 80; front and left side aromatic rings 46, 48, 59, 67, and 82; and the paired aromatics in the right channel, high for c and c_{550} (10 and 97 in tuna numbering) and lower for *R. rubrum* c_2 (20 and 107). But note that this shift in aromatic ring positions is peculiar to *R. rubrum.* All three rhodopseudomonads have the rings in the high position encountered in c and c_{550}.

Leucine-85 in c_{550} (equivalent to tuna 74) is a striking departure from the familiar pattern of aromatic rings, which is not shared by any species of c_2. The possible importance of the almost invariantly basic residues 13 and 79 at top and bottom of the heme crevice will be considered in the section on redox mechanisms. The aromatic rings in positions equivalent to tuna 33 and 36 probably have no essential role since they are absent from many eukaryotes and some c_2's. Tyrosine-52 has been suggested by Salemme *et al.* (*31*) as possibly having a role in electron transfer which could also be played by the invariant asparagine of eukaryotic *c*'s. Its replacement by isoleucine and methionine in c_{550} and two other c_2's makes this unlikely.

Table VIII shows the 12 amino acids which have remained totally invariant throughout the 67 known eukaryotes and the five photosynthetic and respiratory bacterial cytochromes of Table VII. It is a remarkably long list considering that it may represent as much as three eons of independent development, and it must provide evidence for an irreducible minimum necessary for a functioning cytochrome molecule. The list could be enlarged by including obvious functional equivalents such as tyrosine–phenylalanine at position 46, but this shorter list gains in impressiveness by its total constancy.

2. *Amino Acid Sequences of Cytochromes c*

Sixty different complete cytochrome sequences now have been determined from 67 species of eukaryotes ranging from vertebrates and insects to microorganisms and higher plants. Although many people have contributed to this work, the prime movers have been Margoliash and E.

TABLE VIII
TOTALLY CONSERVED POSITIONS AMONG 67 EUKARYOTES, FOUR BACTERIAL c_2, AND c_{550}

Amino acid	Numbering			Role
	c	c_2	c_{550}	
Cys	17	17	18	Heme attachment
His	18	18	19	Heme ligand
Met	80	91	99	Heme ligand
Tyr	48	48	54	Heme hydrogen bond
Trp	59	62	70	Heme hydrogen bond
Leu	32	32	38	Heme neighbor
Phe	82	93	101	Heme neighbor
Gly	6	7	8	Tight region
Gly	29	29	35	Tight region by heme
Gly	34	34	40	Tight region?
Pro	30	30	36	Sharp bend beside heme
Pro	71	73	82	Sharp bend

Smith and their respective co-workers for animals (28 species) and Boulter and associates for plants (24 species). This body of comparative sequence information is without parallel, the closest competitor being hemoglobin. These 60 sequences not only represent the results of extensive chemical research but also become study material for future research in their own right—in protein folding, evolutionary theory and history, and the mechanism of electron transfer. Because these sequences are scattered through the literature and not generally available in a complete and thoroughly checked form, and because some important sequence corrections have recently been made which do not appear in other compilations, we have assembled all the known sequences in Table IX. All have been checked against the original literature and subsequent corrections, which are listed as refs. *35–96*.

35. H. Matsubara and E. L. Smith, *JBC* **238,** 2732–2753 (1963).
36. S. B. Needleman and E. Margoliash, unpublished.
37. J. A. Rothfus and E. L. Smith, *JBC* **240,** 4277–4283 (1965).
38. E. Margoliash, E. L. Smith, G. Kreil, and H. Tuppy, *Nature* (*London*) **192,** 1125–1127 (1961).
39. O. F. Walasek and E. Margoliash, unpublished.
40. L. Gürtler and H. J. Horstmann, *FEBS* (*Fed. Eur. Biochem. Soc.*) *Lett.* **18,** 106–108 (1971).
41. T. Nakashima, H. Higa, H. Matsubara, A. Benson, and K. T. Yasunobu, *JBC* **241,** 1166–1177 (1966).
42. J. W. Stewart and E. Margoliash, *Can. J. Biochem.* **43,** 1187–1206 (1965).
43. S. K. Chan, S. B. Needleman, J. W. Stewart, and E. Margoliash, unpublished (1964).
44. M. Skolovsky and M. Moldovan, *Biochemistry* **11,** 145–149 (1972).
45. A. Goldstone and E. L. Smith, *JBC* **241,** 4480–4486 (1966).
46. R. C. Augusteyn, M. A. McDowall, E. C. Webb, and B. Zerner, *BBA* **257,** 264–272 (1972).
47. M. A. McDowall and E. L. Smith, *JBC* **240,** 4635–4647 (1965).
48. D. J. Strydom, S. J. van der Walt, and D. P. Botes, *Comp. Biochem. Physiol. B* **43,** 21–24 (1972).
49. S. B. Needleman and E. Margoliash, *JBC* **241,** 853–863 (1966).
50. C. Nolan and E. Margoliash, *JBC* **241,** 1049–1059 (1966).
51. S. K. Chan and E. Margoliash, *JBC* **241,** 507–515 (1966).
52. S. K. Chan, J. Tulloss, and E. Margoliash, unpublished (1966).
53. R. C. Augusteyn, *BBA* **303,** 1–7 (1973).
54. S. K. Chan, J. Tulloss, and E. Margoliash, unpublished (1967).
55. S. K. Chan, J. Tulloss, and E. Margoliash, *Biochemistry* **5,** 2586–2597 (1966).
56. O. P. Bahl and E. L. Smith, *JBC* **240,** 3585–3593 (1965).
57. S. K. Chan, O. F. Walasek, G. H. Barlow, and E. Margoliash, unpublished (1967).
58. G. Kreil, *Hoppe-Seyler's Z. Physiol. Chem.* **334,** 154–166 (1963); **340,** 86–87 (1965).
59. T. Nakayama, K. Titani, and K. Narita, *J. Biochem.* (*Tokyo*) **70,** 311–326 (1971).

The principal corrections are to four microorganisms, bringing them into much better homology with other eukaryotes. After correction, *Saccharomyces, Candida, Neurospora*, and *Humicola* all have the sequence

60. L. Gürtler and H. J. Horstmann, *Eur. J. Biochem.* **12,** 48–47 (1970).
61. A Goldstone and E. L. Smith, *JBC* **242,** 4702–4710 (1967).
62. C. Nolan, W. M. Fitch, T. Uzzell, L. J. Weiss, and E. Margoliash, *Biochemistry* **12,** 4052–4060 (1973).
63. R. H. Brown, M. Richardson, D. Boulter, J. A. M. Ramshaw, and R. P. S. Jefferies, *BJ* **128,** 971–974 (1972).
64. C. Nolan, L. J. Weiss, J. J. Adams, and E. Margoliash, unpublished (1968).
65. S. K. Chan and E. Margoliash, *JBC* **241,** 335–348 (1966).
66. S. K. Chan, *BBA* **221,** 497–501 (1970).
67. K. Narita and K. Titani, *J. Biochem.* (*Tokyo*) **65,** 259–267 (1969).
68. Y. Yaoi, *J. Biochem.* (*Tokyo*) **61,** 54–58 (1967).
69. F. Lederer, A. M. Simon, and J. Verdière, *BBRC* **47,** 55–58 (1972).
70. K. Sugeno, K. Narita, and K. Titani, *J. Biochem.* (*Tokyo*) **70,** 659–682 (1971).
71. K. Narita and K. Titani, *J. Biochem.* (*Tokyo*) **63,** 226–241 (1968).
72. F. Lederer, *Eur. J. Biochem.* **31,** 144–147 (1972).
73. J. Heller and E. L. Smith, *JBC* **241,** 3165–3180 (1966).
74. F. Lederer and A. M. Simon, *BBRC* **56,** 317–323 (1974).
75. W. T. Morgan, C. P. Hensley, Jr., and J. P. Riehm, *JBC* **247,** 6555–6565 (1972).
76. K. G. Bitar, S. N. Vinogradov, C. Nolan, L. J. Weiss, and E. Margoliash, *BJ* **129,** 561–569 (1972).
77. G. W. Pettigrew, *FEBS* (*Fed. Eur. Biochem. Soc.*) *Lett.* **22,** 64–66 (1972).
78. G. W. Pettigrew, *Nature* (*London*) **241,** 531–533 (1973).
79. D. K. Lin, R. L. Niece, and W. M. Fitch, *Nature* (*London*) **241,** 533–535 (1973).
80. R. H. Brown and D. Boulter, *BJ* **133,** 251–254 (1973).
81. E. W. Thompson, M. V. Laycock, J. A. M. Ramshaw, and D. Boulter, *BJ* **117,** 183–192 (1970) (see corection in reference *88*).
82. E. W. Thompson, M. Richardson, and D. Boulter, *BJ* 124, 783–785 (1971).
83. M. Richardson, J. A. M. Ramshaw, and D. Boulter, *BBA* **251,** 331–333 (1971).
84. E. W. Thompson, M. Richardson, and D. Boulter, *BJ* **124,** 779–781 (1971).
85. D. G. Wallace, R. H. Brown, and D. Boulter, *Phytochemistry* **12,** 2617–2622 (1973).
86. R. H. Brown and D. Boulter, *BJ* **137,** 93–100 (1974).
87. E. W. Thompson, B. A. Notton, M. Richardson, and D. Boulter, *BJ* **124,** 787–791 (1971).
88. E. W. Thompson, M. Richardson, and D. Boulter, *BJ* **121,** 439–446 (1971).
89. R. Scogin, M. Richardson, and D. Boulter, *ABB* **150,** 489–492 (1972).
90. D. Boulter, private communication (1974).
91. R. H. Brown and D. Boulter, *BJ* **131,** 247–251 (1973.)
92. J. A. M. Ramshaw, E. W. Thompson, and D. Boulter, *BJ* **119,** 535–539 (1970).
93. F. C. Stevens, A. N. Glazer, and E. L. Smith, *JBC* **242,** 2764–2779 (1967).
94. R. J. Delange, A. N. Glazer, and E. L. Smith, *JBC* **244,** 1385–1388 (1969).
95. R. H. Brown, M. Richardson, R. Scogin, and D. Boulter, *BJ* **131,** 253–256 (1973).
96. J. A. M. Ramshaw, M. Richardson, and D. Boulter, *Eur. J. Biochem.* **23,** 475–483 (1971).

Leu-Gln or Ala-Gln at positions 15–16 between the heme attachments, and glutamine-16 becomes an evolutionarily invariant residue. Residues 19–25 of *Neurospora* and *Humicola* also are completely rearranged to bring them into line with other organisms. (Since an X-ray structure analysis of *Neurospora* was once contemplated by the authors solely on the basis of the strange sequence anomalies in the 20's loop, this correction of an erroneous sequence in a way is disappointing.)

The formidable quantity of data contained in Table IX can be reduced to more manageable proportions in Table X, where the frequency of occurrence of each amino acid among the 60 known sequences is tabulated against position along the polypeptide chain. The first row, marked by a dash at the left, indicates no amino acid present at that position. The next four rows are the charged side chains, which are almost invariably on the outside of the molecule: lysine (K), arginine (R), and aspartic (D) and glutamic (E) acids. Following these are the polar but uncharged residues which can exist both at the surface and in the interior: asparagine (N) through cysteine (C), with glycine (G) and alanine (A) included because the price in free energy of bringing them into a hydrophilic environment is small. Last are the larger hydrophobic residues, valine (V) through tryptophan (W), which are found mainly in the interior of proteins. At the very end are the two amide ambiguities: B for aspartic acid or asparagine and Z for glutamic acid or glutamine; and X indicating a trimethylated lysine, found at positions 72 and 86 in most plants.

a. Invariant Amino Acids. The constancy or near-constancy of the heme attachments, hydrogen bonding groups, and aromatics now can be verified. Of the 28 totally invariant residues noted earlier in Table V, seven are glycines, frequently located in tight corners where no side chain can be accomodated. This accounts for the exclusive presence of glycine at positions 6, 29, 34 (a 3_{10} bend; see ref. *12*), 41, 77 (another 3_{10} bend), and possibly 84, but there seems to be no explanation for the invariance of glycine-1, seen also in bacterial c_2 and c_{550}. Perhaps this reflects the manner in which the apocytochrome is folded prior to attachment of the heme or a possible interaction with the mitochondrial membrane or another molecule. Whatever the cause, the complete invariance at position 1 of the chain, even when this chain is elongated by four to nine residues in invertebrates, is quite striking.

The invariance of the groups covalently attached to the heme has already been discussed. Two constant leucines (32 and 68) are in tight van der Waals contact with the heme. Several other such contact positions, although not invariant, are strictly limited to leucine, isoleucine, or valine. In no case is an internal hydrophobic group close to the heme ever replaced by anything other than another hydrophobic residue. The requirement for a hydrophobic environment around the heme is absolute.

Three invariant prolines define sharp bends in the chain at positions 30, 71, and 76. Two of these are retained in both bacterial cytochromes, and the other (76) is not because the bacterial chains wander off in new directions at that point. Threonine-78 is hydrogen-bonded to the outermost heme propionic acid, and possibly to tyrosine-67 also. This position is still hydrogen bonding in bacteria: threonine in c_{550} and serine in c_2. Asparagine-52 is hydrogen-bonded to the inner propionic, but no obvious role can be assigned to the equally invariant asparagine-70. Glutamine-16 probably is invariant because it interacts with cytochrome oxidase or reductase at the edge of the heme crevice, and we will return to this point later.

Positive charges apparently are subject to strong evolutionary constraints. The two arginines in cytochrome *c* at positions 38 and 91 are totally invariant among eukaryotes, although not in bacteria. Three lysines (72, 73, 79) are unchanging, and several others (8, 27, 53, 87, 99) are present in all but four or five species. When one of these five residues is not lysine, another positive charge usually is to be found in the immediate vicinity on the surface of the molecule. It is important to keep in mind that the test of whether a molecule can function or not is conducted on the folded molecule and not on the uncoiled amino acid chain. Two positions widely separated along the sequence may be functionally equivalent if they are neighbors in the completely folded protein. Although no acidic residue is totally conserved, there is conservatism in the location of such negative groups. Many acidic residues are clustered around the upper rear of the molecule as seen in Fig. 7, in the vicinity of residues 2, the early 90's, and the 60's. Negative charge is rigidly excluded from the front face and top—the region of the heme crevice. The possible functional meaning of this will be considered later.

One elèven-amino acid segment of chain, from 70 to 80, was thought for many years to be absolutely essential because it was invariant among all species then known. More recently this invariance has been broken, but in a conservative manner. Tyrosine-74 has been observed as phenylalanine in *Humicola* and *Crithidia,* and isoleucine-75 is valine in *Euglena* and methionine in *Crithidia.* (These two species also have the anomalous alanine at position 14.) Such conservative changes in a small number of distantly related eukaryotic organisms do not detract from the prediction that the left side of the molecule, residues 70–80, must have an important functional role that keeps its sequence so nearly invariant.

b. Hypervariable Regions. In contrast to the foregoing, other parts of the molecule are quite variable in a way that suggests that they have little or no function. These "hypervariable" regions are illustrated in color in reference *14,* in the drawing from which Fig. 6 was adapted. As Table V indicates, 10 different amino acids are encountered at position

TABLE IX
Amino Acid Sequences of Eukaryotic Cytochromes *c*

I. Animals

SPECIES		REF.	1 2 3 4
MAN=CHIMP	1	35,36	@GDVEKGKKIFIMKCSQCHTVEKGGKHKTGPNLHGLFGRKT
RHESUS	2	37	@GDVEKGKKIFIMKCSQCHTVEKGGKHKTGPNLHGLFGRKT
HORSE	3	38	@GDVEKGKKIFVQKCAQCHTVEKGGKHKTGPNLHGLFGRKT
DONKEY=ZEBRA	4	39,40	@GDVEKGKKIFVQKCAQCHTVEKGGKHKTGPNLHGLFGRKT
COW=PIG=SHEEP	5	41-43	@GDVEKGKKIFVQKCAQCHTVEKGGKHKTGPNLHGLFGRKT
CAMEL=GR WHALE	6	44,45	@GDVEKGKKIFVQKCAQCHTVEKGGKHKTGPNLHGLFGRKT
ELEPHANT SEAL	7	46	@GDVEKGKKIFVQKCAQCHTVEKGGKHKTGPNLHGLFGRKT
DOG	8	47	@GDVEKGKKIFVQKCAQCHTVEKGGKHKTGPNLHGLFGRKT
BAT	9	48	@GDVEKGKKIFVQKCAQCHTVEKGGKHKTGPNLHGLFGRKT
RABBIT	10	49	@GDVEKGKKIFVQKCAQCHTVEKGGKHKTGPNLHGLFGRKT
KANGAROO	11	50	@GDVEKGKKIFVQKCAQCHTVEKGGKHKTGPNLNGIFGRKT
CHICKEN=TURKEY	12	51,52	@GDIEKGKKIFVQKCSQCHTVEKGGKHKTGPNLHGLFGRKT
EMU	13	53	@GDIEKGKKIFVQKCSQCHTVEKGGKHKTGPNLNGLFGRKT
KING PENGUIN	14	54	@GDIEKGKKIFVQKCSQCHTVEKGGKHKTGPNLHGIFGRKT
PEKIN DUCK	15	52	@GDVEKGKKIFVQKCSQCHTVEKGGKHKTGPNLHGLFGRKT
PIGEON	16	52	@GDIEKGKKIFVQKCSQCHTVEKGGKHKTGPNLHGLFGRKT
S TURTLE	17	55	@GDVEKGKKIFVQKCAQCHTVEKGGKHKTGPNLNGLIGRKT
RATTLESNAKE	18	56	@GDVEKGKKIFTMKCSQCHTVEKGGKHKTGPNLHGLFGRKT
BULLFROG	19	57	@GDVEKGKKIFVQKCAQCHTCEKGGKHKVGPNLYGLIGRKT
TUNA	20	58	@GDVAKGKKTFVQKCAQCHTVENGGKHKVGPNLWGLFGRKT
BONITO	21	59	@GDVAKGKKTFVQKCAQCHTVENGGKHKVGPNLWGLFGRKT
CARP	22	60	@GDVEKGKKVFVQKCAQCHTVZBGGKHKVGPNLWGLFGRKT
DOGFISH	23	61	@GDVEKGKKVFVQKCAQCHTVENGGKHKTGPNLSGLFGRKT
PACIFIC LAMPREY	24	62	@GDVEKGKKVFVQKCSQCHTVEKAGKHKTGPNLSGLFGRKT
GARDEN SNAIL	25	63	*GZAZKGKKIFTQKCLQCHTVEAGGKHKTGPNLSGLFGRKQ
FRUIT FLY	26	64	*GVPAGDVEKGKKLFVQRCAQCHTVEAGGKHKVGPNLHGLIGRKT
S W FLY	27	54	*GVPAGDVEKGKKIFVQRCAQCHTVEAGGKHKVGPNLHGLFGRKT
SAMIA CYNTHIA	28	65	*GVPAGNAENGKKIFVQRCAQCHTVEAGGKHKVGPNLHGFYGRKT
T H W MOTH	29	66	*GVPAGNADNGKKIFVQRCAQCHTVEAGGKHKVGPNLHGFFGRKT

5 6 7 8 9 10	NOTES	
GQAPGYSYTAANKNKGIIWGEDTLMEYLENPKKYIPGTKMIFVGIKKKEERADLIAYLKKATNE		MAN
GQAPGYSYTAANKNKGITWGEDTLMEYLENPKKYIPGTKMIFVGIKKKEERADLIAYLKKATNE		RHESUS
GQAPGFTYTDANKNKGITWKEETLMEYLENPKKYIPGTKMIFAGIKKKTEREDLIAYLKKATNE		HORSE
GQAPGFSYTDANKNKGITWKEETLMEYLENPKKYIPGTKMIFAGIKKKTEREDLIAYLKKATNE		DONKEY
GQAPGFSYTDANKNKGITWGEETLMEYLENPKKYIPGTKMIFAGIKKKGEREDLIAYLKKATNE		COW
GQAVGFSYTDANKNKGITWGEETLMEYLENPKKYIPGTKMIFAGIKKKGERADLIAYLKKATNE		G WHALE
GQAPGFSYTDANKNKGITWGEETLMEYLENPKKYIPGTKMIFAGIKKTGERADLIAYLKIATKE		ELE SEAL
GQAPGFSYTDANKNKGITWGEETLMEYLENPKKYIPGTKMIFAGIKKTGERADLIAYLKKATKE		DOG
GQAPGFSYTDANKNKGITWGEATLMEYLENPKKYIPGTKMIFAGIKKSAERADLIAYLKKATKE		BAT
GQAVGFSYTDANKNKGITWGEDTLMEYLENPKKYIPGTKMIFAGIKKKDERADLIAYLKKATNE		RABBIT
GQAPGFTYTDANKNKGIIWGEDTLMEYLENPKKYIPGTKMIFAGIKKKGERADLIAYLKKATNE		KANGAROO
GQAEGFSYTDANKNKGITWGEDTLMEYLENPKKYIPGTKMIFAGIKKKSERVDLIAYLKDATSK		CHICKEN
GQAEGFSYTDANKNKGITWGEDTLMEYLENPKKYIPGTKMIFAGIKKKSERADLIAYLKDATSK		EMU
GQAEGFSYTDANKNKGITWGEDTLMEYLENPKKYIPGTKMIFAGIKKKSERADLIAYLKDATSK		PENGUIN
GQAEGFSYTDANKNKGITWGEDTLMEYLENPKKYIPGTKMIFAGIKKKSERADLIAYLKDATAK		PEKIN DUCK
GQAEGFSYTDANKNKGITWGEDTLMEYLENPKKYIPGTKMIFAGIKKKAERADLIAYLKQATAK		PIGEON
GQAEGFSYTEANKNKGITWGEETLMEYLENPKKYIPGTKMIFAGIKKKAERADLIAYLKDATSK		S TURTLE
GQAVGYSYTAANKNKGIIWGDDTLMEYLENPKKYIPGTKMVFTGLSKKKERTNLIAYLKEKTAA		R SNAKE
GQAAGFSYTDANKNKGITWGEDTLMEYLENPKKYIPGTKMIFAGIKKKGERQDLIAYLKSACSK	1	BULLFROG
GQAEGYSYTDANKSKGIVWNNDTLMEYLENPKKYIPGTKMIFAGIKKKGERQDLVAYLKSATS-		TUNA
GQAEGYSYTDANKSKGIVWNENTLMEYLENPKKYIPGTKMIFAGIKKKGERQDLVAYLKSATS-		BONITO
GQAPGFSYTBANKSKGIVWBZZTLMEYLENPKKYIPGTKMIFAGIKKKGERADLIAYLKSATS-	2	CARP
GQAQGFSYTDANKSKGITWQQETLRIYLENPKKYIPGTKMIFAGLKKKSERQDLIAYLKKTAAS		DOGFISH
GQAPGFSYTDANKSKGIVWNQETLFVYLENPKKYIPGTKMIFAGIKKEGERKDLIAYLKKSTSE	3	LAMPREY
GQAPGFAYTDANKGKGITWKNQTLFQYLENPKKYIPGTKMVFAGLKBZTERVHLIAYLZZATKK		SNAIL
GQAAGFAYTNANKAKGITWQDDTLFEYLENPKKYIPGTKMIFAGLKKPNERGDLIAYLKSATK-		FRUIT FLY
GQAAGFAYTNANKAKGITWQDDTLFEYLENPKKYIPGTKMIFAGLKKPNERGDLIAYLKSATK-		S W FLY
GQAPGFSYSNANKAKGITWGDDTLFEYLENPKKYIPGTKMVFAGLKKANERADLIAYLKESTK-		S CYNTHIA
GQAPGFSYSNANKAKGITWQDDTLFEYLENPKKYIPGTKMVFAGLKKANERADLIAYLKQATK-		T H W MOTH

TABLE IX (*Continued*)

II. Microorganisms and higher plants

SPECIES		REF.	1 2 3 4
SACCHAROMYCES	30	67-69	*TEFKAGSAKKGATLFKTRCLQCHTVEKGGPHKVGPNLHGIFGRHS
DEBARYOMYCES	31	70	*PAPYEKGSEKKGANLFKTRCLQCHTVEEGGPHKVGPNLHGVVGRTS
CANDIDA KRUSEI	32	71,72	*PAPFEQGSAKKGATLFKTRCAQCHTIEAGGPHKVGPNLHGIFSRHS
N. CRASSA	33	73,74	*GFSAGDSKKGANLFKTRCAQCHTLEEGGGNKIGPALHGLFGRKT
HUMICOLA	34	74,75	*AKGGSFEPGDASKGANLFKTRCAQCHSVEQGGANKIGPNLHGLFGRKT
USTILAGO	35	76	*GFEDGDAKKGARIFKTRCAQCHTLGAGEPNKVGPNLHGLFGRKS
CRITHIDIA	36	77	@PXAREPLPPGDAAKGEKIFKGRAAQCHTGAKGGANGVGPNLFGIVNRHS
EUGLENA	37	78,79	@GDAERGKKLFESRAAQCHSAQKGV-NSTGPSLWGVYGRTS
NIGELLA	38	80	@ASFDEAPAGNSASGEKIFKTKCAQCHTVDQGAGHKQGPNLHGLFGRQS
MUNG BEAN	39	81	@ASFDEAPPGNSKSGEKIFKTKCAQCHTVDKGAGHKQGPNLNGLFGRQS
CAULIFLOWER=RAPE	40	82,83	@ASFDEAPPGNSKAGEKIFKTKCAQCHTVDKGAGHKQGPNLNGLFGRQS
PUMPKIN	41	84	@ASFNEAPPGNSKAGEKIFKTKCAQCHTVDKGAGHKQGPNLNGLFGRQS
HEMP	42	85	@ASFDEAPPGNSKAGEKIFKTKCAQCHTVGRGAGHKQGPNLNGLFGRQS
ELDER	43	86	@ASFAEAPPGNPKAGEKIFKTKCNQCHTVDKGAGHKQGPNLNGLFGRQS
ABUTILON SEED	44	87	@ASFQEAPPGNAKAGEKIFKTKCAQCHTVEKGAGHKQGPNLNGLFGRQS
COTTON SEED	45	87	@ASFQEAPPGNAKAGEKIFKTKCAQCHTVDKGAGHKQGPNLNGLFGRQS
CASTOR BEAN	46	88	@ASFNEAPPGNVKAGEKIFKTKCAQCHTVEKGAGHKQGPNLNGLFGRQS
TOMATO	47	89	@ASFDEAPPGNPKAGEKIFKTKCAQCHTVEKGAGHKQGPNLNGLFGRQS
MAIZE	48	90	@ASFSEAPPGNPKAGEKIFKTKCAQCHTVEKGAGHKQGPNLNGLFGRQS
ARUM	49	90	@ASFAEAPPGNPKAGEKIFKTKCAQCHTVEKGAGHKQGPNLNGLFGRQS
SESAME SEED	50	88	@ASFDEAPPGNVKSGEKIFKTKCAQCHTVDKGAGHKQGPNLNGLFGRQS
LEEK	51	91	@ATFAEAPPGNQKAGEKIFKLKCAQCHTVEKGAGHKQGPNLNGLFGRQS
ACER	52	86	@ASFAEAPPGNPAAGEKIFKTKCAQCHTVDKGAGHKQGPNLNGLFGRQS
NIGER	53	90	@ASFAEAPAGDAKAGEKIFKTKCAQCHTVEKGAGHKQGPNLNGLFGRQS
SUNFLOWER SEED	54	92	@ASFAEAPAGDPTTGAKIFKTKCAQCHTVEKGAGHKQGPNLNGLFGRQS
NASTURTIUM	55	86	@ASFAEAPAGDNKAGDKIFKNKCAQCHTVDKGAGHKQGPNLNGLFGRQS
PARSNIP	56	86	@ASFAEAPPGDKDVGGKIFKTKCAQCHTVELGAGHKQGPNLNGLFGRQS
WHEAT GERM	57	93,94	@ASFSEAPPGNPDAGAKIFKTKCAQCHTVDAGAGHKQGPNLHGLFGRQS
BUCKWHEAT SEED	58	82	@ATFSEAPPGNIKSGEKIFKTKCAQCHTVEKGAGHKQGPNLNGLFGRQS
SPINACH	59	95	@ATFSEAPPGNKDVGAKIFKTKCAQCHTVDLGAGHKQGPNLNGLFGRQS

5 6 7 8 9 10	NOTES	
GQAEGYSYTDANIKKNVLWDENNMSEYLTNPXKYIPGTKMAFGGLKKEKDRNDLITYLKKACE-	4	SACCHAROMY
GQAQGFSYTDANKKKGVEWTEQDLSDYLENPXKYIPGTKMAFGGLKKAKDRNDLITYLVKATK-		DEBAROMYCE
GQAEGYSYTDANKRAGVEWAEPTMSDYLENPXKYIPGTKMAFGGLKKAKDRNDLVTYMLEASK-	4	CANDIDA
GSVDGYAYTDANKQKGITWDENTLFEYLENPXKYIPGTKMAFGGLKKDKDRNDIITFMKEATA-	4	N. CRASSA
GSVEGYSYTDANKQAGITWNEDTLFEYLENPXKFIPGTKMAFGGLXKNKDRNDLITYLKEATK-	4	HUMICOLA
GTVEGFSYTDANKKAGQVWEEETFLEYLENPKKYIPGTKMAFGGLKKEKDRNDLVTYLREETK-		USTILAGO
GTVEGFAYSKANADSGVVWTPEVLDVYLENPXKFMPGTKMSFAGIKKPQERADLIAYLENLK--		CRITHIDIA
GSVPGYAYSNANKNAAIVWEEETLHKFLENPKKYVPGTKMAFAGIXAKKDRQDIIAYMKTLKD-		EUGLENA
GTVAGYSYSAANKNKAVNWEEKTLYDYLLNPXKYIPGTKMVFPGLXKPQERADLLAYLKESTA-	5	NIGELLA
GTTAGYSYSTANKNMAVIWEEKTLYDYLLNPXKYIPGTKMVFPGLXKPQDRADLIAYLKESTA-	6,7	MUNG BEAN
GTTAGYSYSAANKNKAVEWEEKTLYDYLLNPXKYIPGTKMVFPGLXKPQDRADLIAYLKEATA-	8	CAULIFLOW.
GTTPGYSYSAANKNRAVIWEEKTLYDYLLNPXKYIPGTKMVFPGLXKPQDRADLIAYLKEATA-	9	PUMPKIN
GTTAGYSYSAANKNMAVTWQEKTLYDYLLNPXKYIPGTKMVFPGLXKPQDRADLIAYLKESTA-	10	HEMP
GTTAGYSYSAANKNMAVNWEEKTLYDYLLNPXKYIPGTKMVFPGLXKPQDRADLIAYLKQSTA-		ELDER
GTTPGYSYSAANKNMAVNWGENTLYDYLLNPXKYIPGTKMVFPGLXKPQDRADLIAYLKESTA-	8	ABUTILON
GTTAGYSYSAANKNMAVQWGENTLYDYLLNPXKYIPGTKMVFPGLXKPQDRADLIAYLKESTA-	8	COTTON SD.
GTTAGYSYSAANKNMAVQWGENTLYDYLLNPXKYIPGTKMVFPGLXKPQDRADLIAYLKQATA-	7	CASTOR BN.
GTTAGYSYSAANKNMAVNWGENTLYDYLLNPXKYIPGTKMVFPGLXKPQQRADLIAYLKEATA-	11	TOMATO
GTTAGYSYSAANKNKAVVWEENTLYDYLLNPXKYIPGTKMVFPGLXKPQERADLIAYLKEATA-		MAIZE
GTTAGYSYSAANKNMAVIWEESTLYDYLLNPXKYIPGTKMVFPGLXKPQERADLIAYLKESTA-		ARUM
GTTPGYSYSAANKNMAVIWGENTLYDYLLNPXKYIPGTKMVFPGLXKPQERADLIAYLKEATA-	7	SESAME SD.
GTAAGYSYSAANKNMAVGWEENTLYDYLLNPXKYIPGTKMVFPGLXKPQDRADLIAYLKESTA-	12	LEEK
GTTAGYSYSAANKNMAVNWGYNTLYDYLLNPXKYIPGTKMVFPGLXKPQDRADLIAYLKQSTAS		ACER
GTTAGYSYSAANKNKAVAWEENSLYDYLLNPXKYIPGTKMVFPGLXKPQERADLIAYLKASTA-		NIGER
GTTAGYSYSAANKNMAVIWEENTLYDYLLNPXKYIPGTKMVFPGLXKPQERADLIAYLKTSTA-		SUNFLOWER
GTTAGYSYSAANKNKAVLWEEATLYDYLLNPXKYIPGTKMVFPGLXKPQDRADLIAYLKESTA-		NASTURTIUM
GTTAGYSYSAANKNKAVLWADNTLYDYLLNPXKYIPGTKMVFPGLXKPQDRADLIAYLKHATA-		PARSNIP
GTTAGYSYSAANKNKAVEWEENTLYDYLLNPXKYIPGTKMVFPGLXKPQDRADLIAYLKKATSS		WHEAT GERM
GTTAGYSYSAANKNKAVTWGEDTLYEYLLNPXKYIPGTKMVFPGLXKPQERADLIAYLKDSTZ-		BUCKWHEAT
GTAASYSYSAANKNKAVIWSEDTLYEYLLNPXKYIPGTKMVFPGLXKPQDRADLIAYLKDSTQ-		SPINACH
GTTAGYSYSTGNKNKAVNWGEQTLYEYLLNPXKYIPGTKMVFPGLXKPQERADLISYLKQATSQE	13	GINKGO

Notes to Table IX:

All amino acid residues have been checked against the original literature.

1, 2, 3, 4, . . . at head of column indicate residues 10, 20, 30, 40, ...

See Table VII for explanation of one-letter amino acid symbols. @, acylated amino terminus; *, free H_2N—group at amino terminus.

1. Sequence 88–92 undetermined.
2. Reported as 52 B, 69 Z, 70 B. Amidation assumed (RED) by analogy.
3. 33 Q in Dayhoff is an error of a prepublication sequence.
4. Includes corrections by Lederer *et al.* (*69,72,74*).
5. Amidation of −5, −4, 2, 16, 21, 22, 28, 89, 90, 93 by personal communication.
6. Includes corrections by Thompson *et al.* (*88*).
7. Amidation of −5, −4, 2 by personal communication.
8. Amidation of −5, 2 by personal communication.
9. Amidation of −5 by personal communication.
10. Amidation of −5, −4, 2, 60, 61, 89, 90 by personal communication. 16 E reported, changed (RED) to Q by analogy with other 59 sequences.
11. −5 should be D, and not N as originally reported (personal communication).
12. −5, 7, 58 and amidation state of −4, 2, 3, 60, 61, 62 by personal communication.
13. Amidation of 16, 21, 61, 62, 89, 90 by personal communication.

100 on the back of the molecule (listed as nine alternates to the tuna sequence) and eight at position 103. Nearby on the molecular surface (although far along the sequence), position 3 exhibits nine different side chains. At the top rear, residues 88, 89, and 92 at the final bend before the C-terminal helix have been observed with 8, 9, and 8 different possibilities. Farther down the back of the molecule, positions 58, 60, 62, and 65 show 9, 9, 8, and 8 choices, and 54 on the bottom has 8 alternatives. Almost any type of amino acid can be tolerated at these positions, but hydrophobic groups are less common because most of the sites lie on the surface of the molecule. It is a reasonable deduction from this variability that the back side of the molecule, from top to bottom, cannot be a critical interaction surface with other macromolecules. This will become important later when possible electron transfer mechanisms are discussed.

3. *Evolution of Eukaryotic Cytochromes c*

The evolution of cytochrome *c* and protein molecules in general has been the subject of many papers and reviews, and can be treated relatively briefly here. The starting points must be Anfinson's "Molecular Basis of Evolution" (*97*), based on the early amino acid sequence work, and the discovery that same year by Kendrew and Perutz of the identity of three-dimensional folding of the related proteins myoglobin and hemoglobin (*98,99*). These illustrate the two components of any study of macromolecular evolution: the chemical sequence of the polymer and the folding of the polymer in three dimensions.

Other important reviews dealing entirely or in part with the evolution of cytochrome *c* include Jukes in 1966 (*100*), Margoliash and Schejter in 1966 (*3*), Nolan and Margoliash in 1968 (*101*), the volumes of the Dayhoff Atlas (*102*), Fitch and Margoliash in 1970 (*103*), Jukes and Cantor in 1970 (*104*), Margoliash's Harvey Lecture in 1972 (*105*), Chap-

97. C. B. Anfinsen, "The Molecular Basis of Evolution." Wiley, New York, 1959.
98. J. C. Kendrew, R. E. Dickerson, B. E. Strandberg, R. G. Hart, D. R. Davies, D. C. Phillips, and V. C. Shore, *Nature* (*London*) **185**, 422–427 (1960).
99. M. F. Perutz, M. G. Rossmann, A. F. Cullis, H. Muirhead, G. Will, and A. C. T. North, *Nature* (*London*) **185**, 416–422 (1960).
100. T. H. Jukes, "Molecules and Evolution." Columbia Univ. Press, New York, 1966.
101. C. Nolan and E. Margoliash, *Annu. Rev. Biochem.* **37**, 727–790 (1968).
102. M. O. Dayhoff, "Atlas of Protein Sequence and Structure," vols. 4, 5, and Suppl. Nat. Biomed. Res. Found. Georgetown Univ. Med. Cent., Washington, D. C., 1969, 1972, and 1973 (resp.).
103. W. M. Fitch and E. Margoliash, *Evol. Biol.* **4**, 67–109 (1970).
104. T. H. Jukes and C. R. Cantor, *in* "Mammalian Protein Metabolism" (H. N. Munro and J. B. Allison, eds.), Vol. 3, pp. 21–132. Academic Press, New York, 1969.
105. E. Margoliash, *Harvey Lect.* **66**, 177–247 (1972).

TABLE X

Frequency of Occurrence of Amino Acids among 60 Species at Different Positions along Chain

	-10	-9	-8	-7	-6	-5	-4	-3	-2	-1	1	2	3	4	5	6	7	8	9	10	11	12	13	14	15	16	17	18	19	20
-	60	59	35	35	33	32	26	26	26	26	0	0	0	0	0	0	0	0	0	0	0	0	0	0	0	0	0	0	0	0
K	0	0	0	1	0	0	0	0	1	1	0	0	2	22	34	0	30	54	0	0	30	0	48	0	0	0	0	0	0	0
R	0	0	0	0	1	0	0	0	0	0	0	0	0	0	1	0	0	1	0	0	0	0	12	0	0	0	0	0	0	0
D	0	0	0	0	0	6	0	0	0	1	0	36	0	4	0	0	1	0	0	0	0	0	0	0	0	0	0	0	0	0
E	0	0	0	0	0	1	24	0	4	0	0	0	1	26	0	0	19	0	0	0	1	0	0	0	0	0	0	0	0	0
N	0	0	0	0	0	2	0	0	0	0	0	20	1	0	2	0	0	3	0	0	0	1	0	0	1	0	0	0	0	0
Q	0	0	0	0	0	2	0	0	0	1	0	0	1	0	0	0	0	0	0	0	0	26	0	0	0	60	0	0	0	0
S	0	0	0	19	0	5	1	0	1	0	0	3	6	1	4	0	0	0	0	0	0	1	0	0	9	0	0	0	2	0
T	0	0	0	4	0	1	0	0	0	0	0	0	0	1	1	0	0	2	2	0	2	27	0	0	0	0	0	0	58	0
H	0	0	0	0	0	0	0	0	0	0	0	0	0	0	0	0	0	0	0	0	0	0	0	0	0	0	0	60	0	0
Y	0	0	0	0	0	0	0	1	0	0	0	0	0	0	0	0	0	0	0	0	0	0	0	0	0	0	0	0	0	0
C	0	0	0	0	0	0	0	0	0	0	0	0	0	0	0	0	0	0	0	0	0	0	0	58	0	0	60	0	0	1
G	0	0	0	0	1	1	6	0	0	0	60	0	0	0	0	60	1	0	0	0	0	1	0	0	0	0	0	0	0	1
A	0	0	24	1	0	10	0	23	0	10	0	0	12	5	16	0	9	0	0	0	0	0	0	2	47	0	0	0	0	1
V	0	0	0	0	0	0	0	4	0	0	0	0	24	0	2	0	0	0	3	0	25	0	0	0	0	0	0	0	0	54
L	0	0	0	0	0	0	0	1	0	0	0	0	0	0	0	0	0	0	7	0	0	1	0	0	3	0	0	0	0	2
I	0	0	0	0	0	0	0	0	0	0	0	0	5	0	0	0	0	0	48	0	2	0	0	0	0	0	0	0	0	1
M	0	0	0	0	0	0	0	0	0	0	0	0	0	0	0	0	0	0	0	0	0	3	0	0	0	0	0	0	0	0
P	0	1	0	0	2	0	3	0	28	21	0	0	8	0	0	0	0	0	0	0	0	0	0	0	0	0	0	0	0	0
F	0	0	0	0	23	0	0	5	0	0	0	0	0	0	0	0	0	0	0	60	0	0	0	0	0	0	0	0	0	0
W	0	0	0	0	0	0	0	0	0	0	0	0	0	0	0	0	0	0	0	0	0	0	0	0	0	0	0	0	0	0
B	0	0	0	0	0	0	0	0	0	0	0	0	0	0	0	0	0	0	0	0	0	0	0	0	0	0	0	0	0	0
Z	0	0	0	0	0	0	0	0	0	0	0	1	0	1	0	0	0	0	0	0	0	0	0	0	0	0	0	0	0	0
X	0	0	1	0	0	0	0	0	0	0	0	0	0	0	0	0	0	0	0	0	0	0	0	0	0	0	0	0	0	0

	21	22	23	24	25	26	27	28	29	30	31	32	33	34	35	36	37	38	39	40	41	42	43	44	45	46	47	48	49	50
—	0	0	0	0	1	0	0	0	0	0	0	0	0	0	0	0	0	0	0	0	0	0	0	0	0	0	0	0	0	0
K	0	41	0	0	29	0	58	0	0	0	0	0	0	0	0	0	0	0	32	0	0	0	0	0	0	0	0	0	0	1
R	0	1	0	0	0	0	0	0	0	0	0	0	0	0	0	0	0	60	0	0	0	0	0	0	0	0	0	0	0	0
D	11	0	0	0	0	0	0	0	0	0	0	0	0	0	0	0	0	0	0	0	0	0	0	1	0	0	0	0	0	26
E	44	2	0	1	0	0	0	0	0	0	0	0	0	0	0	0	0	0	0	0	0	0	0	13	0	0	0	0	0	1
N	0	3	0	0	0	5	0	0	0	0	58	0	23	0	0	0	1	0	0	0	0	0	0	0	0	0	0	0	0	5
Q	1	2	0	0	0	0	0	23	0	0	0	0	0	0	0	0	0	0	23	1	0	32	0	2	0	0	0	0	0	0
S	0	0	0	0	0	0	1	0	0	0	1	0	3	0	0	0	1	0	0	29	0	3	0	0	1	0	52	0	27	0
T	0	0	0	0	0	0	0	22	0	0	0	0	0	0	0	0	0	0	2	30	0	25	20	0	0	0	2	0	33	2
H	0	0	0	0	0	55	0	0	0	0	0	0	28	0	0	0	0	0	3	0	0	0	0	0	0	0	0	0	0	0
Y	0	0	0	0	0	0	0	0	0	0	0	0	1	0	0	2	0	0	0	0	0	0	0	0	0	33	0	60	0	0
C	0	0	0	0	0	0	0	0	0	0	0	0	0	0	0	0	0	0	0	0	0	0	0	0	0	0	0	0	0	0
G	2	0	59	35	24	0	1	0	60	0	0	0	0	60	0	0	58	0	0	0	60	0	0	0	59	0	0	0	0	0
A	1	8	1	23	2	0	0	0	0	0	1	0	0	0	0	0	0	0	0	0	0	0	34	23	0	0	6	0	0	24
V	0	0	0	1	0	0	0	13	0	0	0	0	0	0	2	2	0	0	0	0	0	0	6	3	0	0	0	0	0	0
L	0	2	0	0	0	0	0	0	0	0	0	60	0	0	51	0	0	0	0	0	0	0	0	0	0	0	0	0	0	0
I	0	0	0	0	0	0	0	2	0	0	0	0	0	0	5	3	0	0	0	0	0	0	0	0	0	0	0	0	0	0
M	0	0	0	0	0	0	0	0	0	0	0	0	0	0	0	0	0	0	0	0	0	0	0	0	0	0	0	0	0	0
P	0	0	0	0	4	0	0	0	0	60	0	0	0	0	0	0	0	0	0	0	0	0	0	18	0	0	0	0	0	0
F	0	0	0	0	0	0	0	0	0	0	0	0	1	0	2	53	0	0	0	0	0	0	0	0	0	27	0	0	0	0
W	0	0	0	0	0	0	0	0	0	0	0	0	4	0	0	0	0	0	0	0	0	0	0	0	0	0	0	0	0	0
B	0	1	0	0	0	0	0	0	0	0	0	0	0	0	0	0	0	0	0	0	0	0	0	0	0	0	0	0	0	1
Z	1	0	0	0	0	0	0	0	0	0	0	0	0	0	0	0	0	0	0	0	0	0	0	0	0	0	0	0	0	0
X	0	0	0	0	0	0	0	0	0	0	0	0	0	0	0	0	0	0	0	0	0	0	0	0	0	0	0	0	0	0

TABLE X (*Continued*)

	51	52	53	54	55	56	57	58	59	60	61	62	63	64	65	66	67	68	69	70	71	72	73	74	75	76	77	78	79	80
-	0	0	0	0	0	0	0	0	0	0	0	0	0	0	0	0	0	0	0	0	0	0	0	0	0	0	0	0	0	0
K	0	0	58	3	42	0	0	0	0	3	0	6	0	0	0	1	0	0	0	0	0	31	60	0	0	0	0	0	60	0
R	0	0	0	1	1	0	0	0	0	0	0	0	0	0	1	0	0	0	0	0	0	0	0	0	0	0	0	0	0	0
D	0	0	0	1	0	0	0	0	0	2	6	19	1	0	1	22	0	0	0	0	0	0	0	0	0	0	0	0	0	0
E	0	0	0	0	0	0	0	4	0	14	47	12	0	0	0	33	0	0	36	0	0	0	0	0	0	0	0	0	0	0
N	0	60	0	43	0	1	0	6	0	4	2	15	1	0	0	0	0	0	0	60	0	0	0	0	0	0	0	0	0	0
Q	0	0	0	2	0	0	1	2	0	5	2	3	0	0	0	1	0	0	0	0	0	0	0	0	0	0	0	0	0	0
S	0	0	0	5	1	0	0	0	0	1	0	1	1	0	3	0	0	0	0	0	0	0	0	0	0	0	0	0	0	0
T	0	0	0	0	0	0	0	26	0	2	0	0	56	0	0	0	0	0	1	0	0	0	0	0	0	0	0	60	0	0
H	0	0	0	0	0	0	0	0	0	0	0	0	0	0	1	0	0	0	0	0	0	0	0	0	0	0	0	0	0	0
Y	0	0	0	0	0	0	0	0	0	0	1	0	0	0	23	0	59	0	0	0	0	0	0	58	0	0	0	0	0	0
C	0	0	0	0	0	0	0	0	0	0	0	0	0	0	0	0	0	0	0	0	0	0	0	0	0	0	0	0	0	0
G	1	0	0	1	0	35	0	1	0	26	0	0	0	0	0	0	0	0	0	0	0	0	0	0	0	0	60	0	0	0
A	59	0	1	4	4	24	0	1	0	2	0	2	0	0	0	0	0	0	0	0	0	0	0	0	0	0	0	0	0	0
V	0	0	0	0	0	0	27	8	0	0	0	0	1	0	0	2	0	0	0	0	0	0	0	0	1	0	0	0	0	0
L	0	0	0	0	0	0	0	3	0	0	0	0	0	57	1	0	0	60	23	0	0	0	0	0	0	0	0	0	0	0
I	0	0	1	0	0	0	32	5	0	0	0	0	0	0	0	1	0	0	0	0	0	0	0	0	58	0	0	0	0	0
M	0	0	0	0	12	0	0	0	0	0	0	0	0	2	22	0	0	0	0	0	0	0	0	0	1	0	0	0	0	60
P	0	0	0	0	0	0	0	0	0	0	1	1	0	0	0	0	0	0	0	0	60	0	0	0	0	60	0	0	0	0
F	0	0	0	0	0	0	0	0	0	0	0	0	0	1	8	0	1	0	0	0	0	0	0	2	0	0	0	0	0	0
W	0	0	0	0	0	0	0	0	60	0	0	0	0	0	0	0	0	0	0	0	0	0	0	0	0	0	0	0	0	0
B	0	0	0	0	0	0	0	0	0	1	0	0	0	0	0	0	0	0	0	0	0	0	0	0	0	0	0	0	0	0
Z	0	0	0	0	0	0	0	0	0	0	1	1	0	0	0	0	0	0	0	0	0	0	0	0	0	0	0	0	0	0
X	0	0	0	0	0	0	0	0	0	0	0	0	0	0	0	0	0	0	0	0	0	29	0	0	0	0	0	0	0	0

	81	82	83	84	85	86	87	88	89	90	91	92	93	94	95	96	97	98	99	100	101	102	103	104	105	106	107	108	109	110
-	0	0	0	0	0	0	0	0	0	0	0	0	0	0	0	0	0	0	0	0	0	0	1	35	59	60	60	60	60	60
K	0	0	0	0	0	34	58	21	8	0	0	1	0	0	0	0	0	0	55	15	1	2	12	8	0	0	0	0	0	0
R	0	0	0	0	0	0	0	0	0	0	60	0	0	0	0	0	0	0	1	0	0	0	0	0	0	0	0	0	0	0
D	0	0	0	0	0	0	0	1	1	21	0	0	58	0	0	0	0	0	0	7	0	0	1	0	0	0	0	0	0	0
E	0	0	0	0	0	0	0	3	2	38	0	3	0	0	0	0	0	0	1	19	1	0	1	12	1	0	0	0	0	0
N	0	0	0	0	0	0	0	1	4	0	0	6	1	0	0	0	0	0	0	1	0	0	8	0	0	0	0	0	0	0
Q	0	0	0	0	0	0	0	0	24	1	0	5	0	0	0	0	0	0	0	6	0	0	1	1	0	0	0	0	0	0
S	1	0	0	0	0	1	0	1	5	0	0	0	0	0	0	1	0	0	0	6	16	1	11	3	0	0	0	0	0	0
T	0	0	1	0	0	0	0	2	3	0	0	1	0	0	0	6	0	0	0	2	1	54	0	0	0	0	0	0	0	0
H	0	0	0	0	0	0	0	0	0	0	0	0	1	0	0	0	0	0	0	1	0	0	0	0	0	0	0	0	0	0
Y	0	0	0	0	0	0	0	0	0	0	0	0	0	0	0	0	59	0	0	0	0	0	0	0	0	0	0	0	0	0
C	0	0	0	0	0	0	0	0	0	0	0	0	0	0	0	0	0	0	0	0	0	2	0	0	0	0	0	0	0	0
G	0	0	6	60	0	0	0	0	10	0	0	2	0	0	0	0	0	0	0	0	0	0	0	0	0	0	0	0	0	0
A	7	0	28	0	0	0	1	4	3	0	0	40	0	0	0	53	0	0	0	1	39	1	24	1	0	0	0	0	0	0
V	27	0	2	0	0	0	0	0	0	0	0	2	0	0	4	0	0	0	1	0	0	0	0	0	0	0	0	0	0	0
L	0	0	0	0	36	0	0	0	0	0	0	0	0	58	1	0	0	57	1	0	2	0	0	0	0	0	0	0	0	0
I	25	0	0	0	24	0	0	0	0	0	0	0	0	2	55	0	0	0	0	1	0	0	0	0	0	0	0	0	0	0
M	0	0	0	0	0	0	0	0	0	0	0	0	0	0	0	0	0	3	0	0	0	0	0	0	0	0	0	0	0	0
P	0	0	23	0	0	0	0	26	0	0	0	0	0	0	0	0	0	0	0	0	0	0	0	0	0	0	0	0	0	0
F	0	60	0	0	0	0	0	0	0	0	0	0	0	0	0	0	1	0	0	0	0	0	0	0	0	0	0	0	0	0
W	0	0	0	0	0	0	0	0	0	0	0	0	0	0	0	0	0	0	0	0	0	0	0	0	0	0	0	0	0	0
B	0	0	0	0	0	0	1	0	0	0	0	0	0	0	0	0	0	0	0	0	0	0	0	0	0	0	0	0	0	0
Z	0	0	0	0	0	0	0	1	0	0	0	0	0	0	0	0	0	0	1	1	0	0	1	0	0	0	0	0	0	0
X	0	0	0	0	0	25	0	0	0	0	0	0	0	0	0	0	0	0	0	0	0	0	0	0	0	0	0	0	0	0

ter 9 of Lemberg and Barrett in 1973 (*4*), and, of course, E. L. Smith's chapter in Volume I of this edition of "The Enzymes" in 1970 (*106*). In view of this wealth of information, the present section can be held to an outline of the goals and successes of molecular evolution and to more recent developments.

The tools of protein evolutionary studies are the amino acid sequences and three-dimensional structures of the molecules, and the goals are twofold:

(a) to use proteins as characters or traits with which to follow the path of evolution of organisms; and

(b) to use the evolutionary record embodied in protein molecules in order to understand how families of proteins evolve, differentiate, and develop for their individual roles.

In short, to use proteins to study evolution and to use evolution to study proteins.

For cytochrome *c*, the first sequences preceded the detailed structure by 8 years and provided the inspiration for the X-ray analysis. Early papers by Margoliash and Smith (*107–109*) noted the great similarity of sequence of cytochromes from different eukaryotic species and the dependence of this similarity on phylogenetic relatedness of the species, and commented on the probable functional purpose of invariant and highly conserved positions along the polypeptide chain. It became clear that cytochrome *c* was the product of a single gene passed down through the family tree of eukaryotes, and therefore was simpler than the globins and a particularly appropriate protein for use as a character in the study of evolution. [Only baker's yeast (*110*) is known to have polymorphs or isozymes of cytochrome *c*.] Fitch and Margoliash began a continuing quantitative study of the degree of difference between cytochrome sequences from various eukaryotic species and the construction of phylogenetic trees (*111–120*). From this work on animals and eukaryotic micro-

106. E. L. Smith, "The Enzymes," 3rd ed., Vol. 1, pp. 267–339, 1970.
107. E. Margoliash, *Proc. Nat. Acad. Sci. U. S.* **50**, 672–679 (1963).
108. E. L. Smith and E. Margoliash, *Fed. Proc., Fed. Amer. Soc. Exp. Biol.* **23**, 1243–1247 (1964).
109. E. Margoliash and E. L. Smith, *in* "Evolving Genes and Proteins" (V. Bryson and H. J. Vogel, eds.), pp. 221–242. Academic Press, New York, 1965.
110. P. P. Slonimski, R. Archer, A. Péré, A. Sels, and M. Somlo, "Mécanismes de la régulation des activités cellulaires chez les microorganismes," p. 435. CNRS, Paris, 1965.
111. W. M. Fitch and E. Margoliash, *Science* **155**, 279–284 (1967).
112. W. M. Fitch and E. Margoliash, *Biochem. Genet.* **1**, 65–71 (1967).
113. W. M. Fitch and E. Margoliash, *Brookhaven Symp. Biol.* **21**, 217–242 (1968).

organisms has come a confidence in the evolutionary validity of such trees, which is now being exploited by Boulter in the less well-established area of higher plants (*96,121–124*). Because cytochrome *c* is found in all eukaryotes without exception, and because the molecules from different species are so similar, cytochrome is especially well suited for studies of type (a) above—the use of proteins to study evolution. The complementary goal (b) above—the use of evolution to study the development of a family of proteins—becomes more important when the inquiry is enlarged to include bacterial photosynthetic and respiratory cytochromes *c*. This section will be concerned mainly with the use of protein sequence comparisons to study the evolutionary process.

a. Amino Acid Difference Matrices. The point of departure for any study of amino acid sequences and phylogeny is a table of differences between cytochromes from different species. A complete difference matrix for all 60 sequences of Table IX is given in Table XI and merits careful study in its own right. Each matrix entry represents the number of amino acids which differ between the two species coresponding to that row and column in the matrix table; for example, the cytochromes of sunflower and horse differ by 46 amino acids, whereas man and rhesus monkey differ only by a single residue. Absences are counted as a twenty-first kind of amino acid. Since sunflower has eight amino acids at the N-terminus that horse does not have, and horse has an extra residue at position 104, these two chain ends contribute nine differences to the matrix entry. (Such differences matrices are familiar from the Dayhoff Atlas, but their value there has been diminished in the latest edition by the conversion of entries from absolute differences to percent change.)

The matrix of Table XI appears as a sea of 40–60 amino acid differences between distantly related organisms with islands of smaller numbers between closer relatives. The higher plants at the lower right of the table are one such island, with a sudden drop in differences from the

114. E. Margoliash and W. M. Fitch, *Ann. N. Y. Acad. Sci.* **151,** 359–381 (1968).
115. E. Margoliash, W. M. Fitch, and R. E. Dickerson, *Brookhaven Symp. Biol.* **21,** 259–305 (1968).
116. W. M. Fitch, *JMB* **49,** 1–14 (1970).
117. W. M. Fitch, *JMB* **49,** 15–21 (1970).
118. W. M. Fitch, *Syst. Zool.* **19,** 99–113 (1970).
119. W. M. Fitch and E. Markowitz, *Biochem. Genet.* **4,** 579–593 (1970).
120. E. Margoliash and W. M. Fitch, *Miami Winter Symp.* **1,** 33–51 (1970).
121. D. Boulter and J. A. M. Ramshaw, *Phytochemistry* **11,** 553–561 (1972).
122. D. Boulter, J. A. M. Ramshaw, E. W. Thompson, M. Richardson, and R. H. Brown, *Proc. Roy. Soc., Ser. B* **181,** 441–455 (1972).
123. D. Boulter, *Pure Appl. Chem.* **34,** 539–552 (1973).
124. D. Boulter, *Syst. Zool.* **22,** 549–553 (1974).

TABLE XI

	M	R	H	D	C	C	S	D	B	R	K	C	E	P	D	P	T	S	F	T	B	C	D	L	S	F	S	S
Man	0	1	12	11	10	10	12	11	11	9	10	13	13	13	11	12	15	14	18	21	21	18	24	20	30	29	27	3
Rhesus	1	0	11	10	9	9	11	10	10	8	11	12	12	12	10	11	14	15	17	21	21	18	23	20	29	28	26	3
Horse	12	11	0	1	3	5	7	6	7	6	7	11	12	12	10	11	11	22	14	19	18	13	17	16	23	24	22	2
Donkey	11	10	1	0	2	4	6	5	6	5	8	10	11	11	9	10	10	21	13	18	17	12	16	15	23	24	22	2
Cow	10	9	3	2	0	2	4	3	5	4	6	9	10	10	8	9	9	20	11	17	16	11	16	14	25	24	22	2
Camel	10	9	5	4	2	0	4	3	5	2	6	9	9	9	7	8	8	19	11	17	16	11	16	15	26	24	22	2
Seal	12	11	7	6	4	4	0	1	4	6	8	10	10	10	8	9	9	21	12	18	17	11	18	15	24	23	21	2
Dog	11	10	6	5	3	3	1	0	3	5	7	10	10	10	8	9	9	21	12	18	17	11	17	14	24	23	21	2
Bat	11	10	7	6	5	5	4	3	0	5	8	10	10	10	8	8	9	21	13	19	18	13	18	16	24	23	21	2
Rabbit	9	8	6	5	4	2	6	5	5	0	6	8	8	8	6	7	9	18	11	17	17	13	17	17	26	23	21	2
Kangaroo	10	11	7	8	6	6	8	7	8	6	0	12	10	10	10	11	11	21	13	18	18	13	20	17	27	26	24	2
Chicken	13	12	11	10	9	9	10	10	10	8	12	0	2	2	3	4	8	19	11	17	17	15	19	18	24	25	23	2
Emu	13	12	12	11	10	9	10	10	10	8	10	2	0	2	3	4	6	20	11	17	17	14	19	18	25	26	24	2
Penguin	13	12	12	11	10	9	10	10	10	8	10	2	2	0	3	4	8	20	12	18	18	15	20	19	26	26	24	2
Duck	11	10	10	9	8	7	8	8	8	6	10	3	3	3	0	3	7	17	11	17	17	14	17	18	25	24	22	2
Pigeon	12	11	11	10	9	8	9	9	8	7	11	4	4	4	3	0	8	18	12	18	18	15	19	19	24	25	23	2
Turtle	15	14	11	10	9	8	9	9	9	9	11	8	6	8	7	8	0	22	10	18	17	13	19	19	27	24	24	2
Snake	14	15	22	21	20	19	21	21	21	18	21	19	20	20	17	18	22	0	24	26	27	26	26	27	29	31	29	3
Frog	18	17	14	13	11	11	12	12	13	11	13	11	11	12	11	12	10	24	0	15	15	13	20	21	29	22	22	2
Tuna	21	21	19	18	17	17	18	18	19	17	18	17	17	18	17	18	18	26	15	0	2	8	20	19	31	25	24	3
Bonito	21	21	18	17	16	16	17	17	18	17	18	17	17	18	17	18	17	27	15	2	0	7	20	19	32	26	25	3
Carp	18	18	13	12	11	11	11	11	13	13	13	15	14	15	14	15	13	26	13	8	7	0	15	12	27	22	21	2
Dogfish	24	23	17	16	16	16	18	17	18	17	20	19	19	20	17	19	19	26	20	20	20	15	0	16	27	26	25	3
Lamprey	20	20	16	15	14	15	15	14	16	17	17	18	18	19	18	19	19	27	21	19	19	12	16	0	26	29	28	3
Snail	30	29	23	23	25	26	24	24	24	26	27	24	25	26	25	24	27	29	29	31	32	27	27	26	0	30	28	3
Fruit Fly	29	28	24	24	24	24	23	23	23	23	26	25	26	26	24	25	24	31	22	25	26	22	26	29	30	0	2	1
S. W. Fly	27	26	22	22	22	22	21	21	21	21	24	23	24	24	22	23	24	29	22	24	25	21	25	28	28	2	0	1
S. cynthia	31	30	29	28	27	27	25	25	25	26	28	28	28	27	27	27	28	31	29	32	33	27	32	32	30	15	14	
T. H. W. Moth	31	30	28	27	27	27	25	25	25	26	28	28	28	27	27	26	29	33	30	30	31	26	31	33	29	14	12	
Saccharom.	44	44	45	44	44	44	45	44	44	44	45	44	45	43	44	44	47	47	46	45	43	44	48	48	49	44	44	4
Debarom.	45	45	44	43	43	43	43	42	42	43	45	44	45	44	44	44	46	48	46	46	45	42	45	47	45	41	42	4
C. krusei	50	50	50	49	49	49	48	48	48	49	50	49	50	48	49	49	51	51	50	46	45	48	51	54	50	46	46	4
N. crassa	43	42	41	41	41	41	41	41	41	41	44	42	43	43	41	41	44	43	44	43	41	41	44	47	46	36	36	4
Humicola	43	42	43	42	42	42	41	41	41	41	45	41	42	42	41	41	44	44	45	41	41	42	46	47	47	39	39	4
Ustilago	47	47	44	43	43	43	42	42	43	44	47	44	45	45	44	44	45	47	46	43	42	42	47	47	46	42	40	4
Crithidia	59	59	56	56	56	55	55	55	56	56	55	57	56	55	56	56	54	63	56	56	56	54	59	57	60	54	53	5
Euglena	48	48	46	46	46	47	47	47	48	48	47	49	49	49	49	49	47	52	48	49	48	44	50	50	54	53	54	5
Nigella	45	45	48	47	47	46	46	46	46	46	49	48	48	48	46	46	48	45	50	48	47	48	50	52	52	47	45	4
Mung bean	46	47	49	48	48	47	47	47	47	47	47	49	47	49	47	47	47	46	50	51	50	49	51	52	53	50	48	4
Cauliflower	44	44	47	46	46	45	45	45	45	45	46	47	45	47	45	45	45	45	48	49	48	47	50	52	51	48	46	4
Pumpkin	43	44	47	46	46	46	45	45	45	46	45	48	46	48	46	46	46	45	50	50	49	47	51	52	51	50	48	4
Hemp	47	46	49	48	48	47	47	47	47	47	49	49	47	49	47	47	47	47	50	51	50	49	49	53	52	48	46	4
Elder	46	46	50	49	49	48	48	48	48	48	49	49	47	49	47	46	48	47	51	52	51	50	52	52	53	51	49	4
Abutilon	43	43	47	46	45	45	44	44	44	45	45	47	45	47	45	45	45	44	49	50	48	47	50	50	50	50	48	4
Cotton	45	45	49	48	47	46	46	46	46	46	47	48	46	48	46	46	46	45	49	51	49	49	51	52	52	50	48	4
Castor	42	42	46	45	44	43	43	43	43	43	44	46	44	46	43	43	43	44	46	48	46	46	49	51	51	47	45	4
Tomato	43	43	47	46	45	44	44	44	44	44	45	46	44	46	44	44	44	44	47	49	47	47	50	52	51	48	46	4
Maize	42	42	45	44	44	43	43	43	43	43	44	45	43	45	43	43	43	43	46	46	44	44	48	49	49	46	44	4
Arum	43	44	47	46	46	45	45	45	45	45	45	47	45	47	45	45	45	43	48	49	48	47	49	50	51	48	46	4
Sesame	40	41	45	44	43	43	42	42	42	43	42	46	44	46	43	44	43	42	47	48	46	45	49	50	50	48	46	4
Leek	44	44	47	46	46	45	45	45	45	45	46	47	45	47	45	45	45	44	48	49	47	47	49	50	51	48	46	4
Acer	46	46	50	49	48	47	47	47	47	47	48	49	47	49	47	46	47	46	50	51	50	51	50	52	53	51	49	4
Niger	43	43	46	45	45	44	44	44	44	44	45	46	44	46	44	44	44	44	47	48	46	46	48	49	51	46	44	4
Sunflower	42	43	46	45	45	44	44	44	44	44	44	46	44	46	44	44	44	43	47	48	46	46	48	49	52	46	44	4
Nasturtium	44	44	47	46	46	45	45	45	44	45	46	47	45	47	45	45	45	44	48	49	48	47	49	50	52	47	45	4
Parsnip	44	44	47	46	46	45	45	45	45	45	46	47	45	47	45	45	45	44	48	47	46	46	48	51	51	45	43	4
Wheat	43	43	46	45	45	44	45	44	44	44	47	46	46	46	46	46	46	47	48	49	47	47	49	51	51	47	45	4
Buckwheat	40	39	44	43	42	41	41	41	41	40	42	40	38	40	40	40	40	42	43	46	46	45	48	49	50	44	42	3
Spinach	43	44	48	47	47	46	46	46	46	45	45	46	44	46	45	46	45	45	48	48	48	47	51	52	53	47	45	4
Ginkgo	42	42	45	44	43	42	42	42	42	42	45	43	43	43	43	42	43	45	45	48	47	45	51	51	51	47	45	4

AMINO ACID DIFFERENCE MATRIX FOR EUKARYOTIC CYTOCHROMES *c*

T	S	D	C	N	H	U	C	E	N	M	C	P	H	E	A	C	C	T	M	A	S	L	A	N	S	N	P	W	B	S	G
31	44	45	50	43	43	47	59	48	45	46	44	43	47	46	43	45	42	43	42	43	40	44	46	43	42	44	44	43	40	43	42
30	44	45	50	42	42	47	59	48	45	47	44	44	46	46	43	45	42	43	42	44	41	44	46	43	43	44	44	43	39	44	42
28	45	44	50	41	43	44	56	46	48	49	47	47	49	50	47	49	46	47	45	47	45	47	50	46	46	47	47	46	44	48	45
27	44	43	49	41	42	43	56	46	47	48	46	46	48	49	46	48	45	46	44	46	44	46	49	45	45	46	46	45	43	47	44
27	44	43	49	41	42	43	56	46	47	48	46	46	48	49	45	47	44	45	44	46	43	46	48	45	45	46	46	45	42	47	43
27	44	43	49	41	42	43	55	47	46	47	45	46	47	48	45	46	43	44	43	45	43	45	47	44	44	45	45	44	41	46	42
25	45	43	48	41	41	42	55	47	46	47	45	45	47	48	44	46	43	44	43	45	42	45	47	44	44	45	45	45	41	46	42
25	44	42	48	41	41	42	55	47	46	47	45	45	47	48	44	46	43	44	43	45	42	45	47	44	44	45	45	44	41	46	42
25	44	42	48	41	41	43	56	48	46	47	45	45	47	48	44	46	43	44	43	45	42	45	47	44	44	44	45	44	41	46	42
26	44	43	49	41	41	44	56	48	46	47	45	46	47	48	45	46	43	44	43	45	43	45	47	44	44	45	45	44	40	45	42
28	45	45	50	44	45	47	55	47	49	47	46	45	49	49	45	47	44	45	44	45	42	46	48	45	44	46	46	47	42	45	45
28	44	44	49	42	41	44	57	49	48	49	47	48	49	49	47	48	46	46	45	47	46	47	49	46	46	47	47	46	40	46	43
28	45	45	50	43	42	45	56	49	48	47	45	46	47	47	45	46	44	44	43	45	44	45	47	44	44	45	45	46	38	44	43
27	43	44	48	43	42	45	55	49	48	49	47	48	49	49	47	48	46	46	45	47	46	47	49	46	46	47	47	46	40	46	43
27	44	44	49	41	41	44	56	49	46	47	45	46	47	47	45	46	43	44	43	45	43	45	47	44	44	45	45	46	40	45	43
26	44	44	49	41	41	44	56	49	46	47	45	46	47	46	45	46	43	44	43	45	44	45	46	44	44	45	45	46	40	46	42
29	47	46	51	44	44	45	54	47	48	47	45	46	47	48	45	46	43	44	43	45	43	45	47	44	44	45	45	46	40	45	43
3	47	48	51	43	44	47	63	52	45	46	45	45	47	47	44	45	44	44	43	43	42	44	46	44	43	44	44	47	42	45	45
0	46	46	50	44	45	46	56	48	50	50	48	50	50	51	49	49	46	47	46	48	47	48	50	47	47	48	48	48	43	48	45
0	45	46	46	43	41	43	56	49	48	51	49	50	51	52	50	51	48	49	46	49	48	49	51	48	48	49	47	49	46	48	48
1	43	45	45	41	41	42	56	48	47	50	48	49	50	51	48	49	46	47	44	48	46	47	50	46	46	48	46	47	46	48	47
6	44	42	48	41	42	42	54	44	48	49	47	47	49	50	47	49	46	47	44	47	45	47	51	46	46	47	46	47	45	47	45
1	48	45	51	44	46	47	59	50	50	51	50	51	49	52	50	51	49	50	48	49	49	49	50	48	48	49	48	49	48	51	51
3	48	47	54	47	47	47	57	50	52	52	52	52	53	52	50	52	51	52	49	50	50	50	52	49	49	50	51	51	49	52	51
29	49	45	50	46	47	46	60	54	52	53	51	51	52	53	50	52	51	51	49	51	50	51	53	51	52	52	51	51	50	53	51
4	44	41	46	36	39	42	54	53	47	50	48	50	48	51	50	50	47	48	46	48	48	48	51	46	46	47	45	47	44	47	47
2	44	42	46	36	39	40	53	54	45	48	46	48	46	49	48	48	45	46	44	46	46	46	49	44	44	45	43	45	42	45	45
5	46	43	46	42	42	42	54	52	42	45	45	45	44	47	41	43	45	44	43	43	43	43	46	43	44	44	45	45	39	43	46
0	44	42	45	41	41	42	54	54	43	46	44	44	44	46	43	45	43	44	42	44	43	44	46	43	44	45	42	42	40	42	44
4	0	27	26	36	38	35	59	60	50	49	47	48	50	49	46	47	46	47	46	49	48	47	51	45	47	47	47	46	46	48	48
2	27	0	24	38	37	35	57	60	51	50	47	49	50	49	49	50	48	49	48	49	50	48	52	48	49	49	48	46	50	50	50
5	26	24	0	38	35	30	58	59	52	51	49	50	51	52	49	50	50	50	50	50	51	49	54	51	51	51	50	49	53	53	54
1	36	38	38	0	25	32	63	50	49	50	48	49	49	53	49	50	49	49	48	51	50	50	54	49	49	49	49	50	50	52	51
1	38	37	35	25	0	31	56	54	48	48	47	47	47	50	46	47	47	47	47	48	48	48	51	49	48	49	47	47	46	47	48
2	35	35	30	32	31	0	54	53	48	48	48	48	48	50	48	48	50	50	48	49	50	49	53	49	49	48	51	48	50	50	51
4	59	57	58	63	56	54	0	56	54	54	54	54	55	55	53	53	54	54	52	53	53	54	54	52	54	55	55	57	53	57	56
4	60	60	59	50	54	53	56	0	60	58	58	57	60	59	57	58	59	60	58	59	59	58	61	58	57	57	60	60	60	61	63
3	50	51	52	49	48	48	54	60	0	10	10	13	12	13	15	14	17	14	13	13	13	16	15	14	15	14	19	16	16	19	22
6	49	50	51	50	48	48	54	58	10	0	5	6	6	7	9	7	10	9	10	7	7	10	11	13	12	11	16	15	13	16	20
4	47	47	49	48	47	48	54	58	10	5	0	4	6	7	9	7	8	7	6	8	8	10	11	11	14	9	13	10	13	16	18
4	48	49	50	49	47	48	54	57	13	6	4	0	8	8	8	8	8	9	8	8	7	11	12	13	14	11	15	13	15	17	20
4	50	50	51	49	47	48	55	60	12	6	6	8	0	8	8	7	9	8	10	8	10	10	11	13	14	12	15	15	13	17	21
6	49	49	52	53	50	50	55	59	13	7	7	8	8	0	8	7	8	8	9	6	11	9	6	11	11	10	15	13	14	17	18
3	46	49	49	49	46	48	53	57	15	9	9	8	8	8	0	3	6	5	8	7	7	8	8	10	12	12	14	15	12	18	18
5	47	50	50	50	47	48	53	58	14	7	7	8	7	7	3	0	5	6	8	7	7	8	7	10	12	10	14	13	12	16	19
3	46	48	50	49	47	50	54	59	17	10	8	8	9	8	6	5	0	5	7	8	7	9	8	11	12	13	12	13	12	18	16
4	47	49	50	49	47	50	54	60	14	9	7	9	8	8	5	6	5	0	5	6	6	9	8	11	11	13	13	13	12	19	15
2	46	48	50	48	47	48	52	58	13	10	6	8	10	9	8	8	7	5	0	5	8	9	11	8	10	11	12	10	10	17	14
4	49	49	50	51	48	49	53	59	13	7	8	8	8	6	7	7	8	6	5	0	8	7	9	6	7	10	14	14	11	17	17
3	48	50	51	50	48	50	53	59	13	7	8	7	10	11	7	7	7	6	8	8	0	12	11	13	12	14	15	15	12	18	19
4	47	48	49	50	48	49	54	58	16	10	10	11	10	9	8	8	9	9	9	7	12	0	11	11	12	11	15	16	13	16	20
6	51	52	54	54	51	53	54	61	15	11	11	12	11	6	8	7	8	8	11	9	11	11	0	13	12	13	14	12	15	18	18
3	45	48	51	49	49	49	52	58	14	13	11	13	13	11	10	10	11	11	8	8	13	11	13	0	8	9	13	16	13	20	18
4	47	49	51	49	48	49	54	57	15	12	14	14	14	11	12	12	12	11	10	7	12	12	12	8	0	11	13	15	15	18	20
5	47	49	51	49	49	48	55	57	14	11	9	11	12	10	12	10	13	13	11	10	14	11	13	9	11	0	13	15	16	18	21
2	47	48	50	49	47	51	55	60	19	16	13	15	15	15	14	14	12	13	12	14	15	15	14	13	13	13	0	14	17	15	21
2	46	46	49	50	47	48	57	60	16	15	10	13	15	13	15	13	13	13	10	14	15	16	12	16	15	15	14	0	17	15	17
0	46	50	53	50	46	50	53	60	16	13	13	15	13	14	12	12	12	12	10	11	12	13	15	13	15	16	17	17	0	11	14
2	48	50	53	52	47	50	57	61	19	16	16	17	17	17	18	16	18	19	17	17	18	16	18	20	18	18	15	15	11	0	22
4	48	50	54	51	48	51	56	63	22	20	18	20	21	18	18	19	16	15	14	17	19	20	18	18	20	21	21	17	14	22	0

50's to less than 20. The higher animals—insects and vertebrates—are another such island, with the fine structure that one would expect from their known evolutionary history. All animals are separated from other eukaryotes by 36 to 60 amino acid changes, but insects and vertebrates differ only by 21–33 positions. Fish are less sharply differentiated from amphibia, reptiles, and birds, and another break is seen between these and mammals. Strangely enough, the two primate species seem to be sharply set off from other mammals, which differ among themselves by only one to eight amino acids. Birds are well demarcated from reptiles and amphibia, higher fish from dogfish and lamprey, and the especially close relationships within the categories of flies and moths is apparent. Yeasts and fungi (*Saccharomyces* through *Ustilago*) are only slightly set off from the general sea of less related species. The most distinctive sequences of all are the microorganisms *Crithidia* and *Euglena,* which also have the previously mentioned peculiarity of a single covalent heme attachment. Such groupings by relatedness agree with classic ideas and suggest that the cytochrome data might be usable to construct phylogenetic trees.

b. Construction of Phylogenetic Trees. Phylogenetic trees can be generated by using either a matrix approach or a reconstruction of ancestral sequences. In the matrix approach adopted by Margoliash and Fitch (*111,113–115*), a computer search is made for the "best" phylogenetic tree such that for every pair of species, the distance from species 1, down to the common branch point and back up to species 2, will agree with the observed differences in cytochrome sequences between the two species. The absolute lengths of the individual branches need have no physical meaning other than that of sequence differences, but if one yields to the conventional temptation to add a time dimension, then one is tacitly assuming a constant rate of cytochrome evolution. This is a viewpoint which can be defended in certain situations (see below) but does not necessarily follow.

It has been maintained that DNA sequences which code for cytochromes should be compared, rather than the amino acids themselves, since it is the DNA which changes and is passed from one generation to the next. Because of the natural redundancy of the genetic code, one cannot reconstruct DNA sequences from a knowledge only of amino acid sequences. Neither can one be sure how many point mutations in DNA were involved in going from one amino acid to another. If histidine is replaced by arginine, this can come about by one-base changes such as CAC to CGC, two-base changes of the type CAU to CCA, or three-base changes such as CAU to AGA. The minimum number of base changes required to replace histidine by arginine is one, and this is inherently more likely because single events are more probable than multiple events. [However, if two mutations have occurred at separate times at a given codon and we do not have an amino acid sequence corresponding to the intermediate form, then we could interpret two successive one-base changes as a single two-base event. When comparing sequences, a statistical corrrection (*125,126*) must be applied to compensate for these "invisible" multiple events.]

A conversion matrix can be set up to translate observed amino acid changes into minimum required changes in bases in the triplet codons, and a matrix similar to Table XI can be constructed, in which the entries are minimum mutation "distances" between the two sequences instead of total amino acid differences. It is not immediately obvious that this extra trouble brings with it any advantages, since the derived DNA sequence has many ambiguities, and the use of minimum mutation distances rather than the (unknown) actual number of mutations leads to a systematic underestimation of mutation rate. There is nothing particularly sacred about DNA as compared with amino acids, considered solely as traits for use in comparative taxonomy. The zoologist who compares bone structures is not concerned because he has no information about the genomes controlling collagen formation or calcium deposition. Nevertheless, both minimum mutation distances and amino acid differences are in common use for the comparison of sequences, and their results do not differ appreciably.

A phylogenetic tree for cytochromes produced by Fitch and Margoliash using inferred base mutations and the matrix approach is shown in Fig. 11. Although no information was used in constructing it other than cytochrome sequences, the results agree generally well with the tree expected from comparative anatomy and the fossil record. This figure gives us confidence that protein sequences are reasonable traits for use in following the evolutionary process and that inferences drawn in less well-characterized areas such as higher plants will be valid. The immense advantage of proteins as probes for the study of phylogeny is that this entire process can be repeated independently using any one of scores of other proteins shared by the same group of organisms, and the resultant phylogenetic trees can be cross-checked for accuracy. So far this amount of information has been available only for cytochrome *c*, hemoglobin, and the fibrinopeptides because of the great labor involved in amino acid sequencing, but newer automatic methods of sequence analysis undoubtedly will change the situation.

125. M. Kimura and T. Ohta, *J. Mol. Evol.* **1**, 1–17 (1971).
126. R. E. Dickerson, *J. Mol. Evol.* **1**, 26–45 (1971).

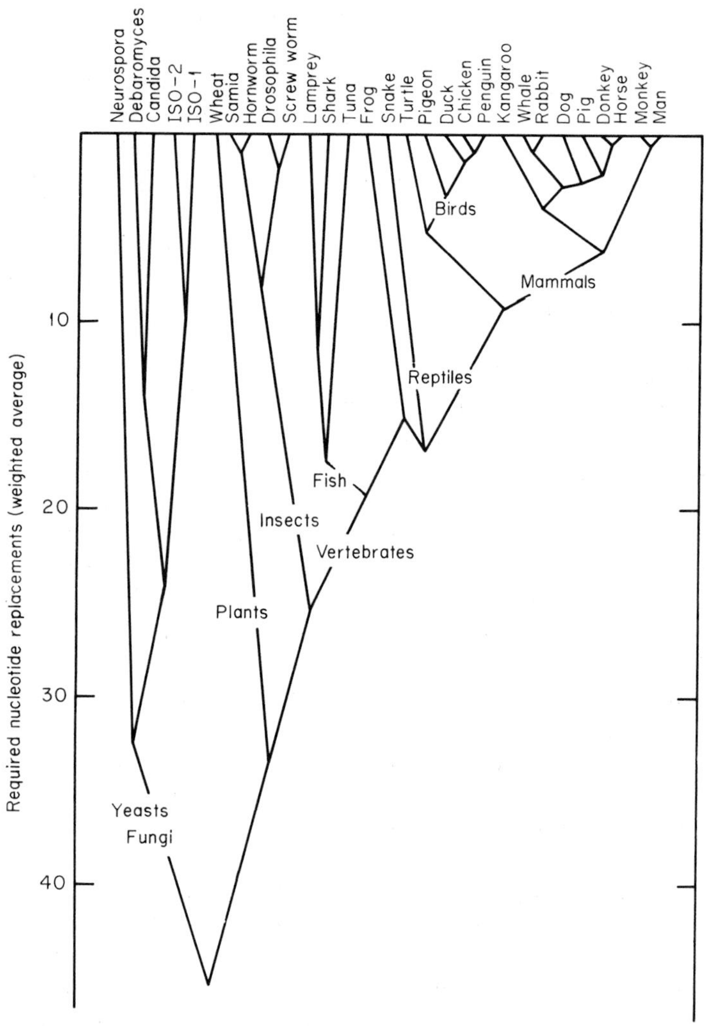

FIG. 11. Phylogenetic tree of the eukaryotes, as deduced from amino acid sequence differences between cytochromes *c*. Adapted from reference *119*.

The one serious anomaly in Fig. 11 is the strange position of rattlesnake, with a reverse bend in the tree. Rattlesnake cytochrome is much less like that of other reptiles and birds than it should be, as the matrix of Table XI confirms. From amino acid sequences alone, rattlesnake might more properly be grouped with the primitive fish. While turtle differs from birds and mammals by 6–15 residues, snake differs by 14–22. It is as different from turtle (22 residues) as from any of the birds and mammals. The problem may lie with the rattlesnake sequence, and the recent correction of errors in sequences of several fungi makes such a hypothesis at least thinkable. The rattlesnake sequence has six positions at which amino acids are reported which are not found there in any of the other 59 sequences. This is atypical, since only one other vertebrate (bullfrog) has as many as two such unique loci and only four other vertebrates have even a single unique position. Two of these unique positions in rattlesnake, lysine-101 and alanine-104, occur in the final chymotryptic peptide of the sequence analysis. If these two residues are interchanged, then only four unique sites are left, and the final peptide falls into a pattern familiar from other closely related vertebrates. A second such interchange, of serine-86 and lysine-89 within another chymotryptic peptide, removes another unique locus in rattlesnake. Replacing asparagine-93 by aspartic acid in that same peptide eliminates a fourth unique position and makes aspartic 93 invariant among all eukaryotes. These five changes—one cannot call them corrections because there is no evidence from published literature that they are in any way justified—would leave rattlesnake with no more unique positions than other vertebrates and would remove most of the difficulties with the matrix and phylogenetic tree. The matter at least is worth a careful reexamination, and other reptilian and amphibian sequences are badly needed to close a gap in the record.

Dayhoff (*102,127,128*) has used the matrix method with amino acid differences instead of minimum mutation distances, but favors generating phylogenetic trees by reconstruction of ancestral sequences at branch points. Boulter (*122–124*) has used similar methods with higher plants. The ancestral sequence method is outlined in refs. *102*, *122*, and *127*, but consists essentially of assuming a topological tree, looking at the amino acids at each chain position in all species on either side of a given branch point in the tree, and then deciding what the ancestral amino acid at that branch point for each chain position probably was. With a set of probable ancestral sequences at all branch points, the pattern of amino acid changes throughout the tree can be recreated. A given phylogenetic tree will require a calculable number of amino acid changes in ancestral sequences in order to produce the set of sequences that we see today. The best phylogenetic tree is the one that requires the smallest number of such changes. It is easy to see why this procedure is carried out on a computer, since thousands of closely related tentative trees may have to be tested in order to find the best one. Figure 12 shows a family tree

127. M. O. Dayhoff, *Sci. Amer.* July, 87–95 (1969).
128. P. J. McLaughlin and M. O. Dayhoff, *J. Mol. Evol.* **2**, 99–116 (1973).

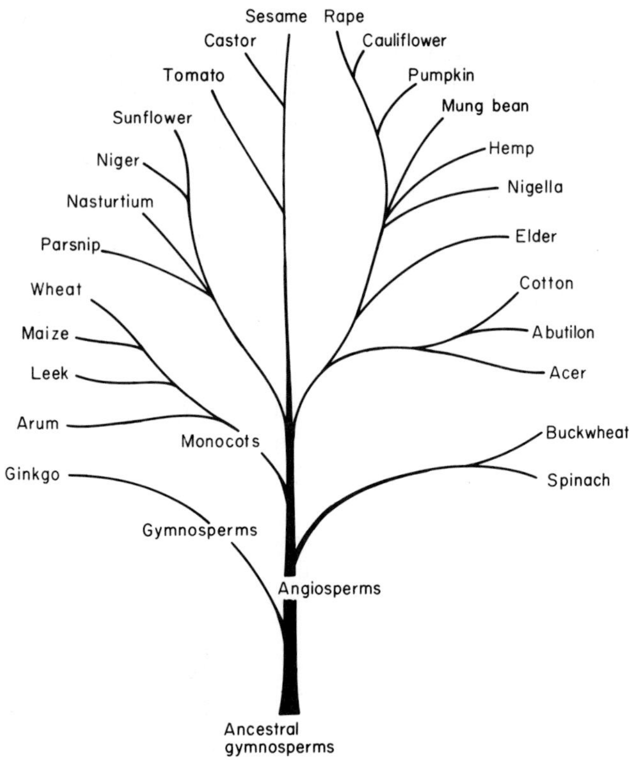

FIG. 12. Phylogenetic tree of the higher plants, deduced from their cytochrome sequences. Adapted from references *122–124*.

of the higher plants obtained by Boulter by the ancestral sequence method. It and the matrix method do not always give identical results, and the differences betwen them in actual practice are discussed in Ref. *122*. Phylogenetic tree building seems to be more difficult for plants than for animals, and varying results have been reported even using the ancestral sequence approach (*123,124*). For the moment these trees must be considered only tentative.

c. Neutral Mutations and the Time Dimension. At one point it was questioned (*104*) and even denied (*129*) that the observed changes between cytochrome sequences had any useful evolutionary or phylogenetic interpretation. Subsequent success in reproducing correct phylogenetic trees from sequence data alone has laid that question to rest. Protein sequences are a valid reflection of the interplay between mutation and

129. G. G. Simpson, *Science* **146,** 1535–1538 (1964).

selection in the organisms that contain them. In comparison with larger scale traits such as anatomical structure, amino acid sequences are closer to the site of mutation (DNA) and farther removed from direct effects of natural selection. This means that they are less affected by fluctuations in selection pressures in the environment. Once a protein has become well adapted for its particular biochemical role, and well suited for the other macromolecules with which it interacts, it may remain virtually unchanged thereafter in the face of gross changes in the whole organism. The cytochromes in the respiratory chain of mitochondria act as evolutionary brakes on one another, since a change in one protein is felt by the proteins before and after it in the chain. The great sequence similarity and common three-dimensional folding in cytochromes *c* of eukaryotes as diverse as bread mold, cauliflower, cow, and snail is evidence for this conservatism. The remarkable finding that any eukaryotic cytochrome *c* of those tested will react equally rapidly with beef heart cytochrome oxidase (*130*) demonstrates the functional chemical identity of all eukaryotic cytochromes *c*. The respiratory machinery of the mitochondrion stabilized slightly over one billion years ago, and selective pressures since then have acted primarily to prevent disruptive changes in a successful system.

The amino acid changes that are observed between species, which can be as many as 63 out of 113, then take on the character of harmless variations in a "standard" molecule. The very earliest papers on cytochrome evolution (*107–109*) recognized the distinction between invariant and variable regions as presumably reflecting essential and nonessential amino acid positions. Regions of evolutionary conservatism or outright invariance become pointers toward the functionally important parts of the cytochrome molecule in a way that would not be true if every amino acid change were to be considered as positively adaptive. Most of the changes observed in cytochromes are there not because they are positively useful to the species, but because they do no harm.

In this picture of cytochrome evolution, point mutations occur at random along the DNA which codes for the amino acid sequence, and the protein produced from this mutated DNA is tested in the organism for its ability to operate as a functioning cytochrome. If the molecule is impaired, then the mutation is deleterious, and if the molecule is ineffective, the mutation is lethal. We never see these lethal mutations in the sequence record because the unfortunate carriers of them are weeded out. Residues such as histidine-18 and methionine-80 are absolutely essential,

130. L. Smith, M. E. Nava, and E. Margoliash, *Oxidases Related Redox Syst. Proc. Symp., 2nd, 1972,* 629–647 (1973).

and point mutations at these positions in the DNA are inevitably fatal. Other residues such as cysteine-14 and tyrosine-74 are strongly favored but not absolutely mandatory, and occasional mutations to other amino acids can be tolerated. Still other regions seem to have few or no restrictions on amino acid substitutions, the hypervariable regions discussed earlier being examples. If a point mutation in one of these unrestricted regions of the molecule occurs, then it has a certain probability of being either lost or fixed in the entire population by the process of genetic drift.

This random mutation and genetic drift argument, although implicit in the cytochrome evolution papers from the beginning, has been presented most convincingly by Kimura (*125,131–134*) and King and Jukes (*135*). (Jukes' term "non-Darwinian evolution" for genetic drift is somewhat overdramatic and has something of the negative emotional impact that "non-Euclidian geometry" had a century ago.) In the simplest version, point mutations in DNA are either advantageous, deleterious, or neutral. More precisely, each possible mutation can be given a selection coefficient or selective advantage describing how much better or worse off the organism is with the new amino acid than the old. Mutations deleting cysteine-14 and -17, histidine-18, and methionine-80 have large negative selection coefficients, although we have seen that the loss of cysteine-14 can be tolerated under certain conditions. At the other extreme, point mutations in the DNA controlling hypervariable positions on the back of the molecule have selection coefficients near zero. The neutral mutation and random drift picture of the divergence of cytochromes would argue that differences in sequence have arisen mainly as the result of gradual fixation of mutations with very low selection coefficients rather than the accumulation within each lineage of changes with high positive selection coefficients for those organisms.

The genetic drift argument was first proposed by Kimura to account for an uncomfortably high rate of observed point mutations in DNA as judged by the standards of classic organismic evolution theory (*131,132*). It since has provided an explanation for another curious fact: the effective constancy of the rate of accumulation of amino acid changes in a protein over long time periods and the existence of different rates of change in different proteins. The concept of a "rate" of evolutionary change in protein sequences goes back to the early hemoglobin work

131. M. Kimura, *Nature (London)* **217,** 624–626 (1968).
132. M. Kimura, *Proc. Nat. Acad. Sci. U. S.* **63,** 1181–1188 (1969).
133. T. Ohta and M. Kimura, *J. Mol. Evol.* **1,** 18–25 (1971).
134. M. Kimura and T. Ohta, *Genetics* **73,** Suppl., 19–35 (1973).
135. J. L. King and T. H. Jukes, *Science* **164,** 788–798 (1969).

(*136–138*). Margoliash and Smith in 1965 defined the unit evolutionary period (UEP) as the time in years required for a one percent difference in observed amino acids to show up between two diverging evolutionary lines, but it was not clear from the data then at hand that such an evolutionary "constant" indeed was constant. The implied sameness of selection pressures from the environment over long time periods argued against it; and even if this could be explained, the constant rate of change of sequence might be better linked to the number of generations elapsed, rather than to time. Dickerson showed in 1971 that this long-term average constancy of evolutionary rate was a fact, at least for cytochrome *c*, hemoglobin, and the fibrinopeptides (*126*), using 30 or more sequences from each protein. The resulting amino acid differences, after correction for multiple mutations at the same locus, are plotted in Fig. 13 against the time since divergence of branches of the phylogenetic tree as determined from the fossil record. Within the limits of variation of the sequence differences, the amino acid changes in these three proteins have been linear with time. Each protein has its characteristic unit evolutionary period: 20 million years per 1% change in cytochrome, 5.8 million years for hemoglobin, and 1.1 million years for the fibrinopeptides. How can this be explained?

It is difficult to account for this behavior on the assumption that all observed differences are the result of positive selection for advantageous mutations. This would require that selection pressures from the environment *shifted at a uniform rate* in all branches of eukaryotes during the past billion years. The explanation in terms of near-neutral mutations and genetic drift is simpler: Each protein has long since stabilized in its present metabolic role, and the strongest selection pressures are against further change in one component of an integrated system. The observed changes are the result of passive fixation of harmless mutations. The observed rate of change for each protein is a function of the intrinsic mutation rate in DNA, and of the fraction of the protein chain that is selectively neutral and available for modification. (This does not imply that certain amino acids are absolutely essential and the rest are completely unimportant. The actual rate of acceptance of mutations would represent an averaging over all amino acid positions in the chain, of varying selection coefficients.) Fibrinopeptides change in sequence nearly 20 times as fast as cytochromes because the constraints on an operative

136. E. Zuckerkandl and L. Pauling, *in* "Horizons in Biochemistry" (M. Kasha and B. Pullman, eds.), pp. 189–225. Academic Press, New York, 1962.
137. E. Zuckerkandl and L. Pauling, *in* "Evolving Genes and Proteins" (V. Bryson and H. J. Vogel, eds.), pp. 97–166. Academic Press, New York, 1965.
138. E. Zuckerkandl, *Sci. Amer.* June, 110–118 (1965).

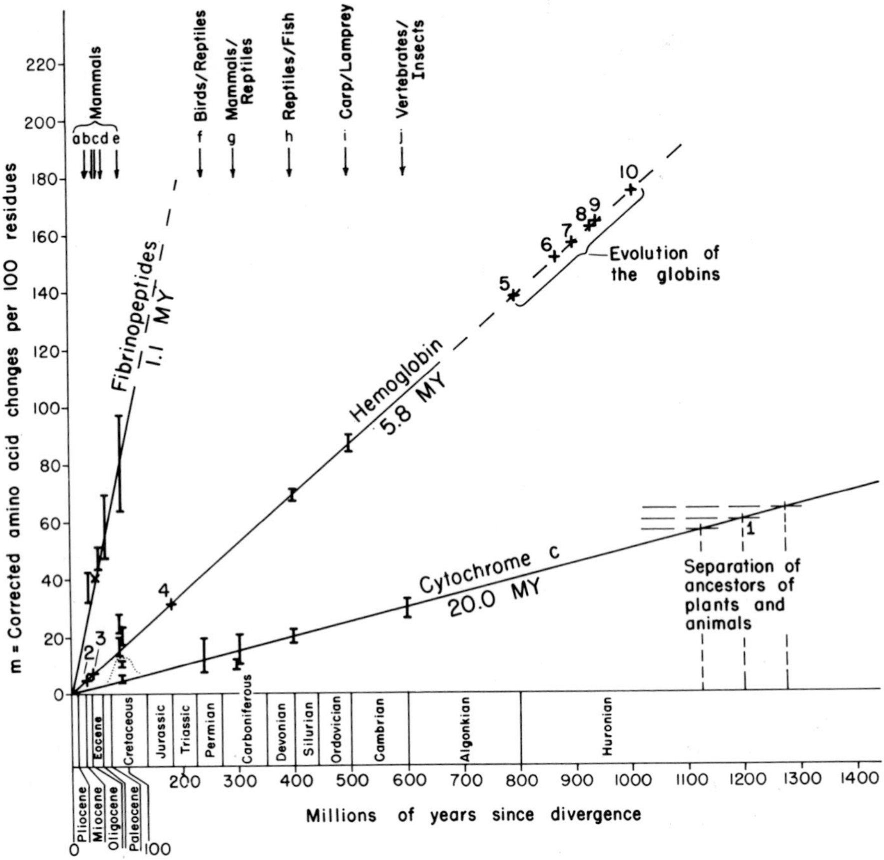

FIG. 13. Approximately linear rates of evolutionary change in the amino acid sequences of three proteins. The time in millions of years (MY) given for each protein is the time required for a 1% change in sequence to occur between two diverging branches of the phylogenetic tree. From reference *126*.

cytochrome molecule are much more severe, and the probability is much lower than a random mutation in DNA will lead to a functioning protein molecule (*126*).

The remaining problem is that of the time scale—whether linearity in rate of change should be by year or by number of generations. Not enough data were available in 1971 to settle the issue, but the wealth of new sequences since then makes new calculations possible. Figure 14 shows a skeletal family tree indicating the relationships between mammals, insects, and plants, and the average differences between their sequences. Although the evolutionary line leading to present day mammals

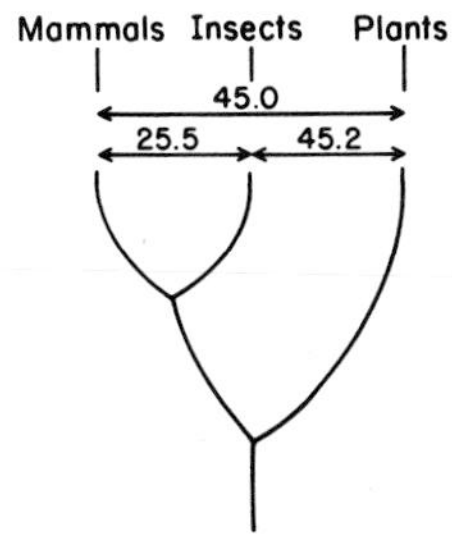

FIG. 14. Average number of differences in amino acids between cytochromes c from mammals, insects, and plants. Mammals and insects have diverged equally far from plants since their common branch point, with elapsed time measured in years.

has included ancestors with short generation times, nevertheless the average generation time is longer in the mammalian leg than the insect line, and more generations of insects than mammals will have arisen since the two lines diverged. If the rate of amino acid change depends on number of generations rather than years, then insects should have diverged farther from plants than mammals have. The evidence in Table XII,

TABLE XII
AMINO ACID DIFFERENCES AMONG CLASSES OF EUKARYOTES

	No. of comparisons	Average no. of differences[a]	
Mammals/higher plants	253	45.0 ± 1.5	
Insects/higher plants	92	45.2 ± 1.9	
Mammals/insects	44	25.5 ± 2.3	
	Average no. of differences[a] when compared with		
Species of microorganism	*Mammals*[b]	*Insects*[c]	*Higher plants*[d]
Saccharomyces	44.3 ± 0.4	44.5 ± 0.8	47.5 ± 1.3
Debaromyces	43.5 ± 0.9	42.0 ± 0.5	49.0 ± 1.0
Candida	49.1 ± 0.7	45.8 ± 0.4	50.7 ± 1.1
Neurospora	41.5 ± 0.8	38.8 ± 2.8	49.8 ± 1.1
Humicola	42.1 ± 0.9	40.2 ± 1.2	47.7 ± 0.9
Ustilago	44.1 ± 1.6	41.5 ± 0.8	49.1 ± 1.0
Crithidia	56.2 ± 1.0	53.8 ± 0.4	54.0 ± 0.9
Euglena	47.1 ± 0.7	53.2 ± 0.8	58.8 ± 1.1
All 8 microorganisms	46.0 ± 3.7	45.0 ± 4.5	50.8 ± 3.0

[a] Values given are mean and mean deviation.
[b] Eleven sequences.
[c] Four sequences.
[d] Twenty-two sequences, excluding ginkgo.

calculated from the 60 sequences of Table IX, shows that this is not so. Mammals and insects differ from plants by exactly the same degree.

Another test is afforded by the data in the second half of Table XII, where the similarities of mammals, insects, and plants to various yeasts, fungi and other microorganisms are compared. Again, if number of generations is the important factor, then mammals should have diverged less from the common ancestor (and hence from the various microorganisms) than insects and plants have, with their shorter generation times. Instead, we see that insects and animals differ from microorganisms to virtually the same extent, whereas plants deviate about 10% more. These data drive us to the conclusion that *the amino acid changes observed in cytochrome c vary at a uniform rate with respect to elapsed time, not number of generations.* This implies that the point mutations in DNA which are responsible for amino acid changes are occurring at a reasonably steady pace with time rather than happening mainly during the copying processes leading in the germ line to reproduction.

The data on plant sequences are just short of permitting a test there of the idea of a constant rate of evolutionary change. One gymnosperm has been sequenced (ginkgo), and 22 angiosperm sequences are known. The average difference between ginkgo and the angiosperms as given in Table XI is 18.5 ± 1.9 residues in all, or 16.4 residues per 100. If 16.4 changes out of 100 amino acid sites are seen, then the correction for multiple mutations (*125,126*) suggests that approximately 18 changes will have occurred. Angiosperms and one particular gymnosperm, ginkgo, differ by an average of 18 residues per hundred.

The problem is to find a reliable estimate of the age of divergence of the two lines of evolution. Angiosperm pollen has been found in the Jurassic era, 180–140 million years ago, and ginkgo ancestors can be traced back to the early Permian, around 270 million years ago. Both of these dates are irrelevant, unfortunately, for what is needed is the time in the past at which the gymnosperm *ancestors* of modern angiosperms and ginkgo separated. If both can be traced back to the pteridosperms or seed ferns of the Carboniferous (*139,140*), then the age of the evolutionary branch point approaches 350 million years. Eighteen sequence differences per hundred residues in 350 million years lead to a unit evolutionary period essentially identical with that calculated from vertebrates, but the fossil record is so poor that the calculation can only be considered approximate.

139. A. Cronquist, "The Evolution and Classification of Flowering Plants." Houghton, Boston, Massachusetts, 1968.
140. R. F. Scagel, R. J. Bandoni, G. E. Rouse, W. B. Schofield, J. R. Stein, and T. M. C. Taylor, "An Evolutionary Survey of the Plant Kingdom." Wadsworth, Belmont, California, 1966.

Fitch and Markowitz have carried out a statistical analysis of the degree of variability of cytochromes of different species, which supports the neutral mutation model *(119,141–143)*. They found that the observed numbers of substitutions at various positions along the protein chain can best be fitted by assuming a certain number of invariant sites, and two different classes of variable sites, one of which changes three times as rapidly as the other. This model is more satisfacory in fitting the observed pattern of variability than a simpler one with only one class of variable sites, and the authors pointed out that their two-state model probably is only a simplification of reality which involves a continuous range of selection coefficient at the variable sites. As successively distant branches of the phylogenetic tree are excluded from the analysis, and a smaller range of species is examined, the size of the invariant group of positions increases, reflecting the greater similarity of more closely related cytochromes. In the extrapolated limit of one species, approximately 90% of the cytochrome sequence falls into the invariant category. The authors concluded that in any one species, only about 10% of the cytochrome sequence can be varied without affecting its function or its interaction with other evolving protein molecules. This restricted group is called the set of "concomitantly variable codons" or "covarions." Different parts of the molecule will fall into the variable 10% at different points on the phylogenetic tree, but the 10% figure represents a maximum "load" of variation that the molecule can tolerate at any one time.

A similar treatment of the fibrinopeptides leads to the conclusion that only one amino acid is invariant and that the molecule contains 18 covarions. Fitch and Markowitz made the remarkable conclusion that although the rate of change of cytochrome *c* and fibrinopeptide per 100 amino acids is quite different, the rate of change per variable site or covarion is nearly the same. If only 10% of the cytochrome *c* molecule is variable at any one time, then the actual rate is 10 times the value obtained by considering the whole molecule. The UEP based on the covarion sites alone is

$$20 \text{ million years} \times 10\% = 2 \text{ million years}$$

The effect of the covarion correction on fibrinopeptide is negligible:

$$1.1 \text{ million years} \times \tfrac{18}{19} = 1.04 \text{ million years}$$

Within the concomitantly variable regions of cytochrome and fibrinopeptides, the rate of change of amino acids is virtually the same.

141. W. M. Fitch, *J. Mol. Evol.* **1**, 84–96 (1971).
142. W. M. Fitch, *Biochem. Genet.* **5**, 231–241 (1971).
143. E. Margoliash, W. M. Fitch, E. Markowitz, and R. E. Dickerson, *in* "Oxidation-Reduction Enzymes" (Å. Åkeson and A. Ehrenberg, eds.), pp. 5–17. Pergamon, Oxford, 1972.

In summary, the picture that has been developed for cytochrome *c* evolution is that of a molecule well adapted for its present metabolic role and essentially interchangeable in function over the entire range of eukaryotes. Some amino acid residues, essential for the proper operation of the molecule and its interaction with other proteins, are virtually fixed; most of the others seem to have few selective constraints. Most if not all of the changes that we observe between eukaryote species represent the results of the random fixation of harmless mutations by genetic drift, the mutations in essential amino acids having been quickly weeded out by the death of their carriers. This model provides an explanation for the observed constant rate of accumulation of changes with time and for the existence of different characteristic rates for different proteins. The explanation of differences in rates of amino acid change in terms of differences in the effective number of essential residues, when carried to its conclusion in the covarion hypothesis, predicts that those residues which can accept changes do so at approximately the same rate in one protein as in another.

It must be added that everything that has been said in this section on cytochrome evolution applies only to proteins which have become stabilized in well-defined metabolic roles as the result of positive selection pressures in an earlier era. When a protein is being adaptively modified for a new role, as lysozyme–α-lactalbumin during the rise of mammals, hemoglobin–myoglobin during the development of metazoans, and cytochromes during the evolution of the respiratory machinery, then one certainly would not expect to find a constant and characteristic average rate of accumulation of amino acid changes with time. It is probably true that we shall be able to use the rate of protein evolution to assign dates to early evolutionary events which are untraceable in the fossil record, but any attempt to push the dating process back into the period in which the subject protein was actively evolving its biochemical role would be misleading.

4. *Molecular Folding and Structural Integrity*

As a result of a fundamental study of the properties of cytochrome *c* (*144*), Theorell proposed in 1941 that the heme group was surrounded and protected from the environment by polypeptide chain, with two histidine side chains coordinated to the iron as the fifth and sixth ligands. The X-ray analysis (*12*) ultimately showed that these two ligands actually were histidine-18 and methionine-80, but methionine had previously been proposed as the sixth ligand by Harbury on chemical grounds (*145*).

144. H. Theorell and Å. Åkeson, *JACS* **63**, 1804–1827 (1941).

Guanidination of all lysines without impairing reaction with oxidase and reductase (*146*), and trifluoroacetylation without alteration of the spectral properties of the molecule (*147*) both ruled out lysine as a heme ligand, and the occurrence of the same spectrum in *Pseudomonas* c_{551}, which has only one histidine, contradicted Theorell's original proposal. Harbury observed that *N*-acetylmethionine and its methyl ester would both bind to a heme octapeptide of cytochrome to produce the hemochromogen spectrum and low-spin iron state that are familiar in intact cytochrome but absent in the simple heme octapeptide in solution. It was also observed that methionine bound more tightly to the reduced iron than to the oxidized, which was a possible contributing factor to the many lines of evidence suggesting that ferrocytochrome *c* is more compact, more stable, and more resistant to denaturation and ligand displacement from the iron than is ferricytochrome (summarized in refs. *3* and *12*).

A characteristic spectral feature of ferricytochrome is a weak absorption band at 695 nm, and it was recognized early that treatment of the molecule which destroyed this band also damaged the redox properties of the cytochrome. The presence of the 695-nm band was used as an indicator of molecular integrity even before its actual origin was known. Schejter and George observed that thermal or solvent denaturation and ligand substitution at the iron eliminated the 695-nm band, and suggested that it was an indicator of protein conformation (*148*). Schechter and Saludjian demonstrated instead that the 695-nm band could be produced simply by complexing methionine with Harbury's cytochrome heme octapeptide (*149*). Although a conformational change in cytochrome *c* follows the breaking of the methionine S—Fe bond, it is actually this ligand bond which is responsible for the 695-nm absorption. Sreenathan and Taylor confirmed that this band was sensitive to the presence or absence of the methionine ligand in the molecule rather than to protein conformation (*150*). Moreover, Folin *et al.* (*151*) showed that methionine-80 could be converted to the sulfoxide without opening up the cytochrome molecule or impairing its redox activity, but that under these conditions the 695-nm band was shifted to 750 nm. This wavelength is almost identical to that seen in the complex of the cytochrome heme undecapeptide with

145. H. A. Harbury, J. R. Cronin, M. W. Fanger, T. P. Hettinger, A. J. Murphy, Y. P. Myer, and S. N. Vinogradov, *Proc. Nat. Acad. Sci. U. S.* **54**, 1658–1664 (1965).
146. T. P. Hettinger and H. A. Harbury, *Proc. Nat. Acad. Sci. U. S.* **52**, 1469–1476 (1964).
147. M. W. Fanger and H. A. Harbury, *Biochemistry* **4**, 2541–2545 (1965).
148. A. Schejter and P. George, *Biochemistry* **3**, 1045–1049 (1964).
149. E. Schechter and P. Saludjian, *Biopolymers* **5**, 788–790 (1967).
150. B. R. Sreenathan and C. P. S. Taylor, *BBRC* **42**, 1122–1126 (1971).
151. M. Folin, A. Azzi, A. M. Tamburro, and G. Jori, *BBA* **285**, 337–345 (1972).

N-acetylmethionine sulfoxide. The 695-nm band, which has been used for years as a marker for the integrity of the cytochrome molecule, appears to measure directly the presence of an intact ligand bond between methionine-80 and the heme iron.

To this point we have been considering the cytochrome *c* molecule as a static object, which is far from correct. As Theorell first reported (*144*), ferricytochrome *c* has five distinct pH-dependent conformational states, separated by the dissociation of single protons:

$$\mathrm{I} \xrightarrow{\;H^+\;} \mathrm{II} \xrightarrow{\;H^+\;} \mathrm{III} \xrightarrow{\;H^+\;} \mathrm{IV} \xrightarrow{\;H^+\;} \mathrm{V}$$

$$pK_a = 0.42 \qquad 2.50 \qquad 9.35 \qquad 12.76$$

These and the three pH states of ferrocytochrome are listed in Table XIII, and the most probable heme-ligand interpretation of the middle three states of ferricytochrome is shown in Fig. 15.

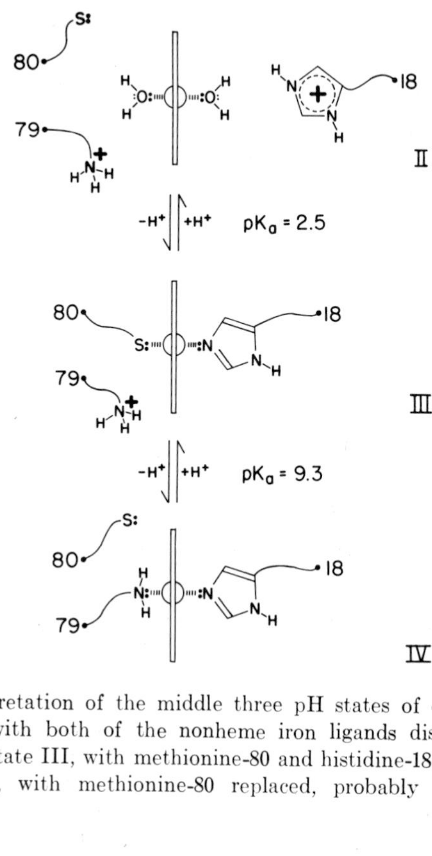

FIG. 15. An interpretation of the middle three pH states of oxidized cytochrome *c*. Top: State II, with both of the nonheme iron ligands displaced. Histidine-18 is ionized. Center: State III, with methionine-80 and histidine-18 as ligands. Bottom: High pH state IV, with methionine-80 replaced, probably by a deprotonated lysine-79.

Ferrocytochrome *c* can be disposed of quite quickly. It remains in a closed conformation from pH 4 to 12, with histidine-18 and methionine-80 as the heme ligands. In this state it is not spontaneously autoxidizable, and carbon monoxide, azide, cyanide, or oxygen are not able to displace

TABLE XIII
CONFORMATIONAL STATES OF CYTOCHROME *c*

Property	State I	State II	State III	State IV	State V
A. Ferricytochrome					
pK_a of transition		0.42	2.50	9.35	12.76
Iron spin	High	High	Low	Low	Low
695 nm band?	No	No	Yes	No	No
Fifth ligand	H_2O	H_2O	His-18	His-18	His-18?
Sixth ligand	H_2O	H_2O	Met-80	Lys-79?	OH^-?
Protein conformation	Completely denatured	Unfolded, open crevice	Folded	Folded	Unfolded
Complexes with					
F^-?		Yes	No	No	
CN^-?		Yes	Slowly	No	Yes
N_3^-?			Yes	No	
NO?			Yes	No	
Reduced by					
Ascorbate			Yes	No	
Ferrocyanide			Yes	No	
Tetrachlorohydroquinone			Yes	No	
Dithionite			Yes	Yes	
Hydrated electrons			Yes	Yes	
Cytochrome c_1			Yes	No	

Property	State I	State II	State III
B. Ferrocytochrome			
pK_a of transition		4	12
Heme crevice	Open	Closed	Partially open
Sixth ligand?	H_2O	Met-80	Met-80?
Protein conformation	Unfolded	Folded	Partially unfolded
Autoxidizable?	Yes	No	Yes
Complexes with CO, N_3^-, CN^-, O_2?	Yes	No	Yes

either ligand and complex with the heme. The integrity of this compact molecule is disturbed below pH 4 or above 12, where the molecule becomes autoxidizable and reactive to monoxide and other external ligands. Because the heme iron has no unpaired electrons, the molecule is always in a low-spin iron state.

In ferricytochrome *c,* state III is the physiologically active one, again with histidine-18 and methionine-80 as the heme ligands. These two strong field ligands force the potentially unpaired iron electrons to pair and give the iron a low spin, and the intact methionine-80 bond produces the characteristic 695-nm absorption. Fluoride ion cannot displace methionine from the iron as a ligand, cyanide displaces methionine slowly, and azide and NO do so rapidly.

As the pH is lowered, ferricytochrome undergoes a transition with a pK_a of 2.50 to a partially unfolded state II with a high-spin iron, and then a second transition with $pK_a = 0.42$ to a completely denatured high-spin form I. The NMR spectrum of state II is that which would be expected from a high-spin iron interacting with weak field ligands such as water molecules (*152,153*). Circular dichroism and viscosity measurements both suggest that state II is unfolded (*154*), but the original low-spin, tightly folded state III molecule can be obtained reversibly by raising the pH again. The 695-nm methionine ligand band disappears in state II, and is replaced by a 620-nm band which is characteristic of a high-spin ferrihemoprotein complex (*155–157*). The transition from state III to II apparently results from the ionization of histidine-18, with an abnormally low pK_a of 2.5 induced by its hydrophobic environment (*154,158–160*). The titration of this group can be followed by proton magnetic resonance and is accompanied by a highly cooperative opening up of the heme crevice. When the pH is raised and the molecule refolds into state III, both the histidine-18 and methionine-80 ligand bonds to the iron must be reformed. Experiments with photooxidation (*151*) and selective carboxymethylation (*154*) of these two groups suggest that his-

152. A. G. Redfield and R. K. Gupta, *Cold Spring Harbor Symp. Quant. Biol.* 405–411 (1971).
153. R. K. Gupta and S. H. Koenig, *BBRC* **45,** 1134–1143 (1971).
154. J. Babul and E. Stellwagen, *Biochemistry* **11,** 1195–1200 (1972).
155. R. Lemberg and J. W. Legge, "Hematin Compounds and Bile Pigments." Wiley (Interscience), New York, 1949.
156. A. S. Brill and R. J. P. Williams, *BJ* **78,** 246–253 (1961).
157. L. S. Kaminsky, M. J. Byrne, and A. J. Davison, *ABB* **150,** 355–361 (1972).
158. E. Stellwagen and R. G. Shulman, *JMB* **75,** 683–695 (1973).
159. J. S. Cohen, W. R. Fisher, and A. N. Schechter, *JBC* **249,** 1113–1118 (1974).
160. J. S. Cohen and M. B. Hayes, *JBC* **249,** 5472–5477 (1974).

tidine plays the more important role of the two in this refolding transition.

Proton resonances are also observed from histidine-26 and -33 in horse cytochrome (*160*). A comparison of spectra with tuna, which has no histidine at position 33, permits the assignment of $pK_a = 6.54$ to histidine-33 and 3.5 to histidine-26. Thus, histidine-33 has the pK_a that would be expected in a polar environment, and the pK_a values of histidine-26 and -18 correlate well with their increasingly shielded hydrophobic surroundings as revealed by the X-ray analysis. The three histidine proton resonances do not become equivalent, indicating complete denaturation and opening up of the cytochrome molecule, until the pH falls to around 0.66. The low pH state I is a totally denatured heme polypeptide, but this denaturation can also be reversed by raising the pH.

Ferricytochrome *c* has two more transistions at high pH, both involving the loss of a proton. In the III to IV transition with $pK_a = 9.35$, the disappearance of the methionine methyl proton resonance from the NMR spectrum indicates that methionine-80 is displaced as the heme ligand (*152*), but another strong field ligand takes its place since the iron remains low spin (*144,152,153*). The fact that not even cyanide or azide ions can displace this new ligand from the iron in state IV suggests that it is a nitrogenous base, and an obvious candidate from both amino acid sequence and X-ray structure is lysine-79, although lysine-72 or -73 cannot be absolutely ruled out. The III to IV transition with $pK_a = 9.35$ probably represents the deprotonation of a lysine side chain, followed by displacement of methionine from the iron by the stronger $—NH_2$ ligand of lysine.

The most striking feature of cytochrome in state IV is that it is not reducible by ordinary reagents such as ascorbate or ferrocyanide. Two forms of cytochrome coexist in equilibrium at alkaline pH: state III reducible and state IV not (*161*). Only the reducible form shows the 695-nm absorption band, which therefore becomes a useful spectroscopic indicator of reducibility (*162,163*). Apparently the intact methionine-80–iron bond is necessary for reduction of the iron in a folded cytochrome molecule, although ascorbate can reduce the heme iron if the molecule is totally denatured with propanol or urea (*164*). This suggests that the protein chain in cytochrome has a protective role: to prevent the iron from being easily reduced except under controlled conditions.

161. C. Greenwood and G. Palmer, *JBC* **240,** 3660–3663 (1965).
162. C. Greenwood and M. T. Wilson, *Eur. J. Biochem.* **22,** 5–10 (1971).
163. M. T. Wilson and C. Greenwood, *Eur. J. Biochem.* **22,** 11–18 (1971).
164. L. S. Kaminsky and V. J. Miller, *BBRC* **49,** 252–256 (1972).

The reduction process just described is represented by a pH-dependent equilibrium between states III and IV, with only the former being reducible:

$$\text{IV} \underset{-\text{H}^+}{\overset{+\text{H}^+}{\rightleftharpoons}} \text{III} \xrightarrow{+e^-} \text{reduced cytochrome}$$

Hess measured the apparent equilibrium constant for reduction of cytochrome by ferrocyanide as a function of pH (*165*), and found that the actual reduction step did not involve a proton and was entirely independent of pH. The free energy of reduction, and hence the reduction potential, change with pH only because the ratio of amounts of state III and IV changes with pH. The observed decrease of cytochrome reduction potential of 60 mV per pH unit above pH 8 is exactly what would be calculated from the simple Nernst equation.

Although the foregoing picture is correct for ascorbate and ferrocyanide, stronger reductants such as dithionite (*166*) or hydrated electrons (*167–171*) can reduce state IV directly. Palmer and co-workers (*166*) have proposed the reduction scheme shown in Fig. 16. At neutrality, both the reduced and oxidized molecules have methionine-80 as the sixth ligand (F and A of Fig. 16), and the oxidized form exhibits the 695-nm absorption band. At higher pH, more and more of the lysines are deprotonated (E and B), and immediately displace methionine in ferricytochrome (C in Fig. 16), although methionine remains as the heme ligand in the reduced molecule (E). State C is not reducible by either ascorbate or ferrocyanide. They bring about reduction at alkaline pH by reducing form A and causing a shift of the equilibrium from C to B to A. In the reverse process, rapid oxidation of ferrocytochrome at pH 10.5 proceeds from states F and E through B to C, and a transient 695-nm band from state B is in fact observed (*166*). In contrast, dithionite and hydrated electrons are strong enough reagents to reduce the lysine-liganded molecules in state C directly, producing unstable form D with a reduced iron

165. K. G. Brandt, P. C. Parks, G. H. Czerlinski, and G. P. Hess, *JBC* **241,** 4180–4185 (1966).
166. D. O. Lambeth, K. L. Campbell, R. Zand, and G. Palmer, *JBC* **248,** 8130–8136 (1973).
167. E. J. Land and A. J. Swallow, *ABB* **145,** 365–372 (1971).
168. I. Pecht and M. Faraggi, *FEBS* (*Fed. Eur. Biochem. Soc.*) *Lett.* **13,** 221–224 (1971).
169. I. Pecht and M. Faraggi, *Proc. Nat. Acad. Sci. U. S.* **69,** 902–906 (1972).
170. J. Wilting, R. Braams, H. Nauta, and K. J. H. Van Buuren, *BBA* **283,** 543–547 (1972).
171. N. N. Lichtin, A. Shafferman, and G. Stein, *Science* **179,** 680–682 (1973).

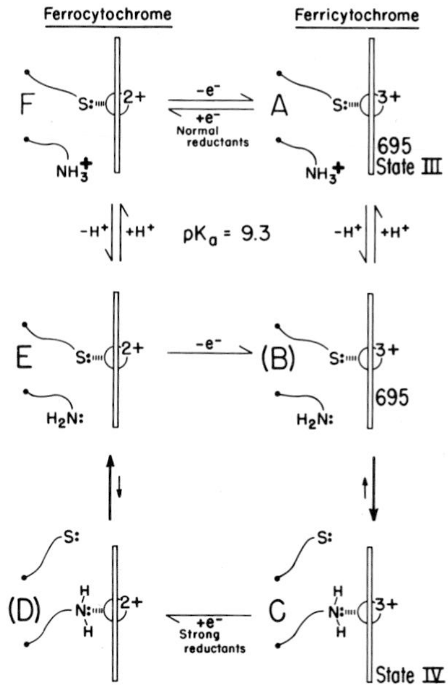

FIG. 16. Explanation, after Lambeth *et al.* (*166*), for the interrelationship between oxidation state, heme liganding, reducibility, and pH in cytochrome *c*. See text for discussion.

liganded to lysine. The stronger binding of methionine in the reduced molecule (*145*) then leads to a rapid ligand exchange, ending at the same states E and F as before. In both the dithionite and hydrated electron experiments at alkaline pH, state D is detectable during rapid reduction as a transient species in the ultraviolet and visible spectrum. The initial reduction with hydrated electrons takes place in nanoseconds, but the subsequent relaxation step (D to E) occurs over several hundred milliseconds and is independent of both protein and hydrated electron concentrations (*169*).

In summary, the most favorable conditions for reduction are with methionine as the sixth iron ligand. Only unusually strong conditions can cause reduction of the lysine-liganded state IV. This is of direct biological interest in view of King and co-workers' report of the pH dependence of the rate constant for reduction of ferricytochrome *c* by ferrocytochrome c_1 (*172*). If these rate constants are plotted against the relative amount of state III cytochrome present, as measured by the 695-nm band intensity (data from ref. *166*), a straight line is obtained from pH 7.5

172. C. A. Yu, L. Yu, and T. E. King, *JBC* **248,** 528–533 (1973).

to 10—evidence that only state III is reducible by physiological agents as well. The dithionite and hydrated electron experiments illustrate an important point to be kept in mind: The ability of a substance to reduce or oxidize cytochrome *c* under laboratory conditions may have no connection with the actual redox process *in vivo*.

The assignment of lysine-79 as the liganding group under alkaline conditions is likely but not quite certainly correct. Modifying all of the lysines in the molecule by guanidination or trifluoroacetylation (*146,147,173*) abolishes the III to IV transition, suggesting the involvement of lysine. Wilgus and Stellwagen (*174*) have made a particularly ingenious use of Corradin and Harbury's discovery (*175–177*) that cytochrome *c* can be cleaved at methionine-65, separated into two peptides, and reconstituted into a well-functioning redox molecule again. They guanidinated each peptide separately, and combined each with an unmodified form of the opposite peptide, producing hybrid molecules which were either guanidinated from residues 1 to 65, or from 66 to 104. Only the first hybrid showed the normal low-spin iron at alkaline pH; blocking lysines in the latter half of the chain led to a high-spin state indicating that a weak field ligand was present instead of lysine. Hence, lysine-13, the next closest to the heme iron after lysine-79, can be ruled out as the high pH ligand, and lysine-79 (or perhaps lysine-73) remains the prime candidate.

Tyrosine-67 has been implicated in holding together the physiologically important state III of cytochrome, since either iodination or nitration of this residue destroys the 695-nm band and shifts the protein to the IV configuration (*178–182*). Skov and Williams (*178*) demonstrated that this actually only represents a shift of the pK_a of the III to IV transition from 9.3 to 6, paralleling the drop in pK_a of the phenol group of tyrosine-67 caused by nitration or iodination. If the pH is reduced below 6, the missing 695-nm band reappears at nearly full strength. Schejter and coworkers extended this work and found that two mononitrotyrosine-67 isomers of cytochrome are produced, probably corresponding to substitutions at the two possible positions ortho to the —OH group on the ring, which is not free to rotate inside the molecule (*181*). One isomer exhibits

173. R. A. Morton, *Can. J. Biochem.* **51**, 465–471 (1973).
174. H. Wilgus and E. Stellwagen, *Proc. Nat. Acad. Sci. U. S.* (in press).
175. G. Corradin and H. A. Harbury, *BBA* **221**, 489–496 (1970).
176. G. Corradin and H. A. Harbury, *Proc. Nat. Acad. Sci. U. S.* **68**, 3036–3039 (1971).
177. J. Babul, E. B. McGowan, and E. Stellwagen, *ABB* **148**, 141–147 (1972).
178. K. Skov, T. Hofmann, and G. R. Williams, *Can J. Biochem.* **47**, 750–752.
179. K. Skov and G. R. Williams, *Can J. Biochem.* **49**, 441–447 (1971).
180. M. Sokolovsky, I. Aviram, and A. Schejter, *Biochemistry* **9**, 5113–5118 (1970).
181. A. Schejter, I. Aviram, and M. Sokolovsky, *Biochemistry* **9**, 5118–5122 (1970).
182. R. A. Morton, *Can. J. Biochem.* **51**, 472–475 (1973).

the III-to-IV-type transition with a pK_a of 5.9, and shows the 695-nm absorption band and is ascorbate reducible in its low pH III state. It is low spin in both III and IV states. The other isomer is high spin and presumably has no sixth ligand up to pH 8.5, where it undergoes a transition to a IV-like form. It is not reducible and has no 695 nm band at any pH. These nitration experiments have been interpreted as indicating that the ionization state of the phenol of tyrosine-67 somehow interacts with lysine-79 and controls its liganding tendency with the heme iron (*182*). A simpler explanation is that a ligand change transition occurs around pH 6 in nitrotyrosyl cytochrome, analogous to that seen at 9.35 in native cytochrome, but that the ligand which displaces methionine-80 is not lysine-79 but is the ionized tyrosine-67 itself. (Salemme has suggested in ref. *31* that tyrosine-67 is the sixth heme ligand at alkaline pH even in normal cytochrome, but this is difficult to harmonize with the lysine-blocking experiments.) When ionized tyrosine-67 is protonated around pH 6, just as when neutral lysine-79 is protonated around pH 9.35 in normal cytochrome, methionine-80 becomes the stronger potential ligand and displaces it from the iron. The 695-nm absorption band reappears, and the molecule becomes reducible by ascorbate.

Acid denaturation of the type that we have seen occurs in two steps: first an opening up of the heme crevice and loss of heme ligands and 695 nm absorption (III → II), and then at lower pH a total uncoiling of the polypeptide chain (II → I). Something similar occurs in denaturation with either temperature or organic solvents. With urea, a crevice-opening step is centered around 3 *M* urea, and a helix-coil transition occurs above 6.5 *M* urea (*183,184*). All of the aromatic residues are in contact with solvent at high urea concentrations, and methionine-80 and histidine-18 are accessible for carboxymethylation reactions. Urea-denatured cytochrome *c* is low spin at neutral pH, changing to high spin with a transition pK_a of 5.2. Babul and Stellwagen have suggested that above pH 5.2, the heme group is liganded by two of the three histidine side chains even though the polypeptide around them is uncoiled (*185*). All of these changes are reversible, and the original folded cytochrome molecule can be obtained when the urea is removed.

If increased temperature is used instead of denaturing solvents, a similar two-state change is produced, with the first step centered at about 55°C, and the second at 82°C (*184*). Sodium dodecyl sulfate (SDS) also causes a two-state change: The 695-nm band is gone and the crevice is opened when 14–20 SDS molecules are bound per molecule of cytochrome, and complete unfolding occurs at higher SDS to protein ratios

183. E. Stellwagen, *Biochemistry* **7**, 2893–2898 (1968).
184. Y. P. Myer, *Biochemistry* **7**, 765–776 (1968).
185. J. Babul and E. Stellwagen, *Biopolymers* **10**, 2359–2361 (1971).

(*186*). Alcohols open the heme crevice as well (*187–189*). In contrast, ferrocytochrome *c* is comparatively resistant to agents which will easily open the heme crevice in ferricytochrome. Ikai and Tanford have proposed that in some reversible refolding of renatured proteins, the observed kinetics can best be explained by the transient existence of an incorrectly folded state, which must uncoil again before it can form the correct molecule (*190,191*). Guanidine hydrochloride has been investigated as a denaturing agent of cytochrome *c* (*192,193*), and it appears that the wrong-fold model may be correct in this case.

One of the more interesting questions for any protein is what factors control the proper folding as the polypeptide chain is spun off the ribosome, what steps the chain goes through in the folding process, and where in the cell this folding takes place. The literature on theories of protein folding is extensive and cannot be discussed here. Cytochrome *c* has an added component: a large, hydrophobic heme group which conceivably could serve as a template for folding. The relative ease of predicting α-helical regions of an amino acid sequence has led to the suggestion that helix formation may act as a primary nucleation process for folding, with the first step being the spontaneous formation of helices at various points along the polypeptide chain, followed by the packing of helices with hydrophobic sides against one another to build a compact molecule. This was an obvious consideration for the entirely α-helical myoglobin and hemoglobin molecules. Moreover, several helices may first build a hydrophobic core which then can serve as a template for the folding of the latter half of the molecule. Sharp turns such as the β or 3_{10} bend (*12,194*) can help to build a compact central unit. Scheraga and co-workers have pursued the theoretical side of this nucleation idea extensively (*195*). Phillips (*196*) has made such a proposal for lysozyme, and Dickerson (*12*) and Salemme *et al.* (*30*) have thought along similar lines for cytochromes.

186. R. K. Burkhard and G. E. Stolzenberg, *Biochemistry* **11**, 1672–1677 (1972).
187. L. S. Kaminsky, F. C. Yong, and T. E. King, *JBC* **247**, 1354–1359 (1972).
188. L. S. Kaminsky, P. E. Burger, A. J. Davison, and D. Helfet, *Biochemistry* **11**, 3702–3706 (1972).
189. L. S. Kaminsky, V. J. Miller, and A. J. Davison, *Biochemistry* **12**, 2215–2221 (1973).
190. A. Ikai and C. Tanford, *Nature* (*London*) **230**, 100–102 (1971).
191. A. Ikai, W. W. Fish, and C. Tanford, *JMB* **73**, 165–184 (1973).
192. R. W. Henkens and S. R. Turner, *Biochemistry* **12**, 1618–1621 (1973).
193. J. A. Knapp and C. N. Pace, *Biochemistry* **13**, 1289–1294 (1974).
194. C. M. Venkatachalam, *Biopolymers* **6**, 1425–1436 (1968).
195. P. N. Lewis, F. A. Momany, and H. A. Scheraga, *Proc. Nat. Acad. Sci. U. S.* **68**, 2293–2297 (1971).
196. D. C. Phillips, *Sci. Amer.* November 78–90 (1966).

A possible sequence of events for the primary folding of eukaryotic cytochrome *c* might be the following (refer to Fig. 7):

(a) Nucleation out of α-helices along the unfolded polypeptide chain: *N*-terminal helix, 50's and 60's regions, and the final C-terminal helix. One side of each of the long helices is covered with polar side chains; the other, with hydrophobic groups.

(b) Attachment of the heme at the end of the N-terminal helix. As Salemme has pointed out, Ehrenberg and Theorell observed as early as 1955 (*197*) that if the first 18 cytochrome residues are built into a helix, the proper stereochemical arrangement is obtained between cysteine-14 and -17 for attachment of the heme. In fact, in *c*, c_2, and c_{550}, the chain from residues 12 through 17 is a passable turn of helix with its axis at about 40° to that of the N-terminal helix. Residues 12–17 may have been a part of this initial helix before the heme was attached.

(c) Anchoring of the loose chain by means of hydrogen bonds to the heme propionic acids, involving tyrosine-46 and -48, tryptophan-59, asparagine-52, and threonine-49. The bending of the chain back to the heme again is facilitated by the formation of sharp 3_{10} bends of the type that are seen in the finished molecule at 20–23 and 75–78. Not all the hydrogen bonds or hydrophobic interactions that are formed during folding need be present in the completed molecule: some bonds may be formed to facilitate the folding process and be ruptured again in subsequent readjustment steps.

(d) Folding of the right side into its final configuration, aided by the packing of hydrophobic groups 30, 32, and 35 against the heme.

(e) Packing of the left side α-helices against the heme, with important hydrophobic contact groups at 64, 67, and 68 in the 60's helix, and 94, 95, and 98 in the C-terminal helix.

(f) Relaxation of the 70's loop, and the 20's loop in c_{550} and some of the c_2's (Table VII) into their final configurations.

In this model, the heme group attached to the initial helix would be the nucleus around which the rest of the molecule would fold. It is easy to see why apocytochrome would be folded differently (see below), and why porphyrin-cytochrome with the iron atom missing could adopt a cytochromelike configuration even without the axial heme ligands. Denaturing solvents and elevated temperatures should simply undo the original folding, and their effects should be reversible within limits. One should expect that, once the heme ligands were broken, the heme crevice could be opened and closed again without general and permanent disruption of the rest of the molecule, and this is what happens in ferricytochrome state II after histidine-18 is protonated. Because the molecule

197. A. Ehrenberg and H. Theorell, *Acta Chem. Scand.* **9**, 1193–1205 (1955).

is held together mainly by hydrophobic interactions with the heme, one would expect it to be neither rigid nor formless, but *flexible*—subject to elastic deformation within limits from a stable minimum energy configuration. The special susceptibility of buried tyrosine-67 to iodination (*198*) and nitration (*178–181*), and the ease with which tryptophan-59 can be formylated (*199*) or oxidized by *N*-bromosuccinimide (*200–202*) support the concept of an elastic polypeptide framework for cytochrome *c*.

Tsong has studied the thermal unfolding and refolding of cytochrome *c* at acid pH (*203*), and found it to be at least a three-state process which can be accounted for in terms of a model very much like the foregoing: initial local nucleation of secondary structure along the chain, a major folding or nucleation step (packing around the heme?), and subsequent less drastic adjustments to yield the stable molecular configuration.

It is interesting to consider what the polypeptide chain in cytochrome *c* would do without the iron atom to hold the axial ligands or without the entire heme group. Stellwagen (*204*) has found that apocytochrome, with the heme group cleaved from the polypeptide chain, is completely unfolded. Tryptophan-59 and all four tyrosines are exposed to the solvent, and both methionines and all three histidines are reactive to carboxymethylation. The heme group apparently is necessary for folding, which is not surprising in view of the way that the polypeptide chain is wrapped around the heme with hydrophobic side chains packed against it (Fig. 7). On the other hand, Fisher *et al.* (*205*) have found that porphyrin-cytochrome, with only the iron missing, is folded into a compact molecule similar to intact cytochrome, although less temperature-stable. The polypeptide chain is capable of folding compactly around the empty porphyrin ring, although we do not know how similar this folding is to that of the intact cytochrome molecule. It would be valuable to polypeptide folding theory to know the X-ray structure of porphyrin-cytochrome *c*, assuming that the molecule was sufficiently regular for crystallization. The iron ligands to histidine and methionine may induce the proper form of compact folding around the porphyrin, or they may do little more than add stability and produce a final fine-structure adjustment to the molecule.

In spite of the attractiveness of an iron-free porphyrin-cytochrome in demonstrating the role of the tetrapyrrole ring as a folding template,

198. E. B. McGowan and E. Stellwagen, *Biochemistry* **9,** 3047–3052 (1970).
199. I. Aviram and A. Schejter, *BBA* **229,** 113–118 (1971).
200. Y. P. Myer, *Biochemistry* **11,** 4195–4203 (1972).
201. Y. P. Myer, *Biochemistry* **11,** 4203–4208 (1972).
202. Y. P. Myer and P. K. Pal, *Biochemistry* **11,** 4209–4216 (1972).
203. T. Y. Tsong, *Biochemistry* **12,** 2209–2214 (1973).
204. E. Stellwagen, R. Rysavy, and G. Babul, *JBC* **247,** 8074–8077 (1972).
205. W. R. Fisher, H. Taniuchi, and C. B. Anfinsen, *JBC* **248,** 3188–3195 (1973).

the process within the living cell apparently proceeds differently. The sequence of cytochrome c is coded in the DNA of the cell nucleus, and the polypeptide chain is spun out on cytoplasmic ribosomes on the endoplasmic reticulum. Kadenbach has presented evidence to show that the apoprotein, which we now can say would be in a random coil configuration, diffuses from the cytoplasm into the mitochondrion, and an iron-containing heme is attached there (*206*). The final folding into the structure of Fig. 7 therefore takes place inside the mitochondrion. It is possible to speculate that an unfolded apocytochrome would find it easier to get through the outer mitochondrial membrane and that attachment of the heme and subsequent folding would also play a control function to help keep the cytochrome molecule where it should be. Sherman and Stewart (*207*) regarded the diffusion of apocytochrome into the mitochondrion prior to heme attachment as less than proven, but in any event the iron is added to the porphyrin prior to its incorporation into the apocytochrome molecule (*208*).

5. *Oxidation–Reduction Mechanisms*

The most important single question that one can ask about the cytochrome c molecule is: "How does it work?" An easy answer to this question has not yet come out of the chemical studies, amino acid sequence comparison, optical and magnetic resonance investigations, or the X-ray structure analyses. The crystal structures have served to define and delimit the areas in which future biochemical studies are likely to be most productive, which perhaps is their proper role. One now is able to eliminate more plausible but incorrect theories than at any time in the past; and the growing body of knowledge about the behavior of the cytochrome molecule leads to the hope that, even barring the sudden intrusion of genius, we may yet arrive at the correct oxidation–reduction mechanism by a rigorous application of the method of Holmes (*209*): ". . . when all other contingencies fail, whatever remains, however improbable, must be the truth."

There are several factors that must be considered in looking for a redox mechanism:

1. details of the changes in protein conformation in going from the oxidized to the reduced state,

2. factors affecting the interaction of cytochrome c with its biological reductase and oxidase *in vivo* and *in vitro*,

206. B. Kadenbach, *Eur. J. Biochem.* **12**, 392–398 (1970).
207. F. Sherman and J. W. Stewart, *Annu. Rev. Genet.* **5**, 257–296 (1971).
208. E. M. Colleran and O. T. G. Jones, *BJ* **134**, 89–96 (1973).
209. S. Holmes, as quoted by J. H. Watson and A. C. Doyle, *in* "The Adventure of the Bruce-Partington Plans," p. 926. London, 1891.

3. effects of systematic modification of the cytochrome molecule, both by chemically altering side chains and by blocking portions of the molecular surface,

4. oxidation and reduction of the molecule with inorganic and other nonbiological reagents, and

5. structure and reactivity comparisons between different cytochrome molecules with comparable roles—both the sequence variations among eukaryotes and the more extensive changes in eukaryotic and bacterial cytochromes.

In the first area we have gone a step backward recently in order to make two steps forward. The reduced tuna cytochrome map at 2 Å resolution is in good shape, the oxidized horse cytochrome map at 2.8 Å is of poor quality, and the oxidized tuna map at 2 Å that will replace it is completed but not interpreted at the time of writing. The opening of the heme crevice at phenylalanine-82 in the oxidized state that was reported from the horse map (*12*) apparently is not correct, and any comments about the finer reorientations of aromatic or other side chains must wait for confirmation from the new tuna map. Much can be said now about areas 2 through 5 above, however, to lay the groundwork for a later discussion of the first area.

a. Oxidase, Reductase, and Modification of Lysines. The one firm statement that can be made about the interaction of cytochrome *c* with its oxidase and reductase is that it is largely ionic and involves positively charged lysine side chains on the cytochrome molecule. The reaction with oxidase (the a/a_3 Complex IV of Fig. 5) is sensitive to changes in ionic strength of the surrounding medium (*210*). Polycations such as salmine or poly-L-lysine are competitive inhibitors, competing with cytochrome *c* for the oxidase surface (*210,211*). The reductase (a particulate preparation from Complexes I–III of Fig. 5) is also subject to competition by polylysine (*212*).

The useful lysine derivatives for cytochrome *c* are shown in Fig. 17. Chemical modifications of the lysine side chains which remove the positive charge or replace it with a negative charge destroy the interaction with cytochrome oxidase (*213–215*). Results with acetylation, succinylation, trinitrophenylation, and guanidination are given in Table XIV. Suc-

210. H. C. Davies, L. Smith, and A. R. Wasserman, *BBA* **85,** 238–246 (1964).
211. B. S. Mochan, W. B. Elliott, and P. Nicholls, *Bioenergetics* **4,** 329–345 (1973).
212. K. A. Davis, Y. Hatefi, F. R. Salemme, and M. D. Kamen, *BBRC* **49,** 1329–1335 (1972).
213. S. Takemori, K. Wada, K. Ando, M. Hosokawa, I. Sekuzu, and K. Okunuki, *J. Biochem.* (*Tokyo*) **52,** 28–37 (1962).
214. K. Wada and K. Okunuki, *J. Biochem.* (*Tokyo*) **64,** 667–681 (1968).
215. K. Wada and K. Okunuki, *J. Biochem.* (*Tokyo*) **66,** 249–262 (1969).

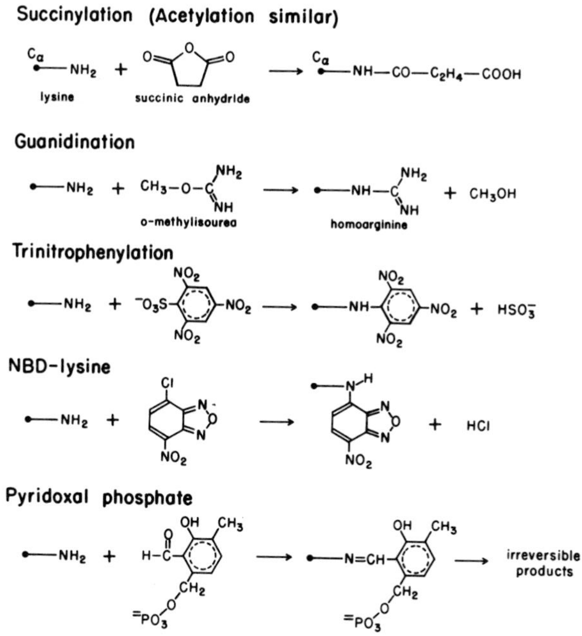

FIG. 17. Chemical modifications of lysine side chains which have proved useful in studying cytochrome *c* mechanisms. See also M. J. Wimmer, M. Foster, K. T. Mo, D. L. Sawyers, and J. H. Harrison, *Fed. Proc., Fed. Amer. Soc. Exp. Biol. Abstr.* **34**, 630 (1975).

cinylation introduces negative charges, acetylation and trinitrophenylation merely eliminate the positive charge, and guanidination replaces lysine by an even stronger base. The elimination of four to six lysine residues is enough to bring the reaction with oxidase to a halt. It would be valuable to know just where all six of these residues are on the surface of the molecule and if the same residues are involved with different blocking reagents. All that is known is that acetylation modifies lysine-22 first and then lysine-13, diminishing the oxidase reaction by 26%; and that trinitrophenylation blocks lysine-13 and then lysine-22 in the course of eliminating 61% of the oxidase reactivity (Table XIV). In each case, modification of lysine-13 causes a larger incremental drop in oxidase activity than lysine-22, and Fig. 7 shows that lysine-13 sits at the upper end of the heme crevice, on the face of the molecule. The identities of the other critical few residues which are involved in destroying oxidase activity during acetylation, succinylation, and trinitrophenylation are badly needed. (The preliminary identification in ref. *216* of lysine-73 as

216. K. Okunuki, K. Wada, H. Matsubara, and S. Takemori, *Oxidases Related Redox Syst., Proc. Symp., 1964* p. 549 (1965).

TABLE XIV
SELECTIVE MODIFICATION OF LYSINES[a]

Type of modification	No. of lysines modified	Residue no., if known	Relative oxidase activity (%)
Acetylation	1	22	90
	2	22 + 13	74
	3	—	61
	6	—	0
Succinylation	4	—	0
Trinitrophenylation	1	13	50
	2	13 + 22	39
	3	—	23
	4	—	14
	5	—	3
Guanidination	9	—	110
	13	—	116
	17	—	123

[a] Data from references *213–215*.

the primary trinitrophenylation site rather than lysine-13 is an unfortunate error, which has proved hard to expunge from the literature.)

In contrast, guanidination of some or all of the 19 lysines *increases* the oxidase activity above that observed with native cytochrome (*146,214*). Presumably the stronger guanidinium bases are more effective binding agents to cytochrome *a* than lysines are. Completely guanidinated cytochrome *c* is also somewhat better than the unmodified protein at restoring oxygen uptake to a particulate succinate oxidase preparation which contains all of the elements of the respiratory chain bound to membrane fragments (*146*). This suggests that the positive charges on cytochrome *c* are involved in both oxidase and reductase binding. No changes are seen in either the reduction potential of guanidinated cytochrome or its spectrum at neutral pH. At alkaline pH, the III to IV transition is eliminated because lysine is not available as an iron ligand. Instead, a high-spin form resembling state V of normal cytochrome appears with a transition pK_a of 9.4 (*173*).

Trifluoroacetylating all of the lysine residues of cytochrome *c* leaves the Soret and visible spectrum of the molecule intact but destroys all activity with the succinate oxidase preparation (*147*). The blocking groups can be gently hydrolyzed away again to recover the intact protein with all lysines free and with 100% activity.

All of the above experiments suffer from nonselectivity. One either modifies all 19 lysine groups or is not sure which ones are being altered if a smaller number react. Margoliash and co-workers (*217*) have modified residue 13 specifically, making the 4-nitrobenzo-2-oxa-1,3-diazole (NBD) derivative of lysine-13 in horse cytochrome (Fig. 17) and the bisphenylglyoxal derivative of arginine-13 in *Candida* (Fig. 18). In agreement with Wada and Okunuki (*215*), they found that the oxidase reactivity is cut in half. They also tested their derivatives against succinate-cytochrome *c* reductase—a particulate succinate oxidase preparation of the type mentioned earlier, but with the cytochrome oxidase poisoned with cyanide and the rate of reduction of cytochrome *c* followed spectrophotometrically. Reductase activity was normal—an important observation because it indicates that the binding sites for oxidase and reductase on the surface of the cytochrome *c* molecule are different.

Aviram and Schejter have provided more evidence for different binding sites by selectively modifying just one or two lysines with pyridoxal phosphate (*218*). Both the singly and doubly substituted derivatives exhibit

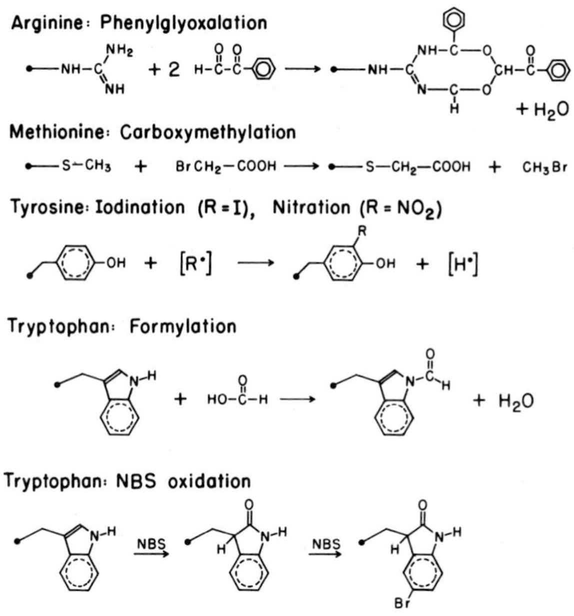

FIG. 18. Useful chemical modifications of other cytochrome *c* side chains.

217. E. Margoliash, S. Ferguson-Miller, J. Tulloss, C. H. Kang, B. A. Feinberg, D. L. Brautigan, and M. Morrison, *Proc. Nat. Acad. Sci. U. S.* **70**, 3245–3249 (1973).
218. I. Aviram and A. Schejter, *FEBS (Fed. Eur. Biochem. Soc.) Lett.* **36**, 174–176 (1973).

a 695-nm absorption band, indicating that the heme crevice structure is intact. The reduced form of both derivatives is not autoxidizable, and the oxidized form shows the normal rate constants for reduction by ascorbate and reaction with cyanide. Care was taken to establish the integrity of the derivatized molecules because of the significant observation that the oxidase reaction is unaffected by pyridoxal phosphate modification, but the reaction with reductase is seriously impaired. Oxidase activity was measured with a purified beef heart cytochrome oxidase preparation, and reductase activity was followed by two assay systems: soluble NADH-cytochrome *c* reductase, and particulate succinate-cytochrome *c* reductase. Both redox processes were monitored even further by the efficiency of restoration of respiratory activity to cytochrome *c*–depleted mitochondria. The singly substituted derivative was partially active, and the doubly substituted molecule was completely inactive.

The location of the two pyridoxal phosphate modification sites is of obvious interest, since they provide a marker for the reductase binding site. In the preliminary report (*218*), three pyridoxal phosphate–containing peptides were found: residues 1–10, 68–74, and 83–94. Candidates for modification therefore include lysines 5, 7, 8, 86, 87, and 88 at the top of the molecule, and 72 and 73 on the left side. The further narrowing down of these groups will be awaited with great interest.

A third line of evidence supporting different binding sites for oxidase and reductase comes from the antibody binding studies of Smith and coworkers (*219*). They found that human cytochrome *c* elicited four different antibodies in rabbit, which they studied in the form of Fab fragments. Two of these have been isolated and purified, and tested for their ability to block the redox reactions of cytochrome. One fraction prevents binding of oxidase but not reductase, and a second component binds to isoleucine-58 in human cytochrome but blocks neither the oxidase nor the reductase. The unseparated mixture of all four components prevents reaction with both oxidase and reductase; thus, one or both of the two unpurified components presumably interferes with the reductase binding site. Smith and colleagues have also reported kinetic evidence for the existence of two different binding sites for oxidase and reductase (*220*).

The conclusions that can be drawn from all this work are few but important:

(a) Both the oxidase and reductase bind to cytochrome *c* with the help of lysine residues on the cytochrome molecule.

219. L. Smith, H. C. Davies, M. Reichlin, and E. Margoliash, *JBC* **248,** 237–243 (1973).
220. L. Smith, H. C. Davies, and M. Nava, *JBC* **249,** 2904–2910 (1974).

(b) Oxidase and reductase have different binding sites on the molecule, although the reaction site for input and outflow of electrons may be shared.

(c) The oxidase complex binds in the vicinity of lysine-13, and the reductase binds either to the top of the molecule as seen in Fig. 7 or to the left side in the vicinity of lysine-72 and -73.

(d) Neither oxidase nor reductase binds at the back of the molecule near residue 58.

It should be kept in mind that the cytochrome *c* molecule is about 30 Å in diameter, and this is approximately the diameter of one of the cylindrical Fab arms of the antibody molecule. A bound Fab fragment probably would block an entire quadrant of the surface of the cytochrome molecule. It could bind to lysine-72 and still block the edge of the heme crevice from approach by oxidase or reductase.

The above points are "boundary conditions" which will have to be observed by any theory of oxidation and reduction of the cytochrome *c* molecule which may be proposed.

b. Other Side Chain Modifications. Modifications of methionine-80, tyrosine-67, and tryptophan-59 produce such similar behavior that they can be treated together (Fig. 18). The general pattern, seen in Table XV, includes displacement of methionine-80 as the sixth heme ligand, loss of the 695-nm absorption band, opening of the heme crevice, loss of reducibility by ascorbate, autoxidizability and CO binding in the reduced state, and impaired reactivity with both oxidase and reductase (*178–182,198–202,217,221–227*). For the sake of brevity the discussion that follows will concentrate mainly on deviations from this basic pattern. It is not at all clear that such extensively damaged derivative molecules can tell us much about the pathways of biological oxidation and reduction since any decrease in reactivity with oxidase or reductase is as likely to have its origin in the general disruption of molecular folding as in the specific modification of the residue in question.

Carboxymethylation of cytochrome *c* with iodoacetate or bromoacetate can be carried out such that only methionine-65, or both methionine-65

221. K. Ando, H. Matsubara, and K. Okunuki, *BBA* **118,** 256–267 (1966).
222. E. Stellwagen, *Biochemistry* **7,** 2496–2501 (1968).
223. A. Schejter and I. Aviram, *JBC* **245,** 1552–1557 (1970).
224. M. Brunori, M. T. Wilson, and E. Antonini, *JBC* **247,** 6076–6081 (1972).
225. R. M. Keller, I. Aviram, A. Schejter, and K. Wüthrich, *FEBS* (*Fed. Eur. Biochem. Soc.*) *Lett.* **20,** 90–92 (1972).
226. M. T. Wilson, M. Brunori, G. C. Rotilio, and E. Antonini, *JBC* **248,** 8162–8269 (1973).
227. K. Wüthrich, I. Aviram, and A. Schejter, *BBA* **253,** 98–103 (1971).

TABLE XV
MODIFICATIONS OF RESIDUES 59, 67, AND 80

Property	Met-65 and Met-80 carboxy-methylation	Tyr-74 and Tyr-67 iodination	Tyr-67 nitration[a]	Trp-59 formylation	Met-65 and Trp-59 NBS[b]
695 nm band?	No	Below pH 5.9	Below pH 5.9	No	Weak
Met-80 ligand?	No	Below pH 5.9	Below pH 5.9	No	?
Heme more exposed?	Yes	Yes	Yes	Yes	—
Fe(III) spin state?	Low	Low	Low	Low	Mixed
Ascorbate reducible?	No	—	Yes	No	—
Fe(II) autoxidizable?	Yes	Yes	Yes	Yes	—
Fe(II) binds CO?	Yes	Yes	Yes	Yes	—
Reaction with					
(a) Cytochrome oxidase?	No	—	—	Slower, same V_{max}	—
(b) NADH-cytochrome reductase?	—	40%	—	—	Somewhat slower
(c) Succinate oxidase?	No	—	—	—	Decreased
(d) Depleted mitochondria?	No	—	Much decreased	8%	Decreased
References	*221–226*	*182,198*	*178–181*	*199,217,227*	*200–202*

[a] Isomer form I, reference *181*.
[b] Three to one *N*-bromosuccinimide (NBS) per protein.

TABLE XVI
EFFECTS OF CHEMICAL MODIFICATIONS ON pK_a

Modification	pK_a of isolated tyrosine or derivative	pK_a of III to IV transition in Tyr-67 modified cytochrome	Ref.
No modification	10.05	9.35	
Diiodination	6.3	5.9	*182*
Mononitration	~7.0	5.9	*178,181*

and -80, are modified (*221–226*). Blocking only residue 65 has little effect on the spectral properties, but carboxymethylating the heme ligand produces the more drastic changes of Table XV. Although the 695-nm band disappears, indicating that methionine no longer is the sixth ligand, some other strong field ligand is evidently found nearby since the iron remains in a low-spin state. The molecule no longer is reducible with ascorbate, fails to react with cytochrome oxidase or succinate-cytochrome *c* reductase (the succinate oxidase particulate preparation poisoned with cyanide), and will not restore respiration to cytochrome *c*–depleted mitochondria.

The results of nonselective iodination of the tyrosines of cytochrome *c* almost exactly parallel the nitration of tyrosine-67 described earlier. Tyrosine-67 is preferentially iodinated, indicating that the polypeptide framework of cytochrome *c* is sufficiently elastic to give KI_3 access to the interior. As with nitration, the pK_a of the state III to state IV ligand transition of normal cytochrome *c* is lowered from pH 9.35 to 5.9, probably as a consequence of the lower pK_a values of iodinated and nitrated tyrosine (Table XVI). In both cases, the new ligand above the transition point probably is the ionized tyrosine itself rather than lysine as in the normal state IV. Results of experiments with acetylation of tyrosines (*228,229*) have been similar but not as conclusive.

Tryptophan-59 has been modified by formylation, with the formyl group replacing the NH proton on the indole ring, and oxidation with *N*-bromosuccinimide (NBS). In both cases the hydrogen bond to the buried heme propionic acid is broken, and the molecule undergoes the change in properties outlined in Table XV. Reductase activity is impaired, and the oxidase reaction is also affected somewhat: The reaction is slower for any given cytochrome *c* concentration short of saturation, but the V_{max} is the same as for unmodified cytochrome *c*. The NBS story

228. J. R. Cronin and H. A. Harbury, *BBRC* **20**, 503–508 (1965).
229. K. M. Ivanetich, J. R. Cronin, J. R. Maynard, and H. A. Harbury, *Fed. Amer. Soc. Exp. Biol. Abstr.* **30**, 1143 (1971).

is similar. The results listed in Table XV are for moderate changes with 3:1 NBS/protein ratios. With higher ratios of NBS, modification of methionine-80 and other side chains occurs, and the molecule is disrupted even more.

The pattern that we have seen with these chemical modifications is remarkably uniform. Carboxymethylation of methionine-80, nitration or iodination of tyrosine-67, and formylation or oxidation of tryptophan-59 all disrupt the molecule, displace methionine-80 as a heme ligand, open up the heme crevice, perturb the reactions with biological oxidases and reductases, and bring about a common pattern of molecular behavior: absence of 695 nm band, nonreducibility of Fe(III) by ascorbate, and ligand binding and autoxidizability of Fe(II). A recent paper (*217*) suggested that the modifications of these three internal residues support their involvement in a specific reduction mechanism of the type that has been proposed in the past by Dickerson (*17*). The balance of the evidence is more easily interpreted as indicating that modification of residues 59, 67, and 80 interferes with the redox properties of the molecule because of the major conformational changes that are induced. In all three cases, the destruction of the 695-nm absorption band is an indicator of trouble. As a guide for future work, it might be assumed that any chemical modification which eliminates the 695-nm band will be of doubtful value in shedding light on the biological redox process.

It must be kept in mind that the interference with redox properties alluded to above may be thermodynamic and not kinetic; that is, the decreased reducibility when residues 67 or 59 are derivatized may arise not from modifications in the direct electron pathway but from a decrease in the reduction potential of the protein. Margalit and Schejter (*230*) have observed that the alkaline state IV of cytochrome *c*, with methionine-80 replaced by another ligand, has a reduction potential which is lowered from the normal +260 mV to less than 90 mV at pH 11.2 or above. If similar factors are at work in the derivatives iodinated or nitrated at tyrosine-67, where the III to IV transition is lowered from pH 9.3 to 5.9, then the decreased reductase activity and ability to restore respiration to depleted mitochondria (summarized in Table XV) could be explained purely on thermodynamic grounds. Kassner (*231,232*) and Dickerson (*18*) have suggested a correlation between reduction potential and the hydrophobicity of the heme environment: a more hydrophobic environment favors the Fe(II) state with no formal charge on the heme, and a more open structure and polar environment stabilizes the Fe(III)

230. R. Margalit and A. Schejter, *Eur. J. Biochem.* **32**, 492–499 (1973).
231. R. J. Kassner, *Proc. Nat. Acad. Sci. U. S.* **69**, 2263–2267 (1972).
232. R. J. Kassner, *JACS* **95**, 2674–2677 (1973).

state with its formal +1 charge. By this hypothesis, any pertubing influence which opened up the heme crevice would tend to lower the reduction potential. In all of the derivatives listed in Table XV, the heme is more exposed than normal, as measured by solvent perturbation methods. One could predict that all would have reduction potentials lower than +260 mV, and that this might be the principal source of their diminished capacity for reduction by NADH-cytochrome *c* reductase, succinate oxidase, or depleted mitochondrial preparations.

One further derivative from ref. *217* should be mentioned in the light of the previous discussion. The enzyme lactoperoxidase catalytically iodinates only tyrosine-74 on the surface of the cytochrome molecule (*233*), and the effects are quite different than with nonselective iodination by KI_3. The iodotyrosine-74 derivative of cytochrome *c* has been stated to have the EPR spectrum indicative of a methionine-80–iron bond (quoted in ref. *217*) and to possess a 695-nm band (E. Margoliash, private communication). It appears to react normally with the oxidase but poorly with reductase (*217*). If only kinetic arguments are considered, this would seem to indicate that the pathway for reduction occurs through tyrosine-74 on the left rear of the molecule. An equally possible explanation, however, is that iodination of this residue has led to a loosening of the molecular folding and a lowering of the reduction potential to the point where the molecule is no longer effective in the assay systems. This could result from the decreased pK_a of iodotyrosine, the increased bulk of the monoiodinated phenyl ring, or both. The reduction potentials of chemically modified cytochrome *c* need to be studied more often than is now the case, in order to distinguish between kinetic and thermodynamic effects. *The reduction potential, like the presence or absence of the 695-nm band, should be routinely reported in discussions of new chemical modifications of cytochrome c* (*233a*).

c. Inorganic Oxidants and Reductants. The reactions of cytochrome *c* have been studied with several inorganic redox agents as possible models

233. M. Morrison and R. E. Gates, *in* "Molecular Basis of Electron Transport" (J. Schultz and B. F. Cameron, eds.), pp. 327–345. Academic Press, New York, 1972.

233a. More recent work in Margoliash' laboratory (*233b*) has led to a classification of the ambiguities regarding monoiodotyrosine-74 horse cytochrome *c*. The iodinated molecule has the properties of the parent cytochrome: normal 695 nm band, reducwith reductase reported in ref. *217* now is attributed to polymerization. These reductase and ascorbate-TMPD-cytochrome *c* oxidase. The diminished reactivity with reductase reported in ref. *217* now is attributed to polymerization. These results definitively remove tyrosine-74 from any involvement with either reduction or oxidation of cytochrome *c*. The only experiments reported in ref. *217* which have not been challenged or retracted are those involving lysine- and arginine-13.

233b. B. A. Feinberg and D. L. Brautigan, *Abstr. FASEB Meet.* (1975).

for the biological processes. These include ferri- and ferrocyanide (*165,166,234–240*), ferrous EDTA (*241*), ruthenium(II) hexammonium ion (*242*), chromous ion (*243–247*), tris(orthophenanthroline) cobalt(III) (*248*), and dithionite (*240,249,250*). Self-exchange between ferricytochrome and ferrocytochrome apparently proceeds by a similar mechanism (*251–255*).

Two fundamentally different mechanisms have been observed in electron transfer reactions between metal complexes. In inner sphere reactions, one of the metal atoms displaces a ligand from the other atom and shares one of its own ligands briefly in a "bridge" which facilitates electron transfer. If the bridging ligand is a single atom, the process is classed as "adjacent," and if a group of atoms is involved, it is "remote." In contrast, outer sphere electron transfer requires no change in the ligands of either metal atom. Electrons pass from one metal, through its ligands, and from there to the ligands of the second metal atom. Outer sphere reactions, of necessity, are remote. They are also usually faster than inner sphere processes because they do not require the displacement of ligands or the formation of a bridged intermediate complex.

234. N. Sutin and D. R. Christman, *JACS* **83,** 1773–1774 (1961).
235. B. H. Havsteen, *Acta Chem. Scand.* **19,** 1227–1231 (1965).
236. G. D. Watt and J. M. Sturtevant, *Biochemistry* **8,** 4567–4571 (1969).
237. R. Margalit and A. Schejter, *FEBS* (*Fed. Eur. Biochem. Soc.*) Lett. **6,** 278–280 (1970).
238. E. Stellwagen and R. G. Shulman, *JMB* **80,** 559–573 (1973).
239. E. Stellwagen and R. D. Cass, in preparation.
240. W. G. Miller and M. A. Cusanovich, *BBA* (in press).
241. H. L. Hodges, R. A. Holwerda, and H. B. Gray, *JACS* **96,** 3132–3137 (1974).
242. R. X. Ewall and L. E. Bennett, *JACS* **96,** 940–942 (1974).
243. A. Kowalsky, *JBC* **244,** 6619–6625 (1969).
244. N. Sutin and A. Forman, *JACS* **93,** 5274–5275 (1971).
245. J. W. Dawson, H. B. Gray, R. A. Holwerda, and E. W. Westhead, *Proc. Nat. Acad. Sci. U. S.* **69,** 30–33 (1972).
246. J. K. Yandell, D. P. Fay, and N. Sutin, *JACS* **95,** 1131–1137 (1973).
247. C. J. Grimes, D. Piszkiewicz, and E. B. Fleischer, *Proc. Nat. Acad. Sci. U. S.* **71,** 1408–1412 (1974).
248. J. V. McArdle, H. B. Gray, C. Creutz, and N. Sutin, *JACS* **96,** 5737–5741
249. C. Cruetz and N. Sutin, *Proc. Nat. Acad. Sci. U. S.* **70,** 1701–1703 (1973).
250. D. O. Lambeth and G. Palmer, *JBC* **248,** 6095–6103 (1973).
251. R. K. Gupta and A. G. Redfield, *Science* **169,** 1204–1206 (1970).
252. R. K. Gupta, S. H. Koenig, and A. G. Redfield, *J. Magn. Resonance* **7,** 66–73 (1972).
253. R. K. Gupta, *BBA* **292,** 291–295 (1973).
254. C. Greenwood, A. Finazzi Agro, P. Guerrieri, L. Avigliano, B. Mondovi, and E. Antonini, *Eur. J. Biochem.* **23,** 321–327 (1971).
255. A. Kowalsky, *Biochemistry* **4,** 2382–2388 (1965).

In the context of cytochrome *c*, inner sphere reactions usually are thought of as those which open the heme crevice or unfold the molecule in some other way in order to give the attacking group a direct approach to the heme iron by displacing the methionine ligand. Measurements of the rate of ligand exchange in cytochrome *c* (*256*) have placed an upper limit on this exchange rate of roughly 60 sec^{-1}. Outer sphere processes would include both the transfer of electrons in and out through the heme edge and other electron flow pathways involving the polypeptide chain and amino acid side chains. The terms "adjacent" and "remote" tend to be used in the cytochrome literature as virtually synonymous with "inner sphere" and "outer sphere."

Second-order rate constants for reduction and oxidation by several inorganic reagents are listed in Table XVII along with enthalpies and entropies of activation where known. The values of the rate constants are remarkably uniform. The general concensus of the discusssions in the literature cited (see also refs. *256* and *257*) is that the transition metal complexes, (a)–(c) and (j)–(k) in Table XVII, reduce or oxidize cytochrome *c* by an outer sphere mechanism, and the assumption is that this is via the heme edge. The conclusion of an outer sphere mechanism follows from the observed kinetics of the reactions, the known nonlability of the ligands on these complexes, and their preference for outer sphere processes in inorganic reactions. Simple Marcus theory (*258*) has been quite successful in predicting reaction rates for these complexes with cytochrome *c* (*242,248*), and equally successful in accounting for the observed rate of exchange between ferri- and ferrocytochrome *c* in solution, suggesting similar outer sphere mechanisms in all cases.

Dithionite reduction occurs by two simultaneous mechanisms: a remote pathway involving $S_2O_4^{2-}$ itself, and an adjacent mechanism involving its dissociation product, SO_2^-. Reduction by chromous ions was expected to be a key reaction (*243*) because the Cr(II) ion is substitution-labile with respect to its ligands, whereas Cr(III) is substitution-inert. The chromous ion should remain firmly bound to protein ligands wherever it found itself at the moment of electron transfer, thereby acting as a marker for the reduction pathway. It was indeed found (*247*) that after reducing cytochrome *c*, the chromium atom was locked between asparagine-52 and tyrosine-67 in the interior of the cytochrome molecule, next to the tyrosine ring and the heme group. As with the iodination and nitration reactions, this once again demonstrates the elasticity of the left side

256. N. Sutin and J. K. Yandell, *JBC* **247**, 6932–6936 (1972).
257. N. Sutin, *Chem. Brit.* **8**, 148–151 (1972).
258. L. E. Bennett, *in* "Current Research Topics in Bioinorganic Chemistry" (S. J. Lippard, ed.), pp. 1–176. Wiley, New York, 1973.

TABLE XVII

SECOND-ORDER RATE CONSTANTS AND ACTIVATION PARAMETERS FOR REDUCTION AND OXIDATION OF CYTOCHROME *c*

Reductant	μ	pH	Temp. (°C)	k_2(M^{-1} sec^{-1})	$\Delta H^{\ddagger}$	$\Delta S^{\ddagger}$	Reference
(a) $Fe(CN)_6^{4-}$	0.18	7.0	25	2.6×10^4	+7.9	−31	*165,238*
(b) $Fe(EDTA)^{2-}$	0.1	7.0	25	2.6×10^4	+6.0	−18	*241*
(c) $Ru(NH_3)_6^{2+}$	0.1	7.0	25	3.8×10^4	+2.9	−28	*242*
(d) Cr^{2+}	0.1	4.1	25	1.9×10^4	+10.0	−7	*241,245,258*
(e) $S_2O_4^{2-}$ (remote)	1.0	6.5	25	1.2×10^4			*249*
(f) $S_2O_4^{2-}$ (adjacent)	1.0	6.5	25	6×10^4			*249*
(g) Ferrocytochrome *c*	0.2	7.0	25	5×10^4			*255*
(h) *Pseudomonas* c_{551}	0.1	7.4	25	8×10^4			*254*
(i) Ferrocytochrome c_1	0.05	7.4	10	3.3×10^6			*172*
Oxidant							
(j) $Fe(CN)_6^{3-}$	0.18	7.0	25	8.1×10^6	+15.2	−0.8	*165,238*
(k) $Co(phen)_3^{3+}$	0.1	7.0	25	1.5×10^3	+11.3	−6.3	*248*

of the molecule and the ease with which ions and molecules can gain access to the interior. Although chemically interesting, this reaction probably has little to do with biological reduction since chromous reduction apparently is an adjacent process requiring the opening of the heme crevice, and one which only occurs at acid pH.

The rates observed for self-exchange reactions between ferrocytochrome and ferricytochrome or between cytochrome *c* from horse and cytochrome c_{551} from *Pseudomonas* suggest that these, like the four metal complexes of Table XVII, are outer sphere reactions in which electrons flow in and out through the exposed edge of the heme at the heme crevice. The recent study using $Co(phen)_3^{3+}$ or tris(1,10-phenanthroline) cobalt (III) as an oxidant of ferrocytochrome *c* is of special interest because the phenanthroline rings (Fig. 19) effectively bury the cobalt atom, just as the iron is buried by the heme and its ligands. If tunneling is not important, then transfer must take place through the π-electron system. Hence, $Co(phen)_3^{3+}$ is a better model compound for heme-containing oxidases and reductases than any of the other complexes of Table XVII.

The actual rate of reduction of cytochrome *c* by cytochrome c_1 (*172*) is two orders of magnitude faster than the self-exchange process or any of these inorganic model reactions. To the extent that the latter are valid model reactions, they represent "proteinless" cytochrome *c*—with the metal-ligand structure but without the surrounding polypeptide chain. The hundredfold increase in rate constant may represent the contribution of the polypeptide chain in terms of specific interaction and binding sites which have evolved in cytochromes *c* and c_1 over the past billion years or so. Since no such specific binding would have been developed for the self-exchange process, it would be expected to resemble the nonspecific inorganic reactions, and it does.

d. Structure and Reactivity Comparisons between Cytochromes. If the differences observed between cytochromes *c* from various eukaryotic species were genuinely adaptive and were the results of evolutionary selection, tailoring them for better interaction with other macromolecules, then one would expect that each eukaryotic cytochrome *c* would function best with its own oxidase and reductase. This seems not to be the case. Apparent differences in reactivities between cytochromes from one species

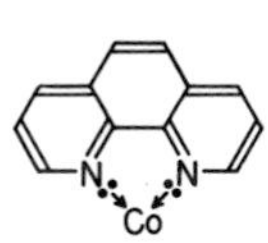

FIG. 19. Orthophenanthroline or 1,10-phenanthroline. Three such molecules supply octahedral ligands to cobalt(III) in $Co(phen)_3^{3+}$.

and oxidases from another (or even within one species) can be observed; but if the cytochromes are exhaustively dialyzed first to remove all ions, then cytochromes from mammals, birds, fish, and insects all react with bovine oxidase at the same rate (*128*). Corresponding information is not yet available about the reductase reaction, but the tentative conclusion is that the oxidase reaction has not been improved appreciably by positive natural selection since the mitochondrial respiratory machinery settled down a billion years or so ago.

In prokaryotes, the molecular function broadens to include photosynthetic electron transport as well as respiration, and the differences in amino acid sequences and three-dimensional folding become more pronounced. Comparative studies have been made of the efficiency with which cytochromes c, c_2, c_{550} and similar proteins interact with the oxidases and reductases of c and c_{550} (*212,259,260*, and L. Smith, private communication). These are summarized in Fig. 20, where the percent-

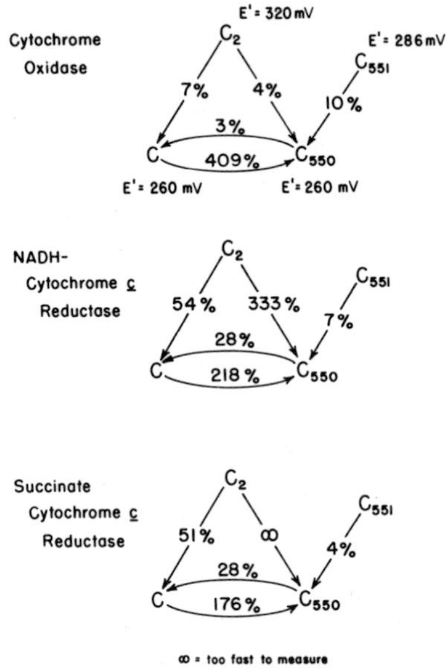

FIG. 20. Patterns of cross-reactivity between cytochromes c (horse), c_2 (*R. rubrum*), c_{550} (*M. denitrificans*), and c_{551} (*P. aeruginosa*), and the oxidases and reductases of c and c_{550}. See text for discussion.

259. M. D. Kamen and L. P. Vernon, *BBA* **17**, 10–22 (1955).
260. L. Smith, N. Newton, and P. Scholes, *in* "Hemes and Hemoproteins" (B. Chance, R. W. Estabrook, and T. Yonetani, eds.), pp. 309–403. Academic Press, New York, 1966.

ages represent magnitudes of first-order rate constants compared with the rate constant of the indicated oxidase or reductase with its own cytochrome as 100%. Each percent figure is for reaction of the cytochrome at the tail of the arrow with the oxidase or reductase belonging to the cytochrome at the head of the arrow. Cytochrome c_{550}, for example, reacts with cytochrome *c* oxidase only 3% as fast as cytochrome *c* itself does. Polylysine interferes with the oxidase and reductase reactions of both eukaryotic and bacterial cytochromes, suggesting that lysine groups on cytochrome *c* are important elements in all of the binding sites.

Several patterns of reactivity can be seen from Fig. 20. Cytochrome *c* reacts well with the c_{550} oxidase and reductase—better, in fact, than c_{550} does. In contrast, c_{550} reacts badly with both the *c* oxidase and reductase, as if the binding of oxidase were more finely structured: stronger for molecules with the proper configuration but less tolerant of altered molecules. Photosynthetic cytochrome c_2 reacts very badly with the two respiratory oxidases but quite well with reductases. Part of this may be thermodynamic rather than kinetic—the reduction potential of c_2 is 60 mV higher than that of *c* or c_{550}, which should correspond roughly to an order of magnitude drop in rate of oxidation. However, the comparable difficulty that c_{550} has with *c* oxidase suggests that reduction potential difference is not the whole story. c_2 also is better adjusted to the bacterial reductase of c_{550} than to the eukaryotic reductase of *c*.

The primary conclusion that can be drawn is that there is a fundamental similarity in all three cytochrome molecules, which permits an appreciable amount of cross-reactivity to occur. We are therefore justified in looking to the common features of these three molecules for the essential regions in terms of molecular mechanism. Four secondary observations can aid our search even more:

1. The oxidase site on these cytochromes *c* and c_{550} is more finely tuned than that of the reductase.

2. Cytochrome *c* interacts easier with the bacterial oxidase and reductase than does c_{550} with the corresponding eukaryotic proteins.

3. Bacterial c_2 and c_{550} behave in much the same way toward the eukaryotic oxidase and reductase.

4. Cytochrome c_2 reacts much better with the oxidase and reductase from bacterial c_{550} than from eukaryotic *c*.

The three-dimensional structures of *c*, c_2, and c_{550} are compared in Figs. 21–23, with the molecules tipped forward by 30° so that the view is more directly into the heme crevice, which runs up the face of the molecule and over the top. If Figs. 7, 9, and 10 are considered as full face views, then Figs. 21–23 are looking at the forehead of the molecule. Of the three, *c* is the "minimal" molecule, and c_2 and c_{550} have extra polypeptide chain at several places which should be recalled from earlier discussions:

Cytochrome c_{550} has extra amino acids in the 20's loop at the right, which are not present in c_2 from *R. rubrum*, but which are shared by the c_2's of two other species of photosynthetic bacteria (see Table VII).

Both cytochromes c_2 and c_{550} have extra material at the bottom (50's helix) and left (70's loop).

Cytochrome c_{550} has more extra material at both ends of the chain, which is folded against the back of the molecule.

All of these alterable regions are less promising candidates for binding sites to oxidase and reductase than are the unchanging front and top—the regions nearest the viewer in Figs. 21–23.

The preference of c_2 for the bacterial oxidase and reductase mentioned in point 4 above can be explained in terms of the greater similarity of structure between c_2 and c_{550} than c. The essential equivalence of the reactions of c_2 and c_{550} with eukaryotic oxidase and reductase (point 3) undoubtedly has the same origin. The greater ease of reaction of c with the oxidase and reductase from c_{550} than the reverse (point 2) may arise from the extra bulk of the c_{550} molecule. If the c_{550} oxidase and reductase evolved to accommodate a large molecule, then the smaller c might be

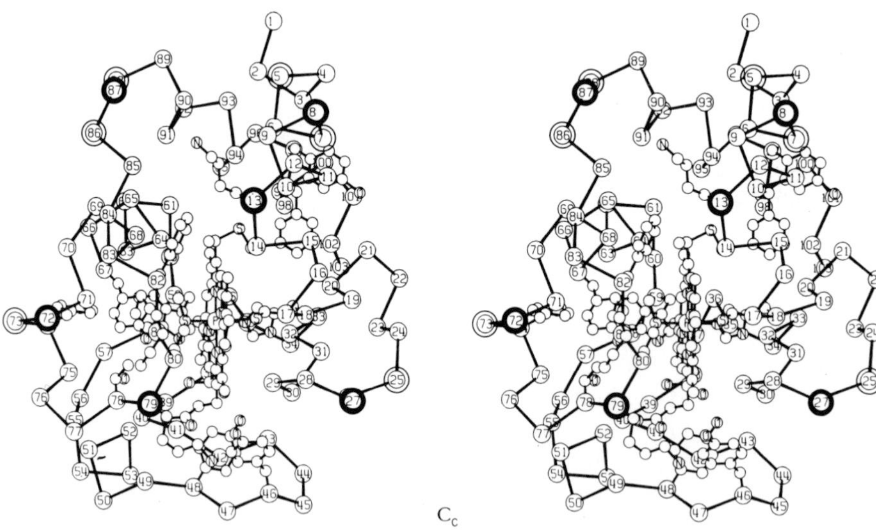

FIG. 21. "Forehead" view of reduced tuna cytochrome c, looking directly into the heme crevice. Lysine or arginine residues on the front face of the tuna cytochrome molecule are indicated by double circles. The six heavy black circles mark front-side residues which are lysine or arginine in 55 or more of the 60 known sequences of Table IX.

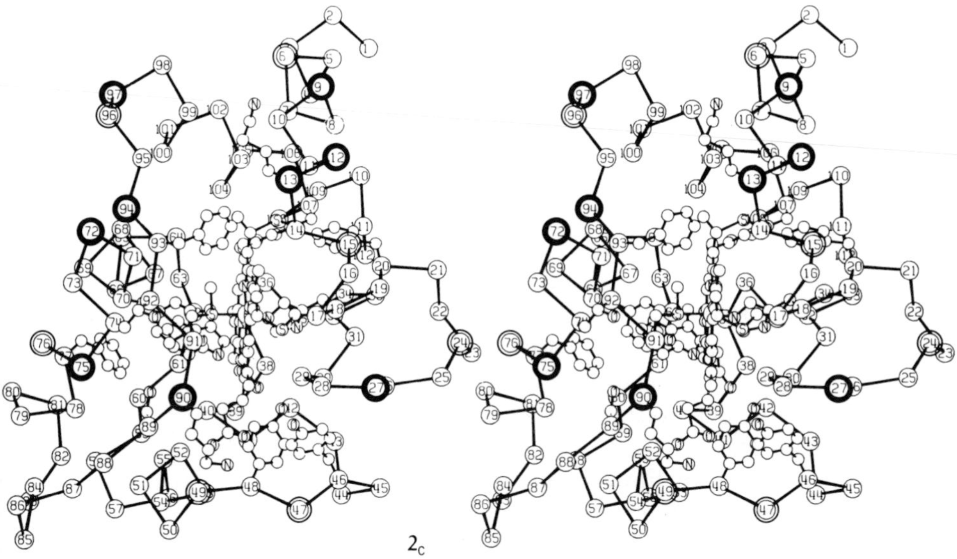

FIG. 22. Corresponding view of *R. rubrum* c_2. Heavy black circles locate lysines and arginines surrounding the heme crevice in *R. rubrum* c_2. Double circles mark front-side lysines and arginines in one or more of the other three c_2's of Table VII. Only groups in the immediate vicinity of the heme crevice are so marked, and lysines on the extra loops not shared with cytochrome *c* are also unmarked.

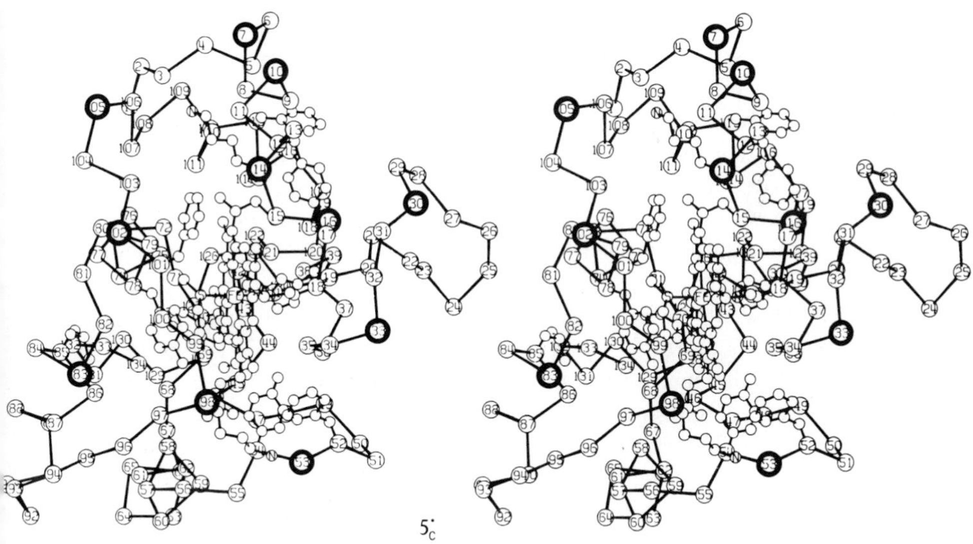

FIG. 23. Corresponding view of *M. denitrificans* c_{550}. Heavy circles indicate lysines and arginines surrounding the heme crevice.

able to slip in and bind as well or even better. Conversely, if the oxidase and reductase of eukaryotic *c* adapted toward a smaller molecule, then the extra loops on the surface of c_{550} might get in the way and diminish or nearly eliminate binding and subsequent reaction.

Where are the binding and electron transfer sites on the surface of the cytochrome molecule? It is important to keep these two aspects separate since biochemists and inorganic chemists sometimes tend to talk at cross purposes: biochemists concentrating on chemical modifications and macromolecular binding sites, and inorganic chemists being more concerned with the mechanics of electron transfer to and from the iron. The chemical modification and antibody binding studies reported in Sections II,A,5,a and b provide clues as to possible oxidase and reductase binding sites but say nothing about electron pathways in and out; and the outer sphere heme edge reactions of Section I,A,5,c suggest a common electron pathway but do not address the issue of shared or separate molecular binding sites. The binding sites of oxidase and reductase evidently are different, and probably neither is to be found primarily in the regions where extra chain occurs in the bacterial proteins. However, these binding sites can be large enough that the oxidase or reductase might "see" any extra material on the perimeter of the site, and bind less well when such material is present. The left side seems not originally to have been an important part of the binding site for either oxidase or reductase. The trimming down of the left side may have been one element in increasing the specificity of binding in eukaryotes, for otherwise it is hard to understand the extreme evolutionary conservatism of the left side in all eukaryotes. Since the most original new function in going from bacterial to eukaryotic cytochromes is integration into the mitochondrial membrane system, we might look to the left side of *c* for that role. The bacterial oxidase and reductase apparently are not troubled by the absence of chain on the left side of *c*, but the equivalent eukaryotic proteins encounter the extra chain c_{550} and c_2 and cannot accommodate it.

These molecular comparisons suggest that one binding site for oxidase or reductase is at the heme crevice on the face of the molecule, and the other one is also next to the heme crevice on the top. There is insufficient evidence at present to say which is which. Blocking residue 13 on the edge of the upper part of the crevice interferes with oxidase binding but not that of reductase, yet the positions that have been implicated in pyridoxal phosphate interference with reductase binding are in the same part of the molecule. It would be of great value to have selective modifications of lysines 79, 5, and 8, and 87–89, and to see the relative effects of each on oxidation and reduction.

e. Toward an Oxidation–Reduction Mechanism for Cytochrome c.

Recent chemical evidence has made it apparent that both of the two detailed oxidation–reduction mechanisms which have been proposed for cytochrome must be abandoned: the Winfield mechanism espoused for eukaryotic *c*, and the heme crevice hydrogen bond network proposed for bacterial c_2. Each mechanism has been undermined by the discovery of unworkable amino acids at key positions in species for which the proposed scheme should hold. A fresh approach is needed.

The Winfield mechanism (*13,16*) was proposed on the basis of the oxidized horse cytochrome structure, and was suggested by the remarkable conservatism of aromatic rings on the left of the heme. It proposed that reduction occurs by the transfer of an electron from the ring of tyrosine-67 to the heme, perhaps through the hydroxyl —OH and methionine S atoms. After the reduced cytochrome structure was solved, an explicit pathway was proposed for the reducing electron from tyrosine-74 to tryptophan-59, to tyrosine-67 after a conformational change, and from there to tetrapyrrole ring 4 of the heme and the iron (*17,18*). A constant objection to this pathway was the high cost in energy of adding an electron to, or removing one from, a tyrosine or phenylalanine side chain. Quantum mechanical tunneling possibilities along a region of stacked aromatic rings were considered as a way of surmounting this difficulty. The discovery that tyrosine-74 and -67 were found as phenylalanines in one or two species was not fatal to the mechanism, but the discovery of leucine at the equivalent of position 74 in c_{550} was. This, plus the radical insertions of new polypeptide chain observed in both of the known bacterial cytochromes, has led the original proponents of the Winfield mechanism to abandon it (*260a*).

The c_2 mechanism proposed by Salemme, Kraut, and co-workers (*30,31*) was designed to take into specific account the fact that a proton is taken up when cytochrome c_2 is reduced, which is not the case for *c*. It postulated a network of hydrogen bonds between serine-89, tyrosine-52, and tyrosine-70. Unfortunately for the generality of this mechanism among bacterial cytochromes, position 52 is isoleucine in c_{550}, and isoleucine, asparagine, and methionine in the other three c_2's from rhodopseudomonads (see Table VII). It is unlikely that this position 52 has any significant functional role.

The one feature of the c_2 proposal that still appears valid—heme crevice approach by both oxidase and reductase—has also been strongly

260a. There are several reasons for naming a mechanism after someone else. One receives credit from colleagues for a becoming modesty. If the mechanism proves to be correct, the name attached to it is unlikely to obscure the source of the papers in which it was developed. On the other hand, if it is wrong, then it is better named after someone else. Out of fairness to M. E. Winfield, it should be stated for the record that it is the Dickerson–Winfield mechanism that has now been abandoned.

advocated by inorganic chemists. The predominance of heme edge, outer sphere mechanisms among the inorganic reagents, including the π-electron heme analog tris(orthophenanthroline) cobalt (III), and the great structural variability of cytochrome molecules everywhere *except* in the vicinity of the heme crevice, inevitably focus attention on the cervice as the pathway for both oxidation and reduction. The chemical modification studies point in the same direction as does the near-invariance of key amino acid sites on the perimeter of the crevice.

Some of the most constant residues around the heme crevice are indicated in Fig. 24, which is a generalized sketch of the molecule seen as in Figs. 21–23 (numbering as in *c*). Six especially constant positive regions surround the crevice in both bacterial and eukaryotic cytochromes, and individual molecules have additional positive sites. The top of the initial helix, around positions 5 and 8, is positively charged in every cytochrome molecule, although the exact positions of the charged residues may vary. Position 12 or 13 is positive in every molecule except one c_2 from *R. spheroides*. Position 27 is lysine in all prokaryotes, and all eukaryotes except *Euglena and Crithidia*. Positive charge is found near 72 and 73 in all cytochromes except for two c_2's; the invariance of lysine 79 is marred only by glycine in *R. capsulata* c_2; and the region around 83–88 always has at least one positive charge. The prokaryotes consistently have a lysine at position 83 in *c* numbering. A study of Tables VII and IX will reveal other positive charges, and it is a firm generalization that all cytochromes *c* whose three-dimensional structures are known have a ring of positive charge around the heme crevice (*31*).

They also have at least one negative charge at the back of the heme crevice around positions 90 and 93, and usually more than one. These may possibly have a directing or orienting influence on the oxidase or reductase, which otherwise is attracted indiscriminately by the collection of positive charges. The bottom of the heme crevice always has an aro-

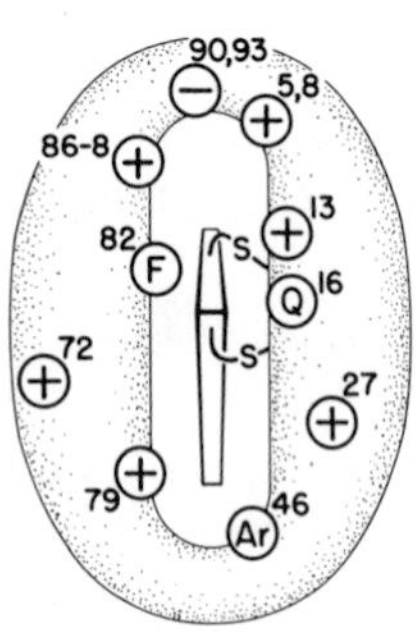

FIG. 24. Schematic view of the heme crevice with highly conservative residues marked (+) basic, (—) acidic, (Ar) aromatic, (Q) glutamine, and (F) phenylalanine. Residue numbering is that of tuna cytochrome.

matic group at position 46, and phenylalanine at position 82 is apparently as vital as are the heme ligands. Glutamine-16 is invariant in eukaryotes but not bacteria, and conceivably could supply a small amount of extra binding energy or directing influence by virtue of its hydrogen bonding ability.

The face of the molecule as shown in Figs. 21–24 is probably the one seen by reductase and oxidase molecules, with entry and egress of electrons through the exposed edge of the heme. This, at least, is the simplest assumption in the absence of compelling evidence to the contrary. Which part of the crevice encounters each of the two molecules—oxidase and reductase—or whether or not they bind in the same region, are questions which can only be answered by selective chemical modification of residues on the surface of the cytochrome molecule or by physical chemical characterization of intermolecular complexes with cytochrome *c*.

f. Quantum Mechanical Tunneling as a Redox Mechanism. In the discussion of possible electron pathways in and out of the cytochrome molecule, the unspoken assumption has been that any acceptable pathway would have to be energetically feasible by classical criteria; that is, that the electron would have enough energy to surmount any potential barrier in its path. This makes it difficult to see how a pathway through adjacent aromatic rings could exist since the ionization energies and electron affinities of tyrosine and phenylalanine side chains could impose a potential energy barrier of 2–3 eV or 50–75 kcal/mole. One possible answer is that the electron does not have this energy but simply leaks through the barrier by a process of quantum mechanical tunneling. The current focus of attention on heme edge transfer makes tunneling mechanisms less attractive, but it is worth establishing for the record that they are not impossible and must be considered.

The initial suggestion of quantum mechanical tunneling came not from respiratory cytochrome *c*, but from the photosynthetic *c* of the bacterium *Chromatium* (*10,261–268*). DeVault, Chance, and co-workers studied the oxidation of *Chromatium* cytochrome *c*, which transfers electrons to the

261. D. DeVault and B. Chance, *Biophys. J.* **6**, 825–847 (1966).
262. B. Chance, *BJ* **103**, 1–18 (1967).
263. B. Chance, D. DeVault, V. Legallais, L. Mela, and T. Yonetani, *Fast React. Primary Process Chem. Kinet., Proc. Nobel Symp., 5th, 1967,* pp. 437–468 (1967).
264. D. DeVault, J. H. Parkes, and B. Chance, *Nature* (*London*) **215**, 642–644 (1967).
265. B. Chance, A. Azzi, I. Y. Lee, C.-P. Lee, and L. Mela, *FEBS Symp.* **17**, 233–273 (1969).
266. B. Chance, T. Kihara, D. DeVault, W. Hildreth, M. Nishimura, and T. Hiyama *Progr. Photosyn. Res., Proc. Int. Congr., 1968* Vol. 3, pp. 1321–1346 (1969).
267. T. Kihara and B. Chance, *BBA* **189**, 116–124 (1969).
268. M. Seibert and D. DeVault, *BBA* **253**, 396–411 (1971).

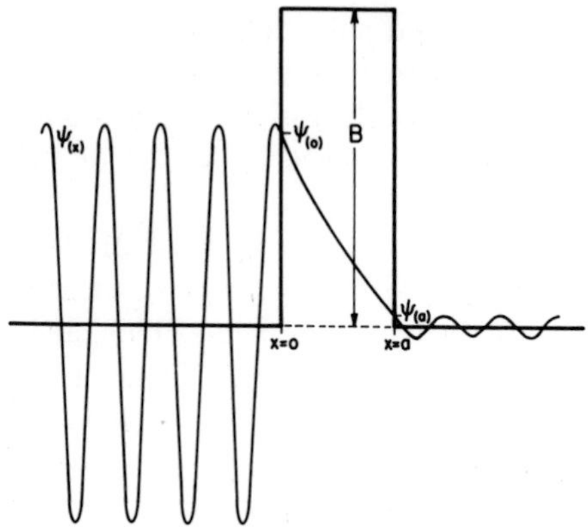

FIG. 25. Schematized wave function for the tunneling from left to right of an electron through a potential barrier of height B above the total energy of the electron, and thickness a. See text for discussion.

bacterial chlorophyll photocenter, immediately after excitation of the photocenter by a short laser pulse in the 10-nsec range. The half-time of oxidation at room temperature is extremely fast, around 2 μsec. This rate falls as the temperature is lowered to 120°K, with an Arrhenius activation energy of 3.3 kcal/mole. Below 120°K the rate of oxidation levels off and becomes nearly temperature-independent, with an Arrhenius activation energy of no more than 80 cal/mole. At these low temperatures the half-time for oxidation is around 2.3 msec. The slowness of the reaction suggests the presence of a significant activation barrier, yet the temperature insensitivity suggests that the electron is not receiving kinetic energy and surmounting the barrier in the classical sense. The alternative is quantum mechanical tunneling from bacterial cytochrome c to chlorophyll.

Simple order of magnitude calculations on tunneling probabilities can be made by assuming a rectangular barrier of height B electron volts and thickness a Ångstroms as diagramed in Fig. 25. More elaborate models quickly become far more complex mathematically, but we can obtain an idea of feasibility from simple model calculations. A representative particle wave function (actually of much too coarse a wavelength) is shown in Fig. 25 at the left of the barrier where the electron originates, and attenuated at the right of the barrier, representing escape by tunneling. To a good approximation, the ratio of the squares of the wave functions on the outside and inside of the barrier represents the probability of escape of an electron as it approaches the barrier from the left:

$$P = \frac{\psi^2(a)}{\psi^2(0)}$$

The falloff of amplitude of the wave function is exponential, such that the escape probability, P, is given by

$$P = \exp\left(-\frac{2a\sqrt{2mB}}{\hbar}\right)$$

where a is the barrier thickness in centimeters, B is the barrier height in ergs, m is the mass of the electron in grams, and $\hbar$ is Planck's constant divided by 2π. If the barrier thickness is expressed in angstroms and the height in electron volts, then this probability of escape simplifies to

$$P = 10^{-0.445aB^{1/2}}$$

If the frequency with which electrons approach the barrier is f electrons per second, then the rate of electron escape in electrons per second is

$$R = fP$$

and the half-time for escape is

$$t = \frac{\ln 2}{R}$$

Bennett and Jones have also made trial calculations demonstrating the feasibility of tunneling as an electron transfer mechanism (*269*). An approximate barrier impact frequency of around 5×10^{15} electrons/sec can be justified from the frequencies of the lowest absorption band of chlorophyll (*261*), from the frequency of an electron in a 7 Å box with 5 eV kinetic energy (*261*), or from $E = h\nu$ and the third ionization energy of iron (*269*). DeVault and Chance considered tunneling as a means of distant transfer of electrons from bacterial cytochrome to chlorophyll through intervening material and showed that penetration over a 30 Å distance through a barrier of 1 eV above the actual energy of the electron leads to a 2.2 msec half-time for reaction in their simple rectangular barrier model, in agreement with observation [(a) of Table XVIII]. The same reaction time can also be obtained by assuming a higher barrier of 4 eV but a shorter tunneling distance of only 15 Å [(b) of Table XVIII]. It proved impossible to fit the entire rate vs. temperature curve for *Chromatium* oxidation by a single mechanism with physically reasonable barrier parameters, and DeVault and Chance concluded provisionally that two mechanisms must be present with different activation energies: tunneling predominating below 120°K and more traditional processes with an activation energy of 3.3 kcal predominating at higher temperatures.

In the model of Bennett and Jones, the barrier to electron flow is not the medium between one macromolecule and another but is the protein

269. L. E. Bennett and W. D. Jones, in preparation.

TABLE XVIII
APPROXIMATE TUNNELING PROBABILITIES

$$P = \exp\left(-\frac{2a\sqrt{2mB}}{\hbar}\right) \qquad k = fP \quad (f = 5 \times 10^{15}\ \text{sec}^{-1}) \qquad t_{1/2} = \frac{\ln 2}{k}$$

a(Å)	B(eV)	$aB^{1/2}$	P	k	$t_{1/2}$
(a) 29.7	1	29.7	6.3×10^{-14}	315 sec^{-1}	2.2 msec
(b) 15	3.9	29.7	6.3×10^{-14}	315 sec^{-1}	2.2 msec
(c) 5	3.0	8.66	1.4×10^{-4}	$7 \times 10^{11}\ \text{sec}^{-1}$	10^{-12} sec
(d) 10	3.0	17.3	2×10^{-8}	$1 \times 10^{8}\ \text{sec}^{-1}$	7×10^{-9} sec
(e) 20	3.0	34.6	4×10^{-16}	2 sec^{-1}	0.35 sec
(f) 30	3.0	52.0	8×10^{-24}	$4 \times 10^{-8}\ \text{sec}^{-1}$	1.7×10^{7} sec

side chain material between the surface of cytochrome c and the heme iron. They also employed the simple rectangular barrier picture of Fig. 25. Known ionization energies and electron affinities of aromatic molecules suggest that a realistic barrier height to transfer of electrons along a set of stacked aromatic side chains might lie in the range of 2.8–4 eV, and trial calculations show that reasonable escape probabilities can be obtained with these and with tunneling distances in the range of 5–15 Å. Lines (c) through (f) of Table XVIII give some representative calculations to illustrate the broad range of reaction half-times that can be accommodated with physically reasonable barrier dimensions.

DeVault and Chance assumed that the energy levels of cytochrome and chlorophyll were properly matched to begin with, perhaps as a result of extensive natural selection on the structures of both molecules; thus, no prior rearrangement of donor and recipient molecules is necessary. Bennett and Jones made explicit allowance for this prior adjustment of energy levels in cytochrome c and a, by assuming that the initial c/a complex may have to undergo an activation step to match energy levels before electron transfer can occur:

$$c^{II} + a^{III} \overset{K_1}{\rightleftharpoons} c^{II}a^{III} \qquad \text{Formation of complex} \qquad (1)$$

$$c^{II}a^{III} \overset{K_2}{\rightleftharpoons} (c^{II}a^{III})^* \qquad \text{Matching of energy levels} \qquad (2)$$

$$(c^{II}a^{III})^* \xrightarrow{k} (c^{III}a^{II})^* \qquad \text{Franck–Condon transfer of electron} \qquad (3)$$

$$(c^{III}a^{II})^* \rightarrow c^{III}a^{II} \qquad \text{Relaxation of complex after transfer} \qquad (4)$$

$$c^{III}a^{II} \rightarrow c^{III} + a^{II} \qquad \text{Dissociation of products} \qquad (5)$$

This is the standard way of formulating an outer sphere inorganic redox mechanism, except that step (3) now represents not a simple electron transfer with a probability of close to 1 but a tunneling process with a probability which will be much less than 1. As in a simple outer sphere reaction without tunneling, step (2) can involve alteration of bond lengths in order to adjust energy levels and can bring with it an activation energy term of its own. The actual observed rate constant for the redox process will be

$$K_{obs} = K_2 \cdot k$$

In effect, DeVault and Chance assumed that for their system, $K_2 = 1$.

A much more elaborate calculation of quantum mechanical tunneling probabilities for the *Chromatium* system has been carried out recently by Hopfield (*270*). In this approach, the frequency factor (assumed earlier to be $f = 5 \times 10^{15}$ electrons/sec) is calculated rather than being taken as constant, and is found to be several orders of magnitude smaller and strongly temperature-dependent. This reduces the permissible thickness for a 1-eV barrier from 30 Å to around 10 Å. Hopfield pointed out that a short tunneling distance and an extreme sensitivity of tunneling probability to distance are both useful in controlling tunneling and keeping it from short-circuiting the entire electron transfer chain. Close physical proximity and proper stereochemical matching of two macromolecules would be as important in regulating electron flow by quantum mechanical tunneling as with more classical electron transfer paths. The most impressive feature of the Hopfield calculations, with their temperature-dependent frequency factor, is that they provide a single explanation of the *Chromatium* rate data over the entire range of temperatures, from 30° to 300°K.

Equations (1) through (5) provide the formalism for classical electron transfer, and quantum mechanical tunneling imposes an additional impediment to reaction in step (3), on top of the normal activation energies. It is not clear why cytochrome c should evolve a tunneling mechanism if a more direct edge transfer method was available, and arguments from simplicity favor the latter. Still, the main conclusion to be drawn from these calculations is that the concept of quantum mechanical tunneling is a reasonable one, using reasonable values for barrier height and tunneling distance. The possibility of such a mechanism cannot be completely excluded.

B. Respiratory Cytochrome c_1

The other mitochondrial respiratory chain cytochrome with a c-type spectrum is cytochrome c_1, and far less is known about its structure and properties. It was discovered independently by Yakushiji and Okunuki

270. J. J. Hopfield, *Proc. Nat. Acad. Sci. U. S.* **71**, 3640–3644 (1974).

(*271,272*) and by Keilin and Hartree (*273*), who called it cytochrome *e* until they realized that it and the Yakushiji–Okunuki protein were identical (*274*). Good background accounts of early work on cytochrome c_1 are to be found in references *4* and *5* and need not be repeated here. Ball and Cooper (*275*) and Estabrook (*276*) established the role of cytochrome c_1 as the reductase for cytochrome *c*, and its predecessor in the mitochondrial electron transport chain as shown in Fig. 4.

It has been difficult to extract cytochrome c_1 from its association with cytochromes *b* and the other components of the mitochondrial succinate oxidase system without damaging it, and this has been the main reason for the relative dearth of information on cytochrome c_1 as a molecular entity. Successful isolation procedures leading to intact and crystallizable protein (two different criteria, it must be added) with the same kinetic behavior in solution as in the unseparated complex have been developed only recently by King and co-workers (*172,277*). Earlier methods using detergents, heat, and denaturing agents tended to be destructive, and the milder methods of King involve extraction of reduced c_1 with β-mercaptoethanol, ammonium sulfate fractionation under anaerobic conditions, and chromatography on a calcium phosphate column. Too little β-mercaptoethanol fails to separate cytochromes *b* and c_1, and too much leads to denaturation of c_1.

Bovine heart cytochrome c_1 has a single heme of the cytochrome *c* type (Fig. 1), per 40,000 molecular weight, attached to the polypeptide chain by thioether links. A small amount of carbohydrate is present (less than 2%) which may be impurity. Gel electrophoresis shows that the 40,000 molecular weight is divided between two subunits of molecular weight ratio 2:1, with the heme attached to the larger subunit. The 40,000 units aggregate in solution: Sedimentation equilibrium measurements indicate an average molecular weight of 210,000 at pH 7.4 (pentamer?) and 93,000 at pH 11. The state of aggregation of c_1 molecules in the mitochondrial membrane is not known. The amino acid composition of the two subunits of c_1 is given in Table XIX, and is not especially remarkable except for the high proline content and dissimilarity to the composition of *c* (*58,78,79,278–280*). The measured isoelectric point of c_1 is 5.5 (T. E.

271. E. Yakushiji and K. Okunuki, *Proc. Imp. Acad. Jap.* **16,** 299–302 (1940).
272. E. Yakushiji and K. Okunuki, *Proc. Imp. Acad. Jap.* **17,** 38–40 (1941).
273. D. Keilin and E. F. Hartree, *Nature* (*London*) **164,** 254–259 (1949).
274. D. Keilin and E. F. Hartree, *Nature* (*London*) **176,** 200–206 (1955).
275. E. G. Ball and O. Cooper, *JBC* **226,** 755–768 (1957).
276. R. W. Estabrook, *JBC* **230,** 735–750 (1958).
277. C. A. Yu, L. Yu, and T. E. King, *JBC* **247,** 1012–1019 (1972).
278. G. Forti, M. L. Bertole, and G. Zanetti, *BBA* **109,** 33–40 (1965).
279. D. B. Knaff and R. Malkin, *ABB* **159,** 555–562 (1973).
280. J. Chiang and T. E. King, unpublished work.

TABLE XIX
AMINO ACID COMPOSITIONS OF TYPICAL CYTOCHROMES *c*

	Photosynthetic		Respiratory			
					Bovine c_1	
Amino acid	Spinach *f*	*Euglena* c_{552}	*Euglena* c_{558}	Tuna *c*	Heme subunit	Subunit 2
Lys	24	8	13	16	15	8
Arg	11	3	4	2	18	8
His	3	2	2	2	10	4
Asx	35	14	9	10	21	10
Glx	37	15	9	9	23	24
Ser	18	9	7	4	16	6
Thr	12	7	5	7	10	9
Tyr	11	5	5	5	14	2
Cys	2	2	1	2	8	4
Gly	28	14	10	13	17	4
Ala	24	15	12	7	18	8
Val	25	9	5	6	15	7
Leu	30	7	5	6	28	15
Ile	20	5	4	4	5	3
Met	5	1	2	2	9	1
Pro	20	4	4	3	25	5
Phe	13	3	3	3	9	5
Trp	6	1	2	2	3	2
Total	324	124	102	103	264	125
Source	*278*	*279*	*78,79*	*58*	*280*	*280*

King, personal communication). As befits its position just before *c* in the electron transport chain, the standard reduction potential of c_1 is slightly lower, +225 mV at pH 7. The standard midpoint potentials as measured in the mitochondria are even closer than the solution values: +235 mV for *c* and +225 mV for c_1 (*7*).

Cytochrome c_1 has many properties similar to those of *c*. The reduced protein shows spectra with α, β, and Soret bands at 552.5, 522.5, and 417 nm, compared with values of 550, 521, and 415 nm for *c*. (The β band in c_1 has shoulders at 530 and 512 nm.) The oxidized molecule is reducible with ascorbate, dithionite, and β-mercaptoethanol; and the reduced molecule is not autoxidizable and does not combine with carbon monoxide between pH 5.5 and 11. There is a strong inference that the geometry of the heme environment is similar in *c* and c_1, with the heme enclosed by polypeptide chain and exposed only at one edge. (The heme edge may be masked in the oligomer of c_1 in solution.) A broad absorption band at 695 nm suggests the presence of a methionine ligand for the iron, as in *c*.

Cytochromes *c* and c_1 form a complex which is stable to Sephadex chromatography (*281*), a fact which raises exciting possibilities for X-ray analysis when larger quantities of c_1 become available. In fluorescence quenching studies with 8-anilino-1-naphthaline sulfonic acid (ANS) bound to cytochrome c_1 by hydrophobic interactions, the formation of a complex of oxidized c_1 with *c* was found to quench the ANS fluorescence half as much as reducing the c_1 molecule did. This has been claimed as evidence that the *c* heme is still exposed to solvent in the complex and evidence against heme edge electron transfer in the complex. But in view of the uncertainties in aggregation state of c_1 in solution and the possibility that some of the effects might result from displacement of ANS molecules from c_1 when *c* binds, these conclusions must be considered only tentative.

The rate of transfer of electrons between c_1 and *c* has been mentioned earlier. The second-order rate constant reaches a maximum of 3.3×10^6 M^{-1} sec^{-1} at pH 7.4 and falls off to 7×10^5 at pH 10. This has been interpreted as a consequence of the decreased positive charge on *c* at higher pH and a lessened attraction for c_1, but a simpler explanation is that less cytochrome *c* is found in reducible state III as the pH increases. A plot of the second-order rate constant from ref. *172* against the amount of state III present as measured by the 695-nm absorbance in ref. *166* is linear from pH 7.5 to 10, as Fig. 26 shows. If a correction is made for the actual amount of state III cytochrome present, then the rate constant for reduction by c_1 appears to have a constant rate of 3.3×10^6 M^{-1}, sec^{-1} at neutral and alkaline pH.

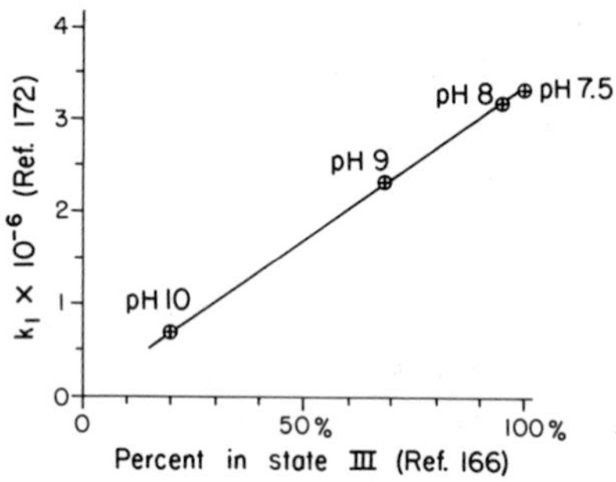

FIG. 26. Linear dependence of the rate constant for reduction of cytochrome *c* by c_1 (k_1) upon the amount of *c* which is present in state III as indicated by absorption at 695 nm. Data from references indicated.

281. L. S. Kaminsky, Y. L. Chiang, C. A. Yu, and T. E. King, *BBRC* **59**, 688–692 (1974).

C. Photosynthetic Cytochromes *f* and c_{553}

Even less is known about the molecular properties of the third *c*-type cytochrome found in eukaryotes: photosynthetic cytochrome $f(=c_6)$ and c_{553} from green plants and algae, respectively. This is not the place for a general discussion of the photosynthetic electron transport chain; for this the reader is referred to several general reviews (*4,282–287*) which, however, must be supplemented by some more recent papers mentioned below. The overall scheme for two-photocenter, water-using and oxygen-liberating photosynthesis as generally accepted is shown in Fig. 27. Light of wavelengths around 700 nm is trapped by various pigments in photocenter I, and used to excite a chlorophyll *a* molecule, designated from its absorption maximum as P-700. This molecule then passes its free energy in the form of reducing electrons to an ill-characterized "fer-

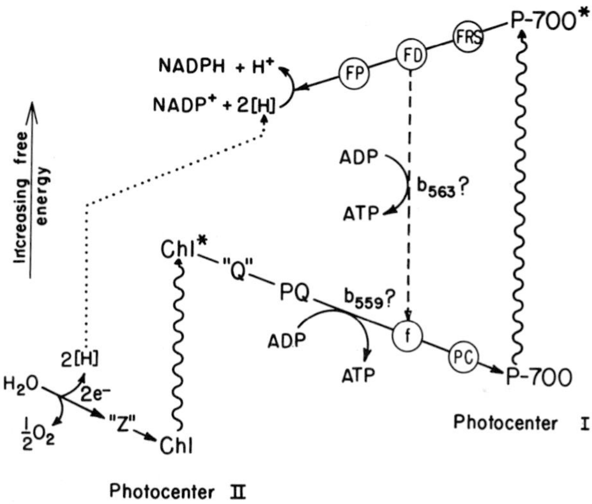

Fig. 27. Electron flow in green plant photosynthesis. Vertical wavy arrows represent excitation of chlorophyll molecules by absorbed light. Reaction intermediates are as follows: "Z", unknown intermediate donating electrons to photocenter II; "Q", unknown intermediate accepting electrons from excited chlorophyll; PQ, plastoquinone, structurally similar to coenzyme Q or ubiquinone of Fig. 4; b_{559}, b_{563}, *f*, cytochrome components; PC, plastocyanin, a copper-containing nonheme protein; FRS, unknown ferredoxin reducing substance; FD, ferredoxin; FP, flavoprotein mediating reduction of $NADP^+$.

282. M. Avron, *Curr. Top. Bioenerg.* **2**, 1–22 (1967).
283. N. K. Boardman, *Advan. Enzymol.* **30**, 1–79 (1968).
284. D. S. Bendall and R. Hill, *Annu. Rev. Plant Physiol.* **19**, 167–186 (1968).
285. N. K. Boardman, *Annu. Rev. Plant Physiol.* **21**, 115–140 (1970).
286. N. I. Bishop, *Annu. Rev. Biochem.* **40**, 197–226 (1971).
287. W. Simonis and W. Urbach, *Annu. Rev. Plant Physiol.* **24**, 89–114 (1973).

redoxin reducing substance," and from there to ferredoxin, to a flavoprotein, and ultimately to $NADP^+$, which is reduced to NADPH as a means of storing chemical free energy. The electron deficiency produced at chlorophyll P-700 is made good by the passage of electrons down another electron transport chain, from an ill-defined substance "Q", through a small plastoquinone molecule, to cytochrome *f*, plastocyanin (a copper protein), and finally to *P*-700. In bacterial photosynthesis these replacement electrons enter the chain from a good reducing agent such as H_2S, but in algal and green plant photosynthesis the source of electrons in H_2O instead. Water is such a poor reducing agent that extra energy must be supplied by a second light absorption at photocenter II before H_2O can act as a source of reducing electrons. The nature of the electron chain from H_2O to photocenter II is unclear, and even the chain connecting photocenters II and I is not firmly established. The order of cytochrome *f* and plastocyanin is reversed by some authors, or the two are placed in parallel; but the arrangement in Fig. 27 is that which is most generally accepted (*288–290*).

The main benefit of photosynthesis as outlined in Fig. 27 is the production of NADPH, which is both a source of stored free energy and of reducing hydrogen equivalents for chemical synthesis. A dividend is the storage of more energy via the synthesis of ATP during the passage of electrons from photocenter II to I. This portion of the electron transport chain resembles the mitochondrial chain, and the structural relatedness of cytochromes *c* and c_2 (bacteria) strongly suggests an evolutionary connection.

The process just described is termed *noncyclic* photophosphorylation, because ATP is synthesized without the cycling of electrons around a closed path. The dashed vertical line in Fig. 27 represents a "short-circuiting" pathway of *cyclic* photophosphorylation, in which electrons can be recycled from ferredoxin to cytochrome *f*, with the synthesis of ATP along the way. The driving force for this process comes only from photocenter I, no water is used, and no NADPH is synthesized.

Two *b*-type cytochromes are present in green plant photosynthesis: b_{559} and b_{563} or b_6. The former has traditionally been placed between plastoquinone and cytochrome *f* in the electron chain, and b_6 has been placed between ferredoxin and cytochrome *f* in cyclic photophosphorylation (*291*). Both of these assignments have been challenged recently (*292–295*), and we must simply note the controversy and drop the issue for the purposes of this chapter.

288. M. Plesnicar and D. S. Bendall, *Eur. J. Biochem.* **34**, 483–488 (1973).
289. J. N. Siedow, V. A. Curtis, and A. San Pietro, *ABB* **158**, 889–897 (1973).
290. V. A. Curtis, J. N. Siedow, and A. San Pietro, *ABB* **158**, 898–902 (1973).
291. H. Böhme and W. A. Cramer, *BBA* **283**, 302–315 (1972).

Cytochrome *f* (for "frons" or leaf) was discovered by Hill and Scarisbrick in 1951 (*296*) as an ethanol-extractable product from green leaves. Davenport and Hill characterized *f* from parsley as having two *c*-type hemes per molecule of 110,000 molecular weight, from sedimentation and diffusion measurements (*297*). Forti and co-workers reported four hemes per 245,000 molecular weight, from Sephadex chromatography (*278*). In a recent and careful purification procedure, Nelson and Racker have obtained the lowest molecular weight yet: one heme per 34,000, determined for spinach *f* by SDS gel electrophoresis (*298*). In line with the normal pattern in biochemistry of a gradual decrease in molecular weight of a new protein as better separation procedures are found, the monoheme protein of molecular weight 34,000 is probably most nearly correct as the prototype of green plant cytochromes *f*.

Cytochrome *f* can be isolated in a particle of molecular weight 103,000 containing two molecules of b_6, one of *f*, nonheme iron, and lipid (*299*)—in many ways similar to mitochondrial Complex III. This particle cannot be dissociated further without destroying the b_6 and producing aggregation of 34,000 molecular weight monomers of *f*. Purified cytochrome *f* apparently exists as an octamer in solution (*298*). This octamer is the size of the solubilized particles reported earlier by Forti and co-workers (*278*), but has twice the heme content.

The absorption spectrum of *f* is similar to that of *c* as shown in Fig. 3. The α, β, and Soret bands occur at 554.5, 524, and 422 nm. Its reduction potential, like that of all high potential photosynthetic cytochromes, is about 100 mV higher than that of respiratory *c*. The potential at pH 7.7 is +365 mV in solution (*297*) and +375 mV in the intact chloroplast membrane (*279,300*). As with cytochrome *c*, the standard reduction potential falls by 60 mV for every increase of one pH unit at alkaline pH, reflecting a proton dissociation to a nonreducible form with a pK of 8.4, which may be analogous to the state III to IV transition in *c*. The molecule is somewhat more sensitive to denaturation by heat and pH extremes than is *c*. The 695-nm absorption band indicative of a methionine ligand in *c* and c_1 apparently has not been looked for in cytochrome *f*.

The amino acid composition of *f* is compared in Table XIX with those

292. N. K. Boardman, J. M. Anderson, and R. G. Hiller, *BBA* **234,** 126–136 (1971).
293. A. W. D. Larkum and W. D. Bonner, *ABB* **153,** 241–248 (1972).
294. A. W. D. Larkum and W. D. Bonner, *ABB* **153,** 249–257 (1972).
295. A. Vermeglio and P. Mathis, *BBA* **314,** 57–65 (1973).
296. R. Hill and R. Scarisbrick, *New Phytol.* **50,** 98–111 (1951).
297. H. E. Davenport and R. Hill, *Proc. Roy. Soc. Ser. B* **139,** 327–345 (1952).
298. N. Nelson and E. Racker, *JBC* **247,** 3848–3853 (1972).
299. N. Nelson and J. Neumann, *JBC* **247,** 1817–1824 (1972).
300. R. Malkin, D. B. Knaff, and A. J. Bearden, *BBA* **305,** 675–678 (1973).

of photosynthetic c_{552} and respiratory c_{558} from *Euglena,* and respiratory c and c_1. [*Euglena* has both a c_{552} typical of photosynthetic algae and a c_{558} which we have seen earlier as the normal eukaryotic respiratory c (see refs. *301–303*).] The amino acid compositions show no particularly striking features. All of these c-type cytochromes have at least two cysteines as heme connections (except for the anomalous *Euglena* c_{558} noted earlier), at least one methionine and two or more histidines, one or more tryptophans, and several other aromatic groups. Unlike respiratory c but like c_1 and the algal and bacterial photosynthetic c's, f is an acidic protein with an isoelectric point around 4.7.

In contrast to green plant cytochrome f, which can only be extracted from the chloroplast membrane by organic solvents, the analogous photosynthetic cytochromes c from eukaryotic and blue-green algae are easily water soluble. From the position of their α-band spectrum they are given the general name of cytochromes c_{553}. They are all monoheme cytochromes with molecular weights 10,000–13,500, and show every indication of being homologous with c and c_2. Yakushiji first discovered such a cytochrome in red algae, *Porphyra tenera,* in 1935 (*304*), and it was subsequently purified and crystallized (*305,306*). Corresponding cytochromes c_{553} have been found in red, brown, green, and blue-green algae, summarized in ref. *307*. The α, β, and Soret bands are observed at 550–555, 521–523, and 425–427 nm, and the standard reduction potentials at pH 7 vary between +340 and +400 eV.

A typical example is the recently purified c_{553} from the green algae, *Kirchneriella obesa* (*308*). It has one heme per 12,000 molecular weight, absorption bands at 553, 522, and 417 nm, an acid isoelectric point, and a rather high standard reduction potential at pH 7 of +400 mV. Oxidized c_{553} from *K. obesa* is easily reduced by ascorbate or dithionite, and the reduced form does not bind CO and is not autoxidized, but can be oxidized by ferricyanide, all being traits familiar from respiratory c. The c-type heme is held to the polypeptide chain by two thioether links.

All of the eukaryotic c cytochromes that we have considered so far have obvious kinships with one another and with bacterial c_2. It would

301. F. Perini, M. D. Kamen, and J. A. Schiff, *BBA* **88,** 74–90 (1964).
302. F. Perini, J. A. Schiff, and M. D. Kamen, *BBA* **88,** 91–98 (1964).
303. M. A. Cusanovich, T. Meyer, S. M. Tedro, and M. D. Kamen, *Proc. Nat. Acad. Sci. U. S.* **68,** 629–631 (1971).
304. E. Yakushiji, *Acta Phytochim.* **8,** 325–329 (1935).
305. E. Yakushiji, Y. Sugimura, I. Sekuzu, I. Morikawa, and K. Okunuki, *Nature (London)* **185,** 105–106 (1960).
306. S. Katoh, *Nature (London)* **186,** 138–139 (1960).
307. Y. Sugimura, F. Toda, T. Murata, and E. Yakushiji, *in* "Structure and Function of Cytochromes" (K. Okunuki, M. D. Kamen, and I. Sekuzu, eds.), pp. 452–458. Univ. of Tokyo Press, Tokyo, 1968.
308. R. M. Nalbandyan, *Biokhimiya* **37,** 1161–1165 (1972).

be astounding if X-ray analysis were to show that algal c_{553} is folded differently from c, c_2, and c_{550}; and it would not be surprising to find that portions of the c_1 and f molecules displayed this same "cytochrome fold." A fascinating story on the rise of eukaryotes and the divergence of respiratory and photosynthetic machinery waits to be told when more structural, sequence, and chemical information is available.

III. Bacterial Cytochromes c

If a certain degree of unity is beginning to appear among the eukaryotic cytochromes, the predominant impression with bacterial cytochromes is one of almost overwhelming diversity. Bacterial cytochromes with a *c*-type spectrum which have been at least partially isolated and purified to date are listed in Table XX. Other *c* cytochromes which have been identified only by their absorption wavelengths in whole cells or extracts are not listed, although some of them will figure in the mechanism discussions that follow. In addition to the specific references *309–356* in the

309. T. Horio and M. D. Kamen, *BBA* **48,** 266–286 (1961).
310. T. Flatmark, K. Dus, H. DeKlerk, and M. D. Kamen, *Biochemistry* **9,** 1991–1996 (1970).
311. R. G. Bartsch, "Methods in Enzymology," Vol. 23, Part A pp. 344–363, 1971.
312. T. E. Meyer, R. G. Bartsch, and M. D. Kamen, *BBA* **245,** 453–464 (1971).
313. P. L. Dutton and B. J. Jackson, *Eur. J. Biochem.* **30,** 495–510 (1972).
314. J. A. Orlando, *BBA* **57,** 373–375 (1962).
315. S. Morita and S. F. Conti, *ABB* **100,** 302–307 (1963).
316. R. G. Bartsch and M. D. Kamen, *JBC* **235,** 825–831 (1960).
317. R. G. Bartsch, T. E. Meyer, and A. B. Robinson, *in* "Structure and Function of Cytochrome" (K. Okunuki, M. D. Kamen, and I. Sekuzu, eds.), pp. 443–451. Univ. of Tokyo Press, Tokyo, 1968.
318. M. A. Cusanovich and R. G. Bartsch, *BBA* **189,** 245–255 (1969).
319. S. J. Kennel and M. D. Kamen, *BBA* **253,** 153–166 (1971).
320. T. E. Meyer, S. J. Kennel, S. Tedro, and M. D. Kamen, *BBA* **292,** 634–643 (1973).
321. T. E. Meyer, R. G. Bartsch, M. A. Cusanovich, and J. H. Matthewson, *BBA* **153,** 854–861 (1968).
322. Y. Shioi, T. Kenichiro, and M. Nishimura, *J. Biochem. (Tokyo)* **71,** 285–294 (1972).
323. R. T. Swank and R. H. Burris, *BBA* **180,** 473–489 (1969).
324. N. P. Neuman and R. H. Burris, *JBC* **234,** 3286–3290 (1959).
325. W. H. Campbell, W. H. Orme-Johnson, and R. H. Burris, *BJ* **135,** 617–630 (1973).
326. T. Yamanaka and S. Imai, *BBRC* **46,** 150–154 (1972).
327. A. M. Lauwers and W. Heinen, *Antonie van Leeuwenhoek; J. Microbiol. Serol.* **38,** 451–455 (1972).
328. K. Miki and K. Okunuki, *J. Biochem. (Tokyo)* **66,** 831–843 (1969).
329. K. Miki and K. Okunuki, *J. Biochem. (Tokyo)* **66,** 845–854 (1969).
330. I. W. Sutherland, *BBA* **73,** 162–164 (1963).

table, other useful general reviews are to be found in references *357* and *358* by Kamen and Horio.

The entries in Table XX are only an experimentally studied subset of all the cytochromes *c* that exist; a complete compilation would very nearly require a minimum of one entry per organism using electron transport in its metabolism. Physiological studies on representative microorganisms throughout the spectrum of bacteria make it safe to generalize that *c*-type cytochromes occur hand in hand with electron transport chains. The apparent exceptions such as *Staphylococcus aureus,* which contains only *b*-, *o*-, and *a*-type cytochromes (*359*), are most likely para-

---

331. G. D. Clark-Walker and J. Lascelles, *ABB* **136,** 153–159 (1970).
332. P. A. Ketchum, H. K. Sanders, J. W. Grycer, and A. Nason, *BBA* **189,** 360–365 (1969).
333. D. A. Tronson, G. A. F. Ritchie, and D. J. D. Nicholas, *BBA* **310,** 331–343 (1973).
334. T. Yamanaka, S. Takenami, N. Akiyama, and K. Okunuki, *J. Biochem.* (*Tokyo*) **70,** 349–358 (1971).
335. G. Milhaud, J.-P. Aubert, and J. Millet, *C.R. Acad. Sci.* **246,** 1766–1769 (1958).
336. P. A. Trudinger, *BJ* **78,** 673–679 (1961).
337. P. A. Trudinger, *BJ* **78,** 680–686 (1961).
338. T. Fujita, *J. Biochem.* (*Tokyo*) **60,** 204–215 (1966).
339. P. B. Scholes, G. McLain, and L. Smith, *Biochemistry* **10,** 2072–2076 (1971).
340. K. Hori, *J. Biochem.* (*Tokyo*) **50,** 480–485 (1961).
341. K. Hori, *J. Biochem.* (*Tokyo*) **50,** 440–449 (1961).
342. T. Horio, T. Higashi, M. Sasagawa, K. Kusai, M. Nakai, and K. Okunuki, *BJ* **77,** 194–201 (1960).
343. T. Horio, *J. Biochem.* (*Tokyo*) **45,** 267–279 (1958).
344. J. Singh and D. C. Wharton, *BBA* **292,** 391–401 (1973).
345. M. A. Cusanovich, S. M. Tedro, and M. D. Kamen, *ABB* **141,** 557–570 (1970).
346. H. Iwasaki and S. Shidara, *J. Biochem.* (*Tokyo*) **66,** 775–781 (1969).
347. R. P. Ambler, *BJ* **89,** 341–349 (1963).
348. T. Kodama and S. Shidara, *J. Biochem.* (*Tokyo*) **65,** 351–360 (1969).
349. K. Furukawa and K. Tonomura, *BBA* **325,** 413–423 (1973).
350. H. Drucker and L. L. Campbell, *J. Bacteriol.* **100,** 358–364 (1969).
351. R. P. Ambler, M. Bruschi-Heriaud, and J. LeGall, *FEBS* (*Fed. Eur. Biochem. Soc.*) *Lett.* **18,** 347–350 (1971).
352. M. Bruschi-Heriaud and J. LeGall, *Bull. Soc. Chim. Biol.* **49,** 753–758 (1967).
353. T. Yagi and K. Maruyama, *BBA* **243,** 214–224 (1971).
354. H. Drucker, E. B. Trousil, and L. L. Campbell, *Biochemistry* **9,** 3395–3400 (1970).
355. J. LeGall, M. Bruschi-Heriaud and D. V. Dervartanian, *BBA* **234,** 499–512 (1971).
356. J. LeGall, M. Bruschi-Heriaud, and N. Forget, *in* "Structure and Function of Cytochrome" (K. Okunuki, M. D. Kamen, and I. Sekuzu, eds.), pp. 467–470. Univ. of Tokyo Press, Tokyo, 1968.
357. M. D. Kamen and T. Horio, *Annu. Rev. Biochem.* **39,** 673–700 (1970).
358. T. Horio and M. D. Kamen, *Annu. Rev. Microbiol.* **24,** 399–428 (1970).
359. H. W. Taber and M. Morrison, *ABB* **105,** 367–379 (1964).

doxes created by our narrow spectral categorization of cytochromes; perhaps a "*b*-type" protein of suitable potential operates in the manner normally ascribed to cytochromes *c*. The only organisms totally free of cytochromes of any type are those subsisting on a purely fermentative metabolism such as *Clostridia* or *Streptococci* (*360*).

Research on bacterial cytochromes has not been directed toward allowing reviewers to publish comprehensive compendia, and Table XX contains many gaps representing properties undetermined, unreported, or unknown to us. Several recent papers have clarified seemingly contradictory or confusing earlier reports, and these must be mentioned first to avoid trouble when reading the older literature.

1. The *c*′ class of cytochromes originally was thought to consist of two groups: a monoheme *c*′ class, and a diheme *cc*′ in which the two hemes had similar properties but the nomenclature emphasized the fact that there were two of them. Kennel and co-workers showed in 1972 that this class distinction does not exist (*361*) and that the reported *cc*′ cytochromes actually are natural dimers of probably identical subunits with molecular weights around 12,000. The traditional *cc*′ nomenclature has been retained in Table XX for ease in referring to the original literature and as a reminder that these cytochromes occur as tightly bound dimers.

2. Among the low potential cytochromes c_3 from the sulfate-reducing *Desulfovibrio*, the reported number of heme groups per peptide chain of molecular weight 13,000 has risen from two, to three, and finally to four. The problem has been the presence of impurities which increase the total ratio of gross protein to heme, detected spectrally or by iron analysis, thus giving low values of the number of hemes per unit of protein. (This may also have been the problem with cytochrome *f*, discussed earlier.) Sequence analyses on cytochromes c_3 (*362*) have shown a uniform pattern of four potential heme binding sites of the form Cys . . . Cys·His, and careful spectroscopic and analytical work has confirmed this (*312*). The figure of four heme groups per molecule of cytochrome c_3 from all strains is now generally accepted (*363*).

3. Any work on bacterial proteins is subject to the danger of misinterpretation arising from an impure mixture of several strains of microorganisms. A striking example was the report of a low potential c_3-like protein from a photosynthetic organism, "*Chloropseudomonas ethylicum*," the

360. R. Y. Stanier, M. Doudoroff, and E. A. Adelberg, "The Microbial World." Prentice-Hall, Englewood Cliffs, New Jersey, 1970.
361. S. J. Kennel, T. E. Meyer, M. D. Kamen, and R. G. Bartsch, *Proc. Nat. Acad. Sci. U. S.* **69**, 3432–3435 (1972).
362. R. P. Ambler, *Syst. Zool.* **22**, 554–565 (1974).
363. J. LeGall and J. R. Postgate, *Advan. Microbial Physiol.* **10**, 81–133 (1973).

TABLE XX
PHYSICAL CHEMICAL PROPERTIES OF REPORTED BACTERIAL CYTOCHROMES[a]

	Absorption bands						
Organism and cytochrome	Reduced	Oxidized	MW	*p*I	E_0' (mV)	Hemes	Ref.
I. Photosynthetic bacteria							
A. Purple nonsulfur (*Athiorhodaceae*)							
Rhodospirillum rubrum (strain S1)							
c_2	550, 521, 415	525, 410	12,500	6.2	+320	1	*309*
cc'	550, 423	495, 390	29,800	5.6	−8	2	*309*
Rhodospirillum molischianum (ATCC 14031)							
c_2	550, 520, 415	529, 410	13,400	10	+288	1	*310*
c_{550}	550, 520, 415	528, 410	10,200	10.5	+381	1	*310*
cc'	555, 426		36,000	7.2		2	*311*
Rhodopseudomonas capsulata (strain SL)							
c_2	550, 416		13,000				*311*
c'	550, 425						*311*
Rhodopseudomonas gelatinosa (ATCC 11169)							
cc'	550, 425		36,000	9.6			*311*
c_{551}	551, 417		12,000	10.5			*311*
c_{552}	552, 418		50,000				*311*
$c_{554,550}$ (split α)	554, 550, 418		27,000	9.6			*311*
Rhodopseudomonas palustris (Van Niel 2.1.37)							
c_2	551, 418		12,700	9.7	+330	1	*311*
c'	552, 426		15,000	9.4	+100	1	*311*
c_{551}	551, 420		13,000			1	*311*

$c_{551.5}$	551, 523, 418	408	16,000	6.1	−150		*312*
c_{554}	554, 418		40,000		−6		*311*
c_{555}	555, 420		12,000		+230	1	*311*
Rhodopseudomonas spheroides (ATCC 11167)							
c_2	550, 416		13,800	5.5	+346	1	*311*
cc'	545, 423		25,000	4.9	−20	2	*311,313*
c_{551}	551, 523, 419	527, 409	21,000	4.3	−254	2	*312*
c_{554}	554, 523, 419	520, 412	44,000	4.1	+120		*311,314*
Rhodomicrobium vannieli							
c_2	550, 521, 415			7	+304		*315*
c_{553}	553, 523, 423				<300		*315*
B. Purple sulfur (Thiorhodaceae)							
Chromatium (strain D)							
cc'	547, 426	490, 395	28,000	4.6	−5	2	*311,316*
Flavin c_{552}	552, 523, 416	530, 410	72,000	4.5	+10	2	*316,317*
$c_{553,550}$ (split α)	553, 550, 523, 417	410	13,000	4.4	+330	1	*318*
Cytochrome complex	555, 552, 520, 423	412	45,000		+340, 0	2	*319*
Thiocapsa pfennigii							
c'	546, 426	490, 398	11,000			1	*320*
$c_{552,545}$	552, 545, 517, 416	524, 407	30,000			2	*320*
$c_{552,550}$	552, 550, 522, 417	522, 412	30,000				*320*
C. Green sulfur (Chlorobacteriaceae)							
Chlorobium thiosulfatophilum (NCIB 8346)							
c_{551}	551, 521, 416	528, 410	45,000	6.0	+135	2	*311,321*
Flavin c_{553}	553, 523, 417	525, 410	50,000	6.7	+98	1	*321*
c_{555} (slightly split α)	555, 523, 418	523, 412	10,000	10.5	+145	1	*311,321*
"*Chloropseudomonas ethylicum*"							
$c_{551.5}$ (c_7)	551, 552, 418	526, 408	11,000	4.1	−194	3?	*312*
$c_{555,549}$	555, 549, 523, 418	525, 412	24,000	5.0	+103	2	*311,322*

TABLE XX (*Continued*)

Organism and cytochrome	Absorption bands: Reduced	Absorption bands: Oxidized	MW	*p*I	E_0' (mV)	Hemes	Ref.
II. Strictly aerobic bacteria							
Azotobacter vinelandii							
c_4	551, 522, 414	409	25,000	4.4	+300	2	*323,324*
c_5	555, 524, 416	414	25,000	4.2	+320	2 (dimer)	*323,324*
c_{555} (denatured subunit of c_5?)			11,800	4.1			*325*
cc' (possible polymer)	550, 423	397	170,000	4.7			*326*
c_{551}	551		12,000	4.6		1	*325*
Bacillus caldolyticus							
c_{550}	550, 520, 416		10,000				*327*
Bacillus subtilis							
c_{550}	550, 520, 414	528, 407	12,500	8.65	+210	1	*328,329*
c_{554}	554, 550, 521, 417	523, 409	14,000	4.44	−80	1	*328,329*
Bordetella pertussis							
c_{550}	550, 520, 418	520, 408			+259		*330*
c_{553}	553, 522, 416	530, 409			+198		*330*
Spirillum itersonii							
c_{550}	550, 522, 416	412	10,411	9.86	+297	1	*331*
III. Aerobic bacteria using inorganic reducing equivalents in respiration							
Nitrobacter agilis							
c_{550}	550, 521, 417	411			+282	1	*332*
Nitrosomonas europaca							
$c_{551.5}$	552, 522, 416		12,600	4.42	+240	1	*333*
c_{552}	552, 523, 462, 417		34,400	4.40			*333*
c_{553}	553, 523, 418		52,000	4.68	+500	2	*333*

Thiobacillus novellus							
c_{550}	550, 520, 414	410	13,270	7.5	+276	1	*334*
c_{551}	551, 522, 416	410		5.2	+260		*334*
Thiobacillus thioparus							
c_{551}					+145		*335*
Thiobacillus X, Trudinger strain							
c_{550}	550, 520, 416	409		>7	+200		*336,337*
c_{553}	553, 523, 418	410		<7	+210		*336,337*
IV. Bacteria using inorganic terminal oxidants in respiration							
Escherichia coli (Yamagutchi)							
c_{552}	552, 523, 420	532, 409	Aggregate	4.4	−200	1 per 12,000	*338*
Micrococcus denitrificans (ATCC 13543)							
c_{550}	550, 520, 418	410	14,890	4.5	+260	1	*339*
Micrococcus strain 203 (halotolerant)							
c_{551}	551, 521, 416	521, 411		<7	+249		*340*
c_{554}	554, 521, 418	525, 414	18,000	3	+180	1	*341*
$c_{554,548}$	554, 548, 521, 418	525, 414	18,000	3.2	+113	2 (dimer)	*341*
Pseudomonas aeruginosa							
c_{551}	551, 521, 416	409	8,100	4.7	+286	1	*342*
c_{554}	554, 525, 420	410			+225		*343*
$c_{556,552}$	556, 552, 521, 420	410	40,500			2	*344*
Pseudomonas denitrificans							
cc'	550, 426	495, 400	27,000	8.95	+132	2	*345*
c_{553}	553, 524, 419	409	45,000		−90, 0	2	*346*
Pseudomonas fluorescens							
c_{551}	551, 520, 416	408	9,000	4.7	+280	1	*347*
Pseudomonas stutzeri (Van Niel)							
c_{552} Type I	552, 523, 418	530, 410	8,300	4.0–6.6	+277	1	*348*
c_{552} Type II	552, 523, 417	525, 410	20,000		+283	2 (dimer)	*348*

TABLE XX (*Continued*)

Organism and cytochrome	Absorption bands: Reduced	Absorption bands: Oxidized	MW	*p*I	E_0' (mV)	Hemes	Ref.
IV. Bacteria using inorganic terminal oxidant in respiration (*Continued*)							
Pseudomonas strain K62							
c_{550}	550, 521, 416	530, 407	26,000			1	*349*
Desulfovibrio gigas (NCIB 9332)							
c_3 or c_3'	552, 522, 418	520, 409	13,000	5.2	−216	4	*350,351*
Desulfovibrio desulfuricans (NCIB 8380)							
c_3	552, 522, 418	520, 409	12,000	10	−205	4	*350,351*
Desulfovibrio vulgaris (Hildenborough, NCIB 8303)							
c_3	552, 522, 418	525, 409	12,000	10.5	−215	4	*309,352*
Desulfovibrio vulgaris (Miyazaki)							
c_3	552, 523, 419	530, 410	13,000	10.6	−290	4	*353*
Desulfovibrio salexigens (NCIB 8403)							
c_3	552, 522, 418	520, 409	13,900	10.8	−205	4	*354*
D. gigas, desulfuricans, vulgaris							
cc_3	552, 522, 418	520, 408	26,000		< +60	4–8	*4,355*
c_{553}	553, 522, 418	409	9,000	8.6	−100 to 0	1	*4,356*

[a] Spectral bands have been rounded to integral values. Some assignments have been made by the present authors from noncalibrated graphs in the original literature. Differences between laboratory instruments can also lead to slightly varying reported wavelengths, and the values reported in the indicated references are used in this table. The oxidized band between 520 and 530 nm is quite broad, and differences of 10 nm are insignificant. Reported molecular weights and isoelectric points in general should be considered only approximate.

first suggestion that this type of cytochrome occurred in anything other than sulfate-reducing bacteria (*312*). However, "*Chloropseudomonas ethylicum*" proved to be a syntrophic mixture of a green photosynthetic bacterium, *Chlorobium*, and a typical sulfate-reducing bacterium (*364*). *Chlorobium* used sulfur in the medium as a source of reducing equivalents in photosynthesis, and the other organism then recycled it by using the oxidized sulfur compounds as terminal electron acceptors in respiration. The low potential cytochrome, which has been called $c_{551.5}$ or c_7 and which appears from amino acid sequence analysis to contain three hemes (*362*), presumably comes from the sulfate-reducing member of the team. Gray and co-workers (*364*) have suggested that other reports of c_3-type cytochromes in photosynthetic purple nonsulfur bacteria (*312*) may also arise from mixed cultures, but this has not been checked.

4. Several cytochromes in Table XX are reported as having more than one α or 550 band. In cases such as *Pseudomonas aeruginosa* $c_{556,552}$, two independent heme groups are present per molecule, but *Chlorobium thiosulfatophilum* $c_{555,551}$ and some others have only one heme. In these latter cases, some peculiar arrangement of ligands evidently is causing a splitting in what otherwise would be a degenerate α-absorption transition.

In the face of the diversity in spectrum, size, heme content, and redox potential seen in Table XX, it would be foolhardy to attempt a comprehensive classification of bacterial cytochromes *c*. Nevertheless, several broad categories do stand out, the most extensive being the proteins similar in size and redox potential to the eukaryotic cytochromes *c*. Within this we can recognize a low molecular weight subclass including *P. aeruginosa* c_{551} (MW 8000), and a "standard" subclass typified by *R. rubrum* c_2 (MW 12,500). The c' class of cytochromes has a peculiar spectrum which resembles that of myoglobin under physiological conditions, suggesting the absence of a sixth heme ligand. Under acid denaturing conditions it reverts to a typical high-spin hemochromogen spectrum, indicating that the heme is the normal *c* type. The split-alpha cytochromes, with their suggestion of a different ligand arrangement, may also be called a separate class. The large cytochromes *c* with bound flavins such as *Chromatium* c_{552} are in a category by themselves, and the low potential multiheme cytochromes c_3 constitute a fifth logical grouping. *Desulfovibrio* c_{553} is anomalous: a single heme protein in the size range of the eukaryotic types, but with a negative reduction potential. These highly varied cytochromes remind one of a collection of vacuum tubes or transistors of different performance ratings, ready to be plugged into electron-moving circuits in microorganisms where needed. It is a reason-

364. B. H. Gray, C. F. Fowler, N. A. Nugent, and R. C. Fuller, *BBRC* **47**, 322–327 (1972).

able assumption that the "circuitry" will be closely similar in closely related microorganisms and that much the same set of cytochromes will be found in all the Rhodopseudomonads, for example, if they are looked for carefully enough.

A. Bacterial Cytochrome Function

1. *Common Methodology in Bacterial Cytochrome Research*

In spite of elegant experimental work dating from the 1950's, the determination of *in vivo* function for most bacterial cytochromes is still in its infancy, as is the detailed tracing of the links in the electron transfer pathways. The number of such pathways is so great in comparison with the situation in eukaryotes that only a few microorganisms have been studied in detail. Even for these, there is a temptation to see analogies with eukaryotic pathways where none may exist, and to report what is probable rather than what is proved. We shall follow this tradition, and attempt to generalize and bring order out of confusion, even where the particular kind of order which results may not be quite right. It is sometimes better to have the wrong scheme of things than no scheme at all, if only to have something to correct later.

In view of these uncertainties, one should take less for granted than usual, and understand the sometimes indirect evidence that is offered in support of proposed electron transfer pathways. Several recent cytochrome reviews contain outlines of methodology (*4,358,365*), with the last-mentioned review being especially well written. Five key methods of studying cytochromes will be mentioned here.

The central tool of cytochrome research since the days of MacMunn has been the spectrophotometer. The wavelengths of the α, β, and Soret bands are the signatures of the cytochrome molecules, and the first step in the study of any new organism is to determine what types (*a, b, c, d, o*, etc.) of cytochrome are present and in what relative amounts. Spectra are measured on whole cell suspensions, extracts, or membrane fragments, at normal or liquid nitrogen temperatures; and in the presence and absence of binding agents such as carbon monoxide or alkaline pyridine which produce characteristic modifications of the spectrum.

With a knowledge of the cytochromes present, the sensitive technique of difference spectroscopy can be used to study the redox state of the individual components as a function of different metabolic conditions; for example, whole cells or an extract can be starved of their source of reducing equivalents such as succinate, NADH, or light, thereby driving

365. D. C. White and P. R. Sinclair, *Advan. Microbial Physiol.* **5**, 173–211 (1971).

all the cytochromes in the chain to a fully oxidized state. Reducing equivalents then can be added quickly to observe which components become reduced and in what order. A similar strategy in reverse can be used by fully reducing the components in an electron transport chain and then rapidly introducing an oxidant and observing the changes.

Inhibitors which selectively block electron transport at certain places are useful in the above experiments. As an example, assume that one has cytochromes of type *b*, *o*, and *c*, and a flavoprotein dehydrogenase, all possibly involved in a chain from succinate to oxygen but with unknown order. The known order in eukaryotes should provide a guide for research, but not a shortcut to conclusions. The intact system may be starved of succinate and 2-*n*-heptyl-4-hydroxyquinoline-*N*-oxide added to block transfer of electrons from *b* to *c*. If, when succinate is readded, the flavoprotein and *b* become reduced while *c* and *o* remain oxidized, this is evidence for an ordering of first flavoprotein and *b*, and then *c* and *o*, along the chain. Blocking cytochrome *o* with cyanide or CO and observing that succinate then could reduce cytochrome *c*, would provide evidence for an ordering within the chain of (flavoprotein, *b*), *c*, and finally *o*. Selective blocking combined with difference spectroscopy provides the most powerful method of establishing the order of components in an electron transport chain.

A danger in this approach is the assumption by analogy that a given inhibitor is capable of the same mode of blocking action in all systems studied. Counter examples are known in which an inhibitor is found to be inactive in a particular system. *Bacillus brevis* is a case in point, producing as a metabolic product large amounts of tyrocidine, which ordinarily is a potent uncoupler of transport-linked oxidative phosphorylation. *Bacillus brevis* survives because it has a variant of mitochondrial-type respiration which has been observed to be relatively unaffected by tyrocidine and other conventional inhibitors (*366*).

The measurement of reduction potential is an important step in establishing the order of chain elements. Reduction potentials vary with relative concentrations of the two redox species, of course, with an order of magnitude increase in ratio of reduced to oxidized form leading to a 60-mV increase in reduction potential in one electron reactions. Differences of a few tens of millivolts between the *standard* reduction potentials of cytochrome *f* and plastocyanin in green plant photosynthesis are not enough to establish their relative order since relative concentrations could reverse the actual reduction potentials. Still, a cytochrome *c* with a standard reduction potential of +50 mV is unlikely to donate electrons to

366. B. Seddon and G. N. Flynn, *Arch. Mikrobiol.* **77**, 252–261 (1971).

another with an E_0' of —200 mV or receive electrons from one with $E_0' = +300$ mV. Standard reduction potentials can be measured on purified cytochromes (if available) or *in situ* by means of optical (*7,367*) or electron paramagnetic (*279,300*) spectroscopy.

The ultimate test, but not one achieved easily, is always the isolation and purification of individual cytochrome components. Cytochrome *c* has been studied more extensively than *b* or *a* cytochromes just because it can be isolated easily, and Table XX is an indication of the success to date. When possible, reconstitution experiments between a purified, extracted component and membrane fragments depleted of that component can provide strong evidence for function. Unfortunately, bacterial systems have proved to be more sensitive to handling then mitochondria, and extraction of cytochrome usually denatures the chain irreversibly (*368*).

A key question to keep asking during any discussion of bacterial cytochrome function, including that which follows, is: To what extent does the primary evidence support the proposed conclusions, and how heavy is the dependence on analogy with known eukaryotic pathways?

2. *Patterns in Bacterial Electron Transport Pathways*

It will be helpful in subsequent discussions of specific transport chains to look first at some of the general patterns of electron flow that are possible. The two main themes in bacterial photosynthesis, as in that of green plants, are cyclic and noncyclic photophosphorylation. If the excited chlorophyll of green plant photocenter II is replaced by a reductant such as H_2S, then Fig. 27 can serve as a rough schematic for bacterial photosynthesis as well. The clearest use for *c*-type cytochromes in bacterial photosynthesis is as electron donors to the bacterial chlorophyll photocenter (without the intervening plastocyanin). These are the high potential cytochromes c_2, entirely analogous to photosynthetic *f* and c_{553} in eukaryotes. Cytochrome *c′* has been suggested as possibly having a photosynthetic role with no analog among eukaryotes, and this will be discussed later.

Respiration provides many more alternative patterns to that found in mitochondria, and some of these are sketched in models I–V of Fig. 28. The mitochondrial pattern of model I is seen in many bacteria. The substrate end of the chain can be branched to provide for different modes of entry into the chain, as with the NADH and succinate entry points in mitochondrial respiration. The acceptor end can also be branched (model II), as in denitrifying bacteria which can use either O_2 or nitrate

367. P. L. Dutton, *BBA* **226**, 63–80 (1971).
368. L. Smith and D. C. White, *JBC* **237**, 1337–1341 (1962).

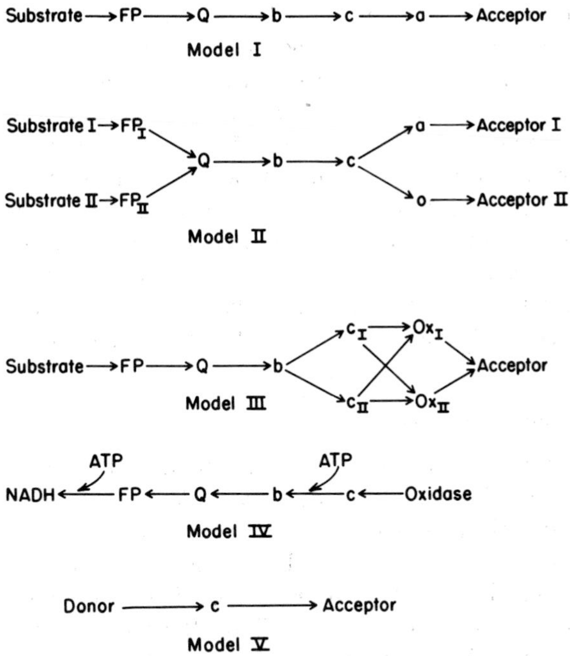

FIG. 28. Five model electron transport chains for bacteria: (I) linear, (II) branched ends, (III) parallel and interconnecting, (IV) reversed flow, and (V) abbreviated chain. Abbreviations: FP, flavoprotein dehydrogenase enzyme; Q, small quinone coenzyme molecule; *b, c, a, o,* cytochromes; and Ox, cytochrome oxidase.

or nitrite as a terminal oxidant. Parallel, possibly intercommunicating pathways as in model III have also been suggested.

Chemoautotrophic bacteria which use inorganic substances instead of carbon compounds as energy sources still need NADH for synthetic purposes but cannot make it by the Krebs cycle as we do. It has been proposed that these autotrophic bacteria might obtain NADH for synthesis by a process of reversed electron flow up an electron transport chain, as in model IV, driven by energy from ATP. Finally, we can imagine more rudimentary electron chains, either primitive or degenerate, in which the free energy span between donor and acceptor is so short that only a few electron "valves" are necessary to regulate flow (model V). These basic patterns, and variations on them, will be seen in the particular examples of the following sections.

B. BACTERIAL PHOTOSYNTHESIS

Only three groups of bacteria are photosynthetic: the green sulfur bacteria or Chlorobacteriaceae, purple sulfur bacteria or Thiorhodaceae, and purple nonsulfur bacteria or Athiorhodaceae. The green and purple sulfur bacteria normally use H_2S as an electron donor in noncyclic photophosphorylation, oxidizing it to elemental sulfur and then all the way to sulfate. Some can also use thiosulfate, H_2, or certain organic molecules as sources of reducing power. Purple nonsulfur bacteria, in contrast, depend exclusively on reduced organic compounds as their source of reducing equivalents. They also have the distinction of being the only bacteria to combine photosynthesis and respiration. They have a Krebs cycle and respiratory electron transport chain, and can obtain energy by oxidizing organic matter in the dark. The sulfur bacteria are obligate anaerobes, unable to tolerate the presence of free O_2, which obviously is not true for the purple nonsulfur bacteria.

1. *Purple Nonsulfur Bacteria*

The purple nonsulfur bacteria are represented by the genera *Rhodospirillum*, *Rhodopseudomonas*, and *Rhodomicrobium*. Cytochromes *c* from all three are described in Table XX. Cytochromes c_2 and c' or cc' seem to be common to all of the Athiorhodaceae, with some species having several other cytochromes *c* of uncertain function. In the absence of light, these bacteria obtain ATP and NADH by the normal Krebs cycle and respiratory chain, using organic molecules as nutrients. The carbon from the nutrients ends as CO_2. Under anaerobic conditions in the light, the story is different. Cyclic photophosphorylation then provides the major part of the ATP needed. Only enough carbon compounds are oxidized by the Krebs cycle to provide the necessary NADH for reduction of other carbon compounds during chemical synthesis of substances needed by the bacterium. In this mode of operation, the bacterium uses the energy of ATP from photosynthesis to dismutate organic matter into more reduced and more oxidized compounds (*360,369,370*).

Rhodospririllum rubrum is the most exhaustively studied of the purple nonsulfur bacteria. The photosynthetic apparatus is located in extensive infolded membrane vesicles called chromatophores. Cytochromes c_2, cc', and $b_{557.5}$ have all been found to be associated with the chromatophores (*371*), with c_2 being the major component. Several lines of evidence, including fast kinetics following the excitation of the bacteriochlorophyll photocenter by laser flash, have suggested the scheme shown in Fig. **29**, where cytochrome c_2 is the immediate source of electrons to the electron-

369. H. W. Doelle, "Bacterial Metabolism." Academic Press, New York, 1969.
370. J. R. Sokatch, "Bacterial Physiology and Metabolism." Academic Press, New York, 1969.
371. T. Kakuno, R. G. Bartsch, K. Nishikawa, and T. Horio, *J. Biochem.* (*Tokyo*) **70,** 79–94 (1971).

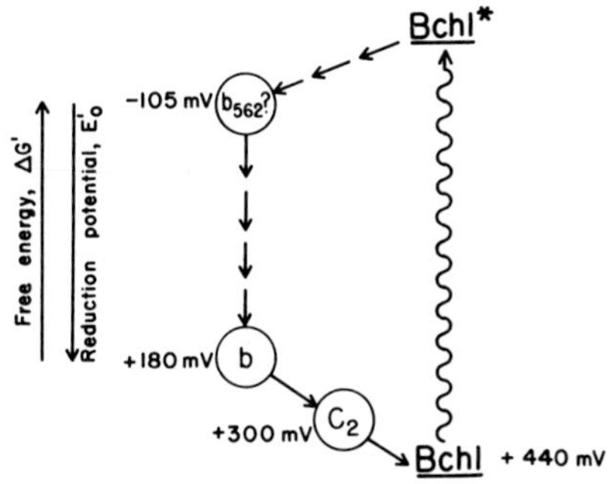

FIG. 29. Cyclic photophosphorylation in *R. rubrum*. Bchl is bacteriochlorophyll, and the vertical wavy arrow indicates excitation by light. b_{562}, b, and c_2 are cytochromes. Multiple arrows indicate the presence of several undetermined intermediate steps. Reduction potentials are given in millivolts. Adapted from reference *313*.

depleted photocenter (*313,358*). Similar cyclic pathways have been found in *Rhodopseudomonas spheroides* (*313,372*), *Rhodopseudomonas capsulata* (*372a–c*), and *Rhodopseudomonas viridis* (*373*). Two c_2-type cytochromes have been seen spectroscopically in the latter organism, c_{558} and c_{553}, and it has been suggested that one may be involved in cyclic and the other in noncyclic photosynthesis

The structure of the respiratory electron chain is very uncertain. The Krebs cycle is useful during both dark and light metabolism as a source of NADH. The respiratory chain, in contrast, is useful only during aerobic respiration in the dark and plays a minor role in the life cycle of the bacterium. In *R. rubrum,* cytochrome c_2 may have nothing to do with the main respiratory chain from substrates NADH and succinate to O_2, and this chain may consist only of *b*- and *o*-type cytochromes. In contrast, there is evidence that in *R. spheroides* (*372*) and *R. capsulata* (*372*a–c) the photosynthetic *b* and *c* are shared as common links with the respiratory chain. In view of the striking sequence and structure resemblance between these photosynthetic c_2's and the purely respiratory c_{550} from *Micrococcus denitrificans* (see Table VII and Figs. 9 and 10), this is an attractive hypothesis.

The functional role of cytochrome *c′* (including *cc′*) in these bacteria

372. J. L. Connelly, O. T. G. Jones, V. A. Saunders, and D. W. Yates, *BBA* **292,** 644–653 (1973).

372a. B. Marrs, C. L. Stahl, S. Lien, and H. Gest, *Proc. Nat. Acad. U. S.* **69,** 916–920 (1972).

372b. B. Marrs and H. Gest, *J. Bacteriol.* **114,** 1045–1051 (1973).

372c. B. Marrs and H. Gest, *J. Bacteriol.* **114,** 1052–1057 (1973).

373. R. J. Cogdell and A. R. Crofts, *FEBS* (*Fed. Eur. Biochem. Soc.*) *Lett.* **27,** 176–178 (1972).

remains obscure. Two conflicting viewpoints have crystallized, and the evidence pro and con is reviewed in references *4* and *358*. One possibility is that *c′* is the terminal cytochrome oxidase for O_2 respiration. This idea arose mainly from action spectra experiments in which O_2 respiration is poisoned by CO binding to some oxidase component as an O_2 mimic, and the CO is then driven off by intense light of variable wavelength. The effectiveness of different wavelengths of light was monitored by its ability to reinstate O_2 uptake. The efficiency spectrum of the incident light was found to match the absorption spectrum of the CO complex of *c′* reasonably well, implying that cytochrome *c′* was the unknown oxidase. Arguing against this idea are the facts that rates and binding constants of CO with membrane fragments and with isolated *c′* are different and the amount of CO bound is much less than the amount of *c′* present. The standard reduction potential for *c′* of around 0 mV also makes it hard to see how *c′* could act as the terminal oxidase. This low *E′* has been claimed to be an artifact introduced by the extraction process, but this is contradicted by the preliminary measurement *in situ* by Dutton and Jackson of —20 mV (*313*).

The second suggestion is that *c′* is involved in photosynthesis. Judged purely by its redox potential, *c′* might be expected to follow succinate in the respiratory chain or to sit somewhere prior to *b* in the cyclic photosynthetic pathway. Pigments with spectral properties similar to *c′* have been observed to become oxidized under illumination in photosynthetically functional cells or fragments. The problem then is to explain the presence of cytochrome *c′* in nonphotosynthetic bacteria such as *Azotobacter* (*326*) or *P. denitrificans* [which may actually be a strain of *Alcaligenes* (*361*)]. It may be that cytochrome *c′* is involved in some transport chain subsidiary to respiration or photosynthesis and hence useful in conjunction with either, perhaps as a bridge or shunt between pathways. It could equally well be part of a biosynthetic pathway which requires reducing equivalents. The roles of *c′* and several of the other *c*-type cytochromes listed under the Athiorhodaceae in Table XX remain to be discovered.

2. *Purple Sulfur Bacteria*

The purple sulfur bacteria are strict anaerobes and photoautotrophs, using inorganic compounds as electron donors for a noncyclic photosynthetic chain. The usual donor is sulfide

$$S^{2-} \rightarrow S^0 + 2\,e^- \rightarrow\ \rightarrow SO_4^{2-}$$

although thiosulfate, H_2, and some organic molecules such as pyruvate and dicarboxylic acids can substitute in some species. Elemental sulfur as an intermediate product can accumulate as oily inclusions within the cell. Photosynthesis provides the sole means of energy production for these autotrophs, and it is interesting to note that no *b*-type cytochromes have ever been detected.

Chromatium is the most intensively studied of the purple sulfur bacteria, and several cytochromes *c* have been found. As in the purple nonsulfur bacteria, these are associated with membrane chromatophores (*374*). Readily buffer-extractable cytochrome *cc′* and an unusual flavin-containing c_{552} were isolated and characterized quite early (*316*). A cytochrome c_{555}, extractable only with difficulty, has been known from spectral studies *in vivo* and referred to as "particle bound cytochrome," although lately Cusanovich and Bartsch (*318*) have succeeded in isolating a pigment identified with this c_{555}. Experiments on whole cells and chromatophore membranes have repeatedly implicated a c_{552} and a c_{555} in photosynthesis (*4,358*). Until 1970 it was assumed that these were the soluble flavin-c_{552} and particle-bound c_{555} because of their wavelength correspondence, and because the undissociated photosynthetic pigments could be titrated with external agents in chromatophore preparations to show redox potentials similar to those of the isolated proteins. More recent experiments have brought this identification into question (*319,375,376*). Kennel and Kamen have succeeded in solubilizing with cholate detergent a $c_{555}/c_{552.5}$ complex from light-active chromatophores, with the same redox potentials expected for the photosynthetic pigments but with amino acid compositions much different from those of the previously known, solubilized cytochromes. Unfortunately, more than one cytochrome component seems to have been hidden behind similar spectral bands and redox potentials.

Figure 30 summarizes the main proposals for cytochrome *c* involvement in *Chromatium* photosynthesis (*377–380*). Cytochromes c_{552} and c_{555} are the components of the cytochrome complex listed in Table XX. There are obvious analogies with green plant and purple nonsulfur bacterial photosynthesis (Figs. 27 and 29), but with less known about intermediate substances. It has been proposed that the high potential c_{555} participates in cyclic photosynthesis and the lower potential c_{552} is involved mainly in substrate-fed noncyclic photosynthesis, but this distinction is muddied because there is some communication between the two cytochromes

374. M. A. Cusanovich and M. D. Kamen, *BBA* **153,** 376–396 (1968).
375. G. D. Case and W. W. Parson, *BBA* **292,** 677–684 (1973).
376. S. J. Kennel and M. D. Kamen, *BBA* **234,** 458–467 (1971).
377. M. Seibert and D. DeVault, *BBA* **205,** 220–231 (1970).
378. J. M. Olson and B. Chance, *ABB* **88,** 40–53 (1960).
379. S. Morita, M. Edwards, and J. Gibson, *BBA* **109,** 45–58 (1965).
380. S. Morita, *BBA* **153,** 241–247 (1968).

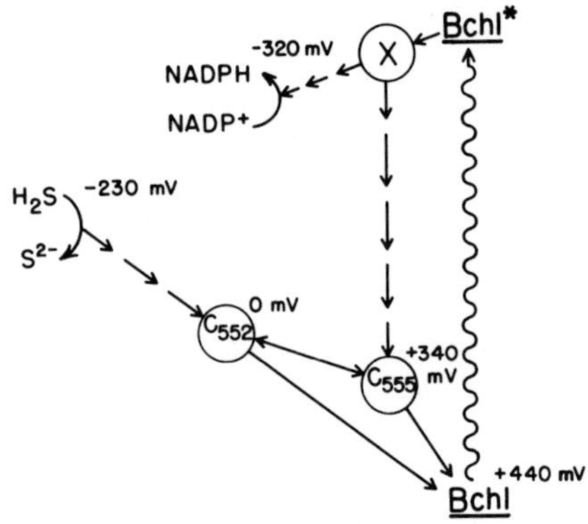

FIG. 30. Proposed scheme of cyclic and noncyclic photophosphorylation in *Chromatium*. Same diagram conventions as Fig. 29. X is an unidentified electron acceptor, probably a ferredoxin. Double-headed arrow represents free interchange of electrons between cytochromes c_{552} and c_{555}. Adapted from information in references *319, 377,* and *379*.

(*377,381*). The possibility has been raised that there are two independent photocenters for cyclic and noncyclic photosynthesis, each with its own cytochrome *c*, but the weight of evidence seems to be against this (*375,377,382*).

The most disappointing loose ends in the *Chromatium* cytochrome story are the lack of clear-cut roles for either cytochrome *cc'* or flavin-c_{552}. For the latter we can only offer the proposal of Kennel and co-workers (*381*) that flavin-c_{552} enhances the rate of reoxidation of the primary photoreductant "X" when readded to chromatophores depleted of their flavin-c_{552}, and thus may function somewhere in the chain between "X" and the c_{552}/c_{555} complex.

3. *Green Sulfur Bacteria*

The final class of photosynthetic bacteria, the "green" sulfur bacteria, are named because of differences in the chemical structure of their bacteriochlorophyll and carotenoids, giving rise to a different set of absorption bands in whole cells from their "purple" counterparts. This has little to do with their actual color but refers to the direction of shift of absorption wavelengths in the infrared. As photoautotrophic, strict anaerobes using sulfide or thiosulfate as an electron source, they resemble the purple sulfur bacteria and have similar electron transport chains. The sulfur that is formed as an intermediate between sulfide and sulfate is deposited out-

381. S. J. Kennel, R. G. Bartsch, and M. D. Kamen, *Biophys. J.* **12**, 882–896 (1972).
382. W. W. Parson and G. D. Case, *BBA* **205**, 232–245 (1970).

side the cell rather than being stored as inclusions within the cell. In both types of purple photosynthetic bacteria, photosynthesis occurs in vesicles formed by infolding of the outer bacterial membrane, sometimes in concentric layers inside the outer membrane, sometimes in stacks connected by tubules like the grana of green plant chloroplasts. The situation in green sulfur bacteria is quite different: The photosynthetic pigments are found in cigar-shaped *Chlorobium* vesicles with their own membranes, just under the outer bacterial membrane but not connected with it.

Three cytochromes *c* are found in *Chlorobium thiosulfatophilium:* a small c_{555}, and much larger c_{551} and flavin-c_{553} (*321*). A rather different set of proteins has been purified from "*Chloropseudomonas ethylicum*" *322,383*), actually another *Chlorobium* strain. Both of the nonflavin cytochromes of *C. thiosulfatophilium* have been assigned roles in photosynthesis (*384,385*), and the assumption of a photosynthetic function for c_{555} seems likely since it bears a strong resemblance to algal cytochromes c_{553} (or *f*), including a characteristic and unusual α peak (*386*).

Kusai and Yamanaka (*387*) have proposed an ingenious scheme, shown in Fig. 31, which ascribes a role to all three *Chlorobium* cytochromes. In this, flavin-c_{553} is also given an enzymic role as the reductase catalyzing the removal of electrons from sulfide and their transfer to c_{555} and from there into the photocenter. This reaction has been amply demonstrated

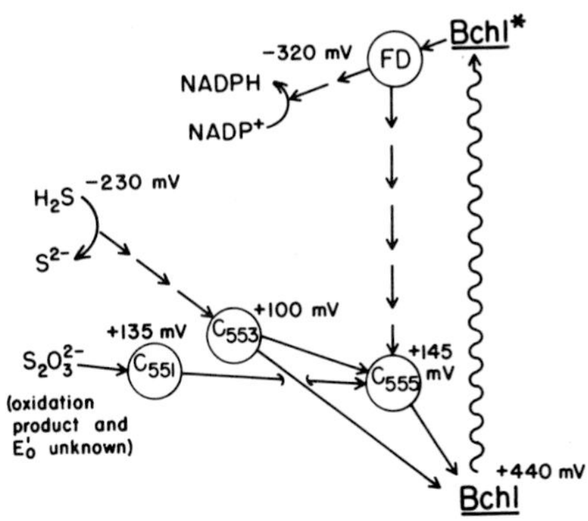

FIG. 31. Possible photosynthetic pathways in *Chlorobium,* after reference *387*. FD, ferredoxin. Transfer from c_{553} directly to bacteriochlorophyll by analogy with *Chromatium.*

383. J. M. Olson and E. K. Shaw, *Photosynthetica* **3,** 288–290 (1969).
384. C. Sybesma, *Photochem. Photobiol.* **6,** 261–267 (1967).
385. C. Sybesma and T. Beugeling, *BBA* **131,** 357–361 (1967).
386. E. Yakushiji, "Methods in Enzymology," Vol. 23, Part A, pp. 364–371, 1971.
387. A. Kusai and T. Yamanaka, *BBA* **325,** 304–314 (1973).

with the purified cytochromes (*388*), but it may be a nonphysiological, artifactual reaction between a good reducing agent, S^{2-}, and a reactive oxidized cytochrome, while some other pathway links sulfide and photosynthesis *in vivo*. If this role for flavin-c_{553} is real, then a comparable role might be considered for the similar flavin-c_{552} in *Chromatium*.

In addition to receiving electrons from other substances and passing them on to the bacteriochlorophyll photocenter, cytochrome c_{555} may have a catalytic role: assisting the transfer of electrons from thiosulfate to c_{551}. Although electrons then flow from c_{551} to c_{555} to the photocenter, the contribution is more than one of a simple electron acceptor. Only two molecules of c_{555} per 40,000 molecules of c_{551} can produce a fivefold increase in the rate of the thiosulfate to c_{551} transfer (*389*).

C. Bacterial Respiratory Chains

Respiration is the passage of reducing equivalents down a potential energy gradient, from external electron donor to a terminal acceptor; respiration is accompanied by conversion of chemical free energy into the stored chemical bond energy of ATP. Within the constraints of demanding a donor–acceptor couple whose free energy span is large enough to generate ATP, bacteria have evolved the ability to use an impressive array of redox reactions to supply their energy needs. The choice of a different chemical couple can lead to the occupying of an otherwise unfilled ecological niche and the utilization of otherwise untapped sources of free energy. Some respiratory chains in bacteria are virtually identical to those found in eukaryotes; others are radically different.

1. *Mitochondrion-Like Respiratory Chains*

Although separated from eukaryotic cells by at least one or two billion years, some bacteria retain electron transport chains almost identical to those of the mitochondria of higher organisms. *Mycobacterium phlei* and *Bacillus subtilis* are good examples (Fig. 32). Being chemoheterotrophs, they require organic compounds in reduced oxidation states as carbon and energy sources just as we do, and they fill most of their energy requirements by oxygen respiration with only an occasional recourse to simple fermentation.

Cohen *et al.* (*390*) have proposed the chain for *M. phlei* shown in Fig. 32 on the basis of spectral studies, inhibitor studies, and other characterizations. For *B. subtilis*, Miki and Okunuki (*328*) have put forth the scheme

388. A. Kusai and T. Yamanaka, *FEBS* (*Fed. Eur. Biochem. Soc.*) *Lett.* **34**, 235–237 (1973).
389. A. Kusai and T. Yamanaka, *BBRC* **51**, 107–112 (1973).
390. N. S. Cohen, E. Bogin, T. Higashi, and A. F. Brodie, *BBRC* **54**, 800–807 (1973).

in Fig. 32 in which the low molecular weight cytochrome c_{554} plays the classical role of eukaryotic c_1, although at a lower reduction potential: −80 mV instead of +220 mV. Other proposals for *Bacillus* are in agreement in all but the c_{554} function (*391*). It seems obvious from the functional role of c_{550} and its size, isoelectric point, and reduction potential that it is a close evolutionary and structural homolog of eukaryotic cytochrome *c*.

2. *Less Mitochondrial-Like Chains*

A chemoheterotrophic metabolism with respiration based on oxygen allows enough thermodynamic "elbow room" to accommodate variations on the mitochondrial type of chain. The *Halobacteria* have been examined primarily with an eye toward how their molecules can function or even survive in the high salt media (*ca.* 4 *M* in NaCl) that make up their natural habitat, but a side benefit has been the elucidation of their respiratory cytochrome chain as a complex branched pattern. Cheah (*392*) has proposed the version shown in Fig. 33 which, while superficially resembling the mitochondrial plan, has both parallel and branching paths.

Haemophilus parainfluenzae has been a favorite organism of study because of its ease of manipulation in the laboratory. It has an intricate respiration scheme which apparently involves only one cytochrome c_{552}, although a small amount of a poorly understood c_{550} is also present. Cytochrome c_{552} seems to be involved in several parallel systems of dehy-

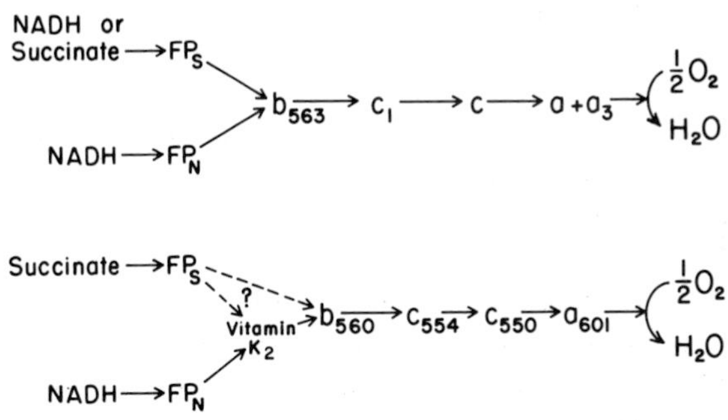

FIG. 32. Mitochondrion-like respiratory chains from *M. phlei* (top) and *B. subtilis* (bottom). FP_S and FP_N are flavoprotein dehydrogenases. Various *b*, *c*, and *a* are cytochromes. Vitamin K_2 is a quinone similar to ubiquinone or coenzyme Q in Fig. 4 and plastoquinone in Fig. 27. Such a quinone would be the expected intermediate between a flavoprotein dehydrogenase and a cytochrome *b*, even though it has not been detected in *M. phlei*. Adapted from references *328* and *390*.

391. J. A. Felix and O. G. Lundgren, *J. Bacteriol.* **115**, 552–559 (1973).
392. K. S. Cheah, *BBA* **216**, 43–53 (1970).

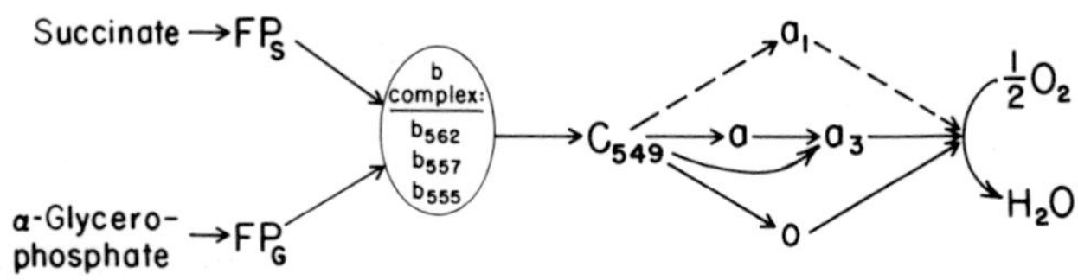

FIG. 33. Proposed electron transport system for *Halobacterium salinarium.* Conventions for component molecules as in previous figures, where o indicates cytochrome *o*. Dashed line to cytochrome a_1 represents a minor pathway. From reference *392*.

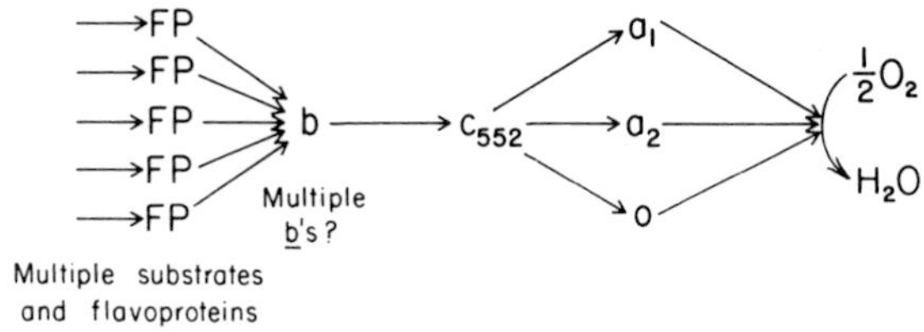

FIG. 34. Multiple and branched pathways in *Haemophilus parainfluenzae,* after reference *365*.

drogenases and oxidases, as is crudely represented in Fig. 34. As with *Halobacterium,* the *c*-type cytochrome evidently serves as a pool of reducing potential which can be diverted in the direction of several different oxidants. (We shall return to the concept of a shared pool of reducing equivalents in the sections on nitrate and sulfate respiration.)

The ecologically important nitrogen fixation by *Azotobacter* has made this a common laboratory organism whose cytochrome system has been studied in detail. Early preparations (*393*) showed two major cytochromes *c*, which unfortunately were given the nonsystematic names c_4 and c_5 (following photosynthetic c_2 and sulfate-respiratory c_3). Two more cytochromes have been found, a c_{551} and a c_{555} which may be a modified subunit of c_5 (*325*). Any proposals for electron transport in *Azotobacter* should take into account the predominance of c_4 as the most plentiful cytochrome in the organism and also the possibility that some of the *c* cytochromes could be involved in nitrogen fixation (*394*). It is doubtful that c_4 is the main respiratory cytochrome, however, since *Azotobacter* membrane fragments containing the cytochrome oxidase system will react with purified, reduced c_5 at twice the rate that they do with pure c_4 (*323*). Moreover, a c_5 shows a smaller Michaelis–Menten constant, K_m. White and Sinclair (*365*) have reviewed all the available spectral and chemical data, and proposed a scheme along the lines of Fig. 35, with possible involvement of cytochromes *c* in nitrogen fixation yet to be demonstrated.

393. A. Tissières and R. H. Burris, *BBA* **20,** 436–437 (1956).

394. I. D. Ivanov, G. I. Matkhanov, M. Y. Belov, and T. V. Gogleva, *Mikrobiologiya* **36,** 205–209 (1967).

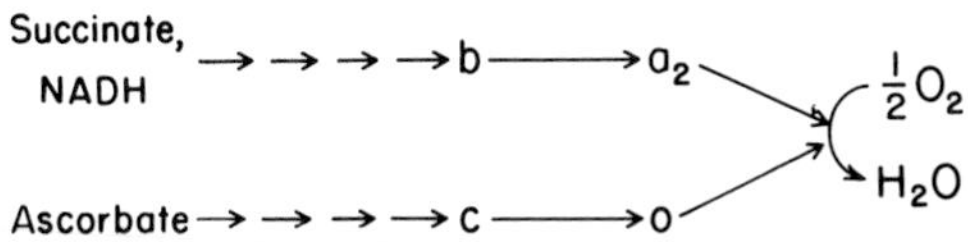

FIG. 35. Parallel chains of electron flow in *Azotobacter,* after reference *365.*

3. *Cytochrome Systems Linking Inorganic Reductants to Molecular Oxygen*

At the next level of diversification of respiratory pathways, reduced organic substrates, entering via specific dehydrogenases or through Krebs cycle conversion to succinate and NADH, are replaced by reduced inorganic molecules as a source of electrons and of useful free energy. The *Thiobacilli* use elemental sulfur, sulfite (SO_3^{2-}), thiosulfate ($S_2O_3^{2-}$), or polythionates ($^-O_3S—(S)_n—SO_3^-$) as electron sources, linking these to oxygen as a terminal electron acceptor or to nitrate in the case of *T. denitrificans.* Their purely chemoautotrophic existence raises the question of how they generate NADH as a source of reducing power for biosynthesis. One answer is the use of ATP to drive a part of the normal respiratory chain in reverse. Peeters and Aleem (*395*) have proposed for *T. denitrificans* the pathway shown in Fig. 36, where we have equated these authors' unspecified cytochrome *c* with the c_{552} reported by Milhaud *et al.* (*335*). At first glance this looks like a normal mitochondrial chain with a T junction entering at the c_{552} level. This junction is the main entry point for *T. denitrificans'* major source of nourishment—thiosulfate ion. Electrons are passed from thiosulfate to c_{552} by a thiosulfate:cytochrome *c* oxidoreductase enzyme, and from there either to a cytochrome *o* or a mitochondria-like $a + a_3$ complex, and ultimately to oxygen, with ATP being generated along the way. Alternatively, a thiosulfate reductase enzyme can dismutate thiosulfate into sulfite and sulfide (an energy-requiring process), which then can enter the electron transport chain at the flavoprotein.

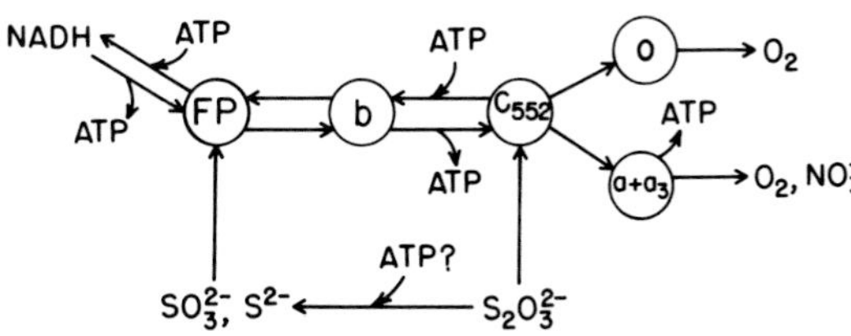

FIG. 36. Forward and reversed electron flow in *T. denitrificans,* after reference *395.* Oxidation of thiosulfate at c_{552} can either generate ATP at cytochrome $a+a_3$, or reverse the normal electron transport chain to generate NADH as a source of reducing power. See text for discussion.

395. T. Peeters and M. I. H. Aleem, *Arch. Mikrobiol.* **71,** 319–330 (1970).

Reducing power in the form of NADH is generated in *Thiobacillus* by running the electron transport chain backward between cytochrome *c* and the flavoprotein. Electrons from thiosulfate need not end on oxygen with the generation of ATP. If extra ATP is available, the chain can be operated "uphill," with electrons going from *c* to *b* to the flavoprotein, and finally being used to make NADH from NAD^+. If all the NADH is not used in synthesis, it can be poured back into the respiratory chain again and its energy recovered. The thiosulfate to sulfite shunt can provide faster or easier synthesis of NADH through the flavoprotein when needed.

Sadler and Johnson (*396*) claimed that reverse electron flow involving cytochrome *c* has not been established for *T. neapolitanus* and *T. thioparus* and proposed the parallel pathways of Figs. 37a and 37b but offered no suggestion as to the source of the NADH. Yamanaka *et al.* (*334*) have found multiple cytochromes in *T. novellus* which they place in series as in Fig. 37c. They proposed that sulfite delivers its electrons to c_{551} by means of a sulfite:cytochrome *c* oxidoreductase enzyme. In all of these *Thiobacillus* species, there is at least one cytochrome *c* with a reduction potential near +280 mV and a molecular weight around 13,000, which probably is an evolutionary homolog of eukaryotic *c*.

In all of the systems discussed so far, the donor enters the electron chain to the acceptor at a point prior to cytochrome *c;* that is, cytochrome *c* has a reduction potential between that of the donor and the acceptor, and hence is capable of acting as a valve for regulating flow.

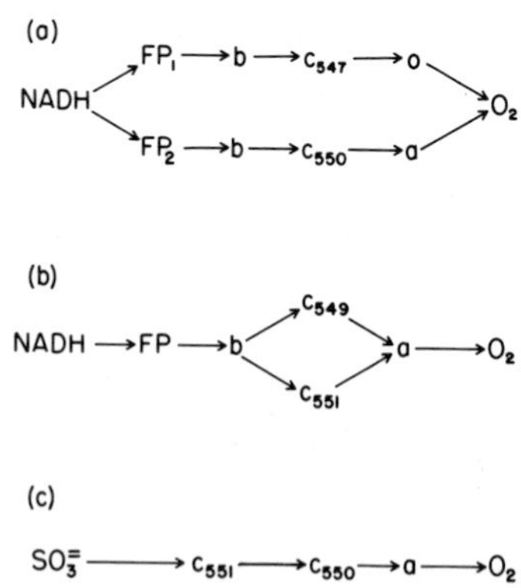

FIG. 37. Proposed respiratory chains in three *Thiobacilli:* (a) *T. neapolitanus* (after reference *396*), (b) *T. thioparus* (after reference *396*), and (c) *T. novellus* (after reference *334*).

396. M. H. Sadler and E. J. Johnson, *BBA* **283**, 167–179 (1972).

For the oxidation of NADH by O_2, the individual reduction potentials are

$$NAD^+ + H^+ + 2\ e^- \rightarrow NADH \qquad E_0' = -320\ mV$$
$$\tfrac{1}{2}\ O_2 + 2\ H^+ + 2\ e^- \rightarrow H_2O \qquad E_0' = +816\ mV$$

and cytochrome *c* clearly falls between these two potential values:

$$c^{III} + e^- \rightarrow c^{II} \qquad E_0' = +260\ mV$$

The obligately respiratory *Nitrobacter* is paradoxical in that its electron-donating reaction, the conversion of nitrite to nitrate, has a higher reduction potential than that of cytochrome *c*:

$$NO_3^- + 2\ H^+ + 2\ e^- \rightarrow NO_2^- + H_2O \qquad E_0' = +421\ mV$$

O_2 certainly can oxidize nitrate in an energy-producing reaction, but cytochrome *c* cannot lie on the electron transport chain between them. Aleem and colleagues circumvented the paradox by assuming that nitrite enters the electron pathway somewhere *after* cytochrome *c*, at a cytochrome *a*-type oxidase, with *c* being involved in reversed, ATP-dependent electron flow to generate NADH (*397–399*). This scheme would resemble Fig. 36 but with the T junction to cytochrome *a* rather than to *c*. A counterproposal is that the nitrogen in nitrite is somehow raised from an oxidation state of +3 to an unspecified reaction intermediate of state +4, perhaps in an energy-requiring process, and then this +4 to +5 (nitrate) couple is used to drive a "normal" electron transport chain involving cytochrome *c* and other components (*332,400*). As Table XX indicates, *Nitrosomonas* may circumvent the E_0' problem since it contains a *c* cytochrome of $E_0' = +500$ mV.

Other organisms oxidizing inorganic substances with O_2, such as *Hydrogenomonas* with H_2 (*401*), have been less well characterized other than to report the spectral detection of a typical complement of *a*, *b*, *c*, and *o* cytochromes.

4. *Cytochrome Systems in Denitrifying Bacteria*

The main alternative to O_2 as a terminal oxidant in respiration is nitrate, which is reduced to N_2, ammonia, or an oxide of nitrogen, depending on the organism. The cytochromes of nitrate respiration cannot be discussed independently of O_2 respiration, for as a general rule nitrate users can also employ molecular O_2 and prefer it if offered a choice. His-

397. M. I. H. Aleem, *BBA* **162,** 338–347 (1968).
398. D. L. Sewell and M. I. H. Aleem, *BBA* **172,** 467–475 (1969).
399. D. L. Sewell, M. I. H. Aleem, and D. F. Wilson, *ABB* **153,** 312–319 (1972).
400. J. C. O'Kelley, G. E. Becker, and A. Nason, *BBA* **205,** 409–425 (1970).
401. U. Bernard, I. Probst, and H. Schlegel, *Arch. Mikrobiol.* **95,** 29–37 (1974).

torically, this has fostered the idea that nitrate respiration is a recent add-on to normal O_2 respiration with only the substitution of a different oxidase or a bifunctional oxidase at the end of the chain. We shall see that this may be an oversimplification.

The role of cytochromes during O_2 respiration in the denitrifiers offers no surprises. Horio *et al.* (*402*) have offered the chain shown in Fig. 38 for *Pseudomonas aeruginosa.* It is essentially the standard mitochondria-like pathway but with quite a bit of "short-circuiting" between components. Since the scheme was deduced from cross-reactions of purified components *in vitro,* some of these minor paths probably are artifacts resulting from disruption of the system. The protein originally proposed as the O_2 "oxidase" has since been found to be the nitrite reductase instead, but a particulate or membrane-bound oxidase which plays the role shown in Fig. 38 has been detected physiologically. Vernon (*403*) has shown that cytochrome c_{550} functions like a mitochondrial *c* during O_2 respiration in *Micrococcus denitrificans,* and Scholes and Smith (*404*) have found normal roles for the other cytochrome pigments.

It is in nitrogen-based respiration that the denitrifiers exhibit novel uses of cytochromes *c.* The reduction of oxides of nitrogen can be carried out for two ends: as an assimilatory process in which the ultimate product, ammonia, is generated as a nitrogen source to fill biosynthetic needs; and as a dissimilatory process of true respiration, coupling the reduction of nitrogen compounds as terminal electron acceptors, to the storage of energy in ATP. The former is typically a flavoprotein-mediated process and does not concern us here, but the latter does involve cytochromes *c.*

In dissimilatory nitrate metabolism the first reduction product, nitrite, in turn becomes a terminal acceptor for another reduction step. So does the product of nitrite reduction, and this ladder continues until gaseous N_2 is reached. Hence, although most denitrifiers use nitrate as their natu-

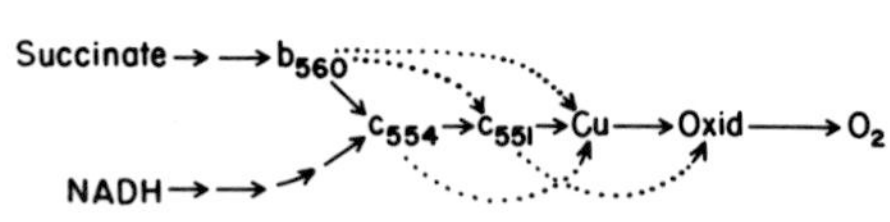

FIG. 38. Electron transport in *P. aeruginosa,* after reference *402*. Multiple arrows represent multiple steps, possibly with undetected intermediates. Abbreviations: Cu, copper-containing protein; Oxid, oxidase enzyme. Dotted arrows indicate side reactions which perhaps are artifacts of extraction.

402. T. Horio, I. Sekuzu, T. Higashi, and K. Okunuki, *in* "Haematin Enzymes" (J. E. Falk, R. Lemberg, and R. K. Morton, eds.), p. 302. Pergamon, Oxford, 1961.
403. L. P. Vernon, *JBC* **222,** 1035–1044 (1956).
404. P. B. Scholes and L. Smith, *BBA* **153,** 363–375 (1968).

ral oxidant, they can also survive on more reduced forms of nitrogen (NO_2^-, NO, and N_2O), since these are natural intermediates in the complete utilization of nitrate. The different steps in reduction of nitrate are shown with their standard free energies and reduction potentials at pH 7 in Table XXI. All of these E_0' values are more than high enough to drive a mitochondrial type electron chain beginning with NADH or succinate.

Oxidized nitrogen compounds thus are good energy surrogates for O_2. All nitrate respirers are also oxygen respirers with competition between the two oxidants. Oxygen usage takes precedence, and even low levels of O_2 will generally inhibit nitrate or nitrite reduction and cause a shift to aerobic respiration. This suggests that nitrate respiration may have evolved from O_2 respiration to provide a viable alternative, perhaps somewhat lower in efficiency, when the organism is faced with anaerobiosis (*405*).

The best understood system of dissimilatory nitrate metabolism is that of *Pseudomonas denitrificans*, with the proposed transport scheme shown in Fig. 39 (*406–408*). The nitrogen intermediates shown between nitrate and N_2 are those which have been isolated and identified with a particular reductase enzyme. There remains the possibility of additional intermediates in the form of transient oxidation states of nitrogen. The single electron chain of O_2 respiration here has been replicated into parallel pathways, each with the general form donor → *b* → *c* → reductase → acceptor. On the N_2O reductase path, the particulate cytochrome c_{552}

TABLE XXI
FREE ENERGIES AND REDUCTION POTENTIALS IN NITRATE RESPIRATION

Reaction	At pH 7	
	$\Delta G'$ (kcal/mole)	E_0' (mV)
$2\,NO_3^- + 4\,H^+ + 4\,e^- \rightarrow 2\,NO_2^- + 2\,H_2O$	−38.82	+420
$2\,NO_2^- + 4\,H^+ + 2\,e^- \rightarrow 2\,NO(g) + 2\,H_2O$	−17.24	+374
$2\,NO(g) + 2\,H^+ + 2\,e^- \rightarrow N_2O(g) + H_2O$	−54.27	+1177
$N_2O(g) + 2\,H^+ + 2\,e^- \rightarrow N_2(g) + H_2O$	−62.35	+1352
$2\,NO_3^- + 12\,H^+ + 10\,e^- \rightarrow N_2(g) + 6\,H_2O$	−172.68	+749

405. K. Knoblock, M. Ishaque, and M. I. H. Aleem, *Arch. Mikrobiol.* **76**, 114–125 (1971).
406. M. Miyata and T. Mori, *J. Biochem.* (*Tokyo*) **64,** 849–861 (1968).
407. M. Miyata, *J. Biochem.* (*Tokyo*) **70,** 205–213 (1971).
408. T. Matsubara, *J. Biochem.* (*Tokyo*) **69,** 991–1001 (1971).

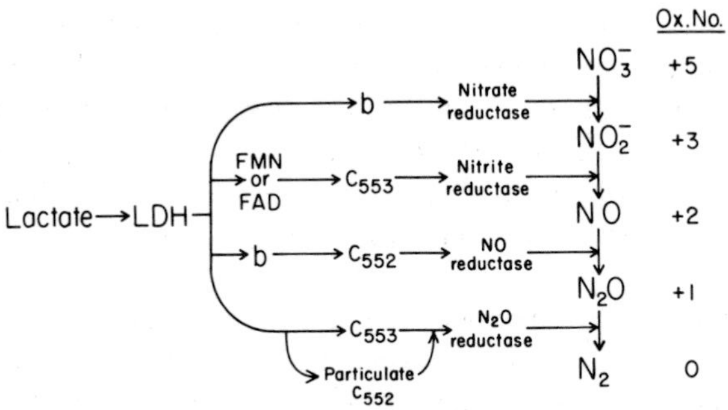

FIG. 39. Proposed parallel electron transport chains feeding the different steps of nitrogen reduction in *P. denitrificans* when it respires with nitrate. Each intermediate reaction step has its own nitrogen compound reductase (or cytochrome oxidase. After references *406–408*.

which can replace c_{553} is firmly attached to membrane particles and quite distinct from the soluble c_{552} of the NO path. A c_{552} also seems to be able to serve in place of c_{553} on the nitrite pathway. Cytochromes *b* in the nitrate and NO branches have not been shown to be definitely different. Other, as-yet undetected components, may participate in the electron transport system. The observation that a soluble c_{552} from *P. stutzeri* with similar physiochemical properties to soluble c_{552} from *P. denitrificans* can substitute for it in the chain suggests that nitrate respiration in *P. stutzeri* is similar (*406*).

Other denitrifying bacteria contain components which at least superficially fit the *P. denitrificans* pattern. Forget (*409*) has purified a nitrate reductase from *Micrococcus denitrificans* and shown it to be a soluble, nonheme iron protein with isoelectric point of 4.2 and molecular weight of around 160,000. It has no flavin, and one iron and one inorganic sulfur per 20,000 molecular weight, with traces of molybdenum. It reacts with $FMNH_2$ or $FADH_2$, but not NADH or NADPH, and can accept ClO_3^- or BrO_3^- as substrate analogs. As in *P. denitrificans,* it is fed electrons by a cytochrome *b* (*410*). A *b* cytochrome similarly reduces the nitrate reductase (and may be part of a complex with it) in *E. coli* (*411,412*). In *Aerobacter aerogenes* it has been suggested that a single *b* cytochrome acts as a branch point for electron transport with one arm leading to nitrate reductase and the other to cytochrome *c* and then to the O_2 oxidase system (*413*). The same proposal has been made for *P. aeruginosa*

409. P. Forget, *Eur. J. Biochem.* **18,** 442–450 (1971).
410. Y. Lam and D. J. D. Nicholas, *BBA* **172,** 450–461 (1969).
411. E. R. Kashket and A. F. Brodie, *JBC* **238,** 2564–2570 (1963).
412. E. Itagaki, *J. Biochem.* (*Tokyo*) **55,** 432–445 (1964).
413. D. L. Knook, J. Van't Riet, and R. J. Planta, *BBA* **292,** 237–245 (1973).

with b_{560} as the branch (*414*). The idea of a common *b* cytochrome for several paths is debatable since identification has come only from difference spectra. New techniques have sometimes shown that presumed single proteins previously identified by difference spectra in whole cells or complex extracts may actually be families of proteins (*415*).

The best characterized terminal reductase enzymes in nitrogen respiration have been the nitrite reductases, and these are found to contain a *c*-type heme. When the nitrite reductase of *P. aeruginosa* was isolated as a soluble protein, it originally was called a cytochrome oxidase since it catalyzed the transfer of electrons from the true respiratory c_{551} to O_2 and was susceptible to typical oxidase inhibitors (*416*). Later work has shown that the most probable function for this enzyme *in vivo* is nitrite reduction, with electrons coming from c_{551} (*417*). The enzyme is present only when the organism is cultured in an anaerobic nitrate medium. It accounts quantitatively for the nitrite reducing ability of whole cells. Aerobic cells contain none of this enzyme, although they must have another functional oxidase which has yet to be characterized. The crystalline nitrite reductase enzyme contains two hemes—a covalently bound *c* and an extractable *d* (formerly called a_2). It reacts rapidly and specifically with cytochrome c_{551}. A nitrite reductase from *M. denitrificans* has similar properties and reacts specifically toward respiratory cytochrome c_{550} (*418,419*). It differs from the *P. aeruginosa* enzyme mainly in reacting less well with oxygen. A soluble nitrite reductase from *P. perfectomarius* has been described as containing a *c*-type cytochrome with a split α peak, perhaps associated with a *d* heme as well (*420*). Cytochrome c_{552} from *E. coli* has been implicated as being involved with a nitrite reductase (*4*), as has c_{550} from *Spirillum itersonii* (*331,421*).

All of the nitrite reductases just described convert NO_2^- to NO in a dissimilatory manner for energy extraction; other nitrite reductases in assimilatory pathways convert NO_2^- directly to NH_4^+ for synthetic purposes. Most of these are nonheme flavoproteins, but one has been found in *Achromobacter fischeri* (*422*) with properties similar to the dissimilatory reductases: two hemes *c* per molecule of weight 95,000. The analogy

414. T. Yamanaka, A. Ota, and K. Okunuki, *BBA* **53,** 294–308 (1961).
415. W. S. Shipp, *ABB* **150,** 482–488 (1972).
416. T. Horio, T. Higashi, T. Yamanaka, H. Matsubara, and K. Okunuki, *JBC* **236,** 944–951 (1961).
417. T. Yamanaka, S. Kijimoto, and K. Okunuki, *J. Biochem.* (*Tokyo*) **53,** 256–259 (1963).
418. N. Newton, *BBA* **185,** 316–331 (1969).
419. Y. Lam and D. J. D. Nicholas, *BBA* **180,** 459–472 (1969).
420. C. D. Cox and W. J. Payne, *Can. J. Microbiol.* **19,** 861–872 (1973).
421. D. K. Gauthier, G. D. Clark-Walker, W. T. Garrand, and J. Lascelles, *J. Bacteriol.* **102,** 797–803 (1970).
422. O. Prakash and J. C. Sadana, *ABB* **148,** 614–632 (1972).

ends here since the electron donor is NADH rather than another *c* cytochrome, but it is interesting that a dissimilatory type enzyme could produce assimilatory products.

The emerging picture of dissimilatory nitrite reductases as diheme proteins is upset by the *P. denitrificans* enzyme: a nonheme, copper-containing protein of molecular weight 150,000 (*423*). It has been suggested that the strain *P. denitrificans* is more appropriately classified with the *Alcaligenes* (see work cited in reference *361*), and, if so, we would be less surprised at finding a difference between its reductase and those of the other *Pseudomonas* strains.

Much less is known about the NO and N_2O reductases. The former is a soluble enzyme with a *c*-type heme, while the latter appears to be membrane-bound (*420*).

5. *Cytochromes in Sulfate-Respiring Bacteria*

Oxygen and nitrate are strong oxidizing agents, and it is not surprising that various organisms have tapped them as means of extracting energy from other molecules. Sulfate, in contrast, is a poor oxidant. This can be illustrated by the standard free energies liberated when one mole of H_2 gas is oxidized by O_2, nitric acid, and sulfuric acid:

$$H_2 + \tfrac{1}{2}\,O_2 \rightarrow H_2O \qquad \Delta G^\circ = -56.7 \text{ kcal/mole}$$
$$H_2 + \tfrac{1}{4}\,HNO_3 \rightarrow \tfrac{1}{4}\,NH_3 + \tfrac{3}{4}\,H_2O \qquad \Delta G^\circ = -37.5 \text{ kcal/mole}$$
$$H_2 + \tfrac{1}{4}\,H_2SO_4 \rightarrow \tfrac{1}{4}\,H_2S + H_2O \qquad \Delta G^\circ = +14.0 \text{ kcal/mole}$$

A smaller amount of free energy is released from oxidation by nitrate than by O_2, and under standard thermodynamic conditions, sulfate oxidation of H_2 requires an *uptake* of free energy. Although living organisms do not operate under standard conditions (1 *M* concentration, pH 0), sulfate respiration always will be less energy producing than nitrate or O_2.

Sulfate respiration has a second disadvantage. Only one pair of electrons is required to reduce an oxygen atom of O_2 to water, and the electron transfer chain can be correspondingly simple: a linear chain from donor to acceptor. Four electron pairs are required for each nitrogen atom when reducing nitrate to ammonia. The respiratory machinery is correspondingly more elaborate, as indicated in Fig. 39, with several steps and reaction intermediates. Four electron pairs are also needed per sulfur atom to take sulfate down to the oxidation level of sulfide; thus, we should not be surprised to see a complex electron transport system in sulfate-respiring organisms. Both nitrate and sulfate respiration share the disadvantage of multiple electron pair transfers and reaction interme-

423. H. Iwasaki, S. Shidara, H. Suzuki, and T. Mori, *J. Biochem.* (*Tokyo*) **53**, 299–303 (1963).

diates, and sulfate has the additional handicap of yielding relatively little energy.

Besides oxygen and nitrate, sulfate is the only other compound that is used as a terminal oxidant in association with cytochrome *c*–containing electron transport chains. It is a common substance, and the end product of *Thiobacillus* respiration; thus, its use as an oxidant by some organism is hardly surprising. The bacteria that respire with sulfate are divided into two groups: the nonsporulating *Desulfovibrio* and the spore-forming *Desulfotomaculum,* a relatively new classification whose members at one time were classed with the totally fermentative *Clostridia.* Since *Desulfotomacula* apparently do not have cytochromes *c,* and depend on cytochromes *b* and ferredoxins for their electron transfer functions, they will not be discussed here. However, one should keep in mind the interesting possibility that *Desulfotomacula* are an evolutionary bridge between the ferredoxin-containing but nonrespiring *Clostridia* and the respiring and cytochrome *c*–containing *Desulfovibrio.*

The *Desulfovibrio* are strict, absolute anaerobes which require sulfate or thiosulfate as terminal acceptors for respiration. Reducing equivalents come from organic molecules such as lactate or from externally supplied hydrogen gas. If deprived of sulfate, the organisms can grow on pyruvate and derive all of their energy needs from substrate level phosphorylation, although growth then is slow:

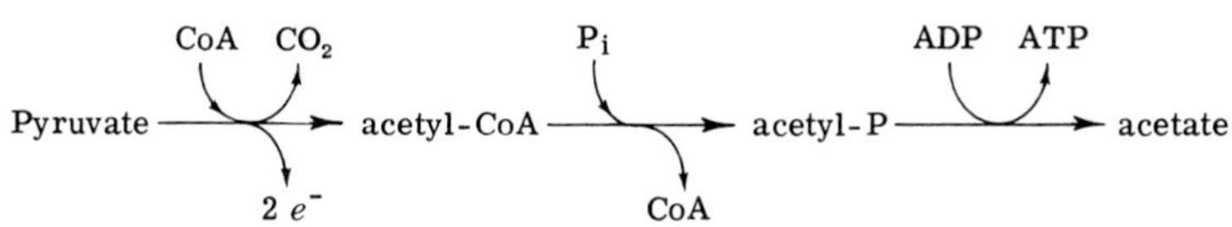

The electrons then are used by a hydrogenase enzyme to produce H_2 gas. *Desulfovibrio* can also convert succinate, fumarate, malate, oxaloacetate, or lactate into pyruvate by reactions which resemble fragments of the Krebs cycle (*360*).

The presence of *c*-type cytochromes in these organisms came as a surprise in 1954. The discovery by Postgate (*424*) in England and Ishimoto and colleagues in Japan (*425*) of a functional cytochrome c_3 in *Desulfovibrio* helped to do away with the idea that cytochromes *c* were confined to oxygen respiration. At the time, c_3 (Table XX) was the only cytochrome that could be detected in *Desulfovibrio,* and this fostered models of short electron transport chains or at least radical alternatives to the standard flavoprotein $\rightarrow b \rightarrow c \rightarrow a$ pattern. Since then, other *c, b,* and

424. J. R. Postgate, *BJ* **56,** xi (1954).
425. M. Ishimoto, J. Koyama, and Y. Nagai, *J. Biochem.* (*Tokyo*) **41,** 763–770 (1954).

d cytochromes have been found in *Desulfovibrio,* but c_3 remains the major cytochrome component.

The involvement of cytochrome c_3 at some level in electron transport was demonstrated by Ishimoto and Yagi (*426*), confirming an earlier proposal by Postgate (*427*). Crude extracts from whole cells, prepared by centrifugation after sonic disruption, were passed through an ion exchange resin to remove all cytochrome. No hydrogen was absorbed in the presence of added sulfite until pure c_3 was supplied, after which sulfite was reduced to sulfide with the expected uptake of three moles of H_2 per mole of sulfite.

Progress from this experiment involving a heterogeneous mixture of proteins to some level of detailed pure component chemistry has been difficult, and no definite answers about the electron transfer network in *Desulfovibrio* are available yet. LeGall and Postgate (*363*) have underscored two problems: The different electron carrier systems may be localized into complexes on specialized membrane areas which are disrupted during cell fragmentation, causing a scrambling of components which normally never would encounter one another; and *in vitro* demonstrations of cytochrome activity, usually consisting at best only of stimulation of electron transport, may be caused by nonphysiological side reactions such as the scavenging of inhibitory substances (O_2 and others) from the system. Another complication is the large number of sulfur oxide intermediates which are formed during sulfate respiration. Each intermediate may require its own transport chain to supply it with reducing equivalents in the manner of nitrate respiration in Fig. 39.

The probable intermediates in the reduction of sulfate to sulfide (*363,428*) are shown in Fig. 40. The structure and even the composition of the four electron transport chains implied by Fig. 40 are almost totally unknown. They presumably diverge from a common pool of reducing power at some point, but whether this is at the initial dehydrogenases, or perhaps at the level of c_3, has yet to be discovered. The standard free energies and reduction potentials under physiological conditions (pH 7) for each of these steps are given in Table XXII. The electron donor—the pyruvate to acetate reaction—is at a low enough reduction potential to make electron transfer possible in each of these steps, but in all except the trithionate reaction the standard potential of the electron acceptor reaction is lower than that of cytochrome c_3. If c_3 is used as an intermediate in the electron transfer chains leading to any of these three reactions,

426. M. Ishimoto and T. Yagi, *J. Biochem.* (*Tokyo*) **49,** 103–109 (1961).
427. J. R. Postgate, *J. Gen. Microbiol.* **15,** 186–193 (1956).
428. G. C. Wagner, R. J. Kassner, and M. D. Kamen, *Proc. Nat. Acad. Sci. U. S.* **71,** 253–256 (1974).

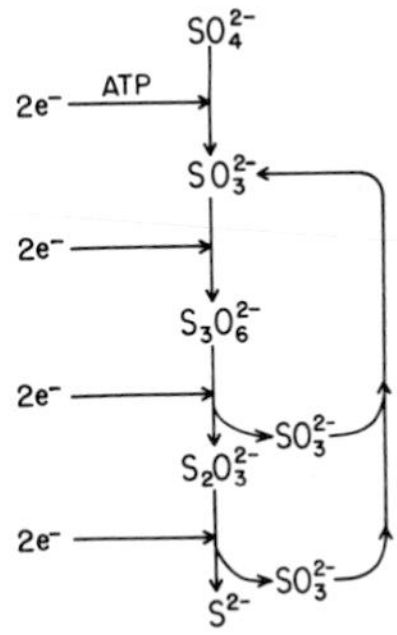

FIG. 40. Probable intermediate steps and dismutation reactions in the reduction of sulfur compounds during sulfate respiration, after references *363* and *428*. The first reduction step, sulfate to sulfite, also requires ATP.

then the majority of the c_3 pool must be present in the reduced form so that its operating reduction potential is depressed below the E_0' of —205 mV by an amount calculable from the Nernst equation.

The first reaction in Fig. 40, reduction of sulfate to sulfite, is the only one about which much is known. This is an energy-priming step in which sulfate combines with ATP to form adenosine 5-phosphosulfate (APS) and pyrophosphate (PP_i). The pyrophosphate breaks down with a re-

TABLE XXII

FREE ENERGIES AND REDUCTION POTENTIALS IN SULFATE RESPIRATION

Reaction	$\Delta G'$ (kcal/mole)	E_0' (mV)
Individual steps[a]		
$SO_4^{2-} + 2\ H^+ + 2\ e^- \rightarrow SO_3^{2-} + H_2O$	+20.39	−442
$3\ SO_3^{2-} + 6\ H^+ + 2\ e^- \rightarrow S_3O_6^{2-} + 3\ H_2O$	+16.37	−355
$S_3O_6^{2-} + 2\ e^- \rightarrow S_2O_3^{2-} + SO_3^{2-}$	−17.57	+381
$S_2O_3^{2-} + 2\ e^- \rightarrow S^{2-} + SO_3^{2-}$	+17.99	−390
Overall reactions[a]		
$2\ SO_3^{2-} + 6\ H^+ + 4\ e^- \rightarrow S_2O_3^{2-} + 3\ H_2O$	−1.20	+13
$SO_3^{2-} + 6\ H^+ + 6\ e^- \rightarrow S^{2-} + 3\ H_2O$	+16.60	−120
$SO_4^{2-} + 8\ H^+ + 8\ e^- \rightarrow S^{2-} + 4\ H_2O$	+37.08	−201
Auxiliary reactions		
Acetate + CO_2 + 2 H^+ + 2 e^- → pyruvate + H_2O	+32.24	−699
Pyruvate + 2 H^+ + 2 e^- → lactate	+8.76	−190
Acetate + CO_2 + 4 H^+ + 4 e^- → lactate + H_2O	+41.00	−444
$2\ H^+ + 2\ e^- \rightarrow H_2$	+19.10	−414
Ferredoxin reduction–oxidation		−310
Cytochrome c_3 reduction–oxidation		−205
Vitamin K_2 (menaquinone) reduction–oxidation[a]		−67

[a] Data from reference *428*.

lease of free energy which pulls the overall reaction toward completion, and APS is reduced to AMP and sulfite:

Reaction	Enzyme
$SO_4^{2-} + ATP \rightarrow APS + PP_i$	ATP-sulfurylase
$PP_i + H_2O \rightarrow 2\ P_i$	Pyrophosphatase
$APS + 2\ e^- \rightarrow AMP + SO_3^{2-}$	APS-reductase

All three enzymes have been isolated and purified, with the APS reductase being especially interesting as a catalyst for a reaction that is not known elsewhere. The APS reductase has a molecular weight of 220,000, 6–8 moles of nonheme iron, and 1 mole of FAD, to which the sulfite product remains covalently attached. The flavin-bound sulfite hence is energetically primed for subsequent reactions in a way that a lone sulfite ion would not be. The physiological electron donor for the third reaction is not known. *In vitro* studies (*363*) use methyl viologen for driving the forward reaction or ferricyanide and AMP for the reverse.

In this initial reduction of sulfate to sulfite by one pair of electrons, the equivalent of two moles of ATP is "invested" for the sake of subsequent energy extraction. (The second ATP is required to bring AMP back up to the level of ADP.) Since the sulfite ion then can use up three more electron pairs on its way to sulfide, this priming of sulfate costs 2 ATP per sulfate reduced to S^{2-}, or 0.5 ATP per electron pair.

An important question is whether *Desulfovibrio* produces any ATP by oxidative phosphorylation in its electron transport chain while SO_3^{2-} is reduced to S^{2-}, or whether its sole supply of ATP is substrate-level phosphorylation between pyruvate and acetate. The short potential span in sulfate respiration as compared with nitrate or O_2 respiration raises the possibility that the electron transport chain may be nothing more than an electron "sink," eliminating reducing equivalents so more substrate-level phosphorylation can take place but contributing no ATP itself. This would be a viable hypothesis if pyruvate were the only nutrient of *Desulfovibrio,* for then the four pyruvates needed to yield four electron pairs to reduce sulfate to sulfide would also yield four ATP, only two of which would be used in priming sulfate. However, lactate is the most common energy source for *Desulfovibrio,* and here the energy arithmetic breaks down. The reduction of two lactate to two pyruvate, and two pyruvates to two acetates, produces eight electron pairs and two ATP by substrate-level phosphorylation, exactly balancing the amounts required for reduction of one sulfate. If *Desulfovibrio* is to obtain any benefit from lactate, then there must be other sites of ATP synthesis within the electron transport chain in spite of the short potential span. The overall free energy is there. Table XXII shows that oxidizing two lactates to acetate produces 82 kcal of free energy, while reducing one sulfate to sulfide requires 37 kcal. The 45-kcal of free energy span would be enough to synthesize as many as three ATP and still have a 50% efficiency of energy conversion, which is not out of line with that in higher organisms. Additional support for the existence of respiratory phosphorylation comes from the observation that *Desulfovibrio* can survive solely on hydrogen as a source of reducing power, oxidizing it with sulfate (*363*). This implies that more than two ATP are synthesized at the respiratory chain level per sulfate reduced, since two are needed for priming.

Where does cytochrome c_3 fit into the electron transport picture? The first identification of molecular components interacting with cytochrome c_3 came from the work of Akagi (*429*). *Desulfovibrio* possesses a hydrogenase enzyme which can transfer electrons to hydrogen ions and liberate H_2 gas if no better acceptor is at hand. Using a crude extract from *D. desulfuricans*, Akagi found the transport chain shown in Fig. 41. The hydrogenase and pyruvate dehydrogenase enzymes were not isolated but contained in a crude extract which was depleted of ferredoxin and c_3. Reduction of c_3 could not be linked to pyruvate oxidation in the absence of ferredoxin, implying the order of components shown. Although ferredoxin could donate electrons directly to the hydrogenase, the rate was much greater when c_3 was present also.

A troublesome aspect of Fig. 41 is the low standard reduction potential of the hydrogen couple at the right. By the criterion of *standard* potentials, c_3 should not be able to donate electrons to H^+. But these standard potentials assume equal amounts of reduced and oxidized c_3, at pH 7, and 1 atm partial pressure of H_2, all of which may be unrealistic. If the ferredoxin and c_3 were predominantly in the reduced state, and if the partial pressure of H_2 was low, then the ferredoxin and c_3 potentials would be shifted down, and the hydrogen potential would be raised. The issue is not whether this electron transport chain works: it does. The problem is only to reconcile this with the standard reduction potentials by an appropriate adjustment of concentrations.

The other side of the coin is the use of this electron transport chain in reverse when *Desulfovibrio* grows autotrophically on H_2 gas, passing

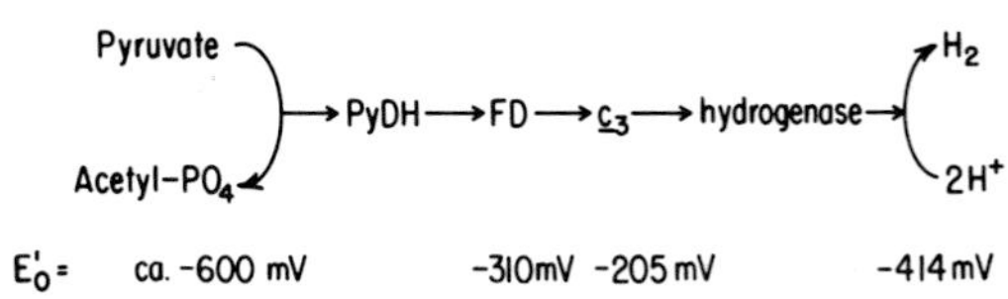

FIG. 41. Experimentally studied electron transport chain from *D. desulfuricans* (*429*). Abbreviations: PyDH, pyruvate dehydrogenase; FD, ferredoxin. Standard reduction potentials at pH 7 are given below, where known.

429. J. M. Akagi, *JBC* **242**, 2478–2483 (1967).

the reducing equivalents to oxidized sulfur compounds, as in the chain in Fig. 42 (*363,429,430*). With a higher concentration of H_2 gas, the hydrogen couple can have a lower reduction potential and can act as a donor of reducing equivalents. Suh and Akagi (*430*) have studied a purified and reconstructed chain in *D. vulgaris,* in which c_3 participates in the passage of electrons from H_2 to a sulfite reductase. Cytochrome c_3 therefore could be a central member of an electron transport chain with a short potential span, capable of running either forward or backward as circumstances and concentrations demanded. c_3 has also been proposed as a carrier of electrons to a formic hydrogenase (*431,432*).

A notable feature of the reduction potentials for the four individual reaction steps in Table XXII is that three of the potentials are as low as that of the hydrogen couple, and the fourth is almost as high as in O_2 or nitrate respiration. A large part of the total free energy of the sulfate-to-sulfide reaction seems to be concentrated in the trithionate ion so it can be released in one step. It is possible to speculate that the three low potential steps—reduction of sulfate, sulfite, and thiosulfate—are only "maintenance level" operations, receiving their electrons from a short chain similar to Fig. 41 which produces no ATP. A longer electron transport chain could operate between approximately —450 and +381 mV to carry electrons to trithionate, and several ATP could be generated within this potential gap.

These ideas are brought together in a hypothetical energy flow diagram for *Desulfovibrio* in Fig. 43. This is an elaboration of a scheme proposed by Vosjan (*433*), with the addition of a common redox pool and two electron transport systems (ETS) feeding reactions of different E_0'. At the left, lactate is oxidized to pyruvate and acetate, with the production of four electron pairs per two lactates, and just enough ATP to prime one sulfate to sulfite. At the right, one sulfate is reduced in a four-step manner, each step requiring one electron pair. From a common pool of reducing equivalents, electrons are passed to two different transport systems. A chain or collection of chains that we have called ETS I operates

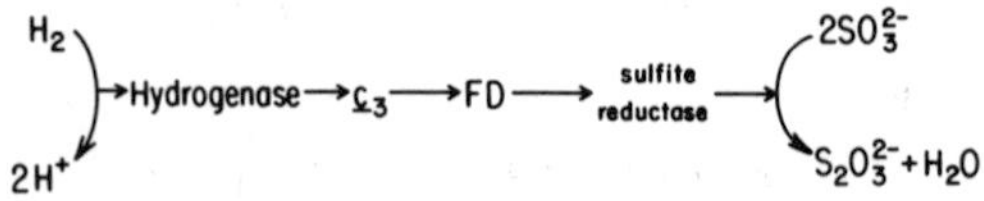

FIG. 42. Reversed flow through the chain of Fig. 41 with a different reaction and enzyme coupled to the ferredoxin (*429, 430*).

430. B. J. Suh and J. M. Akagi, *J. Bacteriol.* **99**, 210–215 (1969).
431. M. Ishimoto, T. Yagi, and M. Shiraki, *J. Biochem.* (*Tokyo*) **44**, 707–714 (1957).
432. J. P. Williams, J. T. Davidson, and H. D. Peck, *Bacteriol. Proc.* pp. 110–111 (1964).
433. J. H. Vosjan, *Antonie van Leeuwenhoek; J. Microbiol. Serol.* **36**, 584–586 (1970).

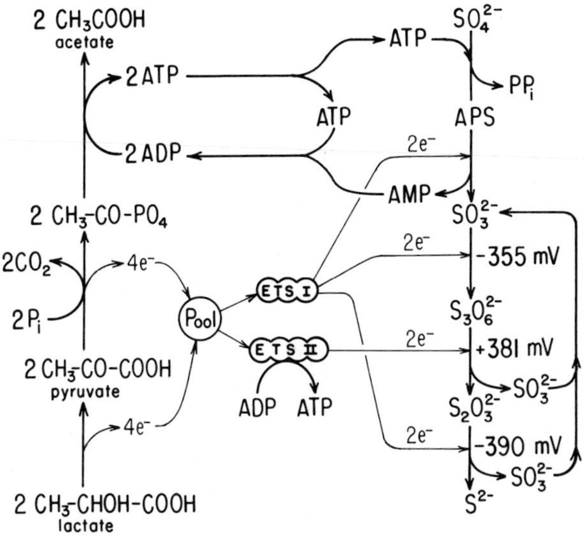

FIG. 43. Hypothetical energy flow diagram for sulfate respiration in *Desulfovibrio*. Left: Oxidation of lactate metabolite to pyruvate and acetate, with generation of reducing electrons, and of ATP by substrate-level phosphorylation. Right: Stepwise reduction of sulfur compounds as seen earlier in Fig. 40. Top center: Stoichiometric balance between ATP generated by substrate-level phosphorylation and ATP used to reduce sulfate to sulfite. Bottom center: Hypothetical electron transfer systems (ETS) drawing on a common pool of reducing equivalents. ETS I operates between the pool and acceptors which have strong negative reduction potentials and generates no ATP. ETS II functions over a larger potential span, and in principle would be capable of generating useful ATP. Freely adapted and extended from ref. *433*.

over a short potential range, supplies electrons to the three reaction steps having negative standard reduction potentials (Table XXII), and generates no ATP. Our postulated ETS II operates between −450 and +381 mV, sends electrons to a trithionate ion, and may be the seat of oxidative phosphorylation. Since, in lactate metabolism, substrate phosphorylation and sulfate priming exactly balance, any useful ATP for *Desulfovibrio* must come from ETS II. Vosjan has measured the overall production of ATP by a growth yield method (*433*) and reported no more than 0.5 net ATP per electron pair, or 2 net ATP per sulfate reduced to S^{2-}. This would imply that every pair of electrons sent down ETS II led to two ATP molecules. However, if *Desulfovibrio* is to grow successfully with only H_2 as a donor, then more than two ATP must be produced per electron pair in ETS II since two ATP are required for sulfate priming.

It must be stressed that our separate ETS I and ETS II are only hypothetical. Electron transport system I could resemble the transport system found by Akagi (*429,430*), and could therefore contain cytochrome c_3.

Cytochrome c_{553} (Table XX) and the *b* and *d* cytochromes reported in *Desulfovibrio* could be components of an energy-storing ETS II. If ETS II is a mitochondria-like chain, it might provide the role for the plentiful menaquinone molecule, which resembles the plastoquinones of photosynthesis or ubiquinone of mitochondrial respiration. Wagner *et al.* (*428*) have suggested this role for menaquinone and pointed out that its reduction potential is compatible with this idea.

The thought that *c*-type cytochromes might serve as a pool of electron donor–acceptor capacity has been advanced before, to explain a multiplicity of oxidases feeding off the same *c* cytochrome (*434*), or bifunctional donor abilities to cytochrome oxidase and to cytochrome *c* peroxidase in yeast (*435*). If the main evolutionary pressure on c_3 were the development and retention of an electron storage capacity, then this might account for the multiplicity of hemes per molecule. The four hemes, which are known to act more or less independently (*363*), then would represent an economical way of building four boxes for electrons in only 13,000 molecular weight of protein. This minimal amount of protein would help solubilize the heme groups, provide specificity in interactions with other macromolecules and membranes, and perhaps help adjust the reduction potential of the heme irons to an appropriate value.

The other cytochrome *c* components of *Desulfovibrio*, c_{553} and cc_3, are even less well understood than c_3. Cytochrome c_{553} has been reported as the natural electron acceptor for a formate dehydrogenase (*436*), and cc_3 has recently been suggested as a transport component in the reduction of thiosulfate in *D. gigas* (*437*). The respiratory metabolism of the sulfate-reducing bacteria is still largely uncharted territory, and the crude map of Fig. 43 still must have regions of imaginary coastline, and blank space with the warning, "Here be monsters."

D. Sequence and Structure

A detailed discussion of the state of knowledge of amino acid sequences of bacterial *c*-type cytochromes has been rendered superfluous by an excellent recent review by Ambler (*362*), from whose laboratory many of these sequences have come. The sequences in that review, identified for reference in Table XXIII, are all that are presently known besides the c_2 and c_{550} sequences of Table VII. With the availability of primary

434. D. C. White, *J. Bacteriol.* **83**, 851–859 (1962).
435. B. Chance, M. Erecinska, D. F. Wilson, P. L. Dutton, and C.-P. Lee, *in* "Structure and Function of Oxidation-Reduction Enzymes" (A. Akesson and A. Ehrenberg, eds.), pp. 263–272. Pergamon, Oxford, 1970).
436. T. Yagi, *J. Biochem.* (*Tokyo*) **66**, 473–478 (1969).
437. E. C. Hatchikian, J. LeGall, M. Bruschi, and M. Dubourdieu, *BBA* **258**, 701–708 (1972).

TABLE XXIII
AMINO ACID SEQUENCE INFORMATION ON BACTERIAL CYTOCHROMES[a]

c_{555}	*Chlorobium thiosulfatophilum* (N-terminus, heme peptide)
	"*Chloropseudomonas ethylica*" (N-terminus, heme peptide)
c_{551}	*Azotobacter vinelandii*
	Pseudomonas aeruginosa
	P. stutzeri
	P. mendocina
	P. fluorescens
	P. denitrificans
c_4	*A. vinelandii* (N-terminus)
	P. aeruginosa (N-terminus)
	P. stutzeri (N-terminus)
	P. mendocina (N-terminus)
c_5	*A. vinelandii* (Heme peptide)
	P. aeruginosa (Heme peptide)
	P. stutzeri (Heme peptide)
	P. mendocina (Entire chain)
	P. fluorescens (Heme peptide)
	P. denitrificans (Heme peptide)
c'	*Alcaligenes*
	Chromatium vinosum (Heme peptide)
c_3	*Desulfovibrio gigas*
	D. vulgaris
	D. desulfuricans
	D. salexigens
"c_7"	"*C. ethylica*"
c_{553}	*D. vulgaris*[b]

[a] From Ambler (*362*).
[b] Reference only. See (*439*).

structure information, the bacterial cytochromes *c* begin to fall into a small number of clear-cut patterns sketched in Fig. 44. Eukaryotic *c* and bacterial *c* with α-band absorption maxima around 550 nm, photosynthetic c_2 and algal *f* or c_{553}, c_5, and the smaller *Pseudomonas* c_{551} and *Desulfovibrio* c_{553} all conform to one basic pattern: a heme attachment by means of the sequence C·x·y·C·H (x and y are various side chains) near the amino end of the chain, and a methionine as a probable heme ligand farther toward the carboxyl end. Three subclasses are distinguished in Fig. 44 according to the number of prolines found near the methionine. These prolines probably are required in individual cases to give the peptide chain the proper geometry to bring the methionine back to the heme as a ligand and to surround that side of the heme with acceptably nonpolar side chains. All of these cytochromes are quite likely to be evolutionary homologs, sharing a common ancestral gene.

Cytochrome c_4 differs in having a second heme attachment farther down the chain, although the initial chain through the first heme site

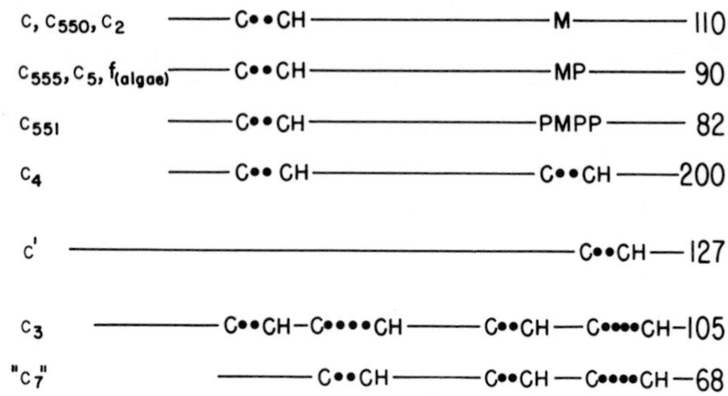

FIG. 44. Schematic representation of amino acid chains of various types of eukaryotic and prokaryotic cytochromes *c*. Cytochromes are listed at the left beside the amino terminus of each chain. Figures at the right, by the carboxyl terminus, are rough numbers of amino acids per chain. C, H, M, and P are cysteine, histidine, methionine, and proline, and large dots represent unspecified amino acid residues between cysteines. The chain lengths as represented by horizontal lines are only approximate. Adapted from reference *362*.

is quite like that of the previous cytochromes. The first 20 or so residues from all these proteins are compared in Table XXIV. Enough common features are visible to encourage ideas of evolutionary relatedness: frequent glycines at position 2 and almost invariant glycines at 6, an aromatic ring at position 10, lysines at 12 or 13, and a general consistency of hydrophobic vs. hydrophilic residues. Ambler observed that some regions of two distantly related cytochromes are similar while other regions in the same two proteins are radically different, and this must be so for c_4 since other cytochromes have no second heme site. Homology hunting must always be carried out with care and skepticism since it is easy to become trapped into seeing false homologies, as has been pointed out in ref. *438*. Nevertheless, the cytochromes that have just been mentioned, with the possible exception of c_4, may have a common genetic ancestry and may show at least traces of the eukaryotic cytochrome folding pattern. In spite of its very low reduction potential, c_{553} from *Desulfovibrio* may still be a member of this class since its sequence resembles that of *Pseudomonas* c_{551} to some extent (*439*) and falls into the general pattern of Fig. 44.

The remaining cytochromes c', c_3, and flavin-*c*, each appear to be distinct from the "eukaryotic" class just discussed. In c' the heme is attached near the carboxyl end of the chain, although the attachment sequence C·x·y·C·H is the same. Two methionines are found earlier in

438. R. E. Dickerson, *JMB* **57**, 1–15 (1971).
439. M. Bruschi and J. LeGall, *BBA* **271**, 48–60 (1972).

TABLE XXIV
N-TERMINAL REGIONS OF VARIOUS *c* CYTOCHROMES[a]

	Tuna numbering:			1				5						10						15				19	
	Invariant in eukaryotes:			*					*					*					(*)		*	*	*		Ref.
c	Tuna			G	D	V	A	K	G	—	K	K	T	F	V	Q	K	—	C	A	Q	C	H	T	*58*
c_2	*R. capsulata*			G	D	A	A	K	G	—	E	K	E	F	—	N	K	—	C	K	T	C	H	S	
	R. rubrum		E	G	D	A	A	A	G	—	E	K	V	S	—	K	K	—	C	L	A	C	H	T	*32*
	R. palustris			E	D	A	A	K	G	—	E	A	V	F	—	K	Q	—	C	M	T	C	H	—	
	R. spheroides	E	E	G	D	P	E	A	G	—	A	K	A	F	—	N	Q	—	C	Q	T	C	H	V	
c_{550}	*M. denitrificans*	N	E	G	D	A	A	K	G	—	E	K	E	F	—	N	K	—	C	K	A	C	H	M	*33*
c_{555}	*C. thiosulfatophilum*			Y	D	A	A	A	G	—	K	A	T	Y	D	A	S	—	C	A	M	C	H	K	*362*
	"*C. ethylica*"		(+9)	Y	D	L	A	N	G	—	K	T	V	Y	D	A	N	—	C	A	S	C	H	A	*362*
c_4	*A. vinelandii*		A	G	D	A	A	—	G	Q	G	K	A	A	—	—	V	—	C	G	A	C	H	G	*362*
	P. aeruginosa		A	G	D	A	A	—	G	Q	A	K	A	A	—	—	V	—	C	G	A	C	H	G	*362*
	P. stutzeri		A	G	D	E	A	—	G	Q	G	K	V	A	—	—	V	—	C	G	A	C	H	G	*362*
c_{551}	*A. vinelandii*						E	T	G	—	E	E	L	Y	K	T	K	G	C	T	V	C	H	A	*362*
	P. aeruginosa						E	D	P	—	E	V	L	F	K	N	K	G	C	V	A	C	H	A	*362*
	P. stutzeri						Q	D	G	—	E	A	L	F	K	S	K	P	C	A	A	C	H	S	*362*
	P. fluorescens						E	D	G	—	A	A	L	F	K	S	K	P	C	A	A	C	H	T	*362*
	P. mendocina						A	S	G	—	E	E	L	F	K	S	K	P	C	G	A	C	H	S	*362*
	P. denitrificans						S	T	G	—	E	E	L	F	K	A	K	A	C	A	A	C	H	S	*362*

[a] See footnote a, Table VII, for explanation of abbreviations.

the chain, but there are reasons to suspect that they are not involved as heme ligands (see below). Cytochrome c_3 from *Desulfovibrio* has four heme attachment sites, alternating between C·x·y·C·H and C·u·v·w·x·C·H. Cytochrome $c_{551.5}$ from the mixed culture "*Chloropseudomonas ethylicum*" (called c_7 by Ambler) appears to be a truncated version of c_3 with one heme site missing. It is the smallest known cytochrome. Cytochrome cc_3 is a large, multiheme cytochrome which is not a dimer of c_3 but may be related to it (*439a*). No sequence information is available yet on flavin-*c*, but it undoubtedly forms a fourth independent class. In addition to its cysteine-linked heme *c*, this protein has a flavin adenine dinucleotide group covalently attached to another cysteine (Fig. 45).

A generally accepted term is needed to designate all those cytochromes with a *c*-type spectrum, which also give evidence of being structural and evolutionary homologs of eukaryotic *c*. Yamanaka (*440*) has proposed calling these "cytochrome *c*" (lower case), and using upper case "cytochrome *C*" for the spectral class as a whole, but this can lead to typographic confusion. For want of a better term, we will use "eukaryotic *c*" to denote any protein which is probably homologous with the *c* from eukaryotes, including those in the first three lines of Fig. 44. The diheme c_4 may or may not turn out to be a "eukaryotic *c*" also.

Four structural classes of *c*-type cytochromes thus are known today: "eukaryotic *c*," *c′*, c_3, and flavin-*c*. Other classes probably will be called for as structural information becomes available on more *c* proteins. These four classes probably represent independent evolutionary convergence on the use of a C . . . CH mode of attaching a heme group rather than divergence from a common ancestral prototype. Within each class, one

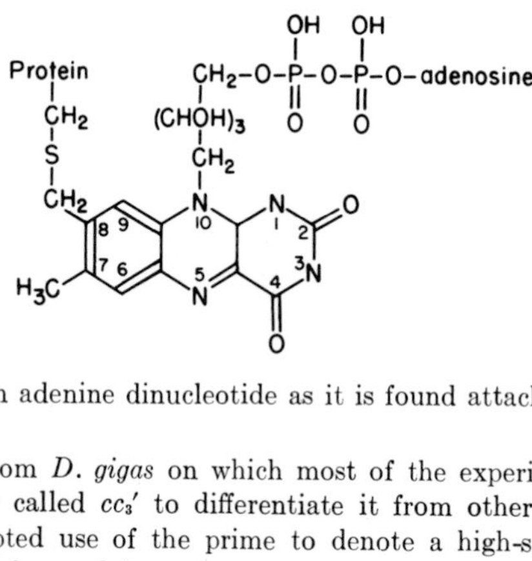

FIG. 45. Flavin adenine dinucleotide as it is found attached to flavin *c*.

439a. The protein from *D. gigas* on which most of the experimental work has been done, was originally called cc_3' to differentiate it from other strains. This conflicts with currently accepted use of the prime to denote a high-spin cytochrome *c*, and the prime should be dropped from cc_3.

440. T. Yamanaka, *Advan. Biophys.* 3, 227–276 (1972).

can hope to carry out the same kind of evolutionary studies that have been done with the globins, serine proteases, and other families of related proteins. It may be possible some day to establish connections between these *c* classes or even with cytochromes *b* and *a*, but nothing that we know today suggests that this will be so.

Only the "eukaryotic *c*" proteins have been studied by X-ray diffraction methods so far. The results with *c*, c_2, and c_{550} have been discussed earlier and need not be repeated. A low resolution electron density map of c_{551} from *P. aeruginosa* by Almassy and Dickerson at Caltech is almost ready at time of writing. A structure for this has been predicted on the basis of the known eukaryotic structure and tentative amino acid sequence homologies, involving the deletion of polypeptide chain at the bottom of the molecule (*438*), and it will be interesting to see if this prediction is borne out. Several crystalline proteins are in early stages of derivative search and structure analysis: *D. gigas* cc_3 by Sieker at Washington, *P. denitrificans* *c′* and *C. thiosulfatophilum* c_{555} by Salemme at Arizona, and algal c_{553} or *f* by Kraut at U.C. San Diego. When these are solved they will do much to extend our understanding of the range of *c*-type cytochromes, especially the *c′* and cc_3 which open up new classes.

The peculiar iron spin state in ferricytochrome *c′* from *Chromatium* has been investigated by Maltempo *et al.* (*441*), who studied the EPR spectrum down to 7°K, and measured magnetic susceptibilities between 4.2 and 1.4°K. They found that *c′* has three pH-dependent states which they named B_1, A, and B_2, with transition pK values of 1.5 and 9.8. Low and high pH states B_1 and B_2 have magnetic properties typical of high-spin ferric heme proteins such as metmyoglobin. The iron at low pH could either be five-coordinated, or could have a weak field sixth ligand such as H_2O. If a sixth ligand is present in state B_2 it cannot be H_2O, since the gradual displacement of water by the stronger OH^- as the pH increased would lead to a change in the B_2 spectrum which was not observed.

The physiological state A is strangest of all. It is a mixture of two states, A_1 and A_2, both of which are quantum mechanical mixtures (not thermal mixtures) of two quantum states with spins $S = \frac{5}{2}$ and $\frac{3}{2}$. Such a situation is unprecedented for biological molecules and clearly distinct from the pure high-spin state found in the globins or the thermal mixture of two pure states in catalase. The possible biological significance of this peculiar magnetic configuration lies in its implications for ligand arrangement. Maltempo *et al.* argued that this spin state mixing is consistent with a mixing of the iron *d* orbitals (in a ligand field approach) between octahedral and square planar symmetry. The latter would imply that only

441. M. M. Maltempo, T. H. Moss, and M. A. Cusanovich, *BBA* **342**, 290–305 (1974).

the four heme-plane ligands were present. In essence this means that the nonheme fifth and sixth ligands are coordinated to the iron very weakly at best. Something in the way the molecule is folded must keep potential ligands away from the faces of the heme.

In line with this exclusion of ligands, anions such as F^-, Cl^-, CN^-, and N_3^- do not have access to the heme iron in oxidized cytochrome *c'*. This could arise from steric hindrance around the heme, from a local concentration of positively charged side chains, or an especially hydrophobic environment for the iron, and the authors favor the latter (*441*). In contrast to this closed-heme pocket picture of the oxidized molecule, and to the behavior of reduced eukaryotic *c*, reduced *c' does* bind carbon monoxide at physiological pH. Either the *c'* molecule opens up slightly upon reduction or the CO molecule can slip into a hydrophobic pocket that charged anions cannot enter.

E. Bacterial Cytochrome Evolution

The amino acid sequences of tuna *c*, *M. denitrificans* c_{550}, and four species of c_2 from *Rhodospirillum* and *Rhodopseudomonas* were presented earlier in Table VII, and their great similarities were noted. The absolute numbers of amino acid differences between all pairs of these six protein sequences are shown in the two matrices in Table XXV. In matrix I, an absence or deletion in one sequence is counted as an evolutionary

TABLE XXV
Amino Acid Differences between Cytochromes of Bacteria and Eukaryotes

Organism	*M. denitrificans*	*R. capsulata*	*R. rubrum*	*R. palustris*	*R. spheroides*	Tuna
I. Counting deletions as a twenty-first amino acid type						
M. denitrificans	0	61	74	85	74	86
R. capsulata c_2	61	0	68	72	63	85
R. rubrum c_2	74	68	0	74	77	84
R. palustris c_2	85	72	74	0	88	87
R. spheroides c_2	74	63	77	88	0	97
Tuna *c*	86	85	84	87	97	0
II. Omitting all deletion regions from the count						
M. denitrificans c_{550}	0	53	58	59	66	59
R. capsulata c_2	53	0	58	54	55	60
R. rubrum c_2	58	58	0	64	65	59
R. palustris c_2	59	54	64	0	66	54
R. spheroides c_2	66	55	65	66	0	68
Tuna *c*	59	60	59	54	68	0

event, and treated as a twenty-first kind of amino acid. In matrix II, positions along the chain where a deletion occurs in one of the two sequences are eliminated from the difference count. The remarkable conclusion from these matrices is how similar all these proteins are and how they differ from one another by roughly the same degree. By the standards of amino acid sequence, just as by three-dimensional folding, c_{550} is a perfectly good member of the c_2 family. If the extra loops not shared by *c* are eliminated (matrix II), then tuna *c* becomes a member of the family also. Eukaryotic *c* differs from c_2's and c_{550}, not by having distinctively different sequences in the shared regions of structure but by lacking certain of the external loops and folds of the other molecules (*441a*).

This is a remarkable state of affairs. We tend to think of photosynthesis and respiration as having had a long independent history, yet in regions of shared structure, respiratory *c* from mammals and respiratory c_{550} from *Micrococcus* are no different from the photosynthetic c_2 of purple nonsulfur bacteria than the individual c_2's are among themselves. Part of our impression that man and *Micrococcus* are farther apart than *R. capsulata* and *R. spheroides* may lie in the eye of the beholder, but we would have expected more distinction between c_2 and the other proteins than is observed.

It could be that the break between respiration and photosynthesis in these bacteria is more recent than we think. Cytochrome c_2 has been suggested to have a respiratory as well as a photosynthetic role in *R. spheroides* (*372*) and *R. capsulata* (*372a–c*) and no alternative respiratory chain has yet been identified in any of the Athiorhodaceae. In some of these organisms a situation may exist as in Fig. 46 with electrons flowing to c^2 both from light-excited bacteriochlorophyll and from external donors, and then from c^2 either to an electron-depleted bacteriochlorophyll or to an oxidase molecule. This would account for the observed control mechanism in the purple nonsulfur bacteria. Under aerobic conditions in the dark, bacteriochlorophyll would not be electron-deficient, whereas the oxidase would be in its oxidized state and capable of accepting electrons from c^2. Under anaerobic conditions, electrons would reduce the oxidase, and further electron transfer down that path would be blocked. Light then would promote electrons away from bacteriochlorophyll and set cyclic photophosphorylation in motion.

Simple arguments from economy make it hard to believe that the respiratory cytochrome *c* from one organism would be virtually identical with the photosynthetic *c* of another, and that the second organism then would

441a. R. Ambler has recently sequenced cytochrome c_2 from *Rhodomicrobium vannielii* (private communication), and observed that it lacks all of the extra loops and insertions of other c_2's. Residue for residue, it is structurally homologous with eukaryotic *c*!

have a totally different respiratory chain. Right or wrong, the shared-cytochrome model is esthetically more pleasing. Respiring bacteria may have evolved from Athiorhodaceae-like ancestors, not by the gain of any new functions, but simply by the loss of the photosynthetic half of Fig. 46. The spur to this development may have been the rise of O_2 levels in the atmosphere to the point where respiration became more efficient in the production of energy than cyclic photophosphorylation. The reason for the similarity of c_2 and c_{550} would then be obvious: They are recent evolutionary homologs. The similarity of eukaryotic cytochrome c to c_2 and c_{550} suggests one more hypothesis: Respiratory bacteria (at least the ancestors of *M. denitrificans* and of eukaryotes) diverged from the purple nonsulfur bacteria by loss of photosynthetic function at about the same time that the line of respiratory bacteria which ultimately led to eukaryotes split off from the others.

Yamanaka has attempted to use the relative cross-reactivities between cytochromes c of one organism and cytochrome oxidases of another to make a quantitative assessment of evolutionary similarity (*440,442,443*, and earlier references cited therein). If this method were successful, it would be a powerful adjunct to sequence and structure comparisons since far less material and labor are required to make an activity measurement than to carry out an amino acid sequence analysis or X-ray diffraction study. One critical requirement, of course, is that only strictly homologous proteins are compared. If noncorresponding proteins in two organisms are compared, a spurious estimate of their unrelatedness would be obtained. It is in this context that Yamanaka emphasized the need for a descriptive term for what we have called "eukaryotic c."

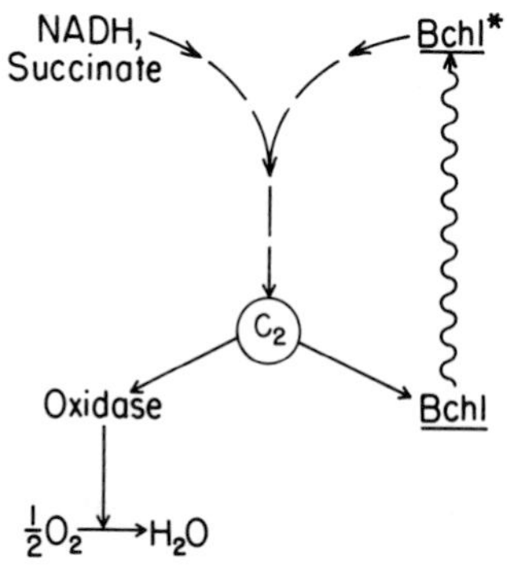

FIG. 46. Possible combination of photosynthetic and respiratory electron pathways in purple nonsulfur bacteria with c_2 being shared by both paths. See text for discussion.

442. T. Yamanaka, *Nature* (*London*) **213**, 1183–1186 (1967).
443. T. Yamanaka, *Space Life Sci.* **4**, 490–504 (1973).

Yamanaka has studied the relative reactivities of presumably homologous cytochrome *c* from 21 bacteria, 6 algae, and 28 higher eukaryotes, with bovine cytochrome oxidase and with *P. aeruginosa* nitrite reductase, a cytochrome oxidase from the nitrate respiration system. Activities are expressed in percent, with the activities of each oxidase with its own cytochrome as 100%. Unfortunately, if a simple linear plot is made of his results instead of the expanded semilogarithmic plot used in the original papers, the only impression one obtains is the not unexpected one that the natural substrates of bovine cytochrome oxidase and bacterial nitrite reductase do not cross-react to a significant extent (Fig. 47). No cytochrome *c* from a higher eukaryote reacts with *P. aeruginosa* nitrite reductase more than 9% as well as *P. aeruginosa* c_{551} does, and only one cytochrome which reacts well with the nitrite reductase is reported as reacting with bovine oxidase at all. Furthermore, the great spread of relative reactivities of *c*'s from higher eukaryotes with bovine oxidase (X's along the ordinate of Fig. 47) is in disagreement with the work of L. Smith and colleagues (*130*), who found that after extensive dialysis to remove all traces of bound ions, cytochromes *c* from five mammals, three birds, tuna, and moth all reacted with bovine oxidase at the same maximal rate. In the absence of these special precautions, an erratic and uncorrelatable spread in rates was observed.

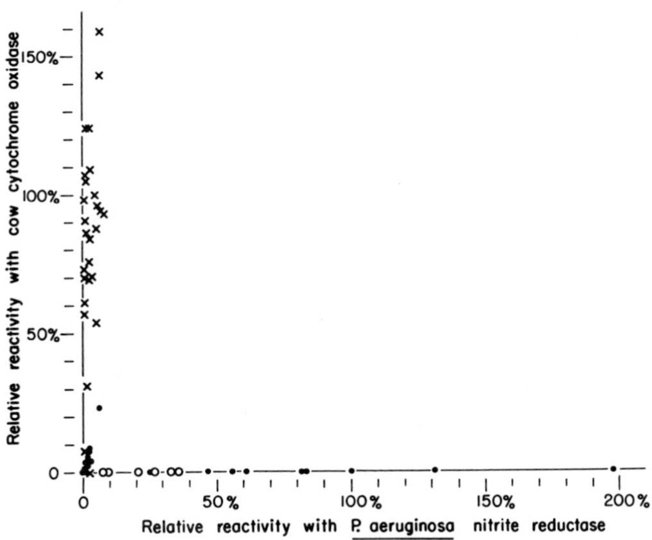

FIG. 47. Cross-reactivity plot of interactions of cytochromes *c* from various organisms with mammalian and bacterial cytochrome oxidases. (●) Bacteria, (○) algae, and (X) fungi and higher eukaryotes. Data replotted from reference *441*.

It was the authors' original intention to conclude this chapter with the presentation of several alternative phylogenetic trees for the bacteria, based on cytochrome *c* utilization, electron transport chain structure, and general energy metabolism. But after considerable reflection, we have finally come around to the viewpoint advocated in slightly different terms by R. Y. Stanier and by J. Kilmer (*443a*). The construction of a detailed evolutionary history, and the very concept of a dentritic phylogenetic tree so useful with eukaryotes may not apply to bacteria. Exchange of genetic information by virial infection and transduction of large segments of the bacterial genome could have blurred the true history of the bacteria beyond present-day recall. It may prove to be easier to draw out a family tree from metabolism than from morphology, but even this may not present a true picture of the evolution of the bacteria.

In the spirit that a distorted picture is better than no picture at all, we should like to close the chapter with a few suggestive ideas on bacterial evolution which have come from the cytochrome study. They are offered less as dogma than as useful targets for future discussion:

1. The four classes of cytochromes mentioned earlier—"eukaryotic *c*," *c′*, c_3, and flavin-*c*—are probably evolutionarily unrelated and represent common convergence on the use of a *c* heme attached by the sequence C . . . CH, which is favored for stereochemical reasons.

The arguments for this have been presented already.

2. Sulfate respiration is probably only a distant relative of the other types of respiration and photosynthesis. The sequence *Clostridium–Desulfotomaculum–Desulfovibrio* represents a probable line of evolutionary development marked by increasing elaboration of cytochrome-using electron transport chains.

The chief argument for unrelatedness is the dissimilarity of the cytochromes of sulfate respirers with those of other organisms.

3. Sulfate respiration may be primitive in the sense of predating oxygen respiration, but nitrate respiration probably is not.

The argument for primitiveness of sulfate respiration hinges on its presence in *Clostridia*-like *Desulfotomaculum* and contains the hidden assumption that present-day *Clostridia* most nearly resemble the ancient fermentative organisms. The widespread but erratic distribution of nitrate respiration and its occurrence only as an alternative to oxygen suggests one of two hypotheses: Either nitrate respiration is an ancient talent which has been superseded in all but a few groups of bacteria which occupy special ecological niches or it is a metabolic innovation which has arisen more than once from the oxygen machinery in response to

443a. "Phyla are made by fools like me,
But only God can make a tree."

special needs. This divergence vs. convergence question might be answered by seeing whether the cytochromes and other enzymes involved only in nitrate respiration in distantly related organisms were more dissimilar than those of O_2 respiration, a situation which would favor independent convergence. The cytochromes c_{551} and nitrite reductases of *Pseudomonas*, *Micrococcus*, and *P. denitrificans* (*Alcaligenes?*) may provide just such a test. The c_{551}'s of all these nitrate respirers are obviously homologous (see ref. *360*), but there are two distinct patterns of nitrite reductases: a diheme protein of molecular weight 95,000 and a nonheme copper protein of molecular weight 150,000. This suggests independent development of nitrite reductase ability.

Modern nitrate deposits, and some sulfate, are the end products of chemoautotrophic, O_2-using bacteria, which could not have been so on the primitive earth. The great solubility and chemical reactivity of nitrates in comparison with sulfates makes it unlikely that large abiotic deposits of nitrate would have remained for long on a water-covered planet with a reducing atmosphere. In contrast, mineral sulfates could have been present then and used by primitive sulfate-respiring organisms. Sulfate would have become plentiful after the appearance of the ancestors of green and purple photosynthetic bacteria, and sulfate respirers may have developed in a symbiotic cooperation with them of the sort that modern biochemists found in the deceptive "*Chloropseudomonas ethylicum.*" Sulfate respiration probably is quite old, but nitrate respiration is more likely to have evolved after the appearance of O^2-using chemoautotrophic nitrifiers such as *Nitrosomonas* and *Nitrobacter*.

4. The Krebs cycle developed first in the ancestors of the purple nonsulfur bacteria, perhaps initially as a means of supplying NADH for synthetic purposes rather than as fuel for a respiratory chain. The possessors of this talent then could eliminate noncyclic photophosphorylation and a need for an external reducing source, and adopt cyclic photophosphorylation instead. This would have marked the break between purple sulfur and nonsulfur bacteria. Respiration followed, perhaps in response to low levels of abiotically produced free oxygen. After the development of O_2-emitting photosynthesis by organisms which may have resembled present-day blue-green algae, there occurred a radiation of oxygen-respiring bacteria from the purple nonsulfur bacteria by loss of a now less-important photosynthetic capability. Among these new oxygen respirers were the ancestors of *M. denitrificans* and of eukaryotes.

These hypotheses are defensible because the Athiorhodaceae resemble the purple sulfur bacteria but are the only bacteria to combine photosynthesis, a Krebs cycle, and respiration; because this Krebs cycle is useful to them even under illuminated anaerobic conditions; because of the suggested dual role for c_2 in some strains; and because of the very great sequence and structural similarity of c^2, c_{550}, and eukaryotic c.

5. The blue-green algae themselves could have arisen from the Athiorhodaceae. The purple sulfur bacteria are either primitive predecessors or specialized offshoots of the Athiorhodaceae, probably the former; and the green sulfur bacteria are more distantly related.

These final hypotheses are supported by the unique position of Athiorhodaceae and blue-green algae among prokaryotes in having both photosynthesis and respiration, and the relative unimportance of the latter to them both; and by the observed similarity of photosynthetic vesicles in purple bacteria and blue-green algae, unlike those in green bacteria.

These proposals have a certain symmetry to them. Only two kinds of photosynthesis are known which use outside sources of reducing equivalents—sulfide-to-sulfate and water-to-oxygen. There also are only two

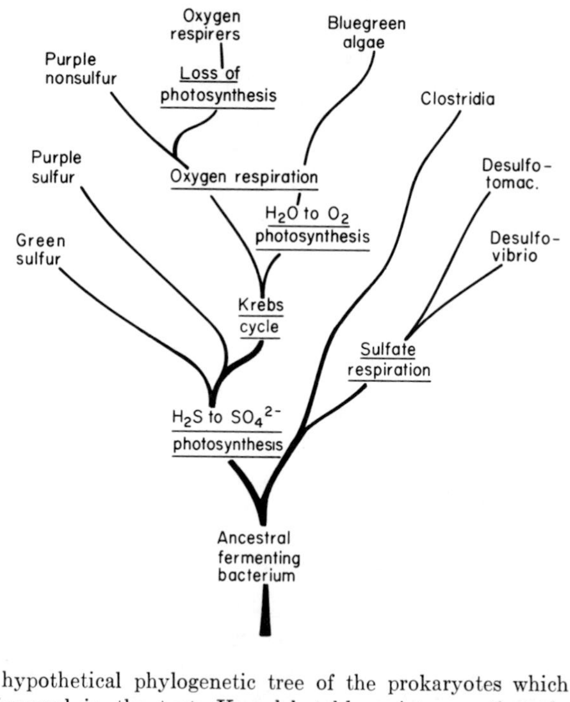

FIG. 48. A hypothetical phylogenetic tree of the prokaryotes which ties together the points discussed in the text. Key debatable points are that the Krebs cycle had a useful history prior to respiration, liberating the Athiorhodaceae from a dependence on H_2S and noncyclic photosynthesis as a source of NAD(P)H, that both the Athiorhodaceae and blue-green algae invented oxygen respiration after the blue-greens had made the planetary atmosphere oxidizing by H_2O-using photosynthesis, and that oxygen-respiring bacteria arose from the Athiorhodaceae by a loss of the photosynthetic function.

patterns of respiration with radically different electron transport components—sulfate-to-sulfide and oxygen-to-water. (Nitrate respiration is considered a late offshoot of oxygen respiration, with many shared chain components.) A sulfide-to-sulfate-to-sulfide syntrophy between ancestors of *Chlorobium* and *Desulfovibrio* could have developed early on the primitive earth in a reducing atmosphere where H_2S was freely available. The water-to-oxygen-to-water syntrophy of present-day plants and animals had to await the development of photocenter II as a means of making a useful electron donor out of a poor reducing agent. When this happened, the redox character of the planetary atmosphere itself was changed. The cytochromes of sulfate respiration diverged from those of photosynthesis quite early, if they did not develop independently. The more familiar cytochromes of oxygen and nitrate respiration, in contrast, separated from those of photosynthesis in more recent times, and show this by their structural and functional similarities.

In spite of the earlier disclaimer, the temptation to end this chapter with a tree is overwhelming. All of the foregoing ideas are incorporated into Fig. 48, which at least manages to avoid some of the chicken vs. egg paradoxes which have plagued other bacterial phylogenies. This tree ultimately may be true or may be ridiculous, but is advanced here only as the most sensible ordering of matters as they appear to two cytochrome chemists.

8

Type b Cytochromes

BUNJI HAGIHARA • NOBUHIRO SATO •
TATEO YAMANAKA

I. Introduction

Type *b* cytochromes are electron-carrying proteins which contain protoheme IX as the prosthetic group; consequently, the position of the α band of their absorption spectra in the ferrous state ranges from 554 to 566 nm. By the pyridine treatment they give the pyridine-protohemochrome with its α band at 556 nm. They are widespread in animals, plants, and various microorganisms. A summary of their occurrence and essential properties is given in Tables I, III, IV, and VI.

Many type *b* cytochromes associated with the classic cytochrome oxidase (a and a_3) show their α peaks at 561–563 nm, and they are usually designated as cytochrome *b*. The name cytochrome b_1 was originally used for the pigments with α peak at 557–560 nm, but it is now generally applied to the cytochrome in the nitrate reductase system. Cytochrome b_2 (α, 557 nm) is the entity of yeast lactate dehydrogenase which contains FMN and protoheme. Cytochromes b_3 (α, 559 nm) and b_5 (α, 556 nm) participate in the microsomal electron transport system in plants and animals, respectively. Both the primary and ternary structures of cytochrome b_5 are known. Cytochromes b_6 (α, 563 nm) and *b*-559 are the components of the photosynthetic electron transport system in plant.

TABLE I
TRIVIAL NAMES OF *b*-TYPE CYTOCHROMES AND THEIR FUNCTIONS

Cytochrome	α band (nm)	Occurrence	Function and remarks
b	561–563	Mitochondria of animals, plants and yeast. Some bacteria.	Electron carrier in respiratory system containing cytochrome aa_3.
b_1	557–560	Bacteria (denitrifying bacteria in a limited sense).	Electron carrier in nitrate respiration system.
b_2	557	Yeast mitochondria	Lactate dehydrogenase (FMN + protoheme)
b_3	559	Plant microsomes	Function, uncertain
b_4			Type *c* cytochrome
b_5	555–556	Microsomes of animals (and plants)	Function, uncertain. Primary and ternary structures are known.
b_6	563	Chloroplasts	Electron carrier in photosynthetic system.
b_7	560	Plant mitochondria, especially in spadix	Function, uncertain. Probably cyanide-insensitive terminal oxidase

Various other type *b* cytochromes are found in invertebrates, protozoa, bacteria, fungi, algae and higher plants and reported under the names of cytochrome b_7, *b*-554, *b*-556, *b*-557, *b*-558, *b*-559, *b*-560, *b*-561, *b*-562, *b*-563, *b*-564, *b*-565, *b*-566, and *b*-567.

Although cytochrome *P*-450 (hemoprotein *P*-450) is also a protohemoprotein, it is a hydroxylating enzyme and is not usually regarded as a cytochrome *b*. In additon, cytochrome *o* contains a protoheme but it acts as a terminal oxidase in various microorganisms. Helicorubin and enterochrome-566 are sometimes included with the type *b* cytochromes.

Recently, a review of type *b* and *b*-like cytochromes was presented by Lemberg and Barrett (*1*).

II. Type *b* Cytochromes in Respiratory Systems

A. Introduction

The presence of different cytochrome *b* components in mitochondrial respiratory chain and their direct role in energy conservation has been postulated for many years. These postulations were based on anomalous kinetic and spectrophotomeric behavior of the *b* cytochromes in intact mitochondria and in phosphorylating submitochondrial particles (*2–13*). During 1970 and 1971 evidence was obtained by several workers pointing

1. F. R. S. Lemberg and J. Barrett, "The Cyochromes," pp. 58–121, 245–264, and 346–360. Academic Press, New York, 1972.
2. B. Chance, *JBC* **233,** 1223 (1958).
3. E. C. Slater and J. P. Colpa-Boonstra, *in* "Haematin Enzymes" (J. E. Falk, R. Lemberg, and R. K. Morton, eds.), Vol. 2, p. 575. Pergamon, Oxford, 1961.
4. A. M. Pumphrey, *JBC* **237,** 238 (1962).
5. B. Chance and B. Schoener, *JBC* **241,** 4567 and 4577 (1966).
6. B. Chance, C.-P. Lee, and B. Schoener, *JBC* **241,** 4574 (1966).
7. J. Kirschbaum and W. W. Wainio, *BBA* **113,** 27 (1966).
8. D. D. Tyler, R. W. Estabrook, and D. R. Sanadi, *ABB* **114,** 239 (1966).
9. M. Klingenberg, *in* "Biological Oxidations" (T. P. Singer, ed.), p. 3. Wiley (Interscience), New York, 1968.
10. W. W. Wainio, J. Kirschbaum, and J. D. Shore, *in* "Structure and Function of Cytochromes" (K. Okunuki, M. D. Kamen, and I. Sekuzu, eds.), p. 713. Univ. of Tokyo Press, Tokyo, 1968.
11. N. Sato and B. Hagihara, *J. Biochem.* (*Tokyo*) **64,** 723 (1968).
12. N. Sato and B. Hagihara, *Cancer Res.* **30,** 2061 (1970).
13. B. Hagihara, N. Sato, K. Takahashi, and S. Muraoka, *in* "Organization of Energy Transducing Membranes" (M. Nakao and L. Packer, eds.), p. 315. Univ. of Tokyo Press, Tokyo, 1973; B. Hagihara, N. Sato, T. Fukuhara, K. Tsutsumi, and Y. Oyanagui, *Cancer Res.* **33,** 2947 (1973).

out the existence of at least two spectroscopically (*14–17*), kinetically (*18*), and potentiometrically (*19,20*), distinct species of *b* cytochromes in intact mitochondria of animals (*14–20*), yeast (*21*), and plants (*22,23*). There is still some uncertainty about the number of the components and their function in electron transport and energy conservation.

B. Different Type *b* Cytochromes in Mammalian Respiratory Systems

The multiplicity of the cytochrome *b* components was first shown by Chance (*2*) in the mitochondrial fragments from beef heart. He suggested the presence of three type *b* cytochromes: The first was characterized as succinate-reducible pigment with an α maximum at 562 nm; the second as succinate-reducible in the presence of antimycin, having an α maximum at 566 nm; and the third as reducible only by dithionite and having a broad α maximum from 556 to 566 nm. Subsequent studies have also shown the existence of at least two forms of cytochrome *b* in mitochondria (*5,7,8,11*) and submitochondrial fragments (*3,4,6,9,10*). Table II summarizes demonstrations of the presence of cytochrome *b* components other than the classic cytochrome *b*.

Slater and Colpa-Boonstra (*3*), and Pumphrey (*4*) proposed the presence of two type *b* cytochromes from the effect of antimycin A on the reducibility and red shift of cytochrome *b*. From studies on the reactivity of cytochrome *b* with substrates in the presence and absence of various inhibitors and chelating agents, Wainio and co-workers (*7,10*) suggested that cytochrome *b* is present in three different compartments or as three moieties in heart mitochondria. Chance and Schoener (*5,6*) reported the presence of a new cytochrome *b* moiety showing an α peak at 555 nm at liquid nitrogen temperature (around 560 nm at room temperature)

14. E. C. Slater, C.-P. Lee, J. A. Berden, and H. J. Wegdam, *Nature* (*London*) **226,** 1248 (1970); *BBA* **223,** 354 (1970).
15. H. J. Wegdam, J. A. Berden, and E. C. Slater, *BBA* **223,** 365 (1970).
16. N. Sato, D. F. Wilson, and B. Chance, *FEBS* (*Fed. Eur. Biochem. Soc.*) *Lett.* **15,** 209 (1971); *BBA* **253,** 88 (1971).
17. M. K. F. Wikström, *BBA* **253,** 332 (1971).
18. B. Chance, D. F. Wilson, P. L. Dutton, and M. Erecinska, *Proc. Nat. Acad. Sci. U. S.* **66,** 1175 (1970).
19. D. F. Wilson and P. L. Dutton, *BBRC* **39,** 59 (1970).
20. P. L. Dutton, D. F. Wilson, and C.-P. Lee, *Biochemistry* **9,** 5077 (1970); *BBRC* **43,** 1186 (1971).
21. N. Sato, T. Ohnishi, and B. Chance, *BBA* **275,** 288 (1972).
22. W. D. Bonner, Jr., and E. C. Slater, *BBA* **223,** 349 (1970).
23. P. L. Dutton and B. T. Storey, *Plant Physiol.* **47,** 282 (1971).

TABLE II

TYPE *b* CYTOCHROMES REPORTED TO EXIST IN MITOCHONDRIAL RESPIRATORY SYSTEMS IN ADDITION TO CLASSIC CYTOCHROME *b* (b_K)

Author	Desig-nation	Position of α peak(s) (nm)	Conditions of reduction	Ref.
Chance		566 556–566	Succinate + AM[a] Dithionite	*2*
Slater and Colpa-Boonstra	*b*′	565	Succinate + AM[a] Dithionite	*3,4*
Chance and Schoener	*b*-555	560 (555 at 77°K)	Succinate + ATP[b]	*5,6*
Wainio *et al.*	*b*′ *b*″		Succinate + AM[a] NADH	*7,10*
Sato and Hagihara	HP-565 HP-559	565 (563 at 77°K) 559 (556 at 77°K)	Dithionite Dithionite	*11–13*
Slater *et al.*	b_i	558 (565 by AM)		*14,15*
Wilson *et al.*	b_T (*b*-565)	565 and 558 (562.5 and 555 at 77°K)	Succinate + ATP	*18,19*

[a] Succinate + AM; the component is reduced with succinate only in the presence of antimycin (AM).

[b] Succinate + ATP; the component is reduced with succinate only in the presence of ATP. The classic cytochrome *b* (=b_K) is easily reduced by succinate alone.

in pigeon heart mitochondria. This absorption band appeared when ATP was added to the mitochondria in the sulfide-inhibited state and disappeared on the addition of uncoupling agents such as dicumarol or dinitrophenol. From the above behavior, this band was proposed to result from the existence of a new type of cytochrome *b*, or a modified form of the classic one, which was tentatively designated as cytochrome *b*-555. Tyler *et al.* (*8*) reported similar observations. Sato and Hagihara (*11–13*) reported the existence of two type *b* cytochromes in addition to the classic cytochrome *b* in various mammalian mitochondria. The above pigments were identified from the difference spectra of mitochondrial and cell suspensions, which were obtained in the presence and absence of dithionite after anaerobiosis resulted from succinate respiration or endogenous respiration, respectively. Slater and co-workers (*14,15*) again proposed the presence of two type *b* cytochromes in beef heart mitochondria and their role in site 2 phosphorylation from the effect of antimycin and ATP.

Evidence for the existence of two potentiometrically distinct cytochrome *b* species was first found by Wilson and Dutton (*19,20*) in rat liver and pigeon heart mitochondria: One of these *b* cytochromes is the classic cytochrome *b* while the other, designated as cytochrome b_T (T for energy transduction), seemed to have an energy-dependent oxidation–reduction midpoint potential. The authors postulated that cytochrome b_T participates in energy transformation of electrochemical potential to chemical potential available for ATP synthesis. Evidence has been obtained by Chance *et al.* for the existence of two kinetically distinct species (*18*). Subsequent papers by Sato *et al.* (*16*), Wikström (*17*), Davis *et al.* (*24*), and Yu *et al.* (*25*) have referred to the existence of at least two spectroscopically distinct species in intact mitochondria, succinate-cytochrome *c* reductase, and complex III derived from heart mitochondria.

C. Absorption Spectra

The classic cytochrome *b*, expressed sometimes as cytochrome b_K, *b*-561, or *b*-562, has a single symmetric α band at 561–562 nm in the reduced minus oxidized difference spectrum at room temperature; at liquid nitrogen temperature (77°K) it has an α band at 558–559.5 nm, a β band at 529 nm, and a Soret band at 428 nm. Cytochrome *b* is readily reduced by succinate and NAD-linked substrate in both coupled and uncoupled mitochondria. This cytochrome is associated with complex III (ubiquinone-cytochrome *c* reductase) (*24*). Figure 1 shows the absorption spectra of reduced cytochrome *b* at different temperatures between liquid helium and room temperature (*26*).

The second cytochrome *b*, designated as cytochrome b_T or as *b*-565 or *b*-566, has a characteristically split α band at 565–566 and 557–559 nm in the reduced minus oxidized difference spectrum at room temperature (cf. footnote *b* of Table III). At liquid nitrogen temperature it has a double α band at 562.5 and 554–555 nm, β bands at 535 and 528 nm, and a γ band at 430 nm (see Fig. 2.). This cytochrome b_T is fully reduced by succinate and NAD-linked substrate in intact mitochondria in the presence of ATP but not fully in uncoupled mitochondria. In the presence of antimycin succinate reduces cytochrome b_T fully in mitochondria and fragmented succinate-cytochrome *c* reductase when oxygen or other electron acceptors are present (*27,28*). The reported value of E_0' is

24. K. A. Davis, Y. Hatefi, K. L. Poff, and W. L. Butler, *BBRC* **46,** 1984 (1972); *BBA* **325,** 341 (1973).
25. C. A. Yu, L. Yu, and T. E. King, *BBA* **267,** 300 (1972).
26. B. Hagihara, R. Oshino, and T. Iizuka, *J. Biochem.* (*Tokyo*) **75,** 45 (1974); B. Hagihara and T. Iizuka, *ibid.* **69,** 355 (1971).
27. J. S. Rieske, *ABB* **145,** 179 (1971).
28. D. F. Wilson, M. Koppelman, M. Erecinska, and P. L. Dutton, *BBRC* **44,** 759 (1971).

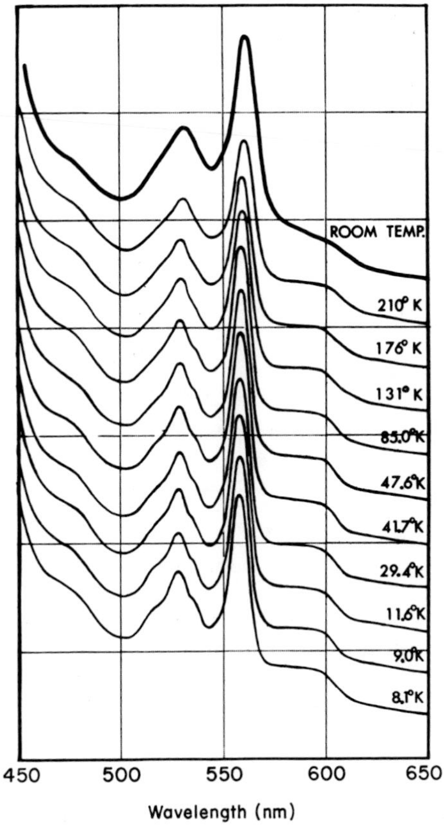

FIG. 1. The absorption spectrum of purified beef heart cytochrome *b* in the reduced form at various temperatures (*26*). The cytochrome was dissolved in 0.01 *M* sodium phosphate buffer containing 1% deoxycholate at pH 7.4 and a trace of sodium dithionite was added for reduction. The optical path was 3 mm in the case of low temperature spectra and 10 mm in the case of the room temperature spectra.

-30 ± 10 mV at pH 7.2 in uncoupled mitochondria, but the observed value is raised to $+245$ mV in the presence of ATP (*19,20*). Wilson and coworkers (*29,30*) postulated, from the phosphate potential ($ATP/ADP \cdot P_i$) dependency of E_0' of cytochrome b_T (*31,32*), that this cytochrome is

29. D. F. Wilson and M. Erecinska, *in* "Mitochondria/Biomembranes" (E. C. Slater, ed.), p. 119. North-Holland Publ., Amsterdam, 1972.
30. D. F. Wilson and P. L. Dutton, *in* "Electron and Coupled Energy Transfer in Biological Systems" (T. E. King and M. Klingenberg eds.), Vol. 1, p. 221. Dekker, New York, 1971.
31. N. Sato, *Med. J. Osaka Univ., Jap. Ed.* **23**, 95 (1971).
32. D. F. Wilson and E. S. Brocklehurst, *ABB* **158**, 200 (1973).

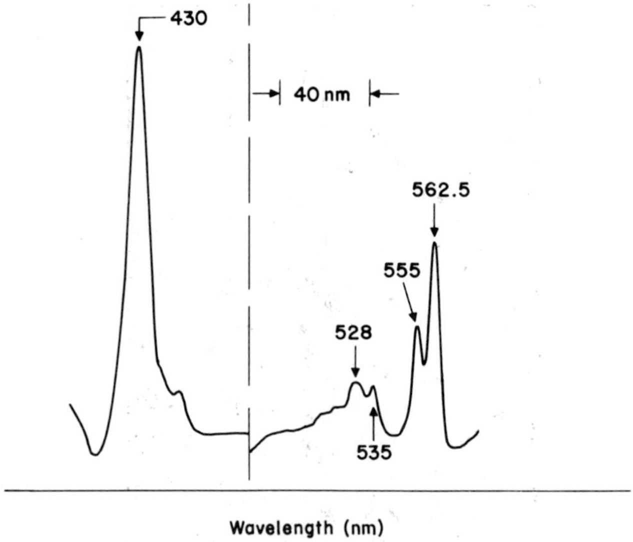

FIG. 2. The absorption spectrum of cytochrome b_T in pigeon heart mitochrondria at 77°K (*16*). The mitochondria were suspended in a medium containing 0.225 *M* mannitol, 0.07 *M* sucrose, 0.2 m*M* EDTA, and 50 m*M* Tris-HCl, pH 7.4, and supplemented with 2.5 μ*M* rotenone, 5 m*M* KCN, 5 m*M* ascorbate, 80 μ*M* TMPD, and 5 m*M* succinate. The suspension was injected into the reference cuvette which had been precooled to 77°K. The suspension was similarly transferred into the measure cuvette after further addition of 1.2 m*M* ATP.

involved directly in primary energy conservation reaction. Cytochrome b_T was reported to be associated with complex III (*24*).

At the wavelength couple 562–575 nm in the reduced minus oxidized spectrum, cytochrome *b* contributes 60–75% while cytochrome b_T contributes 25–40% (*17,20,33*); at the 566–575 nm couple, the contribution to the absorbance is 68% for cytochrome b_T and 32% for cytochrome *b* (*34*).

In mitochondrial fragments and complex III from beef heart, the EPR signals were reported to be at $g = 3.44$ and $g = 3.78$ for cytochrome *b* and cytochrome b_T, respectively, and interpreted as low field resonance of low-spin ferric heme compounds (*35,36*).

33. P. L. Dutton, M. Erecinska, N. Sato, Y. Mukai, M. Pring, and D. F. Wilson, *BBA* **267**, 15 (1972).
34. N. Sato, B. Hagihara, R. Oshino, and B. Chance, *J. Jap. Biochem. Soc.* **44**, 388 (1972).
35. N. R. Orme-Johnson, R. E. Hansen, and H. Beinert, *BBRC* **45**, 871 (1971).
36. D. V. Dervartanian, S. P. J. Albracht, J. A. Berden, B. F. van Gelder, and E. C. Slater, *BBA* **292**, 496 (1973).

The spectral and potentiometric characteristics of type *b* cytochromes in a succinate-cytochrome *c* reductase, prepared from pigeon breast muscle mitochondria using a mixture of ionic and nonionic detergents, have shown a close resemblance to those in intact mitochondria (*28,34,37*). The succinate-cytochrome *c* reductase may be fractionated into two complexes, succinate-ubiquinone reductase and ubiquinone-cytochrome *c* reductase, so-called complexes II and III, respectively. Complex III contains type *b* cytochromes, cytochrome c_1, and nonheme iron protein in a stoichiometry of 2:1:1 (*38*). One of the type *b* cytochromes shows an α peak at 562 nm (559.5 nm at 77°K) by the reduction with succinate and is identified as cytochrome *b*. The other, with the α peak at 566 nm (562.5 and 554 nm at 77°K), is reduced by succinate only in the presence of antimycin or by dithionite and is identified as cytochrome b_T. These two cytochromes do not combine with CO.

Splitting of type *b* cytochromes from these preparations by proteinase or bile salts modified the cytochromes to a form which combines with CO. Gradual disappearance of the double peak of cytochrome b_T was seen with concomitant increase in the absorbance at 559 nm (at 77°K) characteristic of cytochrome *b* (*34,37*).

Complex II from bovine heart mitochondria was first reported by Doeg *et al.* (*39*) to contain a type *b* cytochrome which shows absorption peaks at 562, 530, and 430 nm in the reduced minus oxidized difference spectrum. Davis *et al.* (*24*) have recently reported the presence in complex II of a cytochrome which has termed cytochrome *b*-557.5 based on the α-band position in the low temperature difference spectrum. This cytochrome shows an α peak at 560–561 nm at room temperature, and this peak is split into two bands at 557.5 and 550 nm at 77°K. This cytochrome, *b*-560, is neither reduced nor oxidized during the course of the reaction. The spectral similarity to the cytochrome split from complex III may indicate this cytochrome to be a modified form of cytochrome *b* or b_T (*40*). Cytochrome *b*-560 might possibly be identical to *b*-559 (*11–13,41*) and *b*-558 (*20,42*) found in intact animal mitochondria as well as the type *b* cytochromes found in the cultured cells (*43*), anaerobic yeast

37. M. Erecinska, R. Oshino, N. Oshino, and B. Chance, *ABB* **157,** 431 (1973).
38. J. S. Rieske, S. H. Lipton, H. Baum, and H. I. Silman, *JBC* **242,** 4854 (1967).
39. K. A. Doeg, S. Krueger, and D. M. Ziegler, *BBA* **41,** 491 (1960); D. M. Ziegler and K. A. Doeg, *ABB* **97,** 41 (1962).
40. E. C. Slater, *in* "Mechanisms in Bioenergetics" (G. F. Azzone *et al.,* eds), p. 405. Academic Press, New York, 1973.
41. A. P. Dawson and M. J. Jelwyn, *BBRC* **32,** 579 (1968).
42. J. G. Lindsay, P. L. Dutton, and D. F. Wilson, *Biochemistry* **11,** 1937 (1972).
43. N. Sato, B. Chance, K. Kato, and W. Klietmann, *FEBS* (*Fed. Eur. Biochem. Soc.*) *Lett.* **29,** 222 (1973); *BBA* **305,** 493 (1973).

cells (*43a*), and *Ascaris lumbricoides* (*44,44a*) from the similarity in spectral property and reactivity to substrates.

Dutton and co-workers (*20,42*) detected, in addition to cytochromes b and b_T, a cytochrome b-like species with the α band at 558 nm at room temperature having the E_0' value of about +130 mV at pH 7.2 in beef heart mitochondria and submitochondrial particles from pigeon heart. Berden and co-workers (*45,46*) also reported type b cytochrome with the E_0' of +154 mV at pH 7.2 in beef heart mitochondria, but the absorption maximum was at 562 nm instead of 558 nm. It is not known whether these two are the same species and/or identical with the reported type b cytochrome in complex II or "chromophore 558" in complex III (*24*). A summary of the spectral and potentiometric properties of type b cytochromes proposed to exist in mitochondria is given in Table III.

D. Oxidation–Reduction Potential

A systemic survey of the E_0' values of the cytochromes in mitochondria has recently been carried out by Wilson and Dutton (*19,47*), who simultaneously measured absorption and potential titrations using dyes to mediate between the membrane-bound cytochromes and the platinum electrode. This gives the E_0' of individual cytochrome *in situ* in intact mitochondrial membrane. They revealed the presence of two type b cytochromes in rat liver and pigeon heart mitochondria (*19,20*). These are cytochrome b (b_K) and cytochrome b_T having E_0' values at pH 7.2 of +30 mV and —30 mV, respectively. When ATP is added to the mitochondrial suspension the observed E_0' of cytochrome b_T rises to 245 mV while that of cytochrome b remains unaltered. Similar findings in submitochondrial particles (*20,42*) suggest that this property is intrinsic and cytochrome b_T is postulated to exist in two states, low and high energy forms. The difference of about 300 mV in the two potential states could well be suitable to bridge the potential gap between the redox components on each side of site 2, thereby acting as an effective energy transducer. This hypothesis has been discussed and developed by Chance (*48*) and

43a. K. Ishidate, K. Kawaguchi, K. Tagawa, and B. Hagihara, *J. Biochem.* (*Tokyo*) **65**, 375 (1969).

44. I. Y. Lee and B. Chance, *BBRC* **32**, 547 (1968); K. S. Cheah and B. Chance, *BBA* **223**, 55 (1970).

44a. H. Hayashi, N. Sato, H. Oya, and B. Hagihara, *in* "Dynamics of Energy Transducing Membranes" (L. Ernster, ed.) p. 29. Elsevier, Amsterdam, 1974.

45. J. A. Berden and F. R. Opperdoes, *BBA* **267**, 7 (1972).

46. J. A. Berden, F. R. Opperdoes, and E. C. Slater, *BBA* **256**, 594 (1972).

47. P. L. Dutton, *BBA* **226**, 63 (1971).

48. B. Chance, *FEBS* (*Fed. Eur. Biochem. Soc.*) *Lett.* **23**, 3 (1972).

TABLE III

TYPE *b* CYTOCHROMES REPORTED TO EXIST IN MAMMALIAN MITOCHONDRIA

Cytochrome	Other expression	Localization	Absorption bands (reduced–oxidized) at room temp. α	β	γ	at 77°K α	β	γ	E_0' at pH 7.2 (mV)	Ref.	Remarks
b	b_K, *b*-561, *b*-562	Respiratory chain	562	532	430	559	529	428	30	*16,18–20,24*	[a]
b_T	*b*-565, *b*-566, *b*′, HP-565	Respiratory chain	566 558	538 531	432	562.5 555	535 528	430	−30 250 (+ATP)	*16,18–20,24*	[b]
b-558			558						125	*20,42*	[c]
b-559	HP-559	Respiratory chain(?)	559	529	428	556	527	427		*11–13,41*	[d]
b-560	b_{II}, *b*-557.5	Respiratory chain	560			557.5 550(?)	531 523	422		*24*	[e]
b-562		Respiratory chain(?)	562						154 107 (+antimycin)	*45,46*	[f]
b_5		Outer membrane	556	526	423	557.5 552.5	532 524.5	422		*105,106*	[g]
Sulfite oxidase		Intermembrane space	(555)	(526)	(423)[h]						[h]

[a] The classic cytochrome *b* which was found by Keilin in 1925 (*93*).

[b] The second α peak at 558 nm may result from a separate component (*b*-558) (*40,64,74*).

[c] Twenty to thirty-five percent of total absorbance at 562–575 nm (*20*).

[d] Found relatively abundant in mitochondria of liver, lung, and tumors (*11–13*). Reduced partially by succinate but not by malate (*34,41*).

[e] Found by Davis *et al.* (*24*) in complex II from beef heart mitochondria and designated as *b*-557.5 from the low temperature spectrum.

[f] Twenty percent of total absorbance at 562–575 nm (*45*). Identity of this cytochrome with the classic cytochrome *b* is not clear.

[g] 556.5 and 550 nm at 77°K are also reported (*89,105*).

[h] Extractable by the hypotonic treatment of rat liver mitochondria. Reducible with sulfite in the absence of oxygen. Absorption peaks in the reduced form are at 555–557, 526, and 423 nm (*91,92*).

by Wilson *et al.* (*29,30*). A good correlation has been obtained between the free energy change available on transfer of two electrons in each energy transduction site and free energy consumed in ATP synthesis (*29*).

The pH dependence of the E_0' was reported to be 60 mV/pH between pH 6.2 and pH 8.1 (*49*), while a break in the E_0'-pH slope has recently been reported at approximately pH 6.9 for cytochrome *b* (*50*) and cytochrome b_T (*51*).

E. Reaction with Substrates, ATP, and Antimycin

1. *Kinetics*

The addition of oxygen to the anaerobic coupled mitochondria by an oxygen pulse technique induces biphasic oxidation of type *b* cytochromes (*18*). The fast phase results from the oxidation of a component having a peak around 560 nm (cytochrome *b*) while the slow phase results from a component with a maximum around 564 nm (cytochrome b_T). Under uncoupled conditions the slow phase disappears and the entire cytochrome *b* is rapidly oxidized in a single phase. Thus, the slow phase component which is affected by the energy level of mitochondria was identified as cytochrome b_T. The biphasic response is also induced by addition of oligomycin instead of ATP, and this induction is easily abolished by uncoupling agent and valinomycin plus K^+ (*52*). Addition of durohydroquinone by a reductant pulse technique also induces biphasic reduction of cytochrome *b* (*53*); the fast phase results from the reduction of a component with the α band around 564 nm which is supposed to be cytochrome b_T, and slow phase, cytochrome *b*. These kinetic results as well as the phosphate potential dependency of E_0' suggest that this cytochrome b_T is functioning in the control of the electron transfer.

The fact that NADH ($E_0' = -0.32$ V at pH 7) or NADH-linked substrates reduce cytochrome b_T only partially and very slowly in uncoupled mitochondria (*17*) and submitochondrial particles (*54*), even

49. J. P. Straub and J. P. Colpa-Boonstra, *BBA* **60**, 650 (1962).
50. P. F. Urban and M. Klingenberg, *Eur. J. Biochem.* **9**, 519 (1969).
51. D. F. Wilson, M. Erecinska, J. S. Leigh, Jr., and M. Koppelman, *ABB* **151**, 112 (1972).
52. S. Papa, A. Scarpa, C. P. Lee, and B. Chance, *Biochemistry* **11**, 3091 (1972).
53. A. Boveris, R. Oshino, M. Erecinska, and B. Chance, *BBA* **245**, 1 (1971); A. Boveris, M. Erecinska, and M. Wagner, *ibid.* **256**, 223 (1972).
54. I. Y. Lee and E. C. Slater, *BBA* **256**, 587 (1972); **283**, 223 (1972).

under anaerobic conditions, may suggest the presence of a barrier for redox equilibrium between cytochrome b_T and NADH dehydrogenase. The increase in pH from 7 to 9 allows rapid reduction of cytochrome b_T (*55*). The addition of dyes such as phenazine methosulfate also causes rapid reduction of cytochrome b_T (*54*). It seems likely that these treatments abolish or lower the accessibility barrier.

2. *ATP-Induced Reduction of Cytochrome* b_T

The addition of ATP to anaerobic or terminally inhibited mitochondria or submitochondrial particles containing succinate ($E_0' = 0.03$ V at pH 7) induces reduction of cytochrome b_T (*16,17,55;* see also *6,56*). The original concept of the possible mechanism of this phenomenon described by Wilson and Dutton (*19*) was that the E_0' of cytochrome b_T changes because of the formation of a "high energy derivative" which is the primary intermediate for site 2 energy conservation reaction in oxidative phosphorylation. However, there has been another possible mechanism presented in which ATP can induce reduction of cytochrome b_T by the decrease in the effective redox potential (E_h) of the cytochrome because of reversed electron flow (*57*) or of the abolition of an accessibility barrier between the substrate and the cytochrome (*58*). The former explanation would be favored by the chemical hypothesis of oxidative phosphorylation, while the latter is favorable for the chemiosmotic hypothesis.

Additional factors may complicate understanding of the real mechanism for the ATP-induced reduction of cytochrome b_T: One is the induction of reversed electron flow by ATP with incomplete equilibrium between cytochromes *b* and adjacent components; the other is a modification of the original E_0' of cytochromes because of a change in its structural environment. The isolated cytochrome *b* was reported to have a much lower E_0' than that bound to particles (*59*). It is possible that the modification of the reactivity of cytochrome *b* as well as cytochrome b_T, which occurs during the fragmentation of mitochondrial structure or even by aging alone, results in the change of the ATP response to these cytochromes as well as in the change of the E_0' values.

Reduction of cytochrome b_T is also induced by coupled respiration with ascorbate and TMPD (*17*, cf. *8*) or by an efflux of K^+ from mitochondria by valinomycin (*14*).

55. A. Azzi and M. Santato, *BBRC* **45,** 945 (1971).
56. B. Chance, *Nature* (*London*) **4766,** 719 (1961).
57. A. W. Caswell, *ABB* **144,** 445, (1971).
58. E. C. Slater, *in* "Molecular Basis of Electron Transport" (J. Schultz, ed.), p. 95. Academic Press, New York, 1972.
59. R. Goldberger, A. Pumphrey, and A. L. Smith, *BBA* **58,** 307 (1962).

3. *Effect of Antimycin*

Antimycin is a potent inhibitor of the mitochondrial electron transfer between cytochromes b and c_1 (*60–62*). This antibiotic enhances the rate of reduction of type b cytochromes (b and b_T) (*2*), and, moreover, increases the amount of cytochromes b reducible by succinate (*2–4*). This results from the reduction of cytochrome b_T (*16*), presumably because of the lowering of the effective oxidation–reduction potentials. Antimycin induces a red shift of less than 1 nm in the α peak of reduced cytochrome b but not in cytochrome b_T (*16,33,46,63*). It also causes changes in the absorption spectrum of oxidized cytochrome b (*4,36*). It shifts the line at $g = 3.44$ of oxidized cytochrome in the EPR spectrum to $g = 3.48$, while it has no effect on the line at $g = 3.8$ of oxidized cytochrome b_T (*36*). Antimycin is supposed to affect, directly or indirectly, the binding of one or both of the ligands of the iron in cytochrome b resulting in the change of its absorption and EPR spectra (*64*). Antimycin abolishes the apparent ATP-induced increase in E_0' of cytochrome b_T (*33*), which suggests that antimycin inhibits the energy transfer from ATP to the coupling site 2. The stoichiometry of these interactions is one antimycin per cytochrome b or cytochrome b_T (*2,4,38,65,66*).

Inhibition by antimycin is reversible by extracting it with ether (*67*), by adding serum albumin (*68*), or by freezing in 0.2 M guanidine in complex III (*38*) but only partly in Keilin–Hartree preparation.

2-n-Heptyl-4-hydroxyquinoline N-oxide has a similar action (*69*), although 10 times less effective as antimycin, and induces the reduction of cytochrome b_T by the substrate but has no effect on the spectrum of cytochrome b or b_T (*63*).

4. *Redox Change during Respiration*

As early as 1952 Chance (*70*) observed a complete reduction of cytochrome b (more correctly cytochrome b_T) when oxygen was added to the

60. K. Ahmad, H. G. Schneider, and F. M. Strong, *ABB* **28**, 281 (1950).
61. V. R. Potter and A. E. Reif, *JBC* **194**, 287 (1952).
62. B. Chance and G. R. Williams, *Advan. Enzymol.* **17**, 65 (1956).
63. J. R. Brandon, J. R. Brocklehurst, and C. P. Lee, *Biochemistry* **11**, 1150 (1972).
64. E. C. Slater, *BBA* **301**, 129 (1973).
65. R. W. Estabrook, *BBA* **60**, 236 (1962).
66. J. A. Berden and E. C. Slater, *BBA* **216**, 237 (1970).
67. J. Bryla, Z. Kaniuga, and E. C. Slater, *BBA* **189**, 319 (1969); J. Bryla and Z. Kaniuga, *ibid.* **153**, 910 (1968).
68. A. E. Reif and V. R. Potter, *Cancer Res.* **13**, 49 (1953).
69. J. W. Lightbown and F. L. Jackson, *BJ* **63**, 130 (1956); H. Löw and I. Vallin, *BBA* **69**, 361 (1963).

anaerobic yeast cell suspension supplemented with antimycin. Many workers (*4,28,70–73*), using oxygen or dyes as electron acceptors, confirmed this phenomenon both in the presence and absence of antimycin in beef heart preparation, in succinate-cytochrome *c* reductase, and in complex III. Kinetic study indicated that immediately after the delivery of oxygen into the anaerobic suspension of mitochondria, cytochrome b_T was changed to the reduced form, independent of the energy state of the mitochondrial membrane (*48,73*). Antimycin alone had no effect on the E_0' of cytochrome *b* and b_T (*28,33*,cf.*45*). Antimycin probably acts as a stabilizer of cytochrome b_T either in the reduced state or in the high energy state, induced by the respiration. The oxidant-induced reduction of cytochrome b_T is explained in a manner similar to that for the ATP-induced reduction of cytochrome b_T: (1) The energy released by the electron flow via the cytochrome *b* region induces an increase in the E_0' of cytochrome b_T resulting in easy reduction of this cytochrome by a low potential component (*28,73*) and (2) oxidants induce a decrease in the effective redox potential of the reductant of cytochrome b_T without any change of the E_0' of cytochrome b_T (*74*). An electrostatic potential difference induced within the membrane, or change in local pH (*25*), might also explain the cytochrome b_T reduction. Oxidation of cytochrome c_1 by the added oxidant may be required for the initiation of the effect, because rapid oxidation of cytochrome c_1 precedes the reduction of cytochrome b_T (*73*). Coenzyme Q may be the donor of reducing equivalents to cytochrome b_T since oxidation of coenzyme Q was kinetically compatible with the reduction of cytochrome b_T (*73*). Cytochrome $b_{(k)}$ is another candidate for the donor of reducing equivalents to cytochrome b_T (*13*).

F. Miscellaneous

1. *Purification of Mammalian Cytochrome b*

Several methods have been reported for the purification of cytochrome *b* (*b* or $b + b_T$) from beef heart mitochrondria (*59,75–86*). In most of

70. B. Chance, *Proc. Int. Congr. Biochem, 2nd, 1952* Abstract, p. 32 (1953).
71. L. Kovac, P. Smigan, E. Hrusovska, and B. Hess, *ABB* **139,** 370 (1970).
72. H. Baum, J. S. Rieske, H. I. Silman, and S. H. Lipton, *Proc. Nat. Acad. Sci. U. S.* **57,** 798 (1967).
73. M. Erecinska, B. Chance, D. F. Wilson, and P. L. Dutton, *Proc. Nat. Acad. Sci. U. S.* **69,** 50 (1972).
74. M. K. F. Wikström and J. A. Berden, *BBA* **283,** 403 (1972); M. K. F. Wikström, *ibid.* **301,** 155 (1973).
75. I. Sekuzu and K. Okunuki, *J. Biochem.* (*Tokyo*) **43,** 107 (1956).
76. D. Feldman and W. W. Wainio, *JBC* **235,** 3635 (1960).
77. R. Bomstein, R. Goldberger, and H. Tisdale, *BBRC* **2,** 234 (1960).
78. K. Ohnishi, *J. Biochem.* (Tokyo) **59,** 1, 9, and 17 (1966).
79. H. Shichi and Y. Kuroda, *ABB* **118,** 682 (1967).

these methods cytochrome *b* is extracted with cholate or deoxycholate together with cytochromes c_1 and $a + a_3$, which can be removed from cytochrome *b* by ammonium sulfate fractionation and detergent treatment. Purified cytochrome *b* is hydrophobic and solubilized with detergents such as cholate, Emasol 1130, Triton X-100, and sodium dodecyl sulfate or by proteolytic digestion. It can be further purified by gel filtration or chromatography on DEAE-cellulose. These purified preparations of cytochrome *b* are always autoxidizable and combine with carbon monoxide. Although their α absorption band is usually found at 561–563 nm, it is not exactly the same as that of cytochrome *b* observed in intact mitochondria and is quite different from that of cytochrome b_T. These preparations, therefore, appeared to be significantly modified materials derived from cytochrome *b* or both cytochrome *b* and b_T. The α band of cytochrome b_T is readily shifted to around 562 nm by various treatments (*34,37*).

2. *Cytochrome b-Like Pigment in Invertebrates*

As noted above, cytochrome *b* in the mammalian respiratory system is firmly bound to the mitochondrial membrane and cannot be solubilized without using detergents such as cholate. It is thus interesting that a cytochrome *b*-like pigment, cytochrome *b*-563, is easily extracted from the larvae and pupae of the housefly, *Musca domestica,* with a simple salt solution, but it cannot be extracted from the adult flies under the same conditions. The cytochrome is obtained as a crystalline preparation after purification by the chromatography on DEAE-cellulose (*78,84*). The crystalline cytochrome *b*-563 has a similar spectral property to mammalian cytochrome *b* and shows absorption peaks at 563, 530, and 428.5 nm in the reduced state. Contrary to purified mammalian cytochrome *b,* this cytochrome does not combine with carbon monoxide and exists as monomer species in the solution without detergent.

3. *Other Type b Cytochromes in Mitochondria*

The presence of a few type *b* cytochromes in mitochondria other than those in the respiratory chain has been reported. These are cytochrome

80. J. S. Rieske and H. D. Tisdale, "Methods in Enzymology," Vol. 10, pp. 349 and 353, 1967.
81. S. Yamashita and E. Racker, *JBC* **244,** 1220 (1969).
82. T. E. King, *Advan. Enzymol.* **28,** 155 (1966).
83. T. E. King, "Methods in Enzymology, Vol. 10, p. 202, 1967.
84. T. Yamanaka, S. Tokuyama, and K. Okunuki, *BBA* **77,** 592 (1963).
85. J. S. Rieske, "Method in Enzymology," Vol. 10, p. 488, 1967.
86. H. Baum, H. I. Silman, J. S. Riske, and S. H. Lipton, *JBC* **242,** 4876 (1967).

b_5 in the outer membrane of liver mitochrondia (*87–89*) (see Section III,A,2) and a type *b* cytochrome which is reported to be involved in sulfite oxidase (*90–92*) (see Table III).

III. Cytochrome b_5

A. Introduction

The name cytochrome b_5 is now generally used for a protohemoprotein which exists in high amounts in the endoplasmic reticulum of mammalian liver cells. It shows an asymmetrical α-absorption band with a peak at 556 nm and a shoulder around 560 nm in the reduced state. However, there has been some confusion about its nomenclature. This cytochrome is firmly bound to the membrane structure and reduced with NADH by a flavoprotein (cytochrome b_5 reductase) which is also bound to the membrane. It is solubilized by proteolytic cleavage from the membrane and has been crystallized from several sources. The molecular weight of such preparations are about 11,000. Although both primary and ternary structures of such preparations are known, its biological role is still uncertain. Earlier findings have been fully reviewed by Strittmatter (*94*).

1. *Nomenclature*

The α-absorption band near 560 nm observed in liver tissue was first referred to as cytochrome b_1 (see Section IV,B) by Keilin and Hartree (*95*). The pigment observed in the microsomal fraction of liver was termed cytochrome *b′* by Yoshikawa (*96*) and *m* by Strittmatter and Ball (*97,98*). The name cytochrome b_5 was given by Pappenheimer and Williams to the pigment found in the midgut of larvae of the *Cecropia*

87. G. L. Sottocasa, B. Kuylenstierna, L. Ernster, and A. Bergstrand, *J. Cell Biol.* **32,** 415 (1967).
88. I. Raw, N. Petragnani, and O. C. Nogueira, *JBC* **235,** 1517 (1960).
89. D. F. Parsons, G. R. Williams, W. Thompson, D. F. Wilson, and B. Chance, "Mitochondrial Structure and Compartmentation" (E. Quagliariello *et al.*, eds.), p. 5. Adriatice Editrice, Bari, 1967.
90. R. C. Bray, "The Enzymes," Vol. 7, p. 533, 1963.
91. H. J. Cohen and I. Fridovich, *JBC* **246,** 367 (1971).
92. A. Ito, *J. Biochem.* (*Tokyo*) **70,** 1061 (1971).
93. D. Keilin, *Proc. Roy. Soc., Ser. B* **98,** 312 (1925).
94. P. Strittmatter, "The Enzymes," 2nd ed., Vol. 8, p. 113, 1963.
95. D. Keilin and E. F. Hartree, *Proc. Roy. Soc., Ser. B* **129,** 277 (1940).
96. H. Yoshikawa, *J. Biochem.* (*Tokyo*) **38,** 1 (1951).
97. C. F. Strittmatter and E. G. Ball, *Proc. Nat. Acad. Sci. U. S.* **38,** 19 (1952).
98. C. F. Strittmatter and E. G. Ball, *J. Cell. Comp. Physiol* **43,** 57 (1954).

silkworm (*99*) and by Chance and Williams to that in liver microsomes (*100*). Use of the term cytochrome b_5 in referring to hemoproteins of biological materials other than liver microsomes, simply by similarity in the wavelength of the α peak, may not be desirable unless: (1) there is similarity of the low temperature spectrum in the reduced state (77°K or lower); or (2) there is similarity in amino acid sequence or immunochemical similarity; or (3) a similar reactivity to that of microsomal cytochrome b_5 reductase is shown.

2. *Distribution*

Besides the endoplasmic reticulum of mammalian liver, cytochrome b_5 is contained in the outer membranes (*89,101*) of mitochondria of the same tissue (*88,102–104*). Spectral difference between the above two cytochromes b_5 have been reported (*89,105*), but their similarity has recently been confirmed spectrometrically (*106*) and immunochemically (*107*). Similar cytochromes contained in erythrocytes (*108–110*) and kidney (*96,111*) have been partially purified (*110,111*). Spectrally similar pigments are found in most of the hormones secreting organs such as pancreas (*112*) adrenal glands (*113–117a*), mammary gland (*112,118*), ovary (*96,113*), intestinal mucosa (*118*), and thyroid gland (*118a*).

99. A. M. Pappenheimer and C. M. Williams, *JBC* **209,** 915 (1954).
100. B. Chance and G. R. Williams, *JBC* **209,** 945 (1954).
101. K. A. Davis and G. Kreil, *BBA* **162,** 627 (1968).
102. I. Raw and W. Colli, *Nature* (*London*) **184,** 1798 (1959).
103. I. Raw and H. R. Mahler, *JBC* **234,** 1867 (1959).
104. I. Raw, R. Molinari, D. F. do Amaral, and H. R. Mahler, *JBC* **233,** 225 (1959).
105. H. Wohlrab and H. Degn, *BBA* **256,** 216 (1972).
106. E. Furuya, N. Sato, and B. Hagihara, *J. Jap. Biochem. Soc.* **42,** 60 (1970).
107. A. Ito, *J. Jap. Biochem. Soc.* **45,** 642 (1973).
108. D. E. Hultquist, D. W. Reed, P. G. Passon, and W. E. Andrews, *BBA* **229,** 33 (1971).
109. D. E. Hultquist and P. G. Passon, *Nature* (*London*), *New Biol.* **229,** 252 (1971).
110. P. G. Passon, D. W. Reed, and D. E. Hultquist, *BBA* **275,** 51 (1972).
111. V. M. Mangum, M. D. Klingler, and J. A. North, *BBRC* **40,** 1520 (1970).
112. G. E. Palade and P. Siekevitz, *J. Biochem. Biophys. Cytol.* **2,** 671 (1956).
113. C. A. McMunn, *J. Physiol.* (*London*), **5,** *xxiv* (1884).
114. I. Huszak, *Biochem. Z.* **312,** 330 (1942).
115. M. J. Spiro and E. G. Ball, *Fed. Proc., Fed. Amer. Soc. Exp. Biol.* **17,** 314 (1958).
116. Y. Ichikawa and T. Yamano, *BBRC* **20,** 263 (1965).
117. D. Y. Cooper, J. Narasimhulu, O. Rosenthal, and R. W. Estabrook, "Amherst," Vol. 2, p. 838 1965.
117a. K. J. Ryan and L. L. Engel, *JBC* **225,** 103 (1957).
118. M. Bailie and R. K. Morton, *Nature* (*London*) **176,** 111 (1955).
118a. T. Hosoya and M. Morrison, *Biochemistry* **6,** 1021 (1969).

3. *Biological Role*

The physiological function of cytochrome b_5 is still uncertain, although several possibilities have been suggested. It has been implicated in fatty acid desaturation reactions in the endoplasmic reticulum of liver (*119*) and in some hydroxylase reactions both in the reticulum (*120*) and mitochondria (*121,122*). It may participate in the reduction of methemoglobin in erythrocytes (*109*) since ferrocytochrome b_5 is readily oxidized by methemoglobin as well as ferricytochrome *c*. The half-life of cytochrome b_5 in rat liver microsomes is reported to be 45 hr (*123*).

4. *Isolation*

Microsomal cytochrome b_5 has been solubilized by treatment with pancreatic lipase (*124–129*), proteolytic enzymes (*127–129*), or detergents (*130,130a*). Solubilization by the lipase was suggested to result from the action of proteolytic enzymes contained in the lipase preparation or in the microsomal suspensions. (*128*). All methods for the purification of the solubilized cytochrome b_5 adopt, as main procedures, ammonium sulfate fractionation and chromatography on DEAE-cellulose as introduced by Strittmatter and co-workers (*126, 131,132*). DEAE Sephadex chromatography was also used for the separation of two forms of cytochrome b_5 after treatment with trypsin (*129,130*).

119. N. Oshino, Y. Imai, and J. Sato, *J. Biochem.* (*Tokyo*) **69,** 155 (1971).
120. A. Hildebrandt and R. W. Estabrook, *ABB* **143,** 66 (1971).
121. U. Schmeling, G. Mayer, H. Diehl, V. Ullrich, and Hj. Staudinger, *Hoppe-Seyler's Z. Physiol. Chem.* **350,** 349 (1969).
122. G. Mayer, V. Ullrich, and Hj. Staudinger, *Hoppe-Seyler's Z. Physiol. Chem.* **349,** 459 (1968).
123. H. Greim, *Naunyn-Schmiedebergs Arch. Pharmakol. Exp. Pathol.* **266,** 261 (1970).
124. P. Strittmatter and S. F. Velick, *JBC* **228,** 785 (1957).
125. P. Strittmatter and S. F. Velick, *JBC* **221,** 253 (1956).
126. P. Strittmatter, *JBC* **235,** 2492 (1960).
127. T. Omura, P. Siekevitz, and G. E. Palade, *JBC* **242,** 3389 (1967).
128. T. Kajihara and B. Hagihara, *in* "Structure and Function of Cytochromes" (K. Okunuki, M. D. Kamen, and I. Sekuzu, eds.), p. 581. Univ. of Tokyo Press, Tokyo, 1968.
129. T. Kajihara and B. Hagihara, *J. Biochem.* (*Tokyo*) **63,** 453 (1968).
130. L. Spatz and P. Strittmatter, *Proc. Nat. Acad. Sci. U. S.* **68,** 1042 (1971).
130a. A. Ito and R. Sato, *JBC* **243,** 4922 (1968).
130b. T. Okuda, K. Mihara, and R. Sato, *J. Biochem.* (*Tokyo*) **72,** 987 (1972).
130c. P. Strittmatter, M. J. Rogers, and L. Spatz, *JBC* **247,** 7188 (1972).
130d. K. Enomoto and R. Sato, *BBRC* **51,** 1 (1973).
131. P. Strittmatter and J. Ozols, *JBC* **241,** 4782 (1966).
132. P. Strittmatter, "Methods in Enzymology," Vol. 10, p. 553, 1967.

The first crystallization of cytochrome b_5, obtained from pig liver was reported by Raw and Coli in 1959 (*102*). Kajihara and Hagihara (*129*) obtained three crystalline cytochrome b_5 preparations from rabbit liver microsomes, two from trypsin extracts, and one from Nagarse (subtilisin BPN′) extracts. The three preparations crystallized in entirely different shapes. Calf liver cytochrome b_5 has also been crystallized by Mathews and Strittmatter (*133*). The preparations by the latter two groups have been used for studies on the amino acid sequence and the three-dimensional structure, respectively.

B. Properties of Cytochrome b_5

1. *Chemical Properties*

Cytochrome b_5 preparations obtained by proteolytic solubilization contain approximately 87–97 amino acid residues, and have a molecular weight of roughly 11,000. The detergent-solubilized cytochrome b_5 (detergent-b_5) contains, in addition to the above molecule of 97 residues, a chain of 44 residues which is composed of about 60% hydrophobic amino acids, and has a molecular weight of 16,700 (*130*). This detergent-b_5 exists as an oligomer (octomer) in aqueous solution without detergent (*130a*). The detergent-b_5 forms a functional aggregate with NADH-cytochrome b_5 reductase which is prepared by detergent treatment (detergent-reductase), and the aggregate shows high activity of cytochrome *c* reduction with NADH (*130b*). On the contrary, no aggregate is formed between trypsin solubilized cytochrome b_5 and the detergent reductase, and the mixture shows very low cytochrome *c* reductase activity. The detergent-b_5 is easily incorporated in microsomes and is highly active as electron acceptor in the NADH-cytochrome b_5 reductase system of the membrane, while trypsin-b_5 is poorly incorporated to the membrane (*130c,130d*). The "hydrophobic tail" in the detergent-b_5 is supposed to participate in the above aggregation and membrane incorporation.

Isolated cytochrome b_5 is considerably stable at temperatures below 50° at neutral or alkaline pH. At acidic pH it becomes progressively more unstable. The cytochrome is slightly autoxidizable but does not combine with carbon monoxide or azide. It reacts with cyanide in a biphasic way owing to the formation of mono- and dicyanide complexes (*134*). It is reduced by dithionite, cysteine and reduced indigo di-, tri-, and tetrasulfonates at pH 7–8, and by borohydride at pH 5.5. In the presence of microsomes (or either NADH or NADPH specific reductase), cytochrome b_5 is reduced by NADH or NADPH.

133. F. S. Mathews and P. Strittmatter, *JMB* **41**, 295 (1969).
134. M. Ikeda, T. Iizuka, H. Takao, and B. Hagihara, *BBA* **336**, 15 (1974).

The heme group in the cytochrome b_5 molecule is easily removed by an acid-acetone treatment at low temperature, and the resulting apoprotein rapidly recombines with protohematin. The reconstituted cytochrome shows the same spectral and enzymic properties as the original cytochrome b_5 (*126*). The apocytochrome b_5 readily combines also with other iron porphyrins (*135*).

2. *Physical Properties*

Cytochrome b_5 behaves as a univalent electron carrier, and the E_0' at pH 7 has been estimated to be +20 mV (*136*). A slightly lower value (—12 to —14 mV) has been reported for the enzyme bound in microsomes (*97,137*).

The heme iron in both ferrous and ferric cytochrome *b* is low spin. A small high-spin signal observed in the ferric form (*138*) has been confirmed by Ikeda *et al.* (*134*) to result from contaminated hemeproteins. However, these authors found from their spectral and EPR studies that the ferricytochrome reversibly changes to a high-spin state below pH 4 at a temperature between 20° and 296°K, presumably because of heme dissociation. Above pH 12 it also changes reversibly to another type of low-spin state as a result of distortion of protein structure. Energy for three t_{2g} orbitals calculated in one hole formalism shows a high symmetry of ligand coordination for the low-spin state at pH 6.2 and a lower symmetry for another type at pH 12 (*134*). The heme environment of cytochrome b_5 as well as cytochrome b_2 was also studied by Keller *et al.* (*139,140*) by measuring the proton NMR spectra; the spectrum is readily observed in the case of ferric and ferrous low-spin compounds. Striking structural similarities were observed between the two cytochromes.

3. *Spectral Properties*

The main peaks (α, β, and γ) in the absorption spectrum of the mammalian liver cytochrome b_5 in the reduced state are given in Table IV. The extinction coefficients at the α and γ peaks are 26 and 171 mM^{-1} cm^{-1}, respectively (*125,141*). The absorption spectrum around the α peak (556 nm) is asymmetrical with a shoulder around 560 nm as shown in

135. J. Ozols and P. Strittmatter, *JBC* **239,** 1018 (1964).
136. S. F. Velick and P. Strittmatter, *JBC* **221,** 265 (1956).
137. Y. Kawai, Y. Yoneyama, and H. Yoshikawa, *BBA* **67,** 522 (1963).
138. R. Bois-Poltoratsky and A. Ehrenberg, *Eur. J. Biochem.* **2,** 361 (1967).
139. R. M. Keller and K. Wüthrich, *BBA* **285,** 326 (1972).
140. R. M. Keller, O. Groundinsky, and K. Wüthrich, *BBA* **328,** 233 (1973).
141. D. Garfinkel, *ABB* **77,** 493 (1958).

TABLE IV

CYTOCHROMES b_5 AND SPECTRALLY SIMILAR PIGMENTS ISOLATED FROM VARIOUS ORGANISMS

Cytochromes	Main absorption peak (nm) Reduced α	Reduced β	Reduced γ	Oxidized γ	MW	E_0' at pH 7.0 (mV)	pI	Ref.
Cytochrome b_5 (mammalian liver microsomes)	556	526	423	413	11,000 16,700	200	Acidic	*93,129,131, 145,152,154*
Cytochrome b_5 (erythrocytes)	556	527	423	413	14,600–18,400		Acidic	*108–110*
Cytochrome *b*-555 (housefly larvae)	555	528	424	414	13,700	6	Acidic	*155*
Cytochrome *b*-555 (mung bean seedlings)	555	527	423	413	13,500	−30	Acidic	*157*
Cytochrome *b*-554 (*Aspergillus oryzae*)	554	526	423	413				*158*
Cytochrome *b*-556 (*Neurospora* sp.)	556	526	420				Acidic	*189*
Cytochrome *b*-556 (*Saccharomyces cerevisiae*)	556	525	423	413	12,700	−23		*159*

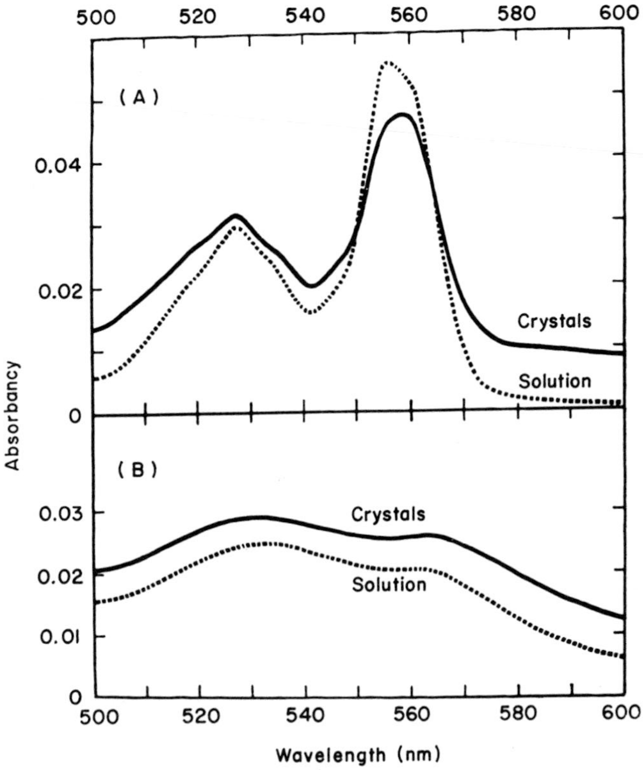

FIG. 3. The absorption spectra in α and β region of (A) reduced and (B) oxidized cytochrome b_5 in solution and crystals (*142*). Equal concentration (approximately 2.2 μM) of cytochrome b_5 was used in all cases.

Fig. 3. This α band becomes symmetrical when the spectrum is measured using the suspension of the crystalline cytochrome or a single crystal (*142*). Estabrook found that the above peak and shoulder are separated into two distinct peaks at low temperature (77°K) (*143*). The spectra at various temperatures from 4° to 273°K (*26*) are shown in Fig. 4.

C. STRUCTURE

The primary structures of ctyochrome b_5 from several mammalian livers have been studied mainly by two groups using the enzyme-solubilized preparations which are supposed to correspond to about 70% of the

142. B. Hagihara and T. Kajihara, *Proc. Jap. Acad.* **44,** 35 (1968).
143. R. W. Estabrook, *in* "Haematin Enzymes" (J. E. Falk, R. Lember, and R. K. Morton, eds., Vol. 2, p. 436. Pergamon, Oxford, 1961.

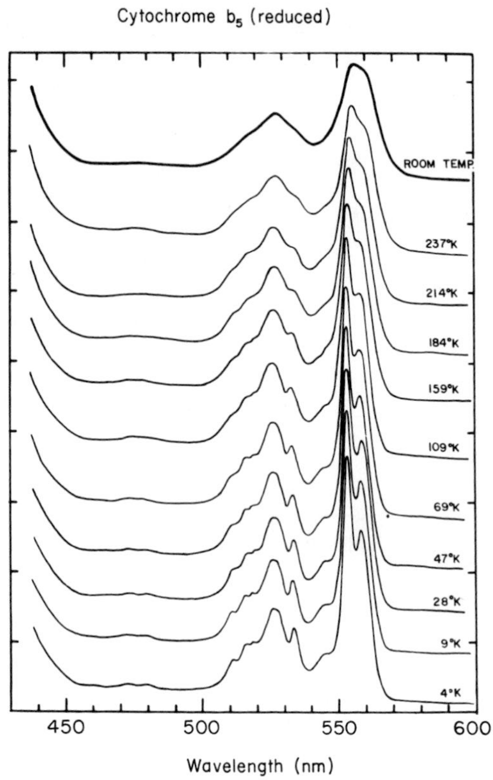

FIG. 4. The absorption spectra of reduced cytochrome b_5 measured at several different temperatures between liquid helium and room temperature (*26*). Conditions are similar to those described in the legend of Fig. 1 except no deoxycholate is contained.

whole molecules. The ternary structure has been determined by Mathews *et al.* (*144–146*) using a cow liver preparation.

1. *Amino Acid Sequence*

The first amino acid sequence of cytochrome b_5 was determined by Ozols and Strittmatter (*147–149*) by using "lipase" extracted calf liver preparation. However, their study contained rather serious errors such

144. F. S. Mathews, M. Levine, and P. Argos, *Nature* (*London*), *New Biol.* **233,** 15 (1971).
145. F. S. Mathews, M. Levine, and P. Argos, *JMB* **64,** 449 (1972).
146. F. S. Mathews, M. Levine, and P. Argos, *Cold Spring Harbor Symp. Quant. Biol.* **36,** 387 (1971).
147. J. Ozols and P. Strittmatter, *Proc. Nat. Acad. Sci. U. S.* **58,** 264 (1967).
148. J. Ozols and P. Strittmatter, *JBC* **243,** 3367 (1968).
149. J. Ozols and P. Strittmatter, *JBC* **243,** 3376 (1968).

as missing nine amino acid residues. Shortly thereafter, Tsugita *et al.* determined the structure of a crystalline rabbit liver preparation obtained by trypsin extraction (*150*). The former authors revised their structure similar to the latter's (*151*). The structures were also determined by Tsugita *et al.* with human, calf, and chicken liver (*152,153*), and by Nobrega and Ozols with porcine, monkey, human, and chicken liver preparations (*154*), all extracted by proteolytic enzymes. The most probable structure of cytochrome b_5 from the above six different sources are shown in Table V.

2. *Ternary Structure*

The X-ray crystallographic studies on calf liver cytochrome b_5 (with 93 amino acid residues) at 2.8 and 2.0 Å resolution were carried out by Mathews *et al.* (*144–146*). Their results show that the cytochrome molecule contains six short helical regions and five short segments of an extended β chain which form a twisted pleated sheet structure. Distribution of these secondary structures, which involve about 75% of the 86 residues in the core part of the molecule, is indicated in Fig. 5.

Like many other proteins, the interior of the molecule is distinctly nonpolar. The heme is buried in a hydrophobic crevice with two vinyl groups lying deep in the interior of the molecule. One of the propionic acid groups lies on the surface of the molecule and is hydrogen-bonded to the γ-oxygen and the peptide nitrogen of Ser-64, and the other projects outward into solution. The walls of the heme crevice are formed by two pairs of roughly antiparallel helices and the floor by the pleated sheet structure. The iron atom is coordinated by His-39 and His-63 which extend from the walls of the crevice. The nitrogens of the two histidines which are in the NH tautomeric form, are hydrogen-bonded to the main chain carbonyl oxygens of Gly-40 and Phe-58, respectively. Furthermore, His-39 is in van der Waals contact with Leu-46 and His-63 lies close and parallel to Phe-58, suggesting a π–π interaction between the latter pair of residues. Thus, the histidine residues are held firmly in place by the rigidity of the backbone structure and by a variety of interactions with the main and side chains.

The first two residues and the last six (1, 2, 87–93) are disordered in the crystal structure. This explains the finding by Strittmatter and Ozols

150. A. Tsugita, M. Kobayashi, T. Kajihara, and B. Hagihara, *J. Biochem.* (*Tokyo*) **64,** 727 (1968).
151. J. Ozols and P. Strittmatter, *JBC* **244,** 6617 (1969).
152. A. Tsugita, M. Kobayashi, S. Tani, S. Kyo, M. A. Rashid, Y. Yoshida, T. Kajihara, and B. Hagihara, *Proc. Nat. Acad. Sci. U. S.* **67,** 442, (1970).
153. M. A. Rashid, B. Hagihara, M. Kobayshi, S. Tani, and A. Tsugita, *J. Biochem.* (*Tokyo*) **74,** 985 (1973).
154. F. G. Nobrega and J. Ozols, *JBC* **246,** 1706 (1971).

TABLE V

AMINO ACID SEQUENCE OF CYTOCHROME b_5 FROM LIVER OF CALF,[a] RABBIT,[b] MAN,[c] MONKEY,[d] PIG,[d] AND CHICKEN[e]

	1 10
Calf	Ser -Lys-Ala -Val -Lys-Tyr-Tyr-Thr-Leu-Glu-Glu-Ile -Glu-Lys-His-Asn-Asn-Ser-
Rabbit	Gln-Ala-Ala-Ser-Asp-Lys-Asp-Val -Lys-Tyr-Tyr-Thr-Leu-Glu-Glu-Ile -Lys-Lys-His-Asn-His -Ser-
Man	Gln-Ala-Ala-Ser-Glx -Glx-Ala -Val -Lys-Tyr-Tyr-Thr-Leu-Glu-Glu-Ile -Glu-Lys-His-Asn-His -Ser-
Monkey	Asx-Ala -Val -Lys-Tyr-Tyr-Thr-Leu-Glu-Glu-Ile -Glu-Lys-His-Asn-His -Ser-
Pig	Ala -Val -Lys-Tyr-Tyr-Thr-Leu-Glu-Glu-Ile -Glu-Lys-His-Asn-Asn-Ser-
Chicken	Gly-Arg-Tyr-Tyr-Arg-Leu-Glu-Glu-Val-Gln-Lys-His-Asn-Asn-Ser-
	20 30
Calf	Lys-Ser-Thr-Trp-Leu-Ile-Leu-His-Tyr-Lys-Val-Tyr-Asp-Leu-Thr-Lys-Phe-Leu-Glu-Glu-His-
Rabbit	Lys-Ser-Thr-Trp-Leu-Ile-Leu-His-His -Lys-Val-Tyr-Asp-Leu-Thr-Lys-Phe-Leu-Glu-Glu-His-
Man	Lys-Ser-Thr-Trp-Leu-Ile-Leu-His-His -Lys-Val-Tyr-Asp-Leu-Thr-Lys-Phe-Leu-Glu-Glu-His-
Monkey	Lys-Ser-Thr-Trp-Leu-Ile-Ileu-His-His -Lys-Val-Tyr-Asp-Leu-Thr-Lys-Phe-Leu-Glu-Glu-His-
Pig	Lys-Ser-Thr-Trp-Leu-Ile-Ileu-His-His -Lys-Val-Tyr-Asp-Leu-Thr-Lys-Phe-Leu-Glu-Glu-His-
Chicken	Glx-Ser-Thr-Trp-Ile -Ile-Val -His-His -Arg-Ile -Tyr-Asp-Ile -Thr-Lys-Phe-Leu-Asp-Glu-His-
	40 50
Calf	Pro-Gly-Gly-Glu-Glu-Val-Leu-Arg-Glu-Gln-Ala-Gly-Gly-Asp-Ala-Thr-Glu-Asp-Phe-Glu-
Rabbit	Pro-Gly-Gly-Glu-Glu-Val-Leu-Arg-Glu-Gln-Ala-Gly-Gly-Asp-Ala-Thr-Glu-Asp-Phe-Glu-
Man	Pro-Gly-Gly-Glu-Glu-Val-Leu-Arg-Glu-Gln-Ala-Gly-Gly-Asp-Ala-Thr-Glu-Asp-Phe-Glu-
Monkey	Pro-Gly-Gly-Glu-Glu-Val-Leu-Arg-Glu-Gln-Ala-Gly-Gly-Asp-Ala-Thr-Glu-Asp-Phe-Glu-
Pig	Pro-Gly-Gly-Glu-Glu-Val-Leu-Arg-Glu-Gln-Ala-Gly-Gly-Asp-Ala-Thr-Glx-Asp-Phe-Glu-
Chicken	Pro-Gly-Gly-Glu-Glu-Val-Leu-Arg-Glu-Gln-Ala-Gly-Gly-Asp-Ala-Thr-Glu-Asp-Phe-Glu-

	60 70
Calf	Asp-Val-Gly-His-Ser-Thr-Asp-Ala-Arg-Glu-Leu -Ser-Lys-Thr-Phe-Ile-Ile-Gly-Glu-Leu-
Rabbit	Asp-Val-Gly-His-Ser-Thr-Asp-Ala-Arg-Glu-Leu -Ser-Lys-Thr-Phe-Ile-Ile-Gly-Glu-Leu-
Man	Asp-Val-Gly-His-Ser-Thr-Asp-Ala-Arg-Glu-Met-Ser-Lys-Thr-Phe-Ile-Ile-Gly-Glu-Leu-
Monkey	Asp-Val-Gly-His-Ser-Thr-Asp-Ala-Arg-Glu-Leu -Ser-Lys-Thr-Phe-Ile-Ile-Gly-Glu-Leu-
Pig	Asp-Val-Gly-His-Ser-Thr-Asp-Ala-Arg-Glu-Leu -Ser-Lys-Thr-Phe-Ile-Ile-Gly-Glu-Leu-
Chicken	Asp-Val-Gly-His-Ser-Thr-Asp-Ala-Arg-Ala -Leu -Ser-Glu-Thr-Phe-Ile-Ile-Gly-Glu-Leu-
	80 90
Calf	His-Pro-Asp-Asp-Arg-Ser -Lys-Ile -Thr-Lys-Pro-Ser- Glu-Ser
Rabbit	His-Pro-Asp-Asp-Arg-Ser -Lys-Leu-Ser -Lys-Pro-Met-Glu-Thr
Man	His-Pro-Asp-Asp-Arg-Pro-Lys
Monkey	His-Pro-Asp-Asp-Lys-Pro-Arg
Pig	His-Pro-Asp-Asp-Arg
Chicken	His-Pro-Asp-Asp-Arg-Pro-Lys-Leu-Arg

[a] Data from Ozols and Strittmatter (*151*) and Tsugita *et al.* (*152*).
[b] Data from Tsugita *et al.* (*150,152*).
[c] Data from Tsugita *et al.* (*152*), Rashid *et al.* (*153*), and Nobrega ad Ozols (*154*).
[d] Data from Nobrega and Ozols (*154*).
[e] Data from Tsugita *et al.* (*152*) and Nobrega and Ozols (*154*).

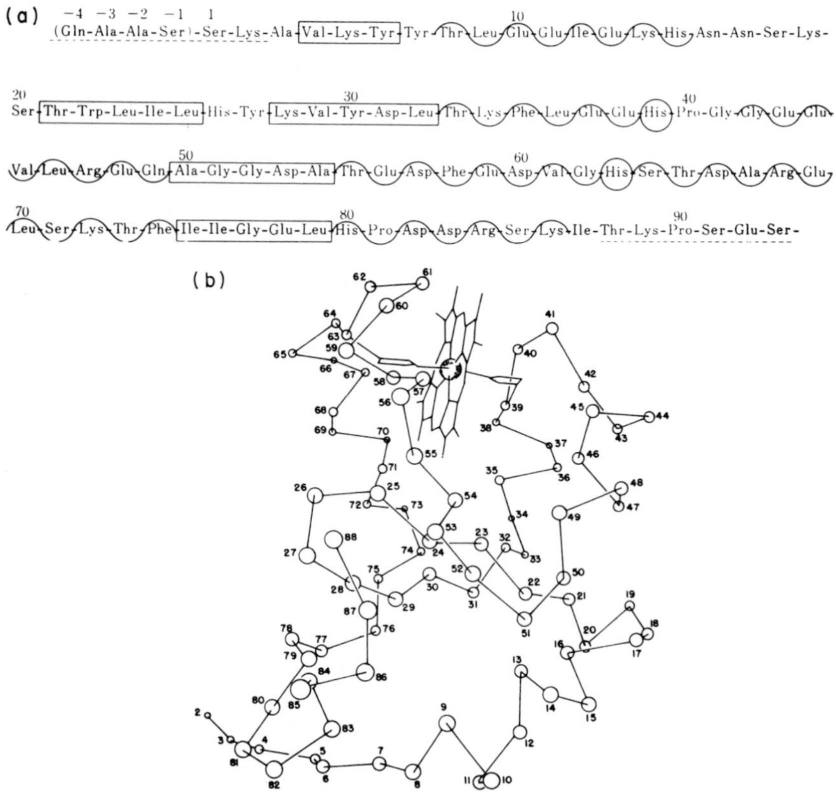

FIG. 5. Structure of calf liver cytochrome b_5 (*145*). (a) Primary and secondary structure. (b) Schematic diagram of the backbone chain of cytochrome b_5. The α-carbon positions are represented by circles whose sizes indicate their relative depth.

that the first two and the last seven residues can be cleaved by trypsin without a loss of activity (*131*). The core part (3–86 residues) contains the heme group at the top, lying in the hydrophobic crevice, and also has a narrow hydrophobic group open to the aqueous environment. The residues principally involved in this latter group are Phe-35, Leu-70, and Phe-74. The site of the action of the NADH-cytochrome b_5 reductase may be this group, and the two phenylalanines in the vicinity of the group may provide a path for an electron to the heme.

D. CYTOCHROME b_5-LIKE HEMOPROTEINS

Although the name cytochrome b_5 is now used for a liver microsomal cytochrome, this name was first given to the pigment of a silkworm larvae (*99*). It is not clear whether this cytochrome is a structurally similar protein to mammalain cytochrome b_5, differing from it only by species-specificity, or an entirely different protein with similar spectrum; but the former case is more probable from the following work. A quite similar pigment to the above is contained also in the larvae and pupae of the housefly, *Musca domestica*. This cytochrome was extracted with a 40%-saturated ammonium sulfate solution together with another cytochrome, *b*-563, then highly purified by DEAE-cellulose chromatography (*155*). The purified cytochrome exhibits closely similar spectral and enzymic properties to mammalian cytochrome b_5. The α band is asymmetrical and is split into two peaks at 552 and 556 nm at liquid nitrogen temperature. The cytochrome is reduced with NADH through NADH-cytochrome b_5 reductase derived from rat liver (*156*).

Cytochrome b_5-like proteins in housefly and presumably other invertebrates are more easily extractable (usually with simple salt solutions) than the liver cytochrome b_5 which is extracted only by the action of proteolytic enzymes or detergents. However, the cytochrome *b*-555 (as well as cytochrome *b*-563) of the housefly is not extracted from the adult flies with the salt solution (*155*).

Similar cytochrome b_5-like proteins are also extracted with a simple salt solution from the seedlings of mung bean, *Phaseolus aureus* (*157*), and *Aspergillus oryze* (*158*). In the case of the fungi, cells are dried by the acetone treatment or lyophilization before the extraction with an appropriate buffer solution. Cytochrome b_5-like hemoprotein can also be obtained from yeast, *Saccharomyces cerevisiae*, by tryptic digestion (*159*). The main properties of the above purified preparations of b_5-like cytochromes are summarized in Table IV, together with those of mammalian cytochrome b_5. The function of the cytochrome b_5-like hemoproteins have not been elucidated. In the case of fungal cytochrome, it may function as the electron donor for nitrate reductase since the nitrate reductase complex with molecular weight of 228,000 isolated from *Neurospora crassa* contains cytochrome *b*-557 (*160*).

IV. Type *b* Cytochromes in Microorganisms

A. Introduction

A large number of type *b* cytochromes have been found in bacteria, fungi, and protozoa, and isolated in partially or highly purified states.

155. T. Yamanaka, S. Tokuyama, and K. Okunuki, *BBA* **77**, 592 (1963).
156. Y. Okada and K. Okunuki, *J. Biochem.* (*Tokyo*) **65**, 581 (1969).
157. H. Shichi and D. P. Hackett, *JBC* **237**, 2955 (1962).
158. T. Yamanaka, T. Nishimura, and K. Okunuki, *J. Biochem.* (*Tokyo*) **54**, 1616 (1963).
159. Y. Yoshida and H. Kumaoka, *BBA* **189**, 461 (1969).
160. R. H. Garrett and A. Nason, *JBC* **244**, 2870 (1969).

These are listed in Table VI (*156,161–192*) with their main spectral and molecular properties as well as their biological roles if they are known. Except in the cases of cytochromes b_1 and b_2, these microbial cytochromes are named by the position of the α peak at room temperature such as cytochromes *b*-554, *b*-563, and *b*-567. Microbial cytochromes including type *b* cytochromes have been recently reviewed by Lemberg and Barrett (*1*) as well as by Yamanaka and Okunuki (*189*). Only highly purified and well-characterized *b* type cytochromes will be described below.

161. S. S. Deeb and L. P. Hager, *JBC* **239,** 1024 (1964).
162. L. P. Hager and S. S. Deeb, "Methods in Enzymology," Vol. 10 p. 367, 1967.
163. L. P. Vernon, *JBC* **222,** 1035 (1956).
164. L. P. Vernon, *JBC* **222,** 1045 (1956).
165. Y. Inoue and H. Kubo, *BBA* **110,** 57 (1965).
166. R. G. Bartsch, T. Kakuno, T. Horio, and M. D. Kamen, *JBC* **246,** 4489 (1971).
167. K. Miki and K. Okunuki, *Seikagaku* **42,** 644 (1970).
168. W. A. Bulen, J. R. LeComte, and S. Lough, *BBRC* **54,** 1274 (1973).
169. H. Iwasaki, *Plant Cell Physiol.* **6,** 199 (1966).
170. J. A. Orlando and T. Horio, *BBA* **50,** 367 (1961).
171. F. J. S. Lara, *BBA* **33,** 565 (1959).
172. E. Itagaki and L. P. Hager, *JBC* **241,** 3687 (1966).
173. L. P. Hager and E. Itagaki, "Methods in Enzymology," Vol. 10, p. 373, 1967.
174. J. G. Hauge, *ABB* **94,** 308 (1961).
175. C. W. Tabor and P. D. Kellogg, *JBC* **245,** 5424 (1970).
176. Y. Birk, W. S. Silver, and A. H. Heim, *BBA* **25,** 227 (1957).
177. T. Mori and K. Hirai, *in* "Structure and Function of Cytochromes"(K. Okunuki, M. D. Kamen, and I. Sekuzu, eds.), p. 681. Univ. of Tokyo Press, Tokyo, 1968.
178. F. L. Jackson and V. D. Lawton, *BBA* **35,** 76 (1959).
179. C. A. Yu, I. C. Gunsalus, M. Katagiri, K. Suhara, and S. Takemori, *JBC* **249,** 94 (1974).
180. C. A. Yu and I. C. Gunsalus, *JBC* **249,** 102 (1974).
181. C. A. Appleby, *in* "Structure and Function of Cytochromes" (K. Okunuki, M. D. Kamen, and I. Sekuzu, eds.), p. 666. Univ. of Tokyo Press, Tokyo, 1968.
182. D. A. Webster and D. P. Hackett, *JBC* **241,** 3308 (1966).
183. L. P. Solomonson and B. Vennesland, *BBA* **267,** 544 (1972).
184. R. Powls, J. Wong, and N. I. Bishop, *BBA* **180,** 490 (1969).
185. Y. Sugimura, F. Toda, T. Murata, and E. Yakushiji, *in* "Structure and Function of Cytochromes" (K. Okunuki, M. D. Kamen, and I. Sekuzu, eds.), p. 452. Univ. of Tokyo Press, Tokyo, 1968.
186. F. Labeyrie and A. Baudras, *Eur. J. Biochem.* **25,** 33 (1972).
187. C. Jacq and F. Lederer, *Eur. J. Biochem.* **25,** 41 (1972).
188. T. Yamanaka, T. Horio, and K. Okunuki, *BBA* **40,** 349 (1963).
189. T. Yamanaka and K. Okunuki, *in* "Microbial Iron Metabolism" (J. B. Neilands, ed.), p. 349. Academic Press, New York, 1974.
190. Y. Yoshida and H. Kumaoka, *BBA* **189,** 461 (1969).
191. R. H. Garrett and A. Nason, *JBC* **244,** 2870 (1969).
192. T. Yamanaka, Y. Nagata, and K. Okunuki, *J. Biochem.* (*Tokyo*) **63,** 753 (1968).

B. CYTOCHROME b_1

The name cytochrome b_1 has been used for the bacterial type *b* cytochromes having the α band near 560 nm since its designation by Keilin (*193*). However, this name is now usually applied in a limited sense to the cytochromes with α band at 558 or 559 nm in the nitrate reductase (nitrate respiration) system, which is a more primitive form of respiration than the mitochondrial respiration system.

1. *Cytochrome b_1 from E. coli*

Cytochrome b_1 from *E. coli* has been purified by several workers after its extraction by digestion with trypsin (*194*) or snake venon (*195*). Deeb and Hager released the enzyme by sonication and crystallized it after purification by chromatography on calcium-phosphate-cellulose columns (*161,162*). The cytochrome shows a peak at 418 nm in the oxidized form and peaks at 567.5, 527.5, and 425 nm in the reduced form. The molecular weight at neutral pH is about 500,000 but the aggregated species, apparently an octomer, is dissociated at higher pH into a monomeric species of molecular weight of approximately 60,000. The crystallized octameric preparation shows the E_0' of —0.34 V at pH 7, while the monomeric form shows that of approximately 0 V. A cruder preparation of *E. coli* cytochrome b_1 also shows the E_0' value of —0.01 V at pH 7, this preparation having absorption peaks at 559, 530, and 427 nm in the reduced state (*196*).

2. *Cytochrome b_1 from Other Organisms*

Denitrifying bacteria generally contain cytochrome b_1 as the electron donor to the nitrate reductase (*197,198*). This cytochrome has been partially purified from *M. dentrificans* (*163,164*), and this preparation (absorption maxima at 559, 528, and 426 nm) was rapidly oxidized by nitrate. A similar cytochrome, *b*-558, was solubilized from pseudomonads (*163,164,199*). Another similar cytochrome, *b*-559, is contained, together

193. D. Keilin, *Nature* (*London*) **133,** 290 (1934).
194. F. R. Williams and L. P. Hager, *BBA* **38,** 566 (1960).
195. S. Taniguchi and E. Itagaki, *BBA* **44,** 263 (1960).
196. T. Fujita, E. Itagaki, and R. Sato, *J. Biochem.* (*Tokyo*) **53,** 282 (1963).
197. F. Egami, M. Ishimoto, and S. Taniguchi, *in* "Haematin Enzymes" (J. E. Falk, R. Lemberg, and R. K. Morton, eds.) Vol. 1 p. 392. Pergamon, Oxford, 1961.
198. H. Takahashi, S. Taniguchi, and F. Egami, *Compr. Biochem.* **5,** 91 (1963).
199. T. Kodama and S. Shimoda, *J. Biochem.* (*Tokyo*) **65,** 351 (1969).

TABLE VI

Some Properties of Purified Type *b* Cytochromes and Related Hemoproteins Derived from Microorganisms

Trivial name	Organism	Main absorption peak (nm)				E_0' at pH 7.0 (mV)	MW	Remarks	Ref.
		Reduced α	Reduced β	Reduced γ	Oxidized γ				
Bacteria									
b_1	*Escherichia coli*	557.5	527.5	425	418	−340	62,000–66,000	Crystallized, see text.	*161* *162*
b_1	*Micrococcus denitrificans*	559	528	426	415	< −50		Similar cytochrome is obtained also from *P. denitrificans*.	*163,164*
b-558	*Streptomyces griseus*	558	527	424				Crystallized. Combines with CO at alkaline pH and autoxidizable.	*165*
b-558	*Rhodospirillum rubrum*	557.5	527	425	417	−204	450,000	pI = 4.6	*166*
b-558	*Bacillus subtilis*	558	528	425	413			Autoxidizable, but does not combine with CO.	*167*
b-558	*Azotobacter vinelandii*	557.5	527		417			Crystallized. Contains nonheme iron but not labile S.	*168*
Hemoprotein-558	*Acetobacter suboxydans*	558		437	402		48,000	Combines with CN^-, N_3^- and CO. CO	*169*

								complex, γ-423 nm.	
b-559	*Rhodopseudomonas spheroides*	559	529	426	419			Autoxidizable but does not combine with CO.	*170*
b-560	*Propionibacterium pentosaceum*	560	530	430	415			Complex with succinate dehydrogenase.	*171*
b-562	*Escherichia coli*	562	531.5	427	418	0.113	12,000	Sequenced. See text. Autoxidizable. pI = 7–8.	*172,173*
b-562	*Bacterium anitratum*	562	532	428	419	0.12–0.14	290,000	Electron acceptor of glucose dehydrogenase.	*174*
Spermidine dehydrogenase	*Serratia marcescens*	562	530	427	414		76,000	Has each one mole of protoheme and FAD.	*175*
b-563	*Streptomyces fradiae*	559	530	428				Autoxidizable	*176*
b-574	*Micrococcus* sp.	574	537	418				Reversibly changes into cytochrome *c*-548, 552	*177*
b-598	*Micrococcus lysodeikticus*	558	530	430				Autoxidizable	*178*
P-450_{cam}	*Pseudomonas putida*	540		411	417		44,000–46,000	Crystallized. CO complex, γ-446 nm.	*179,180*
P-450	*Rhizobuim japonicum*	541		408	414.5		50,000	Autoxidizable. CO complex, γ-446 nm.	*181*

TABLE VI (*Continued*)

Trivial name	Organism	Main absorption peak (nm)				E_0' at pH 7.0 (mV)	MW	Remarks	Ref.
		Reduced α	Reduced β	Reduced γ	Oxidized γ				
Algae									
o	*Vitreoscilla* sp.	553	423		399	−90	22,500	Very autoxidizable and combines with CO.	*182*
b-557	*Chlorella* sp. (Berlin strain)	557	527	423	412			Nitrate reductase. Also from *C. pyrenoidosa*.	*183*
b-558	*Scenedesmus obliquus*	558	527	426	413			Autoxidizable. Not reducible with ascorbate.	*184*
b-562	*Scenedesmus obliquus*	562	532	433	419			Reducible with ascorbate.	*184*
b-563	*Enteromorpha prolifera*	563	530	431				Similar cytochrome also from *Monstroma* sp. and *Ulva pertusa*.	*185*

Fungi									
b_2	*Saccharomyces cerevisiae*	557	528	422	411		58,000	Lactate dehydrogenase. See text.	*186,187*
b-554 (b_5?)	*Aspergillus oryzae*	554	526	423	413			α Peak, with a shoulder around 560 nm.	*158*
b-554, 561	*Sclerotinia libertiana*	561, 554	528	424	413			Double-peaked band.	*188*
b-556 (b_5?)	*Neurospora* sp.	556	526	420				α Peak, with a shoulder around 560 nm.	*189*
b-556 (b_5?)	*Saccharomyces cerevisiae*	556	525	423	413	−0.023	12,700	From the organism anaerobically cultivated.	*159*
b-557	*Neurospora crassa*	557	528	423—424	412—413		228,000	Nitrate reductase	*160*
b-563	*Neurospora* sp.	563	534	428					*190*
Protozoan									
b-560	*Tetrahymena pyriformis*	560	529	424	411			Combines with CO and CN^-. Easily modified.	*192*

with cytochromes *d* and a_1, in a particulate preparation of succinate oxidase from *Pasteurella fularensis* (*200*).

C. Cytochrome *b*-562

1. *Distribution and Preparation*

Cytochrome *b*-562 of *E. coli* has been highly purified and well characterized (*172,173,201*). It has also been purified from *Erwinia aroideae, Erwinia carotovoro*, and *Serratia marcescens* (*202*). Spectrally similar cytochromes are also found in *Xanthomonas phaseoli* (*114*) and *Bacterium anitratum* (*174*). The *Serratia* and *Bacterium* cytochromes have been established to react as an electron acceptor of dehydrogenases (*173,174,203*).

Itagaki and Hager have crystallized cytochrome *b*-562 from *E. coli* (*172*). Purification involved chromatography on calcium phosphate gel and ammonium sulfate fractionation (*172,173*). Both the oxidized and reduced forms of the cytochrome were crystallized.

2. *Properties and Structure of Cytochrome b-562 from E. coli*

The crystalline preparation of *E. coli* cytochrome *b*-562 (*172,173*) shows absorption peaks at 418 nm in the oxidized form and peaks at 427, 531.5, and 562 nm in the reduced form. The cytochrome does not combine with CO in the pH range of 3.0–10.5 but at a higher pH CO causes a small shift of the α maximum. It does not react with cyanide or azide. Its molecular weight is calculated to be 12,000, 11,900–12,700, and 11,954, based on the heme content, ultracentrifugal data, and amino acid analysis, respectively. The amino acid sequence of cytochrome *b*-562 has been determined (*201*). As shown in Table VII this protein is composed of 110 amino acid residues, and lacks CySH and tryptophan. The cytochrome molecule possesses only two histidines, both of which probable coordinate to heme. Therefore, the cytochrome may be said to contain minimal numbers of histidines which are necessary to keep protoheme in the cytochrome molecule. However, it is also possible that one of the two histidines, especially near the carboxyl terminus, does not par-

200. J. H. Felman and R. C. Mills, *J. Bacteriol.* **79,** 800 (1960).
201. E. Itagaki and L. P. Hager, *BBRC* **32,** 1013 (1968).
202. T. Fujita, *J. Biochem.* (*Tokyo*) **60,** 204 (1966).
203. Y. Sugimura, F. Toda, T. Murata, and E. Yakushiji, *in* "Structure and Function of Cytochromes" (K. Okunuki, M. D. Kamen, and I. Sekuzu, eds.), p. 452. Univ. of Tokyo Press, Tokyo, 1968.

TABLE VII

AMINO ACID SEQUENCE OF *E. coli* CYTOCHROME *b*-562[a]

1 10
Ala-Asp-Leu-Glu-Asp-Asp-Met-Gln-Thr-Leu-Asn-Asp-Asn-Leu-Lys-Val-
20 30
Ile-Glu-Lys-Ala-(Asx,Asx,Glx)-Lys-Ala-Asn-Asp-Ala-Ala-Gln-Val-
40
Lys-Leu-Lys-Met-Arg-Ala-Ala-Ala-Leu-Asn-Ala-Gln-Lys-Lys-Ala-Thr-
50 60
Pro-Pro-Lys-Leu-Glu-Asp-Lys-Ser-Pro-Asn-Ser-Gln-Pro-Met-Lys-Asp-
70
Phe-Arg-His-Gly-Phe-Asp-Ile-Leu-Val-Gly-Glu-Ile-Asp-Asp-Ala-Leu-
80 90
Lys-Leu-Ala-Asn-Glu-Gly-Lys-Val-Lys-Glu-Ala-Gln-Ala-Ala-Glu-Ala-
100 110
Gln-Leu-Lys-Thr-Thr-Arg-Asn-Ala-Tyr-Lys-His-Gln-Lys-Tyr-Arg

[a] Data from Itagaki and Hager (*201*).

ticipate as the heme ligand. As in the case of cytochrome *c*, a methionine residue such as Met-35 may be one of the ligands.

D. CYTOCHROME b_2

Cytochrome b_2 is also called yeast lactate dehydrogenase (EC 1.1.2.3, lactate: cytochrome *c* oxidoreductase) (*204,205*). This cytochrome possesses both FMN and protoheme (*206–211*). Although the native enzyme also contains deoxyribonucleic acid, this moiety can be removed by deoxyribonuclease without affecting the enzymic activity of the enzyme (*212*) and is supposed to be a heterogeneous component (*213*). This enzyme completely differs in its heme content from animal lactate dehydrogenase and also from yeast D(—)-lactate dehydrogenase.

204. S. J. Bach, M. Dixon, and L. G. Zerfas, *Nature* (*London*) **149,** 21 (1942).
205. S. J. Bach, M. Dixon, and L. G. Zerfas, *BJ* **40,** 229 (1946).
206. C. A. Appleby and R. K. Morton, *Nature* (*London*) **173,** 749 (1954).
207. C. A. Appleby and R. K. Morton, *BJ* **73,** 539 (1959).
208. R. K. Morton, *Nature* (*London*) **192, 727** (1961).
209. E. Boeri, E. Cutolo, and L. Tosi, *Bull. Soc. Ital. Biol. Sper.* **31,** 1392 (1955).
210. E. Boeri, E. Cutolo, M. Luzzati, and L. Tosi, *ABB* **56,** 487 (1955).
211. A. P. Nygaard, *BBA* **33,** 517 (1959); **35,** 212 (1959).
212. R. K. Morton, *Rev. Pure Appl. Chem.* **8,** 161 (1958).
213. J. F. Jackson, R. D. Kornberg, P. Berg, U. L. Rajehandary, A. Stuart, H. G. Khorana, and A. Kornberg, *BBA* **108,** 243 (1965).

Purification of cytochrome b_2 has been carried out by various workers (*208,211,214–217*). The cytochrome was first crystallized by Appleby and Morton in 1954 (*206*) as a preparation (type I) containing DNA (*214,217*). Another crystalline preparation (type II) which contains no DNA was also obtained after dialysis against strong ammonium sulfate solution or by chromatography on DEAE-cellulose (*218–221*). Another crystalline preparation, which is free from flavin and enzymically inactive, was obtained by Okunuki *et al.* (*222–224*). This cytochrome has a molecular weight of about 20,000 and appears to be similar to but different from the "cytochrome b_2 core" described later.

The crystalline type I preparation shows the α-, β-, and γ-absorption maxima in the reduced state at 556.5, 528, and 423 nm, respectively (*214*), and the γ peak at 413 nm in the oxidized state. The α band is not split into two at the liquid nitrogen temperature (*214*) differing from cytochrome b_5 which shows very similar α-peak position.

Both in solution and in crystals, the cytochrome exists as a tetramer. The molecular weight of monomer cytochrome b_2 is 57,000 which has been reported to be composed of one peptide chain (*187*). It has been reported that when the native cytochrome b_2 was subjected to ultracentrifugation in the presence of 6 *M* guanidine hydrochloride, or to gel electrophoresis in the presence of sodium dodecyl sulfate, the protein split into two peptide chain parts, one with a molecular weight of 21,000 and the other of 36,000 (*225*). However, these two chains are supposed to be produced by proteolytic digestion during the purification and crystallization procedures (*201*). Although there was an assumption that cytochrome b_2 is composed of a hemoprotein and a flavoprotein (*226*), it is now almost certain that both a heme and a flavin are associated to a single

214. C. A. Appleby and R. K. Morton, *BJ* **71,** 492 (1959).
215. E. Boeri and M. Rippa, *in* "Haematin Enzymes" (J. E. Falk, R. Lember, and R. K. Morton, eds.), Vol. 2, pp. 524, 563, and 568. Pergamon, Oxford, 1961.
216. E. Boeri and M. Rippa, *ABB* **94,** 336 (1961).
217. M. Rippa, *ABB* **94,** 333 (1961).
218. J. McD. Armstrong and R. K. Morton, *Aust. J. Sci.* **24,** 137 (1961).
219. R. K. Morton and K. Shepley, *Nature* (*London*) **192,** 639 (1961).
220. R. K. Morton and K. Shepley, *Biochem. Z.* **338,** 122 (1963).
221. R. H. Symons, *BBA* **103,** 298 (1965).
222. J. Yamashita, T. Higashi, T. Yamanaka, M. Nozaki, H. Mizushima, H. Matsubara, T. Horio, and K. Okunuki, *Nature* (*London*) **179,** 959 (1957).
223. T. Yamanaka, T. Horio, and K. Okunuki, *J. Biochem.* (*Tokyo*) **45,** 291 (1958).
224. J. Yamashita and K. Okunuki, *J. Biochem.* (*Tokyo*) **52,** 117 (1962).
225. F. Lederer and A. M. Simon, *Eur. J. Biochem.* **20,** 469 (1971).
226. T. Horio, J. Yamashita, T. Yamanaka, M. Nozaki, and K. Okunuki, *in* "Haematin Enzymes" (J. E. Falk, R. Lemberg, and R. K. Morton, eds.), Vol. 2, pp. 552 and 560. Pergamon, Oxford, 1961.

peptide chain. By tryptic digestion of the native cytochrome b_2 under an appropriate condition, Labeyrie *et al.* (*227,228*) obtained a derivative with a molecular weight of 11,000. This derivative shows absorption spectra very similar to that of the native cytochrome b_2, and is called cytochrome b_2 core. Recently, the structural similarities between cytochrome b_2 core and cytochrome b_5 have been shown to exist by proton magnetic resonance (*140*).

Cytochrome b_2 exists at least partly in mitochondria (*229*) and mediates electron from L-lactate to the respiratory system (*230*).

$$\begin{array}{ll} \text{L-Lactate} \rightarrow b_2 & \\ \qquad\qquad\quad\ \downarrow & \\ \qquad\qquad\quad c_1 \rightarrow c \rightarrow a,a_3 \rightarrow \mathrm{O}_2 & \end{array}$$

V. Type *b* Cytochrome in Plants

A. Introduction

Plants contain type *b* cytochromes participating in photosynthetic electron transport systems in addition to those in mitochondrial and microsomal electron transport systems. There is some confusion about the names of these cytochromes (*231*). Several type *b* cytochromes, including photosynthetic *b*-559 and b_6 and microsomal *b*-555, have been highly purified and fairly well characterized. The last one is described in Section III (see Table IV).

B. Type *b* Cytochromes in Photosynthetic Systems

Two type *b* cytochromes are known to occur in the photosynthetic systems of chloroplasts: Cytochrome *b*-559 (*232*) may participate in the electron transfer between the photosystems I and II, while cytochrome b_6 (*233*) seems to function in the cyclic photophosporylation system (*234*). Both the type *b* cytochromes have been purified from spinach chloroplasts.

227. F. Labeyrie, O. Groudinski, Y. Jacquet-Armand, and L. Naslin, *BBA* **128,** 492 (1966).
228. F. Labeyrie, A. di Franco, M. Iwatsubo, and A. Baudras, *Biochemistry* **6,** 1791 (1967).
229. E. Vitols and A. W. Linnane, *J. Biophys. Biochem. Cytol.* **9,** 701 (1961).
230. T. Ohnishi, K. Kawaguchi, and B. Hagihara, *JBC* **241,** 1797 (1966).
231. B. Chance, W. D. Bonner, and B. T. Storey, *Annu. Rev. Plant. Physiol.* **19,** 295 (1968).
232. N. K. Boardman and J. M. Anderson, *BBA* **143,** 187 (1967).
233. R. Hill and R. Scarisbrick, *New Phytol.* **50,** 98 (1951).
234. N. K. Boardman, *Advan. Enzymol.* **30,** 1 (1968).

Cytochrome b_6 was discovered by Davenport (*235*) and Hill (*236*) and further studied by several workers (*232,237–239*). Ferrocytochrome b_6 shows α, β, and γ-absorption peaks at 563, 536, and 434 nm, respectively. The E_0' at pH 7 was found to be 0–60 mV. Cytochrome b_6 was contained in the lighter fraction obtained by differential centrifugation of digitonin-treated chloroplasts, while the heavier fraction contained cytochrome *b*-559 (*232,240*).

Cytochrome b_6 has been purified to the electrophoretically homogeneous state by Stuart and Wasserman (*241*). This cytochrome is reduced by dithionite but not by ascorbate. The reduced form is autoxidizable and does not combine with CO. Its molecular weight was determined to be 40,000 on the basis of heme content. A similar cytochrome, showing the α band at 561–563 nm, has been observed in several species of green algae (*203,242–251*), *Euglena* (*252,253*), and red algae (*254*).

Chloroplast cytochrome *b*-559 was first described by Lundegårdh (*255–258*) under the name cytochrome b_3 which had been designated as a soluble microsomal cytochrome with the α band at 559 nm (*277*). This cytochrome is tightly bound to chloroplast lamellae (*232,241*). It has been obtained by Garewal *et al.* (*259*) in 95% pure state as judged from

235. H. E. Davenport, *Nature* (*London*) **170,** 1112 (1952).
236. R. Hill, *Nature* (*London*) **174,** 501 (1954).
237. R. Hill and F. Bendall, *Nature* (*London*) **186,** 136 (1960).
238. W. O. James and R. M. Leech, *Proc. Roy. Soc., Ser. B* **160,** 13 (1964).
239. D. S. Bendall, *BJ* **109,** 46 (1968).
240. N. K. Boardman and J. M. Anderson, *Nature* (*London*) **203,** 66 (1964).
241. A. L. Stuart and A. R. Wasserman, *BBA* **314,** 284 (1973).
242. B. Kok, *Nature* (*London*) **179,** 583 (1957).
243. B. Chance and R. Sager, *Plant Physiol.* **32,** 548 (1957).
244. B. Chance, H. Schleyer, and V. Legallais, "Studies on Microalgae and Photosynthetic Bacteria" (Japanese Society of Plant Physiology, ed.), p. 337. Univ. of Tokyo Press, Tokyo, 1963.
245. R. M. Smillie and R. P. Levine, *JBC* **238,** 4058 (1963).
246. A. Müller, B. Rumberg, and H. T. Witt, *Proc. Roy. Soc., Ser. B* **157,** 313 (1963).
247. S. Katoh and A. San Peitro, *BBRC* **20,** 406 (1965).
248. G. Hind and J. M. Olson, *Brookhaven Symp. Biol.* **19,** 188 (1966).
249. D. S. Bendall and R. Hill, *Annu. Rev. Plant Physiol.* **19,** 167 (1968).
250. T. Hiyama, *Fed. Proc., Fed. Amer. Soc. Exp. Biol.* **28,** Abstr. 1500 (1969).
251. R. Powls, J. Wong, and N. I. Bishop, *BBA* **180,** 490 (1969).
252. J. M. Olson and R. M. Smillie, *Nat. Acad. Sci.—Nat. Res. Counc.* **1145,** 42 (1963).
253. F. Perini, M. D. Kamen, and J. A. Schiff, *BBA* **88,** 74 (1964).
254. M. Nishimura, *BBA* **153,** 838 (1968).
255. H. Lundegårdh, *Nature* (*London*) **192,** 243 (1961).
256. H. Lundegårdh, *BBA* **57,** 352 (1962).
257. H. Lundegårdh, *BBA* **88,** 37 (1964).
258. H. Lundegårdh, *Proc. Nat. Acad. Sci. U. S.* **52,** 1587 (1964).
259. H. S. Garewal, J. Singh, and A. R. Wasserman, *BBRC* **44,** 1300 (1971).

the results in polyacrylamide gel electrophoresis. This cytochrome shows α, β, and γ bands at 559, 530, and 429 nm, respectively, in the reduced state, while a peak at 415 nm is seen in the oxidized form. The cytochrome is autoxidizable and completely reducible with ascorbate. Ferrocytochrome *b*-559 does not combine with CO. A similar cytochrome, *b*-560 of spinach, was also purified by Matsuzaki and Kamimura (*260*). The molecular weight of this cytochrome is 30,000. Its E_0' is +130 mV at pH 7. It is rapidly reduced on illumination with red or infrared light in the presence of spinach chloroplasts as in the case of cytochrome *b*-559. However, it is open to question whether or not these two cytochromes are the same protein.

Cytochrome *b*-559 is found in three forms in chloroplasts (*261*); they are distinguishable from one another by their redox potentials although they have the same α peak at 559 nm. The high potential, middle potential, and low potential forms have the E_0' values at pH 7 of 0.330–0.350 V, 0.05–0.080 V, and lower than 0 V, respectively. The high potential form is predominant in freshly prepared chloroplast preparations, and is changed into the middle potential form by treatments such as aging, sonication, and mild heating. The latter form is transformed into the lower potential form by more drastic treatments. The molecular mechanisms of the above change in the redox potential have not yet been made clear.

C. Type *b* Cytochromes in Plant Mitochondria

In plant mitochondria, as in mammalian mitochondria, there are probably three kinds of type *b* cytochromes (*22,262–265*). They are called cytochromes *b*-565, *b*-560, and *b*-556 from the α peak of the reduced minus oxidized spectra at room temperature or, sometimes, b_{562}, b_{557}, and b_{553}, respectively, from those *at the* liquid nitrogen temperature. The E_0' of cytochromes *b*-565, *b*-560, and *b*-556 are −77 mV, +42 mV, and −75 mV, respectively (*266*). An oxygen pulse experiment (*267*) has shown that cytochrome *b*-560 is most rapidly oxidzed, with a half-time

260. E. Matsuzaki and Y. Kamimura, *Plant Cell Physiol.* **13,** 415 (1972).
261. K. Wada and D. I. Arnon, *Proc. Nat. Acad. Sci. U. S.* **68,** 3064 (1971).
262. W. D. Bonner, *in* "Haematin Enzymes" (J. E. Falk, R. Lemberg, and R. K. Morton, eds.), Vol. 2, p. 479. Pergamon, Oxford, 1961.
263. W. D. Bonner Jr., *Proc. Int. Congr. Biochem., 5th, 1961* Vol. 2, p. 50, (1963); *in* "Plant Biochemistry" (J. Bonner and J. E. Varner, eds.), 2nd ed., p. 89. Academic Press, New York, 1965.
264. W. D. Bonner and M. Plesnicar, *Nature* (*London*) **214,** 616 (1967).
265. B. Chance, W. D. Bonner, Jr., and B. T. Storey, *Annu. Rev. Plant Physiol.* **19,** 781 (1968).
266. H. Shichi and D. P. Hackett, *JBC* **237,** 2959 (1962).
267. B. T. Storey, *Plant Physiol.* **44,** 413 (1969).

of around 8 msec, considerably more rapid than that of cytochrome *b* in animal mitochondria. Cytochrome *b*-565 was oxidized with half times in the range of 15–35 msec, and cytochrome *b*-556, most slowly with a half time of 500 msec. Cytochrome *b*-560 is completely reduced by ascorbate plus TMPD in uncoupled mitochondria, while cytochrome *b*-556 is only partially reduced and cytochrome *b*-565 remains largely oxidized (*267*).

Cytochrome *b*-565 is only partially reduced by succinate in anaerobic uncoupled mitochondria and can be fully reduced by succinate in coupled mitochondria (*22*). At steady state this cytochrome is more reduced than other type *b* cytochromes and its reduction is further increased by energization of the mitochondrial membrane. Thus, cytochrome *b*-565 is spectrally and functionally very similar to cytochrome b_T (*16,18–20*) in animal mitochondria. However, in contrast to cytochrome b_T, the E_0' of this cytochrome is not affected by the energy state of mitochondria (*22*), and its α band seems to be a single band with no appreciable shoulder even at 77°K (*265*).

Cytochrome *b*-560 isolated and purified from mung bean seedlings exists in mitochondria of the organism and corresponds functionally to cytochrome *b* in animal mitochondria in that it shows similar spectral properties and standard potential (*266*). Bonner has suggested that cytochrome *b*-560 is identical to cytochrome b_7 which was found in mitochondria from the spadix of *Arum maculatum* (*268*). The E_0' of cytochrome b_7 (*269,270*) was found to be at -30 mV. Both cytochromes *b*-560 and b_7 are autoxidizable but do not combine with CO or cyanide. Oxidation of cytochrome *b*-560 is inhibited by antimycin (*263,267*) while oxidation of cytochrome b_7 is not (*268,271,272*). Thus, the identification of cytochrome *b*-560 with cytochrome b_7 is not quite clear. Cytochrome b_7 seems to function as the terminal oxidase in the cyanide-insensitive respiratory system in the spadix (*269*), and it has been shown that phosphorylation could occur coupled to this electron transfer (*271*).

Cytochrome *b*-556 is fully reduced by succinate in both coupled and uncoupled mitochondria. It is partially reduced by endogenous substrate in the presence of antimycin, thus differing from type *c* cytochrome as assumed by earlier workers (*273,274*). The α band of cytochrome *b*-556

268. D. S. Bendall and R. Hill, *New Phytol.* **55**, 206 (1956).
269. D. S. Bendall, *BJ* **70**, 381 (1958).
270. D. S. Bendall, *BJ* **109**, 46 (1968).
271. S. B. Wilson, *BBA* **223**, 383 (1970).
272. D. P. Hackett, *J. Exp. Bot.* **8**, 157 (1957).
273. H. Lundegårdh, *Nature* (*London*) **181**, 28 (1958).
274. J. T. Wiskich, R. K. Morton, and R. N. Robertson, *Aust. J. Biol. Sci.* **13**, 109 (1960).

lies at 555–556 nm at room temperature and at 552–553 nm at 77°K (*267,275*).

A few respiratory type *b* cytochromes, which have an α band of around 562 nm but are distinct from photosynthetic cytochrome b_6, have been reported to exist in *Euglena* and green algae (*247,249,276*).

D. Type *b* Cytochromes in Plant Microsomes

A microsomal type *b* cytochrome with the α peak at 559 nm is called cytochrome b_3 (*275,277,278*). The molecular weight of cytochrome b_3 obtained from *Vicia faba* leaves (*279,280*) is 28,000. This preparation is reduced by ascorbic acid and does not combine with CO. The E_0' of wheat-root cytochrome b_3 (α band 559 nm), which is identical with or very similar to microsomal cytochrome b_3, is reported to be +40 mV (*233*).

Plant microsomes contains another cytochrome which is very similar to mammalian cytochrome b_5 (*157,275,281–284*). The absorption peaks, molecular weight, and E_0' at pH 7 of cytochrome *b*-555 purified from mung bean seedlings (*157*) are given in Table V. The α band of this cytochrome is split into two peaks at liquid nitrogen temperature in a way similar to mammalian cytochrome b_5 (*143*).

VI. Other Type *b* Cytochromes

A. Introduction

Several cytochromes typical for invertebrates have been reported under the names helicorubin, enterochrome-566, enterochrome-556 (*285,286*), and cytochromes *h* (*287–289*). The former two pigments are type *b* cyto-

275. C. Lance and W. D. Bonner, Jr., *Plant Physiol.* **43,** 756 (1968).
276. J. K. Raison and R. M. Smillie, *BBA* **180,** 500 (1969).
277. E. M. Martin and R. K. Morton, *Nature* (*London*) **176,** 113 (1955).
278. E. M. Martin and R. K. Morton, *BJ* **65,** 404, (1957).
279. H. Shichi, D. P. Hackett, and G. Funatsu, *JBC* **238,** 1156 (1963).
280. H. Shichi, H. E. Kasinsky, and D. P. Hackett, *JBC* **238,** 1162 (1963b).
281. R. K. Morton, *in* "Haematin Enzymes" (J. E. Falk, R. Lember, and R. K. Morton, eds.), Vol. 2, p. 498. Pergamon, Oxford, 1961.
282. H. Shichi and D. P. Hackett, *Nature* (*London*) **193,** 776 (1962).
283. H. E. Kasinsky, H. Shichi, and D. P. Hackett, *Plant Physiol.* **41,** 739 (1966).
284. H. Shichi and D. P. Hackett, *J. Biochem.* (*Tokyo*) **59,** 84 (1966).
285. K. Kawai, *BBA* **43,** 349; **44,** 202 (1960).
286. K. Kawai, *J. Biochem.* (*Tokyo*) **52,** 241 and 248 (1961).
287. J. Keilin, *BJ* **64,** 663 (1956).
288. J. Keilin, *Nature* (*London*) **180,** 427 (1957).
289. J. Keilin and P. Orlans, *Nature* (*London*) **223,** 304 (1969).

chromes while the latter two, which are very similar to each other, are probably hematohemoproteins (*291*). However, pyridine ferrohemochrome of the heme isolated from cytochrome *h* shows the α peak at 550 nm (*291*), while that of hematoheme has it at 546 nm.

In many bacteria and some other microorganisms there exists a type of protohemoprotein called cytochromes *o*. This type of cytochrome is highly autoxidizable and is believed to function as a cytochrome oxidase.

B. Helicorubin and Enterochrome-566

Helicorubin has been designated for a pigment found in snail gut fluid (*290*), and highly purified by Keilin (*291*) from the gut fluid of *Helix pomatia*. The pigment contains a protoheme as a removable prosthetic group and shows absorption peaks at 563, 531, and 428 nm in the reduced form and at 416 nm in the oxidized form. The α band at 563 nm in the reduced form splits into two bands at 562 and 556 nm at the liquid nitrogen temperature. The ferrocytochrome does not react with carbon monoxide and is slowly autoxidizable. The ferric form does not react with cyanide or fluoride. The molecular weight is 12,100 and the E_0', +20 mV (*288,289,291*).

Enterochrome-566 is a very similar pigment to helicorubin and is also found in gastropods as well as pelecypods. It has been purified from the gut fluid of *Hyriopsis schlegelii, Cristaria plicata, Euhadra amaliae,* and *Euhardra sandai* (*286*). Absorption peaks occur at 566, 530, and 430 nm in the reduced form and at 416 nm in the oxidized form. Its E_0' is 200 mV at pH 6.4 and 150 mV at pH 7.8 (*286*). The biological function of both helicorubin and enterochrome-566 is not known.

C. Cytochrome *o*

The existence of CO-binding type *b* cytochromes which behave as a cytochrome oxidase was revealed by Chance *et al.* as the result of spectrophotometric studies; they have been designated as cytochrome *o* (*292–294*). This group of cytochromes is distributed in a wide variety of bacteria (*292–313*) and also in protozoa (*314,315*) as well as in algae

290. C. F. Krukenberg, "Vergleichende Physiologische Studien," Ser. 2, Part 2, p. 63. 1882.
291. J. Keilin, *in* "Structure and Function of Cytochromes" (K. Okunuki, M. D. Kamen, and I. Sekuzu, eds.), p. 691. Univ. of Tokyo Press, Tokyo, 1968.
292. B. Chance, *JBC* **202**, 383 (1953).
293. B. Chance, L. Smith, and L. N. Castor, *BBA* **12**, 289 (1953).
294. L. N. Castor and B. Chance, *JBC* **234**, 1587 (1959).
295. H. Taber and M. Morrison, *ABB* **105**, 367 (1964).
296. L. N. Castor and B. Chance, *JBC* **217**, 453 (1955).

(*316*). The position of their α peak in the reduced minus oxidized difference spectra ranges from 555 to 565 nm. The absorption peaks in their difference spectra between the CO reduced and the reduced state are found at 565–575, 532–540, and 410–421 nm (*292–313*). The prosthetic group of cytochrome *o* has been chracterized as protoheme by the formation of the pyridine hemochrome or by other indentification methods (*295,297,309,316*). Cytochrome *o* of a colorless blue-green alga, *Vitreoscilla* sp., has been highly purified. This preparation shows an absorption peak at 399 nm in the oxidized form and peaks at 423 and 553 nm in the reduced form. Its reduced-CO minus reduced difference spectrum shows absorption peaks at 566, 532, and 418 nm. The cytochrome shows the E_0' value of —0.09 V at pH 7 and is extremely autoxizable (*316*).

The CO complex of a protoheme protein, hemoprotein-558, isolated by Iwasaki (*300,301*) from *Acetobacter suboxydans* shows absorption peaks at 574, 544, and 423 nm, and, as judged from the positions of these peaks, may differ from cytochrome *o*. However, it is noteworthy that the lactate dehydrogenase preparation obtained from *A. suboxydans* acquires a high lactate oxidase activity upon addition of this hemoprotein preparation. A complex of cytochrome $(a + a_3)$ and cytochrome *o* was reported to have been obtained from *M. phlei* (*299*).

297. S. Taniguchi and M. D. Kamen, *BBA* **96,** 395 (1965).
298. M. D. Kamen and T. Horio, *Annu. Rev. Biochem.* **39,** 673 (1970).
299. B. Revsin, E. D. Marquez, and A. F. Brodie, *ABB* **139,** 114 (1970).
300. H. Iwasaki, *Plant Cell Physiol.* **7,** 199 (1966).
301. H. Iwasaki, *Abstr. Int. Congr. Biochem., 7th, 1967* p. 881 (1968).
302. K. Arima and T. Oka, *J. Bacteriol.* **90,** 734 (1965).
303. C. A. Appleby, *BBA* **172,** 88 (1969).
304. M. Dworkin and D. J. Niederpruem, *J. Bacteriol.* **87,** 316 (1964).
305. R. M. Daniel, *BBA* **216,** 328 (1970).
306. J. Biggins and W. E. Dietrich, *ABB* **128,** 40 (1968).
307. J. B. Baseman and C. D. Cox, *J. Bacteriol.* **97,** 1001 (1969).
308. P. L. Broberg and L. Smith, *BBA* **131,** 479 (1967).
309. K. S. Cheah, *BBA* **180,** 320 (1969).
310. N. J. Jacobs and S. F. Conti, *J. Bacteriol.* **89,** 675 (1965).
311. B. Revsin and A. F. Brodie, *JBC* **244,** 3101 (1969).
312. D. C. White and L. Smith, *JBC* **237,** 1032 (1962).
313. T. Sasaki, Y. Motokawa, and G. Kikuchi, *BBA* **197,** 284 (1970).
314. J. S. Perlish and J. H. Eichel, *BBRC* **44,** 973 (1971).
315. H. K. Srivastava, *FEBS* (*Fed. Eur. Biochem. Soc.*) *Lett.* **16,** 189 (1971).
316. D. A. Webster and D. P. Hackett, *JBC* **241,** 3308; *Plant Physiol.* **41,** 599 (1966).

Author Index

Numbers in parentheses are reference numbers and indicate that an author's work is referred to, although his name is not cited in the text.

E

G

H

I

J

M

N

O

P

S

U

V

W

Subject Index

A

B

C

D

E

F

G

I

K

L

M

N

U

A 5
B 6
C 7
D 8
E 9
F 0
G 1
H 2
I 3
J 4